TD 745 .R68 1995
Rowe, Donald R.
Handbook of wastewater
    reclamation and reuse

CORETTE LIBRARY
CARROLL COLLEGE

DISCARD

Handbook of
WASTEWATER RECLAMATION and REUSE

Handbook of
WASTEWATER RECLAMATION and REUSE

Donald R. Rowe
Isam Mohammed Abdel-Magid

LEWIS PUBLISHERS

Boca Raton   New York   London   Tokyo

Library of Congress Cataloging-in-Publication Data

Rowe, Donald R.
  Handbook of wastewater reclamation and reuse / Donald R. Rowe and
Isam Mohammed Abdel-Magid.
     p.   cm.
  Includes bibliographical references and index.
  ISBN 0-87371-671-X
  1. Sewage — Purification. 2. Water reuse. I. Abdel-Magid
  Isām Mohammed. II. Title.
TD745.R68 1995
628.1′62 — dc20   94-46589
                                                                                   CIP

   This book contains information obtained from authentic and highly regarded sources. Reprinted material is quoted with permission, and sources are indicated. A wide variety of references are listed. Reasonable efforts have been made to publish reliable data and information, but the author and the publisher cannot assume responsibility for the validity of all materials or for the consequences of their use.
   Neither this book nor any part may be reproduced or transmitted in any form or by any means, electronic or mechanical, including photocopying, microfilming, and recording, or by any information storage or retrieval system, without prior permission in writing from the publisher.
   CRC Press, Inc.'s consent does not extend to copying for general distribution, for promotion, for creating new works, or for resale. Specific permission must be obtained in writing from CRC Press for such copying.
   Direct all inquiries to CRC Press, Inc., 2000 Corporate Blvd., N.W., Boca Raton, Florida 33431.

© 1995 by CRC Press, Inc.
Lewis Publishers is an imprint of CRC Press

No claim to original U.S. Government works.
International Standard Book Number 0-87371-671-X
Library of Congress Card Number 94-46589
Printed in the United States of America 1 2 3 4 5 6 7 8 9 0
Printed on acid-free paper

Preface

The three Rs in the wastewater reclamation and reuse field are reclamation, recycling, and reuse.

Wastewater reclamation signifies the treatment of wastewater in order to make it suitable for a beneficial or controlled use that would not otherwise occur.

Wastewater recycling generally refers to a specific industrial process where the treated wastewater is recycled in the plant and not discharged to the surrounding environment. The aim is zero discharge.

Wastewater reuse can be direct or indirect. By direct reuse we mean that the treated wastewater is piped directly into a potable water-supply distribution system or a water treatment plant. At present there is no operation where direct potable reuse is practiced in the U.S. However, demonstration projects involving the production of potable water from wastewater has been or is underway at Denver, CO, Tampa, FL, and San Diego, CA. Windhoek, Nambia is the only city in the world that presently practices direct potable reuse.

Unplanned indirect reuse of treated wastewater occurs continuously in the environment. This results when a water supply has a surface or groundwater source that contains wastewater effluents. This type of potable reuse is commonplace and occurs whenever an upstream water user discharges their wastewater effluent into a surface water that serves as a water supply for a downstream user. It has been estimated that the Rhine, Thames, and Ohio Rivers, at minimal flow, contain from 20- to 50% wastewater.

Planned indirect potable reuse is similar to unplanned indirect potable reuse; however, the wastewater effluent is discharged to a water source with the intent of reusing the water, instead of as a means of disposal. This type of indirect potable reuse is becoming more common as water resources become less plentiful and clean water is in short supply.

Less than 1% of the total water on the planet is readily available for human use. Some 26 countries, including 11 in Africa and 9 in the Middle East, already have more people than their water supplies can support. In some cases, a liter of bottled water costs four to five times that of a liter of gasoline. Conflicts in the past have been over such things as oil, land, or money. It is conceivable that water could also be added to this list.

Potential water conflicts currently exist between countries such as the U.S., Canada, and Mexico; Jordan and Israel; and Iraq and Turkey.

Wastewater reclamation and reuse may not be able to solve all these problems, but it clearly could contribute to alleviating some of the water supply shortages.

Water and wastewater treatment technology is available to produce a physically, chemically, and biologically safe water of various levels of quality. This technology can produce water all the way from a high quality potable water supply to a water quality suitable for the biggest water consumer of all, agricultural irrigation.

The Authors

Donald R. Rowe, Ph.D., is President of D.R. Rowe Engineering Services, Inc., Bowling Green, Kentucky. He received his B.S. degree (1948) in Civil Engineering from the University of Saskatchewan, Canada and his M.S. (1962) and Ph.D. degrees (1965) from The University of Texas at Austin.

From 1964 to 1969, Dr. Rowe was Associate Professor, Department of Civil Engineering, Tulane University; and from 1969 to 1981 he was Associate Professor and Professor, Department of Engineering Technology, Western Kentucky University. From 1981 to 1988, he was Professor, Department of Civil Engineering, King Saud University, Riyadh, Saudi Arabia. From 1971 to 1993, he was Vice-President of the Larox Research Corporation. Since 1990 he has been president of his own company.

Dr. Rowe has authored or co-authored over 70 reports and publications in the areas of air pollution, solid waste, water treatment, and wastewater reclamation and reuse. He also co-authored the book entitled, *Environmental Engineering* (McGraw-Hill Publishers).

Dr. Rowe holds, with co-inventors, over ten patents on catalytic conversion removal processes for air contaminants in the United States, Canada, Great Britain, and Japan. In 1980, he was recipient of a Fulbright-Hays Award to Ege University in Izmir, Turkey. While at King Saud University, Dr. Rowe carried out research on wastewater reclamation and reuse.

From 1984 to the present, he has served on Peer Review Evaluation Committees for research projects funded by King Abdulaziz City for Science and Technology (KACST), Riyadh, Saudi Arabia.

**Isam Mohammed Abdel-Magid, Ph.D.,** is Associate Professor in the Department of Civil Engineering of the University of Khartuum, Sudan on secondment to the Department of Civil Engineering at Sultan Qaboos University, Muscat, Sultanate of Oman.

He received his Ph.D. in Public Health Engineering from Strathclyde University, Great Britain in 1982; his Diploma in Hydrology, from Padova University, Italy, in 1978; his Diploma in Sanitary Engineering, from Delft University of Technology in The Netherlands in 1979 (equivalent to M.Sc.), and his B.Sc. in Civil Engineering at the University of Khartoum in 1977.

From 1977 to 1988, he was Teaching Assistant, Assistant Professor, and Associate Professor at the University of Khartoum; from 1988 to 1991, Assistant Professor, University of United Arab Emirates; and from 1991 to the present, Assistant Professor at Sultan Qaboos University.

Dr. Isam Mohammed Abdel-Magid has authored or co-authored over 29 publications, 8 scientific textbooks, and numerous technical reports dealing with water supply, wastewater disposal, water resources, wastewater reclamation and reuse, and solid waste disposal. He has edited or co-edited numerous conference proceedings as well as the proceedings for two international conferences.

Dr. Isam Mohammed Abdel-Magid has supervised numerous research projects leading to various student degrees, such as Diploma, B.Sc., M.Sc., and Ph.D.

The Sudan Engineering Society awarded him the prize for the Best Project in Civil Engineering in 1977. The same year, the Ministry of Irrigation awarded him the Second Best Performance in Civil Engineering and, in 1986, the University of Khartoum awarded him the Honourly Scarf for Enrichment of Knowledge.

Acknowledgments

As with any book the number of individuals who need to be thanked for their contribution is legion.

I would first like to thank the following former students whose help contributed to the development of this book: Joseph P. Abbott, Robert M. Adams, William E. Blacketer, James G. Grunow, Stanley O. Hopkins, Rose M. Hullett, Phillip Porter, and Larry Miller.

I would like to thank Gary Asbury, Ernest U. Earn, Dr. Adnan Gur, Gamar A. H. Mohammed, Tara Clark, and Pamla Wood for information that was included in this book.

I would especially like to thank Melvin L. Newman, General Manager, Clayton County Water Authority (CCWA), Morrow, GA, as well as Neal Wellson and Jim Poff, CCWA staff members, for the information, and the reviewing of the case study entitled, *E. L. Huie, Jr. Land Application Facility.*

I would also like to thank Harry Wojnar and Phillip Porter of the Logan Aluminum Company, Russellville, KY, for the information and the reviewing of the case study entitled *Logan Aluminum Constructed Wetlands Project.*

I wish to gratefully acknowledge the help of Dr. W. G. Lloyd for his direction and assistance relating to any computer work included in this book. Also, I wish to thank Dr. Larry Canter for his encouraging review of the first proposal for this book.

I am indebted to Gary Whittle for the significant contribution he has made in locating the most pertinent literature in the wastewater reclamation and reuse field. I also wish to thank librarians Jean M. Almand and Bettye M. Nichols for their help and direction in locating periodicals, publications, and books that deal with wastewater reclamation and reuse.

I wish to thank Sarah M. Singleton, Manager, and Patti Reagle, Graphic Designer, of Copies Unlimited, Inc., for their skillful work in the preparation of the manuscript.

I wish to express my deep appreciation to James M. Chansler, Director of Public Utilities, Boca Raton, FL, for reviewing this manuscript and making a real contribution with his comments, corrections, and suggested changes as well as providing the information used in the case study, *In-City Reclamation Irrigation System City of Boca Raton, Project IRIS.* It was fortunate that he agreed to evaluate this material.

Lastly, I wish to express my appreciation to Brian Lewis for his understanding and direction in preparing this manuscript.

Donald R. Rowe

I would like to thank Dr. Mohammed Ahmed Osman Ibnouf of the College of Agriculture, Department of Agricultural Economics and Rural Studies, Sultan Qaboos University for reviewing Chapter 7, Economics of Water and Wastewater Reclamation and Reuse Projects. Thanks are to be extended to Dr. Osman Ahmed Hamad of the Chemical Engineering Department, Faculty of Engineering, University of United Arab Emirates, Al-Ain, U.A.E., who contributed to the development of the book. Last, but not least, I would like to acknowledge the assistance, support, patience, and understanding of my family: Layla, Lubna, Hisham, Mohammed, Taghreid, Tasneem, and Ayah.

Isam Mohammed Abdel-Magid

To my daughter, Delaire
and my son, René

Donald Rowe

To my wife, Layla

Isam Mohammed Adel-Magid

Contents

Chapter 1. General Aspects of Wastewater Reclamation and Reuse 1

Chapter 2. Reclaimed Wastewater Quality: Criteria, Standards, and Guidelines 15

Chapter 3. Properties and Characteristics of Water and Wastewater 59

Chapter 4. Health Aspects of Using Reclaimed Water in Engineering Projects 107

Chapter 5. Wastewater Reclamation and Reuse Treatment Technology 165

Chapter 6. Risk Assessment in Wastewater Reclamation and Reuse 261

Chapter 7. Economics of Water and Wastewater Reclamation and Reuse Projects 299

Chapter 8. Reclaimed Wastewater Monitoring Sampling and Analysis 325

Chapter 9. Legal Aspects of Wastewater Reclamation and Reuse 395

Chapter 10. Case Studies 425

Chapter 11. Wastewater Reclamation and Reuse Research 483

Appendix 517

Index 533

Chapter 1

General Aspects of Wastewater Reclamation and Reuse

CONTENTS

    I. Water and Its Properties 1
   II. The Earth's Water Resources 2
  III. U.S. Water Resources — Especially Wastewater Reclamation and Reuse 2
  IV. Other Major Countries Involved in Wastewater Reclamation and Reuse — Especially in Arid Regions 5
      A. Australia 5
      B. Israel 6
      C. Mexico 6
      D. Saudi Arabia 6
      E. South Africa 10
      F. United Arab Emirates (U.A.E.) 10
      G. India 11
   V. Homework Problems 11
References 12

I. WATER AND ITS PROPERTIES

Water constitutes a crucial element for all ecosystems, and its availability has governed the siting of communities and the development and progress of man's economic activities. Water is of indispensable importance for many reasons:[1-14]

- The existence and evolution of plant and animal ecosystems.
- It is an essential constituent of photosynthesis.
- It establishes a means of nutrition for plants, and forms the natural living condition for many living species.
- It acts as a solvent of organic and inorganic materials (universal solvent!).
- It plays a role in metabolism.
- It is a necessary component of body liquids.

It is to be noted that water plays a vital role in the constituents of living organisms. The protoplasm of most living cells contains approximately 80% water and any reduction of this amount can have damaging effects and may be fatal. Likewise, most of the biochemical reactions occurring within the living cell include water and all of these reactions happen in the presence of water. The amount of water contained in the human body represents about 65 to 70% of the body weight, reaching about 90% within the plasma. The amount of body water varies with differences in age, showing a tendency toward reduction with old age (the body weight of a human is made up of 97% water 3 days after birth, 81% 8 days later, and 65 to 75% in old age). Water plays a chief role in the different chemical reactions occurring within the body. The amount of water found within each organ or cell differs — the largest percentage (82 to 94%) is contained within cells of the nervous system and the brain, lesser values (70 to 80%) within internal organs, and the smallest value (22 to 34%) within the bones and fat tissues. It is a well-known fact that the presence of water within the human body governs and regulates the body temperature through perspiration. The latter is affected by atmospheric humidity, work load and duration, and the health conditions of the individual. The role of water also is significant in nutrients and foodstuff. For example, water amounts to about 78 to 97% in vegetables and fruits, while fish contain about 80% water. Meat approaches a water content of 72%. In the case of plants, the needed carbohydrates are produced by making use of carbon dioxide, water, and sunlight. In the presence of needed energy and chlorophyll the following reaction takes place (photosynthesis):

$$6CO_2 + 6H_2O + 688 \text{ kcal} \xrightarrow{\text{(sunlight)}} C_6H_{12}O_6 + 6O_2 \qquad [1.1]$$

Water initiates agricultural and industrial projects:

- It is used for irrigating agricultural sectors, e.g., production of food, crop irrigation.
- It is utilized for the production and generation of energy, e.g., hydropower.
- It enters into the production and manufacturing of industrial products, goods, and commodities.
- It is used for cooling purposes, transfer of heat, quenching of white-hot products, and preparation of baths for different aspects, e.g., electroplating, electrophoresis, etc.
- It is used for scrubbing of gaseous substances.
- It enters into air conditioning and cooling tower operation.
- It is vital for hygienic needs and general amenity aspects.
- It is used in recreational activities, e.g., boating, development of scenic shores and river banks, fishing, swimming, etc.
- It is used for transportation of raw materials via navigational routes by ships, steamers, and the like.
- It is used in disposal of wastes, e.g., sludge whether from domestic, industrial, or agricultural origins.

II. THE EARTH'S WATER RESOURCES

Table 1.1 shows the approximate distribution of water in the hydrosphere, of which only about 0.62% is found in freshwater lakes, rivers, and groundwater supplies available for human consumption.

Water is in a constant state of motion as indicated by its movement through the hydrologic cycle. This signifies the cyclic movement of water evaporated from surface waters or evapotranspirated from plants to the atmosphere. Atmospheric water condenses and precipitates to the earth as rain, snow, or in some other form. On the surface of the earth some of the precipitated water then runs off into streams, lakes, ponds, and the sea. The rest percolates through soil strata to form groundwater aquifers that ultimately flow into surface water bodies.

Such a cyclic order of events does occur, but it is not as simple as that. As outlined by Wilson,[16] the cycle may be short-circuited at several stages; and there is no uniformity in the time a cycle takes. The intensity and frequency of the cycle depend on geography and climate.[16]

III. U.S. WATER RESOURCES — ESPECIALLY WASTEWATER RECLAMATION AND REUSE

As indicated in Table 1.1, only about 5% of the earth's water is classified as freshwater and only about 0.62% is available for human consumption. The source of this fresh water is precipitation, and averages about 71 cm (28 in.) per year over land. In the U.S. the average precipitation is 76 cm (30 in.) per year or about 5,840 cubic kilometers per year (km^3/yr) or 4200 billion gallons per day (bgd). Of this, only 912 km^3/yr or 675 bgd is available for beneficial use.

Figure 1.1 presents the trends in freshwater withdrawals and consumption by use from 1960 to 1985.[17] Freshwater withdrawals decreased from a high of 378 bgd in 1980 to 338 bgd in 1985, and according to an estimate by the U.S. Water Resources Council should decrease to 306 bgd by the year 2000.[18]

The reduction in water withdrawals for 1985 and the estimated reduction in withdrawals by the year 2000 are considered to be due to water quality regulations, water reuse, recycling, and conservation. For

Table 1.1 Distribution of Water in the Hydrosphere

Type of water	% Total water	% Fresh water	% Water available (i.e., fresh and unfrozen)
Total			
Salt	95		
Fresh	5		
Fresh			
Frozen	4	80	
Liquid	1	20	
Fresh liquid			
Groundwater	0.99	19.7	99
Lakes	0.01	0.2	1
Soil	0.002	0.04	0.2
Rivers	0.001	0.02	0.1
Atmospheric	0.001	0.02	0.1
Biological (vegetation cover)	0.0005	0.001	0.005

Adapted from Reference 15.

instance, the manufacturing sector's freshwater withdrawals are estimated to decrease 62% from 1975 to the year 2000. The user-pay principle and pricing of water-related services will also contribute to a reduction in freshwater withdrawals.[17,18]

Figure 1.2 presents the freshwater withdrawals for 1985 for various uses in the U.S. and indicates how much of this water is consumed or returned.[17] By consumption we mean water that is evaporated, transpired, incorporated into vegetation, consumed by people or livestock, and is not available for downstream uses or for groundwater recharge. Irrigation and livestock demands represent the highest consumptive use of water, and in 1985 was responsible for 82% of the total water consumed in the U.S.[17]

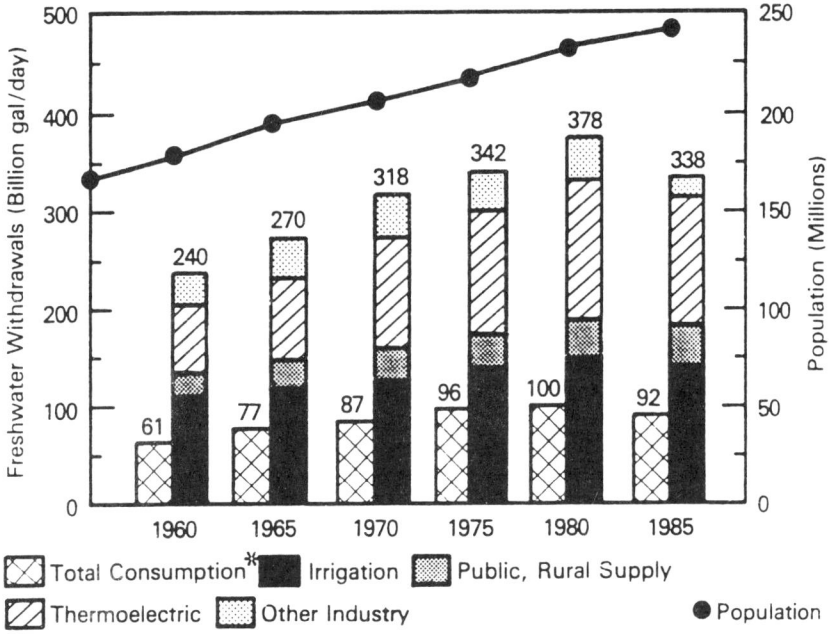

Figure 1.1 Trends in U.S. freshwater withdrawals and consumption by usage — 1960 to 1985.[17] Total consumption is water that is no longer available because it has evaporated, transpired, or has been incorporated into products or crops consumed by humans or livestock, or otherwise removed from the water environment. It is also referred to as water consumed.

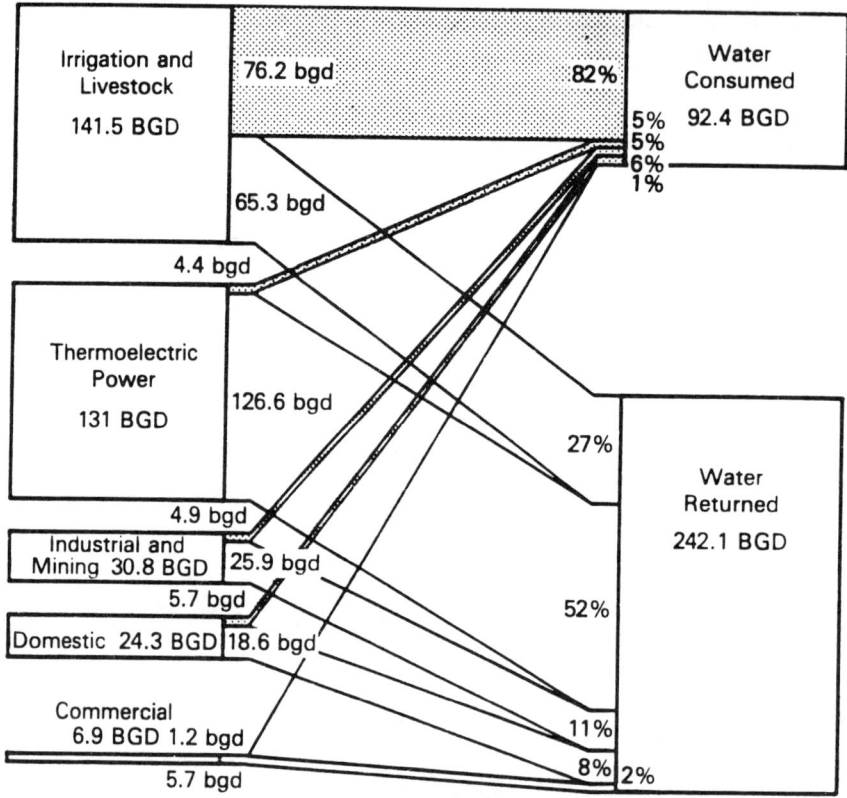

Figure 1.2 Freshwater withdrawals and disposition in 1985. (BDG or bdg = billion gallons per day.)[17]

The two major sources of our freshwater supply are surface waters (lakes, reservoirs, and streams) and groundwater — about 75% comes from our surface waters and 25% from groundwater sources.

As indicated, one reason for the anticipated reduction in freshwater withdrawals by the year 2000 is wastewater reclamation and reuse. Wastewater reclamation, reuse, and recycling will undoubtedly increase.

The most numerous wastewater reclamation projects in the U.S. are in Arizona, California, Colorado, Florida, Georgia, Kansas, South Carolina, and Texas. South Carolina recently reported at least 66 reclamation and reuse projects involving over 57,000 m³/d (15 mgd).[19]

In California, from 1970 to 1987 there has been an overall 35% increase in the volume of wastewater reclaimed and reused, however, even with this increase, only 7 to 8% of the municipal wastewater is reclaimed and put to a beneficial reuse.[20] In 1975 it was reported California had 379 wastewater reclamation and reuse projects involving 727,000 m³/d (192 mgd) while in 1987 there were 854 such projects involving 900,000 m³/d (238 mgd), indicating a 20% increase in volume of reuse from 1975 to 1987.[20,21]

Table 1.2 presents the municipal wastewater reuse in California in 1975 and in 1987 indicating the increase in the number of projects and in the volume of reclaimed water involved.

The last national survey of wastewater reclamation and reuse projects in the U.S. was carried out in 1975 and reported by the Department of the Interior in 1979.[21]

Table 1.3 presents a summary of municipal wastewater reuse projects in the U.S. These 1975 data indicate that irrigation used 62% of the reclaimed water involving 470 irrigation projects. However, the 1,590,000 m³/d (420 mgd) of reclaimed water used for irrigation purposes is only about 0.3% of the total freshwater used for this purpose. This figure should increase in the future in order to help reduce our freshwater withdrawals.

PL 92–500 (the Clean Water Act of 1972) promoted innovative and alternative (I/A) wastewater treatment methods. Incentives for wastewater reuse projects included up to 85% funding, however, by 1984 this had been reduced to 55%, and federal funding for construction projects ended in 1990. The State

Table 1.2 Municipal Wastewater Reuse in California, 1975 and 1987

Category	Number of projects		Reclaimed water (mgd)	
	1975	1987	1975	1987
Agricultural irrigation	232	240	132 (68%)	148 (62%)
Landscape irrigation	104	571	26 (14%)	30 (13%)
Groundwater recharge	5	7	23 (12%)	34 (14%)
Industry	22	12	9 (5%)	6 (2%)
Recreational impoundments	13	4	1 (0.5%)	6 (2%)
Other	3	20	1 (0.5%)	14 (7%)
Total	379	854	192 (100%)	238 (100%)

Compiled from Reference 20 and 21.

Table 1.3 Municipal Wastewater Reuse Projects in the U.S. in 1975

Category	Number of projects	Reclaimed water (mgd)
Irrigation — total	470	420 (62%)
Agriculture	150	199
Landscape	60	33
Not defined	260	188
Industrial — total	29	215 (32%)
Process		66
Cooling		142
Boiler feed		7
Groundwater recharge	11	34 (5%)
Other (recreation, etc.)	26	10 (1%)
Total	536	679

Adapted from Reference 21.

Revolving Loan Fund was phased in and now state agencies are the primary source of funding for municipal wastewater and reclamation projects.

IV. OTHER MAJOR COUNTRIES INVOLVED IN WASTEWATER RECLAMATION AND REUSE — ESPECIALLY IN ARID REGIONS

There is a great deal of interest in the utilization of wastewater in arid regions in the world, especially in such countries as Australia, Israel, Mexico, Saudi Arabia, South Africa, and the United Arab Emirates. There is also a great deal of wastewater being utilized in such countries as India where conservation of their natural resources (water and nutrients) is needed in order to promote agricultural production. A brief account of wastewater reclamation and reuse will be presented here for the above-mentioned countries. However, there are many, many more countries with arid, semiarid, and humid regions that are also involved in wastewater reclamation and reuse. These include Canada, Cyprus, Egypt, France, Germany, Indonesia, Japan, Jordan, Kuwait, Mexico, Poland, Portugal, Syria, and the United Kingdom.

A. AUSTRALIA

The Werribee Farm, 35 km southwest of Melbourne, has been in operation since 1897 and was established as a result of a cholera epidemic in Melbourne. This facility utilizes the treated and untreated wastewater effluent from over 50 municipal facilities.

The mean annual rainfall here is 488 mm (19 in./yr) with an evapotranspiration rate of 1,460 mm/yr (57 in./yr). In 1979 the total area involved was 11,000 hectares (ha) (27,181 acres) (42.5 sections). The average flow of sewage to the Werribee Farm is 440,000 m^3/d (116 mgd) with BOD$_5$ of 570 mg/L and TSS of 620 mg/L. Three methods of treatment are used: land filtration, overland flow (grass filtration), and lagooning. Thus far, lagooning appears to be the safest reclamation technology. The land filtration section of about 4,280 ha (10,576 ac) utilizes mainly raw sewage and provides water for irrigation of pastures. Grazing of cattle and sheep takes place some 10 to 14 days after irrigation; 22,000 cattle and

30,000 sheep are involved. There has been no evidence of unusual veterinary problems. However, careful inspection of slaughtered animals for any infection is practiced. Only about a 0.02% increase of tapeworm *(Taenia saginata)* has been noted and this is easily controlled by cooking and freezing.

About 50 to 60% of the applied water percolates through the soil and is collected in drainage ditches.

Most nutrients and heavy metals are removed during land filtration. The amounts removed include BOD_5 — 99%, SS — 98%, organic C — 94%, N and P — 95%, Zn, Cu, and Pb — 95%, Cr — 90%, Cd and Hg — 85%, and Ni — 75%.[23]

Overland flow is employed during periods of low evaporation when irrigation is not practiced. Lagooning (aerobic and anaerobic) is used the year round during periods of peak daily flows and wet weather.

The health of the farm workers is as good as that in the general community and no epidemics have been reported.[23,24]

Another successful wastewater reclamation and reuse project, also located in the Melbourne area, is the Frankston Vegetable Research Station.

Here, the comparative effect of irrigating with reclaimed water and well water was made. The vegetable crops involved were lettuce, carrots, cabbage, celery, spinach, and tomatoes. A few of the interesting results so far indicate that using the reclaimed water saved 60, 33, and 40% of the N, P, and K fertilizer, respectively, and around 35% in fertilizer costs. No adverse effects of boron or accumulation of Pb, Cd, Cu, Fe, Ni, and Zn were noted.[23]

B. ISRAEL

Israel is a semiarid country with a mean annual rainfall of about 450 mm (18 in./yr). The per capita water consumption in Israel is only 450 m^3/yr (326 gpd) compared with the per capita consumption of 2310 m^3/yr (1672 gpd) in the U.S. In 1984 about 60×10^6 m^3/yr (44 mgd) of treated wastewater effluent was used for agricultural irrigation. The reuse of reclaimed wastewater by 1986 was expected to be 140×10^6 m^3/yr (101 mgd). It was anticipated that the use of reclaimed wastewater will replace about 30% of the total amount of water used today for agricultural irrigation. Only about 1% of the total 250×10^6 m^3/yr (66,050 mil gal/yr) of municipal wastewater in 1984 was used for industrial purposes, and about 7.5% of this quantity for aquifer recharge.[25]

C. MEXICO

The use of sewage water to irrigate land in the Mezquital Valley began in 1886, however, it was not until 1945 that the government established the Number 3 Mezquital Irrigation District in the State of Hidalgo. In 1983 to 1984, 52,175 ha (129,000 ac) were harvested for food crops. The mean annual rainfall for the Irrigation District is 494 mm/yr (19.5 in./yr); the average temperature is 17°C (63°F). The water used for irrigation is a combination of the sewage effluent from Mexico City mixed with rainfall, runoff water, and clean running water from local reservoirs. Most of the 1.125×10^9 m^3 (297,381 mil gal) of irrigation water used in 1983 to 1984 was the sewage water from Mexico City — about 80%. The irrigation water has a turbidity of 104 Nephelometric Turbidity Units (NTUs), a pH of 7.05, total dissolved solids — 1063 mg/L, electrical conductivity of 1470 µmhos/cm, organic carbon — 685 mg/L, and a BOD_5 of 0.66 mg/L. The concentration of the main cations in the water was 120 mg/L Ca^{2+}, 185 mg/L Mg^{2+}, 83 mg/L Na^+, and 274 mg/L K^+, while the main anions concentrations were HCO_3^- 212 mg/L, CO_3^{2-} 56 mg/L (total alkalinity = 268 mg/L as $CaCO_3$), Cl^- 195 mg/L, SO_4^{2-} 116 mg/L, and NO_3^- 258 mg/L.

The concentrations of eight elements were determined in the irrigation water. Table 1.4 presents the limiting tolerable concentrations and the concentrations found in the irrigation water.[26]

The high levels of elements found in the irrigation water and the presence of pathogenic organisms represent a serious problem for the Irrigation District. However, the continuous utilization of this land for production has prevented farmers from abandoning it. The economic advantages, such as an improved standard of living and an increased land value, are felt to exceed the disadvantages, which include health problems.

The methods used in the irrigation of land with sewage effluent has improved over the last 10 years. This, combined with the characteristics of the soil such as good drainage and the calcareous, light-to-medium texture of the soil, has helped to establish a prosperous and stable agricultural region.[26]

D. SAUDI ARABIA

In the Kingdom of Saudi Arabia the policy is to promote wastewater reclamation and reuse. Reclamation and reuse projects are in operation at Riyadh, Madinah, and Qassim, as well as Jubail, Jeddah, Makkah,

Table 1.4 Elements Contained in Irrigation Waters

Element	Limit of tolerated concentrations (mg/L)	Concentration found in irrigation water (mg/L)
As	0.10	0.49
B	0.75	2.83
Cd	0.01	0.03
Cu	0.20	0.35
Cr	0.10	0.21
Fe	5.00	12.91
Mn	0.20	0.32
Mo	0.01	1.44

From Duron, F. S., *Treatment and Use of Sewage Effluent for Irrigation*, Butterworths, London, 1988, 273. With permission.

Table 1.5 Summary of Water Demands for 1990 And 2000 in Saudi Arabia and Estimated Reclaimed Water Available

Category	1990 Mm³/yr	%	2000 Mm³/yr	%
Agriculture	3,684	76	5,119	71
Domestic	1,117	23	1,940	27
Independent industry	74	1	182	2
Total	4,875	100	7,241	100
Wastewater available for reuse	368	8	674	9

Adapted from Reference 27.

Hail, and Dammam. The emphasis on the utilization of reclaimed wastewater is due to the lack of rainfall in this arid region.

The major water consumers in the Kingdom of Saudi Arabia are agriculture, domestic, and industrial users. Table 1.5 presents a summary of water demands for 1990 and the year 2000 and the estimated reclaimed water available.[27]

It was anticipated in the Kingdom of Saudi Arabia Third Development Plan 1980 to 1985, that by 1990 approximately 6% of the water available in the kingdom would come from reclaimed wastewater and that by the year 2000 this would increase to over 11%.[28] The figures presented in Table 1.5 indicate that 8% of the water demand could be met by wastewater reclamation and reuse in 1990, increasing to 9% by the year 2000. This is in fair agreement with the 1980 to 1985 estimates. However this volume of wastewater reuse is still small compared to the water demand for the whole country.

A brief account of wastewater reclamation and reuse will be presented here for Riyadh, Madinah, and the Qassim regions.

The mean annual rainfall in the Riyadh area (capital of Saudi Arabia) is 100 mm/yr (4 in./yr) while the evaporation rate can reach 400 mm/month (16 in./month) in the summer. The monthly average air temperature varies from 14 to 33°C (32 to 91°F) with the mean maximum temperature being 43°C (110°F) in August. The relative humidity is rather low, generally less than 50%.

There are at least 30 wastewater reclamation and reuse projects presently operating in the Riyadh area. Table 1.6a presents a summary of the design criteria for five of these facilities that are producing effluent used for landscape irrigation.[28]

The largest wastewater reclamation and reuse projects in the Riyadh area are at Dirab, Diraiyah, and Ammariyah. Water for these projects comes from the Riyadh Sewage Treatment Plant (Manfouha). The Riyadh Sewage Treatment Plant presently provides full biological treatment and chlorination for 340,000 to 380,000 m³/d (90 to 100 mgd) of wastewater. (See Table 1.6b for typical influent and effluent quality.) Plans have been completed to provide advanced treatment by rapid-gravity sand filtration, however, this was not in operation as of June 1992. Plans have also been executed for the construction of another wastewater facility of 200,000 m³/d (53 mgd) capacity at Manfouha. This plant will be in operation by

Table 1.6a Summary of Design Criteria for Some Major Wastewater Facilities in the Riyadh Area Where Effluent Is Used for Landscape Irrigation

Plant	Population To Be Served	Average Design Capacity (m³/d)	Secondary Treatment	Additional Treatment	Design Water Quality (mg/L)		
						Influent	Effluent
King Saud University Wastewater Treatment Facility	45,000	10,000	Trickling filters	Chlorination	BOD_5	347	24
					SS	408	41
Ministry of Foreign Affairs Wastewater Treatment Facility	4,500	1,135	Rotating biological contactors (3-stage system)	Pressure filtration Chlorination	BOD_5 SS	306 306	10 15
The Riyadh Diplomatic Quarter Wastewater Treatment Facility	31,299	9,270	Extended aeration	Coagulation Flocculation Sedimentation Filtration Chlorination	BOD_5 SS	200 250	10 5
Imam University Wastewater Treatment Facility (constructed but not operational, 1988)	32,000	4,800	Conventional activated sludge	Gravity filtration Pressure filtration Activated carbon adsorption Reverse osmosis	BOD_5 SS	350[a] 400[a]	≤10 ≤10
National Guard Khashm Al-Aan Wastewater Treatment Facility	50,000	10,000	Aerated rotating biological contactors (4 stages)	Gravity Rapid Sand filtration Chlorination	BOD_5 SS	300 300	≤10 ≤10

[a] Estimate

From Rowe, D. R., Al-Dhowalia, K. H., and Whitehead, A., *Reuse of Riyadh Treated Wastewater,* King Saud University, Report. No. 18/1402, November. 1987. With permission.

October 1992. Again, full biological treatment is provided, however, incorporation of a nitrification-denitrification process is included. The advanced treatment consists of rapid sand filtration and chlorination. The present tentative maximum contaminant levels for unrestricted irrigation is BOD_5 10 mg/L, TSS 10 mg/L, and the average fecal coliform standard is set at an MPN of 2.2/100 mL as determined over a 7-day period.

The new plant will meet these standards; however, the BOD_5 for the old plant averages 45 mg/L and the suspended solids 40 mg/L, with the total coliform being 50 to 100 colonies per 100 mL.

The largest portion of the Riyadh treated wastewater goes to Dirab, Dariyah, and Ammariyah for agricultural irrigation. The transmission line to Dirab has a capacity of 120,000 m³/d (32 mgd) with an on-line storage of 300,000 m³ (79 mil gal). The transmission line to Dariyah has a capacity of 80,000 m³/d (21 mgd) with an on-line storage capacity of 200,000 m³ (53 mil gal). The transmission line to the Dirab irrigation site is 55 km (34 mi) in length and 1,000 mm (39 in.) in diameter. The transmission line to Dariyah and Ergah is 50 km (31 miles) in length and 800 mm (31 in.) in diameter. A 32-km (20-mi) transmission line ranging from 500 to 1000 mm (20 to 39 in.) in diameter to the Ammariyah area is now in operation. Here, 61 farms will be irrigated, involving around 4000 ha (9900 ac) and utilizing 60,000 m³/d (21 mgd) of treated wastewater. In the past the farmers at Dirab depended on groundwater for irrigation. However, since 1982 Riyadh treated wastewater has been available to them for use. The volume of treated wastewater being pumped to Dirab has been rapidly increasing and the number of farms using this water has almost doubled since 1982. The maximum amount of treated wastewater pumped to Dirab in 1985 was 92,000 m³/d (24 mgd). In 1981 there was an estimated 850 ha (2100 ac) under cultivation, and by 1985 there was 2000 ha (4940 ac) utilizing the treated wastewater for irrigation. The farms at Dirab are large, generally over 65 ha (161 ac) in size, and are therefore suitable for mechanized

Table 1.6b Typical Composition of Influent and Effluent for the Riyadh Wastewater Treatment Facility (All in mg/L Except for Settleable Solids, pH, and Total Coliform)

Constituent	Concentration	
	Influent	Effluent
Total dissolved solids	1300	1200
Suspended solids	250	40
Settleable solids (mL/L)	3.0	ND
BOD_5, 20°C	250	45
COD	450	100
Ammonia - nitrogen	25	25
Nitrates as Nitrogen	—	≤1
Phosphates	10	7
Chlorides	180	160
Alkalinity	240	200
Grease	100	10
Temperature (°C)	29	27
Free available chlorine	0	0.8
Total chlorine residual	0	≥4.0
pH (standard units)	7.3	7.4
Dissolved oxygen	0	5
ABS[a]	12–20	≤5
Total coliform	millions/ml	50–100/100 ml

[a] Alkyl benzene sulfonates.

From Rowe, D. R., Al-Dhowalia, K. H., and Whitehead, A., *Reuse of Riyadh Treated Wastewater,* King Saud University, Report. No. 18/1402, November. 1987. With permission.

cultivation. The main crops are wheat, fodder, and vegetables. Due to the lack of water in the past, less than half the land available for cultivation was under crop each year.

The pumping of treated wastewater to Dariyah for agricultural irrigation also started in 1982. The maximum amount of treated water pumped to Dariyah in 1985 was 70,400 m³/d (19 mgd). The volume of treated wastewater being pumped to Dariyah and Ergah has increased rapidly and is approaching the transmission lines capacity of 80,000 m³/d (21 mgd). The farms at Dariyah and Ergah are much smaller than at Dirab and average around 15 ha (37 ac) in area. Dariyah and Ergah grow mainly date palms, fruit trees, vegetables, and fodder. In 1981 the estimated area under cultivation at Dariyah was 950 ha (2350 ac) that was irrigated with groundwater. As of 1987 an additional 850 ha (2100 ac) were being irrigated with Riyadh treated wastewater.

An evaluation of Riyadh treated wastewater and the well water used for irrigation at Dirab and Dariyah indicated that, in general, the physical and chemical characteristics of the Riyadh treated wastewater is as good or better than the groundwater, except for turbidity and suspended solids. While the NH_3-N and $PO_4^{\equiv}$ concentrations found in the treated wastewater were well above those found in the groundwater these constituents are essential plant nutrients and contribute to plant growth, and thus are beneficial when the water is to be used for agricultural irrigation. The NO_3-N concentration in the treated wastewater and groundwater is in much the same range and is well below the tentative standard of 10 mg/L.[28,29]

The concentration of boron in the treated wastewater was below that found in the groundwater. The treated wastewater was at or slightly below the tentative standard of 0.7 mg/L.

Of the major trace metals monitored, the concentrations of the metals of greatest concern in the treated wastewater were Cd, Cr, Pb, and Zn, and they were at or below the concentrations found in the groundwater. Only the copper concentration in the treated wastewater tended to be above that found in the groundwater. The Cd, Cr, and Pb concentrations were found in a few samples for both the treated wastewater and groundwater to be at or slightly above the tentative standards. The average Cd, Cr, Cu, Pb, Ni, and Zn concentrations in the treated Riyadh wastewater were found to be at or below the tentative Saudi Arabian Standard for unrestricted agricultural irrigation. The Cd, Cr, Ni, and Zn concentrations in the Dirab soil were in the normal range found throughout the world, while the Cu concentration was lower than normal and the Pb concentration slightly higher.

Grain-size analysis of the Dirab and Dariyah soil indicated it was made up of from 80 to 90% of a uniformly sized sand, the balance being silt with probably little or no clay present.

The Cd, Cr, and Ni concentrations in the plants sampled at Dirab were in the normal range found in plants, however the Pb concentrations in the plants averaged slightly above the normal range. The average Cu and Zn concentrations in the plants sampled at Dirab indicated they had a deficiency of these two trace metals.[28]

The overall concentration factors relating trace metal levels in plants to the levels found in the soil ranged from 0.01 to 0.15. The concentration factors for Cr and Zn indicated they were not as readily taken up by the plants as were Cd, Pb, Ni, and Cu. Cd and Pb gave the highest concentration factors. The concentration factors for the plants and soil at Dirab had the following selective order:

$$Cd \geq Pb \geq Ni \geq Cu \geq Cr = Zn$$

The trace metal sorption by Dirab and Dariyah soils indicated that it would take from 18 to 72 years to saturate the top centimeter of soil with Cd, 39 to 124 years for Zn, and 37 to 78 years for Pb, provided the treated wastewater was applied at the rate of 1 m/m^2/yr and at the trace metal concentration presently found in the Riyadh wastewater.[28]

The General Petroleum and Minerals Organization (Petromin) utilizes 20,000 m^3/d of Riyadh treated wastewater for industrial purposes. The Petromin water reclamation plant produces three grades of water:

1. Utility water for hose stations and fire fighting
2. Process water for crude oil desalting and cooling tower use
3. Boiler feedwater

The treatment of the effluent from the Riyadh Sewage Treatment Plant consists of three stages. In the first stage the incoming flow is equalized in a surge pond before the lime treatment. The second stage involves the use of cooling towers and dual media filtration. The third stage consists of carbon adsorption, reverse osmosis, and ion exchange.

Table 1.7 presents the characteristics of the treated wastewater after primary reverse osmosis and after passage through the ion exchange system.[29]

Another project planned for Riyadh treated wastewater is at Hayer, which is close to Riyadh. A 6-km, (3.7-mi) 450-mm (18-in.) diameter line is to be constructed at this site to irrigate 1200 ha (3000 ac).[28]

In 1987 the wastewater from the Madinah treatment plant was about 27,000 m^3/d (7 mgd), this flow was estimated to increase to 54,000 m^3/d (14 mgd) by 1990 and 141,000 m^3/d (37 mgd) by the year 2000. This treated wastewater is to be used to irrigate 3100 ha (7660 ac).[27]

In the Qassim area, the total quantity of wastewater from the five proposed municipal plants will be 98,700 m^3/d (26 mgd) and will be utilized to irrigate 823 ha (2034 ac) of land.[27]

E. SOUTH AFRICA

The country's average rainfall is about 487 mm (19 in./yr), of which 91% is lost to the atmosphere by evaporation and transpiration and only 9% reaches the rivers. In the early 1970s the wastewater flows from over 20 major cities indicated that 1,230,000 m^3/d of treated wastewater was available for reclamation and reuse. Of this flow, 31.9% was reclaimed and used — 16.1% for irrigation of crops, parks, trees, and sports fields, 8.7% for power station cooling water, and 7.1% for industrial purposes. Windhock Namibia has a full-scale plant where potable water is reclaimed from municipal wastewater. The proportion of reclaimed water in the municipal system generally ranges from 20 to 50%.[30]

F. UNITED ARAB EMIRATES (U.A.E.)

One very successful wastewater reclamation and reuse project in the U.A.E. is at Abu Dhabi. Abu Dhabi is situated in a desert environment typified by hot summers 38 to 50°C (100 to 122°F) from May to October. The winters are mild 20 to 35°C (68 to 95°F) and extend from December to March. The humidity in the summer can reach 90%. Rainfall in Abu Dhabi rarely exceeds 45 mm/yr (2 in./yr) and usually falls in the winter.

The Mafrag Sewage Treatment Works operated by the Abu Dhabi Municipality treats 115,000 m^3 (30 mgd) of wastewater daily. The plant provides full biological treatment followed by chlorination and rapid gravity sand filtration. The plant was designed to achieve

5-Day BOD	≤ 10 mg/L
Suspended solids (SS)	≤ 10 mg/L
Coliform/100 mL	≤ 100 in 80% samples

Table 1.7 Water Quality After Reverse Osmosis and After Ion Exchange

Parameter	Concentration
Average Water Quality from Primary Reverse Osmosis System (All in mg/L except pH which is in pH units)	
pH	5
NH_3-N	7–10
SiO_2	5
COD	3
TDS	200–300
Total hardness as $CaCO_3$	50
Total alkalinity as $CaCO_3$	5
SO_4	45–60
Average Water Quality from Ion Exchange System	
NH_3-N	0
SiO_2	0.5
COD	0
TDS	10
Total hardness as $CaCO_3$	0

Al-Dhowalia, K. H., Rowe, D. R., and Whitehead, A., *Proc. Third Internat. Symp. Water Reuse,* 1984, 1137. With permission.

The present effluent has an average BOD_5 of 1.3 mg/L, SS of 1.5 mg/L, and a MPN for coliform organisms of 10/100 mL.

The treated effluent is returned to Abu Dhabi along a 32-km (20-mi) pipeline for distribution in the city; about 85% is used for landscape irrigation and the other 15% is used at the camel tracks. The treated wastewater has greatly improved the expansion of greenery in this capital.[31]

G. INDIA (GENERALLY HUMID CLIMATE)

Land disposal of wastewater at some sites has been practiced in India for up to 160 years. Sewage farming as a method of wastewater disposal is reported to take place at more than 200 sites. About 10×10^6 m^3 (2642 mgd) of wastewater flow is available for this purpose. The percentage of wastewater given primary treatment is 37%, with 8% of the wastewater flow given secondary treatment. A recent survey conducted by the National Environmental Engineering Research Institute (NEERI) in Nagpur, India showed that many of the environmental and public health problems at these wastewater utilization sites could be attributed to over-application of the untreated sewage. Hookworm, roundworm, whipworm, pinworm, dwarf tape worm, *Entamoeba histolytica,* and *Giardia intestinalis* occurred from 15 to 60% more often in the farm workers than in a control group.

Experimental results show that primary treated and secondary treated wastewaters are superior to untreated sewage in terms of crop yields and soil nutrient utilization efficiency and would, of course, reduce the health problems associated with utilizing raw sewage. Tentative guidelines with regard to using secondary treated and disinfected wastewater indicate it is suitable for all crops without restrictions.[32]

V. HOMEWORK PROBLEMS

1. Define the following terms used in the preface of this book.

    Wastewater reuse
    Wastewater reclamation
    Wastewater recycling
    Direct reuse
    Indirect reuse

2. What is the key to life?
3. Indicate the beneficial ways that water is used by man.

4. What percentage of the world's water is "fresh" or available for human consumption?
5. What percentage of the human body is made up of water?
6. What percentage of the freshwater in the United States comes from surface waters (lakes, reservoirs, rivers, and streams) and what percentage from groundwater?
7. Who are the four principal users of reclaimed wastewater?
8. Indicate three major water quality categories that must be considered in the use of reclaimed treated wastewater and give a prime example in each category.
9. What are some of the major water quality factors associated with wastewater reclamation and reuse?
10. Indicate at least five beneficial uses that can be made of reclaimed wastewater?
11. What three major nutrients present in reclaimed wastewater contribute to plant growth?
12. Public health professionals are concerned over the presence of toxic chemicals such as trace metals present in our water. Which trace metals have been given the most attention in the wastewater and reclamation field?
13. Identify and define the following acronyms or terms that are used in Chapter 1.

    BOD_5
    COD
    TDS
    SS
    NTU
    µmhos/cm

    (If the definitions cannot be located in Chapter 1, they are however presented in following sections of this book.)

14. Prepare a glossary of the following terms used in Chapter 1.

    Photosynthesis
    Metabolism
    Chlorophyll
    Hydrologic cycle
    Evapotranspiration
    Water consumed
    Aerobic
    Anaerobic
    Alkalinity
    Electrical conductivity
    Pathogenic
    Trickling filter
    Rotating biological contactor
    Coagulation
    Flocculation
    Filtration
    Chlorination
    Extended aeration
    Activated carbon adsorption
    Reverse osmosis
    Total hardness as $CaCO_3$

    (If definitions for these terms cannot be located in Chapter 1, they are however presented in following sections of this book.)

REFERENCES

1. **Abdel-Magid, I. M. and B. M. ElHassan,** *Water Supply in the Sudan,* Khartoum University Press, Sudan National Council for Research, Khartoum, 1986 (Arabic).
2. **Abdel-Magid, I. M.,** *Water Treatment and Sanitary Engineering,* Khartoum University Press, Khartoum, 1986 (Arabic).
3. **Barnes, D., et al.,** *Water and Wastewater Engineering Systems,* Pitman International, Bath, U.K., 1981.

4. Camp, T. R., *Water and its Impurities,* Reinhold, New York, 1973.
5. Hammer, M. J., *Water and Wastewater Technology,* 2nd ed., John Wiley and Sons. New York, 1986.
6. Husain, S. K., *Textbook of Water Supply and Sanitary Engineering,* 2nd ed., Oxford and IBH Publications, New Delhi, India, 1981.
7. Lorch, W., Ed. *Handbook of Water Purification,* 2nd ed., McGraw-Hill, London, 1981.
8. Merritt, F. S., *Standard Handbook for Civil Engineers,* McGraw-Hill, New York, 1976.
9. Peavy, H. S., D. R. Rowe, and G. Tchobanoglous, *Environmental Engineering,* McGraw-Hill, New York, 1985.
10. Salvato, J. A., *Environmental Engineering and Sanitation,* 4th ed., Wiley-Interscience, New York, 1992.
11. McGhee, T. J., and E. W. Steel, *Water Supply and Sewerage,* 6th ed., McGraw-Hill, New York, 1991.
12. Tebbutt, T. H. Y., *Principles of Water Quality Control,* Pergamon Press, New York, 1993.
13. Vesilind, P. A., J. J. Peirce, and R. Weiner, *Environmental Engineering,* 2nd ed., Butterworth-Heinemann, Boston, 1988.
14. Walker, R., *Water Supply, Treatment and Distribution,* Prentice-Hall, Englewood Cliffs, NJ, 1978.
15. UNESCO, FAO Working Group on the International Hydrological Decade, Man's influence on the hydrological cycle, FAO, Irrigation and Drainage Paper, Special Issue 17, Food and Agriculture Organization, Rome, 1973.
16. Wilson, E. M., *Engineering Hydrology,* 3rd ed., MacMillan Education, Hong Kong, 1987.
17. Council on Environmental Quality, The 18th Annual Report of the Council on Environmental Quality, U.S. Government Printing Office, Washington, D.C., 1987–88, 116, 117.
18. U.S. Water Resources Council, The Nation's Water Resources 1975–2000, Vol. 1, Summary, Second National Water Assessment, U.S. Government Printing Office, Washington, D.C., 1978, 29.
19. Caughman, G. M., Personal Communication, Domestic Wastewater Division, Bureau of Water Pollution Control, Columbia, SC, March, 1992.
20. California Municipal Wastewater Reclamation in 1987, California State Water Resources Control Board, Sacramento, CA, 1990, 3, 19, 20.
21. Culp/Wesner/Culp, Water Reuse and Recycling, Evaluation of Needs and Potential, OWRT/RU/79, Vols. 1 and 2, Office of Water Research and Technology, Department of the Interior, Washington, D.C., 1979, 44–47.
22. Asano, T., and R. S. Madancy, *Water Reuse,* E. J. Middlebrooks, Ed., Ann Arbor Science Publishers, Ann Arbor, MI, 1982, 282.
23. Feigin, A., I. Ravina, and J. Shalhevet, *Irrigation with Treated Sewage Effluent,* Springer-Verlag, New York, 1991, chap. 4.
24. Messer, J., *Water Reuse,* Middlebrooks, E. J., Ed., Ann Arbor Science Publishers, Ann Arbor, MI, 1982, 549–576.
25. Zoller, V., Detergents in Treated Municipal Wastewaters Reused for Irrigation, Industrial Recycling, and Aquifer Recharge, in *Proc. Water Reuse Symposium III,* AWWA Research Foundation, Denver, CO, 1984, 1163.
26. Duron, N. S., Mexican Experience in Using Sewage Effluent for Large Scale Irrigation, M. B. Pescod and A. Arar, Editors, *Treatment and Use of Sewage Effluent for Irrigation,* Butterworths, London, 1988, 249–257.
27. Kalthem, M. S., and Ahmed M. Jamaan, Plans for reuse of wastewater effluent in agriculture and Industry in the Kingdom of Saudi Arabia, *Treatment and Usage of Sewage Effluent for Irrigation,* M. B. Pescod and A. Arar, Eds., chap. 22, Butterworths, London, 1988, 273–283.
28. Rowe, D. R., K. H. Al-Dhowalia, and A. Whitehead, *Reuse of Riyadh Treated Wastewater,* College of Engineering, Research Center, King Saud University, Final Project Report No. 18/1402, November, 1987.
29. Al-Dhowalia, K. H., D. R. Rowe, and A. Whitehead, Water Reuse for Riyadh, *Proc. Third Internat. Symp. Water Reuse,* American Water Works Association, San Diego, California, August, 1984, 1137–1143.
30. Hart, O. H., and L. R. J. van Vuuren, Water Reuse in South Africa, in *Water Renovation and Reuse,* H. Shuval, Ed., Academic Press, New York, 1977, 355–394.
31. Report, Government of Abu Dhabi, Abu Dhabi Municipality, Mafrag Sewage Treatment Plant, 1991.
32. Shende, G. B., et al., Status of wastewater treatment and agricultural reuse with special reference to Indian Experience and research and development needs, in *Proc. FOA Regional Seminar,* M. B. Pescod and A. Arar, Eds, Butterworths, London, 1988, 185–209.

Chapter 2

Reclaimed Wastewater Quality Criteria, Standards, and Guidelines

CONTENTS

I. Introduction 16
II. Criteria 17
III. Standards 18
IV. Guidelines 20
V. Water Quality Requirements For Wastewater Reclamation and Reuse 20
    A. Reclaimed Wastewater Reuse For Agricultural Irrigation 23
        1. Salinity 23
        2. Exchangeable Cations (Sodium, Calcium, and Magnesium) 25
        3. Boron in Reclaimed Wastewater 27
        4. Trace Metals or Elements in Reclaimed Wastewater 29
    B. Reclaimed Wastewater For Groundwater Recharge 31
    C. Reclaimed Wastewater For Potable Use 33
    D. Reclaimed Wastewater For Industrial Use 35
    E. Reclaimed Wastewater For Recreational Use 36
        1. Elementary Body-Contact Recreational Water 36
        2. Secondary Body-Contact Recreational Water 37
        3. Non-Contact Recreational Water 37
    F. Reclaimed Wastewater For Use in Aquatic Environments 38
    G. Reclaimed Wastewater Used For Livestock and Wildlife Purposes 38
VI. World Health Organization (WHO) Guidelines For Drinking Water Quality 39
VII. Sultanate of Oman Drinking Water Quality Standards 41
VIII. World Health Organization (WHO) Criteria For Wastewater Reuse 50
IX. Developing Countries: Engelberg Report/Guidelines For Wastewater Reuse 51
X. Sultanate of Oman Wastewater Reuse Requirements 52
XI. Homework Problems 54
    A. Discussion Questions 54
    B. Specific Mathematical Problems 55
    C. Suggested Exercises 56
References 56

I. INTRODUCTION

Two major issues of concern for users of reclaimed wastewater are the quantity and quality of this water. The concern regarding a reliable quantity of reclaimed wastewater deals with how dependable the source is for operations, whether they be domestic, commercial, or industrial. The quantity or volume of reclaimed wastewater available must be assured otherwise the user will probably not participate in any wastewater reclamation or reuse project.

The other major concern deals with the quality of the reclaimed wastewater, which in turn depends upon several parameters such as the quality of the original water supply, the sources of the wastewater (domestic, industrial, or commercial — generally, point sources) or storm water runoff, land drainage (non-point sources), infiltration into the collection system, as well as the treatment processes to which the reclaimed water has been subjected.

The major concerns regarding the quality of the reclaimed wastewater include its physical, chemical, biological, and radiological characteristics. These concerns therefore necessitate the formulation of criteria, standards, and guidelines that are appropriate for the users or the consumers of this water.

A first step in developing water quality standards is water quality criteria. Water quality criteria tells us what science thus far has been able to measure of the obvious as well as the insidious effects of contaminants on man and his environment. Criteria documents take into account all of man's previous experience in evaluating an environmental hazard and provide information in making judgments to develop water quality standards. Criteria do not have a legal basis nor do they imply an ideal condition. Water quality criteria reflect the latest scientific knowledge useful in indicating the kind and extent of all identifiable effects on health and welfare which may be expected from the presence of specific pollutants in the water.[1,2]

As indicated, criteria are used as a basis in establishing water quality standards. While criteria are without a legal basis, water quality standards do have a legal basis and spell out specific numbers that can be used by an established authority for administrative action and enforcement. In the U.S. the EPA uses two methods to develop regulatory standards for various pollutants or contaminants. The EPA either establishes a maximum contaminant level (MCL) or, if this is not economically or technically feasible, then a treatment method will be specified for the removal of that contaminant from a water supply.

The maximum contaminant level (MCL) is defined as the maximum permissible level of a contaminant in water at the free-flowing outlet of the ultimate user of a public water system, except in the case of turbidity, where the maximum permissible level is measured at the point of entry to the distribution system.[3] The MCL for a drinking water standard requires a balance between public health benefits and what is technologically and economically feasible.

The development of water quality standards or MCLs involves an intensive technological evaluation that includes assessments of:[1]

- Occurrence in the environment
- Human exposure in specific and general populations
- Adverse health effects
- Risks to the population
- Methods of detection
- Chemical transformations of the contaminant in drinking water
- Treatment technologies and costs

However, before MCLs are set maximum contaminant level goals (MCLGs) are developed. EPA publishes MCLGs for each contaminant, which in the judgment of the EPA administrator may have adverse effects on public health and is known or anticipated to occur in public water systems. The MCLG is nonenforceable and is set at a level at which no known or anticipated adverse health effects in humans occur and which allows for an adequate margin of safety. Factors considered in setting an MCLG include health effects data and sources of exposure other than drinking water.[1,4]

Also, in the U.S., administrative agencies can issue guidelines and can do so without the usual public notice or comment, however, such guidelines are not law and therefore not directly enforceable. At the international level the World Health Organization (WHO) has developed *Guidelines For Drinking-Water Quality* Volumes 1, 2, and 3.[5] These guidelines are intended to help countries develop drinking water standards that ensure an aesthetically pleasing water and do not result in any significant risk to the health of the consumer. These WHO guidelines are used throughout the world and provide all nations with the basic information necessary to set their own water quality standards.[5]

Table 2.1 Contaminants Scheduled for Regulation by 1989 Under the 1986 Amendments to the Safe Drinking Water Act

By 1987 (9 contaminants)		
VOLATILE ORGANIC CHEMICALS		
Benzene	1,2-Dichloroethane	Trichloroethylene
Carbon tetrachloride	1,1-Dichloroethylene	Vinyl chloride
p-Dichlorobenzene	1,1,1-Trichloroethane	
INORGANICS		
Fluoride		
By 1988 (40 contaminants)		
VOLATILE ORGANIC CHEMICALS		
Chlorobenzene	*cis*-1,2-Dichloroethylene	*trans*-1,2-Dichloroethylene
o-Dichlorobenzene		
MICROBIALS AND TURBIDITY		
Giardia	Turbidity	Viruses
Total coliforms		
INORGANICS		
Arsenic	Chromium	Nitrate
Asbestos	Copper	Selenium
Barium	Lead	
Cadmium	Mercury	
ORGANICS		
Acrylamide	1,2-Dichloropropane	Methoxychlor
Alachlor	Epichlorohydrin	PCBs
Aldicarb	Ethylbenzene	Pentachlorophenol
Carbofuran	Ethylene dibromide	Styrene
Chlordane	(EDB)	Toluene
2-4-D	Heptachlor	Toxaphene
Dibromochloropropane	Heptachlor epoxide	2,4,5-TP
(DBCP)	Lindane	Xylene
By 1989 (34 contaminants)		
VOLATILE ORGANIC CHEMICALS		
Methylene chloride	Trichlorobenzene	
(Dichloromethane)		
MICROBIALS AND TURBIDITY		
Legionella	Standard plate count	
INORGANICS		
Antimony	Cyanide	Sulfate
Beryllium	Nickel	Thallium
ORGANICS		
Acrylonitrile	Endothall	Phthalates
Adipates	Endrin	Picloram
Atrazine	Glyphosate	Simazine
Dalapon	Hexachlorocyclo-	2,3,7,8-TCDD (Dioxin)
Dinoseb	pentadiene	1,1,2-Trichloroethane
Diquat	PAHs	Vydate (Oxamyl)
RADIONUCLIDES		
Radium-226	Uranium	Photon radioactivity
Radium-228	Gross α particle activity	
Radon	β Particle activity	

From Calabrese, E. J., Gilbert, C. E., and Pastides, H., *Safe Drinking Water Act, Amendments, Regulations and Standards,* Lewis Publishers, Chelsea, MI, 6, 23, 145, 189.

II. CRITERIA

The United States Drinking Water Criteria Documents (CDs) are formulated for each contaminant before a standard is established. The first step after preparing the CDs is the establishment of MCLGs based on data from such areas as pharmacokinetics research, human exposure studies, acute and chronic toxicity data for both animals and humans, epidemiological reports, as well as dose-response curves for each contaminant.[1]

The 1986 Amendments to the United States Safe Drinking Water Act (SDWA) established an agenda for the regulation of 83 contaminants over a three-year period. The contaminants that were scheduled for regulation by 1987, 1988, and 1989 are summarized in Table 2.1.[1]

The Criteria Documents for each of these contaminants consists or will consist of nine chapters:[1]

- Summary
- Physical and chemical properties
- Toxicokinetics
- Human exposure
- Health effects in animals
- Health effects in humans
- Mechanism of toxicity
- Quantification of toxicological effects (QTE)
- References

The QTE chapter essentially draws together the key health effects for each contaminant and provides the basis for the establishment of the proposed MCLG. The risk assessment/analysis evaluation is performed in this chapter and consists of separate evaluations for both noncarcinogenic and carcinogenic substances. Details as to the procedures for risk assessment and analysis are presented in Chapter 6 of this book.

III. STANDARDS

As designated in Section II, CDs are first formulated in order to establish MCLGs and MCLs. However, on a broad international basis the adoption and implementation of water quality standards also need to be preceded by such activities as:[6,7]

- Surveying and evaluating the current status of the water resources in a particular area or region.
- Registration of the different pollution-producing sectors with a nationally formed body or agency.
- Identification of pollution producing bodies and characterization of pollution streams emanating from each contributor. This is followed by portrayal of pollution streams issuing from each sector.
- Information regarding adoption of appropriate pretreatment or treatment methodologies.
- Locating well-equipped national and regional reference laboratories.
- Training of personnel to fill technical positions.
- Escalating and expanding the community education procedures and programs.
- Estimating origin of pollutants that are likely to invade the human body through the three major routes; namely: water, food, and air.
- Coordinating the different governmental ministries, municipalities, institutions, and organizations. Figure 2.1 outlines a suggested system of coordination that may be followed for optimization and better utilization of resources. A specific authority may be selected to act as a focal point at the national and international levels. One of the major responsibilities of that authority is the coordination between various agencies and institutions.

After the preliminary studies and activities are completed and the development of criteria documents has been carried out, then water quality standards can be promulgated.

Water quality standards may be defined as the limits on the amount of physical, chemical, or microbiological impurities allowed in water that is intended for a particular use. These standards are legally enforceable by governmental agencies, and include rules and regulations for sampling, testing, and reporting procedures.[8]

Water quality standards are necessary and important for various reasons and include:[9-11]

- Protecting public health and welfare.
- Determining the type and degree of treatment required for the removal of the various contaminants.
- Establishing goals for the design of water treatment systems.
- Helping to define and determine if a water quality problem exists.
- Encouraging the measurement and evaluation of the contaminants that may be present in a water supply.
- Preserving fairness in the application of the law.
- Providing a basis for decision making.

While the necessity for water quality standards are indeed important they also lead to other important activities such as:[5]

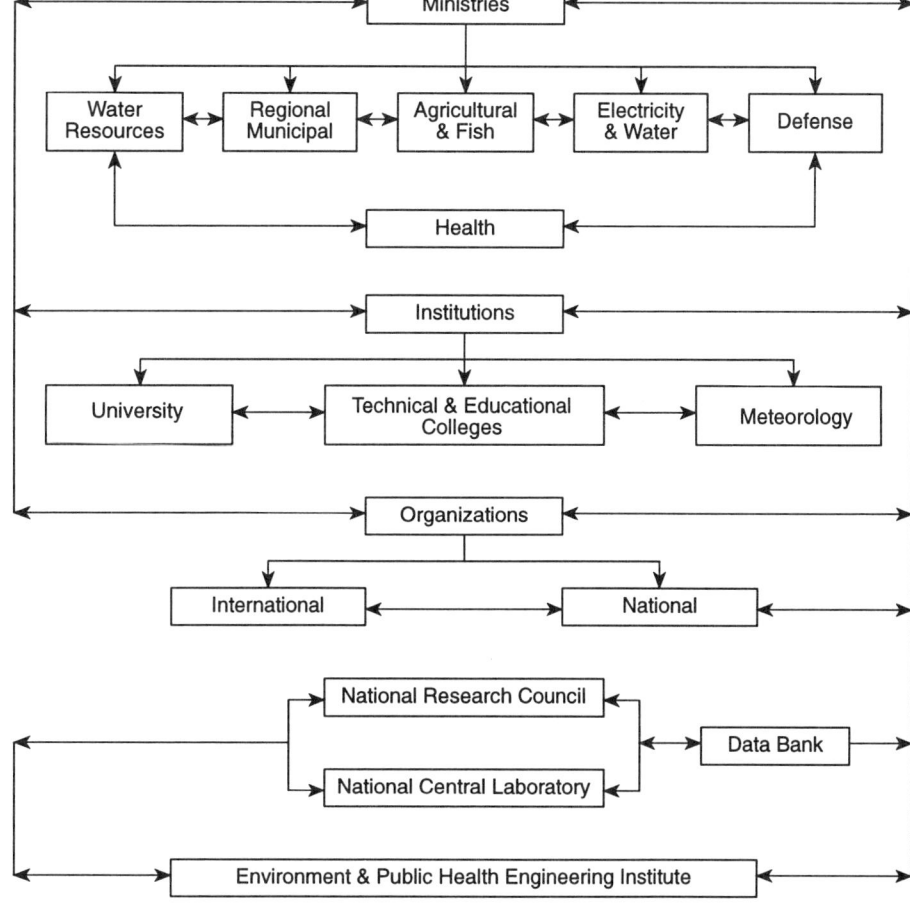

Figure 2.1 An example of a coordinated system for establishing water quality standards.

- Approval of new water supply sources (including privately owned supplies).
- Watershed protection.
- Approval of the construction and operating procedures of water works including:
  - Disinfection of the water and of the distribution system after repair or interruption of supply,
  - Periodic flushing programs and cleaning of water storage facilities,
  - Certification of operators,
  - Regulation of chemical substances used in water treatment,
  - Cross-connection control, back-flow prevention, and leak detection control programs.
- Sanitary surveys.
- Monitoring programs, including provision for central and regional analytical laboratory services.
- Development of codes of practice for well construction, pump installation, and plumbing.
- Inspection and quality control in bottled water and ice manufacturing operations.

In order to implement, evaluate, and follow up on the accepted standards proper surveillance procedures have to be implemented. Therefore, for proper surveillance of the water quality certain components must be present such as:

- An efficient transportation system,
- Optimum required number of trained staff and personnel,
- Continuous availability of laboratory supplies and equipment, and
- Proper supervision.

Water quality standards demand the existence of sufficient and acceptable legislation which can be defended by regulatory codes. That is, to define the water quality that ought to be distributed and employed by the user. The exact nature of the required legislation is dictated by national, constitutional, and other considerations, but some features commonly incorporated in such legislation have been outlined by WHO and include:[5]

- Specification of the scope of authority
- Delegation of powers to administer the law to a specified agency or agencies
- Provision for the establishment and amendment of regulations for the development, production, maintenance, and distribution of safe drinking water; and
- Provision of enforcement.

The surveillance process signifies a policing function on behalf of the public to oversee operations and to ensure the reliability and safety of the water. It is recommended that this process be conducted by a separate agency.[5]

Usually, the standards differ greatly in the way they are presented. Part of the standards may be incorporated in a legislative format, while another part may be furnished under an enabling legislation pattern. The value of this type of legislation depends on:

- The practical method used by the administrator to implement the program,
- The methods used to impose established punishments and penalties, and
- The responsible organization for the prosecution of the standards.

The introduction of appropriate methods and approaches for securing the compliance with discharge standards may be of little value if the pollutant content of the effluent is not regularly monitored.

Figure 2.2 shows the relationship of different water quality standards that can be used for water and reclaimed water flows.

IV. GUIDELINES

It is difficult and probably almost impossible on a global or international basis to develop water quality standards for a large, diverse region or country by establishing just one water quality standard. Therefore, the adoption of guidelines rather than standards is a more realistic approach. The employed guidelines need to be translated into local standards in different regions whenever applicable. This can be done in such a way that flexibility is attained together with variations in social and economic conditions. The register of the formulated standards and the allowable concentrations ought to be frequently evaluated in order to reflect changing conditions.

The development of guidelines need not only follow the procedures outlined in Section III, but also should address the following concerns:[6,7]

- The differing climates within the area or region.
- The cultural and social habits within the various regions or states of the country.
- The type of diet and eating patterns of the population.
- Socioeconomic status of the citizens.
- Opinions, beliefs, taboos, and local habits and customs.
- Availability, reliability, and accessibility of the needed amount of of water.
- Existence of efficient and suitably run water and wastewater treatment plants.
- Methods used for disposal of human, animal, and municipal waste.
- Industries established within the country and their national impact.
- Industrial expansions and potential growth.
- Regulations governing disposal and discharge of industrial and agricultural wastewater.
- Annexation and extension of services and amenities.

V. WATER QUALITY REQUIREMENTS FOR WASTEWATER RECLAMATION AND REUSE

There are no U.S. national standards for the reuse of water. Any potable use of reclaimed water, however, must meet drinking water standards as well as the reliability criteria.[12]

Most states do not have regulations or guidelines that cover all potential users of reclaimed water. Regulations refer to actual rules that have been enacted and are enforceable by state governmental agencies. These state agencies generally issue reuse permits specifying conditions, requirements and

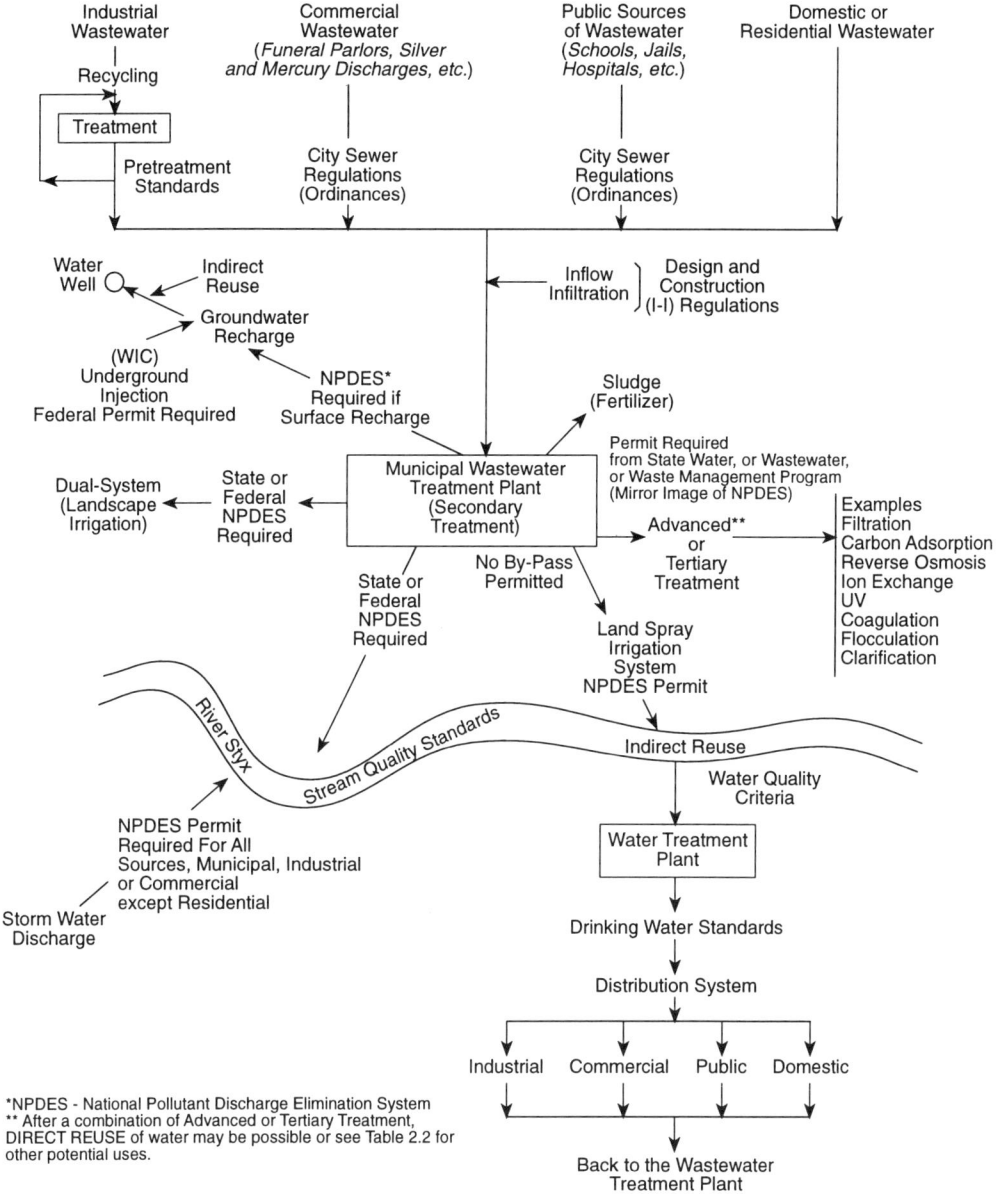

Figure 2.2 Wastewater recycling, direct and indirect reuse, and the various water quality standards and permits required.[11]

limitations, as well as enforcement actions for violation of the permit. Guidelines, on the other hand, are not enforceable but can be used in the development of water reuse regulations.[13,14]

Arizona, California, Florida, Texas, Oregon, Colorado, Nevada, and Hawaii have extensive regulations or guidelines that prescribe requirements for a wide range of end uses of reclaimed water. Other states have regulations or guidelines which tend to focus on land treatment of wastewater effluent, and emphasize additional treatment or effluent disposal rather than beneficial reuse, even though the effluent may be used for irrigation of agricultural sites, golf courses, or public access lands.[13]

As of March 1992, 18 states had adopted regulations regarding the reuse of reclaimed water, 18 states had guidelines or design standards, and 14 states had no regulations or guidelines. In states with no

Table 2.2 Categories of Municipal Wastewater Reuse and Potential Constraints

Wastewater reuse category[a]	Potential constraints
Agricultural and landscape irrigation	
Crop irrigation	Effects of salts on soils and crops.
Commercial nurseries	Public health concerns, surface and groundwater pollution, marketability of crops, and public acceptance.
Parks	
School yards	
Freeway medians	
Golf courses	
Cemeteries	
Greenbelts	
Residential areas	
Industrial reuse	
Cooling	Scaling, corrosion, biological growth, and fouling; public health concerns.
Boiler feed	
Process water	
Heavy construction	
Groundwater recharge	
Groundwater replenishment	Potential toxicity of chemicals and pathogens.
Salt water intrusion	
Subsidence control	
Recreational and environmental	
Lakes and ponds	Health concerns and eutrophication.
Marsh enhancement	
Streamflow augmentation	
Fisheries	
Snowmaking	
Nonpotable urban uses	
Fire protection	Public health, fouling, scaling, corrosion, and biological growth.
Air conditioning	
Toilet flushing	
Potable reuse	
Blending in water supply	Potentially toxic chemicals, public health, and public acceptance.
Pipe-to-pipe water supply	

[a] Arranged in descending order of anticipated volume of use.

From Asano, T. D., et al., *Water Environ. Technol.*, 4, 36, 1992. With permission.

specific regulations or guidelines on water reclamation and reuse, programs may still be permitted on a case-by-case basis.[13]

The reclaimed wastewater can be used either directly or indirectly. Direct reuse indicates planned, deliberate, or intentional use of treated wastewater for some beneficial purpose. An example of the indirect reuse of wastewater occurs when the treated effluent is discharged and diluted in a river and later withdrawn downstream for some beneficial use. There are many beneficial uses for both direct and indirect use of reclaimed wastewater. However, here the concentration will be on the many beneficial uses for properly treated wastewater, which range all the way from potable reuse to agricultural and landscape irrigation. Table 2.2 presents six classified municipal wastewater reuse categories. Other categories of wastewater reuse application have also been developed, such as:[13]

- Urban
- Industrial
- Agricultural
- Recreational
- Habitat restoration/enhancement
- Groundwater recharge
- Augmentation of potable supplies

However, these categories cover essentially the same reuse applications as those presented in Table 2.2.[15]

At present in the U.S. the largest volumes of treated wastewater are used for agricultural irrigation (e.g., 34% for Florida, 63% for California) followed by landscape irrigation, groundwater recharge, and industrial reuse.[13]

A. RECLAIMED WASTEWATER REUSE FOR AGRICULTURAL IRRIGATION

The use of reclaimed wastewater for agricultural irrigation has some advantages as well as some disadvantages.[16,17]

Advantages include:

- Source of additional irrigation water.
- Savings of high quality water for other beneficial uses.
- Low-cost source of a water supply.
- Economical way to dispose of wastewater and prevent pollution and sanitary problems.
- Reliable, constant water source.
- Effective use of plant nutrients contained in the wastewater, such as nitrogen and phosphorus.
- Provides additional treatment of the wastewater before being recharged to the groundwater.

Disadvantages include:

- Wastewater not properly treated can create potential public health problems.
- Potential chemical contamination of the groundwater.
- Some of the soluble constituents in the wastewater could be present at concentrations toxic to plants.
- The treated wastewater could contain suspended solids at levels that may plug nozzles in the irrigation distribution system as well as clog the capillary pores in the soil, however these same solids can provide beneficial humic material required by the soil.
- The treated wastewater supply is continuous throughout the year while the demand for irrigation water is seasonal.
- Major investment in land and equipment.

Regulations, guidelines, and criteria have been developed for the use of reclaimed wastewater for agricultural irrigation and are generally based on the following parameters:[18]

- For farm workers and public health protection, the reclaimed water must pose no bacteriological or virological hazard.
- Salinity (total dissolved solids or TDS) must be low enough to maintain favorable osmotic pressures for plants to take up water.
- Certain ions making up TDS, such as boron, chlorides, and sodium must not be of levels harmful to crops, and sodium must not be at levels harmful to soils.
- Trace levels of certain metals and synthetic organics must be controlled such that crop growth is not adversely affected.
- Concentrations of other heavy metals, such as molybdenum and possibly cadmium, must not be high enough in plants to be toxic to animals eating the plants (which themselves might be unaffected by the substance).
- Suspended solids, chemical precipitates, and algae growth must be controlled to prevent clogging of spray nozzles and drip applicators of irrigation units.

The recommended water quality criteria for long-term agricultural irrigation as well as long-term landscape irrigation are presented in Chapter 9, Table 9.6. Also, the Food and Agricultural Organization (FAO), an agency in the United Nations, has developed guidelines for water quality to be used for agricultural irrigation. The details of these guidelines are presented in Table 2.3.[20]

Several components in the water used for agricultural irrigation are of particular importance and include salinity, exchangeable ions (Na, Ca, and Mg), boron, and trace metals (Cd, Cr, Cu, Hg, Ni, Mo, Pb, Zn).

1. Salinity

It has been stated that salinity is the single most important parameter in determining the suitability of a water for agricultural irrigation.[13,21] Salinity is defined as the total solids in a water sample after all carbonates have been converted to oxides, all bromides and iodides have been replaced by chlorides, and all organic matter has been oxidized, and is a measure of the concentration of dissolved mineral

Table 2.3 Guidelines for Interpretation of Water Quality for Irrigation

Potential irrigation problem	Units	Degree of restriction on use		
		None	Slight to moderate	Severe
Salinity (affects crop water availability)[a]				
EC	dS/m	<0.7	0.7–3.0	>3.0
or, TDS	mg/L	<450	450–2000	>2000
Infiltration (affects infiltration rate of water into the soil. Evaluation using EC and SAR together)[b]				
SAR = 0–3 and EC =		>0.7	0.7–0.2	<0.2
= 3–6 and EC =		>1.2	1.2–0.3	<0.3
= 6–12 and EC =		>1.9	1.9–0.5	<0.5
= 12–20 and EC =		>2.9	2.9–1.3	<1.3
= 20–40 and EC =		>5.0	5.0–2.9	<2.9
Specific ion toxicity (affects sensitive crops)				
Sodium (Na)[b]				
Surface irrigation	SAR	<3	3–9	>9
Sprinkler irrigation	mg/L	<3	<3	
Chloride (Cl)[c]				
Surface irrigation	mg/L	<4	4–10	>10
Sprinkler irrigation	mg/L	<3	>3	
Boron (B)	mg/L	<0.7	0.7–3.0	>3.0
Trace elements (See Table 2.8)				
Miscellaneous effects (affects susceptible crops)				
Nitrogen (NO$_3$-N)[d]	mg/L	>5	5–30	>30
Bicarbonate (HCO$_3$)				
(overhead sprinkling only)	mg/L	<1.5	1.5–8.5	>8.5
pH		Normal range		6.5–8.4

[a] EC = electrical conductivity, a measure of the water salinity, reported in deciSiemens per meter at 25°C (dS/m) or in units millimhos per centimeter (mmho/cm). Both are equivalent. TDS = total dissolved solids, reported in milligram/liter (mg/L).

[b] SAR = sodium adsorption ratio. At a given SAR, infiltration rate increases as water salinity increases. Evaluate the potential infiltration problem by SAR as modified by EC.

[c] For surface irrigation, most tree crops and woody plants are sensitive to sodium and chloride; use the values shown. Most annual crops are not sensitive. With overhead sprinkler irrigation and low humidity (<30%) sodium and chloride may be absorbed through the leaves of sensitive crops.

[d] NO$_3$-N, nitrate nitrogen, reported in terms of elemental nitrogen (NH$_4$-N and organic-N should be included when wastewater is being tested).

From Ayers, R. S. and Westcot, D. W., *FAO*, 7, 11, 54, 69, 1976. With permission.

substances in the water.[21,22] To determine salinity one normally uses indirect methods involving measuring total dissolved solids or conductivity.

Total dissolved solids (TDS) or residue refers to the material left in a vessel after evaporation of a filtered water sample and its subsequent drying in an oven at a defined temperature.[22] The TDS content relates to the conductivity of the water sample. The TDS residue can be estimated by multiplying conductivity (in mmhos per centimeter) by an empirical factor. This factor can vary from 550 to 900 depending upon the ionic components in the solution. A factor most often used for treated wastewater used in agricultural irrigation is 640.

$$\text{TDS}\,(\text{mg/L}) = \text{EC}\,(\text{mmho/cm or dS/m}) \times 640$$

$$[1\,\text{mmho/cm} = 1\,\text{deceSiemens per meter}\,(\text{dS/m})] \qquad [2.1]$$

Table 2.4 General Guidelines for Salinity in Agricultural Irrigation Water[a]

Classification[b]	TDS (mg/L)	EC (mmhos/cm)[c]
Water for which no detrimental effects are usually noticed	500	0.75
Water that can have detrimental effects on sensitive crops	500–1,000	0.75–1.50
Water that can have adverse effects on many crops, requiring careful management practices	1,000–2,000	1.50–3.00
Water that can be used for tolerant plants on permeable soils with careful management practices	2,000–5,000	3.00–7.50

[a] Normally only of concern in arid and semiarid parts of the country.
[b] Crops vary greatly in their tolerance to salinity (TDS or EC).
[c] EC = electrical conductivity.

Adapted from USEPA, Office of Water Program Operations, EPA-430/9-75-001, 1975.

Some general guidelines as to the salinity hazard, TDS content, and electrical conductivity are given in Table 2.4.[23] In this case the conversion factor is 666 rather than 640.

In Saudi Arabia it has been recommended that due to soil conditions the upper conductivity limit be 0.750 mmhos/cm (500 mg/L salinity) for salt-sensitive crops, with an upper limit of 2.25 mmhos/cm electrical conductivity (1500 mg/L salinity) for salt-tolerant plants cultivated on adequately drained soil.[17] Salt-tolerant plants include date palm, certain fruits and vegetables, and field crops such as wheat and barley.

An adequately drained soil can be evaluated by determining the leaching requirement in order to evaluate salt build-up.[13]

$$LR = EC_{iw}/EC_{dw} \times 100 \qquad [2.2]$$

where: LR = leaching requirement (%)
EC_{iw} = electrical conductivity of irrigation water (mmho/cm or dS/m)
EC_{dw} = electrical conductivity of drainage water and is determined by the salt tolerance of the crop to be grown (mmho/cm or dS/m)

The extent of salt accumulation in the soil depends on the concentration of salts in the irrigation water and the rate at which it is removed by leaching.

The deleterious effects of salinity can be augmented by a soil with poor drainage characteristics, high evapotranspiration rates, and the type of crop being grown.

The only suitable way to control a salinity problem is by applying more irrigation water than can be used by the plant and in this way provide excess water that leaches throughout the plant's root zone and carries off excess salt and thus maintains a soil salt concentration at an appropriate level.[18]

2. Exchangeable Cations (Sodium, Calcium, and Magnesium)

The concentration of sodium, calcium, and magnesium ions in water used for agricultural irrigation must be considered. High sodium concentrations not only reduce the clay-bearing soil's permeability, but also affect the soil structure. When calcium is the predominant cation adsorbed in the exchangeable soil complex, the soil tends to have a granular structure and is easily worked and readily permeable. When sodium concentrations are high the clay particles are dispersed and the soil permeability is reduced.[20]

To estimate the degree to which sodium will be adsorbed by a soil from a given water, the sodium adsorption ratio (SAR) has been developed. SAR is defined as follows:

$$SAR = \frac{Na^+}{\sqrt{\dfrac{CA^{2+} + Mg^{2+}}{2}}} \qquad [2.3]$$

where: Na^+ = sodium milliequivalents/liter (meq/L)
Ca^{2+} = calcium (meq/L)
Mg^{2+} = magnesium (meq/L)

For example the SAR for a treated wastewater used for irrigation purposes at Dariyah, Saudi Arabia was found to be

Na^+ = 80 mg/L = 3.48 meq/L
Ca^{2+} = 110 mg/L = 5.50 meq/L
Mg^{2+} = 38 mg/L = 3.11 meq/L
(Equivalent weight Na = 23.0, Ca = 20.0, Mg = 12.2)

$$SAR = \frac{3.48}{\sqrt{\frac{5.50+3.11}{2}}} = \frac{3.48}{2.07} = 1.68$$

For sensitive fruits, the tolerance limit for irrigation water is about 4. For general crops a limit of 8 to 18 is generally considered within a usable range.[24]

A fairly new procedure employs a modification of the SAR and is called the Adjusted Sodium Adsorption Ratio.[20]

The Adjusted Sodium Adsorption Ratio (Adj. SAR) is calculated from the following equations.[20]

$$\text{Adj SAR} = \frac{Na^+}{\sqrt{\frac{Ca^{2+}+Mg^{2+}}{2}}} \left[1+(8.4-pH_c)\right] \quad [2.4]$$

$$pH_c = (pK_2' - pk_c') + p(Ca+Mg) + p(ALK) \quad [2.5]$$

where: $pK_2' - pK_c'$ = empirical constants (see Table 2.5)
$p(Ca^{2+}+Mg^{2+})$ = negative logarithm of the calcium and magnesium ion concentration in moles/liter
$p(ALK)$ = negative logarithm of the total alkalinity in milliequivalents/liter

The adjusted SAR for the treated wastewater used at Dariyah for irrigation purposes would be as follows:

Na^+ = 3.48 meq/L
Ca^{2+} = 5.50 meq/L
Mg^{2+} = 3.11 meq/L
Total = 12.09 meq/L
HCO_3^- = 320 mg/L as $CaCO_3$ = 6.4 meq/L (alkalinity)

Interpolate from Table 2.5

$pK_2' - pK_c'$ = 2.3
$Ca^{2+} + Mg^{2+}$ = 5.50 + 3.11 = 8.61 meq/L

Interpolate from Table 2.5

$p(Ca^{2+} + Mg^{2+})$ = 2.37
or, by calculation,

Ca^{2+} = 110 mg/L = 0.00274 mol/L
Mg^{2+} = 38 mg/L = 0.00156 mol/L
Total = 0.00430 mol/L

(Mole of Ca = 40,100 mg, Mg = 24,300 mg)

$-\log$ of 0.00430 = 2.37
$p(Ca^{2+} + Mg^{2+})$ = 2.37
Alkalinity = 320 mg/L as $CaCO_3$ (equivalent weight $CaCO_3$ = $100/2$ = 50 g)
= 6.4 meq/L

Table 2.5 Table for Calculating pH_c

$(pK'_2 - pK'_c)$ is obtained from using the sum of Ca + Mg + Na in meq/L	
$p(Ca + Mg)$ is obtained from using the sum of Ca + Mg in meq/L	Obtained from water analysis
$p(ALK)$ is obtained from using the sum of $CO_3^{2-} + HCO_3^-$ in meq/L	

Sum of Concentration (meq/L)	$pK'_2 - pK'_c$	$p(Ca + Mg)$	$p(ALK)$
0.05	2.0	4.6	4.3
0.10	2.0	4.3	4.0
0.15	2.0	4.1	3.8
0.20	2.0	4.0	3.7
0.25	2.0	3.9	3.6
0.30	2.0	3.8	3.5
0.40	2.0	3.7	3.4
0.50	2.1	3.6	3.3
0.75	2.1	3.4	3.1
1.00	2.1	3.3	3.0
1.25	2.1	3.2	2.9
1.5	2.1	3.1	2.8
2.0	2.2	3.0	2.7
2.5	2.2	2.9	2.6
3.0	2.2	2.8	2.5
4.0	2.2	2.7	2.4
5.0	2.2	2.6	2.3
6.0	2.2	2.5	2.2
8.0	2.3	2.4	2.1
10.0	2.3	2.3	2.0
12.5	2.3	2.2	1.9
15.0	2.3	2.1	1.8
20.0	2.4	2.0	1.7
30.0	2.4	1.8	1.5
50.0	2.5	1.6	1.1
80.0	2.5	1.4	1.1

Note: pH_c is a theoretical, calculated pH of the irrigation water in contact with lime and in equilibrium with soil CO_2.

From Ayers, R. S. and Westcot, D. W., *FAO*, 7, 11, 54, 69, 1976. With permission.

Interpolate from Table 2.5

$p(ALK) = 2.19$

or by calculation

$ALK = {}^{320}/_{50,000} = 0.0064$
$-\log 0.0064 = 2.19$
$pH_c = 2.3 + 2.37 + 2.19$
$= 6.86$
Adj. SAR $= 1.68 [1 + (8.4 - 6.86)]$ (from Equation 2.4)
$= 1.68 (2.54)$
$= 4.27$

In both cases the treated wastewater SAR and Adjusted SAR are well below the suggested 8 to 18 value suggested for general crop use. In fact the calculated SAR values are in the range suitable for sensitive fruit crops.

3. Boron in Reclaimed Wastewater

The most current toxicity problem for crops irrigated with reclaimed wastewater is from boron.[13] The source of boron in wastewater is usually from household detergents containing perborate — the bleaching

Table 2.6 Relative Tolerance of Crops and Ornamentals to Boron

Tolerant (4.0 mg/L of Boron)	Semitolerant (2.0 mg/L of Boron)	Sensitive (1.0 mg/L of Boron)
Athel	Sunflower, native	Pecan
Asparagus	Potato	Walnut, black and Persian or English
Palm	Cotton, Acala and Pina	Jerusalem artichoke
Date palm	Tomato	Navy bean
Sugarbeet	Sweetpea	American elm
Mangel	Radish	Plum
Garden beet	Field pea	Pear
Alfalfa	Ragged-robin rose	Apple
Gladiolus	Olive	Grape (Sultanina and Malaga)
Broadbean	Barley	Kadota fig
Onion	Wheat	Persimmon
Turnip	Corn	Cherry
Cabbage	Milo	Peach
Lettuce	Oat	Apricot
Carrot	Zinnia	Thornless blackberry
	Pumpkin	Orange
	Bell pepper	Avocado
	Sweet potato	Grapefruit
	Lima bean	Lemon
(2.0 mg/L of boron)	(1.0 mg/L of boron)	(0.3 mg/L of boron)

Note: Relative tolerance is based on the boron concentration in irrigation water at which boron toxicity symptoms were observed when plants were grown in sand culture. It does not necessarily indicate a reduction in yield. Tolerance decreases in descending order in each column.

From Ayers, R. S. and Westcot, D. W., *FAO*, 7, 11, 54, 69, 1976. With permission.

component in the detergent. Other sources can be industrial plants or runoff entering (infiltration) the sewage system where boron fertilizers are used.

The average boron concentration in the earth's crust is 10 mg/kg with an average concentration in the ocean of 5 mg/L. Boron concentrations in freshwater rarely go above 1 mg/L and are generally less than 0.1 mg/L.[20,25]

Boron can accumulate in the upper soil layers in arid regions to levels that are toxic to plants. The amount of boron available to plants in soil is pH dependent. Maximum boron adsorption by soil has been found to be at pH 9.

The increase in boron adsorption with rising soil pH accounts for the lower boron availability in alkaline soils. The increase in borate adsorption with rising soil pH results in less boron being available to plants, and if overliming of a soil takes place it will result in a boron deficiency in crops.[26] Boron toxicity in plants is often associated with arid or semiarid regions where boron levels are frequently high in the soil.

Crops differ in their sensitivity to boron, however it must be kept in mind that boron is very important in crop production and an excess or deficiency can adversely affect crop yields. Also, boron is one of the most import micronutrients for plants necessary to obtain a high quantity and quality crop yield.

A wide range of crops have been tested for boron tolerance. These crops have been grouped as to their relative tolerance and are presented in Table 2.6.[20]

Wastewater treatment plants are not effective at removing boron in their treatment processes unless some form of tertiary treatment such as chemical precipitation is carried out. Typical wastewater influent boron concentration are in the 1.0 mg/L range, while typical effluent concentrations for plants providing secondary treatment and chlorination are around 0.75 mg/L.[27,28]

Guidelines from the Food and Agricultural Organization of the United Nations for boron concentrations in water used for agricultural irrigation indicate that no problem will exist for plants at concentrations less than 0.75 mg/L. However, increasing problems occur between 0.75 and 2.0 mg/L and severe problems result at concentrations above 2.0 mg/L.[20]

Observation at the University of California Agricultural Extension Service indicated that boron below 0.5 mg/L in the irrigation water was satisfactory for all crops. At 0.5 to 1.0 mg/L boron concentrations,

Table 2.7 Limits of Boron in Irrigation Water

Permissible Limits (Boron in milligrams per liter or parts per million)

Class of water	Crop Group		
	Sensitive	Semitolerant	Tolerant
Excellent	<0.33	<0.67	<1.0
Good	0.33–0.67	0.67–1.33	1.0–2.0
Permissible	0.67–1.0	1.33–2.0	2.0–3.0
Doubtful	1.0–1.25	2.0–2.5	3.0–3.75
Unsuitable	>1.25	>2.5	>3.75

From van der Leeden, F., Troise, F. L., and Todd, D. K., *The Water Encyclopedia*, 2nd ed., Lewis Publishers, Boca Raton, FL, 1990, 466.

the water was satisfactory for most crops but sensitive crops would show a lower yield. Between 1.0 and 2.0 mg/L of boron, the water was satisfactory for semitolerant crops, and from 2.0 to 10 mg/L of boron only tolerant crops would produce satisfactory yields.[26]

More detailed guidelines for the permissible limits for the allowable concentration of boron in reclaimed water intended for agricultural irrigation purposes are presented in Table 2.7.[29]

As indicated, the toxicity problems for boron can occur at levels down to 0.5 mg/L. Certain management practices can be used to reduce the toxic effects of boron in the treated wastewater.[20]

These management practices include:

- Irrigate more frequently
- Use additional water for leaching; "rule of thumb" apply 90 cm depth of water through 30 cm depth of soil to remove 80% of the boron
- Change or blend water supplies
- Plant less sensitive crops
- Use additional nitrogen to maximize fertility of the soil for growth of a crop such as citrus

4. Trace Metals or Elements in Reclaimed Wastewater

All wastewaters delivered to treatment facilities contain trace metals or elements. Industrial plants are an obvious source, but wastewaters from private residences can also have high trace metal concentrations. Trace metals in reclaimed water normally occur in concentrations less than a few milligrams per liter, with usual concentrations less than 100 µg/L. Some are essential for plant and animal growth, but all can become toxic at elevated concentrations. The most important trace metals in wastewater include cadmium, chromium, copper, mercury, molybdenum, nickel, lead, and zinc.[13,20,30]

Trace metal concentrations in wastewater are affected by their sources and the wastewater treatment processes provided. The concentration of trace metals in treated wastewater are important in land application situations because they may shorten the lifetime of the site through the accumulation of one metal or a combination of metals in excess of the biological toxicity threshold.

Table 2.8 presents the range and average concentrations of trace elements found in treated wastewater in the U.S. as well as the average concentrations in the effluent from the Riyadh Saudi Arabia Wastewater Treatment Plant (RWTP). The EPA water quality criteria for agricultural irrigation water is also included in this table.[13,17,31]

It appears from Table 2.8 that the average concentration of the major trace metals in the Riyadh treated wastewater are all below suggested water quality criteria used in the U.S. for agricultural irrigation water.

The secondary wastewater treatment processes available vary in their effectiveness at the removal of trace metals.

Table 2.9 presents the removal of the significant trace metals by the activated sludge (AS) process, the trickling filter (TF) process, and the rotating biological contractor (RBC) process.[31]

From the data provided in Table 2.9 it would appear that the RBC process is the most effective at removal of the significant trace metals, except for cadmium which falls between the average removal for the TF and AS processes.

However, advanced wastewater treatment processes such as chemical coagulation and carbon adsorption can in most cases remove over 90% of the trace metals from the influent wastewater.[31]

Table 2.8 Average Concentrations of Trace Elements Found in Treated Wastewater in the U.S. and Riyadh Saudi Arabia WTP and the EPA Water Quality Criteria for Agricultural Irrigation Water

Element	U.S. wastewater effluent (mg/L) Range	U.S. wastewater effluent (mg/L) Median	RWTP[a] effluent (mg/L) Average	EPA recommended limits (mg/L) agricultural irrigation water Long-term use	EPA recommended limits (mg/L) agricultural irrigation water Short-term use
Cadmium	0.005–0.22	0.005	0.01	0.01	0.05
Chromium	0.001–0.1	0.001	0.1	0.1	1.0
Copper	0.006–0.053	0.018	0.01	0.2	5.0
Lead	0.003–0.35	0.008	0.10	5.0	10.0
Mercury	0.0002–0.001	0.0002	—	—	—
Nickel	0.003–0.60	0.004	0.1	0.20	2.0
Zinc	0.004–0.35	0.04	0.01	2.0	10.0

[a] Riyadh Wastewater Treatment Plant Effluent

Compiled from References 13, 17, and 31.

Table 2.9 Average Trace Metal Removal for Various Secondary Treatment Processes

Constituent	Average Removal (%)	Average Reliability 10%	Average Reliability 50%	Average Reliability 90%	Average Effluent Concentration (mg/L)
Secondary Treatment with Activated Sludge (AS) Average Process Train Performance[31]					
Cadmium	71	96	96	20	0.002
Chromium	85	98	87	60	0.025
Copper	88	98	89	62	0.014
Lead	85	98	85	54	0.022
Mercury	23	31	18	0	0.086
Zinc	67	90	67	23	0.138
Secondary Treatment with Trickling Filters (TF) Average Process Train Performance[31]					
Cadmium	38	75	37	5	0.005
Chromium	44	81	49	6	0.094
Copper	78	95	82	24	0.026
Lead	73	97	68	21	0.040
Mercury	11	—	8	—	0.100
Zinc	72	95	75	20	0.117
Secondary Treatment with Rotating Biological Contactors[a] (RBC) Average Process Train Performance[31]					
Cadmium	57				0.003
Chromium	93				0.012
Copper	94				0.007
Lead	98				0.003
Mercury	84				0.018
Zinc	81				0.080

[a] Sedimentation is chemically aided with alum.

A few of the trace metals in wastewater are essential for life, and when applied may enrich the soil. Zinc is the metal most likely to provide an environmental benefit. Large areas of land have too little zinc for the growth of some crops and also the average dietary zinc intake by humans is marginal. Nevertheless, such essential-to-life metals (and other nutrients, too) can accumulate and pose potential long-term hazards to plant growth or to animals or humans consuming the plants. Cadmium, copper, zinc, and nickel are examples of metals that can accumulate in soils and decrease plant growth (phytotoxicity).[30]

While nickel and zinc are of a lesser concern than cadmium, copper, and molybdenum, they have visible adverse effects on plants at lower concentrations than the levels harmful to animals and humans. However, the nickel toxicity is reduced as the pH is increased. Cadmium, copper, and molybdenum can be harmful to animals at concentrations too low to affect plants.

Copper is not toxic to monogastric (stomach, one chamber) animals, but may be toxic to ruminants (stomach, several chambers). In any case the animal's tolerance increases as available molybdenum increases. Molybdenum can also be toxic when available in the absence of copper. Cadmium is of particular concern as it can accumulate in the food chain. Cadmium does not adversely affect ruminants in the small amounts they ingest. Most milk and beef products are also unaffected by livestock ingestion of cadmium because it is stored in the liver and kidneys of the animal rather than the fat or muscle tissues.[13]

Table 2.10 presents EPAs recommended limits for constituents of heavy metals in irrigation water.[13]

B. RECLAIMED WASTEWATER FOR GROUNDWATER RECHARGE

While groundwater recharge in the U.S. is growing rapidly the same cannot be said on an international basis with of course the exception of Israel, where in 1987, 30% of the estimated 70 million gallons per day (mg/d) (3000 L/s) available from reclaimed wastewater was used for groundwater recharge.[13] A main groundwater recharge project in Israel is the Dan Region Wastewater Reclamation project involving about 50 mgd (2200 L/s).[13] In Florida about 11% of the reclaimed wastewater, 30 mgd (1300 L/s), is involved in groundwater recharge, while in California the figure is 14 to 15%; 45 mgd (2000 L/s) of available reclaimed wastewater is used for groundwater recharge.[32-34]

The purposes of groundwater recharge using reclaimed water can be

- To establish saltwater intrusion barriers in coastal aquifers.
- To provide further treatment for future reuse.
- To augment potable or nonpotable aquifers.
- To provide storage of reclaimed water, or to control or prevent ground subsidence.

There are three methods used for groundwater recharge utilizing reclaimed wastewater:[12]

- Surface spreading or percolation, and infiltration.
- Direct injection.
- River bank or stream infiltration as a result of streamflow augmentation.

In surface spreading, treated wastewater percolation and infiltration through the unsaturated zone takes advantage of the subsoil's natural ability for biodegradation and filtration, thus providing additional *in situ* treatment of the wastewater and additional treatment reliability to the overall wastewater management system. Another advantage of groundwater recharge by surface spreading is that it can be carried out in the vicinity of metropolitan and agricultural areas and thus counteract falling groundwater tables.[12,13]

Also, groundwater recharge helps provide a loss of identity between reclaimed water and groundwater. This loss of identity has a positive psychological impact where reuse is contemplated and is an important factor in making reclaimed water acceptable for a wide variety of uses, including potable water supply augmentation.[13]

In direct injection, treated wastewater is pumped under pressure directly into the groundwater zone, usually into a well-confined aquifer. Groundwater recharge by direct injection is practiced, in most cases, where groundwater is deep or where the topography or existing land use makes surface spreading impractical or too expensive. This method of groundwater recharge is particularly effective in creating freshwater barriers in coastal aquifers against intrusion of saltwater. Both in surface spreading and in direct injection, locating the extraction wells at as great a distance as possible from the spreading basins or the injection wells increases the flow path length and residence time of the recharged groundwater, as well as the mixing of the recharged water and the other aquifer contents.[12]

River bank or stream bed infiltration is a means of indirect groundwater recharge and is widely practiced in Europe. Here, groundwater recharge may be used as a treatment scheme in water supply systems where the source is a surface water contaminated by substantial discharges of industrial and municipal wastewater (such as the Rhine River in Germany and the Netherlands). The contaminated

Table 2.10 Recommended Limits for Constituents in Reclaimed Water for Irrigation

Constituent	Long-Term Use (mg/L)	Short-Term Use (mg/L)	Remarks
Trace Heavy Metals			
Aluminum	5.0	20	Can cause nonproductivity in acid soils, but soils at pH 5.5 to 8.0 will precipitate the ion and eliminate toxicity.
Arsenic	0.10	2.0	Toxicity to plants varies widely, ranging from 12 mg/L for Sudan grass to less than 0.05 mg/L for rice.
Beryllium	0.10	0.5	Toxicity to plants varies widely, ranging from 5 mg/L for kale to 0.5 mg/L for bush beans.
Boron	0.75	2.0	Essential to plant growth, with optimum yields for many obtained at a few-tenths mg/L in nutrient solutions. Toxic to many sensitive plants (e.g., citrus) at 1 mg/L. Usually sufficient quantities in reclaimed water to correct soil deficiencies. Most grasses relatively tolerant at 2.0 to 10 mg/L.
Cadmium	0.01	0.05	Toxic to beans, beets, and turnips at concentrations as low as 0.1 mg/L in nutrient solution. Conservative limits recommended.
Chromium	0.1	1.0	Not generally recognized as essential growth element. Conservative limits recommended due to lack of knowledge on toxicity to plants.
Cobalt	0.05	5.0	Toxic to tomato plants at 0.1 mg/L in nutrient solution. Tends to be inactivated by neutral and alkaline soils.
Copper	0.2	5.0	Toxic to a number of plants at 0.1 to 1.0 mg/L in nutrient solution.
Fluoride	1.0	15.0	Inactivated by neutral and alkaline soils.
Iron	5.0	20.0	Not toxic to plants in aerated soils, but can contribute to soil acidification and loss of essential phosphorus and molybdenum.
Lead	5.0	10.0	Can inhibit plant cell growth at very high concentrations.
Lithium	2.5	2.5	Tolerated by most crops at up to 5 mg/L; mobile in soil. Toxic to citrus at low doses — recommended limit is 0.075 mg/L.
Manganese	0.2	10.0	Toxic to a number of crops at a few-tenths to a few mg/L in acid soils.
Molybdenum	0.01	0.05	Nontoxic to plants at normal concentrations in soil and water. Can be toxic to livestock if forage is grown in soils with high levels of available molybdenum.
Nickel	0.2	2.0	Toxic to a number of plants at 0.5 to 1.0 mg/L; reduced toxicity at neutral or alkaline pH.
Selenium	0.02	0.02	Toxic to plants at low concentrations and to livestock if forage is grown in soils with low levels of added selenium.
Tin, Tungsten, & Titanium	—	—	Effectively excluded by plants; specific tolerance levels unknown.
Vanadium	0.1	1.0	Toxic to many plants at relatively low concentrations.
Zinc	2.0	10.0	Toxic to many plants at widely varying concentrations; reduced toxicity at increased pH (6 or above) and in fine-textured or organic soils.

Constituent	Recommended Limit	Remarks
Other Parameters		
pH	6.0	Most effects of pH on plant growth are indirect (e.g., pH effects on heavy metals' toxicity described above).
TDS	500–2,000 mg/L	Below 500 mg/L, no detrimental effects are usually noticed. Between 500 and 1,000 mg/L, TDS in irrigation water can affect sensitive plants. At 1,000 to 2,000 mg/L, TDS levels can affect many crops and careful management practices should be followed. Above 2,000 mg/L, water can be used regularly only for tolerant plants on permeable soils.
Free Chlorine Residual	<1 mg/L	

Adapted from Reference 13.

water percolates not only from the riverbank or streambanks but also from spreading basins to an aquifer and then travels through the aquifer to extraction wells, some distance from the source.

In some cases, the residence time underground is only 20 to 30 days, and there is almost no dilution by natural groundwater. In the Netherlands, dune infiltration of treated Rhine River water has been used to restore the equilibrium between fresh and saltwater in the dunes, while serving to improve water quality and provide storage for potable water systems. Dune infiltration also provides protection from accidental spills of toxic contaminants into the Rhine River.[13] There are four major water quality factors to be considered in groundwater recharge with reclaimed wastewater:[12]

- Pathogens,
- Total minerals,
- Heavy metals, and
- Stable organic substances.

Effluent quality guidelines or criteria are generally more stringent for direct injection than for land spreading. The reason for the more stringent quality requirements is that there is no added protection using direct injection because the water enters the aquifer directly without percolating or filtering through the soil above the aquifer. The water quality requirements vary from region to region depending on the existing groundwater quality and its usage. Suggested water criteria for groundwater recharge by land spreading is presented in Table 2.11.[19]

C. RECLAIMED WASTEWATER FOR POTABLE USE

Up to 1982 there were only two examples for direct reuse of reclaimed wastewater for drinking water: Chanute, Kansas and Windhoek, South Africa.[13]

Table 2.11 Water Quality Criteria for Groundwater Recharge

Parameter	Water Quality Criteria Used for This Evaluation (mg/L)[a]
Arsenic	0.05
Bacteria	
Fecal coliform (MPN/100 mL)	23
Barium	2.0
Biological Oxygen Demand	10.0
Boron	0.02
Cadmium	0.01
Chloride	0.05
Chromium	0.15
Copper	2.0
Cyanide	0.2
Iron	0.10
Lead	0.05
Manganese	0.10
MBAS	0.50
Mercury	0.01
Nitrogen	
Ammonia	5
Nitrate	10
Nitrite	0
Odor	Virtually free
Oil	Virtually free
Oxygen	Aerobic
pH (standard units)	5.0–9.0
Selenium	0.01
Silver	0.10
Solids, suspended	10.0
Zinc	10

[a] All units in mg/L except as noted.

Adapted from Reference 19.

During a drought period from 1952 to 1957 the town of Chanute, Kansas, with a population of 1,200, made direct use of effluent from their wastewater treatment plant which treated domestic as well as industrial wastewater. The Chanute wastewater treatment plant provided preliminary primary and secondary treatment (trickling filters) followed by chlorination, while the Chanute water treatment plant provided prechlorination, alum, lime, and soda ash for water softening, chlorination, recarbonation, sedimentation, rapid sand filtration, postfiltration chlorination, and a 4¼-hour clearwell storage time. This direct use of wastewater project operated for a total of five months. It was estimated that there were approximately ten cycles through the system over the five-month operating period. During this time period, the tap water never failed to meet drinking water standards, though it was pale yellow in color, had a musty unpleasant taste, and foamed in the glass due to its high organic content. The water was high in fluoride, sodium, total solids, and organics. Coliform organisms were found on three different occasions during this time period but were within mandated standards.[35]

Public acceptance of the finished water was poor and sales of bottled water flourished. A limited telephone survey showed 61% of Chanute residents accepted the treated wastewater for potable purposes. Considering the conditions described above this seems like a remarkable acceptance rate.[35]

At present, the only pipe-to-pipe wastewater reclamation project where direct potable reuse is currently practiced in the world is Windhoek, Namibia. This project became operational in 1969 and operates only intermittently, depending upon the rainfall and water demand. The reclaimed water has at times contributed from 15 to 40% of the total daily water supplied to the city, which has a population of 50,000.

The wastewater treatment plant at Windhoek provides traditional biological treatment (trickling filters). The effluent from this plant goes to maturation ponds that serve as polishing units, and provides at least 14 days detention. The Windhoek water reclamation plant (WRP) includes lime treatment, primary settling, sludge drying beds for the settled sludge, NH_3 stripping towers, primary recarbonation, chlorination, secondary settling, secondary recarbonation, sand filtration, breakpoint chlorination, carbon adsorption, and final chlorination if needed. The last modification in the plant was to move the NH_3 stripping operation to the wastewater treatment plant.

The quality of the water from the Windhoek WRP consistently complied with the water quality standards recommended by the World Health Organization.[12,13,35]

In the U.S., the most extensive research and operational projects for direct potable reuse of treated wastewater is being conducted in Denver, CO, Tampa, FL, and San Diego, CA.

The Denver plant has invested over $30 million in a pilot plant with a capacity of 1 mgd (44 L/s) and went into operation in 1984. The raw water supply for the reuse plant is unchlorinated secondary effluent from the metropolitan Denver wastewater treatment facility, and as such, has received biological treatment. The process train includes high-pH lime treatment, single- or two-stage recarbonation, pressure filtration, selective ion exchange for ammonia removal, two stages of activated carbon adsorption, ozonation, reverse osmosis, air stripping, and chlorine dioxide disinfection. Sidestream processes included a fluidized bed carbon reactivation furnace, vacuum sludge filtration, and selective ion exchange regenerant recovery.[12]

In 1986, the water treatment costs were running about $2.50 per 1000 gallons while treating fresh water from mountain sources was around $0.30 per 1000 gallons.[36]

The water produced at this plant was reported to be of better quality than many potable water sources in the region and met all the existing and proposed federal and state drinking water standards[12,13,36] (see Table 2.12).[13]

Despite the generally excellent results achieved in the Denver project, there are no immediate plans to implement potable reuse there.[13]

While direct reuse of wastewater for potable purposes is clearly limited in extent, indirect reuse for potable purposes takes place constantly and on a worldwide basis. The flows in such rivers as the Rhine, Thames, and Ohio are anywhere from 20 to 50% urban and industrial wastewater and these rivers are the water supply source for many large cities. Other examples of indirect reuse of wastewater for potable use in the U.S. is at Whittier Narrows, CA, El Paso, TX, and Occoguan, VA. Indirect potable reuse is more acceptable to the public than direct potable reuse as the water loses its identity as it moves through a river, lake, or aquifer. Indirect reuse, by virtue of the residence time in the water course, reservoir, or aquifer, often provides additional treatment and offers an opportunity for monitoring the quality and taking appropriate measures before the water is ready for distribution. In some instances, however, water quality may actually be degraded as it passes through the environment.[13]

Table 2.12 Denver Demonstration Plant Finished Water Quality

Parameter	After reverse osmosis	After ultrafiltration
Turbidity (NTU)	0.06	0.2
Color	a	a
Alkalinity, mg/L as $CaCO_3$	3	166
Hardness, mg/L as $CaCO_3$	6	108
Total suspended solids, mg/L	a	a
Total dissolved solids, mg/L	18	352
Specific conductance, µmhos/cm	67	648
Dissolved oxygen, mg/L	8.3	6.9
pH	6.6	7.8
Total organic carbon, mg/L	a	0.7
Ammonia nitrogen, mg/L	5	19
Nitrate nitrogen, mg/L	0.1	0.3
Nitrite nitrogen, mg/L	a	a
Total phosphate, P, mg/L	0.02	0.05
Total coliform, count/100 ml	a	a
Fecal strep, count/100 ml	a	a
Fecal coliform, count/100 ml	a	a
Enteric virus	a	a

[a] More than 50% of data were below detection limits.

Adapted from Reference 13.

D. RECLAIMED WASTEWATER FOR INDUSTRIAL USE

Industry represents an important potential market for reuse of reclaimed wastewater. In California in 1987, 2% of the reclaimed wastewater was utilized for industrial purposes while in Florida in 1990, 5% of the reclaimed water went for commercial and industrial purposes.[32,33] Industrial facilities are good candidates for using reclaimed wastewater.

On a national basis, industry in 1985 represented 8% of the total U.S. water demand, while in some states as much as 43% of the state's total water demands was from industry.[13]

Industry can recycle their water within the plant such as is done in the steel mills, breweries, electronics plants, and chemical mineral processing, and in this way conserve water as well as avoid stringent industrial effluent standards and regulations.[12,13] The in-plant recycling processes will not be covered here as this is a complete field in itself.

The other sources of reclaimed wastewater for industry are from publicly owned treatment works (POTWs).

The major factors that influence an industry in using reclaimed wastewater is the availability of the water, the industry's discharge requirements, water quality, volume, economics, and reliability.[12]

The major industrial categories that use reclaimed wastewater include:[13]

- Evaporative cooling water,
- Boiler feedwater,
- Process water, and
- Irrigation and maintenance of plant grounds, fire protection, and dust control.

Water quality criteria, standards, guidelines and requirements vary from industry to industry as well as within a single industry. Specific water quality requirements for many industries have not been established but possible detrimental effects of various components in the reclaimed wastewater on specific processes and equipment must be taken into account.[18]

Of the various industrial users of reclaimed wastewater, cooling water is currently the biggest single application. The cooling water can be a once-through cooling operation or a recirculating cooling system using towers, cooling ponds, or lakes.[12,13]

Quality requirements for cooling water are related to three common problems: scaling, corrosion, and biofouling. Scale-forming constituents found in effluent include calcium carbonate and calcium phosphate. Constituents in effluent known to cause corrosion are total dissolved solids (TDS), including chlorides and ammonia. Ammonia is particularly corrosive to copper alloys commonly used in heat

Table 2.13 Recommended Cooling Water Quality Criteria for Make-Up Water to Recirculating Systems

Parameter[a]	Recommended limit[b]
Cl	500
TDS	500
Hardness	650
Alkalinity	350
pH[c]	6.9–9.0
COD	75
TSS	100
Turbidity	50
BOD[c]	25
Organics[c]	1.0
NH_4-N[c]	1.0
PO_4^c	4
SiO_2	50
Al	0.1
Fe	0.5
Mn	0.5
Ca	50
Mg	0.5
HCO_3	24
SO_4	200

[a] All values in mg/L except pH and turbidity.
[b] Water Pollution Control Federation, 1989.
[c] Methylene blue active substances.

Compiled from Reference 13.

exchange systems. Nutrients in effluent, such as nitrogen and phosphorus, are known to cause biofouling.[18] Table 2.13 lists water quality criteria for cooling water systems.[13]

Recent water quality criteria for industrial boiler feedwater and the pulp and paper and the textile industries can be found in Reference 13.

E. RECLAIMED WASTEWATER FOR RECREATIONAL USE

When reclaimed water is to be employed for recreational use specific criteria, standards, and guidelines may be formulated given the particular use and the degree of physical contact experienced by the user as well as the secondary pollutional sources. The sources of the secondary pollutants in recreational areas, such as bathing places, may include:

- Body discharges such as the mucous from the nose, saliva, sweat, traces of fecal matter, urine, dead skin, etc.
- Air contaminants such as dust, pollens, particulate matter, etc.
- Street and work-area soil which accumulates on the skin.
- Different body creams, ointments, oils, lotions, etc.
- Sewage from domestic, industrial, commercial, institutional, recreation places, hotels, municipal works, etc.
- Cultivated fields, farms, etc.
- Animals.

The criteria, standards, or guidelines for reclaimed water to be used for recreational purposes can be subdivided into the following three classes.

1. Elementary Body-Contact Recreational Water

The reclaimed water used for contact recreational applications include swimming, bathing, waterskiing, etc. This class addresses the situations where there is intimate and prolonged contact between the

Table 2.14 Some Diseases Transmitted by Swimming Pool Water

Disease	Causative agent
Conjunctivitis	Virus
Sinusitis and otitis	*Streptococci* and *Staphylococci* (propagated by nasal mucus)
Certain types of enteritis	Some pathogens or certain viruses ingested with the water
Skin diseases:	
Eczemas	*Koch bacillus*
Granuloma	*Mycobacterium marinum*
Epidermophytosis	Brought about by the fungus that attaches itself to the skin between the toes and is contracted particularly easily when walking on areas around the pool.
Typhoid fever	*Salmonella typhi*
Dysentery	*Entamoeba histolytica, Shigella*
Infectious hepatitis	Virus

Compiled from References 3 and 37.

individual and the water and where there is a great risk of ingesting a large quantity of water which may impose a health threat. For example, both swimmers and non-swimmers may ingest from 10 to 15 ml of water every time they bathe. As such, the bathers may ingest viruses that are to be found in the water.

The routes of transmission of viruses may occur due to ingestion of water, or via the exposed mucous membranes and breaks in the protective skin barrier. Swimming pools have been implicated as the source of adenovirus conjunctivitis and pharyngitis, as well as enterovirus meningitis. Coxsakievirus B5 has been isolated from patients who had been swimming in lakes in which this microorganism was found to be present.[37]

Table 2.14 indicates some of the swimming-pool-water-transmittable diseases.[3,37]

Usually, the criteria, standards, or guidelines that are required for this class of reuse are more stringent than those required for non-body contact sports. Reclaimed wastewater requirements include:

- Reclaimed water needs to be aesthetically attractive.
- The water used must have an acceptable physical quality; this is to be established through the control of parameters such as color, taste, odor, temperature, solids concentration, and turbidity.
- Reclaimed water must be free of toxic compounds and other harmful chemical substances. For example, the reclaimed water must have an acceptable pH level. The pH can range from 6.5 to 8.3. The lacrimal fluid of the human eye has a pH of around 7. The deviation of the pH of the reclaimed water from the normal value may result in irritation to the eyes.
- The reclaimed water must be hygienically safe and free from disease-causing agents.

2. Secondary Body-Contact Recreational Water

This class of reclaimed water includes water utilized for boating, canoeing, camping, fishing, and landscape and golf course irrigation. The quality requirement for this category of reuse is less strict than for elementary body contact.

3. Non-Contact Recreational Water

The reclaimed water used in situations where there is no intimate contact between the human body and the water signifies this subdivision. It includes recreational confined water bodies, fountains, aquaculture, etc. The most significant quality criteria that need to be considered include:

- The furnishing of a reasonable temperature to sustain aquatic life.
- The supply of a suitable concentration of dissolved oxygen.
- The provision of suitable chemical quality aspects with respect to the concentration of trace elements, acidity, alkalinity, pH, pesticides, insecticides, biotoxins, toxic substances, and radionuclides.
- The elimination of nutrients to avoid the development of eutrophic conditions.
- The supply of reclaimed water with reasonable microbiological quality.

Table 2.15 presents suggested criteria for both body-contact and secondary body-contact situations.[18,19]

Table 2.15 Summary of Suggested Water Quality Criteria for Body-Contact and Secondary Body-Contact Recreations

Parameter	Suggested Water Quality Criteria For Body-Contact Recreation	Suggested Water Quality Criteria For Secondary Body-Contact Recreation
Aquatic growth	Virtually free	Virtually free
Bacteria		
Total coliform	2.2 MPN/100 ml	2.2 MPN/100 ml
COD, mg/L	30	60
Floating debris and scum	Virtually free	Virtually free
Odor	Virtually free	Virtually free
Oil	Virtually free	Virtually free
Oxygen dissolved, mg/L	10	Aerobic
pH units	6.5–8.3	6.5–8.3
Solids		
Settleable	Virtually free	Virtually free
Suspended, mg/L	5	—
Temperature		
Minimum °C	15	
Maximum °C	35	
Turbidity (NTU)	1	5

Compiled from References 12, 18, and 19.

F. RECLAIMED WASTEWATER FOR USE IN AQUATIC ENVIRONMENTS

The introduction of reclaimed water to augment flow in rivers and streams can have an impact on the aquatic life. Usually, the water quality in a river or stream is related directly to the quantity of its flow. Generally, the greater the stream flow the more pollutants it may incorporate without violating the water quality standards. This of course is not to say that "dilution is the solution to pollution."

When water quality is managed through control of the concentration of the input waste, the degree of needed treatment ought to be indicated. The required treatment must be as economical and efficient as possible. To accomplish these requirements, the limits for pollutants in receiving water courses must be carefully defined. The basis of the limits first, should be the public health of the community, and second, the environmental health of biological systems within the receiving water.[9]

Discharge of pollutants to surface waters should be controlled if they contain wastes that:

- Will settle or form objectionable deposits.
- Will float or form objectionable debris, oil scum, and other matter.
- Will present objectionable color, odor, taste, and turbidity.
- Will produce undesirable physiological responses in man, fish, and other aquatic life. These materials include radionuclides and toxic substances.[29]

Table 2.16 presents a proposed set of water quality fishing standards.[19]

G. RECLAIMED WASTEWATER USED FOR LIVESTOCK AND WILDLIFE PURPOSES

The term livestock generally applies to farm animals and wild animals. The most important parameter of concern with livestock drinking reclaimed water is salinity. The salts of most concern related to a water's salinity include calcium, magnesium, sodium, sulfates, bicarbonates, and chlorides. Water with a high salinity can cause physiological problems and even death for livestock due to an osmotic imbalance.[19] Total dissolved solids (TDS) of 1000 mg/L (electrical conductivity of 1.5 mmohs/cm or less) is considered safe for both livestock and cattle.[29] Some states such as Hawaii do have proposed regulations governing the watering of non-dairy livestock as well as regulations regarding washing of non-dairy livestock. In the first case the reclaimed water must be[38]

1. Oxidized, then coagulated, then clarified, then filtered, and then disinfected so that at some location in the treatment process the median number of total coliform bacteria determined by multiple-tube

Table 2.16 Proposed Stream Standards for Fishing

Quality parameter	Suggested level of stream standard (mg/L)
Ammonia (NH_3)	<1
Carbon dioxide	<12
Heavy metals	<1
Copper	<0.02
Arsenic	<1
Lead	<0.1
Selenium	<0.1
Cyanides	<0.012
Phenols	<0.02
Dissolved solids	<1000
Detergents	<0.2
Dissolved oxygen	>2
Pesticides	
DDT	<0.002
Endrin	<0.004
BHC	<0.21
Methyl parathion	<0.1
Malathion	<0.16
pH (pH units)	6.5–8.5

From Culp, G., et al., *Wastewater Reuse and Recycling Technology,* Noyes Data Corp., Park Ridge, NJ, 227, 1980.

fermentation does not exceed 2.2/100 mL as determined from the bacteriological results of the last 7 days for which analyses have been completed, and the number of total coliform bacteria does not exceed 23/100 mL in more than one sample within any 30-day period; and

2. Treated under process conditions that have been demonstrated to the satisfaction of the Department of Health to consistently provide a degree of treatment as defined above and so that the number of detectable enteric animal viruses is less than 1/40 L (washing non-dairy livestock).

Table 2.17 presents suggested criteria for acceptable water quality for livestock and wildlife purposes. One set of water quality values for livestock is difficult to establish due to factors such as the nature of fodder consumed, species, age, and productivity of the animals.[29]

VI. WORLD HEALTH ORGANIZATION (WHO) GUIDELINES FOR DRINKING WATER QUALITY

The primary aim of the WHO guidelines for drinking water quality is the protection of public health.

The guidelines are intended for use by countries as a basis for the development of national standards, which, if properly implemented, will ensure the safety of their own drinking water supplies. The levels recommended in the guidelines for water constituents and contaminants are not mandatory limits. In order to define such limits, it is necessary to consider the guidelines in the context of prevailing environmental, social, economic, and cultural conditions.[5] WHO does not promote the adoption of international standards for drinking water quality as there is an advantage provided by the use of a risk-benefit approach to the establishment of national standards and regulations. This policy enables each country to develop and implement their own appropriate drinking water standards.

One of the first steps in providing a safe drinking water supply is the protection of the water source from contamination. While the protection of the water source is almost invariably the best method of ensuring a safe drinking water supply, other factors that must also be considered are treatment efficiencies and reliability, the design, maintenance, and operation of the water distribution system, the monitoring of the water, and of course, the costs involved in developing the public water supply system.

The guidelines prepared by WHO emphasize microbiological safety. This is due to the fact that more than half the world population is still exposed to waters that are not free from pathogenic organisms. The source of these pathogenic organisms is from human and animal wastes which contain bacterial, viral, and protozoan pathogens and helminth parasites.

Table 2.17 Water Quality Criteria for Livestock and Wildlife

Parameter	EPA, October, 1973 (mg/L except as noted)
Alkalinity	30–130[a]
Aluminum	5.0
Aquatic growth	Avoid heavy growth of blue-green algae
Arsenic	0.2
Bacteria	
Total coliform	5000/100 ml
Fecal coliform	1000/100 ml
Beryllium	No limit
Boron	5.0
Cadmium	0.05
Chromium	1.0
Cobalt	1.0
Copper	0.5
Fluoride	2.0
Iron	No limit
Lead	0.1
Manganese	No limit
Mercury	0.001
Molybdenum	No limit
Nitrogen	
Nitrite as N	10
Nitrite + Nitrate as N	100
Oil	No visible floating
Pesticides	[b]
pH, units	6.0–9.0
Radionuclides	[b]
Selenium	0.05
Total dissolved solids	3000
Vanadium	0.1
Zinc	25.0

[a] Fluctuation should be less than 50 mg/L (protection of waterfowl habitat).
[b] Same as for Federal Drinking Water Standards.
From Culp, G., et al., *Wastewater Reuse and Recycling Technology,* Noyes Data Corp., Park Ridge, NJ, 227, 1980.

The assessment of the risks associated with pathogens in a drinking water supply is difficult and controversial because of the lack of epidemiological studies, the number of variables involved, and the changing interrelationships between these variables.

According to WHO, based on a world perspective, chemical contaminants in a public water supply are not normally associated with acute health problems and thus are in a lower priority category than microbial contaminants, the effects of which are usually acute and widespread. The WHO guidelines indicate that it can be argued that chemical standards for drinking water are of secondary consideration in a supply subject to severe bacterial contamination.

The WHO guidelines for drinking water quality indicate that the problems associated with chemical constituents in drinking water arise primarily from their ability to cause adverse health effects after prolonged periods of exposure; and of particular concern are contaminants that have cumulative toxic properties such as heavy metals and substances that are carcinogenic.

They also note that the use of chemical disinfectants in water treatment usually results in the formation of chemical by-products, some of which are potentially hazardous. However, they consider the risks to health from these by-products are extremely small in comparison with the risks associated with inadequate disinfection, and that it is important that disinfection should not be compromised in attempting to control such by-products.[5]

In evaluating the physical quality of a drinking water supply, the consumer generally requires the water to be low in color, turbidity, and suspended solids, and free of undesirable tastes and odors.

The nature of the guideline values for drinking-water quality recommended by WHO are to be interpreted along the following lines:[5,39]

- A guideline value represents the concentration of a constituent that does not result in any significant risk to the health of the consumer over a lifetime of consumption.
- The quality of water defined by the *Guidelines For Drinking Water Quality* is such that it is suitable for human consumption and for all usual domestic purposes, including personal hygiene. However, water of a higher quality may be required for some special purposes, such as renal dialysis.
- When a guideline value is exceeded, this should be a signal: (1) to investigate the cause with a view to taking remedial action; (2) to consult with, and seek advice from, the authority responsible for public health.
- Although the guideline values describe a quality of water that is acceptable for lifelong consumption, the establishment of these guideline values should not be regarded as implying that the quality of drinking water may be degraded to the recommended level. Indeed, a continuous effort should be made to maintain drinking water quality at the highest possible level.
- Short-term deviations above the guideline values do not necessarily mean that the water is unsuitable for consumption. The amount by which, and the period for which, any guideline value can be exceeded without affecting public health depends upon the specific substance involved. It is recommended that when a guideline value is exceeded, the surveillance agency (usually the authority responsible for public health) should be consulted for advice on suitable action, taking into account the intake of the substance from sources other than drinking water (for chemical constituents), the toxicity of the substance, the likelihood and nature of any adverse effects, the practicability of remedial measures, and similar factors.
- In developing national drinking water standards based on these guideline values, it will be necessary to take account of a variety of geographical, socioeconomic, dietary, and other conditions affecting potential exposure. This may lead to national standards that differ appreciably from the guideline values.
- In the case of radioactive substances, screening values for gross alpha and gross beta activity are given, based on a reference level of dose.[5]

Tables 2.18 through 2.22 give a summary of the WHO drinking water guideline values. The values shown denote those which, if found in a water supply system, pose no significant risk to the user. The assumptions made are that the average weight of the consumer is taken to be 60 kg and is ingesting a daily intake of 2 L of water.[5]

In any case, where treated reclaimed water is to be used for potable purposes the water quality will have to be evaluated on the basis of drinking water criteria, standards, and guidelines.[39–41]

Table 2.21 gives the WHO water quality guideline values for radioactive substances. The guidelines are based on an assumed daily intake of drinking water of 2 L and the dose resulting from a given intake of radioactive material has been calculated on the basis of the metabolism of an adult.

The recommended guideline values do not apply to water supplies contaminated during emergencies arising from accidental releases of radioactive substances to the environment.[5]

Table 2.23 presents a summary of the principal health effects of some elements or substances found in drinking water.[42,43]

VII. SULTANATE OF OMAN DRINKING WATER QUALITY STANDARDS

It can be noted that the formulated and applied Omani standards for drinking water quality resemble the WHO standards.[43] Tables 2.24 through 2.29 follow a pattern set in accordance with the classifications established by the Ministry of Commerce and Industry.[44] Drinking water standards were generally grouped into three categories. The first class addresses physical water characteristics such as total solids concentration, taste, odor, color, and turbidity (see Table 2.24).

The second group of drinking water standards deals with the chemical characteristics of the water. These are further subdivided into four main subgroups which include: toxic substances, chemical substances of health significance, chemical substances that affect drinking water quality, and the minimum residual chlorine to be used for effective chlorination. Tables 2.25 through 2.28 summarize these standards.

Table 2.18 World Health Organization Guideline Values for Bacteriological Quality of Drinking Water[a]

Organisms	Guideline value
All Water Intended for Drinking	
E. coli or thermotolerant coliform bacteria[b,c]	Must not be detectable in any 100-ml sample
Treated Water Entering the Distribution System	
E. coli or thermotolerant coliform bacteria[b]	Must not be detectable in any 100-ml sample
Total coliform bacteria	Must not be detectable in any 100-ml sample
Treated Water in the Distribution System	
E. coli or thermotolerant coliform bacteria[b]	Must not be detectable in any 100-ml sample
Total coliform bacteria	Must not be detectable in any 100-ml sample. In the case of large supplies, where sufficient samples are examined, must not be present in 95% of samples taken throughout any 12-month period

[a] Immediate investigative action must be taken if either E. coli or total coliform bacteria are detected. The minimum action in the case of total coliform bacteria is repeat sampling; if these bacteria are detected in the repeat sample, the cause must be determined by immediate further investigation.

[b] Although E. coli is the more precise indicator of fecal pollution, the count of thermotolerant coliform bacteria is an acceptable alternative. If necessary, proper confirmatory tests must be carried out. Total coliform bacteria are not acceptable indicators of the sanitary quality of rural water supplies, particularly in tropical areas where many bacteria of no sanitary significance occur in almost all untreated supplies.

[c] It is recognized that, in the great majority of rural water supplies in developing countries, fecal contamination is widespread. Under these conditions, the national surveillance agency should set medium-term targets for the progressive improvement of water supplies, as recommended in Volume 3 of *Guidelines For Drinking Water Quality*.

From *Guidelines for Drinking Water Quality,* Vol. I: Recommendations, World Health Organization, 2nd ed., Geneva, 1993.

Table 2.19 World Health Organization Water Quality Guideline Values for Chemicals of Health Significance in Drinking Water

A. Inorganic constituents

	Guideline value (mg/L)	Remarks
Antimony	0.005 (P)[a]	
Arsenic	0.01[b](P)	For excess skin cancer risk of 6×10^{-4}
Barium	0.7	
Beryllium		NAD[c]
Boron	0.3	
Cadmium	0.003	
Chromium	0.05 (P)	
Copper	2 (P)	ATO[d]
Cyanide	0.07	
Fluoride	1.5	Climatic conditions, volume of water consumed, and intake from other sources should be considered when setting national standards
Lead	0.01	It is recognized that not all water will meet the guideline value immediately; meanwhile, all other recommended measures to reduce the total exposure to lead should be implemented
Manganese	0.5 (P)	ATO
Mercury (total)	0.001	
Molybdenum	0.07	

Table 2.19 (continued) World Health Organization Water Quality Guideline Values for Chemicals of Health Significance in Drinking Water

A. Inorganic constituents (continued)

	Guideline value (mg/L)	Remarks
Nickel	0.02	
Nitrate (as NO_3^-)	50	The sum of the ratio of the concentration of each to its respective guideline value should not exceed 1
Nitrite (as NO_2^-)	3 (P)	
Selenium	0.01	
Uranium		NAD

B. Organic constituents

	Guideline value (µg/L)	Remarks
Chlorinated alkanes		
Carbon tetrachloride	2	
Dichloromethane	20	
1,1-Dichloroethane		NAD
1,2-Dichloroethane	30[b]	For excess risk of 10^{-5}
1,1,1-Trichloroethane	2000 (P)	
Chlorinated ethenes		
Vinyl chloride	5[b]	For excess risk of 10^{-5}
1,1-Dichloroethene	30	
1,2-Dichloroethene	50	
Trichloroethene	70 (P)	
Tetrachloroethene	40	
Aromatic hydrocarbons		
Benzene	10[b]	For excess risk of 10^{-5}
Toluene	700	ATO
Xylenes	500	ATO
Ethylbenzene	300	ATO
Styrene	20	ATO
Benzo[a]pyrene	0.7[b]	For excess risk of 10^{-5}
Chlorinated benzenes		
Monochlorobenzene	300	ATO
1,2-Dichlorobenzene	1000	ATO
1,3-Dichlorobenzene		NAD
1,4-Dichlorobenzene	300	ATO
Trichlorobenzenes (total)	20	ATO
Miscellaneous		
Di(2-ethylhexyl)adipate	80	
Di(2-ethylhexyl)phthalate	8	
Acrylamide	0.5[b]	For excess risk of 10^{-5}
Epichlorohydrin	0.4 (P)	
Hexachlorobutadiene	0.6	
Edetic acid (EDTA)	200 (P)	
Nitrilotriacetic acid	200	
Dialkyltins		NAD
Tributyltin oxide	2	

C. Pesticides

	Guideline value (µg/L)	Remarks
Alachlor	20[b]	For excess risk of 10^{-5}
Aldicarb	10	
Aldrin/dieldrin	0.03	
Atrazine	2	
Bentazone	30	

Table 2.19 (continued) World Health Organization Water Quality Guideline Values for Chemicals of Health Significance in Drinking Water

C. Pesticides (continued)

	Guideline value (µg/L)	Remarks
Carbofuran	5	
Chlordane	0.2	
Chlorotoluron	30	
DDT	2	
1,2-Dibromo-3-chloropropane	1[b]	For excess risk of 10^{-5}
2,4-D	30	
1,2-Dichloropropane	20 (P)	
1,3-Dichloropropane		NAD
1,3-Dichloropropene	20[b]	For excess risk of 10^{-5}
Ethylene dibromide		NAD
Heptachlor and heptachlor epoxide	0.03	
Hexachlorobenzene	1[b]	For excess risk of 10^{-5}
Isoproturon	9	
Lindane	2	
MCPA	2	
Methoxychlor	20	
Metolachlor	10	
Molinate	6	
Pendimethalin	20	
Pentachlorophenol	9 (P)	
Permethrin	20	
Propanil	20	
Pyridate	100	
Simazine	2	
Trifluralin	20	
Chlorophenoxy herbicides other than 2,4-D and MCPA		
2,4-DB	90	
Dichlorprop	100	
Fenoprop	9	
MCPB		NAD
Mecoprop	10	
2,4,5-T	9	

D. Disinfectants and disinfectant by-products

Disinfectants	Guideline value (µg/L)	Remarks
Monochloramine	3	
Di- and trichloramine		NAD
Chlorine	b	ATO for effective disinfection there should be a residual concentration of free chlorine of ≥0.5 mg/L after at least 30 minutes contact time at pH <8.0
Chlorine dioxide		A guideline value has not been established because of the rapid breakdown of chlorine dioxide and because the chlorite guideline value is adequately protective for potential toxicity from chlorine dioxide
Iodine		NAD

Table 2.19 (continued) World Health Organization Water Quality Guideline Values for Chemicals of Health Significance in Drinking Water

Disinfectant by-products	Guideline value (µg/L)	Remarks
Bromate	25[b] (P)	For 7×10^{-5} excess risk
Chlorate		NAD
Chlorite	200 (P)	
Chlorophenols		
2-Chlorophenol		NAD
2,4-Dichlorophenol		NAD
2,4,6-Trichlorophenol	200[b]	For excess risk of 10^{-5}, ATO
Formaldehyde	900	
MX		NAD
Trihalomethanes		The sum of the ratio of the concentration of each to its respective guideline value should not exceed 1
Bromoform	100	
Dibromochloromethane	100	
Bromodichloromethane	60[b]	For excess risk of 10^{-5}
Chloroform	200[b]	For excess risk of 10^{-5}
Chlorinated acetic acids		
Monochloroacetic acid		NAD
Dichloroacetic acid	50 (P)	
Trichloroacetic acid	100 (P)	
Chloral hydrate (trichloroacetaldehyde)	10 (P)	
Chloroacetone		NAD
Halogenated acetonitriles		
Dichloroacetonitrile	90 (P)	
Dibromoacetonitrile	100 (P)	
Bromochloroacetonitrile		NAD
Trichloroacetonitrile	1 (P)	
Cyanogen chloride (as CN)	70	
Chloropicrin		NAD

[a] (P) — Provisional guideline value. This term is used for constituents for which there is some evidence of a potential hazard but where the available information on health effects is limited; or where an uncertainty factor greater than 1000 has been used in the derivation of the tolerable daily intake (TDI). Provisional guideline values are also recommended: (1) for substances for which the calculated guideline value would be below the practical quantification level, or below the level that can be achieved through practical treatment methods, or (2) where disinfection is likely to result in the guideline value being exceeded.

[b] For substances that are considered to be carcinogenic, the guideline value is the concentration in drinking water associated with an excess lifetime cancer risk of 10^{-5} (one additional cancer per 100 000 of the population ingesting drinking water containing the substance at the guideline value for 70 years). Concentrations associated with estimated excess lifetime cancer risks of 10^{-4} and 10^{-6} can be calculated by multiplying and dividing, respectively, the guideline value by 10.

In cases in which the concentration associated with an excess lifetime cancer risk of 10^{-5} is not feasible as a result of inadequate analytical or treatment technology, a provisional guideline value is recommended at a practicable level and the estimated associated excess lifetime cancer risk presented.

It should be emphasized that the guideline values for carcinogenic substances have been computed from hypothetical mathematical models that cannot be verified experimentally and that the values should be interpreted differently than TDI-based values because of the lack of precision of the models. At best, these values must be regarded as rough estimates of cancer risk. However, the models used are conservative and probably err on the side of caution. Moderate short-term exposure to levels exceeding the guideline value for carcinogens does not significantly affect the risk.

[c] NAD — No adequate data to permit recommendation of a health-based guideline value.

[d] ATO — Concentrations of the substance at or below the health-based guideline value may affect the appearance, taste, or odor of the water.

From *Guidelines for Drinking Water Quality,* Vol. I: Recommendations, World Health Organization, 2nd ed., Geneva, 1993.

Table 2.20 World Health Organization Water Quality Guideline Values for Substances and Parameters in Drinking Water that May Give Rise to Complaints from Consumers

	Levels likely to give rise to consumer complaints[a]	Reasons for consumer complaints
Physical parameters		
Color	15 TCU[b]	Appearance
Taste and odor	—	Should be acceptable
Temperature	—	Should be acceptable
Turbidity	5 NTU[c]	Appearance; for effective terminal disinfection, median turbidity ≤1 NTU, single sample ≤5 NTU
Inorganic constituents		
Aluminum	0.2 mg/L	Depositions, discoloration
Ammonia	1.5 mg/L	Odor and taste
Chloride	250 mg/L	Taste, corrosion
Copper	1 mg/L	Staining of laundry and sanitary ware (health-based provisional guideline value 2 mg/L)
Hardness	—	High hardness: scale deposition, scum formation Low hardness: possible corrosion
Hydrogen sulfide	0.05 mg/L	Odor and taste
Iron	0.3 mg/L	Staining of laundry and sanitary ware
Manganese	0.1 mg/L	Staining of laundry and sanitary ware (health-based provisional guideline value 0.5 mg/L)
Dissolved oxygen	—	Indirect effects
pH	—	Low pH: corrosion High pH: taste, soapy feel, preferably <8.0 for effective disinfection with chlorine
Sodium	200 mg/L	Taste
Sulfate	250 mg/L	Taste, corrosion
Total dissolved solids	1000 mg/L	Taste
Zinc	3 mg/L	Appearance, taste
Organic constituents		
Toluene	24–170 µg/L	Odor, taste (health-based guideline value 700 µg/L)
Xylene	20–1800 µg/L	Odor, taste (health-based guideline value 500 µg/L)
Ethylbenzene	2–200 µg/L	Odor, taste (health-based guideline value 300 µg/L)
Styrene	4–2600 µg/L	Odor, taste (health-based guideline value 20 µg/L)
Monochlorobenzene	10–120 µg/L	Odor, taste (health-based guideline value 300 µg/L)
1,2-Dichlorobenzene	1–10 µg/L	Odor, taste (health-based guideline value 1000 µg/L)
1,4-Dichlorobenzene	0.3–30 µg/L	Odor, taste (health-based guideline value 300 µg/L)
Trichlorobenzenes (total)	5–50 µg/L	Odor, taste (health-based guideline value 20 µg/L)
Synthetic detergents	—	Foaming, taste, odor
Disinfectants and disinfectant by-products		
Chlorine	600–1000 µg/L	Taste and odor (health-based guideline value 5 mg/L)
Chlorophenols		
2-Chlorophenol	0.1–10 µg/L	Taste, odor
2,4-Dichlorophenol	0.3–40 µg/L	Taste, odor
2,4,6-Trichlorophenol	2–300 µg/L	Taste, odor (health-based guideline value 200 µg/L)

[a] The levels indicated are not precise numbers. Problems may occur at lower or higher values according to local circumstances. A range of taste and odor threshold concentrations is given for organic constituents.
[b] TCU, time color unit.
[c] NTU, nephelometric turbidity unit.

From *Guidelines for Drinking Water Quality,* Vol. I: Recommendations, World Health Organization, 2nd ed., Geneva, 1993.

Table 2.21 World Health Organization Water Quality Guideline Values for Radioactive Constituents in Drinking Water

	Screening value (Bq/L)	Remarks
Gross alpha activity	0.1	If a screening value is exceeded, more detailed radionuclide analysis is necessary.
Gross beta activity	1	Higher values do not necessarily imply that the water is unsuitable for human consumption

From *Guidelines for Drinking Water Quality,* Vol. I: Recommendations, World Health Organization, 2nd ed., Geneva, 1993.

Table 2.22 World Health Organization Water Quality Guideline Values for Chemicals Not of Health Significance at Concentrations Normally Found in Drinking Water

Chemical	Remarks
Asbestos	U
Silver	U
Tin	U

Note: U — It is unnecessary to recommend a health-based guideline value for these compounds because they are not hazardous to human health at concentrations normally found in drinking water.

From *Guidelines for Drinking Water Quality,* Vol. I: Recommendations, World Health Organization, 2nd ed., Geneva, 1993.

Table 2.23 The Principal Health Effects of Some Elements or Substances[5,42,43]

Contaminant	Principal health effects
Aldrin/dieldrin	Central nervous system effects
Arsenic	Dermal and nervous system toxicity effects, cancer
Asbestos	Gastric leiomyosarcoma, malignancies in animals. Cancer in humans
Barium	Circulatory system effects. Toxic, muscle stimulant, cardiovascular diseases
Beryllium	Rhinitis, pharyngitis, pneumonitis, pulmonary edema, regarded carcinogen
Cadmium	Kidney effects. Itai-Itai disease (a bone disease), hypertension
Carbon tetrachloride (CCl_4)	Poisoning, liver/kidney injury, hepatic complications
Chlordane	Vomiting
Chlorobenzenes	Irritating to respiratory system, central nervous system depressant.
Chloroform	Central nervous system depressant, liver/kidney effects, unconsciousness, toxic, carcinogenic effects
Chromium	Liver/kidney effects. Toxic, cancer
Cyanide	Higher exposures may be fatal
DDT	Toxic, nervous system effects, affects kidney
Endrin	Nervous system/kidney effects
Fluoride	Skeletal damage (fluorosis), dental caries, mottling of teeth, toxic
Hardness	Urolithiasis, cardiovascular risk
Hexachlorobenzene	Toxic, carcinogen
Lead	Central and peripheral nervous system damage; kidney effects; highly toxic to infants and pregnant women
Lindane	Nervous system/kidney effects
Mercury	Central nervous system disorders; neurological and renal disturbances
Methoxychlor	Nervous system/kidney effects
Nickel	High doses cause minimal toxic effects, dermatitis, renal problems
Nitrate and nitrite	Methemoglobinemia (blue-baby syndrome)
Pentachlorophenol	Poisoning, cardiac arrest
Selenium	Gastrointestinal effects
Silver	Discoloration of skin, hair, and fingernails (argyria)
Sodium	Excessive amounts cause vomiting and the elimination of much of the salt
Tetrachloroethane	Central nervous system depressions, tumor initiator
Trichloroethane	Central nervous system depressions, anesthetic
Trihalomethanes (THMs)	Cancer risk
2,4-D	Liver/kidney effects
1,2-Dichloroethane	Kidney/liver damage

The first subgroup of chemical substances points to inorganic constituents of health significance together with the fluoride and nitrates (mentioned in the second subgroup) and are presented in Tables 2.25 and 2.26.

Table 2.24 The Sultanate of Oman Drinking Water Quality Standards for Physical Characteristics

Substance or characteristic	Highest desirable level	Maximum permissible level
Color	5[a]	50[a]
Odor	Unobjectionable	Unobjectionable
Taste	Unobjectionable	Unobjectionable
Turbidity	5[b]	25[b]
Total dissolved solids	500 mg/L	1500 mg/L

[a] On the platinum-cobalt scale.
[b] Turbidity units (NTU).

From Ministry of Commerce and Industry, Sultanate of Oman Standards: Drinking Water, No. 8, 1978, General Department of Standards and Weights, Muscat, Oman.

Table 2.25 The Sultanate of Oman Drinking Water Standards for Toxic Chemical Substances

Characteristic	Maximum permissible level (mg/L)
Lead (as Pb)	0.1
Selenium (as Se)	0.01
Arsenic (as As)	0.05
Cadmium (as Cd)	0.01
Cyanide (as CN)	0.05
Mercury (as Hg)	0.001

Note: The concentration of indicated substances must not be greater than the values indicated herein.

From Ministry of Commerce and Industry, Sultanate of Oman Standards: Drinking Water, No. 8, 1978, General Department of Standards and Weights, Muscat, Oman.

Table 2.26 The Sultanate of Oman Drinking Water Standards for Chemical Substances of Special Health Effects

Characteristic	Maximum permissible level (mg/L)
Fluoride (as F)	0.8
Nitrate (as NO_3)	45

Note: The concentration of indicated substances must not be greater than the values indicated herein.

From Ministry of Commerce and Industry, Sultanate of Oman Standards: Drinking Water, No. 8, 1978, General Department of Standards and Weights, Muscat, Oman.

The third subgroup of chemical substances of concern are presented in Table 2.27.[44]

The third set of standards deals with the biological and bacteriological parameters for both untreated and treated water for drinking water purposes (see Table 2.29).[44]

Two major water quality parameters for drinking water have not been included, such as organic constituents other than phenols and radioactive substances.

Table 2.27 The Sultanate of Oman Drinking Water Standards for Chemical Substances that Affect Drinking Water Quality

Characteristic	Highest desirable level (mg/L)	Maximum permissible level (mg/L)
Total dissolved solids	500	1500
Copper (as Cu)	0.05	1.5
Iron (as Fe)	0.1	1.0
Magnesium (as Mg)	Not more than 30 mg/L if there are 250 mg/L of sulfate; if there is less sulfate, magnesium up to 150 mg/L may be allowed	
Manganese (as Mn)	0.05	0.5
Zinc (as Zn)	5.0	15.0
Calcium (as Ca)	75	200
Chloride (as Cl)	200	600
Sulfate (as SO_4)	200	400
Phenolic compounds (as phenol)	0.001	0.002
Total hardness (as $CaCO_3$)	100	500
pH range	(7–8.5)	(6.5–9.2)

From Ministry of Commerce and Industry, Sultanate of Oman Standards: Drinking Water, No. 8, 1978, General Department of Standards and Weights, Muscat, Oman.

Table 2.28 The Sultanate of Oman Drinking Water Standards for Minimum Residual Chlorine

The minimum concentration of residual chlorine for effective chlorination should not be less than 0.2–0.5 mg/L in treated water.

From Ministry of Commerce and Industry, Sultanate of Oman Standards: Drinking Water, No. 8, 1978, General Department of Standards and Weights, Muscat, Oman.

Table 2.29 The Sultanate of Oman Drinking Water Standards for Bacteriological Characteristics

Treated Water

No sample should contain *E. coli* in 100 mL
No sample should contain more than 10 viable microorganisms of the coliform group per 100 mL
Throughout any year, 95% of all samples should not contain any microorganisms in 100 mL

Untreated Water

No sample should contain *E. coli* in 100 mL
No sample should contain more than 10 viable microorganisms of the coliform group per 100 mL

From Ministry of Commerce and Industry, Sultanate of Oman Standards: Drinking Water, No. 8, 1978, General Department of Standards and Weights, Muscat, Oman.

Table 2.30 Suggested Treatment Processes to Meet the Given Health Criteria for Wastewater Reuse for Agricultural Irrigation

Unit treatment process	Type of agriculture reuse		
	Crops not for direct human consumption	Crops eaten cooked	Crops eaten raw
Primary treatment	+++	+++	+++
Secondary treatment		+++	+++
Sand filtration or equivalent polishing methods		+	+
Disinfection		+	+++
Health criteria	A + F	D + F	

Note: +++ = Essential.
  + = May sometimes be required.
  A = Freedom from gross solids; significant removal of parasite eggs.
  D = Not more than 100 coliforms per 100 mL in 80% of samples.
  F = No chemicals that lead to undesirable residues in crops.

Compiled from Reference 45.

Table 2.31 Suggested Treatment Processes to Meet the Given Health Criteria for Wastewater Reuse in Recreational Facilities

Unit treatment process	Type of recreation	
	No contact	Contact
Primary treatment	+++	+++
Secondary treatment	+++	+++
Sand filtration or equivalent polishing methods		+++
Disinfection	+	+++
Health criteria	B	D + G

Note: +++ = Essential.
  + = May sometimes be required.
  B = Freedom from gross solids; significant removal of parasite eggs plus a significant removal of bacteria.
  D = Not more than 100 coliforms per 100 mL in 80% of samples.
  G = No chemicals that lead to irritation of mucous membranes and skin.

Compiled from References 41 and 45.

VIII. WORLD HEALTH ORGANIZATION (WHO) CRITERIA FOR WASTEWATER REUSE

To allow for safe reuse of reclaimed water for agricultural irrigation the WHO criteria for reuse of effluents recommend that the irrigation of crops consumed raw be biologically treated plus disinfected in order to produce a coliform count of not more than 100 coliform organisms per 100 ml in 80% of the samples tested. Table 2.30 illustrates the recommended wastewater treatment to meet these recommended health criteria.[45]

The World Health Organization indicates that if reclaimed wastewater is to be used for recreational purposes, the kind and degree of treatment is governed by the potential for human contact. Table 2.31 presents the WHO suggested criteria and treatment processes that should be applied for both body contact and non-body contact sports.[41,45]

Table 2.32 Recommended Microbiological Quality Guidelines for Wastewater Use in Agriculture[a]

Category	Reuse conditions	Exposed group	Intestinal nematodes[b] (arithmetic mean no. of eggs per liter[c])	Fecal coliforms (geometric mean no. per 100 ml[c])	Wastewater treatment expected achieve the required microbiological quality
A	Irrigation of crops likely to be eaten uncooked, sports fields, public parks[d]	Workers, consumers, public	≤1	≤1000[d]	A series of stabilization ponds designed to achieve the microbiological quality indicated, or equivalent treatment
B	Irrigation of cereal crops, industrial crops, fodder crops, pasture and trees[e]	Workers	≤1	No standard recommended	Retention in stabilization ponds for 8–10 days or equivalent helminth and fecal coliform removal
C	Localized irrigation of crops in category B if exposure of workers and the public does not occur	None	Not applicable	Not applicable	Pretreatment as required by the irrigation technology, but not less than primary sedimentation

[a] In specific cases, local epidemiological, sociocultural, and environmental factors should be taken into account, and the guidelines modified accordingly.
[b] *Ascaris* and *Trichuris* species and hookworms.
[c] During the irrigation period.
[d] A more stringent guideline (≤200 fecal coliforms per 100 ml) is appropriate for public lawns, such as hotel lawns, with which the public may come into direct contact.
[e] In the case of fruit trees, irrigation should cease two weeks before fruit is picked, and no fruit should be picked off the ground. Sprinkler irrigation should not be used.

Compiled from References 46 and 47.

IX. DEVELOPING COUNTRIES: ENGELBERG REPORT/GUIDELINES FOR WASTEWATER REUSE

The review meeting of environmental specialists and epidemiologists held at Engelberg, Switzerland formulated a model for the health risks that are linked to the use of untreated excreta and wastewater in agriculture and aquaculture. The model advocated that the significance of diseases be in a sequential order as follows: intestinal nematode infections such as *Ascaris, Trichuris,* and hookworms, first; excreted bacterial infections such as dysentery and typhoid, and excreted viral infections such as rotavirus and infectious hepatitis A, second. The model has been adopted for the formulation of excreta and wastewater use guidelines. The recommended guidelines on the health risks from wastewater or excreta use has been based on the analysis of credible epidemiological studies that have demonstrated the health effects from such use. Likewise, the model is based on the factors which affect the potential transmission of different wastewater and excreta pathogenic agents.[46–49]

Table 2.32 presents the Engelberg meeting recommendations for the microbiological quality of treated wastewater to be used for agricultural irrigation in tropical and subtropical countries.[46,47]

The strict helminth standard recommended was selected as an indicator for all of the easily settleable pathogens including some of the protozoans such as *Schistosoma, Amoeba,* and *Giardia*.[49]

The Engelberg report advocates the employment of waste stabilization ponds for the treatment of wastewater. However, the extensive land area needed for the ponds may restrict their application in certain localities.

Table 2.33a Wastewater Reuse — Areas of Application of Standards A and B

	A	B
Crops	Vegetables likely to be eaten raw	Vegetables to be cooked or processed
	Fruit likely to be eaten raw and within 2 weeks of any irrigation	Fruit if no irrigation within 2 weeks of cropping Fodder, cereal and seed crops
Grass and Ornamental areas	Public parks, hotel lawns, recreational areas	Pastures
	Areas with public access	Areas with no public access
	Lakes with public contact (except places which may be used for praying and hand washing)	
Aquifer recharge	All controlled aquifer recharge	
Method of irrigation	Spray or any other method of aerial irrigation not permitted in areas with public access unless with timing control	
Any other reuse applications	Subject to the approval of the Ministry	

Adapted from Reference 50.

The report does not include the reuse of reclaimed water originating from municipal or industrial treatment systems. Likewise, the report does not include the reuse of reclaimed wastewater in other reuse applications apart from agricultural and aquacultural fields.

X. SULTANATE OF OMAN WASTEWATER REUSE REQUIREMENTS

In case of wastewater sludge treatment and disposal, the Omani Regulations have been divided into two broad areas. The first area marks parameters to be considered for wastewater reuse and discharge, and sludge application to land.[50] This is in accord with the Ministerial Decision 145/93 concerning regulations for wastewater reuse and discharge. Table 2.33 (see also Chapter 10, Tables 10.29 and 10.30) gives an outline of the Sultanate of Oman Regulations concerning wastewater reuse and discharge, while Table 2.34 presents limits for disposal of treated sludge.

As outlined in Table 2.33, emphasis was placed on chemical substances. Radioactive materials are covered in a separate regulation incorporated with hazardous wastes.[51]

Other articles of the Ministerial decision 145/93 include:[51]

- The discharge to the environment of any wastewater or sludge in whatever form or condition is prohibited without a Permit to Discharge issued by the Ministry. The Permit to Discharge may be amended by the Ministry at any time after giving reasonable notice of any change to the owner.
- The final point(s) of discharge of wastewater to the environment shall only be at the point(s) marked on the drawing(s) listed in the Permit to Discharge.
- Wastewater quality shall at all times be within the limits that are set out in Table 2.33 as they relate to the permitted method of discharge or as may be modified and supplemented by any other limits that might be included in any specific Permit to Discharge.
  The soil on which sludge may be applied shall be tested by the owner for the metals listed in Table 2.34 and for pH value, prior to any initial application; and the sludge quality and application constraints shall at all times be within the limits that are set out in Table 2.34 as they relate to the permitted method of sludge reuse, or as may be modified and supplemented by any other limits that might be included in any specific Permit to Discharge.
- Any sludge having concentrations of metals greater than the limits prescribed in Table 2.34 shall be disposed of in sanitary landfills or to other facilities but only with the prior approval of the Ministry.
- Facilities and equipment shall be provided and maintained by the owner to the requirements and satisfaction of the Ministry for sampling, measuring, and recording the quantity and rate of discharge of the wastewater and for determining its characteristics.
- Samples and readings shall be taken by the owner at intervals stated in the Permit to Discharge, or as required by the Ministry. All data shall be recorded and submitted at the end of each month to the Ministry in an approved format.

Table 2.33b Wastewater — Maximum Quality Limits

Parameter	Standards (See Table 10.29)	
	A	B
	(mg/L except where noted)	
Biochemical oxygen demand (BOD) (5 d @ 20°C)	15.000	20.000
Chemical oxygen demand (COD)	150.000	200.000
Suspended solids (SS)	15.000	30.000
Total dissolved solids (TDS)	1500.000	2000.000
Electrical conductivity (EC) (µS/cm)	2000.000	2700.000
Sodium absorption ration (SAR)	10.000	10.000
pH (within range), pH units	6–9.000	6–9.000
Aluminum (as Al)	5.000	5.000
Arsenic (as As)	0.100	0.100
Barium (as Ba)	1.000	2.000
Beryllium (as Be)	0.100	0.300
Boron (as B)	0.500	1.000
Cadmium (as Cd)	0.010	0.010
Chloride (as Cl)	650.000	650.000
Chromium (total as Cr)	0.050	0.050
Cobalt (as Co)	0.050	0.050
Copper (as Cu)	0.500	1.000
Cyanide (total as CN)	0.050	0.100
Fluoride (as F)	1.000	2.000
Iron (total as Fe)	1.000	5.000
Lead (as Pb)	0.100	0.200
Lithium (as Li)	0.070	0.070
Magnesium (as Mg)	150.000	150.000
Manganese (as Mn)	0.100	0.500
Mercury (as Hg)	0.001	0.001
Molybdenum (as Mo)	0.010	0.050
Nickel (as Ni)	0.100	0.100
Nitrogen: Ammoniacal (as N)	5.000	10.000
Nitrate (as NO_3)	50.000	50.000
Organic (Kjeldahl, as N)	5.000	10.000
Oil and grease (total extractable)	0.500	0.500
Phenols (total)	0.001	0.002
Phosphorus (total as P)	30.000	30.000
Selenium (as Se)	0.020	0.002
Silver (as Ag)	0.010	0.010
Sodium (as Na)	200.000	300.000
Sulfate (as SO_4)	400.000	40.0000
Sulfide (total as S)	0.100	0.100
Vanadium (as V)	0.100	0.100
Zinc (as Zn)	5.000	5.000
Fecal coliform bacteria (per 100 mL)	200.000	1000.000
Viable nematode ova (/L)	<1.000	<1.000

Adapted from Reference 50.

- Wastewater or sludge shall not be dicharged sacrificially except in an exceptional circumstance where no form of wastewater reuse is possible.
- No wastewater or sludge shall be transported from the site of its origin without the prior approval of the Ministry. Approval shall be subject to conditions that will include the obligation for all transport movements to be recorded in a manner defined in the approval.
- The Ministry shall have the absolute right to inspect and/or monitor any wastewater treatment plant and to take samples of any wastewater, sludge, or soil at any time and place.
- These Regulations shall not apply to discharges from septic tanks or to discharges of wastewater to the marine environment, or discharges of wastewater or sludge which contain radioactive matter, which are subject to separate legislation.

Table 2.34 Reuse of Sludge in Agriculture — Conditions for Application to Land

Metal	Maximum Concentration (mg/kg of dry Solids)	Maximum Application Rate (kg/ha/yr)*	Maximum Permitted Concentration in Soil (mg/kg of dry solids)
Cadmium	20	0.150	3
Chromium	1000	10	400
Copper	1000	10	150
Lead	1000	15	30
Mercury	10	0.100	1
Molybdenum	20	0.100	3
Nickel	300	3	75
Selenium	50	0.150	5
Zinc	3000	15	300

After the spreading of sludge there must be a minimum period of three weeks before grazing or harvesting of forage crops.

Sludge use is prohibited:
- on soils while fruit or vegetable crops, other than fruit trees, are growing or being harvested.
- for six months preceding the harvesting of fruit or vegetables which grow in contact with the soil and which are normally eaten raw.
- on soils with a pH < 7.0.

* Based on a 10-year average and a soil pH > 7.0.

Ministry of Regional Muncipalities and Environment, Sultanate of Oman, Regulations for the Management of Hazardous Waste, Ministerial Decision 18/93 dated 2 February, 1993.

XI. HOMEWORK PROBLEMS

A. DISCUSSION QUESTIONS

1. Define criteria, standards, guidelines, and regulations.
2. What are the major concerns that an organization must address before using reclaimed wastewater?
3. What are the two methods used by EPA to develop water quality standards?
4. Explain the difference between an MCLG and an MCL.
5. What are the major contents of a criteria document?
6. Adoption, development, and enforcement of criteria, standards, quidelines, and regulations for drinking water quality must be preceded by certain activities. What are at least six of these activities?
7. Give the reasons why it is important to have established water quality standards.
8. Enumerate the advantages and disadvantages in setting water quality standards.
9. Explain the difference between the terms direct and indirect use of reclaimed wastewater.
10. What are at least seven categories into which the use of reclaimed wastewater may be classified?
11. What are the advantages and disadvantages in using reclaimed wastewater for agricultural irrigation purposes?
12. What are the major parameters used to evaluate the suitability of reclaimed wastewater for agricultural irrigation?
13. What management practices can be used to reduce the toxic effects of boron in reclaimed wastewater used for agricultural irrigation?
14. What are at least five trace metals of major concern in using reclaimed wastewater for agricultural irrigation?
15. Which method in the secondary treatment processes appears to be the most effective at removing trace metals from the reclaimed wastewater?
16. What are the three methods used for groundwater recharge using reclaimed wastewater?
17. List the five categories included in the World Health Organization guidelines for drinking water quality.
18. Explain the effect of pH on the availability of boron in soil for plant growth?
19. What major water quality characteristics must be considered in using reclaimed wastewater for groundwater recharge?
20. At this time, how many direct reuse (potable) of reclaimed wastewater projects are in operation in the world?

21. Which of the industrial categories at present uses the largest volume of reclaimed wastewater?
22. What is the single most important component in reclaimed wastewater that must be considered if this water is to be used for livestock watering?
23. List five physical characteristics that affect the aesthetic quality of drinking water?
24. Indicate the principle health effects of excess levels of the following elements or substances in drinking water: beryllium, cadmium, lead, nickel, mercury, nitrates.
25. What method of wastewater treatment would be recommended for reclaimed wastewater used for irrigating crops that are to be consumed raw?
26. Discuss the model formulated by the Engelberg group for the health risks which are linked to the use of untreated excreta and wastewater in agriculture and aquaculture.
27. Matching (use only one number for each item)

    ___ a. Conductivity (salinity)
    ___ b. *Ascaris, Trichuris,* hookworms
    ___ c. Turbidity
    ___ d. SAR
    ___ e. Color
    ___ f. Alkalinity
    ___ g. Boron
    ___ h. Pathogen
    ___ i. Wadis
    ___ j. Dysentery

    1. Seasonal streams
    2. Disease-causing organism
    3. $CO_3^{2-} + HCO_3^-$ in meq/L
    4. Bleaching component in detergent
    5. NTUs
    6. Intestinal nematodes
    7. Platinum-cobalt standard
    8. *Shigella*
    9. Total dissolved solids (TDS)
    10. $\dfrac{Na}{\sqrt{\dfrac{Ca+Mg}{2}}}$

B. SPECIFIC MATHEMATICAL PROBLEMS

1. A chemical analysis of the effluent from a wastewater treatment plant is as follows:

    SiO_2 = 10 mg/L
    Fe = 0.57 mg/L
    Ca^{2+} = 3.8 mg/L
    Mg^{2+} = 1.9 mg/L
    Na^+ = 280 mg/L
    K^+ = 5.0 mg/L
    HCO_3^- = 609 mg/L as HCO_3^-
    CO_3^{2-} = 35 mg/L as CO_3^{2-}
    SO_4^{2-} = 53 mg/L
    Cl^- = 113 mg/L
    F = 0.6 mg/L
    B = 0.26 mg/L
    TDS = 703 mg/L
    pH = 8.4

    BOD = 15 mg/L
    SS = 40 mg/L
    $NH_3 - N$ = 25 mg/L
    $NO_3 - N$ = 0.4 mg/L
    Total P = 7 mg/L
    Total chlorine residual = 1.5 mg/L
    Dissolved oxygen = 5 mg/L

    Evaluate this water for its suitability for agricultural irrigation. Determine the SAR and Adj. SAR. Estimate the salinity (electrical conductivity) at 25°C. Also determine the hardness and alkalinity as mg/L $CaCO_3$.

2. Which of the following reclaimed wastewaters whose characteristics are given below would be most suitable for agricultural irrigation?

	Water #1	Water #2
Ca^{2+}	30 mg/L	40 mg/L
Mg^{2+}	6.1 mg/L	24.4 mg/L
Na^+	46 mg/L	92 mg/L
HCO_3^-	122 mg/L	122 mg/L
CO_3^{2-}	44 mg/L	ND (nondetectable)
pH	8.4	7.6
Cl^-	180 mg/L	250 mg/L
NO_3^{2-}	5 mg/L	12 mg/L
TDS	550 mg/L	1000 mg/L
Electrical conductivity	0.86 mmhos/cm	1.6 mmhos/cm

3. Evaluate the salinity (electrical conductivity) of drainage water if reclaimed wastewater is used for agricultural irrigation and a 90-cm-depth of water is applied through a 30-cm-depth of soil to achieve a leaching requirement of 20%. Use electrical conductivities for the irrigation water presented in Problem 2. Also estimate the total dissolved solids (TDS) in the drainage water.

C. SUGGESTED EXERCISES

1. Conduct a library and literature search for a wastewater reuse project in your community or region, then visit the project and prepare a report regarding this project.

   The report or case study should contain some or all of the following elements:

   1. Introduction
   2. Background and Development of Project
   3. Description of the Facility
   4. Treatment Processes Used
   5. Plant Operation and Performance
   6. Monitoring Techniques
   7. Analytical Techniques
   8. Cost of Construction and Operation
   9. Compliance with Regulations
   10. Conclusions and Discussion
   11. References

2. Contact the agency in your state or area responsible for regulations governing the reuse of reclaimed wastewater. Ask for information as to criteria, standards, or regulations pertaining to wastewater reclamation and reuse. Ask for information regarding the number of existing water reuse projects in your state or area and the categories included, such as agricultural irrigation, landscape irrigation, industrial use, groundwater recharge, recreational use, stock watering, cooling water, fire fighting, and street cleaning, and the volume of water involved. Also ask for brochures, references, or publications regarding specific projects in which they are involved.

REFERENCES

1. **Calabrese, E. J., C. E. Gilbert, and H. Pastides,** *Safe Drinking Water Act, Amendments, Regulations and Standards,* Lewis Publishers, Chelsea, MI, 6, 23, 145, 1989.
2. Air Quality Criteria for Carbon Monoxide, U.S. Department of Health, Education and Welfare, AP-62, U.S. Government Printing Office, Washington, D.C., 1970.
3. *Glossary Water and Wastewater Control Engineering,* 3rd ed., American Public Health Association, Washington, D.C., 225, 1981.
4. **Goldfarb, W.,** *Water Law,* 2nd ed., Lewis Publishers, Chelsea, MI, 259, 1988.
5. Guidelines for Drinking Water Quality, Volume 1: Recommendations, World Health Organization, 2nd ed., Geneva, 1993.
6. **Abdel-Magid, I. M. and El-Hassan, B. M.,** Reflections on drinking water quality guidelines for the Sudan, *J. Water International,* 12(1), 33, 1987.
7. **Abdel-Magid, I. M. and El-Zawahry, A.** Formulation of Omani Standards for Water and Wastewater, A paper presented at the seminar on Transfer of Technology held at the College of Engineering, Sultan Qaboos University, March 1st, 1992.
8. **Nathanson, J. A.,** *Basic Environmental Technology: Water Supply, Waste Disposal and Pollution Control,* Prentice-Hall, Englewood Cliffs, NJ, 1986.
9. **Waite, T. D., with chapters by Quon, J. E. and Freeman, N. J.,** *Principles of Water Quality,* Academic Press, New York, 1984.
10. **McGauhey, P. H.,** *Engineering Management of Water Quality,* McGraw Hill, New York, 1968.
11. **Hammer, M.,** *Water and Wastewater Technology,* Prentice-Hall, Englewood Cliffs, NJ, 1986.
12. WPCF, *Water Reuse,* Manual of Practice SM-3, 2nd ed., Water Pollution Control Federation, Alexandria, VA, 78, 201, 1989.
13. USEPA, Manual — Guidelines For Water Reuse, EPA/625/R-92/004, Office of Water, Office of Wastewater Enforcement and Compliance, U.S. Environmental Protection Agency, Washington, D.C., September, 76, 77, 81, 105, 123, 1992.
14. **Chansler, J. M.,** Personal communication, 1993.

15. **Asano, T., D. Richard, R. W. Crites, and G. Tchobanoglous,** Evolution of Tertiary Treatment Requirements in California, *Water Environ. Technol.,* 4, 36, 1992.
16. **Shuval, H. I.,** *Water Renovation And Reuse,* Academic Press, New York, 6, 46, 1977.
17. **Rowe, D. R., K. Al-Dhowalia, and A. Whitehead,** *Reuse of Riyadh Treated Wastewater,* Project No. 18/1402, King Saud University, The College of Engineering Research Center, Riyadh, Saudi Arabia, 1988.
18. **Lieuwen, A.,** *Effluent Use In The Phoenix And Tucson Metropolitan Area,* Water Resources Research Center, University of Arizona, Phoenix, 20, 1990.
19. **Culp, G., et al.,** *Wastewater Reuse And Recycling Technology,* Noyes Data Corp, Park Ridge, NJ, 227, 1980.
20. **Ayers, R. S. and D. W. Westcot,** Water Quality For Agriculture, Food and Agriculture Organization of the United Nations, Rome, 7, 11, 54, 69, 1976.
21. **USPHA,** *Standard Methods For The Examination of Water and Wastewater,* 18th ed., American Public Health Association (AWWA, WEF), Washington, D.C., 2–46, 1992.
22. **USPHA,** *Standard Methods for the Examination of Water and Wastewater,* 15th ed., American Public Health Association, Washington, D.C., 90, 1980.
23. **USEPA,** Evaluation of Land Application Systems, Office of Water Program Operations, EPA-430/9–75–001, U.S. Environmental Protection Agency, Washington, D.C. 20460, 25, 1975.
24. **FWPCA,** Report of the Committee on Water Quality Criteria, Federal Water Pollution Control Administration, U.S. Department of the Interior (Green Book), Washington, D.C., 164, 1968.
25. **Kemmer, F. N.,** *The Nalco Water Handbook,* McGraw-Hill, New York, 6–10, 1979.
26. **Mengel, K. and E. A. Kirby,** *Principles of Plant Nutrition,* International Potash Institute, Bern, Switzerland, 533, 534, 1982.
27. **USEPA,** Process Design Manual For Land Treatment of Municipal Wastewater, U.S. Environmental Protection Agency, Office of Water Programs Operations, Washington, D.C., October, 7-7, 7-22, 7-52, 1977.
28. **AWWA,** *Municipal Wastewater Reuse News,* AWWA Foundation, Denver, CO, October, 19, 1977, February, 9, 21, 1981.
29. **Van der Leeden, F., Troise, F. L., and Todd, D. K.,** *The Water Encyclopedia,* 2nd ed., Lewis Publishers, Boca Raton, FL, 466, 1990.
30. **USEPA,** Process Design Manual For Land Treatment of Municipal Wastewater, EPA 625/1–77–008, E1, E2 U.S. Environmental Protection Agency, Washington, D.C., October, 1977.
31. **Culp/Wesner/Culp,** Water Reuse And Recycling Volume 2, Evaluation of Treatment Technology, U.S. Department of the Interior OWRT/RU-79/2, Washington, D.C., 70, 110, 1979.
32. **York, D.,** Reuse in Florida — An Update, *Florida Water Resources Journal,* 43, 28, 1991.
33. California Municipal Wastewater Reclamation In 1987, California State Water Resources Control Board, Office of Water Recycling, Sacramento, 4, 5, 1990.
34. **Cook, J., T. Asano, and M. Mellor,** Groundwater Recharge with Reclaimed Water in California, *Water Environment and Technology,* 8, 1990.
35. **Olson, B. H. and W. Bruvola,** Influence of Social Factors on Public Acceptance of Renovated Wastewater, and K.D. Linstedt and M.R. Rothberg, Potable Water Reuse, *Water Reuse,* E.J. Middlebrooks, Ed., Ann Arbor Science Publishers, Ann Arbor, MI, 56, 199, 1982.
36. Denver Residents Plan to Drink Their Sewage, *Environmental News Record,* 20, December 18, 1986.
37. **WHO,** Report of a WHO Scientific Group, Human Viruses in Water, Wastewater and Soil, TRS 639, WHO, Geneva, 1979.
38. Hawaii State Department of Health, Wastewater Branch, Proper Guidelines For Use of Water Reclamation, Honolulu, April, 1991.
39. **Gorchev, H. G. and Ozolins, G.,** WHO Guidelines for Drinking Water Quality, a paper presented at the International Water Supply Association Congress, Zurich, September, 6–10, 1982.
40. **Pontius, F. W.,** Ed., 4th ed., *American Water Works Association, Water Quality and Treatment: A Handbook of Community Water Supplies,* McGraw-Hill, New York, 1990.
41. **Hillman, P. J.,** Health Aspects of Reuse of Treated Wastewater for Irrigation, *Treatment and Use of Sewage Effluent for Irrigation,* Pescod, M.B. and Arar, A., Eds., Proceedings of the FAO Regional Seminar on the Treatment and Use of Sewage Effluent for Irrigation, Nicosia, Cyprus, 52–73, October 7–9, 1985, Butterworths, London 1988.
42. **Masters, G. M.,** *Introduction to Environmental Engineering and Science,* Prentice-Hall, Englewood Cliffs, NJ, 1991.
43. **WHO,** International Standards For Drinking Water, World Health Organization, 3rd ed., Geneva, 1971.
44. Ministry of Commerce and Industry, Sultanate of Oman Standards: Drinking Water, No. 8, 1978, General Department of Standards and Weights, Muscat, Oman, 1978.
45. WHO Meeting of Experts, Reuse of Effluents: Methods of Wastewater Treatment and Health Safeguards, WHO Technical Report Series Number 517, World Health Organization, Geneva, 1973.
46. Engleberg Report, Health Aspects of Wastewater and Excreta Use in Agriculture and Aquaculture, Report of a Review Meeting of Environmental Specialists and Epidemiologists, Engelberg, Switzerland, IRCWD, Duebendorf, Switzerland, July, 1–4, 1985.
47. **WHO,** Health Guidelines for the Use of Wastewater in Agriculture and Aquaculture, Technical Report Series 778, World Health Organization, Geneva, 39, 1989.

48. **Shuval, H. I., Adin, A., Fattal, B., Rawitz, E., and Yekutiel, P.,** Integrated Resource Recovery: Wastewater Irrigation in Developing Countries: Health Effects and Technical Solutions, World Bank Technical Paper Series Number 51, UNDP Project Management Report Number 6, World Bank, Washington, D.C., 1986.
49. **Shuval, H.,** Rationale for Engelberg Guidelines, International Reference Center for Waste Disposal, WHO Collaboration Center for Waste Disposal, IRCWD News Number 24/25, World Health Organization, Geneva, 18–19, May, 1988.
50. Ministry of Regional Municipalities and Environment, Sultanate of Oman, Regulations for Wastewater Reuse and Discharge, Ministerial Decision 145/93 dated 13th June, Muscat, 1993.
51. Ministry of Regional Muncipalities and Environment, Sultanate of Oman, Regulations for the Management of Hazardous Waste, Ministerial Decision 18/93 dated 2 February, 1993.

Chapter 3

Properties and Characteristics of Water and Wastewater

CONTENTS

- I. Introduction ... 60
- II. Physical Characteristics ... 60
  - A. Temperature ... 60
  - B. Turbidity ... 61
  - C. Taste ... 63
  - D. Odor ... 64
  - E. Color ... 64
  - F. Conductivity ... 66
  - G. Salinity ... 66
  - H. Solids Content ... 67
  - I. Density ... 68
- III. Radioactivity ... 68
- IV. Rheological Properties ... 70
  - A. Moisture Content of Sludges ... 72
- V. Chemical Characteristics ... 73
  - A. General Considerations ... 73
  - B. Important Chemical Determinations ... 73
  - C. Hydrogen Ion Concentration (pH) ... 75
  - D. Alkalinity ... 75
  - E. Acidity ... 76
  - F. Hardness ... 76
  - G. Dissolved Oxygen (DO) ... 79
  - H. Oxygen Demand ... 80
  - I. Dissolved Gases ... 80
  - J. Chloride ... 81
  - K. Nitrogen ... 81
  - L. Toxic Metals ... 82
  - M. Nutrients ... 82
  - N. Proteins ... 82
  - O. Oil and Grease ... 82
  - P. Carbohydrates ... 82
  - Q. Phenols ... 83
  - R. Detergents ... 83
  - S. Biochemical Oxygen Demand (BOD) ... 83
    1. BOD Kinetics ... 83
    2. Chemical Oxygen Demand (COD) ... 88

VI. Biological and Bacteriological Characteristics 88
    A. Environmental Microbiology 88
    B. Nature of Biological Growth 90
        1. Culture Systems 92
            a. Batch Type 92
            b. Continuous Culture System 93
            c. Complete Mixed System 94
    C. Bacteria 96
    D. Fungi or Molds 97
    E. Algae 98
        1. Control of Algae 99
    F. Animals 99
        1. Protozoa 99
        2. Rotifers 99
        3. Crustaceans 99
        4. Worms and Larvae 100
    G. Viruses 100
VII. Homework Problems 100
    A. Discussion Problems 100
    B. Matching Questions 101
    C. Specific Mathematical Problems 102
    D. Suggested Exercise 105
References 105

I. INTRODUCTION

Knowledge about the properties and characteristics of water or wastewater discharges is of indispensable value for a number of reasons, such as:

- Assessment of degree of pollution or contamination introduced in the past, present, or in the future
- Governing discharges
- Formulation of guidelines and standards
- Installation of needed treatment units
- Efficient evaluation of treatment and applicability of methods and procedures
- Application of suitable surveillance and monitoring systems
- Initiation and development of operational, maintenance, and training centers
- Engineering management of environmental quality
- Gaining a true picture of the nature of a particular sample

Characteristics can generally be grouped as physical, chemical, biological, and radiological.

II. PHYSICAL CHARACTERISTICS

Physical parameters signify those characteristics of a liquid that are the result of the application of physical forces. Some knowledge and understanding of physical properties of water and wastewater is of significance for adequate performance of treatability, supply, or disposal. Physical characteristics are easy to measure and quantify. The most important physical parameters are discussed below:

A. TEMPERATURE

Temperature is one of the physical parameters that reveals a great deal of information about the water's source and its state. Changes in temperature may be due to seasonal or daily variations or disposal of hot water, or due to disposal of wastes from industrial processes and/or power stations (thermal pollution). The notable effects of a temperature increase include:

- Speeding up and motivation of chemical reactions and reaction rates. The temperature may affect the reaction rate of microorganisms to the extent of doubling it for each 10°C increase.
- Reduction of dissolved oxygen concentration.
- Reduction in the solubility of gases.

- An increase of biochemical oxygen demand (BOD).
- An increase in the rate of corrosion of substances.
- An increase of sensitivity of aquatic animals for toxic dissolved substances in the aquatic environs. An increase in temperature can cause a change in the species of fish that can exist in the receiving water body if significantly large quantities of heated water are discharged to it. Furthermore, higher temperatures can support the growth of undesirable aquatic plants and wastewater fungus.
- An increase in taste and odor.
- Affecting the time required for substantially complete digestion, which depends on the temperature at which the digester is operated.

The most important parameters that affect the temperature include:

- Latitudinal variations.
- Influence of land and water masses.
- Effects of vegetation. During the day the temperature of air near the ground surface is lower than that just above the treetops. At night, the treetops act as radiating surfaces, and the soil beneath is protected from excessive heat losses. On the other hand, the shading effect of trees tends to keep the daily maximum temperature lower.
- Orographic influences (temperature decrease with elevation).
- Effects of cities. In a large city, the amount of heat produced annually is roughly equal to a third of the solar radiation reaching an equivalent area.

Continuous monitoring of temperature is a prerequisite that needs to be conducted for water sources and wastewater effluent.

Example 1.1
Convert 50, 104, and –22°F to degrees Celsius.

Solution

1. Use the relationship of the Celsius temperature to the Fahrenheit temperature

$$°C = [5/9](°F - 32) \qquad [3.1]$$

2. Find the equivalent desired temperature values

$$C_{50} = [5/9](50 - 32) = 10°C$$

3. Repeat Steps 1 and 2 to find other values as:

$$C_{104} = 40°C, \text{ and } C_{-22} = -30°C$$

B. TURBIDITY

Turbidity is a measure of the suspended matter that interferes (absorption or scattering) with the passage of light through water. Thus the factors that affect the scattering of light would affect its measurement.[1,2] The influential factors include:

- Number, size, and shape of the particles,
- Refractive index of the particles,
- Wavelength of the incident ray from the instrument,
- Characteristics and quality of the measuring device.

Turbidity measurements are of indispensable importance as a guide to quality as well as an essential parameter for proper control and operation of treatment plants.

Because colloidal particles have dimensions greater than the average wavelength of white light, they interfere with the passage of light. Light which strikes them may be reflected. As a result, a beam of light passing through a colloidal suspension is visible to an observer who is at or near a right angle to the beam of light. This phenomenon is called the Tyndall effect. This test is often used to prove the presence of a colloid, as true solutions and coarse suspensions do not produce this phenomenon. The Tyndall effect is used as one basis for determining very low turbidities in filtered water (nephelometric method).

Turbidity in water is caused by suspended matter such as clay, silt, rock fragments, finely divided organic and inorganic matter, soluble colored organic compounds, metal oxides from the soil, vegetable fibers, and plankton and other microscopic organisms. Wastewater contains a wide variety of turbidity-producing materials; such discharges of wastes may augment the turbidity of natural bodies of the receiving water.

Turbidity is an expression of the optical property that causes light to be scattered and absorbed rather than transmitted in straight lines through the sample.

The method in which ordinary white light transmitted through a finely divided suspension is compared with that transmitted by a standard suspension is known as Turbidimetry.

In water treatment, turbidimeters have been standardized so that the light scattered at an angle of 90° to an incandescent lamp source is measured. This is a fairly recent development. Previously, a variety of instruments measuring scattered and transmitted light were used, including Jackson Candle Turbidimeters or Standard Bottles for measurement of turbidity.

Measurement of turbidity can be made easily and rapidly, and the instrumentation is relatively inexpensive. Turbidity measurements do not give complete information on size, number, mass or type of particles. Small particles (e.g., maximum dimension <0.1 micrometer, μm) do not scatter much visible light. Therefore, a water containing asbestos fibers, viruses, or humic substances may have a large particle number concentration, but a low turbidity. Large particles such as clays or plankton, which have particle diameters approximately the length of visible light, scatter light more effectively and thus yield higher turbidities.

Turbidity is also measured by comparing the water with standard suspensions of silica, kaolin, or formazin polymer. Measurements of turbidity in water can be defined in two ways: the turbidity resulting from 1 mg/L of Fuller's Earth suspended in water, or the depth of the column of water that just obscures the image of a burning standard candle viewed vertically through the sample (Jackson Candle Unit, JCU).

$$1 \text{ nephelometric turbidity unit (NTU)} \simeq \text{JCU} \qquad [3.2]$$

The absorption of light is related to the depth or thickness of colored liquids through Lambert's law (Bouguer's law). This law states that each layer of equal thickness absorbs an equal fraction of the light which transverses it. Thus, when a ray of monochromatic light passes through an absorbing medium its intensity decreases exponentially as the length of the medium increases.

$$\text{Tr} = I/I_o = 10^{-kL} \qquad [3.3]$$

or

$$\text{Ab} = \log I_o/I = kL \qquad [3.4]$$

where:
- I_0 = intensity of light entering solution
- I = intensity of light leaving solution
- L = length of absorbing layer
- k = constant for particular solution
- Tr = transmittance of solution
- 100 Tr = percentage transmittance of solution
- Ab = absorbance, or optical density of solution

Light absorption is related to solution concentration through Beer's law. This law states that the intensity of a ray of monochromatic light decreases exponentially as the concentration of the absorbing medium increases.

$$\text{Tr} = I/I_o = 10^{-kC} \qquad [3.5]$$

or

$$\text{Ab} = \log I_o/I = kC \qquad [3.6]$$

where: k = constant for a particular solution
       C = concentration of solution

Combined Lambert and Beer's Law

$$Tr = I/I_o = 10^{-kLC} \qquad [3.7]$$

or

$$Ab = \log I_o/I = kLC \qquad [3.8]$$

If light of some intensity enters two different solutions and adjustments of depths are made so that emerging beams are of the same intensity, then the transmission are the same and

$$C_1L_1 = C_2L_2 \qquad [3.9]$$

Measuring the turbidity of a clear water sample or of a treated wastewater sample does not usually present a problem, however, a cloudy wastewater sample may require dilution in order to properly evaluate the turbidity.

Example 1.2

In a turbidity determination test a sample containing 0.1 mg/L of nitrogen compound produces 80% transmittance. Determine the amount of transmission that would be produced by another sample that contains 0.3 mg/L of the same nitrogen compound (assume Beer's Law to be valid).

Solution

1. Given: concentration of first sample, $C_1 = 0.1$; transmittance for first sample, $Tr_1 = 0.8$ (80%); concentration for second sample, $C_2 = 0.3$.
2. Use Beer's law ($Tr = 10^{-kc}$ [Equation 3.7]) to obtain the relationship between transmission of the two samples as: $\log (Tr_2/Tr_1) = C_2/C_1$.
3. Substitute given values (see 1 above) in this equation to obtain: $\log Tr_2/\log 0.8 = 0.3/0.1 = 3$ or $\log Tr_2 = -0.291$. The antilog of -0.291 can be determined from logarithmic tables, for instance, $9.709 - 10.000 = -0.291$ gives $Tr_2 = 0.512$, or the inverse log of $9.709 - 10.000$ on a pocket calculator also gives 0.512. This yields a transmission for the second sample of: $Tr_2 = 51.2\%$.

C. TASTE

Usually, drinking water must be almost tasteless to please the consumer. Taste and odor are subjective properties which are rather difficult to measure. Presence of taste may be due to some dissolved impurities that have found their ways into the water. These substances may be of organic or inorganic origin. Examples of organic substances are phenols, chlorophenol, oil, fats and grease, and unsaturated hydrocarbons. Inorganic substances include dissolved salts, iron, manganese, chlorides, and gaseous substances such as hydrogen sulfide (H_2S) that are produced by the decomposition of organic matter by microorganisms such as fungi, algae, protozoa, bacteria, etc. Taste and odor can be caused by decaying aquatic vegetation, as well as by decaying leaves, weeds, grasses, etc. Most organic and some inorganic chemicals contribute some taste or odor. These chemicals may originate from municipal and industrial waste discharges, from natural sources such as decomposition of vegetable matter, or from associated microbial activity.

Measurement of taste is rather difficult and the classifications generally include: Salty, Bitter, Sour, or Sweet.

While it is relatively easy and safe for a person to evaluate the taste of a drinking water supply, no one would be anxious to taste wastewater before or after conventional treatment. However, there have been cases where officials drank a glass of conventionally treated wastewater in order to convince their constituents that a fine wastewater treatment facility has just been put into operation. Nevertheless, there is at least one case in Denver, Colorado where treated wastewater does become potable. In this case secondary unchlorinated effluent is given lime treatment, recarbonation, filtration, ion exchange, carbon adsorption, ozonation, another carbon adsorption treatment, reverse osmosis, and finally chlorine dioxide treatment. While costly, the water produced is certainly drinkable.

Table 3.1 Major Classes of Offensive Odors

Class of odor	Responsible compounds
Ammoniacal	NH_3, ammonia
Decayed flesh	Diamines, algae, protozoa
Fecal	Skatole ($C_8H_5NHCH_3$)
Fishy	Amines, algae, protozoa
Grassy	Algae
Rotten eggs	Hydrogen sulfide (H_2S)
Rotten cabbage	Organic sulfides
Skunk	Mercaptan

Compiled from Reference 3.

D. ODOR

A wholesome supply of water is normally odor free. Existence of odors in water may be due to a number of reasons, such as:

- Biodegradation of organic and inorganic compounds of nitrogen, phosphorus, and sulfur.
- Decomposition of algae and other microorganisms.
- Generation of substances such as ammonia, sulfides, chlorine, cyanide, and hydrogen sulfide.

It is easy to check for an odor associated with a drinking water supply.

Offensive odors are generally associated with a raw wastewater that has been in the sewerage system too long and anaerobic conditions have developed. The sewage turns black and gives off hydrogen sulfide.

The effects of this unpleasant odor can include:

- Psychological stress
- Headache, nausea, vomiting
- Mental depression and blurred vision
- Fatigue, loss of appetite
- Impaired respiration
- Irritation of eyes
- Loss of sleep
- Reduction in production and work efficiency

If an odor problem extends beyond the wastewater facility it can result in a decline of market and rental values, tax revenues, and a decline in sales in the area.

Measurements of odors concentrate on intensity, character, hedonics, detectability, and adaptability. Table 3.1 presents the major classes of offensive odors.

Odors are of public concern, and although odor thresholds (a smell limit below which a substance cannot be detected by the human nose) differ greatly from one person to another, thresholds can be decided by a group of trained odor panelists. The threshold is taken as the lowest concentration at which half the panel members recognize the odor.

Some methods that are used to reduce objectionable tastes and odors include:

- Elimination or control at the source of the odor
- Activated carbon adsorption
- Aeration
- Coagulation and filtration
- Ozonation or chlorination
- Chlorine-ammonia treatment
- Chlorine dioxide treatment

E. COLOR

Sources of color include:

- Natural sources such as extracts from organic debris (leaves, wood, peat).
- Industrial origins such as mine wastes, textile industry, paper industry, and dye industry.
- Domestic sewage.

True color in natural waters is caused by large organic molecules. Pure water is colorless. Color in water may result from the presence of natural metallic ions such as iron oxides (causes red color) and manganese oxides (causes brown or black color). Other sources are humus and peat materials, plankton, weeds, and industrial wastes (e.g., textile and dye operations, pulp and paper production, food processing, chemical manufacturing, mining, refining, and slaughterhouse operations). The greatest contributors of color by plants are the humic acids which produce a yellow-brownish color together with tannin and humate from the decomposition of lignin. The lignin derivatives are highly colored and resist biological degradation.

Color is classified as:

1. Apparent color (due to substances in solution + suspended matter).
2. True color: true color units (TCUs) — water from which turbidity has been removed by methods such as filtration or centrifugation where the color was due to vegetable or organic extracts that are colloidal.

True color is normally caused by substances in solution, while organic color is mostly caused by suspended organic matter. Also, color can be caused by many different chemicals such as dyes, dissolved minerals (iron and manganese), and by plant and animal by-products and decomposition products. One of the major colorations is due to humic acids. These are materials produced by plants and they are polymeric compounds with carboxylic and phenolic functional groups. They may exist as large molecules or as colloids.

Color intensity increases with an increase in pH (color is pH dependent). Color can be determined by:

- Visual comparison of the sample with known concentrations of colored solutions (Nessler tubes), or with special properly calibrated glass color disks.
- Spectrophotometric methods.
- Field kit.

Units of color (TCUs) are based on a standard solution containing potassium chloroplatinate and cobalt chloride dissolved in a hydrochloric acid solution.

The following solution has been adopted as the color standard: 1246 mg Pt (potassium chloroplatinate K_2PtCl_6) together with 1000 mg Co (Cobaltons chloride $CoCl_2$) dissolved in 1000 ml of water — this represents 500 CUs (sometimes called Hazen units). For drinking water the upper limit for color is 15 CUs but 5 CUs is desirable. Many industrial wastes are highly colored and some contain colored substances that are quite resistant to biological degradation. An example of this is the waste from pulping of wood, which contains lignin derivatives.[4]

Fresh wastewater is usually gray in color. However, when organic matter is being biodegraded by microorganisms and dissolved oxygen is depleted, the color changes to black (septic or stale).[1]

Example 3.1

The following results reveal percent absorbance for ammonia samples and an unknown one. Determine the ammonia concentration of the unknown sample.

Ammonia concentration in sample (mg/L)	% Absorbance
Blank sample	0.0
1.0	7.5
1.5	11.0
2.0	15.0
2.5	18.5
3.0	22.5
Unknown sample	26.0
4.0	30.0

Solution

1. Assume that Beer's law is valid (color intensity and hence light absorbance is directly proportional to concentration of the unknown ion).
2. Plot ammonia concentration of standards vs. percent absorbance.
3. From the graph find the concentration of the unknown sample corresponding to an absorbance of 26% (3.5 mg/L).

F. CONDUCTIVITY

Conductivity is a numerical expression of the tendency of an aqueous solution to carry an electric current. This ability depends on:

- Presence and type of ions.
- Total concentration of ions.
- Mobility, valence, and relative concentration of ions.
- Temperature of the solution.

Solutions of most inorganic acids, bases, and salts are relatively good conductors. Electrical resistance, R (in ohms) of a conductor can be found as:

$$R = R_s \cdot l/A \qquad [3.10]$$

where: R_S = resistivity of conductor (Ωm)
  R = resistance (Ω)
  l = length of conductor (m)
  A = cross-sectional area of conductor (m²)

The reciprocal of the resistance is conductance expressed in reciprocal ohms or mhos:

$$R_d = 1/R_s = a_c/R_m \qquad [3.11]$$

where: R_s = specific resistance (resistance of a cube 1 cm on an edge)
  a_c = cell constant
  R_m = measured resistance

Conductivity may be defined as "the electrical conductance of a conductor of unit length and unit cross-sectional area" and is commonly expressed in micromhos/cm (μmhos/cm or microSiemens/cm).

Freshly distilled water has a conductivity of 0.5 to 2 μmhos/cm, increasing after a few weeks of storage to 2 to 4 μmhos/cm. The increase is due mainly to absorption of atmospheric carbon dioxide (CO_2) and, to a lesser extent, ammonia.

Pure water is normally not a good conductor of electricity. The increase of dissolved salts in water increases its conductivity. As such, the conductivity of the water is sometimes used for indicating the degree of its purification or pollution. The conductivity value is proportional to the concentration of dissolved solids.

$$a \cdot EC = TDS \qquad [3.12]$$

where: EC = electrical conductivity
  TDS = total dissolved solids
  a = constant

Example 3.2

A water sample has a total dissolved solids concentration (TDS) of 1200 mg/L and electrical conductivity (EC) of 1800 μmohs/cm. Calculate the electrical conductivity of another water sample which has a TDS of 7200 mg/L.

Solution

1. Compute the constant a from Equation 3.12: $a \cdot EC = TDS$: a = 1200/1800 = 0.667
2. Find the electrical conductivity (EC) for the other sample as: EC = TDS/a = 7200/0.667 = 10,800 μmohs/cm

G. SALINITY

Salinity is the total dissolved solids in water after all carbonates have been converted to oxides, all bromide and iodide have been replaced by chloride, and all organic matter has been oxidized.

$$\text{Salinity (g/kg)} = 0.03 + 1.805 \cdot \text{chlorinity (g/kg)} \qquad [3.13]$$

Table 3.2 Solids Classification System[6,9]

Solids	Size (× 10^{-6} m)
Settleable	>100
Supercolloidal	1–100
True colloidal	0.001–1
Dissolved	<0.001

Compiled from References 6 and 9.

The sources of chlorides in natural waters can be from:

- Leaching of chloride-containing rocks and soils,
- Saltwater intrusion (coastal areas),
- Agricultural, industrial, and domestic wastewater.

H. SOLIDS CONTENT

Solids content is defined as the matter that remains as residue upon evaporation and drying at 103 to 105°C.[5] Solids can be classified as:

1. **Dissolved Solids (DS).** In potable water these consist mainly of inorganic salts, and small concentrations of organic matter. Waters with high dissolved solids generally are of inferior palatability and may induce an unfavorable physiological reaction in the consumer. Highly mineralized waters are also unsuitable for many industrial applications.
2. **Suspended Solids (SS).** In water these solids may be of inorganic particles such as clay, silt, and other soil constituents; or they may be of organic origin such as plant fibers, or biological solids like algae, bacteria, etc. These are the solids that can be filtered out by a fine filter paper. Waters high in suspended solids may be aesthetically unsatisfactory for such purposes as bathing, and they provide adsorption sites for chemical and biological agents.
3. **Volatile and Fixed Solids (VS + FS).** They give a measure of the amount of organic matter present in a sample. The test is carried out by burning organic matter to convert it to carbon dioxide and water, at a controlled temperature of 550°C, to prevent the decomposition and volatilization of inorganic substances.
4. **Settleable Solids.** These are solids in suspension that can settle in quiescent conditions under the influence of gravitational attraction.

According to particle size, wastewater solids have been classified as indicated in Table 3.2.

The lower size limit of $1 \cdot 10^{-6}$ m for true colloidal solids has been used as it represents the size above which the behavior of most particles tends to be controlled by chemical composition rather than size and surface area.

Most solid measuring procedures are gravimetric tests involving the mass of residues as related to volume.

Example 3.3

Given the following data:

- Weight of crucible dish = 46.6225 g
- 100 mL of sample is placed in the dish and evaporated.
- Constant weight of dish plus solids dried at 104°C = 46.6475 g.
- Dish is placed in 550°C furnace, then cooled. Weight = 46.6305 g.

Find the total and the volatile and fixed solids.

Solution:

1. Determine mass of total solids:

    mass of dish + solids = 46.6475 g
    mass of dish = 46.6225 g
    mass of solids = 0.0250 g = 25 mg

2. Determine the concentration of total solids:

    = mass of solids (mg)/volume of sample (L)
    = 25·1000/100 = 250 mg/L

3. Determine mass of fixed solids:

mass of dish + fixed solids	= 46.6305 g
mass of dish	= 46.6225 g
mass of fixed solids	= 0.0080 g

4. Determine concentration of fixed solids:
   = 8·1000/100 = 80 mg/L
5. Determine concentration of volatile solids:
   = concentration of total solids − concentration of fixed solids
   = 250 − 80 = 170 mg/L

I. DENSITY [ρ]

The density of a fluid is defined as its mass per unit volume

$$\rho = m/V \qquad [3.14]$$

For water at standard pressure 760 mmHg and at 4°C the density is 1000 kg/m^3.

The reciprocal of the density $(1/\rho)$ is termed the specific volume. It is defined as the volume of the fluid occupied by a unit mass of it. The ratio of the weight (or density, ρ_s) of a substance to the weight (or density, ρ_w) of an equal volume of water at standard conditions is denoted as the specific gravity (s.g.)

$$\text{s.g.} = \rho_s/\rho_w \qquad [3.15]$$

Since molecular activity and spacing increase with temperature, fewer molecules exist in a given volume of fluid as the temperature is increased. Therefore, density decreases with an increase in temperature. The application of pressure forces a larger number of molecules into a given volume. This results in an increase in density.

III. RADIOACTIVITY

Radiation is characteristic of unstable atoms; therefore, the approach to understanding radioactivity should begin at the level of the atom.

Atomic structure — Atoms are made up of three primary subatomic building blocks — protons, neutrons, and electrons — structured into a nucleus and orbital electrons. The simplest atom is that of hydrogen with a nucleus of one proton and a single orbital electron. Next in order of simplicity is helium, with two protons and two neutrons in the nucleus and two orbital electrons. Next is lithium with three protons, four neutrons, and three orbital electrons. Electrons have an electrically negative charge, protons a positive charge, and neutrons have no charge. An electrically neutral atom has an equal number of electrons and protons.

Radioactivity — Is the property of certain nuclides of spontaneously emitting particles or gamma radiation or X radiation following orbital electron capture, or of undergoing spontaneous fission.

Artificial radioactivity — Manmade radioactivity produced by particle bombardment or electromagnetic irradiation, as opposed to natural radioactivity.

Induced radioactivity — Radioactivity produced in a substance after bombardment with neutrons or other particles. The resulting activity is "natural radioactivity" if formed by nuclear reactions occurring in nature, and "artificial radioactivity" if the reactions are caused by man.

Natural radioactivity — The property of radioactivity exhibited by more than 50 naturally occurring radionuclides.[10]

Atoms are said to be unstable when the ratio of protons and neutrons in their nuclei differs from well-defined ratios for stable atoms. Unstable atoms undergo spontaneous nuclear rearrangement resulting in the liberation of energy in the form of either particulates or electromagnetic radiation. Particulate radiations include alpha and beta emissions (particles), and electromagnetic radiations include those energy particles called gamma emissions (see Figure 3.1):

Alpha particles (α) — These are relatively slow-moving particles of small charge-to-mass ratio. They are helium (He) atoms which have lost their two orbital electrons, and thus have a net positive charge. They are ejected with nearly the same velocity (10% of the speed of light) from a radioactive substance.

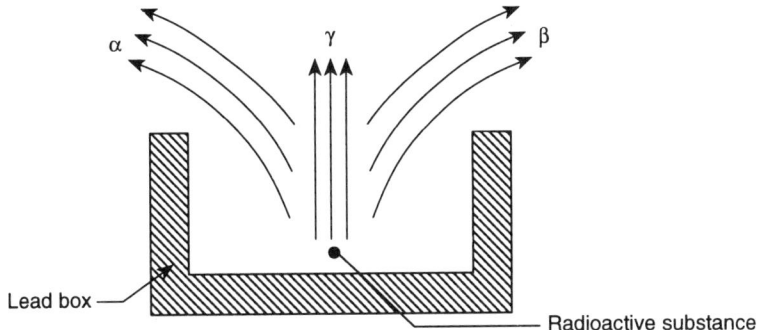

Figure 3.1 Radioactive emissions.

Their range is several centimeters in air and most are stopped by a very thin sheet of aluminum foil or by an ordinary sheet of paper. Alpha particles have extremely high ionizing action within their range.

Beta particles (β) — These are fast-moving particles (speeds from 30 to 99% of the speed of light) and have a great charge-to-mass ratio. Beta particles are streams of high-energy electrons emitted from a radioactive source with variable velocities approaching that of light (3×10^8 m/s). The more energetic beta particles are able to penetrate an aluminum sheet several millimeters thick. The ionizing action of the β particles is much less than that of α particles. When a radioactive atom emits an α or β particle it changes its chemical nature. When an α particle is emitted, the mass number of the atom decreases by 4 and the atomic number decreases by 2.

$$^{Z}_{A}X \rightarrow ^{Z-4}_{A-2}Y + ^{4}_{2}He \qquad [3.16]$$

where: X = parent nucleus
          Y = daughter nucleus
          Z = mass number = sum of number of protons and neutrons in the nucleus of the atom
          A = atomic number = number of protons or number of electrons of an atom

Example 3.4

$$^{226}_{88}Ra \rightarrow ^{222}_{86}Rn + ^{4}_{2}He$$

where Ra = radium, Rn = radon, He = the helium nucleus or the omitted α-particle.

The loss of a beta particle involves no change in the mass number of an atom but causes an increase of 1 in its atomic number.

$$^{Z}_{A}X \rightarrow ^{Z}_{A+1}Y + ^{0}_{-1}e \qquad [3.17]$$

This is an example of emission of a beta particle (−1 indicates that it has a single negative charge).

Example 3.5

$$^{239}_{92}U \rightarrow ^{239}_{93}Np + ^{0}_{-1}e$$

where U = uranium isotope, Np = neptunium.

If the nucleus of an atom is bombarded by a certain particle such as an α particle ($^{4}_{2}He$), a β particle ($^{0}_{-1}e$), a neutron ($^{1}_{0}n$), a proton ($^{1}_{1}H$), or a deuteron ($^{2}_{1}D$), the bombarding particle may be captured by the nucleus. As a result, the atom is converted into a new element or an isotope of the original element. The capture of the particle may or may not be accompanied by the expulsion from the nucleus of some other particle.

Example 3.6

Upon bombarding nitrogen (N) with an α particle, the nitrogen nucleus captures one α particle. During this capturing process it also ejects a proton, H_1^1. The product is an atom of oxygen

$$^{14}_{7}N + ^{4}_{2}He \rightarrow ^{17}_{8}O + ^{1}_{1}H$$

This represents an example of an α and a β transformation.

Gamma rays (γ) — These are true electromagnetic radiations that travel with the speed of light. These rays occupy a band among X-rays with a shorter wave length and thus have greater penetrating power. The highest-energy rays are very penetrating and proper shielding from them demands several centimeters of lead.

Energies of radiations can be evaluated through Einstein's energy mass equivalence formula, namely:

$$E = mv^2 \qquad [3.18]$$

where: E = energy, (g·cm/s, ergs)
m = mass of the particle (g)
v = velocity of light = 3×10^8 m/s

Generally, the unit of energy used in particle physics is the electron volt (eV), which is defined as the energy acquired by an electron (or proton) in moving through a unit potential difference.

$$1\,eV = 1.6 \times 10^{-12}\,ergs$$

Each radioactive substance is designated by a period known as its half life. The half life is defined as the time taken for half the atoms in any given sample of the substance to decay. These half life periods are different for each element as shown in Table 3.3.

The unit of radioactivity is the curie (c), which is defined as the number of disintegrations occurring per second in 1 g of pure radium (3.7×10^{10} disintegrating atoms per second, dps) and is applicable to all radioactive nuclides. For convenience, the curie may be extended to the megacurie or reduced to the picocurie.

1 megacurie (Mc) = 3.7×10^{16} dps = 10^6 c
1 curie (c)   = 3.7×10^{10} dps
1 millicurie (mc) = 3.7×10^{7} dps = 10^{-3} c
1 microcurie (μc) = 3.7×10^{4} dps = 10^{-6} c
1 picocurie (pc) = 3.7×10^{-2} dps = 10^{-12} c

The rate of decay of any nuclide is represented by the equation:

$$\log(n_t/n_o) = -k_t/2.303 \qquad [3.19]$$

where: n_0 = number of atoms (fraction of the total atoms) present at zero time (starting [initial] time)
n_t = number of atoms (or the fraction of the initial total atoms) present at time t
k_t = decay constant for the particular decay reaction

$$k_t = 0.693/t_{0.5} \qquad [3.20]$$

where $t_{0.5}$ = half life of the particular nuclide.

IV. RHEOLOGICAL PROPERTIES

Rheology is the science of deformation and flow of matter. The basic instrument for the investigation of this property in a liquid is the viscometer. Viscosity is one of the physical parameters which is comparatively easy to measure.

Table 3.3 Some Common Radioactive Isotopes

Element	Symbol	Half life	Particle emitted
Bromine	^{78}Br	6.4 minutes	β,γ
Carbon	^{14}C	5730 years	β
Cobalt	^{60}Co	5.3 years	β,γ
Cesium	^{137}Cs	30 years	β
Fluorine	^{131}F	8 days	β,γ
Hydrogen	^{3}H	12.3 years	β
Lead	^{214}Pb	26.8 minutes	β
Potassium	^{40}K	1.28 × 10^9 years	β
Sodium	^{24}Na	15 hours	β,γ
Phosphorus	^{32}P	14.3 days	β
Polonium	^{218}Po	3.05 minutes	α
Radium	^{226}Ra	1600 years	α
Strontium	^{90}Sr	28.1 years	β
Thorium	^{234}Th	24.1 days	β
Uranium	^{238}U	4.51·10^9 years	α
Uranium	^{234}U	2.48 × 10^5 years	α

Newton found that when water and similar liquids were subjected to shearing forces, the resisting shear, τ, was directly proportional to the shearing rate, G (see Figure 3.2).

$$\text{Force } F = -\mu \cdot A \cdot dv/dy \qquad [3.21]$$

$$\tau = F/A = -\mu \cdot dv/dy = -\mu \cdot G \qquad [3.22]$$

where τ = shearing stress, dv/dy = shearing rate = velocity gradient, and μ = constant of proportionality, termed the fluid's viscosity

The coefficient of dynamic viscosity, μ, is defined as the shear force per unit area required to drag one layer of fluid with unit velocity past another layer through a unit distance in the fluid and is reported in Newton-seconds per square meter (N-s/m^2).

The coefficient of kinematic viscosity, ν, is defined as the ratio of dynamic viscosity to mass density and is reported in square meters per second (m^2/s).

$$\nu = \mu/\rho \qquad [3.23]$$

where ν = kinematic viscosity, m^2/s
  μ = dynamic (absolute) viscosity, N-s/m^2
  ρ = density, kg/m^3
  Poise = 0.1 N-s/m^2, 1 centistoke = 10^{-6} m^2/s.

Viscosity for a given liquid at a given temperature is a characteristic physical constant. However, for many liquids the stress-rate relationship is not a simple ratio. Such liquids are broadly classed as being non-Newtonian, e.g., sewage sludge.

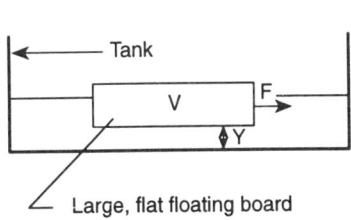
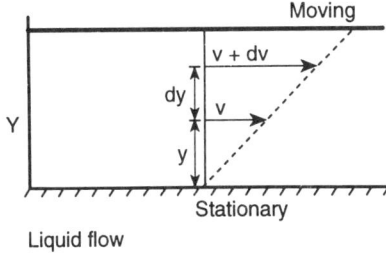

Figure 3.2 Shearing forces in fluids.

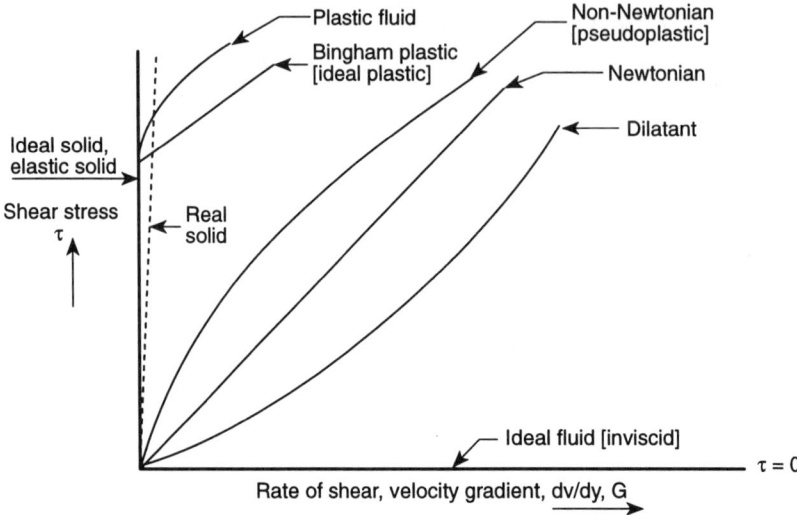

Figure 3.3 Types of fluids based on their rheological properties.

The factors that influence viscosity include:

- Temperature variations: viscosity of liquids decreases with an increase in temperature, however, for gases the viscosity increases with an increase in temperature.
- Rate of variation of shearing stress.
- Flow conditions.
- Characteristics of the fluid.

According to their rheological properties, fluids can be classified as shown in Figure 3.3.

Newtonian fluids — Newtonian fluids exhibit no internally connected structure, therefore shearing commences instantaneously with the application of stress and viscosity is independent of shear rate.

Time-independent fluids — This type of fluid behavior may be subdivided into several groups:

1. Pseudoplastic (shear thinning): are those which become increasingly less viscous as the rate of shear is increased.
2. Dilatant (shear-thickening): dilatance is the opposite of pseudoplasticity. This phenomenon is not common and is associated with suspensions of repelling particles.
3. Viscoplastic or Bingham plastics: in this case a finite yield stress is required to initiate flow when the solid phase of a suspension is present in sufficient concentration to form a continuous but disoriented structure.

Time-dependent fluids — This type of fluid exhibits time-dependent flow effects — effects that can be either spontaneously reversible or irreversible. Further subdivisions of these fluids reveals the following classes:

1. Thixotropic fluids: they possess a structure which breaks down with time when sheared at a given rate until equilibrium is reached. At equilibrium the internal forces tending to rebuild the structure are equal to the applied force. In this group hysteresis is seen with an increase of shear rate to a maximum and then a reduction to a minimum.
2. Rheopexy or antithixotropy: this fluid behaves in an opposite manner to that of a thixotropic fluid.

Viscosity measurements of digested sewage sludge reveal that it is not an ideal non-Newtonian liquid. It exhibits behavior characteristics of both pseudo- and Bingham plastics with only a slight thixotropic behavior.[11–13]

A. MOISTURE CONTENT OF SLUDGES

Normally, wastewater and sludges are of low solids content. Rarely does the solids concentration of raw sludges exceed 6%; in most secondary treatment units sludges are about 1% solids.[11] Thus any costs incurred in transportation of a sludge will be mainly due to the water content.[14]

There are various classes of water in sludges. For example Taylor[14] classified the water in sludges as:

1. Free water: water which will drain readily from the sludge, with or without the assistance of vacuum or pressure.
2. Bound water: water which is a part of the particle structure by virtue of the charge upon the particle, and which is released by neutralization of that charge.
3. Contained water: water which is an essential part of the structure of the particle and which requires drastic methods such as drying to effect removal.

The water in a granular bed can be divided into three classes:[15]

1. The pendular form, in which the water is held as a discrete ring at the points of contact with the particles.
2. The funicular form, where the rings have increased to the point where many coalesce and give a continuous network of liquid through the bed interspersed with air.
3. The capillary form, where all the pore spaces are completely filled with water.

As a result of their studies on the filtration and drying characteristics of sewage sludges, Coackley and co-workers classified the moisture as:[12,16,17]

1. Free or readily drained moisture, the lower limit of which is defined as a first critical moisture content.
2. The capillary water, which is the bulk of the water held between the first and the second critical moisture content.
3. Floc or particle moisture, which is the water held internally by osmotic and hydration forces within the body of the floc.
4. Chemically bound water, this being water held at moisture contents somewhere below the equilibrium moisture content. Moisture in the form of 2 and 3 above is removed solely by evaporation. The moisture content of the sludge determines the magnitude of the interparticle force system on the particles. Therefore, it determines the degree of compression and reduction of pore space.

The moisture content (m.c.) of a sample can be computed as:

$$\text{m.c.} = \{(\text{weight of water})/(\text{weight of wet sludge})\} \cdot 100(\%) \qquad [3.24]$$

Table 3.4 presents a summary of the units used in describing the physical properties of sludges.

V. CHEMICAL CHARACTERISTICS

A. GENERAL CONSIDERATIONS

The following review presents some basic aspects of chemistry that will help to understand the technology involved in the water and wastewater fields.[2,4,18]

- An element is a substance which cannot be changed into any simpler substances.
- An atom is the smallest unit of each of the elements. The atom consists of a positively charged nucleus containing protons and neutrons. Electrons occupy the extranuclear space. The nucleus is about 10^{-15} m in radius. The radius of the atom is about 10^{-10} m; thus the atom is mostly an empty space. Since the electron has a mass of 1/1837 of that of the proton, most of the mass of the atom is in its nucleus.
- Atomic weight (AW) of an element refers to the relative weight of the atoms of an element. Elements do not have atomic weights that are whole numbers because of the existence of isotopes.
- Gram atomic weight (GAW) of an element denotes the quantity of element in grams corresponding to the atomic weight.
- Atomic number of an atom is the number of protons in its nucleus.
- Mass number of an atom signifies the number of protons and neutrons.
- Molecular weight (MW) refers to the sum of the atomic weights of the atoms of a molecular formula.
- Molecular formula signifies a formula that describes both the number and types of atoms in a molecule.

B. IMPORTANT CHEMICAL DETERMINATIONS

- Molality = mass of solute/1000 g of solvent [mol/kg]
- Molarity = moles of solute/liter of solution

$$M = (m/MW)/V \qquad [3.25]$$

Table 3.4 Units of Analysis for Wastewater and Sludges

Analysis	Formula	Unit
Color		TCUs
Density	Mass of solution/unit volume	kg/m^3
Mass ratio	Mass (mg)/mass·10^6 (mg)	mg/L
Percent by mass	Mass of solute ·100/total mass of solute and solvent	%
Percent by volume	Volume of solute ·100/total volume of solution	%
Solids content	Mass/volume	mg/L, g/m^3
Temperature	°C = (°F − 32) × 5/9 K = °C + 273	°C, °F, K
Turbidity		NTU
Viscosity	Dynamic, μ Kinematic, $\nu = \mu/\rho$	N-s/m^2 m^2/s
Volume ratio	Volume (mL)/volume (L)	ml/L

where: M = molarity (mol/L; M)
        m = mass of solute (g)
        MW= molecular weight
        V = volume of solution (L)

- Normality = equivalents of solute/liter of solution

$$N = (m/EW)/V \qquad [3.26]$$

where: N = normality (equivalent/L; N)
        EW = equivalent weight = MW/a$_v$
        a$_v$ = valency
        V = volume of solution (L)

Example 3.7

For the reaction A + B = C + D, the equilibrium constant is 121 at a particular temperature. If 0.6 mol of each of reactants A and B are placed in a 2-L flask at that temperature, determine the concentrations of all species that will be present at equilibrium?

Solution

1. Accordingly, from the solubility product constant and the reaction equation:

$$k_s = [C][D]/[A][B] = 121$$

2. Find the concentrations of reactants remaining:
    If x is the number of moles of each of the reaction products C and D at equilibrium (resulted from the dissociation), then (0.6/2 − x) mol/L will remain of reactants A and B.
3. Compute the concentrations at equilibrium:

$$x \cdot x/(0.3-x)(0.3-x) = 121; \text{ or } x/(0.3-x) = 11$$

$$x = 0.3 \cdot 11/12 = 0.275$$

At equilibrium: [A] = [B] = 0.3 − 0.275 = 0.025 M
                 [C] = [D] = 0.275 M

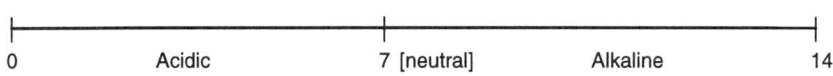

Figure 3.4 The pH scale.

C. HYDROGEN ION CONCENTRATION (pH)

The pH is a measure of the acid or alkaline nature of a solution, and affects the quality of a water or wastewater.

$$\text{pH} = -\log(\text{H}^+) = \log 1/[\text{H}^+] \qquad [3.27]$$

where: (H$^+$) = concentration of hydrogen ions.

The pH ranges from 0 to 14, with 7 as neutrality; below 7 being acidic and above 7 being alkaline (see Figure 3.4). The pH is an important parameter for both natural waters and wastewaters. The concentration range suitable for the existence of most biological life is narrow and critical. Wastewaters with an adverse concentration of pH are difficult to treat by biological means, and if the concentration is not altered before discharge the wastewater effluent may change the pH in natural waters.[3]

One of the best controls for biological growth is pH. At a low pH the hydrogen ion concentration causes denaturation of the key enzyme proteins. Most microorganisms cannot survive below pH 4, but a few sulfate-oxidizing bacteria can exist at a pH of 1. The same is true of the hydroxyl ion concentration. As the pH rises over 9.5, the hydroxyl ions begin to exert a toxic effect. Few, if any, microorganisms can survive above a pH of 11. Control of pH at either a high or low range can be used to prevent decomposition of stored matter until desired. Actually, pH control is the most significant economic control the sanitary microbiologist has over the growth and death of microorganisms.[19]

Optimum pH for methane-forming bacteria is between 6.8 and 7.2, although the organisms can operate over a slightly wider range of pH 6.6 to 7.6.[3]

pH can be adjusted by the addition of an acid or alkaline compound. In water treatment the following compounds are usually used: H_2SO_4, HCl, CO_2, $Ca(OH)_2$, Na_2CO_3, and $NaOH$.

Example 3.8

Which is more strongly acidic: a solution with a pH of 3, or a solution containing 0.03 g of H$^+$ per liter?

Solution:

1. The pH of first solution = 3
2. Find pH of second solution:
   since concentration of the other solution = 0.03 M, then ph = $-\log$ [H$^+$] = $-\log$ 0.03 = 1.5
3. Since the pH of the second solution is lower, then it is more acidic than the first solution.

Example 3.9

Calculate the OH$^-$ concentration in grams of OH$^-$ per liter of a solution containing $1 \cdot 10^{-10}$ mol of H$^+$ per liter.

Solution:

1. The dissociation constant for water is
   [H$^+$][OH$^-$] = 10^{-14}
2. Find the concentration of hydroxide as:
   [OH$^-$] = $1 \cdot 10^{-14}/1 \cdot 10^{-10}$ = 10^{-4} M
   = $10^{-4} \cdot$MW = $10^{-4} \cdot 17$ = $1.7 \cdot 10^{-3}$ g OH$^-$/L.

D. ALKALINITY

Alkalinity is a measure of the buffering capacity of water. Alkalinity is caused primarily by chemical compounds dissolved from the rocks and soil and is mainly due to the presence of hydroxyl (OH$^-$), carbonate (CO$_3^{2-}$) and bicarbonate (HCO$_3^-$) ions. These compounds are mostly the carbonates and bicarbonates of sodium (Na), potassium (K), magnesium (Mg), and calcium (Ca). Other ions that may contribute to alkalinity but are generally found at low concentrations are $H_2PO_4^-$, HPO_4^{2-}, PO_4^-, $HSiO_3^-$, $H_2BO_3^-$, and HS^-. Normally wastewater is alkaline. In the anaerobic digestion process sufficient alkalinity has to be present to ensure that the pH will not drop below 6.2, the methane bacteria cannot function below that point. When digestion is proceeding satisfactorily, the alkalinity will normally range from 1000 to 5000 mg/L as $CaCO_3$.

Alkalinity in a water is determined by titrating a sample of water with 0.02 N, H_2SO_4 solution. Total alkalinity (T) is found by titrating to pH 4.5 (the methyl orange end point) with a color change from orange to pink. The phenolphthalein alkalinity (P) is determined by titration to a pH of 8.3 if, of course, the water sample has a pH above 8.3, or with a color change from pink to colorless. See Chapter 8 for detailed laboratory procedures for alkalinity determinations.

The alkalinity is reported in mg/L as $CaCO_3$:

$$\text{Alkalinity, mg/L as CaCO}_3 = \frac{(A-B) \times N \times 50{,}000}{\text{mL sample}}$$

where: A = mL standard acid used for sample
B = mL standard acid used for blank
N = normality of acid (0.02 N)

Figure 3.5 presents both a graphical and tabular relationship for determining the various forms of alkalinity.[2,4]

E. ACIDITY

Acidity is usually attributed to samples with a pH below the value of 7. In unpolluted water acidity comes from dissolved CO_2 or organic acids leached from the soil. Atmospheric pollution also may cause acidity. Acid waters corrode metal or concrete.

The acidity of water is determined by titrating a water sample with 0.02 N NaOH to pH 8.3.
The calculation for acidity is as follows:[34]

$$\text{Acidity, as mg CaCO}_3/L = \frac{[(A-B) \times C] - [(D \times E)] \times 50{,}000}{\text{mL sample}}$$

where: A = mL NaOH titrant used for sample
B = mL standard NaOH titrant used for blank
C = actual normality of standard NaOH titrant (0.02 N)
D = mL standard H_2SO_4 used (note: this term may be zero if pH of sample is ≤4.0.)
E = actual normality of standard H_2SO_4 (0.02 N)

If the analyst is to report the acidity in millequivalents per liter, divide the acidity (mg $CaCO_3$/L) values by 50:

$$\text{Acidity as meq/L} = \frac{\text{Acidity as mg CaCO}_3/L}{50}$$

F. HARDNESS

Hardness in a water will prevent the formation of a soap lather, and is usually due to divalent metallic cations such as calcium, Ca^{2+}, magnesium, Mg^{2+}, strontium, Sr^{2+}, ferrous ions, Fe^{2+}, and manganous ions, Mn^{2+} (see Table 3.5).

When hardness is numerically greater than the sum of carbonate and bicarbonate alkalinity, that amount of hardness equivalent to the total alkalinity is called carbonate hardness. The amount of hardness in excess of this is called noncarbonate hardness. When the hardness is less than or equal to total alkalinity, all hardness is carbonate hardness and noncarbonate hardness is absent (see Figure 3.6).

$$\text{Hardness, [mg equivalent CaCO}_3/L] = 2.497\, Ca^{2+}[mg/L] + 4.118\, Mg^{2+}[mg/L]^{\text{Ref. 5}} \quad [3.28]$$

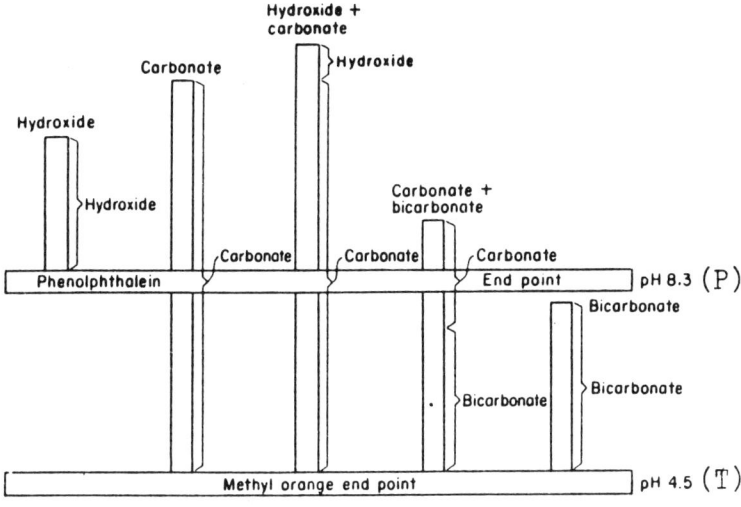

Result of Titration	Hydroxide Alkalinity as CaCO$_3$	Carbonate Alkalinity as CaCO$_3$	Bicarbonate Alkalinity as CaCO$_3$
P = 0	0	0	T
P < ½ T	0	2P	T - 2P
P = ½ T	0	2P	0
P < ½ T	2P - T	2(T - P)	0
P = T	T	0	0

Figure 3.5 Graphical and tabular relationships for various forms of alkalinity.[2,4]

when:

$$\text{Alkalinity} < \text{total hardness; carbonate hardness} = \text{alkalinity [mg/L]} \qquad [3.29]$$

$$\text{Alkalinity} > \text{total hardness; carbonate hardness} = \text{total hardness [mg/L]} \qquad [3.30]$$

Impacts of hardness include:

- Economic losses to water users through consumption of soap.
- Formation of precipitates on hot water appliances, boilers, kettles and domestic appliances, bath tubs, sinks, dishwashers and wash basins.
- Staining of clothes, dishes, and other household utensils.
- Residues of the hardness-soap precipitate may remain in pores of the skin making it feel rough and uncomfortable.
- Development of a laxative effect on new consumers, especially with the presence of magnesium sulfates.

Table 3.5 Principal Cations Causing Hardness in Water and the Major Anions Associated with Them

Cations	Anions
Ca^{2+}	HCO_3^-
Fe^{2+}	NO_3^-
Mg^{2+}	SO_4^{2-}
Mn^{2+}	Cl^-
Sr^{2+}	SiO_3^{2-}

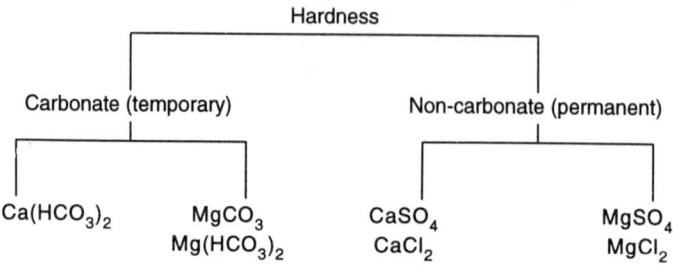

Figure 3.6 Classification of hardness.

The merits that are to be gained from usage of hard water include:

- Aiding in growth of teeth and bones,
- Reduction of toxicity to man by poisoning with lead oxide (PbO) from pipelines made of lead (precipitation of lead carbonate ($PbCO_3$), plumbosolvancy).
- Soft waters are suspected to be associated with cardiovascular diseases.

Table 3.6 presents a descriptive evaluation of hardness as it relates to the numerical hardness reported in mg/L as $CaCO_3$.

The total hardness of a water sample can be easily determined by the EDTA titrimetric method.[5]

Example 3.10

Analysis of a sample of water revealed the following results:

mg/L Cations	mg/L Anions
Na^+ = 20.2	Cl^- = 40
Ca^{2+} = 16	SO_4^{2-} = 16
Mg^{2+} = 10	Alkalinity = 50 mg $CaCO_3$/L
Sr^{2+} = 2.6	

1. Convert concentrations of substances to milliequivalents/liter.
2. If an experimental error of 10% is accepted, should the analysis be considered complete?
3. Draw a bar diagram in milliequivalents per liter of this water.
4. Calculate the total, carbonate, and noncarbonate hardness of the water sample.
5. List the hypothetical chemical combinations of positive and negative ions from the bar graph.

Solution:

1. Find the concentration of different substances in terms of meq/L by dividing given concentrations, in mg/L, by the equivalent weight of each substance.
2. Convert the concentrations expressed as meq/L calculated in 1, above, to mg/L as $CaCO_3$. This is to be done by multiplying the concentrations of equivalent/L by the equivalent weight of $CaCO_3$. EW of $CaCO_3$ = (40 + 12 + 16 · 3)/2 = 50. Results are as tabulated below:

Constituent	Col. I EW mg/meq	Col. II Concentration mg/L	Col. III meq/L [II/I]	Col. IV [mg/L $CaCO_3$] [III·50]
Cations:				
Ca^{2+}	20	16	0.8	40
Mg^{2+}	12.2	10	0.82	41
Sr^{2+}	43.8	2.6	0.06	3
Na^+	23	20.2	0.88	44
			2.56	
Anions				
HCO_3^-	61	61	1	50
SO_4^{2-}	48	16	0.33	17
Cl^-	35.5	40	1.13	56.3
			2.46	

Table 3.6 Degree of Hardness

mg/L as $CaCO_3$	Degree of hardness
0–75	Soft
75–150	Moderately soft
150–175	Moderately hard
175–300	Hard
>300	Very hard

From Berger, B. B., Ed., *Control of Organic Substances in Water and Wastewater*, Noyes Data Co., Park Ridge, NJ, 1987. With permission.

3. Calculate hardness:
    1. In this case only divalent cations, Ca^{2+}, Mg^{2+}, and Sr^{2+} cause hardness. Thus, total hardness = Ca^{2+} + Mg^{2+} + Sr^{2+} = 0.8 + 0.82 + 0.06 = 1.68 meq/L = 1.68 · 50 = 84 mg/L as $CaCO_3$.
    2. Since alkalinity < total hardness, then carbonate hardness = alkalinity [Equation 3.29] = 50 mg/L $CaCO_3$.
    3. Noncarbonate hardness = total hardness – carbonate hardness. Thus, noncarbonate hardness = 84 – 50 = 34 mg/L as $CaCO_3$.
4. Calculate percent experimental error as:
   experimental error = (cations – anions)/cations = (anions – cations)/anions = [(2.56 – 2.46)/2.56] · 100 = 3.9%, which is <10%, therefore analysis can be accepted.
5. Plot the bar diagram as shown below:

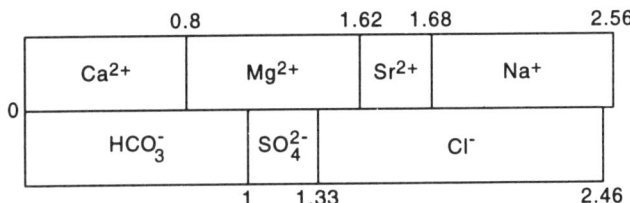

6. Determine the hypothetical chemical combinations as illustrated in the figure below:

0.8	0.2	0.3	0.29	0.06	0.78
$Ca(HCO_3)_2$	$Mg(HCO_3)_2$	$MgSO_4$	$MgCl_2$	$SrCl_2$	$NaCl$

G. DISSOLVED OXYGEN (DO)

Oxygen dissolved in sewage or water is needed for the maintenance of aerobic conditions, but the solubility of oxygen in water is low.

Drinking water saturated with oxygen has a pleasant taste, while water lacking dissolved oxygen has an insipid taste.

$$C_g = P_g \cdot MW/R_u T \qquad [3.31]$$

where: C_g = gas concentration in gas phase [g/m³]
P_g = partial pressure of the respective gas in the gas phase $\left[Pa = N/m^2\right]$

$$P_g = x_g \cdot k_H \qquad [3.32]$$

$$x_g = \text{mole fraction of gas} = n_g/[n_g + n_w] \qquad [3.33]$$

n_w = moles of water
n_g = moles of gas
k_H = Henry's constant
MW = molecular weight of the gas
R_u = universal gas constant = 8.3143 J/K.mol
T = absolute temperature (K)

$$C_s = K_D C_g \qquad [3.34]$$

where: C_s = saturation concentration [g/m³]
K_D = distribution coefficient

Oxygen is slightly soluble in water. The actual quantity of O_2 that can be present in solution is governed by:

- Solubility of the gas
- Partial pressure of the gas in atmosphere
- Temperature
- Purity (salinity, suspended solids, etc.) of water

Example 3.11

Calculate the saturation concentration of dissolved oxygen in a water sample with a temperature of 24°C and a dissolved oxygen concentration of 8.15 mg/L when the atmospheric pressure is 730 mmHg. Assume sample salinity is zero mg/L.

Solution:

1. Given: T = 24; C′ = 8.15; P = 730; salinity = 0
2. Use the equation that relates oxygen solubility concentration to pressure as:

$$C' = C_s (P - p_w)/(760 - p_w) \qquad [3.35]$$

where: C′ = solubility of oxygen at barometric pressure P and given temperature, mg/L
C_s = saturation concentration at given temperature, mg/L
P = barometric pressure, mm
p_w = pressure of saturated water vapor at temperature of water, mm

3. Substitute given values from 1 in Equation 3.35. Therefore: 8.15 = C_s (730 – 22)/(760 – 22) [from Table 2 in the Appendix, for T = 24, p_w = 22]. This yields C_s = 8.5 mg/L.

H. OXYGEN DEMAND

Oxygen demand is the amount of O_2 needed to stabilize organic matter.

1. Biochemical oxygen demand (BOD) is a measure of the amount of pollution by organic substances in water.
2. Permanganate value (PV) is the chemical oxidation of a water sample using a potassium permanganate solution ($KMnO_4$).
3. Chemical oxygen demand (COD) is the chemical oxidation of a water sample using a mixture of concentrated H_2SO_4 and potassium dichromate $K_2Cr_2O_7$

$$PV < BOD < COD \qquad [3.36]$$

I. DISSOLVED GASES

Natural waters contain dissolved gases with varying concentrations depending upon their solubility in water. When the water is anaerobic and there is microbial activity, free ammonia, hydrogen sulfide, and methane may exist. In the latter case the water needs to be oxygenated before use. From the point of view of water purity, the most important gases are oxygen and carbon dioxide.

J. CHLORIDE
Sources of chlorides in natural waters are

- Leaching of chloride from rocks and soils.
- Salt water intrusion (coastal areas).
- Agricultural, industrial, domestic and human wastewater.
- Infiltration of groundwater into sewers adjacent to salt water.

Chloride, in the form of the Cl^- ion, is one of the major inorganic anions in water and wastewater. In potable water, the salty taste produced by chloride concentration is variable and depends on the chemical composition of the water. Some waters containing 250 mg/L Cl^- have a detectable salty taste if the cation involved is Na^+. On the other hand, the typical salty taste may be absent in waters containing as much as 1000 mg/L Cl^- when the predominant cations are calcium and magnesium.[5]

The chloride concentration is higher in wastewater than in raw water because NaCl is a common part of the diet and passes unchanged through the digestive system.

Along the sea coast, Cl^- may be present in high concentrations because of seepage of salt water into groundwater sources or into sewerage systems. A high chloride content is harmful to metallic pipes and structures, as well as to growing plants.

When chlorine dioxide is used in water treatment (disinfection) chlorite ion is formed as a by-product. Chlorite is known to cause methemoglobinemia (a condition in which hemoglobin of the blood is oxidized to a metabolically inactive [ferric] state).

Example 3.12

In a titration test, 6.5 mL of a solution of silver nitrate was used for a water sample of volume 100 mL. If 1 mL of the silver nitrate solution is equivalent to 1/1000 g of chloride ion, what is the chloride content of water?

Solution:

1. The amount of silver nitrate used = 0.001·6.5 = 0.0065 g Cl^-
2. The chloride concentration = 0.0065·1000 mL in a liter/100 mL water = 0.065 g/L = 65 mg/L

K. NITROGEN

In waters and wastewaters nitrogen exists in four main forms, and biological treatment cannot proceed unless some of these forms are present.

- Organic nitrogen, N, is organically bound in the tri-negative oxidation state. Analytically organic N and NH_3-N can be determined together (Kjeldahl nitrogen). Organic nitrogen includes such natural materials as proteins, peptides, nucleic acids, urea, and numerous synthetic organic materials.

Total oxidized nitrogen = nitrite nitrogen + nitrate nitrogen:

$$\text{Organic N} + O_2 \rightarrow NH_3 - N + O_2 \rightarrow NO_2 - N + O_2 \rightarrow NO_3 - N \qquad [3.37]$$

- Ammonia, NH_3-N, is present naturally in surface and wastewaters. Its concentration generally is low in groundwater because it adsorbs to soil particles and clays and is not leached readily from soils. It is produced largely by de-aeration of organic nitrogen-containing compounds and by hydrolysis of urea.
- Nitrite, NO_2-N, is an intermediate oxidation state of nitrogen. It can enter a water supply system through use as a corrosion inhibitor in industrial process water. Nitrite is the actual etiologic agent of methemoglobinemia. Nitrous acid, which is also formed from nitrite under acidic conditions, can react with secondary amines (RR′NH) to form nitrosamine (RR′N-NO), many of which are known to be carcinogens.
- Nitrate, NO_3-N, is derived from the oxidation of ammonia. High concentration of nitrate (greater than 10 mg/L [NO_3-N]) in water can cause cyanosis in infants (methemoglobinemia: an illness especially affecting infants less than 6 months old). Nitrates generally occur in trace quantities in surface water but may attain high levels in some groundwater. Nitrate is an essential nutrient for many photosynthetic autotrophs and in some cases has been identified as the growth-limiting nutrient.

L. TOXIC METALS

Toxicity is the adverse effect a substance has on a test organism exposed to that substance. Toxicity is the result of a concentration and time exposure test, modified by variables such as temperature, chemical form, and availability. Toxicity may be

1. Acute (short-term lethal).
2. Chronic (long-term effects that may be related to changes in appetite, growth, metabolism, reproduction, and even death or mutations).

The degree of toxicity depends upon the compounds involved such as copper, lead, silver, chromium, arsenic, and boron. These metals have to be taken into consideration when designing biological treatment systems. The presence of other trace metals such as nickel, manganese, and mercury at high concentration will also interfere with wastewater treatment processes. Toxic anions such as cyanide and chromates, often found in industrial wastewaters, also hinder biological treatment and should be removed by pretreatment at the source before being discharged to the municipal sewerage system.

M. NUTRIENTS (BIOSTIMULANTS)

Nitrogen and phosphorus are essential growth factors together with other trace elements like iron, potassium, magnesium, calcium, cobalt, copper, sulfur, and zinc. If wastewaters are to be treated by biological processes, the nutrient balance has to be considered in order to establish optimum operating conditions.

N. PROTEINS

Proteins are nitrogenous organic substances of high molecular weight found in the animal kingdom and to a lesser extent in the plant kingdom. The amount present varies from a small percentage in watery fruits (e.g., tomatoes) and in the fatty tissues of meat, to quite a high percentage in beans or lean meats. Proteins consist wholly or partially of very large numbers of amino acids united by peptide links. They contain carbon, hydrogen, oxygen, nitrogen, sulfur, and sometimes phosphorus (see Table 3.7).

It has been shown that proteinaceous materials constitute a large part of the wastewater sludges, and that the sludge particles, if they do not consist of pure protein, will be covered by a layer of protein which will govern their chemical and physical behavior.[12,16,17] Coackley found that the protein content ranges between 15 to 30% of the organic matter present for digested sludge, and 28 to 50% in the case of activated sludge. This layer of protein can account for the hydrophilic (water loving) properties of the sludge.

Under the influence of microorganisms proteins undergo decomposition, giving end products which often have objectionable foul odors. For example, in sewage treatment proteins may be hydrolyzed to polypeptides and then to amino acids, which may be further degraded to ammonia, hydrogen sulfide, and to simple organic compounds.

O. OIL AND GREASE

Oil and grease compounds are insoluble in water but dissolve in such organic substances (solvents) as petroleum, chloroform, ether, etc. They are esters of alcohol or glycerol and fatty acids. Fats are among the more stable of organic compounds and are not easily decomposed by bacteria. However, they can be attacked by mineral acids resulting in the formation of glycerin and fatty acid.

When grease is encountered in sufficient quantities it causes clogging of filters, nozzles, and sand beds.[20] Grease can coat the walls of sedimentation tanks and decompose and increase the amount of scum. If grease is not removed before discharge of the wastewater, it can interfere with the biological processes in the surface waters and create unsightly floating matter. Both trickling filters and the activated sludge process are adversely affected by grease which can coat the biological forms sufficiently to interfere with oxygen transfer from the liquid to the interior of the living cells.[4]

P. CARBOHYDRATES

Carbohydrates are organic substances that include starch, cellulose, and sugars; they contain carbon, hydrogen, and oxygen. Carbohydrates are widely distributed in nature (plants and animals).

Carbohydrates may be grouped as simple sugars (monosaccharides), or complex sugars (disaccharides and polysaccharides). The primary function of carbohydrates in higher animals is to serve as a source of energy. Bacteria utilize carbohydrates for the synthesis of fats and proteins as well as for energy. The majority of carbohydrates in wastewater are in the form of large molecules that cannot penetrate the cell membrane of microorganisms.

Table 3.7 Protein Constituents in Percentages

Carbon	51–55
Hydrogen	6–7
Oxygen	20–24
Nitrogen	15–18
Sulfur	0–2.5
Phosphorus	0–1

Compiled from Reference 4.

Bacteria break down the large molecules into diffusible fractions for assimilation into the cell in order to metabolize high molecular mass substances. The first step in bacterial decomposition of organic compounds is hydrolysis of carbohydrates into soluble sugars and proteins and to amino acids and fats. Further aerobic decomposition results in formation of water and carbon dioxide. In the absence of oxygen the end products are organic acids, alcohols, as well as gases such as CO_2, CH_4, and H_2S. It should be noted that formation of organic acids in large quantities can overtax the buffering capacity of the wastewater resulting in a drop in pH and a cessation of biological activity.

Q. PHENOLS

Phenols are a group of aromatic compounds with one or more hydroxyl groups attached to a benzene ring. Phenols can be recovered from coal tar while greater amounts are manufactured synthetically. Phenols in wastewater may be industrial in origin, such as from coal, gas, or petroleum operations. Phenols cause taste problems in drinking water, particularly when the water is chlorinated. This is due to formation of chlorophenol.

R. DETERGENTS

Detergents are large organic molecules. They are slightly soluble in water, and may cause foaming in wastewater treatment plants and in the surface waters into which the wastewater effluent is discharged. They can also seriously reduce the oxygen uptake in biological treatment processes. Synthetic detergents are classified as anionic, cationic, or nonionic due to their electrical charge or lack of one when they dissolve in water.[3,4] Synthetic detergents are used in households and industry. In comparison with soaps, they are more effective cleansing agents; they are not changed in cleansing action by acid, alkali, or alkaline earth metals. However, some of their desirable properties as detergents affect wastewater treatment processes adversely. They lower the surface, or interfacial, tension of water and increase its ability to wet surfaces with which they come in contact; emulsify grease and oil, deflocculate colloids; induce flotation of solids and give rise to foams; and may kill useful bacteria and other living organisms.

S. BIOCHEMICAL OXYGEN DEMAND (BOD)

BOD determination involves the measurement of the dissolved oxygen consumed by microorganisms in the biochemical oxidation of organic matter. The test determines the approximate quantity of oxygen that will be required to biologically stablize the organic matter present. The test's advantages include:

- Determination of the size of waste treatment facilities.
- Measurement of the efficiency of some treatment processes.
- Determination of the approximate quantity of O_2 needed for stablization of organic matter present.

Biological oxidation is a slow process and theoretically takes an infinite time to go to completion. Within a 20-day period, the oxidation is about 95 to 99% complete, and in the 5-day period used for the BOD test, oxidation is from 60 to 70% complete. The 20°C temperature used is an average value for slow-moving streams in temperate climates and is easily duplicated in an incubator. Different results would be obtained at different temperatures because biochemical reaction rates are temperature dependent. The test requires exclusion of light during the incubation period to prevent oxygen formation by algae in the sample.

1. BOD Kinetics

The exertion of BOD is considered to be a first-order reaction kinetics and may be expressed as:

$$dL_t/dt = -k'L_t \qquad [3.38]$$

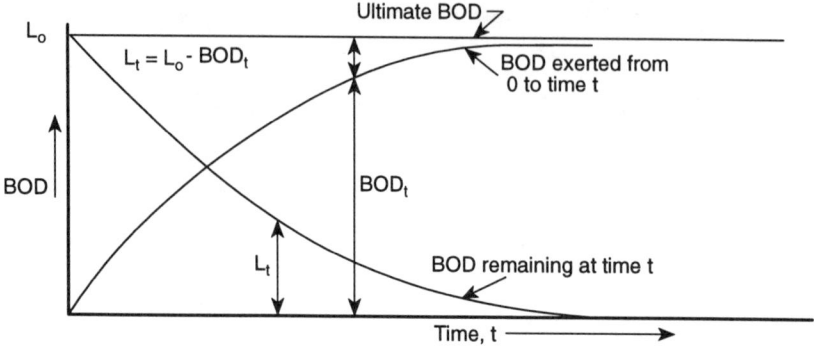

Figure 3.7 Formulation of the first-stage BOD curve.

where: L_t = amount of 1st stage BOD remaining in the sample at time t
t = time
k′ = rate constant

Integrating Equation 3.38

$$L_t/L_o = e^{-k't} = 10^{-k_1 t} \quad [3.39]$$

$$k_1 = k'/2.303 = 0.4343\, k' \quad [3.40]$$

where L_o = BOD remaining [at time t = 0] = total or ultimate first stage BOD initially present.
Amount of BOD exerted at any time t equals:

$$BOD_t = L_o - L_t \quad [3.41]$$

$$= L_o - L_o \cdot 10^{-k_1 t} = L_o\left(1 - 10^{-k_1 t}\right) \quad [3.42]$$

$$BOD_5^{20} = L_o\left(1 - 10^{-5k_1}\right) \quad [3.43]$$

The relationship is shown in Figure 3.7.
The values for k′ and L_o may be determined from a series of BOD measurements, e.g., the Thomas method. The Thomas method is based on the similarity of two series functions [expansions for $(1 - e^{-k't})$ and $k't(1 + k't/6)^{-3}$]. Thomas developed the approximate formula:

$$BOD = L_o\, k't(1 + k't/6)^{-3} \quad [3.44]$$

which can be linearized as:

$$(t/BOD)^{1/3} = (k'\, L_o)^{-1/3} + \left(\left[k'^{2/3}\right] \cdot t / \left[6\, L_o^{1/3}\right]\right) \quad [3.45]$$

where: BOD = BOD that has been exerted in time interval t
k′ = reaction rate constant [to base e]
L_o = ultimate BOD

Equation 3.45 has the form of a straight line

$$x = a + bt \quad [3.46]$$

in which

$$x = (t/BOD)^{1/3},\ a = (k'\, L_o)^{-1/3},\ b = k'^{2/3}/6L_o^{1/3} \quad [3.47]$$

Table 3.8 Wastewater Strength in Terms of BOD and COD

Strength	BOD_5 (mg/L)	COD (mg/L)
Weak	<200	<400
Medium	200–350	400–700
Strong	351–500	701–1000
Very strong	>750	>1500

Compiled from References 3, 21, and 22.

x can be plotted as a function of t, and the slope b and the intercept a of the line of best fit of the data can be used to calculate k′ and L_0

$$k' = 6b/a \text{ and } L_o = 1/k'\, a^3 \qquad [3.48]$$

or

$$(t/BOD)^{1/3} = (2.3\, k_1\, L_o)^{-1/3} + \left[k_1^{2/3}\right]\cdot t/\left[3.43\, L_o^{1/3}\right] \qquad [3.49]$$

$$k_1 = 2.61\, b/a \text{ and } L_o = 1/2.3\, k_1\, a^3 \qquad [3.50]$$

The higher the concentration of waste matter in wastewater, the stronger it is said to be. Wastewater strength is most often judged by its BOD_5 or COD, as indicated in Table 3.8.

Limitations in the biochemical oxygen demand test include:

- A high concentration of active, acclimated seed bacteria is required.
- Pretreatment is needed when dealing with toxic wastes, and the effects of nitrifying organisms must be reduced.
- Only the biodegradable organics are measured.
- The test does not have stoichiometric validity after the soluble organic matter present in solution has been used.
- An arbitrary long period of time is required to obtain results.

Perhaps the most serious limitation is that the 5-day period may or may not correspond to the point where the soluble organic matter that is present has been reduced. This reduces the usefulness of the test results.

Example 3.13

A BOD test is carried out on a sample of wastewater which has a rate constant k_1 value of 0.15/day. Compute the value of the BOD_5 as compared to the ultimate BOD.

Solution:

1. Use the first order reaction rate equation, i.e., Equation 3.43

$$BOD_5^{20} = L_o\left(1 - 10^{-k_1 t}\right)$$

where BOD_5^{20} = amount of biochemical oxygen demand removed over 5 days at 20°C. Then L_o = initial BOD at zero time = ultimate BOD and k_1 = rate constant (/day) = time (days) = 0.15/day [given].

2. Substituting given values in the above equation gives:
$BOD_5^{20}/L_o = 1 - 10^{-k_1 t} = 1 - 10^{-0.15 \cdot 5} = 0.82$

Thus, the 5-day BOD amounts to about 82% of the ultimate BOD.

Example 3.14

The 5-day BOD of a wastewater is 270 mg/L. The ultimate BOD is reported to be 390 mg/L. Determine the rate at which the wastewater is being oxidized.

Solution:

1. Use Equation 3.42: $BOD_5^{20} = L_o(1 - 10^{-k_1 t})$.
2. Substitute given values of $BOD_5 = 270$ mg/L; and $L_o = 390$ mg/L in the equation to find the reaction rate constant:

$270 = 390(1 - 10^{-5K_1})$.

This yields a value for k_1 equal to 0.102/day.

Example 3.15

A sample of wastewater was incubated for 2 days and its BOD was found to be 200 mg/L at a temperature of 20°C. Compute its 5-day BOD, assuming that the rate constant is 0.1/day.

Solution:

1. Use Equation 3.42: $BOD = L_o(1 - 10^{-k_1 t})$
2. Substitute given values, then ultimate BOD will be

$L_o = 200/(1 - 10^{-0.1 \cdot 2}) = 541.9$ mg/L.

Thus, $BOD_5 = 541.9(1 - 10^{-0.1 \cdot 5}) = 371$ mg/L.

Example 3.16

The following BOD results were obtained on a sample of wastewater at a temperature of 20°C.

t (days)	BOD (mg/L)
0.5	50
1	91
1.5	120
2	142
2.5	158
3	169
3.5	177
4	184
4.5	187
5	190

1. Draw a BOD vs. time graph.
2. Compute the reaction rate constant.
3. Find ultimate 1st stage BOD using the Thomas method.
4. Determine the value of k_1 (to base 10).

Solution

1. Find the value $(t/BOD)^{1/3}$ using the Thomas method [Equation 3.45] as tabulated below:

t (days)	BOD (mg/L)	$(t/BOD)^{1/3}$
.5	50	0.215446
1	91	0.222324
1.5	120	0.232079
2	142	0.241498
2.5	158	0.25105
3	169	0.260862
3.5	177	0.270416
4	184	0.279092
4.5	187	0.288707
5	190	0.297444

2. Plot a graph of BOD vs. t.
3. Plot a graph of $(t/BOD)^{1/3}$ vs. t.
4. Find, from the graph drawn in 3, the slope of the straight line which equals b = 0.018598.
5. Use Equation 3.47:

$$0.018598 = k'^{2/3}/6L_o^{1/3} \qquad (1)$$

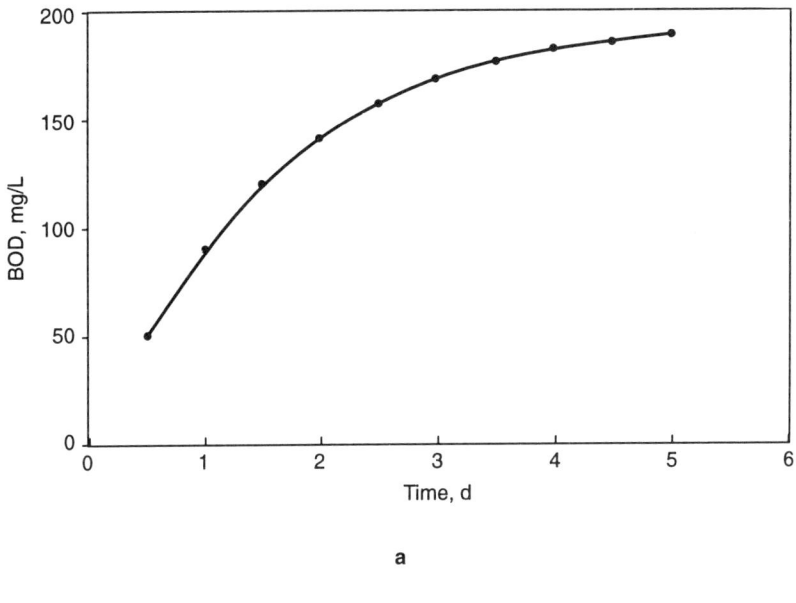

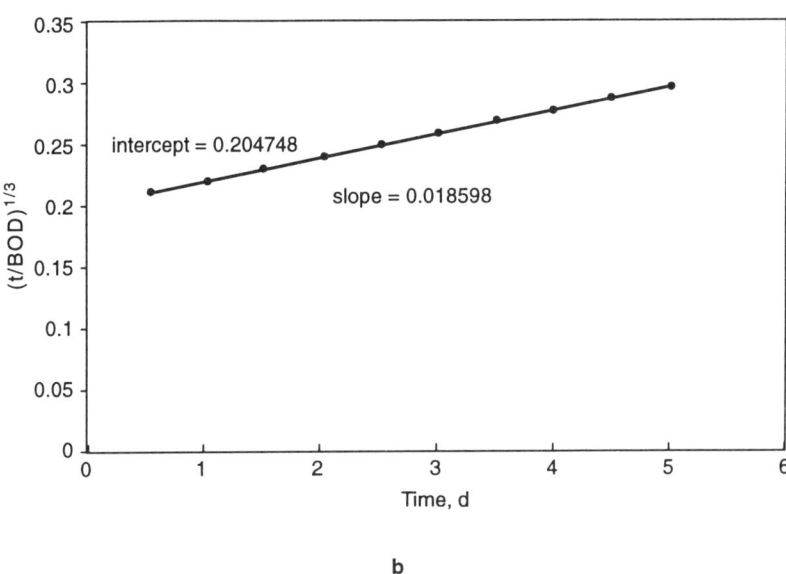

Example 3.16

6. Determine intercept from graph drawn in 3 which is

$$\text{Intercept a} = 0.204748 = (k' \cdot L)^{-1/3} \quad (2)$$

From the above equation

$$L^{1/3} = k'^{-1/3}/0.204748 \quad (3)$$

Substitute (3) in (1), then

$$0.018598 = \left(k'^{2/3}_t \times 0.204748\right)/6k'^{-1/3}$$

Thus, the reaction rate constant can be found to be

$$k' = 0.018598 \times 6/0.204748 = 0.545/day \text{ (to base e)}.$$

The ultimate first stage BOD can be determined as:

$$L_o = 1/(0.204748^3 \, k') = 1/(0.204748^3 \times 0.545) = 214 \text{ mg/L}$$

7. Reaction rate constant to base 10 can be found to be

$$k_1 = 0.4343 \, k' = 0.4343 \times 0.545 = 0.24/day$$

Example 3.17

The 5-day BOD of a wastewater at 20°C is 200 mg/L. Assuming $k_1 = 0.17$, what is the ultimate BOD?

Solution:

1. Use Equation 3.42: $BOD_t = L_o (1 - 10^{-k_1 t})$
2. Find ultimate BOD after substituting given values as $200 = L_o (1 - 10^{-0.17 \times 5})$, thus yielding $L_o = 233$ mg/L.

Example 3.18

A sample of wastewater was incubated for 2 days and the BOD of the sample was observed to be 165 mg/L at 20°C. Determine its 5-day BOD. Assume $k_1 = 0.1/day$.

Solution:

1. Use Equation 3.42: $BOD_t = L_o (1 - 10^{-k_1 t})$
2. Substitute given values: $165 = L_o (1 - 10^{-0.1 \times 2})$, thus $L_o = 447$ mg/L
3. Compute the 5-day BOD as:
   $BOD_5 = 447 (1 - 10^{-0.1 \times 5}) = 306$ mg/L

2. Chemical Oxygen Demand (COD)

The COD test involves an acid oxidation with potassium dichromate. A measured amount of dichromate is added, the acidified sample is boiled for 2 hours, cooled, and the amount of the dichromate remaining is measured by titration with a 0.25 N solution of ferrous ammonium sulfate (FAS), using the Ferroin indicator for end-point determination. COD results are generally higher than BOD values since the test will oxidize materials such as fats and lignin which are only slowly biodegradable.[2]

VI. BIOLOGICAL AND BACTERIOLOGICAL CHARACTERISTICS

A. ENVIRONMENTAL MICROBIOLOGY

Environmental microbiology is a branch of biology that deals with microorganisms commonly found in water and soil. Some groups of microbes have properties in common with plants, while others, to some extent, resemble animals.

Table 3.9 presents a highly abbreviated form of a five-kingdom classification for the living world.[3]

Many biologists classify all microbes in a kingdom of their own, that of the Protista. This contains all unicellular and simple multicellular organisms giving rise to some properties similar to plants as well as animals (see Table 3.10).

Applied microbiology is a rapidly growing and strongly diversified field of technology that is playing important roles, for instance, in the:

- Food industry (baker's yeast, cheese, yogurt, vinegar, etc.).
- Chemical industry (ethanol, antibiotics, enzymes, biopolymer). Antibiotics are substances produced by microbes, which in low concentrations [around 1 mg/L] may prevent growth of certain other microbes. Their therapeutic value is based on the fact that they do not damage the cells of human and animal tissues. Their action can be bacteriostatic (cells are prevented to grow but not killed) or bactericidal (cells are also killed). Antibiotics are used:

Table 3.9 The Simplified Five-Kingdoms of Microorganisms

Plantae	Fungi	Animalia
Photosynthetic plants with somatic cell and tissue differentiation	Absorptive or osmotrophic nutrition, flagellated stages generally absent, multinucleate, and mycelial	Ingestive or phagotrophic nutrition, advanced multicellular organization with tissue differentiation
Tracheophyta (vascular plants)	*Basidomycetes* (club fungi)	Many well-recognized phyla in animal kingdom
Bryophyta (mosses and liverworts)	*Ascomycetes* (sac fungi)	*Porifera* (sponges)
Rhodophyta (red algae)	*Deuteromycetes* (fungi imperfecti)	
Phaeophyta (brown algae)	Phycomycetes (lower fungi)	
Charophyta (stoneworts)	Myxomycetes (slime molds)	

Protista

Eucaryotic unicellular organisms not included in other groups, principally Algae and Protozoa

Monera

Procaryotic blue-green algae, bacteria, and related forms

From Frobisher, M., et al., *Fundamentals of Microbiology,* 9th ed., W. B. Saunders, Philadelphia, PA, 1974, 120, 152.

Table 3.10 Protista in the Living World

Unicellular or multicellular without differentiation	Multicellular with extensive differentiation	
	Plantae	Animalia
Protista	Chlorophyll present	Chlorophyll absent
	Cell wall present	Cell wall absent
	Growth mode: open	Growth mode: closed
	Energy source: solar radiations	Energy source: organic matter
	Movement absent	Active movement

Compiled from References 19, 25, and 26.

1. In treatment of infectious diseases in man and animals.
2. As a supplement in nutrition of poultry and cattle to increase growth rate.
3. For conservation in the food industry.

- Agriculture field (biological N_2-fixation through seed and soil inoculation, biological pest control).
- Water treatment field (slow sand filtration bacteria).
- Waste treatment field (activated sludge, trickling filter, oxidation pond, anaerobic digestion).
- Waste stabilization field (production of fertilizer, fuel: methane production, fodder, and food).

Metabolism or the chemical processes in a living cell can be separated into two categories:

1. Dissimilation or catabolism: this comprises energy-producing reactions.
2. Assimilation or anabolism: this comprises reactions that lead to formation of cell material with the aid of energy liberated in dissimilation.

In both groups, chemical transformation proceeds by way of many successive step-reactions, each of which is catalyzed by a specific enzyme (organic catalysts with a protein-based structure). Enzymes accelerate biochemical reactions without alteration in themselves. Enzymes may pass through an intermediate stage, in combination with a substance. Thus, they convert the substance to an end product that finally releases the enzyme. Enzymes exist in enormous variety throughout life and each one has its specific biochemical function, either outside a cell (extracellular or exoenzyme) or inside one (intracellular enzymes or endoenzyme). The action of enzymes is strongly influenced by:

- Temperature: the rate of enzyme reactions is approximately doubled with each 10°C rise in temperature up to 35°C. Above 35°C the protein fraction of the enzyme undergoes denaturation, resulting in the destruction of the enzyme.[19]

- pH: pH exerts a pronounced effect on enzyme reactions. The effect of pH differs with the particular enzyme in question. Some enzymes are optimum at low pH, while others are optimum at high pH.
- Toxic substances: the nucleus is the source of the enzymes within the cell.

Constitutive enzymes are produced continuously by the nucleus, regardless of the biochemical environment in which the cell is growing.

Adaptive enzymes are produced by the cell when it is placed in a nutrient medium not normal for its growth. Therefore, adaptive enzymes are temporary enzymes. The cell ceases to produce them once the stimulus for the adaptive enzyme has been removed.

Enzymatic reactions are all reversible chemical reactions which are driven in accordance with a decrease in energy. The reaction will be affected by concentration of reactant, concentration of enzyme, and concentration of end product.

The basic enzyme reaction is

$$\text{Reactant} + \text{enzyme} \to \text{reactant-enzyme complex} \to \text{enzyme} + \text{end-product} \quad [3.51]$$

An example of this reaction is the homolactic acid fermentation for energy production:

$$C_6H_{12}O_6 \xrightarrow{\text{enzyme}} 2CH_3.CHOH.COOH + \text{energy} \quad [3.52]$$

where $C_6H_{12}O_6$ = glucose, and $CH_3 \cdot CHOH \cdot COOH$ = lactic acid

Tables 3.11, 3.12, and 3.13 present the manner in which bacteria can be grouped, the differences between higher animals and plants, and the general classification of bacteria according to sources of energy.

Certain microbial metabolic processes have been used by man since prehistoric times for the preparation of food, drink, and textiles. In many cases, these processes have become controlled and perfected to an astonishing degree by purely empirical methods. Outstanding examples of such traditional microbiological processes are

- Leavening of bread: an essential step in the production of bread is the alcoholic fermentation by yeast. The yeasts used for all baking belong to the species *Saccharomyces cerevisiae*.
- Using yeast as a food supplement for animal and food stuffs.
- The manufacture of vinegar by acetic acid bacteria.
- Manufacture of such milk products as butter, cheese, and yogurt which involves the action of microorganisms, among which the lactic acid bacteria are particularly important.
- Stripping of flax and hemp to free the fibers used in the making of linen. The anaerobic butyric acid bacteria attack the plant pectin, loosening the stem structure and freeing the fibers.
- The acetone-butanol fermentation by using *Clostridium acetobutylicum*.
- Production of chemotherapeutic agents:
  - Treatment of bacterial infection by sulfonamide.
  - Production of antibiotics, e.g., penicillin (*Penicillium* spp), streptomycin *(Streptomyces griseus)*, and tetracycline *(Streptomyces aureofaciens, S. rimosus)*.
- Production of organic acids, e.g., citric and gluconic acids (produced by *Aspergillus niger*), itaconic acid (produced by *Aspergillus itaconicus*), and lactic acid (produced by lactobacilli).
- Production of some enzymes, e.g., streptokinase by *Streptococcus hemolyticus*.

B. NATURE OF BIOLOGICAL GROWTH

Microbial growth is a rapid process when conditions are optimal. Multiplication is generally by binary fission. The process proceeds in a logarithmic fashion. Two cells are simultaneously formed from one parent cell, each growing at about the same rate as the parent cell does. Furthermore, the generation time (i.e., the period of time needed by a growing bacterial population to double its number) can be quite short, e.g., of the order of 15 to 20 minutes. The number of cells formed from one cell after a number of divisions = n divisions is equal to 2^n. In practice, of course, growth is always soon restricted by exhaustion of nutrients, and the rising concentration of toxic waste products.

The resulting growth follows a pattern that is qualitatively the same for all microbes. This occurs in the presence of a limited amount of sterile, liquid nutrient medium that is inoculated with a microbe that can grow in it. It is important to keep the culture under fixed conditions. The growth is best represented

Table 3.11 Distinctions between Eukaryotic, Prokaryotic and Akaryotic Bacterial Groups

Group	Cellular organization	Differentiation	Organisms
Eukaryotic	Multicellular	Extensive	Higher plants, (including ferns, mosses, liverworts) and vertebrate and invertebrate animals
	Unicellular, coenocytic, or mycelial	Little or none	Protists (algae, fungi, and protozoa)
Prokaryotic	Unicellular	Little or none	Bacteria (including cyanobacteria)
Akaryotic	Not cellular	—	Viruses

From Sterritt, R. M. and Lester, J. N., *Microbiology for Environmental and Public Health Engineers,* E. and F. N. Spon, London, 1988. With permission.

Table 3.12 Main Differences between Higher Animals and Higher Plants[19,25,26]

	Plants	Animals
Cell walls	Present	Absent
Chlorophyll	Present	Absent
Energy source	Photosynthesis	Organic matter
Mode of growth	Continuous	Fixed for the adult
Movement	Absent	Present
Principal reserve food	Starch	Glycogen, fat

Table 3.13 Classification of Living Organisms to Sources of Energy

	Electron Donor	
Energy Source	Inorganic Substance (Lithotroph)	Organic Substance (Organotroph)
I Oxidizable Substance (Chemotrophs)	*Chemolithotrophs* H_2 bacteria Colorless sulfur bacteria Nitrifying bacteria Iron bacteria	*Chemoorganotrophs* Most bacteria Fungi Protozoa Animals, including *Homo sapiens*
II Light (Phototrophs)	*Photolithotrophs* Green plants Algae Purple sulfur bacteria Green sulfur bacteria	*Photoorganotrophs* Purple nonsulfur bacteria

From Frobisher, M., et al., *Fundamentals of Microbiology,* 9th ed., W. B. Saunders, Philadelphia, PA, 1974, 120, 152.

by the so-called growth curves (see Figures 3.8 and 3.9). This curve is obtained by plotting the logarithm of the number of living cells in the culture against time. In this semilogarithmic plot a straight line is obtained for the period in which exponential growth occurs. In general, a typical growth curve can be divided into the following different phases.

Lag phase — In this phase the cells of the inoculum have to adapt their enzymatic systems to the new medium they are confronted with. The length of the lag phase in a given medium depends on the previous history of the cells in the inoculum. This length is zero if the cells are taken directly from cultures growing exponentially in a medium of the same composition. It is appreciable when the cells have been taken from an old culture or from an exponentially growing culture in a different medium. In the first case, enzymes and/or key intermediates in metabolism have dropped to suboptimal concentrations in the cells. Thus, enzymes have to be restored to normal levels before exponential growth can resume. In the second case, when there is a change of medium, new enzymes very often have to be synthesized in response to the altered conditions. This important process is called induced enzyme synthesis.

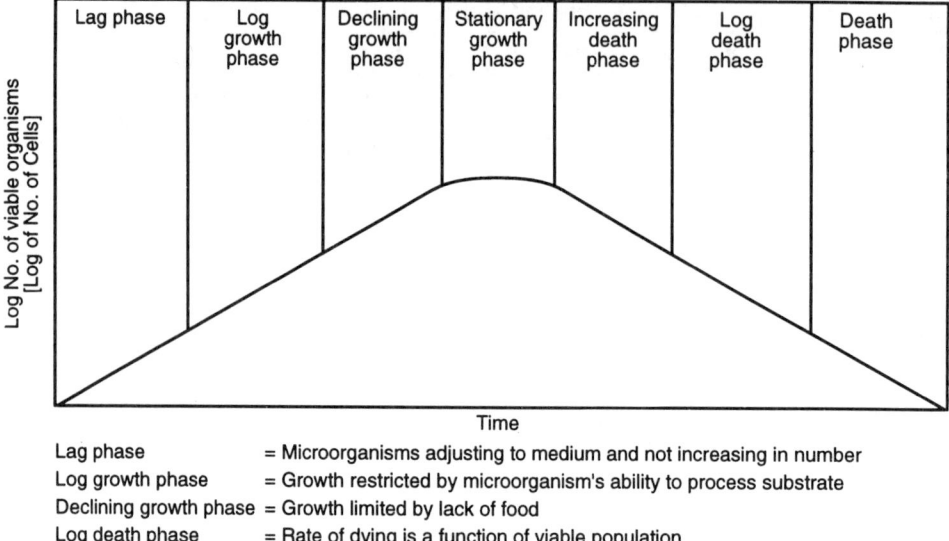

Lag phase = Microorganisms adjusting to medium and not increasing in number
Log growth phase = Growth restricted by microorganism's ability to process substrate
Declining growth phase = Growth limited by lack of food
Log death phase = Rate of dying is a function of viable population

Figure 3.8 Growth pattern based on the number of microorganisms.

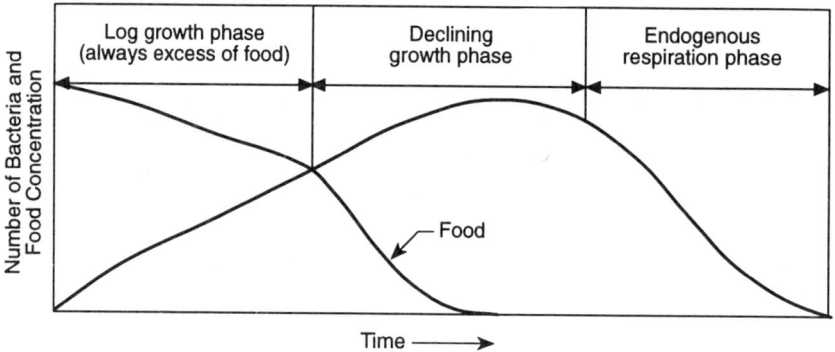

Figure 3.9 The biological growth curve.

Phase of exponential growth — Here the cells divide at a constant rate. In this phase the cells are more sensitive to adverse conditions than they are in the lag phase. The growth rate during this phase is expressed as the generation time — the time required for a population to double its number.

1. Culture Systems

Culture systems for microorganisms can be of the batch type or of the continuous culture type:

a. Batch Type

In this system a fixed amount of medium is inoculated and during cultivation no change of solid or liquid material with the environment is allowed to take place. Under constant conditions, with full mixing by stirring, the mass of cells as a function of time is represented by the growth curve. During the logarithmic growth phase, the logarithm of the microbial mass X increases linearly with time (t)

$$dX/dt = u_s \cdot X \qquad [3.53]$$

where X = microbial mass (number of cells), t = time, and u_s = specific growth rate (hr^{-1}). Integrating Equation 3.53 from X_1 at t_1 to X_2 at t_2

$$\ln X_1 = u_s t_1 + \text{constant} \qquad [3.54]$$

$$\ln X_2 = u_s t_2 + \text{constant} \qquad [3.55]$$

Solving these two Equations 3.54 and 3.55 yields:

$$\ln 2 \cdot X/X = \ln 2 = u_s(t_2 - t_1) \qquad [3.56]$$

The division time t_d of one cell is found by integration from X to 2X and it amounts to:

$$t_d = \ln 2/u_s = 0.69/u_s \qquad [3.57]$$

Usually u_s is determined from the slope of a logarithmic plot, and represents the maximum growth rate $[u_s]_{max}$ (or growth rate constant) the organism can have under the conditions chosen. If the growth rate declines because of depletion of the medium, one of the components of the medium will be the first to limit growth. The relationship between the concentration of the limiting substrate component (s*) and u_s is given by the empirical relationship found by Monod.

$$u_s = [u_s]_{max}(s*/K_s + s*) \qquad [3.58]$$

where: u = specific growth rate
        u_{max} = growth rate constant (maximum growth rate)
        s* = concentration of limiting substrate
        K_s = saturation constant, represented by the substrate concentration at which $u_s = 1/2\ [u_s]_{max}/2$; generally, K_s is very small, $[10^{-4}\ M$ or less].

b. Continuous Culture System

The most simple system is that of a fully mixed culture. In this form a constant flow of culture medium enters and an equal flow of culture liquid flows out. The dilution rate DR is expressed as:

DR = number of times liquid volume in flask is replaced per hour
   = (mL liquid passing through per hour)/(mL liquid volume in the flask)

If DR is fixed at a certain value and the flask is inoculated, a steady state will establish itself. This steady state is characterized by constant cell density in the culture, constant chemical composition of the culture liquid, and constant growth rate u_s = DR.

If all parameters are kept constant, such a steady state can be maintained indefinitely.

The increase in cell mass per mL is given at all times by net increase = production of cells − amount of cells washed away

$$dX/dt = u_s \cdot X - DR \cdot X \qquad [3.59]$$

where X = microbial mass, t = time, u_s = specific growth rate, and DR = dilution rate. The increase in concentration of limiting substrate is given at all times by net increase = amount flowing in − amount washed out − amount consumed:

$$ds*/dt = DR \cdot S_R^* - DR \cdot S* - u_s \cdot X/Y \qquad [3.60]$$

where: s* = concentration of limiting substrate
        t = time
        DR = dilution rate
        S_R^* = substrate concentration in the system
        Y = yield factor, i.e., the weight in mg of cells formed per mg of limiting substrate
           = u_s*X/substrate consumed to form u_s*X cell mass.

The above equations show that there are many values of u_s, DR, and s^* at which a steady state is possible. These are found in the relationship obtained by substituting the value of zero for dX/dt and ds*/dt.

Steady states occur in all situations that meet the following equations.

$$u_s = DR = [u_s]_{max} s^*/(K_s + s^*) \qquad [3.61]$$

$$s^* = K_s(DR/([u_s]_{max} - DR)) \qquad [3.62]$$

$$X = Y(s_R^* - s^*) \qquad [3.63]$$

The equations clearly show the self-regulatory character of this type of continuous culture, called a chemostat.

For a non-steady (= transient) state if:

1. $u_s > DR$, dX/dt is positive, X increases; ds*/dt is negative and s^* decreases. This would continue until $u_s = DR$, i.e., steady state is obtained.
2. $u_s < DR$, dX/dt is negative and X decreases; and ds/dt is positive and s^* increases until $u_s = DR$, i.e., steady state is obtained.
3. DR approaches $[u_s]_{max}$, s^* increases. As soon as s^* reaches the value s_R^*, X = 0 and the culture has been washed out. Since K_s is very much smaller than s_R^*, then for $s^* = s_R^*$, DR becomes equal to $[u_s]_{max}$.

All parameters of the system can accurately be predicted for known values of $[u_s]_{max}$, K_s, and Y for the cultural conditions, the organism, and the limiting substrate under study.

It is worthwhile to note that s^* does not depend on s_R^*. If s_R^* is changed, only X changes. This is of great importance when a substrate is used which is toxic in high concentrations (e.g., phenol), provided a steady state is maintained and no other factor (e.g., O_2 concentration) becomes limiting. The value of s^* can always be kept below the toxic value even if s_R^* is at a toxic concentration. The cell density X, however, has to be kept at a high level. This can be achieved by decreasing DR, i.e., increasing the detention time. If the rate of increase is proportional to the number of cells present at any time, then $dX/dt = u_s*X$, and $\ln X = u_s t + a$.

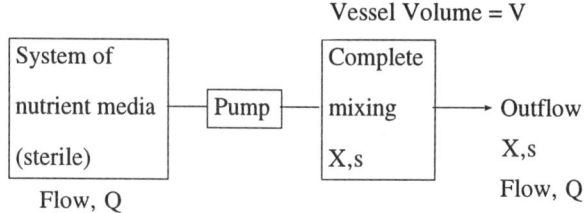

For the vessel not to drain and for steady state conditions:

$$\text{Outflow = Inflow} \qquad [3.64]$$

$$u_s = [u_s]_{max} s^*/(K_s + s^*) \qquad [3.65]$$

$$\text{Dilution rate DR} = Q/V - 1/DR \qquad [3.66]$$

$$\text{Detention time } t = V/Q = 1/DR \text{ [hr]} \qquad [3.67]$$

c. Complete Mixed System

The change in concentration of cells/mL in a vessel is given by:

$$dX/dt = u_s X = DR * X$$

$$0 = u_s X - DR * X \qquad [3.68]$$

where $u_s = DR$ for all values of $DR < [u_s]_{max}$.

For the nutrient concentration (culture vessel)

$$dS^*/dt = DR \cdot s_R^* - DR \cdot s^* - u_s X/Y \quad [3.69]$$

where Y = yield constant.

$$\text{When } ds^*/dt = 0 \quad [3.70]$$

$$\text{Then } DR \cdot s_R^* - DR \cdot s - u_s^* X/Y = 0 \quad [3.71]$$

but X is constant $u_s = DR$

$$DR \cdot s_R^* - DR \cdot s^* - DR \cdot X/Y = 0 \quad [3.72]$$

or

$$X = Y(s_R^* - s^*) \quad [3.73]$$

$$\text{Recycling ratio } R_r = (1 - r_l)/(1 - r_m) \quad [3.74]$$

where: r_l = fraction of liquid volume that is returned
r_m = fraction of cell mass that is returned
$dX/dt = u_s X - DR_r \cdot X/R_r$
$0 = u_s X - DR^* X/R_r$

for steady state condition

$$DR/R_r = u_s \text{ or } DR = u_s R_r \ [R_r > 1] \quad [3.75]$$

then

$$ds^*/dt = DR \cdot s_R^* - DR \cdot s^* - u_s X/Y$$
$$= DR \cdot s_R^* - DR \cdot s^* - [DR/R_r] \cdot [x/y] \quad [3.76]$$

Steady state $0 = DR - s_R^* - DR \cdot s^* - DR \cdot X/R_r \cdot Y$

$$X = R_r \cdot Y(s_R^* - s^*) \quad [3.77]$$

The main factors determining the generation time include natural aspects, concentration of chemical compounds in the medium (nutrients or toxic substances including gases like O_2 and CO_2), temperature (maximum growth at optimum temperature for psychrophiles $T_{opt} < 20°C$; for mesophiles $T_{opt} = 20$ to $45°C$, and for thermophiles $>45°C$), pH, osmotic pressure, radiation, hydrostatic pressure, and nature of the organism.

In general, a compound present in a medium may act as a nutrient, or it may be indifferent to the microbe in question, or it may be toxic to it. In the latter case the effect may be twofold: bacteriostatic (preventing growth without killing); or bactericidal (bacteria-killing). Table 3.14 presents elements and compounds needed for bacterial growth.

Table 3.14 Inorganic Substances Needed by Most Microorganisms

Substantial amounts	Trace amounts
Ca^{2+}, Cl^-, HCO_3^-, K^+, Mg^{2+}, Na^+, PO_4^{3-}, SO_4^{2-}	B^{3+}, Co^{2+}, Cu^{2+}, Fe^{2+}, I^-, Mn^{2+}, Mo^+, Se^{2-}, V^{2+}, Zn^{2+}

Compiled from References 19, 25, and 26.

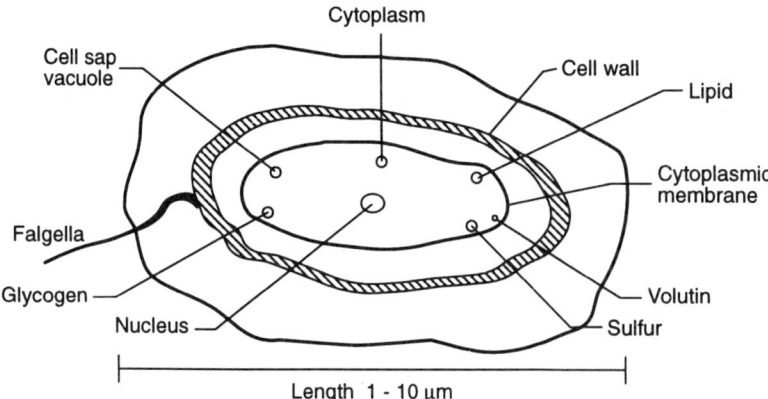

Most bacteria multiply by binary fission. This process can be subdivided
1. Nuclear division
2. Cellular elongation
3. Cellular division
4. Cellular separation

Figure 3.10 Schematic diagram of a typical bacterium.

Stationary Growth Phase — During this phase the growth rate levels off. After this, the cells start dying frequently as a result of partial saturation or accumulation of toxic materials. Nevertheless, the death rate is still counterbalanced by some growth. In such a case the death rate and growth rate are equal.

Death Phase — In this phase the death rate has increased to its maximum value and the number of cells declines exponentially. The death rate in some cases may decrease as a result of utilization of nutrient materials released by dead cells. Autolysis may be encountered in this phase, i.e., digestion of cellular material by enzymes.

C. BACTERIA

The word bacteria (singular: bacterium) comes from the Greek word meaning "rod" or "staff", a shape characteristic of many bacteria. Bacteria are single-celled microscopic organisms that multiply by splitting in two (binary fission). In order to multiply they need carbon obtained from carbon dioxide (CO_2) if they are autotrophs, or from organic compounds (dead vegetation, sewage, meat) if they are heterotrophs. Their energy comes either from sunlight if they are photosynthetic or from chemical reactions if they are chemosynthetic. Bacteria are present in air, water, earth, rotting vegetation, and the intestines of animals. Bacteria are fundamental to all biological processes, especially in the degradation of organic matter which takes place in trickling filters, activated sludge processes, and sludge digestion.

Bacteria average about 1 μm in size but can vary from 0.5 to 10 μm and are classified by shape into the following groups (Figure 3.10 shows the typical schematic diagram for a bacterium):

1. Spherical or ovoid cells generally referred to as coccus (or cocci for plural). The cells are spherical or occasionally very slightly elongated at right angles to the axis of each pair, and the opposed surfaces and are flat or sometimes concave. Cells are differentiated according to their reaction to Gram's staining method.
2. Cylindrical cells, referred to as rods, are bacillus. They are relatively straight, rod-shaped, and non-flexuous cells.
3. Curved cylindrical cell, showing less than one spiral, often with the appearance of a comma, referred to as vibrio. Some vibrio cause cholera *(Vibrio cholerae)*.
4. Helical, spiral-shaped cylindrical cells are referred to as spirillum.

Under ideal conditions bacteria may divide (generation time) every 20 minutes. Nevertheless, they take up food so quickly that they are likely to be limited by shortages of food or oxygen or water. Most bacteria multiply by binary fission. This process can be subdivided into: nuclear division, cellular elongation, cellular division, and cellular separation.

Examination of the dry protoplasm of bacteria indicate an approximate concentration of major elements as outlined in Table 3.15. Other elements are in trace quantities only.

Table 3.15 Concentration of Major Elements Found in Bacteria

Element	% of cell dry weight
Carbon (C)	49
Hydrogen (H)	6.0
Oxygen (O)	27.0
Nitrogen (N)	11.0
Phosphorus (P)	2.5
Sulfur (S)	0.7
Sodium (Na)	0.7
Potassium (K)	0.5
Calcium (Ca)	0.7
Magnesium (Mg)	0.5
Iron (Fe)	0.1

Compiled from Reference 19.

Table 3.16 General Role of Primary Elements in Bacterial Growth

Element	General physiological function
Calcium	Component of cell; cofactor for some enzymes (e.g., proteinase)
Carbon	Component of cell
Cobalt	Constituent of vitamin B_{12} and its coenzyme derivatives
Copper	Component of special enzymes
Hydrogen	Constituent of cell water and organic cell
Iron	Constituent of cytochromes and proteins; cofactor for some enzymes
Magnesium	Component of cell, cofactor for many enzymes; component of chlorophyll
Manganese	Cofactor for some enzymes
Molybdenum	Component of special enzymes
Nitrogen	Constituent of proteins, nucleic acids, and coenzymes
Oxygen	Component of cell water, important in respiration for aerobes
Phosphorus	Component of nucleic acids, phospholipid, coenzyme
Potassium	Constituent of cell, cofactor for some enzymes
Sulfur	Constituent of proteins; component of some coenzymes
Zinc	Component of special enzymes

Compiled from References 5, 19, and 26.

Table 3.16 presents the general role of the primary elements in bacterial growth.

Examination of many different bacteria grown under many different environmental conditions indicate that bacteria are about 75 to 80% water and 20 to 25% dry matter. The dry matter is about 90% organic and 10% inorganic. The organic fraction gives an approximate empirical formulation of $C_5H_7O_2N$.

D. FUNGI OR MOLDS

Fungi (singular: fungus) are tiny aerobic, heterotrophic Protista containing no chlorophyll. They can tolerate drier and more acidic conditions than most bacteria and also are often many-celled. They live in the earth, fresh water, and sea water. Often they can grow so large that they can be seen with the naked eye [mushrooms]. Many grow as filaments and may be seen in polluted rivers, trickling filters, or activated sludge. The optimum pH for most types is between 2 and 9. Because fungi are wholly aerobic, they can, in animals, exist only on the skin or in the bloodstream or lungs. Consequently, there are relatively few fungi that cause disease in humans. Nevertheless, many fungi cause disease in plants. An example of a disease caused by fungi is potato blight, which has sometimes caused famines in certain regions in the world. Many organic substances can be attacked by fungi such as cellulose, phenols, and hydrocarbons. Attacked organic compounds are converted into simple compounds which can be used as nutrients by other organisms. Spoilage of food and deterioration of paper, wood, or leather are often caused by fungi. Even electrical as well as optical instruments may be rendered useless by fungal growth, particularly in hot and humid climates. Spoilage of seeds, fruits, and vegetables results in release of

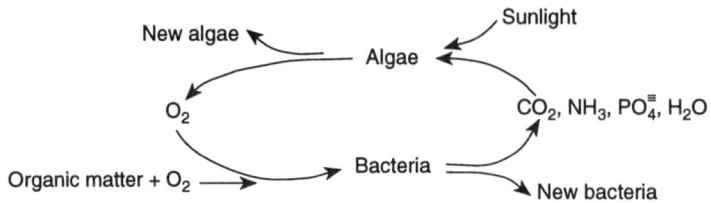

Figure 3.11 Bacteria-algae symbiosis (synergistic reaction).

enzymes that destroy tissue or produce toxins that are harmful to man and animal (e.g., aflatoxin). Certain fungi are used in making cheese and bread; others serve as food (like mushrooms), and still others are used as a source of drugs (such as antibiotics).

The main groups of fungi are Lower fungi *(Phycomycetes);* and Higher fungi *(Ascomycetes, Basidiomycetes,* and *Fungi-imperfecti).*

The vegetative body of a fungus is composed of thread-like structures called hyphae. A mass of a hyphae in turn is called mycelium. Reproduction of fungi can be asexual (by means of asexual spores) such as conidia produced on specialized hyphae named conidiophores or sporangiophores. Reproduction can also be by sexual reproduction resulting in sexual spores. An approximate empirical formulation for fungi is $C_{10}H_{17}O_6N$.

E. ALGAE

Algae (singular: alga) form a large group in the Protista. Being photosynthetic, they are classified as plants. (Photosynthetic species utilize sunlight to release oxygen and they use carbon dioxide (CO_2) from the air or water as the source of carbon for making their cells.) Algae are single-celled or multicellular autotrophs. (Autotrophs utilize inorganic compounds for their protoplasm.) At night some algae react by chemosynthesis, consuming oxygen. Thus a water containing algae has a diurnal variation in dissolved oxygen; however, during sunlight the CO_2 concentration falls. The CO_2 originates from the symbiotic relationship with bacteria (see Figure 3.11), or from bicarbonates releasing hydroxyl ions (OH^-) which tend to raise the pH of the water:

$$HCO_3^- \rightarrow CO_2 + OH^-$$

Photosynthetic reaction (during daytime) can be represented by the following reaction:

$$CO_2 + H_2O + \text{sunlight} \rightarrow CH_2O \text{ (cell material)} + O_2$$

This process utilizes CO_2. The utilization of CO_2 during the day may lead to a considerable rise of pH and results in a softening of the water due to the precipitation of $CaCO_3$, as represented by the following reaction:

$$Ca(HCO_3)_2 \rightarrow CaCO_3 + H_2O + CO_2$$

For most waste stabilization ponds to function properly, algae such as *Chlamydomonas, Chlorella,* and *Euglena* are needed to supply oxygen to aerobic heterotrophic bacteria that consume and oxidize the organic matter in sewage. (Heterotrophs need organic matter as an energy source.) The most important classes of freshwater algae are

1. Green algae, *Chlorophyta,* including *Chlorella.*
2. Motile, bright green flagellate unicellular *Euglenophyta* including *Euglena.*
3. Yellow-green or golden brown, usually unicellular *Chrysophyta.*
4. Blue-green algae, *Cyanophyta.*

Freshwater algae are microscopic, although they can mat together to cover a large water surface. In drinking water, algae are troublesome because they clog filters, may leave a taste when they die, and produce toxins that can poison cattle.

Algae in reservoirs can be reduced by oxygenating the water and reducing its CO_2 content, or by adding an algicide such as copper sulfate ($CuSO_4$) or potassium permanganate ($KMnO_4$), or by destratification of the lake or reservoir.

Algal growth in the catchment area of a water supply system may lead to obnoxious taste and odor problems that are hard to remove and may require treatment with activated carbon. Algae may cause fish kills by excreting toxic materials, but also by causing oxygen depletion. Algae present in raw water are known to interfere with floc formation in coagulation plants. This phenomenon can be indirectly due to changes in pH, alkalinity, or hardness.

1. Control of Algae

1. Prevention

   - Reduce discharges of industrial and domestic waste.
   - Minimize runoff containing organic matter and fertilizers.
   - Clear impounding reservoirs from organic matter as much as possible before they are filled.
   - Control vegetation in reservoirs with pronounced changes of water level and shallow banks.

2. Applying an algicide such as copper sulfate ($CuSO_4$) with a dose of 0.1 to 2 mg/L of ($CuSO_4 \cdot 5H_2O$). Toxicity can be lowered by:

   - Presence of organic matter that combines with Cu^{2+} ions and renders them ineffective.
   - High temperature (promotes toxic action).
   - Low alkalinity.
   - Low pH (at pH values above 8.5, $CuSO_4$ is no longer effective as the hydroxide and/or carbonate precipitate).

3. Chlorination in certain situations (doses of up to 1 mg/L Cl_2).
4. Using cations such as Ni^{2+} and Co^{2+} ions.

F. ANIMALS
1. Protozoa

Protozoa (singular: protozoan) are mobile, single-celled Protista ranging from two to several hundred microns in length. Most are aerobic and heterotrophic. They move by cilia (hair-like feelers), flagella, or pseudopodium. Main groups of the protozoa are

1. Mastigophora: motility by flagella.
2. Sarcodina (pseudopodia: false foot).
3. Ciliata (free swimming [short hair]).
4. Ciliata (stalked).
5. Suctorial.
6. Sporozoa (e.g., causative agent of malaria found in the genus *Plasmodium*).

Protozoa may indicate, by their type, the condition of activated sludge. They are also important in the operation of trickling filters. They feed on the bacteria, and some utilize algae. Most protozoa are harmless, only a few cause illness in humans — *Entamoeba histolytica* (amebiasis), *Giardia lamblia* (giardiasis), *Trypanosoma* (trypanosomiasis: sleeping sickness), and *Plasmodium* (malaria) being some of the exceptions.

Because protozoa are easily seen under the optical microscope, they are valuable indicators of river conditions. In streams with high sewage pollution *Paramecium* and *Colpidium* are often found in the zooplankton, while *Vorticella* and *Opercularia* are found in the bottom mud. Often, the population densities of protozoa are more important than the species present. Empirical formulation of protozoa protoplasm indicate $C_7H_{14}O_3N$.

2. Rotifers (Rotifera: Wheel Animal)

Rotifers are tiny aerobic creatures ranging 50 to 250 µm in length. Rotifers are the simplest of the multicell invertebrate animals. They have cilia around their mouths and can swallow bacteria or other organic matter. Their presence in an effluent indicates highly efficient aerobic biological treatment.

3. Crustaceans

Crustaceans mainly are water animals that use oxygen, consume organic substances, and have a hard body or crust. They are an important food for fish. Crustaceans are not normally found in biological treatment processes. Usually, they are an indicator of clean water. They include crabs, lobsters, shrimps, and prawns, and are important predators on microorganisms. The metabolic complexity of the crustaceans limits their growth to relatively stable streams and lakes.

4. Worms and Larvae

Worms and larvae are the normal inhabitants in organic mud and biological slime. They have aerobic requirements but can metabolize solid organic matter not readily degraded by other microorganisms. The common organisms used in stream pollution studies as indicators of pollution are the worm, *Tubifix* (found in very polluted streams), and the midge fly larva, *Chironomidae* (found after the zone of active decomposition as the stream begins recovery).

G. VIRUSES

A virus is an entity that carries the information needed for its replication, but does not possess the machinery for such replication.[26] Thus, all viruses are obligately parasitic and they are unable to reproduce outside a host cell. As such, viruses exist in two fundamental states: (1) as intracellular entities, and (2) as extracellular particles. Viruses typically range in size from 5 to 25 nanometer (nm) in diameter and up to 800 nm in length. Viruses are not cells, but particles composed of a protein sheath surrounding a nucleic acid core. Viruses are classified according to the type of nucleic acid (DNA or RNA), entire geometry (capsid), and presence or absence of an envelope. Viruses of particular interest in drinking water are hepatitis A, Norwalk-type viruses, rotaviruses, adenoviruses, enteroviruses, and reoviruses.[27] Testing for viruses in water is a difficult process because:

- Viruses are small in size,
- They are of low concentrations in natural waters,
- There are numerous varieties (more than 100 different human enteric viruses are recognized),
- They are unstable,
- There are limited identification methods available.

VII. HOMEWORK PROBLEMS

A. DISCUSSION PROBLEMS

1. What are the most important water quality requirements for a good drinking water supply?
2. Outline five important physical quality characteristics of a good water supply.
3. Name five chemical parameters of concern in water-quality management.
4. Discuss taste and odor in water supplies. Outline suitable methods for their removal.
5. What are the units used in a color measurement?
6. What are the names of the major color-causing compounds?
7. How can the color of a water be reduced?
8. What objections would there be if dissolved solids, chlorides, fluorides, and hardness are present in excess in a water?
9. Describe the origin and nature of alpha, beta, and gamma radiation.
10. Describe methods available for the disposal of radioactive material.
11. Describe factors that affect the pH of waters, wastewater, and sludges.
12. Define hardness of water, note two broad classifications of hardness, and discuss their sources and impact.
13. What are the advantages and disadvantages of softening hard water?
14. Indicate two methods to be used for water softening, with illustrations of the reactions involved.
15. Would you recommend water softening for a village in a developing country? Give your reasons.
16. Discuss the factors that influence the saturation concentration of a gas in water.
17. Outline the processes involved in the degradation of fats under aerobic and anaerobic conditions.
18. Explain why not all of the organic matter in wastewater is converted to carbon dioxide and water in biological treatment.
19. Discuss problems as well as the precision and accuracy associated with the 5-day BOD determination.
20. What is the significance of the BOD in the characterization of wastewater?
21. Thomas' method for determining the BOD exerted at time t is based on the following equation:

$$(t/y)^{1/3} = (k_1 L)^{-1/3} + \left([k_1]^{2/3} \cdot t/6L^{1/3}\right)$$

Define the parameters shown in the equation. Indicate advantages of this method.
22. Why is the BOD test adopted as the standard used to determine the pollution strength of wastewater?
23. Compare and contrast BOD (Biochemical Oxygen Demand) with alternatives such as COD (Chemical Oxygen Demand) or TOC (Total Organic Carbon).
24. Why are samples incubated at 20°C in the dark for 5 days in the BOD test?
25. List three main reasons why testing for viruses in a water sample is a difficult process.
26. What is the significance of bacteria of the coliform group with respect to drinking water?
27. Define autotrophic bacteria and discuss the modeling of the growth kinetics for a pure culture of microorganisms in the presence of a substrate.
28. In terms of kinetics equations, explain the shape of the biomas time curve for a batch culture.
29. Discuss the relationship between growth of bacteria and the temperature of their environment.
30. Does the presence of fecal coliform in water conclusively prove pollution by human wastes? Explain.

B. MATCHING QUESTIONS

1. For each of the headings or phrases in group (A), choose one of the words or phrases in group (B) that is mostly associated with it.

Group (A)	Group (B)
1. Aerobe	a. Proteins
2. Amino acids	b. Fats
3. Galvanized steel	c. Silver
4. Bacteria	d. Suspended solids
5. Cation	e. Mixing
6. Flocculation	f. Anaerobic digestion
7. Polymers	g. Positively charged
8. E. coli	h. Binary fission
9. Hygrometer	i. Electrical conductivity
10. Molarity	j. Moles/gram
11. Molality	k. Zinc
12. Discolors laundry	l. Feces
13. Biogas	m. Moles/liter
14. Lipids	n. Long chain molecules
15. Softness	o. Molecular oxygen
16. Turbidimeter	p. Specific gravity
17. Entamoeba histolytica	q. Bacteria
18. Shigella dysenteriae	r. Manganese
19. Darkening of skin	s. Soap lathering
20. Dissolved salts	t. Protozoa

2. Match the following:

1. Ferroin indicator	a. NTUs
2. Mesophilic range for bacteria	b. Produce CO_2, H_2O, SO_4, NO_3, PO_4, NH_3
3. Beer's Law	c. Platinum-cobalt standard
4. Color	d. Incubate at 35°C
5. Anaerobic microorganisms	e. Protein
6. Fungi	f. COD test
7. Total coliform	g. HCO_3^-, CO_3^{2-}, OH^-
8. Protozoa	h. 20 to 40°C
9. Turbidity	i. Produce CO_2, H_2O, H_2S, CH_4, NH_3, N_2
10. Psychrophilic range for bacteria	j. Multicellular nonphotosynthetic plants
11. Aerobic microorganisms	k. Divalent metallic ions
12. BOD_5	l. Virus
13. Alkalinity	m. 0 to 20°C
14. Thermophilic range for bacteria	n. $Ab = \log Io/I = KC$
15. Hardness	o. Single-celled animals
16. Hepatitis	p. Incubate at 20°C
17. Enzyme	q. 40 to 60°C

C. SPECIFIC MATHEMATICAL PROBLEMS

1. At what temperature will the readings on Celsius and Fahrenheit thermometers coincide?
2. Relate heat lost by an individual who drinks 500 ml of cold water at a temperature of 0°C to that lost by another who sweats out 500 ml of water. (Take specific heat of water = 4180 J/kg.°C; heat of vaporization of water = 2410 J/g at 37°C.)
3. In a water displacement experiment for collecting oxygen gas of pressure 720 mmHg, the barometer indicated a pressure of 730 mmHg. Determine the water temperature.
4. When the color in a sample of drinking water was looked at through a depth of 10 cm it corresponds in intensity to a standard solution having a color of 30 color units. Determine the corresponding color units for the same sample when viewed through another depth of 25 cm.
5. The following results were acquired for a sample: total solids 550 mg/L; suspended solids 600 mg/L; volatile suspended solids 250 mg/L; and fixed suspended solids 150 mg/L. Which of the numbers is doubtful? Justify your computations.
6. Determine the total dissolved solids concentration of a sample given the following data: weight of evaporating dish = 45.6505 g; volume of water evaporated = 250 mL; and weight of the dish and evaporated dry material = 45.8570 g.
7. What is the difference between the density of water at room temperature and the density of water at 60 to 70°C? Explain the difference.
8. What will the emission of an alpha particle from the nucleus of $^{226}_{88}Ra$ produce?
9. A radioactive nuclide has a half-life of 14.3 days. How long would a sample containing 15 μc of this nuclide have to be stored in order to reduce it to 0.3 μc?
10. A radioactive nuclide was found to have decayed 60% in 24 hours. Compute its half life.
11. Thorium, $^{234}_{90}Th$, decays by beta radiation with a half life of 24.1 days. Calculate its decay constant.
12. The radioelement $^{24}_{11}Na$ decays with a half life of 15 hours according to the reaction: $^{24}_{11}Na \rightarrow\ ^{24}_{12}Mg + e^0_{-1}$. Find how many hours will it take for 40% of the element to decay? What fraction of the sample will remain after 6 days?
13. The c.g.s. unit of viscosity is the poise (1 g/cm-sec). Prove that one poise is equivalent to 0.1 N-s/m² (Newton-second/square meter).
14. The viscosity of water is 0.9608 centipoise. Compute:
    a. the absolute viscosity in Pa-s (Pascal-second).
    b. the value of the kinematic viscosity (in m²/s) if the relative density of water is 0.9978.
15. Calculate the equivalent weight of sulfuric acid (H_2SO_4), carbonate ion (CO_3^{2-}), and nitrate ion (NO_3^-).
16. At a certain temperature the equilibrium constant of the reaction between hydrogen and carbon dioxide is one. The original concentration of the reactants were $[H_2] = 0.6$ mol/L, and $[CO_2] = 0.3$ mol/L. Determine the concentration of all the four substances when equilibrium is established. [Hint: the reaction equation is $H_2 + CO_2 = CO + H_2O$].
17. Substance A reacts with substance B according to the following reaction equation: A + B = C + D. Calculate the percentage of substance A that has reacted given that: the initial concentration of substance B = 4 mol/L, equilibrium constant = 1, and initial concentration of substance A = 2 mol/L.
18. Hydrogen iodide, HI, dissociates into iodine, I_2, and hydrogen, H_2, when heated. At a certain temperature, the equilibrium constant of this reaction is 0.0015. Calculate the percentage of hydrogen iodine that has dissociated at the given temperature if its initial concentration is 0.4 mg/L.
19. Nitrogen dioxide dissociates into nitric oxide and oxygen at a certain temperature [$2NO_2 \rightarrow 2NO + O_2$]. At equilibrium the following concentrations are found: $[NO_2] = 0.05$ mol/L, $[NO] = 0.2$ mol/L, $[O_2] = 0.01$ mol/L. Find the equilibrium constant.
20. How many grams of calcium carbonate salt are found in 1 L of a saturated solution? Take the solubility product of the salt to be $1.7 \cdot 10^{-8}$.
21. Given the solubility product of barium sulfate [$BaSO_4$] to be $1 \cdot 10^{-10}$, determine its concentration in a saturated aqueous solution.
22. Determine the weight in grams of soda, $Na_2CO_3 \cdot 10H_2O$, required to prepare 400 mL of a 0.1 N solution.
23. An analysis of a water source indicated the following results: carbon = 62, nitrogen = 1, phosphorus = 0.01 mg/L. Which element would be the limiting factor during the growth of algae? Outline your reasons. (Assume algae to have the following chemical composition by weight: $C_{106}H_{181}O_{45}N_{16}P$.)

24. What are the percentages of C, H, N, and O in the compound $NH_2C_6H_4C_6H_3(NH_2)NO_2$? What weights of carbon dioxide, water, and nitrogen would be produced by burning 5 mg of this compound in aerobic conditions?
25. Commercial H_2SO_4 is 96.4% sulfuric acid by mass and its density is 1.84 g/ml. Calculate the molarity of commercial sulfuric acid.
26. A sample shows a pH of 7.6. What is its hydrogen ion concentration?
27. Calculate the concentration of ions and pH in a 0.5 M hydrochloric acid solution.
28. How many grams of commercial hydrofluosilicic acid H_2SiF_6 with a purity of 30% should be added per cubic meter of water to increase the fluoride ion concentration from 0.1 mg/L to 1 mg/L?
29. Raw water contains 0.3 mg/L of fluoride ion. After treating 250 L of the water each second, the fluoride concentration was increased to 1 mg/L. This increase was achieved by dosing a solution of hydrofluosilicic acid (H_2SiF_6) which contains 0.2 kg of the acid per liter. Calculate the rate of the feed of hydrofluosilicic acid required.
30. Analysis of water revealed the following: Na^+ = 20 mg/L, K^+ = 30 mg/L, Ca^{2+} = 5 mg/L, Mg^{2+} = 10 mg/L, Sr^{2+} = 2 mg/L, Cl^- = 40 mg/L, HCO_3^- = 67 mg/L, $SO_4^=$ = 5 mg/L, and NO_3^- = 10 mg/L. What is the total hardness and carbonate and noncarbonate hardness in mg/L as $CaCO_3$?
31. Tests for common ions are run on a sample of water and the results are as shown below. If a 10% error in the balance is acceptable, should the analysis be considered complete?

Constituents (mg/L)

Ca^{2+}	120	HCO_3^-	244
Mg^{2+}	48.6	SO_4^{2-}	96
Na^+	92	Cl^-	71

Determine the carbonate, noncarbonate, and total hardness. List the hypothetical chemical combinations of ions.

32. A surface water has the following analysis:

Calcium	70 mg/L	Chloride	missing
Magnesium	55 mg/L	Alkalinity	250 mg $CaCO_3$/L
Sodium	23 mg/L	Sulfate	144 mg/L

   1. Assuming negligible experimental error, with the exception of the chloride ion, and that the alkalinity is due to bicarbonate, compute the missing chloride concentration.
   2. Calculate total and noncarbonate hardness of the water expressed as mg/L $CaCO_3$.
   3. Draw a bar diagram in milliequivalents per liter of this water.
   4. List the hypothetical chemical combinations of positive and negative ions from the bar graph.
   5. What conclusions can you draw from your calculations?

33. An analysis of hard water revealed the following data:

Total hardness	250 mg/L as CaO
Calcium hardness	130 mg/L as $CaCO_3$
Alkalinity	200 mg/L as $CaCO_3$
Acidity	0.3 meq/L

   1. Construct the bar diagram of this water.
   2. Compute the amount of lime (in mg/L) needed for softening this water.
   3. Indicate whether this water is aggressive to asbestos-cement pipes (use an appropriate corrosion potential and assume the hydrogen ion concentration to be $2.5 \cdot 10^{-7}$ mol/L).

34. The following water analysis is submitted for your review:

Mg^{2+} as Mg	30.1 mg/L
Alkalinity as $CaCO_3$	135 mg/L
Ca^{2+} as Ca	90 mg/L
Sodium as Na	72.2 mg/L
K^+ as K	6.3 mg/L
Sulfate as SO_4^{2-}	missing, (y; mg/L)
Chloride as Cl^-	120 mg/L
Carbon dioxide as CO_2	3 mg/L
Temperature	25°C

   1. Assuming that all of the constituents with the exception of the sulfate ion (SO_4^{2-}) have been analyzed correctly and that the alkalinity is due solely to bicarbonate, determine the sulfate concentration.

2. Determine the total and noncarbonate hardness (express results in mg/L as $CaCO_3$).
3. Draw a bar diagram in milliequivalents per liter of the water.
4. List the hypothetical chemical combinations of positive and negative ions from the bar graph.

35. What is the saturation value of dissolved oxygen in a water containing 1000 mg/L of chloride ion at a temperature of 22°C and at a pressure of 720 mmHg?
36. At what depth of the water is the partial pressure of oxygen 34 kPa (kiloPascals)?
37. Determine the saturation concentration of oxygen in pure water at 10°C exposed to air at a pressure of 101.3 kPa. Assume that dry air contains 20.948% oxygen by volume.
38. Air of atmospheric pressure contains 21% of oxygen by volume. Compute the saturation concentration in water at a depth of 6 m and 25°C.
39. Compute the saturation value of dissolved oxygen in water exposed to water-saturated air containing 20.9% oxygen under a pressure of 732 mm of mercury. Assume a temperature of 20°C and a chloride concentration in water of 8000 mg/L.
40. The dissolved oxygen (DO) of a stream is 8 mg/L. A city in the vicinity discharges its wastewater into the stream. The waste discharged has a DO content of zero. The ratio of stream flow to that of the wastewater is 7:1. Find the DO of the mixture.
41. The BOD_5 of a wastewater is determined to be 150 mg/L at 20°C. The k (to base e) value is known to be 0.23 per day. What would the BOD_5 be if the test were run at 15°C?
42. The BOD value of a wastewater was measured at 2 and 8 days and found to be 125 and 225 mg/L, respectively. Determine the 5-day value using the first-order rate model.
43. A sample of sewage was incubated for 2 days and the BOD of the sample is observed to be 175 mg/L at 20°C. Determine its 5-day BOD (assume $k_1 = 0.1$).
44. Assuming that each person contributes a BOD load of 0.055 kg/d, and a suspended solids load of 0.08 kg/d, find the daily domestic pollution load of a town of 40,000 inhabitants. If the DWF is 150 L/capita·d, compute the concentration of BOD and suspended solids in the sewage.
45. A waste was found to have an ultimate BOD of 400 mg/L, and was found to follow first order reaction kinetics with a k value of 0.2 per day. Compute the 5-day BOD value.
46. A small sewage works serves a population of 15,000 whose per capita BOD_5 generation is 0.06 kg/d. Water consumption is metered at 275 L/capita·d. A small industry in the neighborhood discharges an effluent of 20 L/s over a period of 4 hours per day. The BOD_5 of the wastewater from this industry was determined by using a 2% dilution (6 mL of wastewater in a 300-mL BOD bottle) and incubated for 5 days at 20°C. At the end of the 5-day incubation period 4 mg/L dissolved oxygen has been depleted in the bottle. The first stage ultimate oxygen demand for the sewage works was 500 mg/L. Determine:

   1. The rate at which the waste is being oxidized (assume a temperature of 20°C). At this rate, a 75% demand will be exerted in how many hours?
   2. The oxidation rate at 25°C, given that:

   $$(k_1)_T = (k_1)_{20} \cdot (1.024)^{T-20}$$

   where $(k_1)_{20}$ = reaeration rate at 20°C.

47. Three factories discharge their combined wastewater into a neighboring stream. The characteristics of their wastewater are as tabulated below:

Factory	Amount of wastewater discharge (m³/s)	BOD_5 (mg/L)
1	0.4	300
2	0.2	450
3	0.3	150

Compute the BOD_5 of the mixture being discharged to the river.

48. A town of 25,000 inhabitants is to discharge treated domestic sewage to the neighboring stream which has a minimum flow of 0.25 m³/s and a 5-day BOD of 3 mg/L. Water consumption in the town is 175 L/capita·d, and the BOD contribution is 0.06 kg/person·d. If the BOD in the stream below the sewage outfall is not to exceed 4 mg/L, determine the maximum effluent BOD permissible.

49. In a 5-day BOD test on a diluted sample of wastewater the following results emerged:

Time (days)	BOD (mg/L)
1	7
2	12
3	16
4	19
5	21
6	22

1. Draw the BOD vs. time curve.
2. Compute the reaction rate constant.
3. Determine the ultimate first-stage BOD of the sample using Thomas' method.
4. Find the value of the reaction rate constant to base 10.

50. The results of a presumptive coliform test were 3 of 5 tubes of 10-mL portions positive, 4 of 5 tubes of 1-mL portions positive, and 0 of 5 tubes of 0.1-mL portions positive. What is the MPN per 100 mL?
51. Suppose you run a multiple-tube coliform test and got the following results: 10 mL samples — all three positive, 1 mL samples — all three positive, and 0.1 mL samples — two of three negative. Estimate the concentration of coliform bacteria.
52. The results of a coliform test were 10 mL samples — all three tubes positive, 1 mL sample — one tube positive, 0.1 mL sample — one tube positive. Estimate the concentration of coliform bacteria.

D. SUGGESTED EXERCISE

Make a field trip to a local wastewater treatment facility and prepare a report regarding the plant's operation.

The report should contain the following information:

1. Design population
2. Design average flow (m^3/day)
3. Peak design flow (m^3/day)
4. Present flow (m^3/day)
5. Design calculations (giving dimensions) for:
   - Preliminary treatment,
   - Primary treatment,
   - Secondary treatment, and
   - Advanced or tertiary treatment
6. Design influent BOD_5 (mg/L)
7. Design influent suspended solids (mg/L)
8. Design effluent BOD_5 (mg/L)
9. Design soluble effluent BOD_5 (mg/L)
10. Design suspended solids effluent (mg/L)
11. Flow diagram of the plant showing the various wastewater treatment processes involved, as well as flow measuring devices.
12. Present a table showing the various tests being made at the plant on the influent and effluent and the concentrations for each item.
13. Indicate where the wastewater effluent is being discharged and whether or not it is being utilized for a beneficial use.
14. Discussion and conclusions regarding the trip.

REFERENCES

1. **Berger, B. B.**, Ed., *Control of Organic Substances in Water and Wastewater,* Noyes Data Co., Park Ridge, NJ, 1987.
2. **Adams, V. D.**, *Water and Wastewater Examination Manual,* Lewis Publishers, Chelsea, MI, 1990, 54–56, 75, 79.
3. **Frobisher, M. et al.**, *Fundamentals of Microbiology,* 9th ed., W. B. Saunders, Philadelphia, PA, 1974, 120, 152.
4. **Sawyer, C. N., McCarty, P. L., and Parking, G. F.**, *Chemistry for Environmental Engineering,* McGraw-Hill, New York, 1994.
5. *Standard Methods for the Examination of Water and Wastewater,* 18th ed., American Public Health Association, Washington, D.C., 1992.

6. **Mills, E. V.,** Studies on the nature and amount of the colloids present in sewage. I. A historical survey. II. The physical and chemical analysis of sewage, *J. Soc. Chem. Ind.,* 51, 1932, 255T, 349T.
7. **Rickert, D. A. and Hunter, J. V.,** Rapid flocculation and material balance of soils fractions in wastewater and wastewater effluent, *J. Water Pollut. Control Fed.,* 39(2), 1967, 1475–86.
8. **Rudolfs, W. and Gehm, W. H.,** Colloids in sewage and sewage treatment. I. Occurrence and role. A critical review, *Sewage Works J.,* 11(5), 1939, 727.
9. **Rudolfs, W. and Palmat, J. L.,** Colloids in sewage. I. Separation of sewage colloids with the aid of the electron microscope, *Sewage Ind. Wastes J.,* 24(3), 1952, 247–56.
10. *Radiological Health,* U.S. Department Health Education And Welfare, U.S. Printing Office, Washington, D.C., 1970, 413–441.
11. **Nebiker, J. H.,** Dewatering of sewage sludge on granular materials, *Environmental Engineering Report, No. 8–68–3,* University of Massachusetts, Amherst, 1968.
12. **Coackley, P.,** Development in our knowledge of sludge dewatering behavior, 8th Public Health Engineering Conference held in the Department of Civil Engineering, University of Technology, Loughborough, 1975, 5.
13. **Abdel-Magid, I. M.,** The influence of additives on the rheological properties of sewage sludges, *Sudan Eng. Soc. J.,* 26, 1984, 31.
14. **Taylor, G.,** Some humus sludge conditioning experiments, *Institute of Sewage Purification, The Journal and Proceedings,* 1957, 242.
15. **Newitt, D. M., T. R. Oliver and J. F. Pearse,** The mechanism of the drying of solids, *Transactions of the Institute of Chemical Engineers,* 27, 1949, 1.
16. **Coackley, P. and R. Allos,** The drying characteristics of some sewage sludges, *Institute of Sewage Purification, The Journal and Proceedings,* 6, 1962, 557.
17. **Coackley, P.,** The theory and practice of sludge dewatering, *J. Inst. Publ. Health Eng.,* 64(1), 1965, 34.
18. **Whitten, K. W., K. D. Gailey, and R. E. Davis,** *General Chemistry with Qualitative Analysis,* W.B. Saunders, Philadelphia, 1988.
19. **McKinney, R. E.,** *Microbiology For Sanitary Engineers,* McGraw-Hill, New York, 1962.
20. **Gilcreas, F. W., W. W. Sanderson, and R. P. Elmer,** Two new methods for the determination of grease in sewage, *Sewage Ind. Wastes,* 25, 1953, 1379.
21. **Abdel-Magid, I. M.,** *Selected Problems in Wastewater Engineering,* Khartoum University Press, National Research Council, Khartoum, 1986.
22. **Mara, D.,** *Sewage Treatment in Hot Climates,* John Wiley & Sons, London, 1980.
23. **White, J. B.,** *Wastewater Engineering,* Edward Arnold, London, 1978.
24. **Wilson, F.,** *Design Calculations in Wastewater Treatment,* E. and F. N. Spon, London, 1981.
25. **Stanier, R. Y., M. Doudoroff, and E. A. Adelberg,** *General Microbiology,* 3rd ed., Macmillan, New York, 1970.
26. **Sterritt, R. M. and J. N. Lester,** *Microbiology for Environmental and Public Health Engineers,* E. and F. N. Spon, London, 1988.
27. **Pontius, F. W.,** *Water Quality and Treatment: A Handbook of Community Water Supplies,* 4th ed., McGraw-Hill, New York, 1990.
28. **Whipple, G. C. and M. C. Whipple,** Solubility of Oxygen in Sea Water, *J. Am. Chem. Soc.,* 33, 1911, 362.

Chapter 4

Health Aspects of Using Reclaimed Water in Engineering Projects

CONTENTS

I. Introduction	108
A. Waterborne Diseases	109
B. Water-Washed Diseases	110
C. Water-Based Diseases	110
D. Water-Related Insect Vectors	112
E. Infections Primarily Due to Defective Sanitation	113
II. Pathogens Found in Reclaimed Wastewater	117
A. Bacteria	119
B. Helminths (Worms)	120
C. Protozoa	123
D. Viruses	123
III. Removal of Pathogens By Various Wastewater Treatment Processes	123
A. Preliminary and Primary Treatment Methods	123
B. Secondary Treatment Methods	124
C. Tertiary (Advanced) Treatment Methods	124
IV. Toxic Chemicals Found in Reclaimed Wastewater	125
A. Inorganic Chemicals That Pose a Health Hazard	126
B. Organic Chemicals That Pose a Health Hazard	127
C. Boron	127
D. Nitrates	128
E. Pesticides	128
V. Radioactivity in Reclaimed Water	128
VI. Removal of Chemical Constituents By Various Treatment Processes	130
A. Preliminary Treatment Methods	130
B. Primary Treatment Methods	130
C. Secondary Treatment Methods	131
D. Tertiary (Advanced) Treatment Methods	131
VII. Important Health Concerns in Engineering Projects	132
A. Introduction	132
B. The Environment and Habitat of Vectors	133
C. Vector Control	135
D. Chemical Methods For Vector Control	137

        1. Major Groups of Insecticides 137
           a. Pyrethrin 137
           b. Organochlorine Compounds (Chlorinated Hydrocarbons) 137
           c. Organophosphate Compounds 138
           d. Carbamate Compounds 138
           e. Pyrethroids Compounds 138
        2. Classes of Pesticides 138
    E. Natural Vector Control Methods 139
        1. Natural Enemies 140
        2. Biological Control Agents (Biological Toxoids) 140
    F. Genetic Vector Control Methods 140
    G. Environmental Management Methods For Vector Control 140
    H. Permanent Vector Control Methods 141
        1. Storage Impoundments 141
        2. Mosquito Problems in Impoundments 142
        3. Irrigation 142
    I. Temporary Methods For Vector Control — Environmental Methods 150
    J. Modification of Human Habitation and Behavior For Vector Control 151
    K. Integrated Vector Control Methods 152
    L. Community Awareness in Engineering Developmental Projects 154
VIII. Homework Problems 158
    A. Discussion Type Problems 158
    B. Specific Mathematical Problems 159
    C. True and False Statements 160
    D. Multiple Choice Questions 160
    E. Matching Questions 162
References 162

I. INTRODUCTION

The reuse of reclaimed wastewater needs to satisfy specific conditions to safeguard the public health and welfare of a community. The basic elements required to assure success of a water reuse project include safety (water quality), adequacy (quantity), dependability (continuity), and convenience. Figure 4.1 outlines the parameters that must be considered in order to guarantee a community that its public water supply is both safe and reliable.

If reclaimed wastewater is to be used for a beneficial use then the quality and quantity of this water must be assured. To assure these two basic elements, careful attention must be paid to planning, design, construction, operation, and maintenance of any water reuse project.

A safe water supply is basic to the management and control of waterborne diseases. Approximately 80% of all diseases in the world are considered to be related to the use of contaminated water.

The history reveals that the connection between drinking contaminated water and disease dates back to ancient times. For instance, the relationship between water and health was postulated by Hippocrates. Hippocrates, the father of medicine, suggested that there was an association between wet marshy places and fevers. In 1645, Anton Van Leeuwenhoek pointed to the correlation between bacteria, protozoa, and the various diseases. Dr. John Snow, a public health worker in London, in 1854, described the connection between cholera and the consumption of water from a well on Broad Street. Shortly thereafter, Budd suggested that the spread of typhoid was related to drinking contaminated water. Manson, in 1877, related filariasis and the consumption of contaminated water. Pasteur and other scientists of the time developed the germ theory for diseases and also related consumption of contaminated water with various infections diseases.

Human health may not only be affected by microorganisms in water but also by chemicals in the water. For example, a high fluoride content in a groundwater source may introduce an adverse effect on the growth of bones. However, at low concentrations in water, around 1 mg/L, fluorides provide protection from dental cavities.

Nevertheless, the chemical health risks are generally not as serious or widespread as that brought about from microbial contamination of the water (except, of course, for lead).

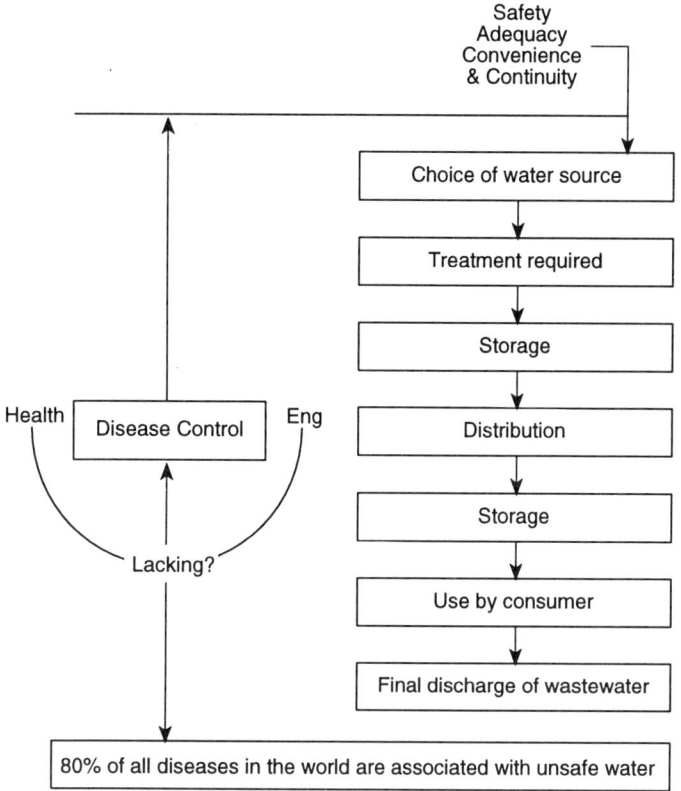

Figure 4.1 Water supply and human health.

Generally, the diseases that are associated with water and/or reclaimed wastewater may be grouped into five main categories. These categories include: waterborne diseases, water-washed diseases, water-based diseases, water-related insect vectors, and infections due to poor sanitation. These categories are briefly summarized here.

A. WATERBORNE DISEASES

Waterborne diseases are those infections that may be spread through a water supply system. Water acts exclusively as a passive vehicle for the pathogen that cause the diseases.

Examples of some of these diseases include: typhoid fever, cholera, giardiasis, dysentery, leptospirosis, tularemia, paratyphoid, and infective hepatitis.

Typhoid and cholera are both due to a relatively fragile microbe whose sole reservoir is man. Typhoid causes a high severe fever. The typhoid victim may recover clinically but may continue passing the pathogen in the feces for a period of time that can extend for months or years. Typhoid carriers are regarded as the principal source of this infectious disease.

Cholera is caused by a toxin produced by the bacterial agent *Vibrio cholerae* growing in the small intestine of humans. The ability of this bacteria to attach itself to the intestinal lining and to multiply in this area is regarded as an important feature in producing the disastrous effects of this disease.[1]

Giardiasis is brought about by the cysts or trophozoites of *Giardia lamblia*. This parasite lives in the upper small intestine and the trophozoites of this microorganism may also be found in the jejunal area (middle part of the small intestine). Giardiasis is responsible for gastrointestinal disturbances, flatulence, diarrhea, and discomfort. Likewise, giardiasis is emerging as a major waterborne disease.[2]

Bacillary dysentery is usually caused by *Shigella flexneri* and *Shigella sonnei,* while the protozoan *Entamoeba histolytica* is responsible for amoebic dysentery and amoebic hepatitis. Amoebic dysentery is normally present with a bloody diarrheal stool, lower abdominal pain, and mild fever. Usually, the disease exhibits a subacute or a chronic sickness.

Recently, an outbreak of a waterborne disease, cryptosporidiosis, affected from 200,000 to 400,000 people in Milwaukee, Wisconsin. The protozoan responsible for this disease is *Cryptosporidium*. The usual symptoms are watery or mucoid diarrhea, lasting 3 to more than 14 days, with vomiting, anorexia, and abdominal pain. Milwaukee shut down one of its water treatment plants and has now taken steps to ensure a safe water supply to the community.[3,4]

B. WATER-WASHED DISEASES

Water-washed diseases are those caused by a shortage of water for personal hygiene. The main infections caused by this shortage of water are those that affect the body exterior surfaces, the eyes, and the skin.

Examples of the water-washed diseases include: bacillary dysentery, skin sepsis and ulcers, conjunctivitis, trachoma, scabies, yaws, leprosy, tinea, louse-borne fevers, diarrheal diseases, and ascariasis.

The eye infections are of special concern in arid and semiarid regions, as the eye conjectiva is adversely affected by a dry atmosphere, suspended dust particles, as well as dust storms. Trachoma may cause inflammation of the eye, rotation of the eyelashes, and the resulting opacity of the cornea. These conditions may lead to partial or complete blindness.

Salmonella pathogens are generally found in municipal wastewaters. The *Salmonella* group includes a great number of species that may initiate infections in man and his domestic pets. Salmonellosis in man may be classified as: enteric fevers (e.g., typhoid fever), septicemia, and acute gastroenteritis.

Diarrhea is a condition that produces watery or bloody stools that leads to a loss in weight. The infection usually includes one or more of the following symptoms: fever, vomiting, cramping, loss of appetite, and malabsorption of nutrients. Diarrhea is usually due to the ingestion of many enteric organisms that cause disease. Diarrhea may also be caused by other non-enteric infections such as measles or even be caused by intestinal abnormalities not related to infection such as lactose intolerance. These enteric organisms are usually transmitted through excreta (human and animal) and may be ingested through contaminated drinking water, food, hands, or other objects. The organisms can go directly from person-to-person, person-object-person, or person-water-person. In some instances self reinfection may continue in an individual, such as in the case with *Giardia*.[5] The pathogenic organisms that may cause diarrhea may be bacteria, viruses, or parasites. Some of the major organisms are *Campulobacter jejune*, *Escherichia coli*, *Shigella* spp., *Vibro cholerae*, rotavirus, *Cryptosporidium*, *Entamoeba histolytica*, and *Giardia lamblia*. The transmission of diarrhea-causing pathogens may be reduced by improved excreta disposal, increasing the amount of water for personal and domestic hygiene, improving the water quality, promoting breast-feeding by mothers, and providing a safe food supply. Vaccine developments are expected to reduce diarrheal diseases.[5]

Diarrhea is considered the main cause of the high morbidity and mortality rates for children less than 5 years old. Diarrhea may lead to death in severe cases, especially in children suffering from malnutrition. The incidence of diarrhea per child in developing countries is estimated at 3.5 cases per year. Some children suffer more than others, and they may have ten or more episodes of diarrhea per annum. Several million children die each year from this disease.[5]

C. WATER-BASED DISEASES

Water-based diseases are considered as infections transmitted through an aquatic invertebrate host, usually an animal. An essential part of the life cycle of the infecting organism takes place in this aquatic animal.

Examples of the water-based diseases include: schistosomiasis, guinea worm, and filariosis.

Schistosomiasis or bilharziasis (swimmer's itch) is a disease that is caused by the infection of the venous system by trematodes of the genus *Schistosoma*. The disease is transmitted through the skin and it may be accompanied by inflammation and itching. Hematuria or bloody urine among children is the classical clinical sign of urinary schistosomiasis.[6] Urinary infection is caused by *S. haematobium* which affect the bladder. Intestinal infection is caused by *S. japonicum* and *S. mansoni* in the portal venous system that transports blood from the intestines to the liver. *S. intercalatum* also causes intestinal infections which tends to be localized to West Africa.[7] In most endemic communities the prevalence of the infection is highest in 10- to 14-year-old children. In many communities of Africa, over 70% of village school children are infected.[6]

Schistosoma mansoni infections are caused by contact with contaminated fresh water in rivers, lakes, streams, ponds, and irrigation canals. People rarely get infected by drinking contaminated water. Any contact with contaminated waters can result in infections. These contacts include water collection for home use, hand and utensils washing, clothes washing, washing animals, washing after defecation, bathing, and swimming. Any activity that requires human-water contact with contaminated water such as irrigation canal and drainage ditch construction, canal cleaning, and fishing can result in contracting schistosomiasis.[6]

The infection is not spread directly from person to person but rather through snails. If the snails are not present, schistosomiasis transmission will not occur.[6] The adult worms produce eggs or ova. When the eggs come in contact with fresh water they hatch and release another form of larvae, denoted *miracidium;* these larvae search for and contaminate snails. Consequently, the snail releases the cercaria larvae. Many of the schistosomiasis victims suffer from lethargy, abdominal pain, and intermittent diarrhea. The enormous number of fresh microscopic eggs produced daily impairs the body's functions and results in the loss of blood through the urine and feces. The eggs are kept in the bladder wall, or transferred to the liver, with devastating results. The eggs deposited in the bladder wall may die and become calcified. Damage to the liver, spleen, kidneys, bladder, as well as to the central nervous system occur in those patients who are most severely infected. Much of the chronic damage to the organs is irreversible, and in severe cases the individual may die from internal bleeding.[6] This disease has a negative impact not only on self-productivity but also on socioeconomic growth.

Schistosoma mansoni infections are endemic to many areas of Africa, the Caribbean, the Middle East, as well as South America. The escalations of land usage through the development of irrigation projects can result in an increased incidence of *S. haematobium* by transmission through *Bulinus* snails breeding in the irrigation canals. *S. haematobium* is widespread in Africa and the Middle East and in certain places in Europe and Southeast Asia. The main genera of snails that act as hosts, besides *Bulinus,* are *Biomphalaria* for *S. mansoni, Oncomelania* for *S. japonicum,*[8] and *Tricula* for *S. mekongi.*[9]

The three common parasites of schistosomiasis in man, *S. haematobium, S. japonicum,* and *S. mansoni,* have a similar life cycle. The parasitic eggs are passed in the urine in the case of *S. haematobium* and in the feces in the case of *S. japonicum* or *S. mansoni.* After their release the eggs hatch in ponds, lake edges, streams, or canals. From the eggs the *miracidia* hatch in the water where they invade the snails. In the snails the *miracidia* develop two generations of sporocysts, the second of which manifests the fork-tailed *cercariae,* a macroscopic larval form released by certain species of freshwater snails. The *cercariae* enters the skin of a new host who is in contact with the contaminated water. When the *cercariae* penetrates the skin they discard their tails and become *Schistosomulae.* The *Schistosomulae* move through the tissues and some enter the liver. In the liver they multiply. From the liver they migrate to the bladder or the colon where spiny eggs are laid. Most of the eggs are discharged from the colon and some from the bladder.

The snails will not become infected unless discharges of urine and feces from infected persons enter the surface waters. The sanitary disposal of excreta from infected humans will clearly help prevent the transmission of this disease. Thus, the two approaches in reducing the transmission of schistosomiasis is preventing contact and preventing contaminated discharges from entering surface waters.[6]

Guinea worm, or dracunculiasis, is found in regions of Africa, Brazil, India, Pakistan, the Middle East, or other localities where the water supply comes from contaminated surface waters or wells. The disease is only transmitted by drinking water infected with the microscopic immature parasite. The disease has no natural reservoir other than human beings. The larvae multiply in the intestines and migrate to and grow in the hosts' subcutaneous tissues. In humans the larvae mature and multiply in about three to four months, after which the males die. However, several months later the mature adult female worm develops. These female worms measure up to a meter in length and excrete a toxin. The toxin causes a feeling of severe local burning below the skin and raises blisters. Through the blisters the female inevitably appears. The appearance of the female enables her to discharge hundreds of thousands of larvae into the surrounding environment. Dracunculiasis is a debilitating disease, victims of which are crippled for weeks or months by the emergence of the long worms through their skin. The worms are inclined to emerge on the lower legs, ankles or feet, but they may also show through the skin or mucous membranes anywhere on the body. The worms commonly emerge during the harvest or planting season, thus the infection has an adverse effect on both the health and economic welfare of an endemic community.[9]

The adults of the hookworms *Necator americanus* and *Ankylostoma duodenale* are attached to the walls of the jejunum by the buccal capsule. The females lay an enormous number of eggs that are discharged with the feces. The eggs mature to larvae which hatch in the soil and feed on bacteria. The larvae grow and manifest the filariform or the infective larvae. Usually, the larvae enter the skin of a new host through the feet. From here they travel into the venule and consequently invade the heart and the lungs. Here they multiply and penetrate the capillaries of the alveoli, trachea, pharynx, and then are ingested and transferred to the small intestine. In the small intestine they develop into adults.

One of the methods used for the extraction of the adult female worm is by gradually winding it round a match-stick as it appears from the subcutaneous tissues. The use of chemotherapy has also decreased the health risks from this disease.

Ascariasis is a disease that is caused by *Ascaris lumbricoides,* an intestinal roundworm sometimes found in man. The worms survive in the small intestine where they lay a large number of eggs that are discharged with the feces of the infected person. The eggs may contaminate vegetables and crops grown on soil where sludges or nightsoil are used as a fertilizer or as a soil conditioner. If raw vegetables and crops are eaten the parasite is transferred to the consumer. The eggs then penetrate the jejunum where the larvae hatch, then enter the mucosa, and pass through the hepatic circulation to the heart and the lungs of the host. In the heart and lungs the worm larvae develop, and gradually reach the alveoli which are located deep in the lungs. They further invade the stomach via the trachea and esophagus, and proceed to the small intestine. In the small intestine the larvae grow to maturity.

The mosquitoes *Culex pipens* complex are the vectors of the nematode worm *Bancroftian filariasis (Wuchereria bancroftii),* especially in Asia. In other areas the vector for the disease may be *Anopheles* or *Aedes* mosquitoes. The mosquitoes breed mostly in highly polluted water such as stagnant open drains, poorly maintained ponds, pit latrines, septic tanks, and soakaway pits.[8] Filariasis is an infection that is conveyed by nematode worms which have different forms. The worms survive in the lymphatic ducts of man, and the disease in an infected victim may block the lymphatic vessels. This condition ultimately leads to swelling of the arms, legs, or genitalia, leading to deformity or elephantiasis.

D. WATER-RELATED INSECT VECTORS

Water-related insect vectors are the infections that are spread by insects which rely on or live near a surface water system.

Examples of the water-related insect vector diseases include trypanosomiasis (sleeping sickness), yellow fever, dengue, onchocerciasis, and malaria.

Trypanosomiasis is a disease caused by parasites of the genus *Trypanosoma.* It includes sleeping sickness in Africa and chagas disease in Central and South America. African trypanosomiasis is known as sleeping sickness in humans and nagana in livestock. Cattle trypanosomiasis is a main cause of economic loss in endemic areas. Trypanosomiasis is detected in all tropical areas of Africa where tsetse flies are found. The disease is due to many *Trypanosoma* species that attack humans, cattle, and a variety of other domestic as well as wild animals. Typical symptoms of the disease include a decrease in the number of red blood cells (anemia), fever, irritation of the inner lining of the eyelid (conjunctivitis), nervous symptoms, paralysis, and death.

The parasitic protozoa are transferred from animals, that serve as reservoirs for the disease, to humans and domestic animals by the tsetse fly. The males and females of the tsetse fly feed exclusively on the larger vertebrates, hunting their prey over extensive areas to take a blood meal every few days. This regular supply of a protein-rich food fits well with the reproductive strategy of tsetse, which is distinct from that of most insects and somewhat like that of mammals. The female lays one egg every nine days, retaining it in her uterus where it develops into a fully grown larva by feeding on a milk-like secretion from the mother. The larva is then deposited in a sandy place where it burrows into the soil to pupate. It emerges as an adult about a month later. This method of reproduction guarantees the abundance and stability for generations of the tsetse fly.[10] The tsetse fly is found in 37 countries and extends over 10 million square kilometers.

Gambiense trypanosomiasis is transmitted by the riverain species of *Glossina* requiring optimum shade and humidity. Shady trees near lakes, rivers, and pools of water are ideal habitats. Man-fly contact is often made when many people gather at a surface water source to get water for household purposes. *Glossina tachinoides* is second in importance to *G. palpalis* as a vector of sleeping sickness. *Rhodesiense trypanosomiasis* can occur in scrub savannah country because the *Glossina* vectors are less dependent on moisture. Moreover, in such terrain wild animals and domestic cattle provide alternative feeding opportunities for the fly. Trypanosomiasis is a serious disease of domestic animals, causing great economic losses. Normally, *T. vivax* and *T. congolense* are the prevailing pathogens. The common vectors of *T. b. gambiense* are *Glossina palpalis* and *G. tachinoides* in West Africa. *T. b. rhodesiense* is associated with *G. morsitans, G. swynnertoni,* and *G. pallidipes.* Other secondary vectors are to be found in certain localities.

A reservoir of *T. rhodes* may be found in wild animals such as the bush-buck *(Tragelaphus scriptus).* If not treated, the person infected with gambiense becomes unconscious and comatose, and sleeps almost continuously. Frequently, the Rhodesiense sickness results in death.

Onchocerciasis, or river blindness, is a disease caused by the parasitic filarial nematode *Onchocerca volvulus*. The disease is endemic throughout the sub-Sahara; Central, West, and East Africa; Yemen, and Central America where it is the chief cause of blindness. The vector of the disease is the black fly, or the Simuliid. The simuliid is a family of biting flies including the important genus *Simulium*.

The crucial ecological environment for the vector is provided by fast-moving, highly oxygenated waters found in streams, rivers, or near waterfalls.

With onchocerciasis there is intense itching, which usually starts in one limb, shoulder or buttock, and continues gradually over the remainder of the body. Itching is normally accompanied by papular urticaria which differs in its severity. Prolonged exposure to the disease causes the skin to become thickened and rough. The skin sags and becomes excoriated from the endless scratching. When the microfilaria die intense inflammation of the skin results.

Yellow fever is an acute, often fatal, disease caused by an arbovirus. The disease causes severe headache and fever as well as aching of the bones. The fever is followed by jaundice, internal hemorrhages, and vomiting. *Aedes aegypti* is the vector for yellow fever, dengue, and dengue hemorrhagic fever. The vector breeds in almost any situation where water is held in either small or large containers, such as small pits or large cisterns.

Malaria is a disease of humans caused by blood parasites of the species *Plasmodium falciparum, P. vivax, P. ovale,* or *P. malariae*. The typical symptoms of the disease involve fever, headaches, and myalgia. The symptoms vary during the first stages of the disease but once the infection develops they tend to follow a tertian pattern.

In cerebral malaria the first symptoms include severe headache, drowsiness, and confusion. If an appropriate treatment is not prescribed at an early stage, the infected person may quickly enter into coma and die.

Figure 4.2 illustrates the possible interactions that may develop between man, water, vector, and parasite within a particular environment.[11] The outcome of this interaction can result in either schistosomiasis, typanosomiasis, onchocerciasis, filariasis, or malaria.

E. INFECTION PRIMARILY DUE TO DEFECTIVE SANITATION

Infections primarily due to defective sanitation are those infections that spread within a community usually due to the absence of suitable sanitation facilities.

Some examples of these diseases include hookworm, round worm, and ascariasis.

Human hookworm penetrates the skin from a damp and contaminated soil, or through ingestion of water or through the consumption of food. The hookworm lives in the small intestines where it causes major blood losses. This situation may result in anemia, especially for those people with a diet low in iron.

Enteric pathogens, disease-causing microorganisms, are able to live for long periods of time on crops or in water or in the soil when they find a suitable environmental. The factors that govern and control the survival of these pathogens include: number and type of the pathogen; conditions affecting the rate of growth of microorganisms such as temperature, humidity, pH, and the nutrients present; the organic matter content of the soil; intensity and duration of precipitation; sunlight; protection provided by foliage; and the presence of other competitive microorganisms.

Table 4.1 shows examples of the survival times of some disease-causing microorganisms and the related media in which the particular organism grow and multiply.[12]

Table 4.2 illustrates a general summary of the classification of the aforementioned groups of infective diseases in relation to the water supply systems.[13]

The incubation period for disease-causing organisms and their associated diseases are presented in Table 4.3.

The transmission of diseases by pathogenic microorganism from one person to another occurs in many different ways. For example these routes may include direct contact between one person and another, by ingestion, by the exposure to contaminated objects, by intermediate vectors, or by animals and birds that harbor the pathogens. Drinking or swimming in contaminated water or the inhalation of a pathogen can result in an infectious disease. Table 4.4 and Figures 4.3 and 4.4 present a summary of the transmission routes for many diseases.[15]

Generally, the disease transmission routes can be grouped broadly as follows:

1. *Mechanical transmission*. Mechanical transmission is the simple transport of pathogens on the feet, body, surface, or proboscis of the vector. The vector deposits the pathogenic organisms in its feces, or it may regurgitate the pathogen onto a surface such as human flesh. In the mechanical transmission of

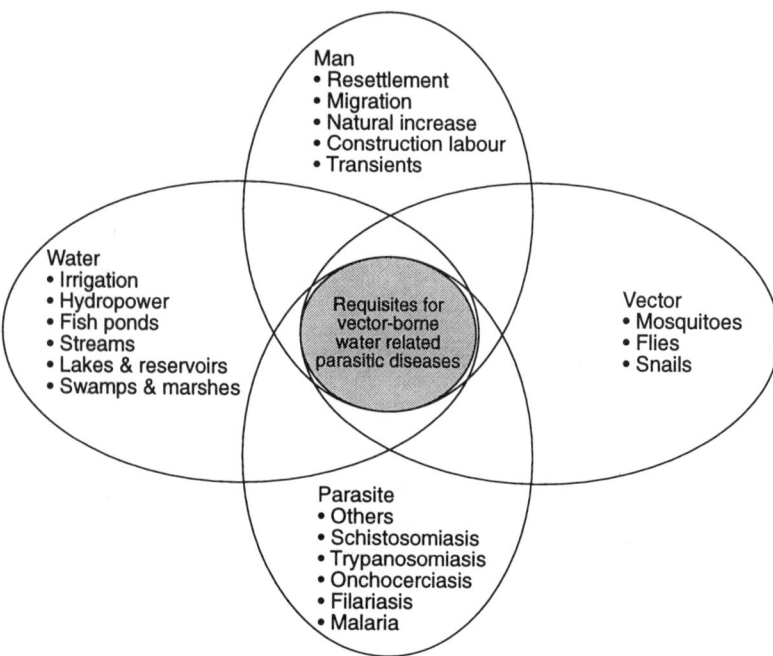

Figure 4.2 The interactions between man-vector-parasite and water.[11]

Table 4.1 Survival Times of Organisms

Organism	Media	Survival time (days)
Ascaris ova	Vegetables	27–35
	Soil	730–2010
Entamoeba histolytica	Vegetables	3
	Soil	6–8
	Water	60
Salmonella (spp.)	Vegetables	3–40+
	Soil	15–280
	Pasture	200+
	Grass	100+
Salmonella typhi	Vegetables	10–53
	Lettuce	18–21
	Soil	2–120
	Water	87–104
Shigella (spp.)	Vegetables	7
	Grass	42
Streptococcus faecalis	Soil	26–77
Vibrio cholera	Vegetables	5
Vibrio comma	Water	32
Poliovirus	Water	20

From Crook, J., *J. Am. Water Works Assoc.*, 77(7), 1985, 60. With permission.

diseases there is no biological cycle or development of the pathogen in or on the vector. Examples of diseases transmitted via this route include trachoma, yaws, and conjunctivitis.

2. *Biological transmission.* The biological transmission route signifies the transmission of diseases by bloodsucking vectors. Some examples of the diseases transmitted via this route include sleeping sickness, malaria, sandfly fever, and yellow fever.
3. *Transmission from an animal host.* Examples of diseases transmitted from animals to man by vectors include plague and typhus fever. Rats involved in the transmission of these diseases include *Rattus norvegicus* (also known as the brown, sewer, or Norway rat) and *Rattus rattus* (also known as the roof,

Table 4.2 Classification of Infective Diseases in Relation to Water Supplies

Category	Examples	Causative organism	Incubation period	Relevant water improvement	Suggested reduction by water improvement (%)
1. Water-borne Infections					
a. Classical	Typhoid Fragile organisms whose sole reservoir is man	*Salmonella typhosa*[b]	7–12[d] Av.–14[d]	Microbiological sterility	80
	Cholera	*Vibrio cholera, vibrio comma*[b]	few hr — 5[d] (3 days)	Microbiological sterility	90
b. Non-classical	Infective hepatitis	Viruses Unknown[V]	10–50[d] Av. 30–35[d]	Microbiological sterility	10?
2. Water-washed infections					
a. skin & eyes	Scabies	Miscellaneous, sarcoptes scabiei[m]	Several days to weeks	Greater volume available	80
	Trachoma	virus-like microbe		Greater volume available	60
b. Diarrheal	Bacillary dysentery (shigellosis)	*Shigella*[b]	1–7[d] usually <4[d]	Greater volume available	50
3. Water-based Infections					
a. Penetrating skin	Schistosomiasis	Schistosoma[H], others	4–6 Weeks or longer	Protection of source	60
b. Ingested	Guinea worm (dracantiasis)	Cyclops,[H] Dracunculus medinensis, minute crustacean	About 12 months	Protection of source	100
4. Infections with water-related insect vectors					
a. Biting near water	Trypanosomiasis (sleeping sickness)	*Trypanosoma gambiense*	2–3 Weeks	Water piped from source	80
b. Breeding in water	Yellow fever	Virus of yellow fever carried by mosquito: (*Aedes aegypti*)	3–6 Days	Water piped from source	10?
5. Infections primarily of defective sanitation	Hookworm Round worm (Ascaris)	*Ascaris lumbricolides*	About 2 months	Sanitary fecal disposal through cooking of food or by sanitation	? ?

Notes: (b) = bacteria, (V) = viruses, (m) = mites, (H) = helminth, and d = days.
Adapted from Reference 13.

Table 4.3 Diseases-Causing Agents and Their Incubation Periods

	Disease	Specific agent	Incubation period	
			Max. range	Usual range
Bacteria	Salmonellosis (Salmonella infection)	*Salmonella typhimurium*	6–48 Hours	12–24 Hr
	Typhoid fever	*Salmonella typhosa*	3–38 Days	10–14 Days
	Paratyphoid fever	*Salmonella paratyphi A*		1–10 Days
	Shigellosis (bacillary dysentery)	*Shigella*	1–7 Days	Less than 4 days
	Cholera	*Cholera, vibrio, vibrio comma*	Few hours–5 days	3 Days
	Tuberculosis	*Mycobacterium tuberculosis*	4–6 Weeks	
	Gastroenteritis	Unknown	8–12 hour (variable)	
Viruses	Infectious hepatitis	Unknown	10–50 Days	30–35 Days (average)
Protozoa	Amoebiasis (ameobic dysentery)	*Entamoeba histolytica*	5 Days or longer	3–4 Weeks average
Pirachetes	Leptospirosis (Weil's disease)	*Leptospira icterohaemorrhagiae*	4–19 Days	9–10 Days (average)
Helminths	Schistosomiasis (Bilharziasis) (Blood flukes)	*Schistosoma haematoblum, S. mansoni S. japonicum*	4–6 Months or longer	1–3 Months or more
	Ascariasis (Intestinal round worms)	*Ascaris lumbricoides*	About 2 months	8 Weeks

From Salvato, J.A., *Environmental Engineering and Sanitation*, 4th ed., John Wiley & Sons, New York, 1992. With permission.

ship, gray, black, Alexandrian, or English rat), which live in close proximity to man. The house mouse, *Mus musculus* plays a very minor role in the transmission of disease, but it may transmit diseases such as *Rickettsia alkari* which causes rickettsial pox in humans, with symptoms similar to that of chicken pox.[16]

Table 4.5 outline some diseases that are transmitted from an animal host to man by vectors.[16]

Health concerns regarding use of reclaimed wastewater relates to the degree of human contact, the effluent quality, and the reliability of the wastewater treatment systems.[17]

For a person to develop a sickness from contact with treated wastewater the following conditions must be present:

1. The infectious agent (disease-causing organism) must be present in the community that produces the wastewater,
2. The disease agents must survive all the wastewater treatment processes to which they are subjected,
3. The person must either directly or indirectly come in contact with the wastewater effluent, and
4. The disease agents must be present in sufficient numbers at the time of contact to cause illness.[18]

The factors that influence the occurrence and spread of a certain disease include: pathogenic dose, infective dose or number of pathogens needed to initiate the disease, pathogenicity or the potential of the pathogen for infection, virulence or the ability of the pathogen to infect a host, and the relative degree of susceptibility of the host.

It is estimated that around 25,000 persons per day, or the equivalent of 9.1 million persons per year, die of preventable, waterborne diseases, and this of course is in combination with malnutrition.[19]

Table 4.4 Common Transmission Routes of Certain Diseases

Disease	Causative agent	Possible transmission route
Diarrhea	*Salmonella* spp., *Shigella* spp., *Escherichia coli,* Viruses, etc.	Man — feces — flies, food, water — man
Cholera	*Vibrio cholerae*	Man — feces, water, food — man
Typhoid	*Salmonella typhi*	Man — food, water — man
Leptospirosis (enteric fever)	*Leptospira* spp.	Animal — oral, skin, eye — man
Infectious hepatitis	Hepatitis virus A	Man — feces, water, food — man
Yellow fever	Yellow fever virus	Man — mosquito — man
Amoebic dysentery	*Entamoeba histolytica*	Man — feces, flies, food, water — man
Bacillary dysentery	*Shigella*	Man — feces, flies, food, water — man
Giardiasis	*Giardia lamblia*	Man — feces, food, water — man
Trypanosomiasis	*Trypanosoma gambiense*	Man — tsetse fly — man
Ascariasis	*Ascaris lumbricoides*	Man — soil, food, water, vegetables, dust, etc. — man
Schistosomiasis	*Schistosoma mansoni, S. haematobium, S. japonicum*	Man — snail — man

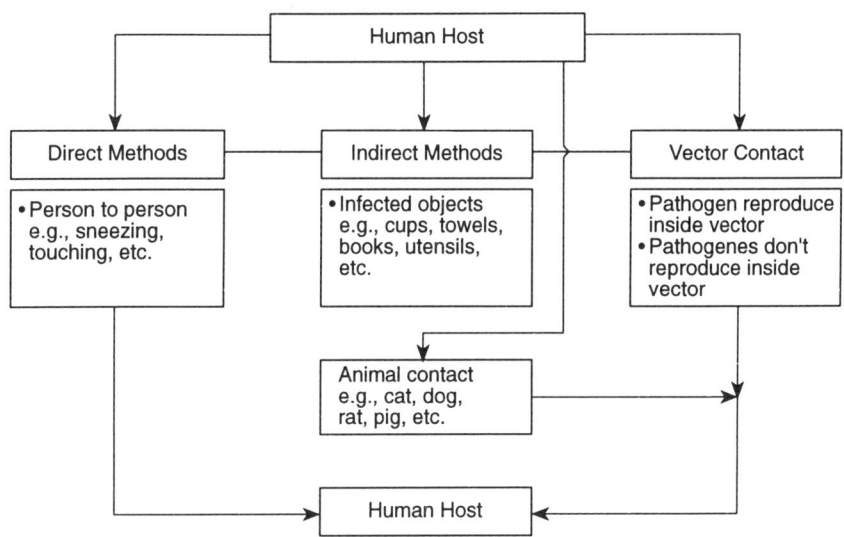

Figure 4.3 Possible transmission routes of pathogenic microorganisms.

II. PATHOGENS FOUND IN RECLAIMED WASTEWATER

Raw or reclaimed wastewater include all of the pathogens that are present in a community. Human excrement is regarded as a primary source for the transmission of many communicable diseases. The concentration of microorganisms found in wastewater has a wide variation. Pathogenic microorganisms are numerous and types of diseases they cause are a hazard to the public health and welfare of a community.

Some of the essential factors that must be considered in a public health risk assessment for diseases include: concentration of the infectious agent, amount of microbes which entered the body, duration of exposure to the parasitic agent, and properties of the exposed microbial cells.[20] Pathogen survival in domestic sewage depends on many complex factors some of which include demography, location, precipitation, season, and temperature.

Table 4.6 outlines the relative health risks due to the use of untreated human excrement and reclaimed wastewater in agriculture and aquaculture schemes.[21]

Eventually, all pathogens die and, of course, do so rapidly when exposed to a hostile environment. The die-off of viable pathogenic microbial organisms is considered to be exponential. This indicates a rapid decrease in the number of pathogens following the first few hours or days of being excreted. However,

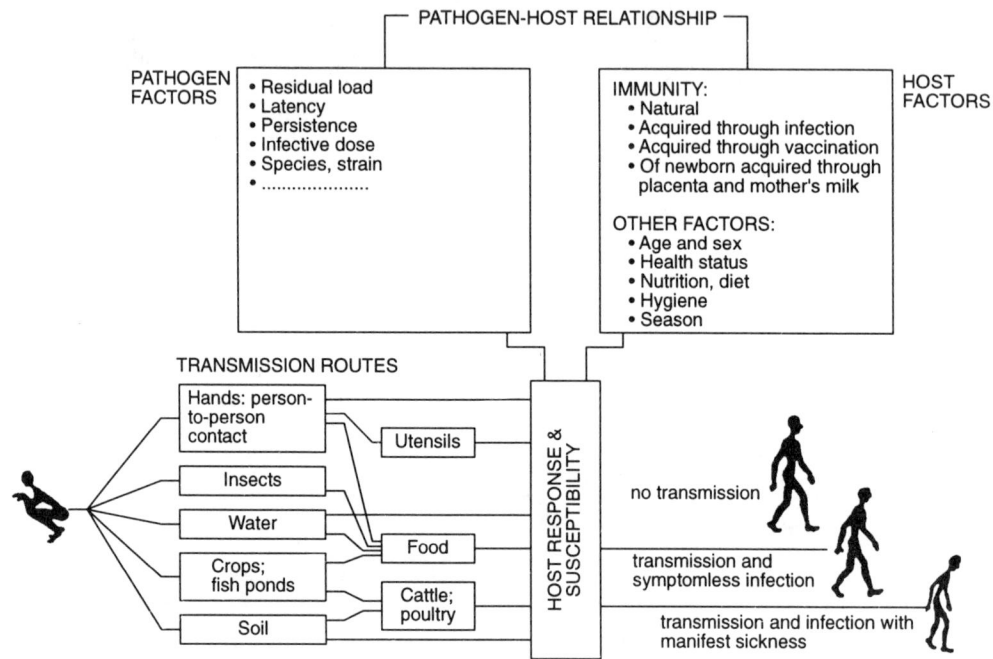

Figure 4.4 The pathogen-host relationship and possible transmission route for excreta-related infections. (From *IRCWD News* No. 23, December, World Health Organization, Geneva, 1985, 1. With permission.)

Table 4.5 Selected Diseases Transmitted from Animals to Man by Insects

Disease	Animals	Insect and its action
Plague:		
Bubonic	Roof rat, Norway rat	Flea regurgitates on biting
Sylvatic	Squirrel, bandicoot, prairie dog, other wild rodents	Flea regurgitates on biting
Typhus:		
Endemic, murine	Roof rat, Norway rat	Flea defecates during biting
Encephalitis:		
Equine	Horse, cow, rabbit, ground squirrel	Mosquito bite
St. Louis	Wild birds, domestic hen	Mosquito bite
Western	Wild birds, domestic hen, horse, deer, squirrel	Mosquito bite
Eastern	Small wild birds, duck, turkey, pheasant	Mosquito bite
Hemorrhagic fevers:		
Some types	Wild rodents	Tick bite
Leishmaniasis:		
Three types	Dog, other canines, wild rodents	Sand fly bite
Rickettsial pox	House mouse, wild rodents	Mite bite
Tick fevers:		
Several types	Small wild rodents	Tick bite
Trypanosomiasis:		
African	Cow, sheep, horse, hog, antelope	Tsetse fly bite
American or Chagas' disease	Dog, cat, monkey, bat, opossum, armadillo, squirrel, other rodents	Fecal material of triatomid bug
Tsutsugamushi or scrub typhus	Field rats and mice, voles, swamp birds, parrot, monkey, bush hen	Mite bite

From Chanlett, E.T., *Environmental Protection*, McGraw-Hill, New York, 1979. With permission.

Table 4.6 Relative Health Risks from Use of Untreated Excreta and Wastewater in Agriculture and Aquaculture

Class of pathogen	Relative amount of excess frequency of infection or disease

1. Intestinal nematodes:
  Ascaris
  Trichuris
  Ancylostoma
  Necator
	High

2. Bacterial infection:
  Bacterial diarrheas (e.g., cholera)
  Typhoid
	Lower

3. Viral infections:
  Viral diarrheas
  Hepatitis A
	Least

4. Trematode and cestode infections:
  Schistosomiasis
  Clonorchiasis
  Taeniasis
	From high to nil, depending upon the particular excreta use practice and local circumstances

Compiled from Reference 21.

some organisms do not follow this rapid die-off pattern, for instance *Salmonella* spp. and most helminths and trematodes (e.g., *Schistosoma* and *Clonorchis sinensis* [Chinese liver fluke]) can survive for long periods of time outside of the body. The major environmental factors that influence excreta pathogen die-off are presented in Table 4.7.[22]

The factors that affect the survival time of pathogens in soil include: level of wastewater treatment, soil moisture content, temperature variations, sunlight intensity and duration, pH level, antibiotics content, toxic substances, competitive organisms, nutrients value, organic matter content, method and time of application of wastewater, and soil type.[23] Figure 4.5 illustrates the survival of pathogens in the soil vs. vegetative growth in warm climates.[22]

Microorganisms do not penetrate the healthy undamaged surfaces of vegetables and crops, they die off swiftly on crop surfaces which are exposed to the sunlight. Pathogens are likely to survive for longer periods inside leafy vegetables or in the protected cracks or stems.[20] Figure 4.6 presents the survival time for pathogens on crops as compared to the vegetable growth periods in warm climatic conditions.[22]

Table 4.8 also gives an outline of the survival times of pathogens found in excreta, night soil, sludges, sewage, soil, and crops.[8]

In the reuse of reclaimed wastewater, potential diseases can be put in two categories. Communicable diseases which are related to the microbiological quality of the reclaimed wastewater, and noncommunicable diseases which are associated with the chemical constituents of the wastewater.

Table 4.9 presents the pathogens that have to be considered in the design of a wastewater reclamation and reuse engineering project.[23] Also, Table 4.9 indicates the pathogenic organisms that are expected to be important globally, and other pathogens that are only significant in limited geographical areas.

Raw wastewater contains many different pathogenic microbial species including bacteria, protozoa, viruses, and helminths. Usually, these pathogens are excreted by individuals who suffer from, or by those who carry, that particular disease-causative agent.

A brief description of the important microorganisms found in reclaimed wastewater are discussed below.

A. BACTERIA

Some of the most important bacterial diseases that could be caused by improperly treated wastewater include bacillary dysentery, cholera, enteritis, typhoid, paratyphoid, and salmonellosis.

Table 4.10 gives examples of the pathogenic bacteria that may cause diseases in man either by drinking contaminated water, eating contaminated food, or breathing air in which these microorganisms are suspended.[8]

Table 4.7 Main Environmental Factors Influencing Pathogen Die-Off

Environmental factor	Effect on pathogen die-off or survival
Temperature	Accelerated die-off with increasing temperature, longer survival at low temperature
Moisture content (of foods or soils or in waste products.)	Generally longer survival in moist environment and under humid weather conditions, rapid die-off under conditions of desiccation
Nutrients	Accelerated die-off if essential nutrients are scarce or absent
Competition by other microorganisms	Longer survival in an environment with few microorganisms competing for nutrients or acting as predators
Sunlight (ultraviolet radiation)	Accelerated die-off if exposed to sunlight
pH	Neutral to alkaline pH tends to prolong survival of bacteria; acid pH tends to prolong survival of viruses

From Strauss, M., IRCWD News No. 23, WHO, 1985, 4. With permission.

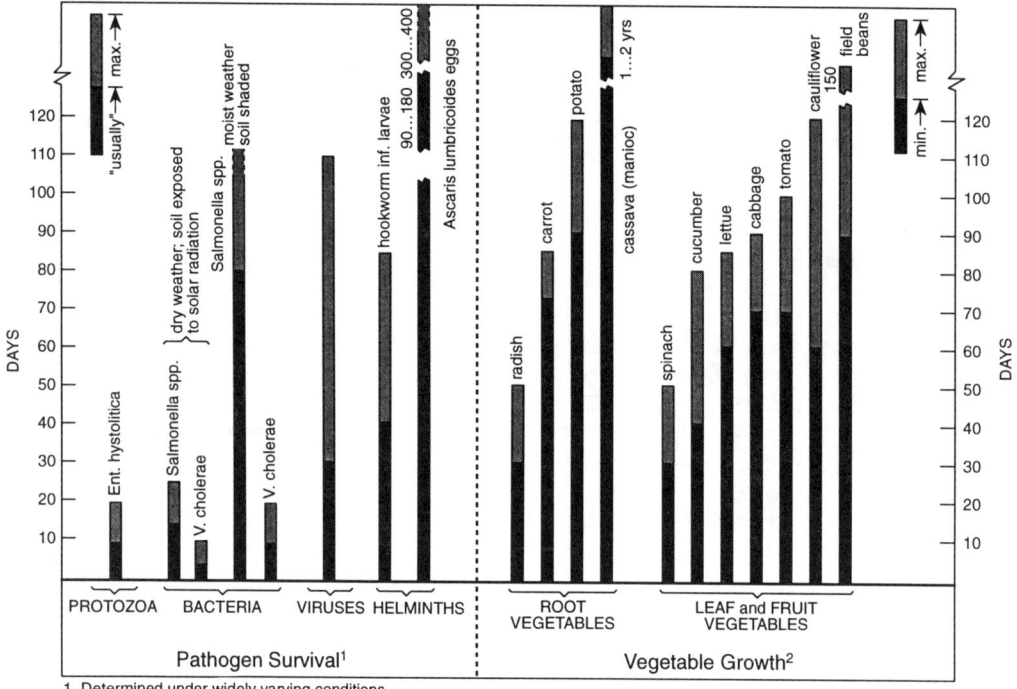

Figure 4.5 Pathogen survival in soil vs. vegetable growth periods in warm climates.[22] (From Strauss, M., *IRCWD News* No. 23, WHO, 1985, 4. With permission.)

B. HELMINTHS (WORMS)

Many of the helminths that can initiate diseases in man and animal are found in wastewater. Helminths may be divided into two main groups, roundworms or nematodes and flatworms. The flatworms are further subdivided into tapeworms or cestodes, and flukes or trematodes. The tapeworms are segmented while the flukes have flat unsegmented bodies. The route of transmission of flukes is normally via the eggs (ova), which may be deposited in the soil, water, or on vegetables, and transferred to the victim by either direct or indirect means. The ova, or sometimes the larvae, are generally discharged with feces. Exceptions to this pattern are the eggs of *Schistosoma haematobium* (the agent of urinary schistosomiasis), which are discharged with urine, and the eggs of guineaworm, which are released through the ruptured skin of the infected individuals. The eggs can survive for prolonged periods of time. Many types of parasitic worms may infect the intestine of man and, with time and repeated infections, can result in impaired function of the intestines and other organs in the body.[8]

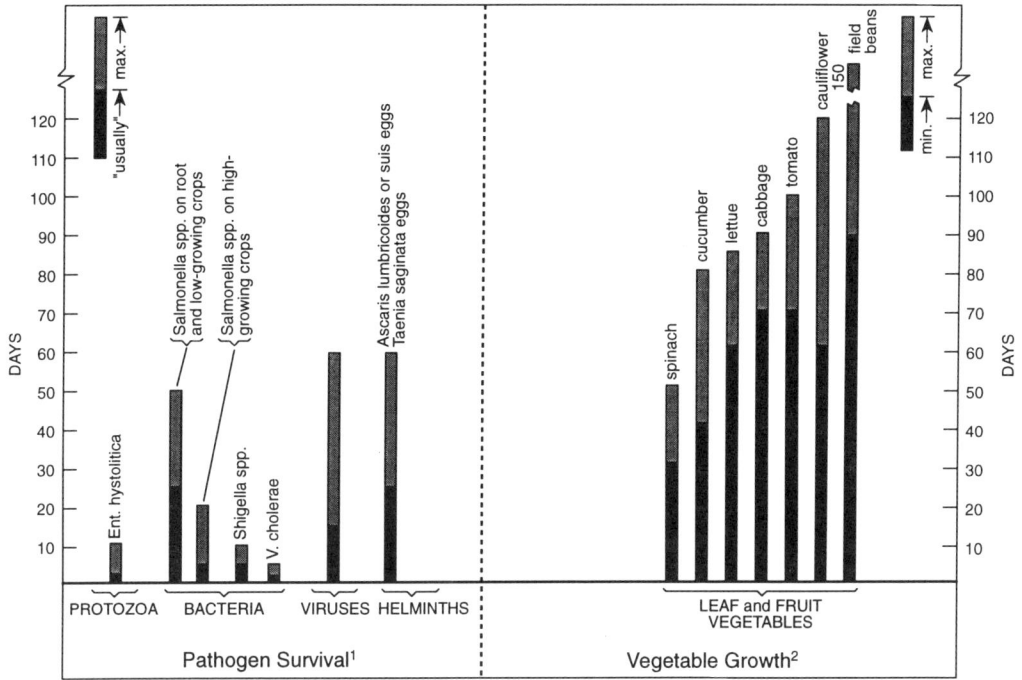

Figure 4.6 Pathogen survival in crops vs. vegetable growth periods in warm climates. (From Strauss, M., *IRCWD News* No. 23, WHO, 1985, 4. With permission.)

Table 4.8 Survival of Excreted Pathogens (at 20–30°C)

Type of pathogen	Survival times (Days)			
	In feces, night soil, and sludge	In fresh water and sewage	In the soil	On crops
Viruses				
Enteroviruses	<100 (<20)	<120 (<50)	<100 (<20)	<60 (<15)
Bacteria				
Fecal coliforms	<90 (<50)	<60 (<30)	<70 (<20)	<30 (<15)
Salmonella spp.	<60 (<30)	<60 (<30)	<70 (<20)	<30 (<15)
Shigella spp.	<30 (<10)	<30 (<10)	—	<10 (<5)
Vibrio cholerae	<30 (<5)	<30 (<10)	<20 (<10)	<5 (<2)
Protozoa				
Entamoeba histolytica cysts	<30 (<15)	<30 (<15)	<20 (<10)	<10 (<2)
Helminths				
Ascaris lumbricoides eggs	Many months	Many months	Many months	<60 (<30)

Note: The values in parentheses show the usual survival times.

From Feachem, R. G., et al., *Sanitation and Disease: Health Aspects of Excreta and Waste Water Management,* John Wiley & Son, 1983. With permission.

Examples of the major diseases transmitted by helminths include tapeworm infections, hookworm infections, roundworm infections, schistosomiasis, and ascariasis.

Table 4.11 presents some of the helminths that are found in the feces of infected individuals.[7,8,24]

Table 4.9 Excreted Pathogens that Should Be Considered in the Design of Wastewater Re-Use Projects

Biological class	Pathogen	Associated disease	Distribution
Viruses	Poliovirus	Poliomyelitis	Global
	Hepatitis A virus	Infectious hepatitis	Global
	Rotavirus	Diarrhea	Global
Bacteria	*Campylobacter jejuni*	Diarrhea	Global
	Escherichia coli	Diarrhea	Global
	Salmonella spp.	Diarrhea/typhoid	Global
	Shigella spp.	Diarrhea/dysentery	Global
	Vibrio cholerae	Diarrhea	Global
Protozoa	*Entamoeba histolytica*	Diarrhea/dysentery	Global
	Giardia lamblia	Diarrhea	Global
Helminths			
Nematodes	*Ascaris lumbricoides*	Roundworm infection	Warm climates
	Trichuris trichiura	Whipworm infection	Warm climates
	Ancylostoma + Necator	Hookworm infection	Warm climates
Trematodes	*Schistosoma* spp.	Schistosomiasis	Local
	Clonorchis spp.	Liver fluke infection	Local
Cestodes	*Taenia* spp.	Tapeworm infection	Local

From Feachem, R. G. and Blum, D., *Reuse of Sewage Effluents,* Thomas Telford, London, 1985, 237. With permission.

Table 4.10 Bacterial Pathogens Excreted in Feces

Bacterium	Disease	Can symptomless infections occur?	Reservoir
Campylobacter fetus spp. jejuni	Diarrhea	Yes	Animal and man
Pathogenic *Escherichia coli*[a]	Diarrhea	Yes	Man[b]
Salmonella			
S. *typhi*	Typhoid fever	Yes	Man
S. *paratyphi*	Paratyphoid fever	Yes	Man
Other salmonellae	Food poisoning and other salmonelloses	Yes	Animal and man
Shigella spp.	Bacillary dysentery	Yes	Man
Vibrio			
V. *cholerae*	Cholera	Yes	Man
Other vibrios	Diarrhea	Yes	Man
Yersinia enterocolitica	Diarrhea and septicemia	Yes	Animals and man[c]

[a] Includes enterotoxigenic, enteroinvasive, and enteropathogenic *E. coli*.
[b] Although many animals are infected by pathogenic *E. coli,* each serotype is more or less specific to a particular animal host.
[c] Of the 30 or more serotypes identified so far, a number seem to be associated with a particular animal species. At present, there is insufficient epidemiological and serological evidence to determine whether distinct serotypes are specific to primates.

From Feachem, R. G., et al., *Sanitation and Disease: Health Aspects of Excreta and Waste Water Management,* John Wiley & Son, 1983. With permission.

Table 4.11 Helminthic Pathogens in Feces

Common name	Helminth	Disease
Fish tapeworm	*Diphyllobothrium latum*	Diphyllobothriasis
Hookworm	*Ancylostoma duodenale*	Hookworm
Hookworm	*Necator americanus*	Hookworm
Roundworm	*Ascaris lumbricoides*	Ascariasis
Schistosome	*Schistosoma haematobium*	Schistosomiasis
Schistosome	*S. japonicum*	Schistosomiasis
Schistosome	*S. mansoni*	Schistosomiasis
Whipworm	*Trichuris trichiura*	Trichuriasis

Compiled from References 7, 8, and 24.

Table 4.12 Protozoal Pathogens in Feces

Protozoan	Disease	Can symptomless infections occur?	Reservoir
Balantidium coli	Diarrhea, dysentery, colonic ulceration	Yes	Man and animals, especially pigs and rats
Entamoeba histolytica	Colonic ulceration, amoebic dysentery, liver abscess	Yes	Man
Giardia lamblia	Diarrhea and malabsorption	Yes	Man

From Feachem, R. G., et al., *Sanitation and Disease: Health Aspects of Excreta and Waste Water Management*, John Wiley & Son, 1983. With permission.

C. PROTOZOA

Untreated wastewater may contain cysts, which are considered to be the more resistant stage of the protozoan development. The cysts may infect the intestinal tract of man and animals and transmit diseases.

Some examples of the diseases that can be caused by protozoa in improperly treated reclaimed wastewater include diarrhea, amoebiasis, amoebic dysentery, and giardiasis. Table 4.12 presents some of the pathogenic protozoa that can be found in the feces of infected persons.[8]

D. VIRUSES

Pathogenic viruses are present in wastewater effluent. More than 100 different viruses have been detected in human feces. The viruses normally infect the alimentary tract and are excreted in large numbers by infected persons.[7] Viruses that are excreted by an infected individual commonly range from 1,000 to 10,000 infective units per gram of feces. The number of viruses in wastewater are also quite seasonal and are most frequently isolated during the summer and early autumn.[18]

Some examples of viral diseases that can be caused by improperly treated wastewater include infectious hepatitis, poliomyelitis, enteric diseases, diarrheal disease, respiratory infections, gastroenteritis in children, and eye diseases.

Many factors are involved in the survival of viruses in the soil or on vegetation and crops. Some of the survival factors include organic content of the soil, pH, temperature, moisture content, and sunlight. In the soil the viruses are adsorbed onto the surface of soil particles. Generally, the survival of viruses on vegetation and crops is shorter than their survival in the soil due to the absence of moisture and the exposure to sunlight. Although viruses cannot multiply outside a living host cell, some viruses can survive for weeks in the environment, especially if the temperature is less than 15°C.[7,24]

Enteric viruses are those that multiply in the intestinal tract and are released in the feces of infected individuals. Over 100 different enteric viruses capable of producing disease are excreted by humans.[18] Table 4.13 presents examples of pathogenic viruses that are contained in the human feces.[24]

Dispersion of aerosolized enteric pathogens can occur around wastewater treatment facilities such as aeration processes, trickling filters, and sprinkler and spray operations involving reclaimed wastewater and sludge.[25] Aerosols may be defined as the particles that are suspended in the air and have diameters ranging between 0.01 and 50 μm.[18] Viruses and most pathogenic bacteria are reported to be in the respirable size range.[26,27] The spread of diseases via this path may rely on many factors such as the concentration of the pathogen in the reclaimed wastewater, wind speed, wind direction, relative humidity, biodegradation patterns, distance of host from source, rate of dispersion in the air, temperature, light, and droplet size. Other routes for the transmission of an infection by aerosols include indirect means such as the deposition of the aerosol on food, vegetation, or clothes.

III. REMOVAL OF PATHOGENS BY VARIOUS WASTEWATER TREATMENT PROCESSES

The degree of removal of pathogens by different treatment unit operations and processes varies according to the method of treatment applied, wastewater characteristics, environmental conditions, and the concentration and type of disease-causing agents.

A. PRELIMINARY AND PRIMARY TREATMENT METHODS

Primary treatment will remove by sedimentation some of the larger and heavier organisms, such as the eggs of helminths, cysts, and protozoa. Also, primary treatment by sedimentation may remove particulate

Table 4.13 Viral Pathogens Excreted in Feces

Virus	Disease	Can symptomless infections occur?	Reservoir
Adenoviruses	Numerous conditions	Yes	Man
Enteroviruses			
Poliviruses	Poliomyelitis, paralysis and other conditions	Yes	Man
Echoviruses	Numerous conditions	Yes	Man
Coxsackieviruses	Numerous conditions	Yes	Man
Hepatitis A virus	Infectious hepatitis	Yes	Man
Reoviruses	Numerous conditions	Yes	Man and animal
Rotaviruses, agent and other viruses	Norwalk Diarrhea	Yes	Probably man

From Shuval, H. I., et al., World Bank Technical Paper Series No. 51, Washington, D.C., 1986. With permission.

matter associated with the microorganisms. Somewhere between 50 to 90% of the parasitic eggs and cysts are removed by primary settling, while only 24% of the bacteria may be removed during this process.[18,28] The primary treatment mechanism is not efficient for the reduction of bacteria and viruses in wastewater.[29,30]

B. SECONDARY TREATMENT METHODS

Generally, bacterial concentrations are greatly reduced by wastewater treatment techniques which include secondary treatment. Nevertheless, large numbers of bacteria are still present even after biological treatment.[31] Research indicates that a series of waste stabilization ponds with a total detention time of around 11 days will accomplish good bacterial removal.

Mara and Cairncross reported that a series of waste stabilization ponds with a total retention time of around 22 days will achieve excellent helminth removal.[31] However, secondary wastewater treatment is not capable of removing all the *Taenai saginata,* and *ascariasis* microorganisms. These organisms can have serious effects on cattle fed on fodder grown on land irrigated with a secondarily treated effluent containing these parasites.[32]

The conventional treatment processes of sedimentation, followed by activated sludge or biological filtration, will achieve a reduction in the cysts but still probably will not provide an effluent free of these parasites. The much longer retention times and natural processes of biodegradation within waste stabilization ponds provides a much greater removal of protozoa.[33]

Table 4.14 presents the efficiency for removal of some of the major pathogenic bacteria by primary and biological treatment processes.[8] The biological treatment processes include trickling filtration, activated sludge, and waste stabilization ponds. Generally, the activated sludge process is reported to be more effective in reducing bacteria and viruses than the trickling filtration process. Activated sludge typically removes more than 90% of the bacteria and 80 to 90% of the viruses. The trickling filter system removes 50 to 90% of the bacteria and viruses.[18]

Some secondary treatment processes claim a removal efficiency of more than 90% for the coliform indicator organisms, which should then relate to the removal of the pathogenic organisms.[18]

C. TERTIARY (ADVANCED) TREATMENT METHODS

The eggs of the *Taenai saginata* worms may be removed from the wastewater by a combination of sedimentation and tertiary treatment methods. However, the eggs of Ascaris may still be found in digested sewage sludges. The removal of these eggs may be accomplished by application of heat treatment for a period of 2 hours or more at a temperature of 55°C or through prolonged exposure to sunlight.[32] Usually, the pathogenic protozoan cysts are removed by tertiary treatment methods such as sand filtration or lagooning.

Reverse osmosis is reported to be very effective at removing viruses and virtually all larger organisms.[18] The efficiency for viral removal by this method is subjected to the factors that influence the reverse osmosis process as outlined in Chapter 5.

Chlorination, as a disinfection method, provides a means for prevention of the spread of waterborne diseases. Appropriate chlorination is reported to be highly efficient in the elimination of bacteria from

Table 4.14 Microorganism Removal by Wastewater Treatment Processes

Type of microorganisms	Percentage removal	
	Primary	Biological
Salmonella spp.	15	96–99.999
Mycobacterium	48–57	Slight–99.9
Amoebic cyst	Limited removal	0–99.9
Helminth ova	72–98	0–76
Viruses	3–extensive	0–84

From Feachem, R. G., et al., *Sanitation and Disease: Health Aspects of Excreta and Waste Water Management,* John Wiley & Son, 1983. With permission.

wastewater. Unless the wastewater has a very low turbidity, there is a high probability that the disinfected wastewater will not be completely free of bacterial or viral pathogens.[18] Chlorination is effective as a bactericide provided there is no interaction with the turbidity, organic matter, and ammonia present in the wastewater. Therefore, the use of chlorination to treat reclaimed wastewater may have some shortcomings. A problem is the formation of chloramines, simple and complex chlorinated organics, and free chlorine residuals which may be toxic, mutagenic, or carcinogenic.[33]

A heavy dose of chlorine (45 mg/L) in the treatment of reclaimed wastewater with a contact time of 1 hour will inactivate protozoan cysts.[34]

Normally the concentration of viruses is not as great as that of bacteria found in wastewater. Nonetheless, viruses are more resistant to the treatment processes. Viruses are not readily removed even by an efficient disinfection processes such as chlorination. However, it is reported that the viruses normally survive for no more than around 7 days in the wastewater effluent.[32]

Secondary treatment methods followed by tertiary treatment such as rapid sand filtration and liming have been reported to have the following effects:[33]

- Killing of bacteria and viruses due to the high pH levels.
- Precipitation of heavy metals.
- Increased electrical conductivity of the effluent at the same time as lowering of the sodium adsorption ratio (SAR).[20] The calculations for the SAR are presented in Chapter 2.

Table 4.15 presents a summary of the degree of removal of pathogens by different wastewater treatment unit operations and processes.[8,20,31]

Table 4.16 gives a summary of the removal efficiency for various enteric pathogens by wastewater treatment unit operations and processes.[24]

IV. TOXIC CHEMICALS FOUND IN RECLAIMED WASTEWATER

Diseases may also be caused by the presence of chemical substances in improperly treated reclaimed wastewater. The chemical substances may be found alone or in combination. Industrial discharges are often the major sources of chemical substances in raw wastewater. If the concentration of the chemicals are not reduced to acceptable levels they can enter the human food chain and negatively affect the public health and welfare of a community.

The buildup of toxic metals, such as cadmium, copper, nickel, lead, and zinc, in the soil or in plants may be increased if the reclaimed wastewater contains toxic elements or metals that are in excess of prescribed limits.

The availability, uptake, and accumulation of toxic metals in crops and soil depends on many interacting parameters. These parameters include climatic conditions, soil characteristics, type of soil, type of plants, degree of plant sensitivity to the chemical, plant age, seasonal variations, type of chemical, and the wastewater's characteristics.

The toxic effects produced by a certain chemical or a group of chemicals depends on many interacting and complex factors such as characteristics of the chemical, lethal dose, average daily intake, ability of the chemical to reach and accumulate in the tissues, and routes of exposure (air, water, food).

Table 4.15 Treatment Units and the Removal of Pathogens

Treatment unit	Pathogen removed
Conventional: Primary	Inefficient for the removal of excreted pathogens[8]
	Inadequate for the removal of heavy metals and salinity[20]
	Ineffective for the removal of helminth eggs[31]
Secondary + tertiary (chlorination + rapid sand filtration)	Eliminates pathogens
Waste stabilization ponds	Eliminate protozoal cysts and helminth eggs
	Reduce concentration of excreted bacteria and viruses[8]
Secondary + tertiary (lime + rapid sand filtration)	Kills and removes bacteria and viruses
	Precipitates heavy metals in solution

Compiled from References 8, 20, and 31.

Table 4.16 Enteric Pathogen Removal Efficiencies by Wastewater Treatment Processes (in Log_{10} Units) (i.e., $4 = 10^{-4} = 99.99\%$ Removal)

Treatment unit	Viruses	Bacteria	Protozoa	Helminths
Primary sedimentation	0–1	0–1	0–1	0–1
Septic tank	0–1	1–2	1–2	1–2
Trickling filter	0–1	0–2	0–1	0–1
Activated sludge	1–2	2–3	1–2	1–2
Stabilization ponds (20 day, 4 cells)	2–4	4–6	4–6	4–6

From Shuval, H. I., et al., World Bank Technical Paper Series No. 51, Washington, D.C., 1986. With permission.

A. INORGANIC CHEMICALS THAT POSE A HEALTH HAZARD

Heavy metals may be found in high concentrations when the reclaimed wastewater includes significant amounts of industrial and trade effluents.

Metals such as copper, chromium, zinc, and nickel are usually related to electroplating, printing, and other heavy metal industries such as steel foundries, motor vehicle and aircraft industries, as well as recirculating cooling water systems.

Mercury is used in the production of many industrial and consumer products such as caustic soda, electrical equipment, chlorine, paint, cosmetics, paper, pulp, fungicides, and the chloroalkali industry. Arsenic may be present in wastewater from metallurgical industries, glassware and ceramic industries, tannery operations, dye manufacturing, pesticide manufacturing, and petroleum refining. Barium is associated with the paint and pigment industry, metallurgical industry, glass and ceramics industries, dye manufacturing, and the rubber industry. Cadmium results from metallurgical alloying, ceramics, electroplating, photography, pigment works, textile printing, and the chemical industry. Lead is associated with storage-battery manufacturing. Silver enters the wastewater in the form of silver nitrate, the source of which can be from the photographic, electroplating, porcelain, and ink manufacturing industries.

Reclaimed wastewater used in aquaculture, such as fish farming, has to be carefully monitored as heavy metals can accumulate in the flesh of the fish. This condition, of course, is a health hazard to the consumer.

Usually, chemicals that pose a health hazard or a significant health risk exist in wastewater at very low concentrations. Thus, health problems may result when the chemical is ingested over prolonged periods. Under acidic conditions which may be brought about in the soil, the discharge or reuse of reclaimed wastewater favors the uptake and buildup of some heavy metals such as cadmium or selenium in plants or crops. This is one of the means by which the heavy metals can enter the food chain.

Table 4.17 gives examples of the chemical substances that may be present in reclaimed wastewater together with the associated disease the chemicals may cause.[32] The action and health implications of a combination of these chemicals still merits research.

The significance of the chemical constituents and the related diseases that may develop due to their presence in high concentrations has been outlined by Tate and Trussel and are presented in Table 4.18.[35]

Table 4.17 Chemical Contaminants in Wastewater and Associated Diseases

Substance	Disease
Lead	Lead poisoning
Nitrate	Methemoglobinemia
Sodium	Hypernatremia
Organic halogens	Cancer
Polynuclear aromatic hydrocarbons	Cancer

From Cowan, J. P. and Johnson, P. R., *Reuse of Sewage Effluents*, Thomas Telford, London, 1985, 107. With permission.

Table 4.18 Significance of Some Major Chemical Constituents

Constituents	Effect
Arsenic	Has been linked with skin cancer and black foot disease; recognized as a carcinogen.
Barium	Muscle stimulant; toxic to heart, blood vessels, and nervous system.
Cadmium	Causes nausea and vomiting, accumulates in the liver and kidney, recognized as a carcinogen.
Chloride	Imparts taste at concentrations above 400 mg/L, no known health effects.
Chromium	Nausea, ulcers after long-term exposure, trivalent form harmless.
Copper	Disagreeable taste above 1 mg/L, therefore, ingestion unlikely.
Cyanide	Toxic gas released at pH values below 6, high concentrations affect nervous system.
Fluoride	About 1 mg/L reduces decay in teeth, especially in children; above about 4 mg/L causes mottled teeth; greater than 15–20 mg/L causes fluorosis.
Iron	High levels impart an unattractive appearance and taste, no health effects.
Lead	Accumulates in bones, constipation, loss of appetite, anemia, abdominal pain, paralysis.
Manganese	Disagreeable taste, discolors laundry, not considered health hazard in water because of unpleasant taste and other dietary sources.
Mercury	Highly toxic to man, gingivitis, stomatitis, tremors, chest pains, coughing.
Selenium	Widely believed to have symptoms similar to arsenic poisoning, has been associated with increased dental carries.
Silver	Fatal at very high concentrations, at low concentrations causes a darkening of the skin.
Zinc	A stringent taste above 5 mg/L, higher concentrations give milky appearance and form a greasy film upon boiling, very high concentrations associated with nausea and fainting.

Compiled from Reference 35.

B. ORGANIC CHEMICALS THAT POSE A HEALTH HAZARD

Many of the organic chemical compounds found in wastewater are stable and persistent. Many of these compounds are believed to be not only carcinogenic but also have mutagenic effects. Some examples of these substances include chlorinated alkanes, chlorinated phenolic compounds, herbicides, pesticides, and polychlorinated biphenyls (PCBs). The sources of phenolic compounds are from the distillation of coal or wood, oil refineries, and other specific chemical plants. If reclaimed wastewater from these plants were used for the production of fish a health hazard exists not only for the consumers but also for the fish, as these compounds impart a disagreeable taste to fish meat. Chlorophenols are reported to taint fish flesh even at extremely low concentration levels.[1,24]

Chemicals in wastewater used for agricultural irrigation may pose negative effects on the health of farmers or consumers. Nevertheless, if properly treated wastewater is used for a beneficial purpose the problems associated with the chemical constituents present can be controlled.

C. BORON

Boron usually originates from domestic sources or from the discharge of industrial effluents. The household source of boron is usually from detergents such as borax, which is used in many laundry powders and in other domestic cleaning compounds. Other sources of boron are from industrial discharges that stem from the food processing industry and may be at concentration above desirable levels. Boron occurs naturally in most rivers. It can be found in fertilizers and in wastewaters, such as those from citrus washing.

Boron is considered to be an essential plant nutrient at very low concentrations. It is phytotoxic to the plant at levels only slightly higher than required for good plant growth. An upper limit from 0.75 to 1.0 mg/L has been suggested for irrigation water.[24] At concentration in excess of 1.0 mg/L, boron is toxic to many boron-sensitive crop species. Examples of the agricultural crops that are very sensitive to low concentrations (in the range below 0.5 mg/L) of boron include lemon *(Citrus limon)* and blackberry *(Rubus* spp.), while crops that are very tolerant (in the range of 6 to 15 mg/L) to the chemical include cotton *(Gossypium hirsutum)* and asparagus *(Asparagus officinalis).*

Reclaimed wastewater contains enough boron to make up for any boron deficiencies in the soils.[36] Symptoms of excess boron include leaf tip and marginal burn, leaf cupping, chlorosis (yellowing leaves), anthocyanin (blue and red leaves), rosette spotting, premature leaf drop, branch dieback, and reduced growth.[36] The boron tolerance varies depending upon certain factors that include climatic conditions, soil conditions, and crop varieties.

Boron adsorption by the crops is slight and it may rapidly filter through the soil with the leaching water. Boron content is a vital parameter that warrants consideration when reclaimed wastewater is to be used for agricultural irrigation.

D. NITRATES

High levels of nitrates have been associated with methemoglobinemia (in infants) as well as diarrheal diseases. In infants the growth of nitrate-reducing bacteria is promoted due to the low acidity of the intestinal tract. The microorganisms convert the nitrate to nitrite which is absorbed into the bloodstream. The nitrite reacts with the oxygen receptor sites on the hemoglobin fraction of the blood and reduces the oxygen-carrying capacity of the blood. This state of oxygen deprivation changes the color of the infant to blue. This nitrate poisoning of infants has been referred to as the blue-baby syndrome, cyanosis, or methemoglobinemia. After the age of 3 to 6 months the flora of the intestinal tract of an infant becomes fully developed, and then nitrate conversion is seldom a problem.

Incidents of infants methemoglobinemia are rare where mothers breast-feed their babies. Methemoglobinemia is much more common in ruminant animals than in humans. Its occurrence in animals is usually associated with high nitrate concentrations in forage rather than in drinking water.[37]

Irrigation with reclaimed wastewater enriched with nitrogen or nitrogen compounds can supply an excess of nutrient nitrogen to growing plants. This condition can result in a significant loss in crop yields, especially rice, through lodging, failure to ripen, and increased susceptibility to pests and diseases through over-luxuriant growth.[38]

E. PESTICIDES

The pesticides that are used in agriculture are not likely to be a major problem in reclaimed wastewater reuse. Chlorinated hydrocarbons can be absorbed into the body of humans through the lungs, gastrointestinal tract, and skin and stored in the fatty tissue. Table 4.19 gives some examples of the hazards that are associated with the use of certain insecticides or pesticides. The use of chlorinated hydrocarbons has been banned in the U.S. and many other industrial nations.

The severity of the adverse effects of exposure to pesticides depends on the dose, rate of exposure, type of pesticide, degree of absorption of the chemical, biodegradability of the pesticide, accumulation rate of the pesticide in the body, and the health status of the person.

V. RADIOACTIVITY IN RECLAIMED WASTEWATER

Usually, the pathways through which human exposure to radioactive elements may develop is governed by the movement of the radionuclide through the biosphere. This movement in the highest trophic levels, usually man, determines the environmental radioactive impact. The radioelement movement may have two pathways, the liquid path and the gaseous or terrestrial pathway. The former involves transport or entry into the food chain through water (see Figure 4.7). The latter pathway introduces the radioactive element into the biosphere through atmospheric dispersion (see Figure 4.8).[39,40]

The health hazards associated with radioactive substances depend on many factors such as the source of radiation as related to the body, type of emissions, radiation dose, period of exposure to the radioactive substance, energy absorbed, and the nature of the decay products.

When a body is subjected to a high dosage of radiation symptoms of acute radiation injury develop a few hours after the exposure. The developed symptoms may include vomiting, malaise and fatigue, loss of appetite, loss of hair, nausea, epilation, sore throat, hemorrhage and infection, purpura, petechiae and

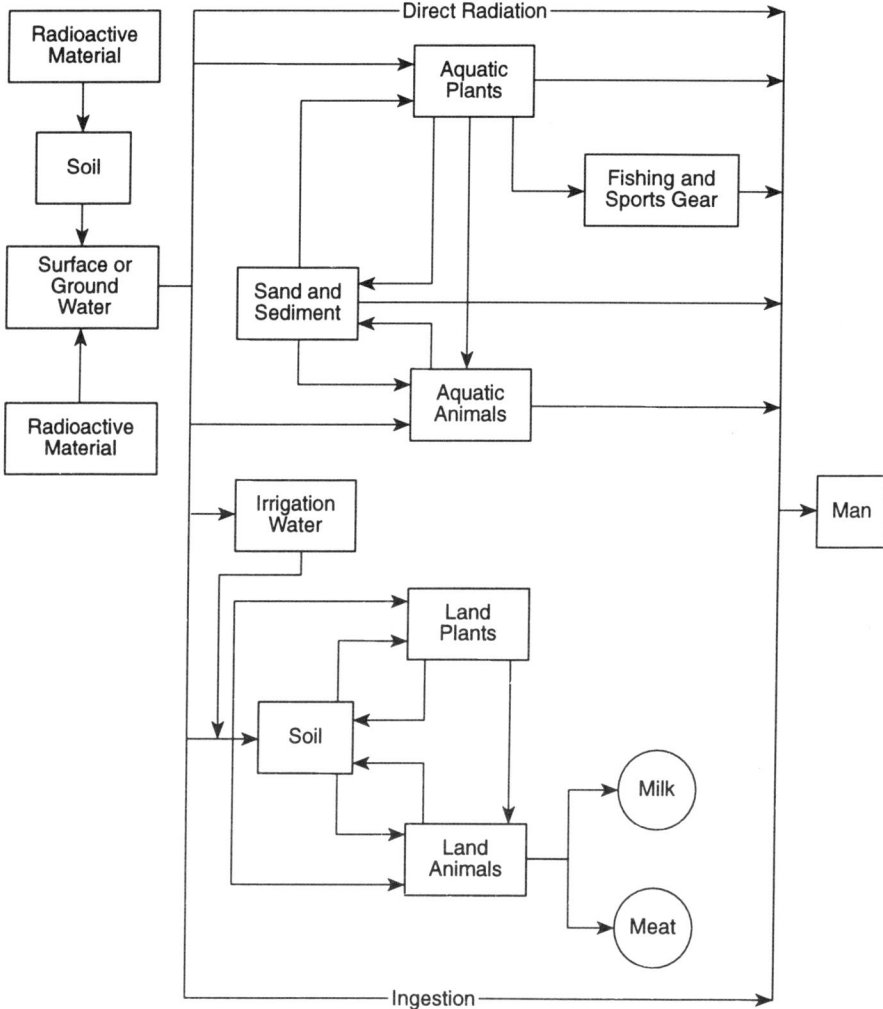

Figure 4.7 Generalized water pathways to man. (From References 39 and 40. With permission.)

diarrhea, impairment to the blood-forming organs, injury to the gastrointestinal tract, and damage to the central nervous system. When a body is subjected to high dosages of radiation, or to relatively small dosages repeated over extended time periods, delayed effects of radiation may arise after prolonged periods of time.

The effects due to the exposure to radiation that are manifested in an exposed person are referred to as somatic effects. The effects that may develop in the descendants of the exposed person and are results of changes transmitted by hereditary mechanisms are referred to as genetic or hereditary effects.

Some examples of somatic health effects observed in irradiated populations include bone cancer, leukemia, lung cancer, malignancies, thyroid cancer, life span shortening, and cataracts. Cancers are most frequent in the skin, bone, and the thyroid gland of exposed persons. Bone cancer develops when radium or other radioelements that are chemically similar to calcium are ingested. These radioelements when deposited in the bones are eliminated very slowly. Cataracts (impairment of vision) result from exposure of the eye lens to relatively high dosages of X-rays, gamma rays, beta particles, or neutrons. The growing cells in the eye lens are affected by developing abnormal fibers in the lens. After several years of exposure the lens becomes opaque. Lung cancers may be attributed to the dispersion of radon into the atmosphere as, for instance, in the case or underground mines. Leukemia is a cancer of the blood-forming organs with an excessive production of malfunctioning white blood cells and also a concurrent shortage of hemoglobin.

The pollution of soil and aquatic environs by radioactive elements may be initiated either from the deposition of material originally introduced into the atmosphere, or from waste products discharged

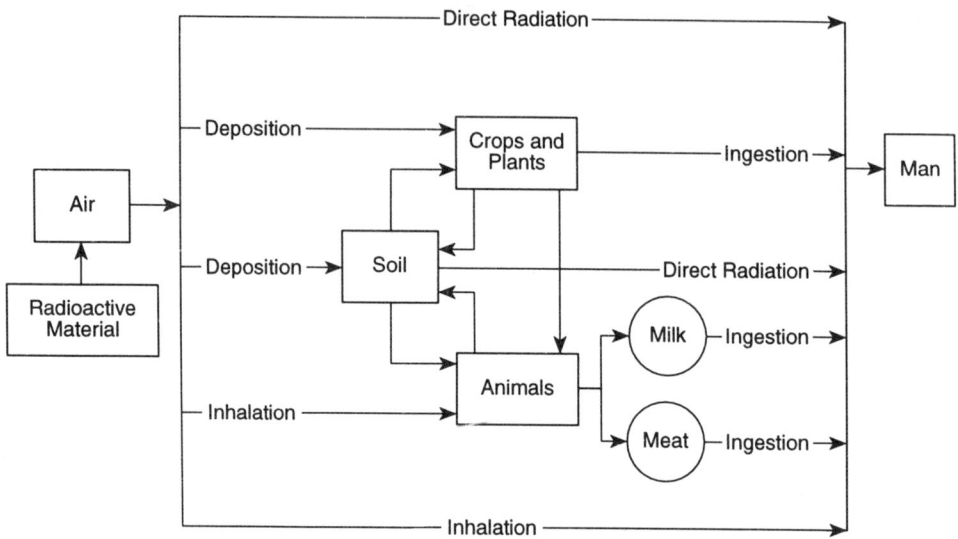

Figure 4.8 Generalized gaseous pathways to man. (From References 39 and 40. With permission.)

Table 4.19 Health Problems Associated with Certain Pesticides or Insecticides

Chemical	Effects
Endrin	Danger to the central nervous system of humans and higher animals.
Lindane	Considered a carcinogen.
Methoxychlor	Of very low mammalian toxicity.
2,4-D and 2,4,5-T	Carcinogenic and teratogenic even in trace amounts.

directly into surface or subsurface waters or buried in the ground, from where they can eventually be picked up by groundwater.[39] Radioactive elements can enter the human food chain through water, milk, plants, or aquatic animals. See Chapter 8 for details regarding terminology, units of measurement, and standards for radionuclides.

VI. REMOVAL OF CHEMICAL CONSTITUENTS BY VARIOUS TREATMENT PROCESSES

Figure 4.9 shows a generalized flow sheet for wastewater treatment unit operations and processes that consist of a combination of physical, chemical, and biological units designed to prepare reclaimed wastewater for reuse.[17]

A. PRELIMINARY TREATMENT METHODS

The removal of chemical constituents by different treatment processes depends on many parameters including the characteristics of wastewaters to be treated, the degree of chemical pollution within the wastewater, the concentration of pollutants, and the types and efficiency of the treatment units. Preliminary treatment contributes very little to the removal of chemical constituent from wastewater.

B. PRIMARY TREATMENT METHODS

Primary settling removes some of the metals which are either insoluble or adsorbed onto particulate matter. The basic purpose of primary treatment is the removal of settleable organic substances, inorganic solids, and scum by the action of sedimentation and flotation forces. The removability of these constituents is in the order of 50 to 70% of the total suspended solids, 25 to 50% of the incoming BOD_5, 35 to 59% of the COD, and around 65% of the oil and grease. During primary sedimentation some organic nitrogen, organic phosphorus, and heavy metals are removed.[41]

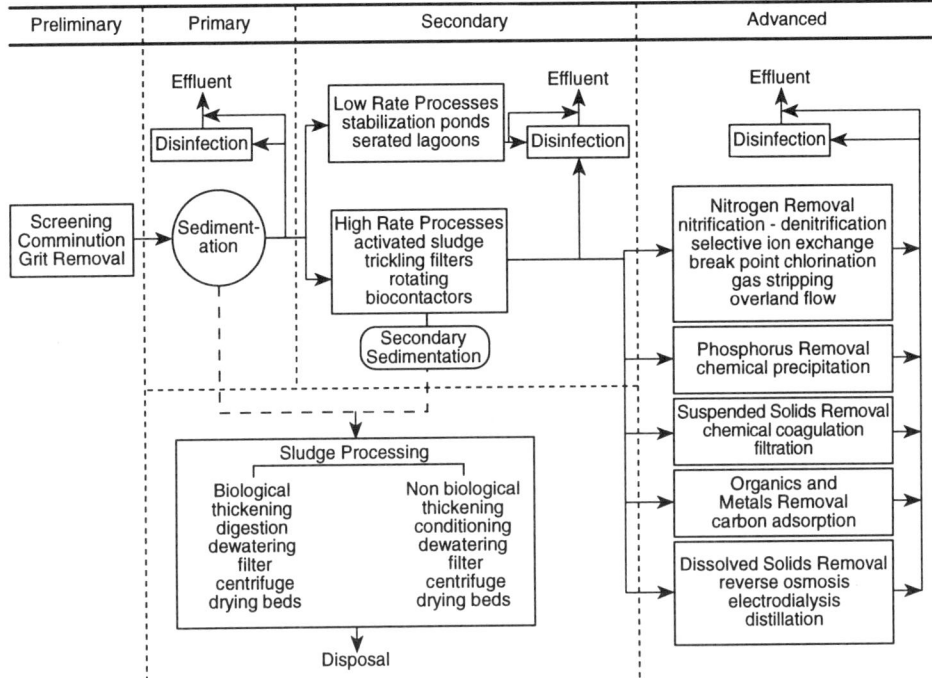

Figure 4.9 Generalized flow sheet for wastewater treatment.[17]

Sedimentation (precipitation) can be used to remove heavy metals such as arsenic, barium, cadmium, and copper. These metals are precipitated as hydroxides by the addition of lime or caustic soda to a pH of minimum solubility.[42]

Conventional wastewater treatment units are not generally designed for the removal of trace elements. Nevertheless, the metals are removed from the effluent by the settling of the suspended solids to which trace elements are attached.[37]

C. SECONDARY TREATMENT METHODS

Although primary settling removes a portion of the metals, further metal removal occurs in the secondary biological stages of the wastewater treatment. Usually, this occurs through the adsorption of dissolved metals or fine particulate metals onto the sludge flocs.

The wastewater which incorporates high concentrations of heavy metals may be treated in an anaerobic pond, whereby a significant percentage of the metals tend to be concentrated in the sludge.[7]

Under normal operating conditions, the trace elements concentrations are reduced by 70 to 90% by secondary treatment processes.[43]

D. TERTIARY (ADVANCED) TREATMENT METHODS

Granular activated carbon is often used to remove organics and, to a lesser degree, heavy metals and viruses from a treated wastewater.

In nitrified wastewater the nitrates are converted anaerobically to nitrogen gas through the denitrification biological process. Under anaerobic conditions the nitrate serves as the electron acceptor, or hydrogen donor, for the oxidation of organic matter by the facultative heterotrophic bacteria. This process is restricted to bacteria of the chemoorganotrophic genera. The denitrification process can be carried out in a packed bed or fluidized reactor followed by a clarifier. An organic carbon source is needed to act as a hydrogen donor and to act as a carbon source for biological synthesis. Examples of substances used as a carbon source include methanol, acetic acid, acetone, ethanol, and sugar. Methanol is used as the carbon source in case of water systems because of its availability, ease of application, and also due to the fact it does not have a residual BOD_5 in the effluent. The detention time required for denitrification of a domestic wastewater is usually in the range of 2 to 4 hours, depending on the nitrate loading and temperature. A carrousel system or a plug flow type of activated sludge system with an anoxic chamber is capable of carrying out denitrification.

Table 4.20 Crops that Include a Manufacturing Process

Crop	Use
Grass, reed	Energy production
Sugar beet	Production of sugar
Camomile, sunflower, castor	Manufacture of oil
Datura	Manufacture of heart drug

Compiled from References 20 and 32.

Nitrates may also be removed by an ion exchange system. The most effective exchanger is a strong basic anion resin that uses sodium chloride for regeneration. The process depends on the amount of sulfate present in the flow, and the exchange capacity of the resin.[44]

There is no simple and economical method for the removal of boron. The best solution is to isolate the chemical at the source and prevent it from being added to the wastewater to be used for irrigation purposes.[24] It may make good sense to grow crops that require a manufacturing process in order to produce a commercial end product. This could eliminate the need for tertiary treatment for the reclaimed wastewater used for irrigation purposes. Tertiary treatment can represent as much as 50% of the total treatment costs. Thus, extensive savings could be realized. Some examples of crops to be considered in this regard are presented in Table 4.20.[20]

VII. IMPORTANT HEALTH CONCERNS IN ENGINEERING PROJECTS

A. INTRODUCTION

Health is defined by the Word Health Organization (WHO) as "A state of complete physical, mental and social well-being and not merely the absence of disease or infirmity." In order to establish this state of health certain environmental factors must be evaluated at both the individual and community levels. One of these environmental factors is the presence of vectors that can transmit infectious diseases.

A vector is an agent that can transmit a pathogen and includes both true vectors and animal reservoirs of the infectious agent. An example of vectors are biting insects. An example of an animal reservoir is the snail, which acquires a suitable ecosystem to survive. An ecosystem involves the interaction between the entire living members in a particular area and their surroundings.

The major ecological factors that influence the life style and pattern of disease-causing vectors include the following:[13,14,16,46,49]

1. *Water movement.* Vectors vary in their choice of a suitable habitat for breading purposes. For example, the *Anopheles* female mosquito, the vector for malaria and other diseases, favors fairly clean, stagnant, or slowly moving stretches of water such as irrigation ditches or impoundments that have still surfaces and are protected from the effects of winds and waves. While the *Anopholes* mosquito likes calm water for reproducing, the *Simuliid* black fly, which transmits onchocerciasis, prefers moving water for its breeding purposes. Following are measures that can be used to help in vector control in the aquatic environment:
   - Control of aquatic vegetation.
   - Control of hydraulic systems. Examples of this type of control include intermittent discharges of reclaimed wastewater as well as water level fluctuations.
   - Control of flow rates. Different species of mosquitoes require water that has a minimum flow rate for the development of their larval and pupal stages.
   - Control of the depth of water. Some vectors can be detected at a considerable water depth. The survival of vectors in deep water impoundments depends upon certain factors such as availability of a food supply, abundance of oxygen, absence of vector predators, presence of floating vegetation which may create a second false bottom, as well as the surrounding turbulence. The development of the vector from an egg to an adult requires about a week, thus, good breeding conditions necessitate appropriate water conditions for that period.

2. *Chemical characteristics of the reclaimed wastewater.* The major chemical characteristics of concern when using reclaimed wastewater include:
   - Water salinity. Salinity varies in its effects on vectors. A high salinity is fatal to most vectors except coastal species such as *Anopheles melas* (vectors of malaria in West Africa) and *Anopheles sundaicus* (vectors of malaria in India), which breed in saltwater habitats.[10,45]

Table 4.21 Vector Relationship with Water

	Insect vectors					
Location	Culicine mosquito	Anopheline mosquito	Simuliid blackfly	Tabanid horsefly	Phlebotomine sandfly	*Glossina* or tsetse fly
Breeds in water	+	+	+			
Breeds in wet ground				+		
Breeds in damp ground					+	+
Lives near water	+	+	+			+
Lives elsewhere					+	+

	Intermediate hosts			
Location	*Cyclops*	*Oncomelania* snails	*Bismphalaria* snails	*Bulinus* snails
Found in drinking water	+			
Lives in various kinds of water		+	+	+

Compiled from Reference 46.

- Organic levels. A high level of organic matter favors the growth of *Culex quinquefasciatus* (vector of *Bancroftian filariasis*) which prefers breeding in septic tanks and polluted waters.
- Oxygen levels. Access to the air at the water surface for breathing is required by all vector genera except Mansonia.

3. *Sunlight.* The incidence of sunlight has an impact on water bodies and impoundments in various ways such as:
   - Light has a direct influences on the behavior of vectors. The factors that affect the influence of light on vectors include the kind of vector, number of species, change in temperature, and availability of food.
   - Light rays incident on the water induce temperature variations in the water.
   - Light influences the chemistry of the water body or the impoundment through its influence on photosynthetic growth of green plants such as algae.
4. *Temperature.* Temperature has a significant effect on the survival and development of the vector, the life cycles of the vector, and the number of vectors present.
5. *Suitable aquatic vegetation.* The presence of aquatic vegetation may establish a stagnant and a calmer water surface. Likewise, plants provide a food supply and shelter for the vector, thus, aquatic plants assist in the growth and development of the vectors. However, in some cases the aquatic plants interfere with the hatching of the larvae, this is partly due to the lack of sunlight.

Table 4.21 illustrates the relationship of vectors with their intermediate hosts and with water. An intermediate host or secondary host is a host in which the parasite is in a larval or a sexual state. Table 4.21 also presents the preferred location for the vectors involved in water-related disease.[46]

B. THE ENVIRONMENT AND HABITATS OF VECTORS

Vectors and vector-borne diseases are not distributed uniformly over a particular geographical zone. Vectors can develop in separate or independent areas. These areas are governed by factors such as the vector habitat, vector growth, and vector development. Engineering projects involved in reclaimed wastewater reuse and other hydraulic and water engineering schemes can modify the development of a vector. These engineering projects can even introduce a brand new vector to an area.

All vectors have separate and preferred habitats. The factors involved in habitats for different vectors include:

1. *Inhabited areas.* Some vectors favor breeding adjacent to, or far from man-inhabited areas. The vectors that choose to reproduce near man can clearly affect the people in that vicinity.
2. *Breeding location and time of breeding.* The breeding location and time of breeding for vectors can be during daytime, during twilight, at night, inside or outside homes, near grassland, as well as adjacent to the human dwelling. Table 4.22 presents information regarding some vectors, their intermediate hosts, their reproductive potential, and their preferred behavior.[13]

Table 4.22 Generalized Biological Information Concerning the Vectors of Some Diseases

Disease	Vector or intermediate host	Reproductive potential				Preferred behavior			Flight dispersal range[a]
		Number of eggs	Egg-to-egg cycle	Number of broods	Life span (weeks)	Feeding time	Resting place	Source of blood	
Malaria	*Anopheles* mosquitoes	200	10–14 days	6–10	20	Night	Indoors and outdoors	Man and animals	1.5 km
Filariasis; viral diseases	*Culex* and *Aedes* mosquitoes	200	8–10 days	6–10	20	Night and day	Indoors, outdoors	Man and animals	0.1–8 km
Onchocerciasis (river blindness)	Black flies (*Simulium damnosum*)	400	2–3 weeks	3–4	1–2	Day	Outdoors, animals	Man and animals	4–8 km
Infantile diarrhea	Houseflies	150	7–14 days	2–3	3	Day	Indoors, outdoors	—	4 km
Schistosomiasis	Aquatic snails	45	30 days	10–20	50	—	Outdoors	—	10–30 m
African trypanosomiasis (sleeping sickness)	Tsetse flies	1 pupa	60 days	10	3–12	day	Outdoors	Man and animals	2–4 km
Chagas disease (American trypanosomiasis)	Triatomid bugs	200	52 weeks	1–2	50	Night	Indoors	Man	10–20 m
Leishmaniasis	Sandflies	50	6–8 weeks	2	12	Night	Outdoors	Man and animals	50 m
Plague	Rats	8 per liter	12 weeks	4	32	Night and day	Indoors, outdoors	—	50–80 m
	Fleas	12	8 weeks	10	15	Night and day	Indoors outdoors	Man and animals	—

Note: The figures given are only indicative and illustrate the major factors affecting transmission. They vary widely from species to species and in different environments.

[a] Under normal static conditions.

Compiled from Reference 18.

Table 4.23 The Main Animal Hosts of Vector-Borne Diseases

Group	Disease	Pigs	Birds	Rodents	Monkeys	Large herbivores	Carnivores	Human is principal host
Arboviruses	Dengue				+			+
	Yellow fever				+			
	Other	+	+	+	+	+	+	
Dracunculiasis					+		+	+
Filariasis	Bancroftian							+
	Brugian				+			+
	Loiasis				+			+
	Onchocerciasis							+
Leishmaniasis	Cutaneous			+			+	
	Visceral			+			+	
Malaria								+
Schistosomiasis	Mansoni			+	+			+
	Haematobium							+
	Japonicum	+		+		+	+	
Trypanosomiasis	Rhodesian	+				+	+	
	Gambian							+

Compiled from Reference 13.

3. *Infective stage.* An infective stage of a vector indicates the period of time during which the vector can transmit the disease. The vectors that are capable of performing this transmission are those that are capable of living through their incubation period.
4. *Resting periods.* Some vectors prefer resting in homes. Others rest outside of dwellings, and prefer resting on plant leaves or beneath vegetation.
5. *Seasonal differences.* Different vectors vary in their breeding behavior and growth patterns depending on seasonal changes. This means that when the vectors are detected in abundance in a particular season then the associated parasitic agents may be transferred at that time.
6. *Human blood.* Certain vector types prefer a human blood meal. Other vector species attack and bite humans when their normal host animal is missing or absent. Table 4.23 presents examples of the main animal hosts for vector-borne diseases.[46]
7. *Domesticated animals.* Vectors that prefer feeding on the blood of domestic animals may easily reach people who own pets or animals.

The majority of vectors are particular in the choice of their habitat. For example, some vectors prefer specific water bodies such as lakes, rivers, or streams with certain quality aspects such as temperature level, water velocity, and availability of shadow.

Other vectors like the aquatic types choose either freshwater, contaminated water, or brackish water source. The selection of the preferred habitat is confined to the particular vector with no overlap in the selection process.

Table 4.24 illustrates some examples of the principal diseases associated with water in relation to the principal habitat of the vectors.[46]

C. VECTOR CONTROL

Vectors that transmit diseases need to be controlled in order to safeguard the health of a community.

Vector control is used against such vectors as mosquitoes, flies, fleas, lice, roaches, rats, mice, and snails. The methods used in vector control serve many purposes, such as:

- The prevention, occurrence, and spread of vector-borne diseases.
- The control of endemic disease.
- The reduction, or prevention of outbreaks of epidemics.
- A decrease in the occurrence of bites by vectors such as insects and rats.
- The reduction in economic losses which are caused by vectors that eat and destroy food.
- The improvement of the aesthetic quality of life. For example seeing insect parts or rat hairs in food is truly disturbing.

Table 4.24 The Principal Diseases Associated with Water in Relation To the Principal Habitats of the Vectors

Group	Principal disease	Arid and semiarid land	Rain forests	Riverain vegetation	Savanna woodlands	High altitudes	Irrigation ditches canals	Lakes and ponds	Wetland rice cultivation	Rivers and streams	Human settlements	Savanna grasslands	Coastal plains
Arboviruses	Dengue										+		
	Yellow fever		+								+		+
Dracunculiasis								+					
Filariasis	Bancroftian		+				+				+		+
	Brugian		+								+		
	Loiasis		+										
	Onchocerciasis						+						
Leishmaniasis	Cutaneous	+	+										
	Visceral	+	+										
Malaria			+	+			+	+	+	+	+		
Schistosomiasis	Mansoni						+	+	+	+	+		
	Haematobium						+	+	+	+	+	+	
	Japonicum								+	+			
Trypanosomiasis (African)					+								

Compiled from Reference 46.

The methods used for vector control apply a combination of techniques. The methods used rely on the technology, equipment, and personnel available, the species composition, and the biological and ecological peculiarities of the vectors present.

The usual vector control and management patterns include:

1. *Breeding sites.* The destruction of the vector's breeding sites and its habitat are a first step in vector control. Examples of such control measures include filling and drainage of impounded waters as well as the application of larvicides or insecticides.
2. *Chemical action.* Chemicals may be used to kill insects, pupae, larvae, and adult vector.
3. *Biological control.* Biological control methods may be used to control vectors. An example of this method is the introduction of larvivorous fish to decrease the number of vector larvae in a vector habitat.
4. *Personal protection.* Personal protection measures include home screening, bed netting, spraying with insecticides, as well as the use of vector repellents.
5. *Community education and participation.* Community education and training are essential to motivate people to participate in the vector control program.

Vector control methods include the following:

- Chemical control methods.
- Biological control measures.
- Genetic control considerations.

D. CHEMICAL METHODS FOR VECTOR CONTROL

Chemical vector control methods are frequently used in vector management techniques. Chemical compounds are applied to the resting areas of the adult vectors or to their breeding sites. The chemical compounds that are used need to be applied for an appropriate period of time by trained personnel.

The advantages in using chemical compounds for vector control include:

- Chemical compounds have a very potent effect in controlling vector-borne disease epidemics.
- Chemical compounds are fast acting and provide optimum control of the vectors.
- Chemical compounds are economical to use in a vector control program.

Two of the disadvantages in using chemical compounds to control vectors is the development of vector resistance to the various chemicals used. Also, these chemicals can be hazardous to the health of humans.

1. Major Groups of Insecticides

An insecticide may be defined as any substance or a mixture of substances used for preventing, killing, or controlling an insect pest. A pest is defined as any animal, plant, or pathogen which causes damage or annoyance to man, his animals, crops, or possessions.

The major insecticides are either chemical compounds or biological compounds. The chemical insecticides act either as a systemic poison or as a contact poison. The chemical insecticides are further subdivided into the following groups:

a. Pyrethrin

These are one of the earliest known insecticides and are of plant origin. One of the best insecticides of this group is pyrethrum which is derived from chrysanthemum flowers. Pyrethrum consists of pyrethrin and cinerins ($C_{20}H_{23}O_3$ and $C_{21}H_{28}O_5$). It is effective against a wide range of insects but is virtually harmless to humans and other mammals. Its residual activity is minimal, and the active ingredients break down in sunlight.[47] The chemical is costly as it requires a specific variety of flower for its production.

b. Organochlorine Compounds (Chlorinated Hydrocarbons)

In the past, the most frequently used insecticides of the organochlorine group were DDT (dichlorodiphenyltrichloroethane; $C_{14}H_9Cl_5$), BHC (the gamma isomer of benzene hexachloride, also known as HCH and commercially as lindane), chlordane, and dieldrin. These compounds are now banned in the U.S. The method of application of these compounds was in solution, emulsion, or as a suspension in water. The water suspension form was most often used as water was generally readily available.

Organochlorine compounds not only have a long residual time, especially on the walls of mud houses, but they also can accumulate in humans.

While being effective against many insects, organochlorine insecticides have a high mammalian toxicity that can last for many years. These adverse health effects have led to the banning of these chemicals except for the treatment of timber.[47]

c. Organophosphate Compounds

Organophosphate compounds are synthetic organic substances which may be used as contact or systemic toxins. Many of the organophosphate compounds have a short residual time as they are biodegradable. This biodegradation aspect is important as it decreases the residues left on food crops treated with these chemical compounds.

Examples of the organophosphorus compounds include malathion (s-1,2-*bis*(ethoxycarbonyl)ethyl,O,O-dimethylphosphorodithioate, $C_{10}H_{19}O_6PS_2$); and fenitrothion (O,O-dimethyl O-4-nitro-m-tolylphosphirothioate, $C_9H_{12}NO_5PS$). Malathion is a nonsystematic insecticide of low toxicity to mammalian animals (having milk-secreting organs in females for feeding of young), and has a short residual time. Malathion is employed in antimalarial control programs as well as for the control of flies and other domestic insects. Fenitrothion is a contact insecticide, with a longer residual time than malathion. However, it is expensive. Fenitrothion is practically insoluble in water but soluble in most organic solvents.

d. Carbamate Compounds

Carbamate compounds are synthetic organics that contain carbon, hydrogen, nitrogen, and sulfur. They are similar in their biological action to that of the organophosphorus insecticides. They have a short residual time. Some examples of the carbamate compounds include propoxur, $C_{11}H_{15}NO_3$, aldicarb (2-methyl-2-(methylthio)propionaldehyde O-methylcarbamoyloxime, $C_7H_{14}N_2O_2S$), and bendicarb ($C_{11}H_{13}NO_4$). Propoxur, a methyl carbamate, is a very expensive chemical compound that is highly toxic to mosquitoes. Aldicarb is a soil-applied pesticide that is employed to eradicate mites, nematodes, and insects.

e. Pyrethroids Compounds

Pyrethroids are synthetic compounds of a chemical composition that corresponds to natural pyrethrin. The active constituents of the insecticide embodies permethrin, allethrin, resmethrin, and other compounds. Most of the chemicals are useful against a broad range of insects and possess a low mammalian toxicity.

2. Classes of Pesticides

A pesticide is a chemical or physical agent used for the killing or the controlling of pests on crops, forests, grass, gardens, lawns, soils, houses, pets, and livestock. Pesticides are used to control arthropods, nematodes, algae, bacteria, fungi, weeds, rodents, and occasionally birds and fish. The word "cida" is a Latin word which means to cut or kill. The most common categories of pesticides may be grouped according to the following:

1. *Type of pest.* Pesticides may be classified according to the type of the pest they control. Examples of these pesticides include algicides (used to kill algae), arboricides (used for control of trees, bushes, and shrubs), bactericides (used to kill bacteria), fungicides (used to kill fungi), herbicides (used to kill weeds), insecticides (used to kill insects and sometimes other related pests such as ticks and spiders), and larvicides (used for the control of larvae).

    Larviciding is an effective method to control adult vectors. A larvicide may be used in spraying the breeding sites for the vectors. Larvicides have limited residual effect and last only for a few days. Some examples of larvicides that are utilized for mosquito control include arsenicals, crude kerosene, chlorpyriphos miticide, or acaricide, distilled petroleum oils; Paris green which is a complex of copper meta-arsenite and copper acetate; temephos; fenthion, molluscoid, and nematicide. Temephos, also known as abate ($C_{16}H_{20}O_6P_2S_3$), is a colorless, crystalline solid used to control mosquito larvae, blackfly, moth, sand flies, and lice.

2. *Effect on pest.* Pesticides may be classified according to the effect they impose on pests. Some examples of types of pesticides include the following:

    - Anti-feedant. An anti-feedant limits the feeding of the vectors and they starve to death.
    - Anti-transpirant. Anti-transpirant reduces the transpiration rate of the vector (reduces the giving off of moisture).
    - Attractant. An attractant draws or attracts the pests to the control area. An example of an attractant is the use of pheromones.
    - Chemosterilant. A chemosterilant prevents the reproduction of the pest.
    - Defoliant. A defoliant alters any unwanted development within the vector without instantly initiating its death.
    - Desiccant. A desiccant dries up vectors.

- Growth regulator. A growth regulator stops, speeds up, or otherwise changes the ordinary growth pattern of the vector.
- Repellant. A repellant drives the vectors away.

3. *Chemical composition.* Pesticides may be classified according to their chemical composition into two main subdivisions: inorganic and organic compounds.
   - Inorganic pesticides. Inorganic pesticides do not include carbon in their composition. These pesticides are prepared from natural elements. Some varieties of this subdivision are persistent and some contain arsenic, mercury, and thallium, which can be cumulative poisons. Some examples of elements used in pesticides include arsenic, boron, copper, lead, mercury, sulfur, tin, and zinc. Specific inorganic compounds used as pesticides include: boracic acid, calcium cyanide, copper oxychloride, copperas oxide, copper sulfate, lead arsenate, mercuric chloride, mercuric oxide, sodium aluminum fluoride, sodium arsenite, sodium chlorate, sodium cyanide, sodium fluoride, sodium fluoroacetate, and thallium sulfate.
   - Organic compounds. Organic compounds may be divided in two major categories, namely: natural compounds and synthetic compounds. The latter category include compounds such as DDT, endrin, dieldrin, and chlordane. Natural compounds are referred to as botanic compounds, and are produced from plant extracts. These insecticides are inclined to be unstable and short-lived. However, they are efficient when used to control vectors. Also, these pesticides are generally of a low mammalian toxicity, with the exception of nicotine. Some examples of natural insecticidal compounds include nicotine, pyrethrum, rotenone, and sabadilla. Nicotine is derived from *Nicotiana tabacumthe* which are plant leaves. Pyrethrum is obtained from the flowers of *Chrysanthemum cinerariaefolium,* and red squill which is from the bulbs of *Urginea maritima.* Pyrethrum is unstable when exposed to sunlight and rapidly hydrolyzes under alkaline conditions. Pyrethrum is a nonsystemic contact insecticide and causes rapid paralysis or death of the vector. Rotenone is made from the roots of the *Derris* plant and *Sabadilla* is made from the plant *Schoenocaulon officinale*.

4. *Mode of action.* Pesticides may be classified according to their mode of action. The pests may die or develop a change in behavior such as growth pattern, metabolism, or asexual development. Some examples of this type of pesticide include:
   - Contact toxicant. Generally, a contact toxicant enters the body of the vector due to its contact with a surface treated with the pesticide. Once the pesticide penetrates the vector it damages its nervous and respiratory systems.
   - Fumigant. A fumigant is a volatile substance which enters the body of the vector in a gaseous form through its respiratory system.
   - Stomach toxicant. A stomach toxicant orally penetrates the body of a vector where it is absorbed through the digestive tract. Stomach poisons are applied during the feeding period of the vector.
   - Suffocation. Suffocation results due to the obstruction of the breathing tract of a vector. A typical example of this type of pesticides is oil.
   - Systemic toxicant. A systemic toxicant applied to a plant can reach a vector when the vector feeds on that plant.

Some of the potentially negative impacts that may be attributed to pesticides include:
- The persistence of pesticidal residues and their entrance into the food chain.
- Exposure of humans to toxic pesticides during the operation.
- Development of resistant vector strains.
- Environmental risks during the transportation, handling, storage, usage, and ultimate disposal of the pesticides.
- Ecological impacts that cannot be predicted.

E. NATURAL VECTOR CONTROL METHODS

The natural vector control methods have developed as an integral part of the approach to vector control. Natural or biological control methods involve the introduction of natural predators or infectious agents that will kill the vectors. Examples of such natural methods include the use of bacterial agents; parasites of larvae; fish and top-feeding minnows that eat the mosquitoes; or cats and ferrets that will kill the rats; or any other predators or pathogenic organisms that will destroy the vector.

To achieve efficient control of vectors the following information is needed: the vector's life span, frequency of biting humans or other animals, and its population density. Following are some of the biological methods that are used for the control of vectors.

1. Natural Enemies

The natural enemies of vectors that are frequently used include:

1. *Plant and invertebrate predators.* Certain plants and invertebrates are predators. An example of an invertebrate predator is *Toxorhynchites.* This invertebrate predator alters the natural development of the vector population. For mosquito vector control another mosquito of the genus *Toxorhynchites* can be used. The success of this method is due to the following properties of the predator mosquito:
   - The female mosquito never bites humans or other animals.
   - The mosquito larvae possesses predatory habits.
   - This mosquito can be mass-produced.
   - This mosquito is very effective in controlling the *Aedes* mosquitoes.

2. *Natural insecticides.* Many plants produce chemicals that have insecticidal properties and can be used to produce pesticides. Pesticides that are produced in this way are safer and less environmentally damaging than synthetic chemicals. Plants with such properties include the Neem tree *(Azadirachata indica).* Neem has long been used as an insecticide and a medicine. The leaves of the Neem has been used to protect clothes and stored grain from insect attack, and are also plowed into the soil as a fertilizer.

3. *Protozoa and fungi.* Protozoal and fungal agents are being developed as vector control agents.

4. *Vertebrate animals.* Vertebrate animals are those that have a spinal column, such as frogs, fish, and ducks. These animals can be used for vector control as they eat the mosquito eggs and larvae, or they can ingest the vectors while feeding. Larvivorous fish such as the *Gambusia affinis, Aplocheilus, Oryzias,* and *Poecilia* have been used in mosquito control programs. The larvivorous fish belongs to the fish species that feed specially on mosquito larvae. The *Gambusia affinis* fish is a voracious eater of mosquito larvae. When the fish is present in adequate numbers in ponds, pools, lakes, and marshes, it will destroy vast number of mosquito eggs, larvae, and pupae. This fish is also tolerant of low dissolved oxygen concentrations in the water. Another larvivorous fish that is often used for mosquito vector control is the guppy *(Poecilia reticulata).* The fish *Oreochromis spilurus (Tilapia melanopleura)* is effective in controlling malaria vectors in brackish water as this fish eliminates floating and semisubmerged aquatic macrophytes which are breeding sites for malaria-transmitting *Anopheles* mosquitoes.[7]

Fish used in biological vector control programs need to have the following characteristics:
- Hardy and small in size.
- Have a high rate of reproduction.
- Adjust readily to a confined habitat.
- Capable of reaching the vector's refuge.
- Survive in either the presence or absence of the mosquito vectors larvae.
- Flourish in shallow waters.
- Not valuable as a food source.

2. Biological Control Agents (Biological Toxoids)

An example of this type of vector control is the production of toxic compounds by bacteria. These toxins are fatal to mosquito larvae and harmless to most other nontarget aquatic organisms and vertebrates. The toxins produced by the bacteria *Bacillus thuriniensis* spp. *israelensis* (serotype H-14, often called "Bti") and *Bacillus sphaericus* have to be consumed by the larvae in order to kill them.[10] Difficulties can be experienced when using these toxins or toxoids as other aquatic animals may also be killed.

F. GENETIC VECTOR CONTROL METHODS

The genetic methods for vector control utilize the mate-seeking behavior of the males. This aspect of genetic vector control is specific, and nontarget organisms will not be harmed. Numerous procedures for genetic control of vectors is now under review. These methods include:

- Use of sterile males. The distribution and release of sterile males is carried out to produce nonproductive matings in a selected population.
- Cytoplasmic incompatibility. This results in the transfer of damaged genetic material that interferes with the development of the vector.

G. ENVIRONMENTAL MANAGEMENT METHODS FOR VECTOR CONTROL

Environmental management for vector control encompasses a broad spectrum of planning, organizing, and monitoring activities. Figure 4.10 illustrates the general environmental management methods and

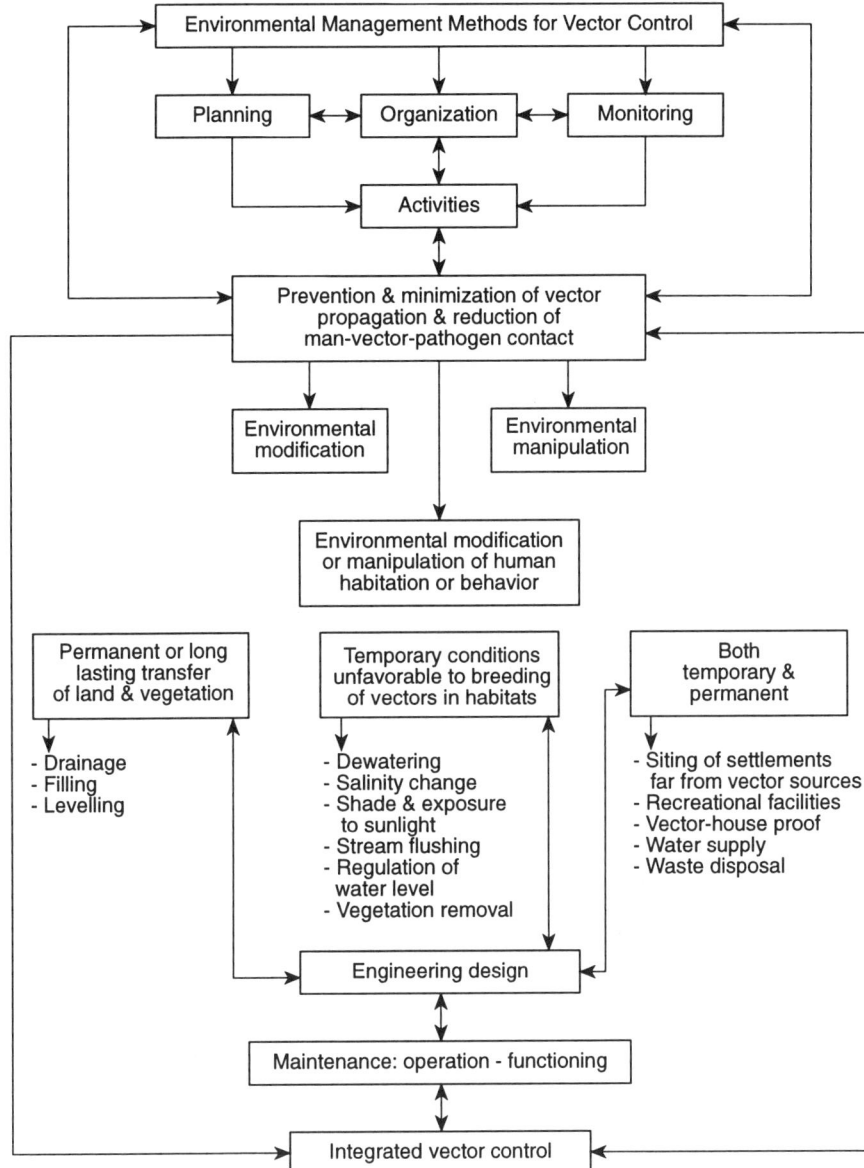

Figure 4.10 Environmental management methods for vector control.

procedures used in the control of disease-transmitting vectors. Some of these operations are discussed briefly in the following section.

H. PERMANENT VECTOR CONTROL METHODS

Permanent control of vectors and the vectors hosts involves physical environmental changes.

These permanent environmental changes are long-lasting and involve changes in the land, water, and vegetation. The physical changes are directed at prevention, elimination, or reduction of the habitat of the vectors and result in a limited impact on the overall environment.

Permanent environmental changes include filling and draining areas of impounded water, alterations in impoundment boundaries, as well as brush clearing.

1. Storage Impoundments

An impoundment can be defined as a reservoir used for the storage of runoff water which collects behind a man-made dam. The function of the dam is to store the excess runoff water during periods of high

Table 4.25 Marsh Potential for Certain Situations

Situation	MP (/m)	Remarks
Shallow run-of-the river (reservoir with mean length [V/A]) or where number and size of bights and indentations caused by streams and ravines produce a very long shore line	18–20	Much greater potential for creating mosquito problem
Deep reservoir with steep slopes in a mountainous area	2–3	

Compiled from Reference 48.

rainfall or the storage of flood waters in order to avoid damage to structures downstream. The collected water can be used for power generation, water supply, agricultural irrigation, recreational activities, as well as for other beneficial purposes.

The impoundments can be designed and constructed in such a way as to contribute to the control of vector breeding sites along its shores.

2. Mosquito Problems In Impoundments

The storage of water behind a dam can provide mosquito breeding sites. Flooding of these sites is an effective way to control vector breeding such as mosquitoes. Vector breeding sites are hard to isolate and treat effectively with insecticides.

Mosquitoes breed in sheltered bays, zigzag shorelines, and naturally protected areas along the banks of an impoundment. The presence of shallow water, aquatic plants, and floating debris provide ideal conditions for vector breeding. Such surroundings provide the mosquito larvae with shelter from such hazards as turbulent currents, high wind, and intense wave action. The larvae may easily find food and obtain shelter from their natural enemies under such conditions.

Sites where mosquitoes cannot find good breeding conditions are in reservoirs or impoundments and include deep water, areas far from the shoreline, areas free of vegetation, and steep shorelines that are exposed to intense wave action.

The scale of the mosquito problem around impoundments or reservoirs is indirectly proportional to the length of the marshy shoreline. To evaluate the mosquito potential in impoundments or reservoirs for mosquito breeding sites, a marsh potential has been formulated.[49] The marsh potential is a parameter that is based on the length of the shoreline and the area and volume of the impoundment, as presented in Equation 4.1.

$$MP = \left(L \cdot (A)^{0.5}\right)\Big/V \qquad [4.1]$$

where: MP = marsh potential, /m
L = shoreline length, m
A = reservoir area, m^2
V = reservoir volume, m^3

Table 4.25 presents marsh potential values that correspond to certain situations involving impoundments or reservoirs.[49]

Figure 4.11 presents a schematic diagram outlining the impact and implications of the building of a dam and its possible impact with regard to vectors and vector-borne diseases.

3. Irrigation

Irrigation is the application of water or reclaimed wastewater to the soil with the objective of providing moisture content for plant growth and increased crop yield.

The operation and management of an irrigation project has an impact on the agricultural activities of the area and also on the health of the surrounding communities.

The conveyance and distribution of the irrigation water often utilizes open channels, ditches, or canals.

These channels or ditches need to be reasonably stable in order to preserve their shape, and thus maintain their water-carrying capacity. Deposition of sediments or erosion of the canals must be avoided. The cross-sectional area of an earth canal is often trapezoidal in shape, with side slopes varying from a ratio of 1:1 to 1:3.

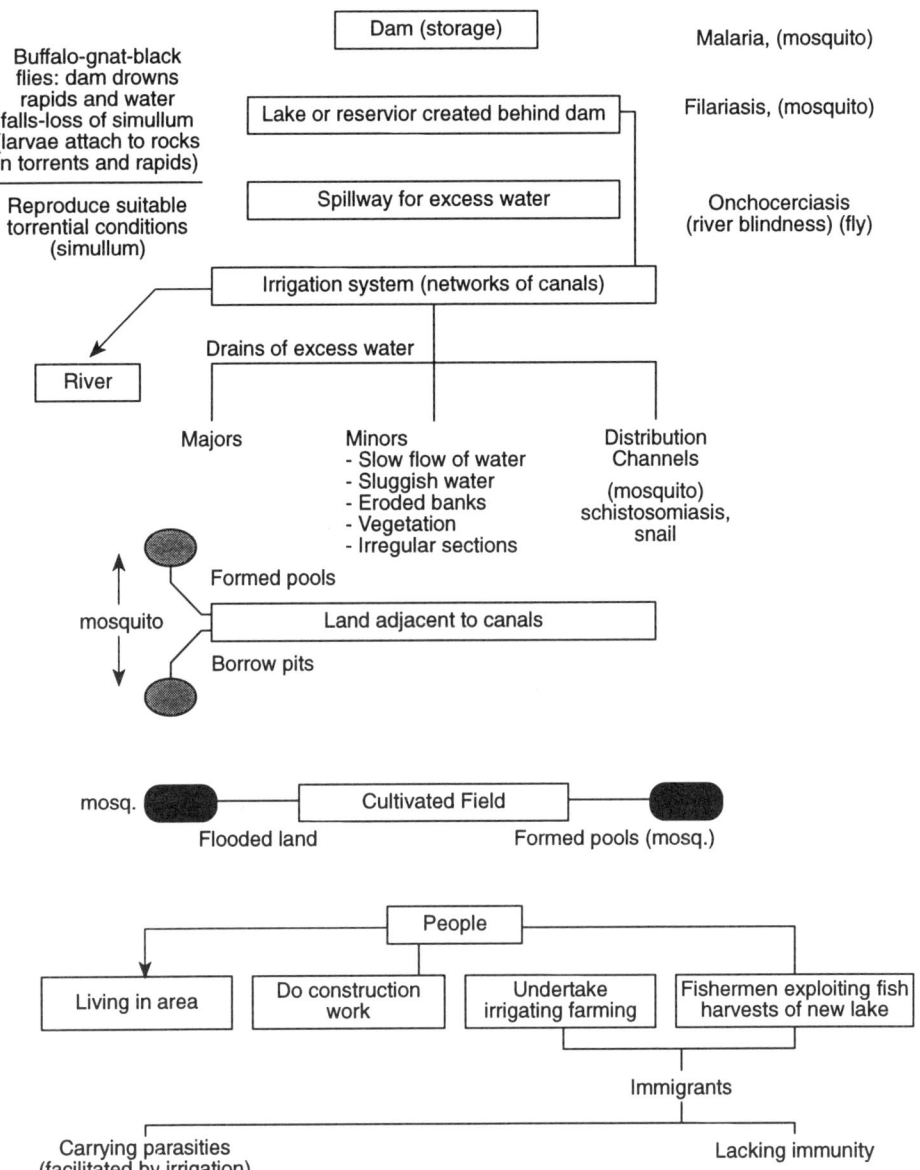

Figure 4.11 Impoundments and the environment.

The important elements to be considered for an open channel water flow system include the following:

- Low initial cost.
- Direct simple construction procedures.
- Easy, effective maintenance at minimum cost.
- Control of vector breeding sites.

Some of the disadvantages associated with earthen canals are as follows:

- Aquatic plants. Excessive growth of aquatic plants can develop in the canals or along their banks. Problems of excessive growth of vegetation in canals result in a decrease in quantity of flow, and an increase in maintenance costs.
- Land area. A large land area is required for the canals or ditch systems. This is particularly true if low flow velocities are to be maintained.
- Water loss. Water losses can be high due to seepage and evaporation.

- Bank breeching. Embankments may be destroyed by overflowing, weathering, erosion, as well as by animal burrowing.
- Bridge demands. Bridges are needed in order to avoid damage to the canals.

Many of the above problems can be reduced or controlled through the use of lined canals.

In some irrigation schemes health problems associated with vectors and their breeding habits can necessitate an increase in water flow velocities. Water flow velocity in the canals can influence the settleability and multiplication of snails and the breeding of mosquitoes. Certain irrigation techniques can contribute to some diseases such as schistosomiasis and malaria.

The greatest potential for vector breeding is to be found in the smaller distribution channels, where the flow is slow, or along channel sections that are irregular. The vector problems can also develop along eroded canal banks or where there is abundant plant growth.

For canal design open channel flow equations are used. The flow under open channel flow conditions can be uniform or nonuniform, steady or unsteady. Under uniform flow conditions the velocity does not change in its direction and magnitude across the channel section. For normal flow the total energy line is the same as the slope of the canal bed. Thus, the specific energy is constant along the channel. A normal flow occurs when the work done against friction is exactly equal to the loss of potential energy. As such, the total energy line, liquid surface, and channel bed (or base) are parallel to each other. Other prerequisites for normal flow conditions demand a constant channel section, slope of the bed, and roughness of the channel walls. The flow is termed steady or unsteady according to whether the velocity and depth at a particular point in the channel varies with time.

If the depth of flow is constant, the velocity of flow is also constant and is given by one of the following equations:

Chezy's formula:

$$V = C \cdot (R_h)^{0.5} \cdot (j)^{0.5} \qquad [4.2]$$

where: V = velocity of flow, m/s,
C = Chezy's coefficient, $m^{1/2}/s$,
R_h = hydraulic radius = hydraulic mean depth cross-sectional area/wetted perimeter = A/w_p, m,   [4.3]
A = cross-sectional area of the channel, m^2,
w_p = wetted perimeter, m,
j = slope of the energy line = hydraulic gradient.

The Chezy's constant, C, depends on the mean velocity of flow, the hydraulic mean depth, the kinematic viscosity, and a relative roughness factor. The values of C can be determined using the following empirical equation.

$$C = \{23 + [0.0015/j] + [1/n]\} \Big/ \Big\{1 + (23 + 0.0015/j) \cdot n \Big/ \big[(R_h)^{0.5}\big]\Big\} \qquad [4.4]$$

where: j = bed slope
n = Manning's or Kutter's coefficient of roughness for the channel lining which increases with increasing roughness, $s/m^{1/3}$
R_h = hydraulic mean depth, m

Manning-Kutter (Strickler's) formula:

$$V = \big([R_h]\big)^{2/3} \cdot j^{0.5} \Big/ n \qquad \text{(Metric units)} \qquad [4.5]$$

$$V = \big(1.49 \cdot [R_h]^{2/3} \cdot j^{0.5}\big) \Big/ n \qquad \text{(English units)} \qquad [4.6]$$

where: V = mean velocity of flow = Q/A, m/s
Q = the flow, or the rate of discharge, or the carrying capacity of an open channel, m^3/s

Table 4.26 Values of n in Manning's Formula

Canal Design	Minimum	Maximum
Earth, straight and uniform	0.017	0.025
Dredged earth, smooth	0.025	0.033
Rock cuts, smooth	0.025	0.035
Rock cuts, jagged and regular	0.035	0.045
Rough beds with weeds on sides	0.025	0.04

Compiled from References 49 and 50.

A = cross-sectional area of the channel, m²
R_h = hydraulic radius = A/w_p, m
w_p = wetted perimeter, m
j = hydraulic gradient, slope
n = Manning's or Kutter's coefficient of roughness depending upon the channel lining, s/m$^{1/3}$
1/n = strickler coefficient, m$^{1/3}$/s

C and n are functions of the depth of flow and are related as shown in Equation 4.7.

$$C = [R_h]^{1/6}/n \qquad [4.7]$$

Table 4.26 presents some selected values for Manning's coefficient of friction.[49,50]

The Manning's formula is the most often used equation as it is simple and reasonably accurate.

Equation 4.2 indicates that the mean velocity under open channel flow conditions can be increased by increasing the hydraulic gradient, or through a reduction in the roughness coefficient, or by increasing the hydraulic radius.

Under open channel conditions, the slope of the hydraulic gradient indicates a loss of energy per unit length. The hydraulic gradient slope, in a canal, can be taken as the slope of the bed when measured between two distant points. The slope of the hydraulic gradient of a canal relies on the topography of the area in which the canal is constructed. If the land to be irrigated is very flat it is very difficult to ensure proper flow conditions.

Equation 4.2 illustrates that the velocity of flow is directly proportional to the square root of the bed slope. This means that if the slope is doubled, the velocity will only be increased by a factor of around 41%.

Equations 4.2 and 4.3 indicate that for a given slope and roughness, the velocity of flow increases with the hydraulic radius. Therefore, for a given cross-sectional area the rate of flow is a maximum when the hydraulic radius is also a maximum. In other words, the wetted perimeter is a minimum. Such a section is referred to as the most efficient cross section for a given area. For a given flow, the cross-sectional area is a minimum when the hydraulic radius is a maximum, and the wetted perimeter is a minimum. Thus, this cross-sectional area is regarded as the most efficient one for a given rate of flow. The most efficient section requires the least amount of lining material or surface area and tends to be the least costly.

Of all geometric shapes the circle has the smallest perimeter for a given cross-sectional area. The hydraulic radius of a semicircle is the same as that of a circle. Therefore, a semicircular open channel allows more flow than other cross-sectional shapes, assuming the same area, surface roughness, and slope.

From a practical point of view the simplicity and ease of constructing the canal must be considered when selecting cross-sectional shape. Canals excavated in earth generally have a trapezoidal cross-sectional area, with the side slopes less than the angle of repose of the saturated bank material. Although the optimum side slopes can be determined, nonetheless, the nature of the soil dictates the slope that is utilized. Unlined earth banks normally do not stand at slopes greater than 1.0 vertical to 1.5 horizontal. In a sandy soil the side slopes may be as flat as 1 to 3. In rock, canals can be constructed with vertical side walls. Canals can also be constructed from such materials as timber, concrete, or metal.

The roughness of a channel surface affects the velocity of flow due to friction.

A practical way to increase the velocity of flow is to reduce the roughness of the canal surface. This can be achieved by lining the canal.

The benefits that may be gained by canal lining can be summarized as follows:

- Water saving. Water may be saved, by canal lining, through a decrease in the rate of water seepage and a decrease in the conveyance losses.
- Structural protection. Canal lining offers structural protection against deformation and destruction of the canal embankments and side slopes.
- Reduction in maintenance work. The lining of the canal reduces the amount of maintenance work that has to be carried out and prevents canal scouring, deposition of silt, and the growth of vegetation.
- Land saving. Lining of a canal reduces the land required for the minor channels and drainage ditches.
- Cost benefits. The cost benefits that can be gained through canal lining is a reduction in the canal size as well as smaller secondary structures.
- More crop yield. Canal lining prevents water leakage, and the waterlogging of the soil, as well as the buildup of the salt content in the soil.
- Elimination of mosquito breeding zones. Lining deters the collection of standing water, and thus eliminates breeding places for the vectors.
- Removal of rooted plants. Lining of the canal and its proper maintenance are factors that help control aquatic vegetation. Eggs of vectors and mosquito larvae can not readily find secure shelter under such conditions.

The lining of a canal may be done by using a paved or hard surface, or by employing membranes. Examples of paved or hard surfaces include asphalt, concrete, brick, Portland cement, and stones. Examples of membranes include butyl rubber sheeting, polymers, polyvinyl or polyethylene. Sandy clay soil or resins may also be used as canal linings.

The water velocity flow in a canal, with any type of earth lining, must not exceed a velocity that will cause erosion. The permissible velocity depends on the nature of the material used and is usually less than 0.9 m/s.

A velocity of around 0.65 m/s at the top of the canal removes snails that are the intermediate hosts of schistosomiasis.

The larvae of *Simulium domnosum,* black-fly vectors of onchocerciasis, can breed with water velocities varying between 0.5 and 2 m/s.

Since the snails settle and grow mainly in quiescent and slow-flowing water, and black-fly larvae need a strong current to reach maturity, an increase in water velocity to displace snails or a reduction in velocity to interrupt black-fly larvae development may actually lessen the numbers of these vectors. However, such measures are inappropriate when both classes of vectors are found in the same locality.

The main engineering techniques that may be used to prevent and control vector breeding, such as mosquitoes in an irrigation canal include:

- Use of closed conduits or lined canals for water conveyance.
- Use of irrigation systems such as sprinklers, or drip irrigation.
- Improved canal design. This may be achieved by:
  - Using appropriate canal hydraulic gradients that ensure flow velocities that control vectors.
  - Providing for good channel alignment, with few sharp curves or bends.
  - Application of appropriate canal maintenance. This includes, for instance, repair of embankments, vegetation control, and channel flushing.
  - Implementation of education programs for farmers regarding irrigation practices.

Table 4.27 lists the principal engineering methods that may be adopted for vector control canal design.[51]

Example 4.1

Wastewater collects in a pond which becomes a mosquito breeding site. The problem is addressed by discharging the water in an amount Q, m³/s using a ditch of a trapezoidal cross-sectional area. The canal section has a width of B, central depth of h meters, and has side slope angle of ε with the horizontal (see the accompanying figure). Determine the depth of the ditch that provides the minimum amount of excavation and yields the maximum wastewater flow.

Solution

1. Given: flow = Q, central depth = h, width = B, angle = ε.

Table 4.27 List of Principal Engineering Methods for Vector Control Canal Design

Canal Design
  Straight canals to eliminate standing pools
  Mechanical screening of water intakes to control snails
  Bridged crossing points
  Built-in chemical dispensers at strategic points
  Vegetation clearance
  Seepage control
Reservoir Design
  Minimum night storage reservoirs
  Periodic drawdown
  Vegetation clearance
  Spillways
  Inundation of breeding sites
  Steep, regular banks
Irrigation and Drainage Design
  Increase water velocity, but no channel scouring
  Desilting and vegetation clearance to prevent slow water flow
  Proper field drainage
  Maintenance of field drains
  Lining major water contact points
  Sprinkler irrigation if feasible
  Filling surface water collection areas
  Intermittent irrigation
Settlement Design
  Properly sited villages
  Piped water supplies
  Properly designed and located latrines to ensure effective use and the prevention of sewage entering the water system
  Fencing and zoning
  Children's swimming pools and other recreation sites
  Communal laundries
  Waste disposal
  Home construction materials and design
  Pathways and bridges
  Domestic animal pens
Earthworks
  Diking
  Drainage
  Grading and infilling

Compiled from Reference 51.

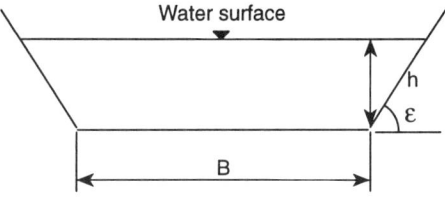

2. Determine the area of the section outlined in the shown figure as:

$$A = B \cdot h + 2([h \cdot h \cdot \cot \varepsilon]/2)$$

$$= B \cdot h + h^2 \cdot \cot \varepsilon \qquad (i)$$

3. Find the wetted perimeter w_p as:

$$w_p = B + 2(h/\sin\varepsilon) = B + (2h/\sin\varepsilon) \qquad (ii)$$

4. Find the width B in terms of the area of the cross section from Equation (i) as:

$$B = (A - h^2 \cdot \cot\varepsilon)/h \qquad (iii)$$

5. Substitute Equation (iii) in Equation (ii), then:

$$w_p = A/h - h \cdot \cot\varepsilon + 2 \cdot h/\sin\varepsilon \qquad (iv)$$

6. From the continuity equation: the flow is related to the velocity and cross-sectional area as: $Q = v \cdot A$
From Manning's equation, $v = ([R_h]^{2/3} \cdot j^{0.5})/n$, then

$$Q = \left\{\left([R_h]^{2/3} \cdot j^{0.5}\right)/n\right\} \cdot A$$

Since the hydraulic radius $R_h = A/w_p$, then:

$$Q = \left\{\left([A/w_p]^{2/3} \cdot j^{0.5}\right)/n\right\} \cdot A = [A]^{5/3} \cdot j^{0.5}/\left(n \cdot [w_p]^{2/3}\right)$$

This equation signifies that the maximum discharge is obtained when the wetted perimeter is minimum. From the same equation upon rearrangement, the area of the cross section may be determined as:

$$A = \left\{\left(Q \cdot n \cdot [w_p]^{2/3}\right)/j^{0.5}\right\}^{0.6}$$

This illustrates again that the minimum area for a given flow is obtained when the wetted perimeter is minimum.

7. From Step 6, the minimum amount of excavation (minimum area) and the minimum wetted perimeter are required for optimum conditions. This condition may be fulfilled when the derivative of Equation (iv) is negligible, i.e., $dw_p/dh = 0$. Therefore, differentiating Equation (iv) with respect to depth, h yields:

$$-(A/h^2) - (\cot\varepsilon) + (2/\sin\varepsilon) = 0$$

$$-(A/h^2) + ([2 - \cos\varepsilon]/\sin\varepsilon) = 0$$

which gives:

$$h^2 = (A \cdot \sin\varepsilon)/(2 - \cos\varepsilon) \qquad (v)$$

Equation (v) gives the central depth for the most economical cross section of the ditch in meters.

Example 4.2

Determine the best angle of the slope of the banks corresponding to the minimum amount of excavation for the channel that is outlined in Example 4.1.

Solution

1. From a practical point of view the angle of the side slope (ε) is usually taken as the steepest angle in order to attain stability. The depth of the ditch is as computed in Equation (v) as: $h^2 = (A \cdot \sin\varepsilon)/(2 - \cos\varepsilon)$.
2. To achieve the criteria mentioned in Step 1, then the derivative of Equation (v) with respect to the side slope angle of the section needs to be negligible, i.e., $dh/d\varepsilon = 0$. Thus, put $u = A \cdot \sin\varepsilon$ and $v = 1/(2 - \cos\varepsilon)$

$$du = A \cdot \cos\varepsilon \quad \text{and} \quad dv = -\sin\varepsilon/(2-\cos\varepsilon)^2 \tag{vi}$$

Therefore, $d(uv) = u \cdot dv + v \cdot du$ yields:

$$\begin{aligned}
d(uv) &= [(-A \cdot \sin\varepsilon \cdot \sin\varepsilon)/(2 - \cos\varepsilon)^2] + [[1/(2 - \cos\varepsilon)] \cdot A \cdot \cos\varepsilon] \\
&= [(-A \cdot \sin^2\varepsilon)/(2 - \cos\varepsilon)^2] + [(A \cdot \cos\varepsilon)/(2 - \cos\varepsilon)] \\
&= A\{(2 - \cos\varepsilon) \cdot \cos\varepsilon - \sin^2\varepsilon\}/(2 - \cos\varepsilon)^2 \\
&= A\{2\cos\varepsilon - \cos^2\varepsilon - \sin^2\varepsilon\}/(2 - \cos\varepsilon)^2
\end{aligned}$$

From trigonometry, $\sin^2\varepsilon + \cos^2\varepsilon = 1$, thus:

$$d(uv) = A\{2\cos\varepsilon - 1\}/(2 - \cos\varepsilon)^2$$

Substituting Equation (vi) in Equation (v) then:

$$2h \cdot dh/d\varepsilon = A\{2\cos\varepsilon - 1\}/(2 - \cos\varepsilon)^2 \tag{vii}$$

From Equation (v) the depth may be determined as:

$$h = \{(A \cdot \sin\varepsilon)/(2 - \cos\varepsilon)\}^{0.5} \tag{viii}$$

Substituting Equation (viii) into Equation (vii), then:

$$dh/d\varepsilon = \left\{A \cdot [2\cos\varepsilon - 1]/\left(2 \cdot [2 - \cos\varepsilon]^2\right)\right\} \cdot \{(2 - \cos\varepsilon)/A \cdot \sin\varepsilon\}^{0.5} = 0 \tag{ix}$$

3. Equation (ix) is valid only when $2\cos\varepsilon - 1 = 0$ or, $\cos\varepsilon = 0.5$. This yields $\varepsilon = 60°$.

Therefore, the best angle of the slope of the banks corresponding to the minimum amount of excavation for the ditch is equal to 60 degrees with the horizontal.

Example 4.3

Show that the hydraulic radius of the most efficient trapezoidal cross section of a channel is one for which the hydraulic radius is equal to one-half its central depth.

Solution

1. Assume a cross section has dimensions as indicated in Example 4.1 with h = depth of section, B = bottom width of the section, and ε = side slope of the channel walls.
2. The wetted perimeter is given by Equation (iv):

$$w_p = B + (2h/\sin\varepsilon) \tag{iv}$$

The area is as determined by Equation (i):

$$A = B \cdot h + h^2 \cdot \cot\varepsilon \tag{i}$$

The width of the channel is determined by Equation (iii):

$$B = (A - h^2 \cdot \cot\varepsilon)/h \tag{iii}$$

The differentiation of the wetted perimeter and equating to zero yields the depth of the cross section as found in Equation (v):

$$h = \{(A \cdot \sin\varepsilon)/(2 - \cos\varepsilon)\}^{0.5} \tag{v}$$

3. Determine the hydraulic radius as given in Equation 4.3:

$$R_h = A/w_p \tag{4.3}$$

From Equation (iii) the width is given by:

$$B = (A/h) - (h \cdot \cot\varepsilon) \tag{iii}$$

But from Equation (v) and upon rearrangement, the area of the cross section may be obtained:

$$A = h^2(2 - \cos\varepsilon)/\sin\varepsilon \tag{x}$$

Therefore, substituting the new form of Equation (x) in Equation (iii) yields:

$$B = \{h^2(2-\cos\varepsilon)/\sin\varepsilon\}/h - h \cdot \cos\varepsilon/\sin\varepsilon$$
$$= h(2 - 2\cos\varepsilon)/\sin\varepsilon = 2h(1-\cos\varepsilon)/\sin\varepsilon \tag{xi}$$

Substitute Equation (xi) in Equation (iv) to obtain an expression for the wetted perimeter:

$$w_p = \{2h(1-\cos\varepsilon)/\sin\varepsilon\} + \{2h/\sin\varepsilon\}$$
$$= 2h(2-\cos\varepsilon)/\sin\varepsilon \tag{xii}$$

Substitute Equations (x) and (xii) in Equation 4.3 to determine the hydraulic radius as:

$$R_h = \{h^2(2-\cos\varepsilon)/\sin\varepsilon\}/\{2h(2-\cos\varepsilon)/\sin\varepsilon\} = h/2 \tag{xiii}$$

4. Equation (xiii) indicates that the hydraulic radius of the most efficient trapezoidal cross section of a channel is one for which the hydraulic radius is equal to one-half its central depth.

I. TEMPORARY METHODS FOR VECTOR CONTROL — ENVIRONMENTAL METHODS

An environmental vector control program must include an environmental management component. A vector control program signifies the periodic and seasonal killing of vectors, hosts, and pests. The control methods involve conditions unfavorable for the breeding of the vectors in their habitats. The objective of a control program is to reduce the number of vectors in a certain locality. Temporary methods are necessary in cases where disease outbreaks, population explosions, or an aroused public demand immediate action or when the people perceive a real or imagined threat is present.

Examples of environmental vector control activities include water salinity changes, stream flushing, water level changes in reservoirs, dewatering or flooding of swampy regions, vegetative growth removal, shading or, just the opposite, exposure to sunlight.

Because of the lack of comprehensive scientific knowledge, environmental vector control programs may have to be conducted on a case by case basis. The problem solving process often utilizes past experiences in order to carry out an effective program.

The most commonly used environmental vector control measures include water management and vegetation control.

Water management can create a biological imbalance for the vector and results in its destruction. The water management of an impoundment or reservoir can be carried out in various ways such as:

1. *Water level fluctuation.* The effect of changing the water surface level of a reservoir can adversely affect not only the vector's breeding site but also interfere with undesirable plant growth.
2. *Sluicing and flushing.* Periodic sluicing and flushing has proved effective in the control of *Anopheles minimus* and *Anopheles maculatus* mosquitoes.
3. *Vegetation control.* Vegetation control affects the biology of many vector species. This type of control can be subdivided into aquatic vegetative control and terrestrial vegetative control. The methods adopted for the control of aquatic plants include water management, cutting programs, and use of biological agents. Biological agents are used to prevent invasion of water bodies by deep-water plants. The biological agents used include herbivorous fish, such as the grass crap. This fish is used in rice fields, and also serves as a source of protein for human consumption. For terrestrial vegetative control there are two main control measures, namely:

   - Vegetation removal. The complete or the selective removal of vegetation is used to eliminate the vector habitat. The disadvantages with this technique include the possibility of removal of economically important plants, and detrimental effects such as erosion or sedimentation.
   - Vegetation change. Changing the present plant species found in bodies of water can deter the breeding of vectors. Where high water tables exist water-loving plants such as eucalyptus trees can be planted.

Figure 4.12 presents a summary of the most important environmental management techniques that can be used for control of vectors.

J. MODIFICATION OF HUMAN HABITATION AND BEHAVIOR FOR VECTOR CONTROL

The modification of human habitation or human behavior is a form of environmental management. This method of vector control is designed to reduce the man-vector-pathogen contact.

Examples of this method include:

- Locating settlements far from vector habitats.
- Mosquito-proofing of houses.
- Personal protection.
- Hygienic measures against vectors.
- Provision of barriers and fences for treatment works such as water supply facilities, wastewater facilities, excreta disposal systems, as well as laundry, bathing, and recreational sites. In this way human-vector-pathogen contact is discouraged or prevented.
- Land use restrictions. Land use restrictions involves the establishment of constraints that will prevent access to infected areas at hazardous times. This procedure can be used in the control of malaria, onchocerciasis, trypanosomiasis, and schistosomiasis. The restrictions of land use may include resettling of people, shifting of people from buffer areas, and establishment of new settlements.
- Household and individual protection. Houses should be vector proof. This is essential in the case where vectors feed during the night. Examples of such measures include:

  - Use of mosquito screens and nets. Screens and nets need to be made of durable fabric. White cloth material is desirable in order to see the mosquitoes resting on it. A thin weave is needed so as to obstruct admission of the vectors. Also the material must provide for good ventilation. Examples of materials used in netting and screens include cotton and synthetic textiles.
  - Closure of cracks. The cementing and closure of cracks and holes in buildings will prevent the entrance of rodents, triatomid, and insects. In this way the human-disease-causing vector contact is reduced.
  - Construction of bridges. The provision of footbridges over streams and canals will contribute to the control of vector-borne diseases such as schistosomiasis.
  - Insect repellents. The use of insect repellents must fulfill certain conditions such as being safe for humans, efficient against mosquito species of the area, have no unpleasant odors, not stain clothes

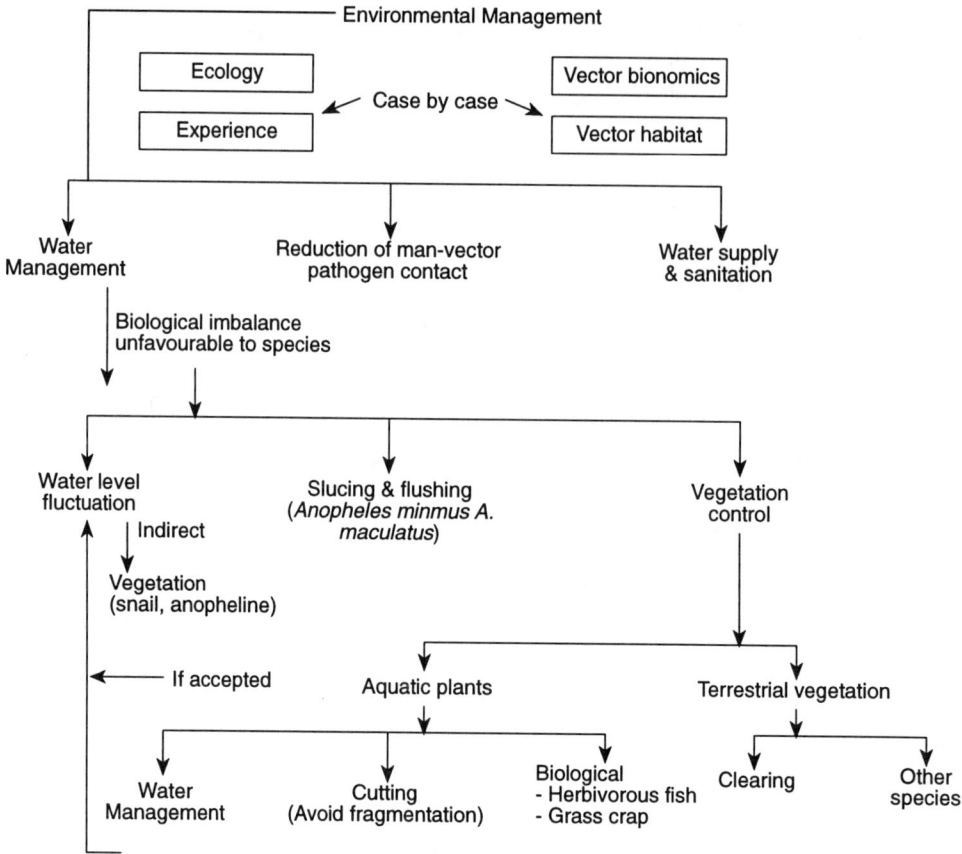

Figure 4.12 Management techniques for vector control.

and be stable. Examples of such repellents include dimethylphthalate, diethyltoluamide, and aerosol dispensers.

- Animals. Wild or domestic animals (zooprophylaxis) are not the usual reservoirs for a given disease. They can, however, be used to divert the blood-seeking vectors away from human hosts. Examples of the use of zooprophylaxis as an environmental technique include horses, cattle, sheep, monkeys, and other domestic animals.

Table 4.28 outlines some characteristics of rural and urban areas in relation to the factors that affect water supply and sanitation in developing countries such as political, social, financial, and technical parameters.[52]

The provision of a safe and convenient water supply and facilities for laundry and excreta disposal are recognized as essential in the control of vector-borne diseases. The effectiveness of water supply and sanitation facilities depends on their proper design, use, and maintenance.

Table 4.29 gives estimates of the decrease of water-related diseases when a potable and palatable water is supplied to a community.[53]

K. INTEGRATED VECTOR CONTROL METHODS

Integrated vector control may be regarded as the implementation of all appropriate technological and management techniques that will result in efficient vector control and is also cost effective.

The integrated control approach has been used successfully in the control of schistosomiasis in the irrigation projects at Rahad and New Halfa in the Sudan.

Table 4.30 gives a summary of the ecological factors which affect the snail hosts and the schistosome.[16]

Tables 4.31 and 4.32 illustrate the environmental management techniques that can be applied for control of vectors and the interventions that are designed for efficient control of different groups of vectors.[13,46]

Table 4.28 Some Characteristics of Rural and Urban Areas in Relation to Factors Affecting Water Supply and Sanitation in Developing Countries

Factors affecting water supply and sanitation	Rural			Urban		
	Scattered population	Nucleated communities	Squatters (unauthorized)	Slums (authorized)	Urban developments	
Political pressure	Very low	Low	Medium-to-low	Medium-high	High	
Political awareness	Very low	Medium	Medium-low	Medium	High	
Manpower availability						
High-level staff	Unlikely	Very limited	Limited	Limited	Available (limited)	
Medium-level staff	Unlikely	Very limited	Limited	Limited	Limited (available)	
Voluntary labor	Possibly available	Often available	Usually not free	Usually not free	Not free	
Maintenance arrangements	Extremely difficult	Very difficult	Very difficult	Difficult	Reasonably difficult	
Repair skills	Low level	Low/medium	Medium	Medium/high	Medium/high	
Spare parts	Not available	Generally absent	—	Sometime available	Sometime available	
Back-up support		Very limited	—	Sometime available	Sometime available	
Income						
Potential for revenue collection	Very low	Difficult but existing	Low and difficult	Possible	Existing and often applied	
Non-public options available						
For water supply	Unprotected surface impoundments and shallow wells	Unprotected surface impoundments and shallow wells	Water vendors	Water vendors	Water vendors and public wells	
For sanitation	Open field defecation	Open field defecation	Street/open areas	Limited/street	limited	
Public land utility						
Availability	Available	Often available	Not available	Very limited	Limited	
Quality	—	—	Low	Low	—	
Required user-involvement in decision making and implementation	High	High	High	Moderate	Limited	
Required government input	Back-up support and advice	Back-up support, advice and coordination	Back-up support, advice and coordination	Coordination/ organization	Coordination	
Power supply	Not available	Limited available	Limited available	Usually available	Usually available	
Road and communication	Difficult	Possible with major time lag	Possible but difficult	Relatively manageable	Relatively easy	
Legislation and control	Very difficult	Difficult	Very difficult	Difficult	Possible	
Industrial development	None	Very limited	Very limited	Considerable	Substantial	
Literacy level	Low	Low	Low	Low	Medium/high	

Compiled from Reference 52.

Table 4.29 Estimates of Water-Related Disease Reduction Through the Supply of a Potable Water Supply

Disease	% Reduction
Guinea worm	100
Typhoid	80
Schistosomiasis	80
Trypanosomiasis gambiense	80
Scabies	80
Yaws	70
Inflammatory eye disease	70
Trachoma	50
Bacillary dysentery	50
Amoebiasis	50
Dysentery, unspecified	50
Tinea	50
Gastroenteritis	50
Skin and subcutaneous infections	50
Paratyphoid and other Salmonella	40
Ascariasis	40
Louse-borne relapsing fever	40

From White, G. F., Bradley, D. J., and White, A. W., *Drawers of Water, Domestic Water Use in East Africa,* University of Chicago Press, IL, 1972. With permission.

Table 4.30 Environmental Factors that Affect Snails and Schistosome

Environmental Factors	Snails		Schistosome	
	Favorable	Adverse	Favorable	Adverse
Temperature, °C	22–26	Below 0, above 40	25–30	Below 20, above 35
Light	Moderate sun, microflora growth	Darkness	Phototrophic, active in daylight hours	Darkness
Seasons	Rainy for breeding	Dry	Rainy	Dry
Water movement	Slow current gentle waves	Swift and violent		
Turbidity	Low to moderately rich in organics	High, silty		
Salinity	Biomphalaria can tolerate gradual increases	Above 1000–2500 mg/L for *Bulinus*		No hatching over 5000 mg/L
Dissolved O_2	Necessary		Necessary	<6.5 mg/L
pH	6–8	Below 5, above 10	5–8	Below 5, above 8

From Chanlett, E. T., *Environmental Protection,* McGraw-Hill, New York, 1979. With permission.

Figure 4.13 gives a summary of the components and the methods that may be considered in an integrated mosquito control program.[49]

L. COMMUNITY AWARENESS IN ENGINEERING DEVELOPMENTAL PROJECTS

Community awareness is vital to motivate participation in a vector control program and to ensure its success. Community participation also helps management collect the needed finances and to allocate human resources.

Community participation may be defined as a continuous, two-way communication process whereby individuals, families, and other community members are involved from the commencement of the project including planning, implementation, and execution of the vector control program.

The general objectives of public participation include:

- Information, education, and liaison.
- Identification of problems, needs, and important values.

Table 4.31 Environmental Management for Vector Control

Vector or intermediate host	Diseases transmitted	Environmental modification						Environmental manipulation						Modification or manipulation of human habitation or behavior				
		Drainage (all types)	Total earth filling	Deepening and filling	Land draining	Velocity alteration	Impoundment	Clearing and burning of terrestrial vegetation	Shading or exposure to sunlight	Water level fluctuation	Sluicing for flushing	Aquatic vegetation control	Salinity regulation	Water supply and sewerage	Screening and bednets	Refuse collection and disposal	Land-use restriction	Improved housing
Anopheles mosquito	Malaria	++	++	++	++	+	–	+	+	+	+	+	++	+	+	+	+	+
Aquatic snails	Schistosomiasis	+	+	++	++	++	–	–	–	+	+	+	+	++	–	–	+	–
Culex and *Aedes* mosquitoes	Filariasis; viral and other diseases	++	+	+	–	+	–	+	+	+	+	+	+	+	+	+	+	+
Blackflies	Onchocerciasis	–	–	–	–	+	++[a]	–	–	–	+	–	–	++	+	–	–	–
Houseflies	Infantile diarrhea	–	–	–	–	–	–	–	–	–	–	–	–	–	–	++	–	+
Tsetse flies	African trypanosomiasis	–	–	–	–	–	–	++	–	–	–	–	–	–	–	–	–	–
Triatomid bugs	Chagas' disease	–	–	–	–	–	–	+	–	–	–	–	–	–	+	–	+	+
Rat fleas	Plague	–	–	–	–	–	–	–	–	–	–	–	–	–	–	++	–	++
Cyclops	Dracontiasis	–	–	–	–	–	–	–	–	–	–	–	–	++	–	–	–	–

Note: – Little or no directly demonstrated value, or not applicable.
+ Partially effective (some species).
[a] small dams = adverse effect; large dams = good effect
++ Primarily effective (most species).

Compiled from Reference 13.

Table 4.32 Examples of Interventions Designed to Control Vectors and the Vector Groups Which Are Affected

Modification	Type of intervention	Anopheline mosquito	Culicine mosquito	Simuliid blackfly	Tabanid horsefly	Phebotomine sandfly	Glossina or tsetse	Cyclops	Aquatic snails
Environmental modifications	Drainage	*	*		*				+
	Earth filling	*	+		*				+
	Deepening and filling	*	+						*
	Land grading	*	+						+
	Velocity alteration	+	+	+		*			+
	Small impoundment			−					
	Large impoundment			*					
Environmental manipulation	Clearing terrestrial vegetation	+	+				*		
	Shading or exposing	+	+						+
	Water level fluctuation	+	+						+
	Sluicing/flushing	+	+						+
	Clearing aquatic vegetation	+	+						+
	Salinity regulation	*	+						*
Modification of human habitation	Water supply/sewerage	+	*	+				*	
	Screening	+	+		*				
	Refuse collection	+	*						
	Zoning	*	+	+		*	*		*
	Improved housing	*	+						

Note: * Primarily effective.
+ Partially effective.
− Detrimental.

Compiled from Reference 13.

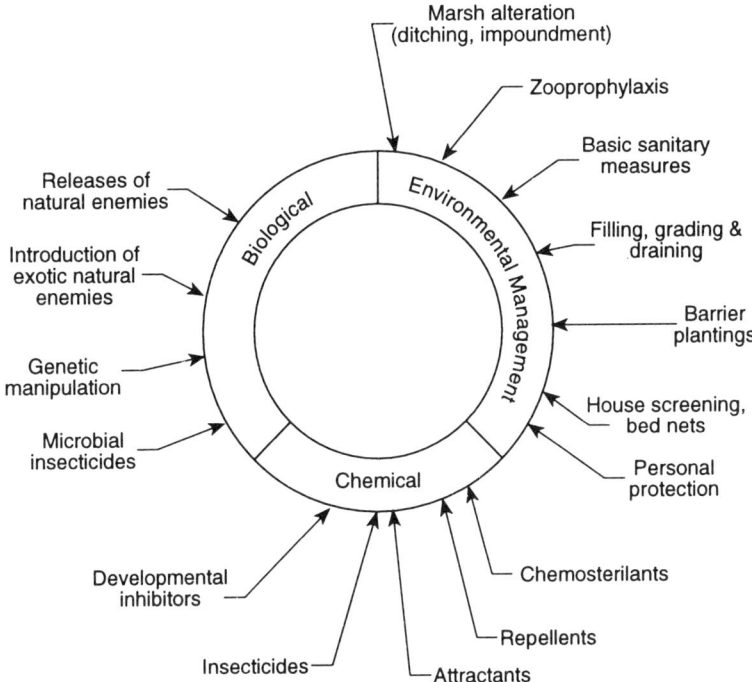

Figure 4.13 Diagram of the components for an integrated approach to mosquito control.[49]

- Idea generation and problem solving.
- Reaction and feedback on proposals.
- Evaluation of alternatives.
- Conflict resolution and consensus.

The following steps may be recommended in order to encourage community enrollment, motivation, and participation in a vector control program:

- Preparatory measures. Preparatory measures and community inspections should to be undertaken to analyze the opinion, belief, and attitude of the inhabitants towards health, vectors, and disease transmission and control. Information concerning the infrastructure and leadership within the community should also to be gathered.
- Specialized policy formulation. The identification of people's demands should be addressed as they relate to health policies and vector control. The formulated program should be presented to the community in an easily understandable format.
- Program approach. The program should include the ideas of the community members and their leaders. The plan should also include the project's objectives, the community's responsibilities, and the techniques to be used to implement and execute the program.
- Budget layout. The plan must incorporate the needed financial support and details regarding expenditures.
- Community education. Community education must be conducted, and include the components explaining the what, why, when, where, and how of the vector control program.
- Follow up and evaluation assessment. The plan must include an evaluation of the project during and after its completion.

Some problems that may be encountered in the community participation process include the following:

- Lack of experience by the community members.
- Confusion of issues by the community.
- Lack of knowledge as to the program's objectives, layout, and method of implementation.
- Erroneous information (rumors and hearsay).
- Inadequate community motivation and lack of interest.

- Budgetary problems.
- Project delays due to social and political issues.
- Tendency of community members to lose interest and thus affect the continuity of the process of community participation throughout the different project phases.
- Uncertainty of the results of the program.

The community participation process can be enhanced by using the analysis and results involved in an environmental impact assessment (EIA).

An environmental impact assessment signifies a study which is conducted to predict the environmental input of a developmental project such as an irrigation scheme or a hydropower dam project. The environmental impact methodology has been incorporated into environmental planning in order to help make decisions that safeguard the technical, social, health, and financial resources of the community.

Usually, the EIA program involves the following activities:

- Identification and assessment of the impacts.
- Explanation and importance of the impacts.
- Presentation of the conclusions of the impact's assessment.
- Selection of suitable monitoring programs.
- Transfer of information to beneficiaries, community leaders, and community members.

The impacts can be both detrimental and beneficial. An EIA focuses on difficulties, conflicts, or natural resources constraints that can influence the feasibility of a project. The continuous environmental assessment process enables appropriate measures to be taken to minimize problems and also outline techniques that can be used to solve such problems.

VIII. HOMEWORK PROBLEMS

A. DISCUSSION TYPE PROBLEMS

1. What is a pathogen?
2. Name the five main categories into which water-related diseases can be divided. Give two examples of each disease in each category.
3. Name the common waterborne bacterial, protozoan, viral, and helminthic diseases.
4. What are the major nutrients required for bacterial growth?
5. Define methemoglobinemia and indicate the mode of action of the component in the water that causes methemoglobinemia.
6. What are the general modes of transmission of vector-borne diseases?
7. What are the major environmental factors that affect the survival time of pathogens?
8. What are the major trace metals in reclaimed water that pose a health hazard.
9. What diseases are associated with the trace metals found in water?
10. What wastewater treatment methods are most effective in removing pathogenic organisms and which treatment methods are least effective?
11. Define health according to the World Health Organization (WHO).
12. What environmental factors influence the reproduction and growth of disease-causing vectors?
13. What is the purpose of a vector control program?
14. What are the major factors or activities used in a vector management control programs? Give examples of these activities.
15. What are some of the problems encountered when chemical methods are used in a vector control program?
16. What are the various types of insecticides used in vector control programs and indicate the advantages and disadvantages of each?
17. Into what two broad categories can pesticides be classified?
18. Compare the chemical and biological methods used in vector control programs.
19. What are the environmental management measures that can be taken to reduce man-vector-pathogen contact in an irrigation scheme in a developing country?
20. Why are there fewer mosquito larvae in dams then in most other bodies of water?
21. What methods of mosquito control constitute environmentally safe procedures?
22. How can you increase the velocity of flow in an open channel? What are the reasons for increasing the flow velocity from a vector control point of view?

23. What are the essential factors that govern the design of an irrigation system?
24. What are the major practical engineering methods that can be used to prevent and control mosquito breeding in an irrigation system? What are the practical methods that can be implemented in an irrigation scheme in a developing country?
25. What hydraulic shape of a canal section is the most efficient for transporting water?
26. Give a mathematical relationship between mosquito breeding and the length of a shoreline for a body of water.
27. What are the major disadvantages of using earth canals for irrigational purposes? How can you reduce or overcome these shortcomings?
28. What are the major disadvantages of using pipe conduits for water conveyance structures in irrigation projects?

B. SPECIFIC MATHEMATICAL PROBLEMS

1. Prove that the shape of the most efficient trapezoidal section, which makes the water discharge a maximum, corresponds to a rectangle whose depth is one-half its width. Also indicate that the most efficient trapezoid is a half-hexagon.
2.a. A rectangular ditch 0.6 m wide at the bottom and a slope of 1 vertical to 500 horizontal is used to convey wastewater from a small pond. Using Chezy's formula, compute the mean velocity of flow and the rate of flow for a water depth of 0.5 m. Assume Chezy's coefficient is 50.
2.b. What volume of water will be drained from this pond in one hour?
3. An open channel has a cross-sectional shape of a trapezoid with a bottom width of 1 m, central depth of 0.75 m, and a wall side slope of 1 to 1.5. Assuming a channel slope of $8 \cdot 10^{-4}$ and a roughness coefficient of 0.03 find the rate of flow in the channel using both the Manning-Kutter's and Chezy's equations.
4. For the channel described in Problem 1 determine the required bed slope to give a flow rate of 0.5 m³/s at an average velocity of 0.75 m/s. Take Manning's coefficient to be 0.025.
5. A trapezoidal channel with a bottom width of B, a central flow depth of h, and a side wall slope of 1 to m, is used to convey water. For a maximum discharge corresponding to a given area prove that the width is given by: $B = 2h \cdot \{(m^2 + 1)^{0.5} - m\}$.
6. A ditch with a top width of 1.2 m carries storm water at a central water depth of 0.6 m. The ditch has side walls with a slope of 1:1. The slope of the bed of the ditch is 8 in 10,000. Using a Chezy's coefficient of 49 determine:
   i) the hydraulic mean depth,
   ii) the velocity of flow, and
   iii) the rate of storm water flow.
7. In Problem 6 find the storm water depth that would occur in the ditch if the storm water flow is doubled.
8. Compute the flow rate of the storm water through the ditch in Problem 6 if the drain has a semicircular cross-sectional area with a radius equal to the central flow depth.
9. A channel with a trapezoidal cross section and side slopes of 60° is required to carry a flow of 1800 m³/hour at a bed gradient of 1 in 1000. Find the depth and width of the cross-sectional area that corresponds to the minimum area. Use Chezy's formula and take the coefficient C to be equal to 60.
10. Rainfall in an area resulted in the creation of a pond with a volume of 50,000 L and a center depth of 50 cm. Mosquitoes are observed to be breeding in the pond. It was decided to drain the water to a stream 0.5 km away. The difference in elevation between the bottom of the pond and the surface water in the stream is 15 cm. Using a Chezy's coefficient of 50 find a suitable width and depth for a ditch that would solve the mosquito problem. (Assume that the local Anopheline vectors will require approximately two weeks to develop from eggs to the adult stage.)
11. A watershed with a 70-ha area diverts its rainfall water to a pond. The average rainfall has an intensity of 10 cm/hour and a duration of 2 hours. The accumulated water is observed to support mosquito breeding, therefore, it is decided to drain the water to a stream 5 km away. The water surface in the stream is 5 m below the bottom of the pond. Using a Manning's coefficient of 0.03 design a suitable drain that would control the breeding of the mosquitoes in the pond. (Take 0.4 as the value of the runoff coefficient for the rational formula; and assume that the local Anopheline vectors would require approximately 10 to 14 days to develop from egg to the adult stage.)

C. TRUE AND FALSE STATEMENTS
Indicate whether the following statements are true, [T], or false, [F].

___ 1. The presence of *E. coli* in a water sample is sufficient proof that fecal contamination exists.
___ 2. All coliforms inhabit the human intestinal tract?
___ 3. A healthy individual can excrete disease-causing organisms?
___ 4. *E. coli* can be a pathogen.
___ 5. *E. coli* is always found in groundwater.
___ 6. Nitrobacter is a bacteria responsible for the oxidation of ammonia to nitrite.
___ 7. The larvae of *Simulium damnosum* need a strong water current in order to reach maturity.
___ 8. Relatively clean, aerated, and fast flowing water favors breeding of Simuliid vectors.
___ 9. An increase in water velocity in canals allow schistosomiasis snails to settle and multiply.
___ 10. Fast currents and an unstable stream bed are favorable sites for snails.
___ 11. Deep water deters mosquito and snails from breeding.
___ 12. Yellow fever is a disease that is related to water and can be transmitted by a mosquito.
___ 13. The mosquito *Aedes aegypti* transmits sleeping sickness (trypanosomiasis).
___ 14. The longer the length of a shoreline the lower will be the potential for mosquito breeding.
___ 15. Vectors may live at a considerable water depth if the water surface is calm and there are no waves or storm activity.
___ 16. With regard to hydraulic radius, the most efficient of all possible hydraulic shapes of a canal section is a semicircle, open at the top and flowing full.
___ 17. A practical way to increase the flow velocity in a canal is to increase the roughness of the canal surface.
___ 18. The principal disadvantage of pipe conduits, used as water conveyance structures for irrigation, is the need to control mosquito breeding in them.
___ 19. Natural courses and canals carrying muddy water have a lesser problem with the growth of vegetation.
___ 20. Most of the organophosphorus compounds which are now used in the chemical control of vectors have a short residual time.
___ 21. Agricultural insecticides used on a large scale enable vectors to develop resistance to a wide range of the chemicals.
___ 22. Use of biological vector control agents (parasites and predators) necessitates periodic new releases of the agents.

D. MULTIPLE CHOICE QUESTIONS
(CIRCLE ONE CORRECT ANSWER FOR EACH QUESTION)

1. Vectors
   a) are pathogens.
   b) are eaters of mosquito larvae.
   c) include animal reservoirs of diseases.
   d) are usually associated with water courses.
   e) utilize human blood as a principal source of food.

2. The bacteriological analysis of a sample of surface water shows that the number of *Escherichia coli* is very high. The result will signify that:
   a) fecal pollution is present.
   b) organic pollution is present.
   c) pathogenic organisms are absent.
   d) filariasis is present.
   e) mottling of teeth will accompany consumption.

3. Bacillary dysentery is a water-washed disease, the most probable causative organism is:
   a) schistosoma.
   b) *Entamoeba histolytica.*
   c) Shigella.
   d) *Salmonella typhi.*
   e) *Sarcoptes scabiei.*

4. Yellow fever is a disease that is related to water and is transmitted by a:
   a) fly.
   b) mosquitoes.
   c) bacteria.
   d) virus.
   e) snail.

5. Which one of the following diseases is transmitted by a mosquito or fly?
   a) tuberculosis.
   b) leprosy.
   c) yellow fever.
   d) typhus.
   e) leptospirosis.

6. Which one of the following diseases is transmitted by a mosquito?
   a) leishmaniasis.
   b) onchocerciasis.
   c) schistosomiasis.
   d) yellow fever.
   e) trypanosomiasis.

7. Which one of the following mosquito control methods constitute an environmentally safe larvicidal program?
   a) land levelling.
   b) water salinity changes.
   c) DDT.
   d) bacterial toxins.
   e) vegetation removal.

8. Triatomid bugs are vectors that transmit the following diseases:
   a) American trypanosomiasis.
   b) schistosomiasis.
   c) river blindness.
   d) infantile diarrhea.
   e) plaque.

9. Which one of the following insect vectors breed in damp grounds?
   a) Culicine mosquito.
   b) Phlobotomine sandfly.
   c) Tabanid horsefly.
   d) Simuliid black fly.
   e) Anopheline mosquito.

10. Drainage of agricultural land is effective in controlling which of the following vectors?
    a) Anopheline mosquito.
    b) black flies.
    c) horseflies.
    d) tsetse flies.
    e) cyclops.

11. When groundwater contains high levels of nitrates, its consumption will:
    a) result in methemoglobinemia in infants.
    b) not quench the thirst.
    c) cause a darkening of the skin.
    d) reduce decay in teeth.
    e) not produce any ill-health effects.

12. Viruses are extremely small, and when inactive they behave much like stable chemical molecules. These microorganisms are peculiar in that they:
    a) use organic or inorganic compounds during growth.
    b) reproduce in absence of a host cell.

c) do not use directly organic or inorganic compounds during growth and reproduce by invading a host cell to redirect the reproductive processes so that more viruses are produced.
d) none of the above.

E. MATCHING QUESTIONS

1. For each of the following questions i), ii), iii), iv) indicate the appropriate A, B, C, or D plan or activity.
   A) Environmental vector control.
   B) Modification or vector control through changes in human behavior.
   C) Environmental impact.
   D) Environmental modification.
   ____ i) Making permanent changes in the environment with the objective of controlling the multiplication of vectors
   ____ ii) Making temporary changes in the environment with the objective of controlling the multiplication of vectors
   ____ iii) Sum of total effects, both beneficial and detrimental, that an intended action exerts on a particular component of the environment
   ____ iv) A form of environmental management that reduces man-vector-pathogen contact.

2. For each of the items in group (A) choose one of the words or phrases in group (B) that is most closely associated with it.

   Group (A) ____ 1) Fly.   ____ 2) Guppy.
              ____ 3) DDT.   ____ 4) Snails.
              ____ 5) Drainage.   ____ 6) Culicine mosquito.
              ____ 7) Malathion.   ____ 8) Stream flushing.
              ____ 9) Sterile males.   ____10) *Simulium damnosum*.

   Group (B) A) Genetic control.
              B) Yellow fever.
              C) Black fly.
              D) Environmental modification.
              E) Organochlorine.
              F) Organophosphate.
              G) Environmental manipulation method.
              H) Biological control of vectors.
              I) Leishmaniasis.
              J) *S. japonicum.*

REFERENCES

1. **Alabaster, J. S. and Lloyd, R.,** *Water Quality Criteria for Freshwater Fish,* Butterworths, London, 1980.
2. **Crook, J.,** Health and Regulatory Considerations, Pettygrove, G. S. and Asano, T., Eds., *Irrigation with Reclaimed Wastewater: A Guidance Manual,* Lewis Publishers, Chelsea, MI, 1985, chap 10.
3. **Parmelee, M. A.,** Milwaukee takes steps to ensure water quality, *AWWA Mainstream,* Vol. 37, No. 5, American Water Works Association, Denver, CO, 1993.
4. **Benenson, A. S., Ed.,** *Control of Communicable Diseases in Man,* 14th ed., American Public Health Association, Washington, D.C., 1985.
5. **Esrey, S. A.,** Report on International Drinking Water Supply and Sanitation Decade (IDWSSD) Impact on Diarrheal Disease, IDWSSD Steering Committee for Cooperative Action, Document prepared for WHO with support from United Nations Development Program on behalf of IDWSSD, World Health Organization, Geneva, July 1990.
6. **Miller, F. D.,** Report on International Drinking Water Supply and Sanitation Decade (IDWSSD) Impact on Schistosomiasis, IDWSSD Steering Committee for Cooperative Action, Document prepared for WHO with support from United Nations Development Program on behalf of IDWSSD, World Health Organization, Geneva, July 1990.
7. **Edwards, P.,** *Reuse of Human Wastes in Aquaculture: A Technical Review,* UNDP/World Bank Water and Sanitation Program, The World Bank, Washington, D.C., 1992.
8. **Feachem, R. G., Bradley, D. J., Garelick, H., and Mara, D. D.,** *Sanitation and Disease: Health Aspects of Excreta and Wastewater Management,* Published for the World Bank by John Wiley & Sons, Chichester, 1983.

9. **Hopkins, D. R.**, Report on International Drinking Water Supply and Sanitation Decade (IDWSSD) Impact on Dracunculiasis, IDWSSD Steering Committee for Cooperative Action, Document prepared for WHO with support from United Nations Development Program on behalf of IDWSSD, World Health Organization, Geneva, July 1990.
10. **Curtis, C. F.,** Ed., *Appropriate Technology in Vector Control,* CRC Press, Boca Raton, FL, 1990.
11. **McJunkin, F. E.,** Water, Engineers, Development and Disease in the Tropics, Agency for International Development, Department of State, Washington, D.C., 1975, 18.
12. **Crook, J.,** Water Reuse in California, *J. Am. Water Works Assoc.,* 77(7), 1985, 60.
13. WHO Expert Committee on Vector Biology and Control, Environmental Management for Vector Control, 4th Report, Technical Paper Series 649, World Health Organization, Geneva, 1980, 16, 22.
14. **Salvato, J. A.,** *Environmental Engineering and Sanitation,* 4th ed., John Wiley & Sons, New York, 1992, 39.
15. IRCWD, Health Aspects of Nightsoil and Sludge Use in Agriculture and Aquaculture, International Reference Center for Waste Disposal, Collaboration Center for Waste Disposal, IRCWD News Number 23, Dec., World Health Organization, Geneva, 1985, 1.
16. **Chanlett, E. T.,** *Environmental Protection,* McGraw-Hill, New York, 1979.
17. **Pettygrove, G. S. and Asano, T.,** Eds., *Irrigation With Reclaimed Municipal Wastewater: A Guidance Manual,* Lewis Publishers, Chelsea, MI, 1985.
18. Manual — Guidelines for Water Reuse, EPA/625/R-92/004, Office of Water, U.S. Environmental Protection Agency, Washington, D.C., 1992.
19. **Dale, J. T.,** World Bank Shifts Focus on Third World Sanitation Projects, *J. Water Pollut. Control Fed.,* 51, 1979, 662.
20. **Pescod, M. B. and Alka, U.,** Urban Effluent Reuse for Agriculture in Arid and Semi-arid Zones, in *Reuse of Sewage Effluents,* Proceedings of the International Symposium Organized by the Institution of Civil Engineers, held in London, England, October 1984, Thomas Telford, London, 1985, 93.
21. WB/UNDP/WHO/IRCWD, Health Aspects of Wastewater and Excreta Use in Agriculture and Aquaculture: The Engelberg Report, International Reference Center for Waste Disposal, Collaboration Center for Waste Disposal, IRCWD News Number 23, Dec., World Health Organization, Geneva, 1985, 11.
22. **Strauss, M.,** Survival of Excreted Pathogens in Excrete and Faecal Sludges, Part II, International Reference Center for Waste Disposal, Collaboration Center for Waste Disposal, IRCWD News Number 23, Dec., World Health Organization, Geneva, 1985, 4.
23. **Feachem, R. G. and Blum, D.,** Health Aspects of Wastewater Reuse, in *Reuse of Sewage Effluents,* Proc. Internat. Symp. Inst. Civil Eng. London, October 1984, Thomas Telford, London, 1985, 237.
24. **Shuval, H. I., Adin, A., Fattal, B., Rawitz, E. and Yekutiel, P.,** Integrated Resource Recovery: Wastewater Irrigation in Developing Countries: Health Effects and Technical Solutions, World Bank Technical Paper Series Number 51, UNDP Project Management Report Number 6, World Bank, Washington, D.C., 1986.
25. **Napolitana, P. J. and Rowe, D. R.,** Microbial Content of Air Near Sewage Treatment Plants, *Water & Sewage Works,* Vol. 113, December 1966, 480.
26. **Sorber, C. A., Bausum, H. T., Schaub, S. A., and Small, M. J.,**, A Study of Bacterial Aerosols at a Wastewater Irrigation Site, *J. Water Pollut. Control Fed.,* 48(10), 1976, 2367.
27. **Bausum, H. T., Schaub, S. A., Bates, R. E., McKim, H. L., Schumacher, P. W., and Brockett, B. E.,** Microbiological Aerosols from a Field-source Wastewater Irrigation System, *J. Water Pollut. Control Fed.,* 55(1), 1983, 65.
28. **Fair, G. M., Geyer, J. C., and Okun, D. A.,** *Water and Wastewater Engineering,* Volumes 1 and 2, John Wiley & Sons, New York, 1968.
29. **Clarke, N. A., Stevenson, R. E., Chang, S. L., and Kobler, P. W.,** Removal of Enteric Viruses from Sewage by Activated Sludge Treatment, *J. Am. Public Health,* 51 (8), 1961, 1118.
30. **Mack, W. N., Frey, J. R., Riegle, B. J., and Mallman, W. L.,** Enterovirus Removal by Activated Sludge Treatment, *J. Water Pollut. Control Fed.,* 34(11), 1962, 1133.
31. **Mara, D. and Cairncross, S.** Guidelines for the Safe Use of Wastewater and Excreta in Agriculture and Aquaculture: Methods for Public Health Protection, Published by the WHO in Collaboration with the United Nations Environment Program, and the IRCWD News Number 24/25, May 1988, 4, World Health Organization, Geneva, 1989.
32. **Cowan, J. P. and Johnson, P. R.,** Reuse of effluent for Agriculture in the Middle East, in *Reuse of Sewage Effluents,* Proceedings of the International Symposium Organized by the Institution of Civil Engineers, held in London, October 1984, Thomas Telford, London, 1985, 107.
33. **Pescod, M. B. and Alka, U.,** *Guidelines for Wastewater Reuse in Agriculture,* Pescod, M. B. and Arar, A., (Eds.) Proceedings of the FAO Regional Seminar on the Treatment and Use of Sewage Effluent for Irrigation, held in Nicosia, Cyprus, October, 1985, Butterworths, London, 1988, 52, 64.
34. WHO Meeting of Experts, Reuse of Effluents: Methods of Wastewater Treatment and Health Safeguards, WHO, Technical Report Series Number 517, World Health Organization, Geneva, 1973.
35. **Tate, C. H. and Trussel, R. R.,** Developing Drinking Water Standards, *J. Am. Water Works Assoc.,* 69, 1977, 486.
36. **Westcot, P. W. and Ayers, R. S.,** Irrigation Water Quality Criteria, *Irrigation with Reclaimed Municipal Wastewater: A Guidance Manual,* Pettygrove, C. S., and Asano, T., Eds., Lewis Publishers, Chelsea, MI, 1985, 3.1.
37. **Broadbent, F. E. and Reisenauer, H. M.,** Fate of Wastewater Constituents in Soil and Groundwater: Nitrogen and Phosphorous, *Irrigation with Reclaimed Municipal Wastewater: A Guidance Manual,* Pettygrove, G. S. and Asano, T., Eds., Lewis Publishers, Chelsea, MI, 1985, chap. 12.

38. **Morishita, T.,** Environmental Hazards of Sewage and Industrial Effluents on Irrigated Farmlands in Japan, in *Treatment and Use of Sewage Effluent for Irrigation,* Pescod, M. B. and Arar, A., Eds., Proceedings of the FAO Regional Seminar on the Treatment and Use of Sewage Effluent for Irrigation, held in Nicosia, Cyprus, 7 9 October 1985, Butterworths, London 1988, 64.
39. **Eisenbud, M.,** *Environmental Radioactivity: From Natural, Industrial and Military Sources,* Academic Press, Orlando, FL, 1987.
40. **Kathren, R. L.,** *Radioactivity in the Environment: Sources, Distribution, and Surveillance,* Harwood Academic, Chur, Switzerland, 1991.
41. **Asano, T., Smith, R. G., and Tchobanglous, G.,** Municipal Wastewater: Treatment and Reclaimed Water Characteristics, *Irrigation with Reclaimed Municipal Wastewater: A Guidance Manual,* Pettygrove, G. S. and Asano, T., Eds., Lewis Publishers, Chelsea, MI, 1985, chap. 2.
42. **Eckenfelder, W. W.,** *Industrial Water Pollution Control,* Series in Water Resources and Environmental Engineering, McGraw-Hill, New York, 1989.
43. **Chen, K. Y., Young, C. S., Jan, T. K., and Rohatgi, N.,** Trace Metals in Wastewater Effluents, *J. Water Pollut. Control Fed.,* 45, 1974, 2663.
44. **Hammer, M.,** *Water and Wastewater Technology,* Prentice-Hall, Englewood Cliffs, NJ, 1986.
45. **Service, M. W.,** *Blood Sucking Insects: Vectors of Disease,* Institute of Biology, Studies in Biology, Number 167, Edward Arnold, London, 1986.
46. WHO/FAO/UNEP Panel of Experts on Environmental Management for Vector Control, Guidelines for Forecasting the Vector-borne Disease Implications in the Development of a Water Resource Project, VBC/86.3, World Health Organization, Geneva, 1987.
47. **Burgess, N.R.H.,** Public Health Pests, *A Guide to Identification, Biology and Control,* Chapman and Hall, London, 1990.
48. Drinking Water Health Effects Task Force, *Health Effects of Drinking Water Treatment Technologies,* Lewis Publishers, Chelsea, MI, 1989.
49. WHO, Manual on Environmental Management for Mosquito Control, with special emphasis on malaria vectors, WHO Offset Publication number 66, World Health Organization, Geneva, 1982.
50. **Douglas, J. F., Gasiorek, J. M., and Swaffield, J. A.,** *Fluid Mechanics,* Longman Scientific & Technical, New York, 1994.
51. WHO Scientific Group, Vector Control in Primary Health Care, WHO Technical Report Series 755, World Health Organization, Geneva, 1987.
52. WHO Study Group, Technology for Water Supply and Sanitation in Developing Countries, Technical Report Series 742, World Health Organization, Geneva, 1987.
53. **White, G. F., Bradley, D. J., and White, A. W.,** *Drawers of Water, Domestic Water Use in East Africa,* University of Chicago Press, Chicago, IL, 1972.

Chapter 5

Wastewater Reclamation and Reuse Treatment Technology

CONTENTS

```
   I. Introduction ............................................................................................................. 167
      A. Sources and Classification of Wastewater ..................................................... 167
      B. Quantity of Wastewater ................................................................................. 170
      C. Population Equivalent ................................................................................... 171
  II. Wastewater Treatment Processes ....................................................................... 171
 III. Reasons For Treatment ....................................................................................... 171
  IV. Advanced or Tertiary Treatment ......................................................................... 174
      A. Filtration ........................................................................................................ 174
         1. Introduction ............................................................................................. 174
         2. Mechanisms of Filtration ........................................................................ 175
            a. Mechanical Sieving and Straining ..................................................... 175
            b. Sedimentation or Precipitation .......................................................... 176
            c. Adsorption Action .............................................................................. 176
            d. Chemical Action ................................................................................ 176
            e. Biological Aspects ............................................................................. 177
         3. Classification of Filters ........................................................................... 177
         4. Hydraulics of Filtration ........................................................................... 179
         5. Filter Clogging ........................................................................................ 183
         6. Rapid Filtration ....................................................................................... 186
            a. Introduction ........................................................................................ 186
            b. Application of Rapid Sand Filters ..................................................... 186
            c. Operation of a Filter .......................................................................... 188
            d. Filter Control Systems ....................................................................... 189
            e. Design Considerations in Filters ....................................................... 189
            f. Cleaning a Filter ................................................................................ 191
            g. Hydraulics of Backwashing a Filter .................................................. 191
         7. Slow Sand Filtration ............................................................................... 193
            a. Introduction ........................................................................................ 193
            b. Principles of Operation of a Slow Sand Filter .................................. 193
      B. Adsorption Process ........................................................................................ 194
         1. Introduction ............................................................................................. 194
         2. Adsorption Isotherms .............................................................................. 194
      C. Desalination ................................................................................................... 197
         1. Introduction ............................................................................................. 197
         2. Classes of Desalination Processes .......................................................... 198
```

- 3. Distillation 198
  - a. Introduction 198
  - b. Principle of the Distillation Process 200
  - c. Double Distillation Systems 202
  - d. Multistage Flash Evaporators 202
  - e. Multiple-Effect Distillation 202
  - f. Solar Distillation 205
    - i. Introduction 205
    - ii. Basic Principles of Solar Distillation 205
- 4. Osmosis 207
  - a. Introduction 207
  - b. Reverse Osmosis 209
  - c. Advantages of the Reverse Osmosis Process 209
  - d. Reverse Osmosis Flow Process 209
  - e. Selection of Membranes 211
  - f. Pretreatment of the Feedwater to an Osmosis Plant 212
  - g. Reverse Osmosis Plants 212
- 5. Electrodialysis 213
  - a. Introduction 213
  - b. Electrodialysis Technology 213
  - c. Electrodialysis Process Disadvantages 217
- D. Ion Exchange Resins 218
- E. Disinfection Process 220
  - 1. Introduction 220
    - a. Physical Methods 220
    - b. Chemical Methods 220
  - 2. Requirements of a Disinfectant 220
  - 3. Chlorination 221
    - a. Introduction 221
    - b. Breakpoint Chlorination 222
    - c. Residual Chlorine 224
    - d. Kinetics of Chlorination 224
- V. Sludge Treatment and Disposal 225
  - A. Sludge Digestion 225
  - B. Sludge Dewatering 226
    - 1. Introduction 226
    - 2. Sludge Dewatering Methods 226
      - a. Drying Beds 227
      - b. Vacuum Filtration 229
      - c. Pressure Filtration 229
      - d. Centrifugation 229
      - e. Elutriation 231
    - 3. Factors Influencing Sludge Dewaterability 231
      - a. Presence of Fine Particles 231
      - b. Solids Content 231
      - c. Protein Content 232
      - d. pH and Particle Charge 232
      - e. Water Content 232
      - f. Shearing Strength 232
      - g. Anaerobic Digestion 232
      - h. Filter Aids 232
    - 4. Filtration of Sludge 233
      - a. Introduction 233
      - b. Filterability Determination 233
      - c. Specific Resistance Concept 233

    d. Specific Resistance Test 235
    e. Capillary Suction Time (CST) 237
    f. Filter Leaf Test 237
   5. Centrifugation 238
  C. Dilution 242
  D. Disposal Into Natural Waters 242
   1. Introduction 242
   2. Oxygen Renewal in Streams (Reaeration) 245
   3. Oxygen Depletion in Streams 246
    a. Biodegradation of Organic Matter 246
    b. Oxygen Depletion Due to the Oxidation of Sludge Deposits 247
   4. Sag Curves in Streams 247
VI. Homework Problems 251
  A. Discussion Problems 251
  B. True and False Statements and MCQs 252
  C. Specific Mathematical Problems 253
  D. Suggested Excercises 257
References 258

I. INTRODUCTION

Wastewater may be defined as "A combination of the liquid or water-carried wastes removed from residences, institutions, and commercial, and industrial establishments together with such groundwater, surface water, and storm water as may be present."[1]

Wastewater or sewage is a complex solution which may contaminate the human environment which includes air, water, food, land, and shelter. Therefore, appropriate treatment and disposal is a prerequisite in order to avoid development of such conditions that may endanger the public health and welfare.

Generally, the problems associated with improper wastewater discharges include the following:[1-15]

- Wastewater affects natural water quality through the production of taste, odor, and malodorous gases. The gases that may be produced include carbon dioxide (CO_2), hydrogen sulfide (H_2S), methane gas (CH_4), ammonia (NH_3), and other trace gases such as hydrogen (H_2) and nitrogen (N_2).
- Wastewater contains pathogenic microorganisms that cause many diseases (see Chapter 4). Likewise, wastewater may introduce other public health effects, especially long-term physiological effects such as those developed from newly introduced organic substances.
- Wastewater sludges may introduce highly persistent detergents, pesticides, and other toxic wastes and compounds.
- Massive quantities of solids may produce objectionable and dangerous levels of sludge on the bottom of water bodies or along their banks. These solids add to the chemical, biological, and physical degradation of natural water courses.
- Wastewater containing grease and oils render bathing sites unusable, present extra problems for treatment works, produce unsightly conditions, and interfere with the processes of biodegradation.
- Wastewater may produce eutrophication or the enrichment of water by plant nutrients, etc. (see Figure 5.1).

For the aforementioned reasons, the immediate and nuisance-free removal of wastewater from its sources of generation, succeeded by proper treatment and final disposal, is not only desirable but also necessary for appropriate environmental sanitation.

A. SOURCES AND CLASSIFICATION OF WASTEWATER

The sources of origin and generation of wastewater may be grouped as follows (see Figure 5.2):

- Domestic (sanitary) wastewater. Domestic wastewater includes discharges from residences and commercial, institutional, and similar facilities.
- Industrial wastewater. Industrial wastewater signifies the industrial wastes generated from industrial localities. It varies with the type and size of the industry and other factors affecting production and processes.

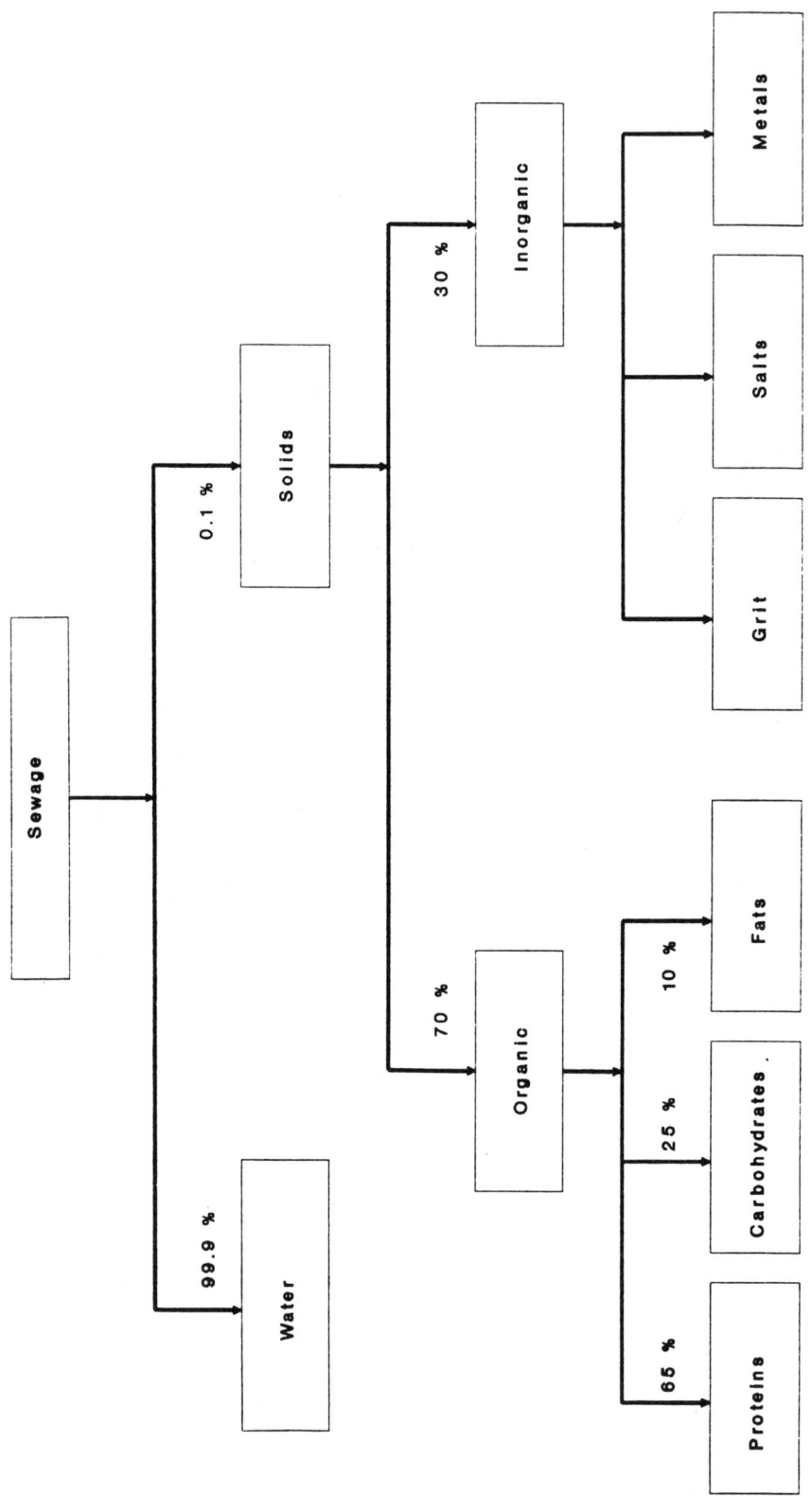

Figure 5.1 Composition of sewage. (From References 7 and 16. With permission.)

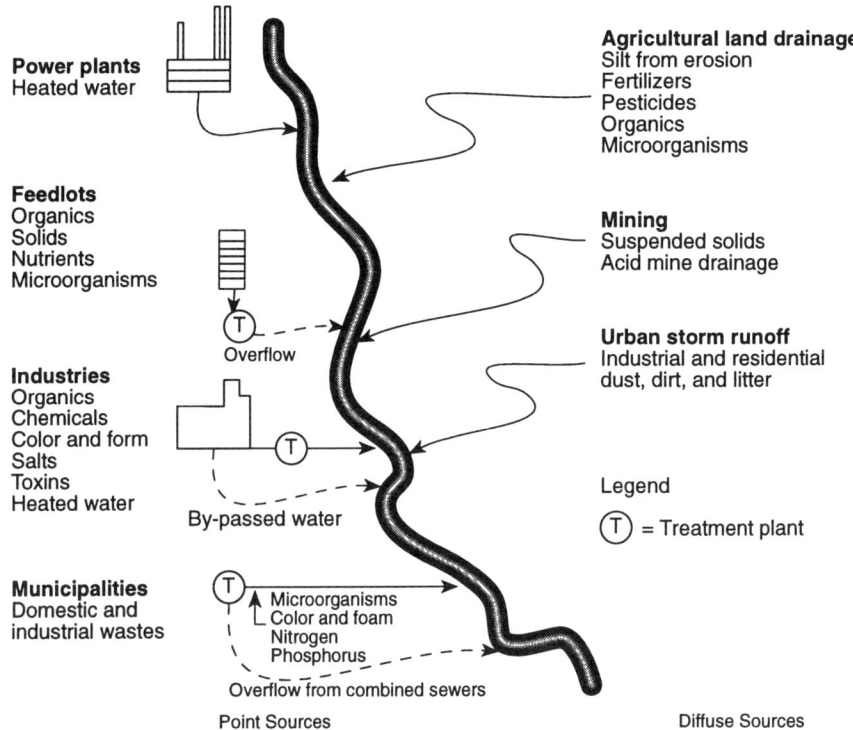

Figure 5.2 Principal sources of water pollution. (From Hammer, M. J., *Water and Wastewater Technology*, John Wiley & Sons, New York, 1977. With permission.

- Infiltration/inflow. Infiltration as a class of wastewater addresses extraneous water that enters the sewer system from the ground through various means. Likewise, it includes storm water that is discharged from sources such as roof leaders, foundation drains, and storm sewers.
- Storm water. Storm water refers to the water that results from precipitation runoff.

Calculations of domestic sewage flows are usually based on the concept of dry weather flow (DWF). DWF is the total average discharge of sanitary sewage and is the normal flow in a sewer during the dry weather. It depends on parameters such as population size, density, and pattern of growth; type and size of area served; rate of water supply and usage; groundwater infiltration; and affecting climatic factors such as the rate of evaporation.

DWF may also be defined as the average daily flow in the sewer after several days during which rainfall has not exceeded 2.5 mm in the previous 24 hours.[17] Equation 5.1 illustrates the mathematical formula for estimating the value of the DWF for a certain locality.

$$DWF = (POP \cdot Q) + I_r + T_w - E_v \qquad [5.1]$$

where   DWF = dry weather flow, L/d
   POP = population served by the sewer. It is usually determined from census data with adjustments made for any future growth, expansions, or development
   Q = average daily per capita water consumption, L/c/d
   I_r = average infiltration into the sewer owing to poor joints or previous material, L/d. Usually, I_r varies between 0–30% of DWF. I_r can be estimated as L/d per kilometer length of sewer for different ages of the sewer
   T_w = average trade waste discharge, L/d. T_w may be estimated from local figures for metered industrial water supplies
   E_v = rate of evaporation; in hot climates the rate of evaporation may reach 30 to 50% of the water consumption

In warm climates only 70 to 80% of the domestic supply is assumed to enter the sewers, the remainder being lost by evaporation or in garden watering, etc.

The DWF may be estimated from data of per capita water consumption as follows:

$$\text{DWF} = (80 \text{ to } 90\%) \cdot \text{water consumption} \cdot \text{population} \quad [5.2]$$

The Babbit formula gives a reasonable estimate of the maximum daily wastewater flow, as outlined in Equation 5.3.

$$\text{Maximum daily wastewater flow} = (2 \text{ to } 4) \cdot \text{DWF} \quad [5.3]$$

The minimum wastewater flow could be taken as:[18]

$$\text{Minimum flow} = (30 \text{ to } 50\%) \cdot \text{average flow } [\text{for small communities}]$$

$$= (66 \text{ to } 80\%) \cdot \text{average flow } [\text{for large communities;}$$

$$\text{i.e., greater than } 100{,}000] \quad [5.4]$$

The ratio of maximum flow to average flow is referred to as a peaking factor (pf).[19] As such:

$$\text{pf} = Q_{max}/Q_{av} \quad [5.5]$$

where: pf = peaking factor
Q_{max} = maximum flow, m³/d
Q_{av} = average flow, m³/d

This factor is estimated as follows:

$$\text{Peaking factor} = \text{pf} = 5/(\text{POP})^{0.167}$$

$$[\text{for a population size of 200 up to 1000}] \quad [5.6]$$

where: POP = design population, in thousands.

Example 5.1

A sanitary sewer is to serve an area with a population of 5000. The average daily sewage flow is 0.5 m³ per capita. Daily infiltration in the area is estimated to be 80 m³ per kilometer length of sewer. Taking a total length of the sewer to be 5 km, find the dry weather flow.

Solution

1. Given: POP = 5000, Q = 0.5 m³/d.c, I_r = 80 m³/length of sewer, l of sewer = 5 km.
2. Use Equation 5.1, DWF = (POP · Q) + I_r + T_w − E_v.
3. Find the average dry weather infiltration into the sewer as: I_r = 80 m³ · 5 km of sewer length = 400 m³/d.
4. Assume both the average trade waste discharge, T_w, and the evaporation rate, E_v, to be negligible.
5. Substitute the given information of Step 1 in Equation 5.1 to find the DWF as:

$$\text{DWF} = 5000 \cdot 0.5 + 400 = 2900 \text{ m}^3/\text{d}$$

B. QUANTITY OF WASTEWATER

The factors that affect the quantity and flow pattern of a wastewater include population increase, population density, and population density change; water use; water demand; water consumption;

industrial and commercial requirements; expansion of service geographically; groundwater geology of the area; and topography of the location.

C. POPULATION EQUIVALENT

Industrial wastewaters are often expressed in terms of the population equivalent concept.

The population equivalent (PE) of a sewage is an expression of some characteristic such as biochemical oxygen demand (BOD) or suspended solids (SS) as a per capita contribution of the sewage as compared to some characteristic of the per capita contribution of some standard sewage. The domestic sewage was usually taken as the standard sewage.

Thus the PE of any sewage would be the number of persons who would be responsible for the sewage that has the same characteristics of BOD or SS as the standard sewage.

Equation 5.7 gives a general expression for the population equivalent:

$$PE = BOD_5 \text{ of waste} \cdot \text{flow rate/BOD of standard sewage} \qquad [5.7]$$

In the U.K. it is assumed that the average person exerts a BOD_5 load of 60 g/day, while in the U.S. the assumption is that the average person exerts a BOD_5 load of 80 g/day.

Advantages of the population equivalent are

- It is a useful index of the strength of industrial sewage for the purposes of treatment at a municipal sewage treatment plant.
- It is of value as a measure for assessing changes of waste treatment against industries instead of only considering the volume of the sewage.
- It aids the assessment of charges for industries.

Example 5.2

Compute the population equivalent (PE) of a certain factory that produces $15 \cdot 10^5$ L of wastewater each day. The wastewater has a 5-day BOD of 400 mg/L.

Solution

1. Given: wastewater flow, $Q = 15 \cdot 10^5$ L/d, BOD = 400 mg/L.
2. Use Equation 5.7 as:
   $PE = BOD_5$ of waste · flow rate/BOD of standard sewage
3. Assume the average person exerts a 5-day BOD load of 0.06 kg/d.
4. Substitute gained information in the equation mentioned in 2 above to find population equivalent as:

$$PE = \left(400 \cdot 10^{-3} \text{ g/L} \cdot 15 \cdot 10^5 \text{ L/d}\right) / \left(0.06 \cdot 10^3 \text{ g/d}\right) = 10,000$$

II. WASTEWATER TREATMENT PROCESSES

Wastewater treatment units can be classified according to their capacity (see Figure 5.3):

1. *Small wastewater treatment plants.* Small wastewater treatment units address wastewater treatment as applied to individual households, or small communities. Usually they are on-site treatment and disposal units.
2. *Large wastewater treatment units.* Large wastewater treatment plants are wastewater works that govern the discharge and treatment of large population sectors. Sewage is collected from different localities and diverted to a central treatment plant.

Table 5.1 gives a quick review of the most essential unit processes involved in water and wastewater treatment plants.

III. REASONS FOR TREATMENT

The major reasons for wastewater treatment may be summarized as follows:

- Reduction in the spread of communicable diseases — to be achieved through the elimination or reduction of pathogens in the sewage.

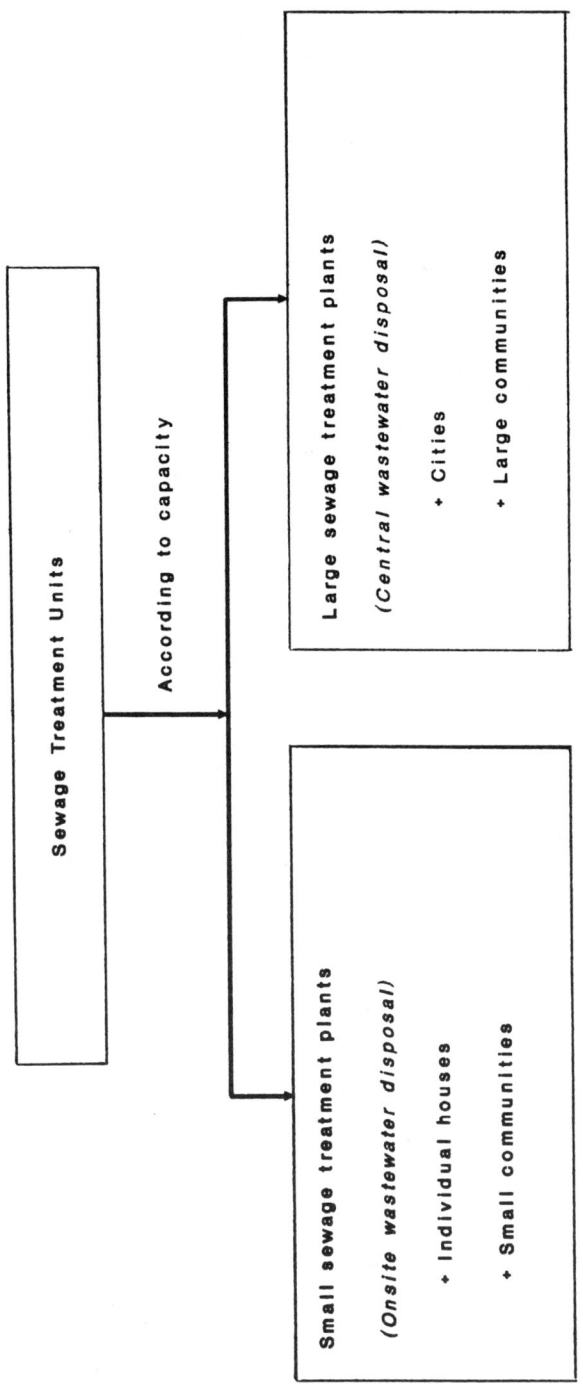

Figure 5.3 Sewage treatment units.

Table 5.1 Unit Processes in Water and Wastewater Treatment

Water Treatment

Screening	Bars, mesh, or strainer for the removal of large coarse suspended and floating matter
Flocculation	Mechanical mixing to favor agglomeration of solid particles
Coagulation	Addition of chemicals to coagulate suspended and colloidal solids
Sedimentation	Settling out of flocculated suspended substances under gravitational forces
Flotation	Removal of particles with a mass density lower than that of the surrounding fluid
Sand filtration	Filtering out remaining suspended matter
Disinfection	Addition of disinfectant to kill harmful microorganisms
Sludge treatment	Thickening by gravity and then disposing
Other units	Water softening, pH adjustment, fluoridation, etc

Wastewater Treatment

Preliminary Treatment

Screening	As above
Grit removal	Removing grit and inorganic material (e.g., sand) but not organic matter
Storm overflow	Diverting sewage in excess of capacity of treatment plant to stormwater holding or storage tanks

Primary Treatment

Primary sedimentation	Settling of suspended solids (only 40 to 60% are removed), no chemicals are added

Secondary Treatment

Oxidation of organic matter	Biodegrading organic matter through the action of microorganisms in a biological treatment unit such as activated sludge, trickling filter plant, waste stabilization ponds, etc
Nitrification	Conversion of inorganic materials by bacterial cultures, or the oxidation of ammonium ion
Secondary sedimentation	Settling out of sludge containing microorganisms to produce a treated effluent

Tertiary Treatment (Effluent Polishing)

Finalizing treatment	Polishing of effluent by systems such as sand filters, microstrainers, etc

Sludge Treatment

Anaerobic digestion	Decomposing thickened sludge under anaerobic conditions
Gravity thickening	Thickening of primary and secondary sewage sludges
Mechanical dewatering	Removing water from sewage sludges by methods such as centrifuges, pressure or vacuum filters, etc
Drying beds:	Drying sewage sludge in open atmosphere

Sludge Disposal

Composted to be used as a soil conditioner (digested only)
Dumped at sea (undigested)
Incinerated (normally undigested but thickened)
Landfilled (preferably digested and dewatered or dried)
Used for stabilizing sand dunes

Compiled from References 1, 2, 4–7, 10, 12, 18, 20, 21, 29, and 33–35

- Prevention or reduction of pollution that may enter the surface or groundwater sources.
- Stabilization of sewage without causing any odors or nuisances and without endangering the public health.
- Water reuse aspects, or for waste by-product recovery.

Figure 5.4 outlines a recommended layout for a wastewater treatment system in relation to the source of the wastewater and the ultimate disposal of this wastewater.

Figure 5.5 gives an outline of a recommended sequence of treatment unit operations (controlled by physical forces) and unit processes (governed by biological and or chemical reactions) in a treatment work.

Primary and preliminary treatment methods are beyond the scope of this book. More information may be gathered in References 1–3, 5, 6, 8–14, 17, 19–27, and 29–32.

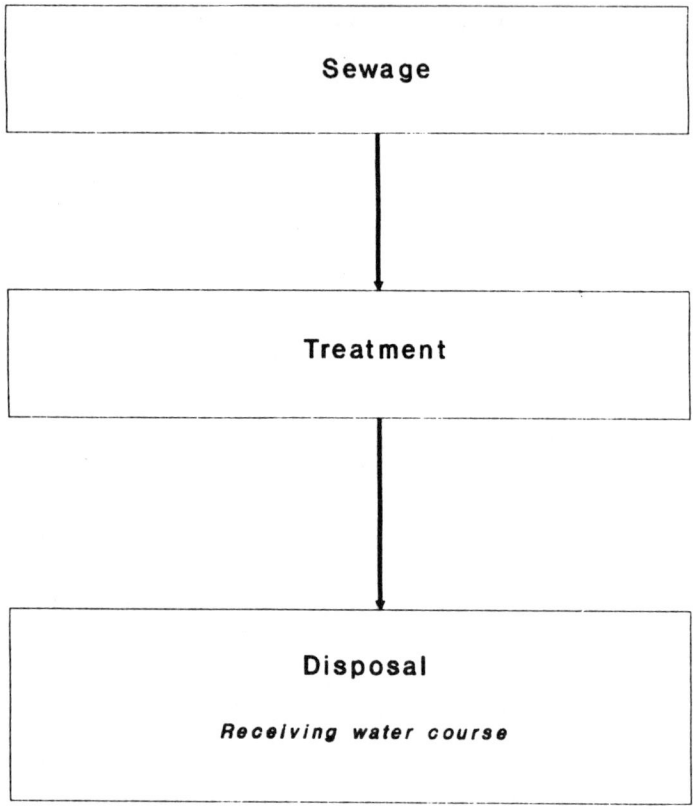

Figure 5.4 Sewage treatment.

IV. ADVANCED OR TERTIARY TREATMENT
A. FILTRATION
1. Introduction

Filtration is defined as the separation of liquid from solid particles contained in them through a partition having pores of such a size that it obstructs the passage of the particles, while allowing the egress of the liquid.

In the field of water purification, the process of filtration improves water quality by:

- Partial removal of suspended solids and colloidal particles.
- Alteration of the chemical properties of the constituents.
- Reduction in the number of pathogenic bacteria and other microbes.
- Removal of color, tastes, odor, iron, and manganese.

The porous filter media to be utilized ought to possess roughly the following properties and characteristics:

- The media need to be low in cost.
- The media are readily available and in sufficient amounts.
- The media to be used in the filters must be of an inert material.
- The selected media must be easily cleaned for reuse.
- The media must be able to withstand the existing pressures.

Many materials fulfilling the above-mentioned specifications have been employed in filtration processes. These filter media materials include sand, anthracite, crushed stone, glass, plastics, porous concrete, cinders, diatomaceous earth, etc. Sand is one of the excellent elements often used in the filtration process.

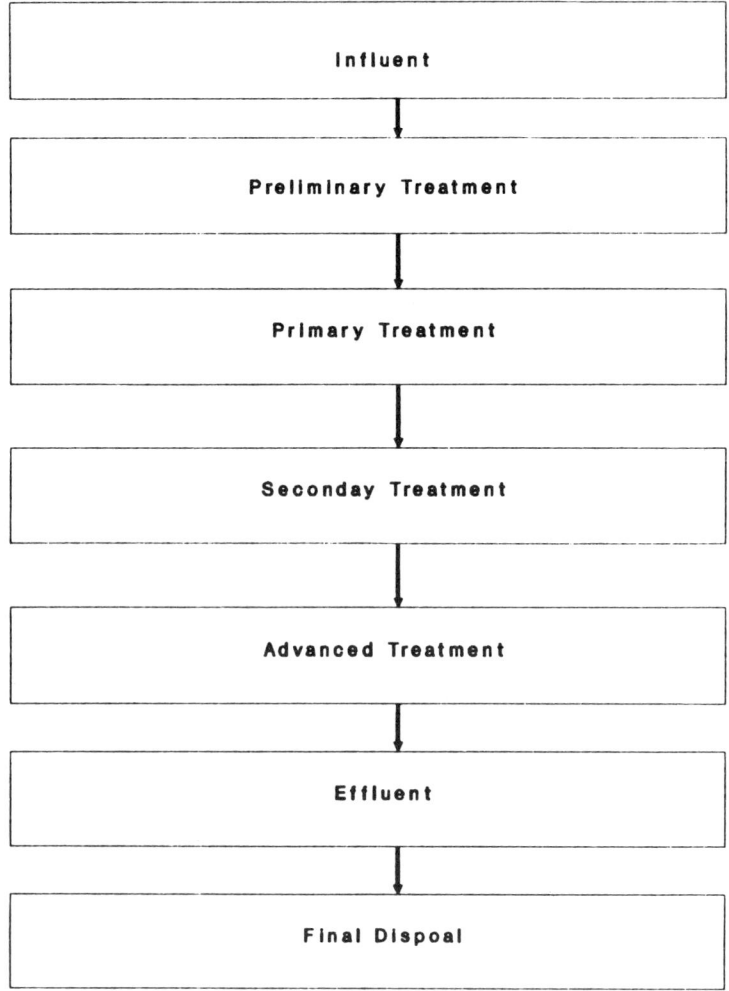

Figure 5.5 Sequence of unit operations and processes in a wastewater treatment facility.

2. Mechanisms of Filtration

The improvement in water quality by filtration is achieved through the reduction of impurities. This improvement is brought about by the action of different mechanisms such as mechanical sieving or straining, precipitation, adsorption, chemical actions, and biological processes.[1,6,14,29,30,35–40]

a. Mechanical Sieving or Straining

Mechanical sieving is a surface phenomenon that pertains to the mechanical segregation of suspended solids. Solids that are to be separated have a diameter bigger than the size of the opening between the sand grains. Straining action is encountered in the first few centimeters of the top of the filter medium. With the commencement of filtration straining only retains solid particles in the water that are big enough to be captured in the pore spaces of the sand particles (see Figure 5.6). The captured particles then start to form a screen (mat) that has smaller pore openings than the original filter medium. In this way, smaller particles suspended in the water are trapped by the formed mat. The newly trapped particles immediately begin acting as a section of the created screen. Therefore, the removal efficiency due to straining tends to increase with the time of filtration.

The parameters that influence the straining action include velocity of filtration, time, specific gravity of particles, water velocity gradients, nature of the particles, and nature and characteristics of the filter bed.

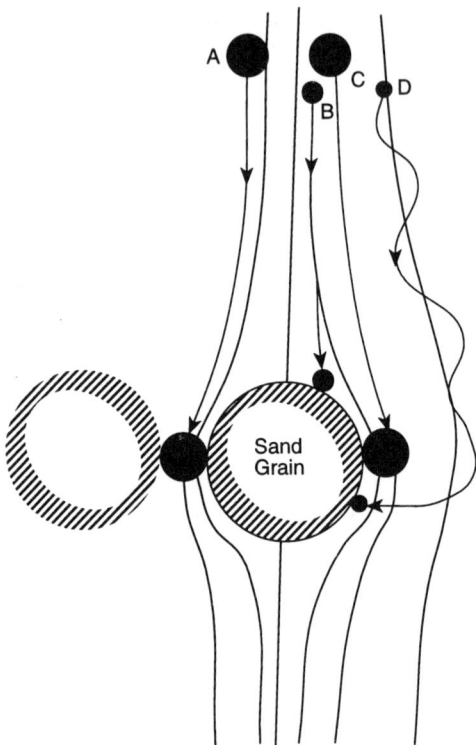

Figure 5.6 Filter bed mechanisms for solids removal: (A) straining, (B) sedimentation, (C) interception, (D) diffusion. (From Hammer, M. J., *Water and Wastewater Technology*, John Wiley & Sons, New York, 1977. With permission.)

b. Sedimentation or Precipitation

Sedimentation has its effect along the water surface to enable the withdrawal of large and heavy suspended solids. The settling particles do not obey the trajectory followed by the fluid streamline but settle on the sand grains.

Settlement resembles that in sedimentation basins. The factors that influence the straining action include velocity of filtration, settling velocity of particles (a function of the specific gravity of particles, gravitational attractive forces, viscous action, diameter of the particles, temperature, degree of flocculation), depth of filter, scouring action, degree of turbulence and flow conditions, together with the purifying capacity of the filter.

c. Adsorption Action

Adsorption is an effect associated fundamentally with the surface properties of the solid. During filtration the adsorption process facilitates the removal of finely divided suspended matter as well as colloidal and molecular dissolved substances.

Adsorption forces play a role over extremely small distances of not more than 0.01 to 1 μm, while the water film surrounding the filter grain has a much greater thickness. Nevertheless, adsorption forces constitute the greatest purifying action in the filter.[29]

Efficient removability of particles is achieved due to the action of other controlling parameters. These parameters cause impurities to move to the neighborhood of the filter grain surfaces. The controlling parameters include inertia, gravity, turbulence, and diffusion. Inertia encompasses particles heavier than water to follow their original direction of movement. Gravity tends to cause particles to migrate resulting in their downward movement. Diffusion addresses the random motion of particles as caused by collisions with the surrounding molecules and hydrodynamic forces.

d. Chemical Action

The chemical processes permit dissolved impurities to be removed by the aforementioned mechanisms. This is achieved basically by the breaking down of the impurities to insoluble and/or less harmful compounds. After these chemical changes other mechanisms such as straining, sedimentation, and adsorption may remove the particles from the flowing water.

Factors that govern the chemical action include water characteristics, dissolved oxygen concentration, microorganisms concentration, and species and presence of catalytic compounds.

In the presence of oxygen organic matter can be converted aerobically:

$$C_5H_7O_2N + 5O_2 \rightarrow H_2O + 4CO_2 + NH_4^+ + HCO_3^-$$

Ammonia is further oxidized with the help of bacteria to form nitrate:

$$NH_4^+ + 2O_2 \rightarrow H_2O + NO_3^- + 2H^+$$

Soluble ferrous compounds are converted into insoluble ferric oxide hydrates:

$$4Fe^{2+} + O_2 + 10H_2O \rightarrow 4Fe(OH)_3 + 8H^+$$

For the removal of manganese the reactions involved are as indicated below:

$$2Mn^{2+} + O_2 + 2H_2O \rightarrow 2MnO_2 + 4H^+$$

e. Biological Aspects

Biological reactions signify the biodegradation of organic matter and the utilization of certain inorganic substances by microorganisms which multiply selectively throughout the filter bed.

The factors of concern include availability of nutrients, dissolved oxygen concentration, temperature variation, filtration time, and filter depth.

Since many viruses, bacteria, and fine clay particles are about 1 µm in size, filtration may not be an effective tool for their removal. Rapid filters cannot produce a safe water from a bacteriological point of view, the reduction of the *E. coli* level being a factor of 2 to 10 only. This makes a preceding coagulation, a subsequent slow sand filtration, or postchlorination a necessity for this purpose.[29,30,35]

Example 5.3

A treatment plant includes aeration and rapid filtration units to treat groundwater that contains 5 mg/L Fe^{2+} and 1.5 mg/L ammonia. The oxygen concentration increased to 8.5 mg/L after aeration. Assuming that the rapid filter is capable of oxidizing the Fe^{2+} and the ammonia, determine the oxygen concentration of the effluent emerging from the filter.

Solution

1. Given: C_{iron} = 5 mg/L, $C_{ammonia}$ = 1.5 mg/L, oxygen concentration entering filter = C_i = 8.5 mg/L.
2. For biological oxidation of the compounds in the filter the following values of oxygen concentrations are needed:
    O_2 required to oxidize 1 mg of iron II = 0.17 mg
    O_2 required to oxidize 1 mg of ammonia = 3.6 mg
3. Using the information presented in 2 above, then:
    amount of oxygen required to oxidize 5 mg of iron II = 0.17 · 5 = 0.85 mg
    amount of oxygen required to oxidize 1.5 mg of ammonia = 3.6 · 1.5 = 5.4 mg.
4. Determine the effluent concentration of oxygen:
    C_i – amount used for oxidation of substances = 8.5 – 0.85 – 5.4 = 2.25 mg/L.
    This demands further aeration to increase the oxygen concentration to acceptable values.

3. Classification of Filters

Sand filters may be classified as outlined herein:

- According to the method and interval of cleaning and the velocity of filtration, sand filters are classified as slow sand filters (SSF) and rapid sand filters (RSF).

Table 5.2 Comparison of Rapid and Slow Sand Filters

Parameter	RSF	SSF
Reason for filtration	Removal of SS, reduction of pathogens.	Reduction in pathogens, finalizing treatment
Location in treatment plant	After coagulation or sedimentation	Without or after coagulation, or after RSF
Efficiency	Depends on raw water quality and design parameters	Depends on raw water quality and design parameters
Raw water turbidity needed	High	Moderate(<15 NTU)
Design period	10–15 years	10–15 years
Life span	Relatively long	Relatively long
Filtration rate ($m^3/m^2/hr$)	5–15	0.1–0.2
Total area (A_t)	Flow rate/filtration velocity	Flow rate/filtration velocity
Area of each filter	($A_t/n - 1$), ($A_t/n - 2$)	($A_t/n - 1$), ($A_t/n - 2$)
Dimensions[37]		$1 = (2A_t/[n+1])^{0.5}$
		$B = (n+1)L/2n$
Effective grain size (mm)	0.4–3	0.15–0.35
Uniformity coefficient	>1.2–1.5	<3–5 (2.5)
Bed thickness (m)	0.6–3	0.8–1.2
Supernatant water level (m)	1–1.5	1–1.5
Minimum depth before resanding (m)	Depends on treatment	0.5
Number of filters[29]	$12(Q)^{0.5}$ (Q in m^3/s)	$15(Q)^{0.5}$
	{Minimum of two filters}	
Operation period	24 hr/d (intermittent not recommended)	24 hr/d (intermittent not recommended)
Interval between successive cleanings	12–72 hr	20–60 d or more
Filter bed resistance	1.5–4 m	
Method of cleaning	Backwashing (water and /or air)	Scraping top 0.5–2 cm layer
Sludge removal	Manual, mechanical, hydraulic	Manual, mechanical, hydraulic
Filter material	Concrete, brick, plastics, etc.	Concrete, brick, plastics, etc.
Maintenance	Continuous	Continuous
Hazards	Algal growth, change in water quality, clogging	Algal growth, change in water quality, clogging
Control measures	Head loss, flow rate, turbidity	Head loss, flow rate, turbidity
Important quality parameters to be analyzed	Turbidity, bacteriological quality	Turbidity, bacteriological quality

- According to filter bed media, sand filters are grouped as single medium, dual media, tri-media, and multimedia filters. Dual and tri-media filters are used in wastewater treatment because they permit solids penetration into the bed and because they offer more storage capacity. Therefore, dual and tri-media filters increase the required time between backwashings. Multimedia filters tend to spread the head loss buildup over time and thus permit longer filter runs.
- According to the water movement, filters are classified as gravity and pressure filters.

Table 5.2 gives a general comparison between rapid and slow sand filters.

Example 5.4

Water is introduced to a treatment plant at an hourly rate of flow of 900 m^3. The effluent from the sedimentation units is introduced to the rapid filters at the filtration rate of 8.5 $m^3/m^2/hr$. Determine the number of filters to be used, the filtration area, and the required area of each filter.

Solution

1. Given: Q = 900 m^3/hr, v_f = 8.5 m/hr.
2. Determine the number of rapid sand filters needed by using the empirical equation:
   $n = 12(Q)^{0.5} = 12(900/[60 \cdot 60])^{0.5} = 6$ filters
   Take 8 filters (two filters are to be used as standby filters).
3. The total surface area of the filters can be computed from the equation:
   $A = Q/V_f = 900/8.5 = 106$ m^2.
4. Determine the unit filter area as:
   $A_n = A/(n - 2) = 106/(8 - 2) = 17.7$ m^2.

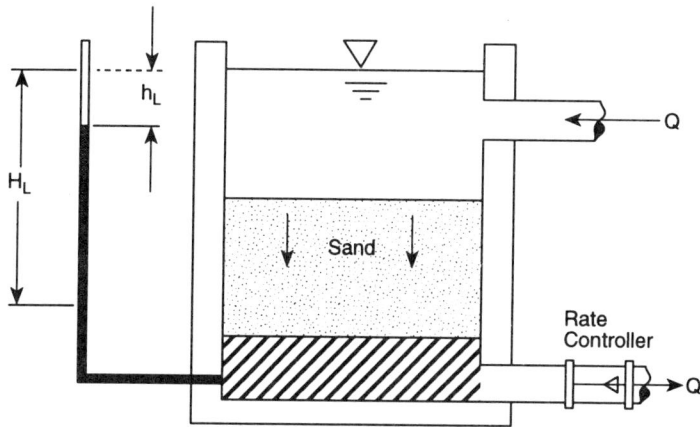

Figure 5.7 Schematic diagram of the head loss in a filter. (From Hammer, M. J., *Water and Wastewater Technology*, John Wiley & Sons, New York, 1977. With permission.)

4. Hydraulics of Filtration

The head loss through the filter media is one of the chief factors to be considered in the design of a filter. As the sand grains become clogged the head loss increases. Figure 5.7 shows a simplified representation of the head loss in a filter.

The head loss in a clean sand filter can be estimated by using several different methods and empirical equations. One of the earliest equations used for prediction of the head loss in a filter is that developed by Carman-Kozeny.

A head loss in a clean sand filter while filtering can be estimated by first considering the filter to behave as a mass of pipes, in which the Darcy's head loss equation is valid. Equation 5.8 gives the mathematical formula for the headloss.

$$H_1 = f \cdot l \cdot v^2 / 2 \cdot D \cdot g \qquad [5.8]$$

where: H_1 = head loss through the filter, m
f = friction factor
l = filter depth, m
v = velocity in pipe, m/s
D = pipe diameter, m
g = gravitational acceleration, m/s^2

Since the pipes or channels through the sand are not straight, and their diameter changes, then the concept of the hydraulic radius can be used as outlined in Equation 5.9:

$$D = 4 \cdot R_h \qquad [5.9]$$

where: D = pipe diameter, m
R_h = Hydraulic radius, m (= area/wetted perimeter) = $(\pi D^2/4)/\pi D = D/4$. [5.10]

Thus, by substituting Equations 5.10 and 5.9 in Equation 5.8:

$$H_1 = f \cdot l \cdot v^2 / 8 \cdot R_h \cdot g \qquad [5.11]$$

The velocity of the water approaching the sand filter is

$$v_a = Q/A \qquad [5.12]$$

and the velocity through the bed is

$$v = v_a / e \qquad [5.13]$$

where: v_a = velocity of water approaching sand grains, m/s
 v = filtration velocity, m/s
 A = surface area of the sand bed, m²
 e = porosity of bed, fraction of pore spaces in the sand.

The total channel volume = porosity of bed · total volume
$$= e \cdot V$$

$$\text{Total solids volume} = (1-e) \cdot V = n \cdot V_p \quad [5.14]$$

where: e = porosity of bed, dimensionless
 V = total volume of bed, m³
 n = number of particles, dimensionless
 V_p = volume occupied by each particle, m³.

Thus, total volume = $n \cdot V_p/(1 - e)$ and total channel volume = $e[n \cdot V_p/(1 - e)]$.
Total wetted surface area = $n \cdot A_p$, where A_p = surface area of each particle.
Therefore, R_h = area/wetted perimeter ∝ volume/area

$$= e[n \cdot V_p/(1-e)]/n \cdot A_p = [e/(1-e)](V_p/A_p) \quad [5.15]$$

For spherical particles:

$$V_p/A_p = (\pi \cdot d^3/6)/(\pi d^2) = d/6 \quad [5.16]$$

For particles that are not true spheres:

$$V_p/A_p = \varphi \cdot (d/6) \quad [5.17]$$

where φ = shape factor = (surface area of equivalent volume of sphere/actual surface area).
Typical values of φ are given in Table 5.3

$$R_h = [e/(1-e)](\varphi \cdot d/6) \quad [5.18]$$

This yields

$$H_l = f_l \cdot l \cdot (1-e) \cdot v_a^2 / \varphi \cdot d \cdot e^3 \cdot g \quad [5.19]$$

Equation 5.19 can be approximated by estimating the friction factor as shown in Equation 5.20.

$$f = [150(1-e)/R_e] + 1.75 \quad [5.20]$$

where: f = friction factor, dimensionless
 R_e = Reynold's number, dimensionless

$$R_e = \varphi \cdot \rho \cdot v_a \cdot d/\mu \quad [5.21]$$

 φ = shape factor, dimensionless
 $v_a = v \cdot e$ = velocity of water approaching sand grains, m/s
 d = diameter of particle, m
 μ = viscosity coefficient, N·s/m²
 ρ = particle density, kg/m³.

Table 5.3 Typical Values of the Particle Shape Factor

Material	φ
Angular sand	0.73
Crushed glass	0.65
Mica flakes	0.28
Spherical sand	1.00
Worn sand	0.89

Compiled from References 5, 6, and 29.

The above equation is relevant to a filter bed that is composed of particles of one size. For a bed composed of nonuniform sand grains the diameter may be determined as:

$$d = (6/\varphi) \cdot (V/A)_{av} \qquad d = (6/\varphi) \cdot (V_{av}/A_{av}) \qquad [5.22]$$

where: φ = shape factor, dimensionless
V_{av} = average volume of all particles, m³
A_{av} = average area of all particles, m²

This yields an approximation for the head loss as presented by Equation 5.23:

$$H_l = \left(f \cdot l \cdot (1-e) \cdot v_a^2\right) / \left(6 \cdot e^3 \cdot g \cdot [V/A]_{av}\right) \qquad [5.23]$$

and $\{A/V\}_{av}$ can be estimated as $(6/\varphi)\Sigma(x/d')$ where x = the weight fraction of impurities held between any two sieves, and d' = the geometric mean diameter between the sieves.
Therefore, the headloss may be determined as:

$$H_l = \left\{f_l \cdot l \cdot (1-e) \cdot V_a^2 / \varphi \cdot e^3 \cdot g\right\} \cdot \left\{\Sigma(x/d')\right\} \qquad [5.24]$$

Equation 5.24 is valid for nonstratified sand beds, for example, conditions similar to those found in a slow sand filter.[13] This is due to the fact that the friction factor does not change with the filter depth.

In the case of stratified particles, such as those found in a rapid sand filter, Equation 5.24 can then be rewritten as:

$$dH_l/dl = \left\{[1-e] \cdot v_a^2 / \varphi \cdot e^3 \cdot g\right\} \cdot \left\{f/d'\right\} \qquad [5.25]$$

The total head loss may be determined by integrating the above equation. Since dl = l·dx (where dx = proportion of particles of size d), then:

$$H_l = \left\{[1-e] \cdot v_a^2 / \varphi \cdot e^3 \cdot g\right\} \cdot l \cdot \left(\int_{x=0}^{x=1} f \cdot dx/d'\right) \qquad [5.26]$$

and if the particles between adjacent sieve sizes are assumed uniform, then the head loss is as given in Equation 5.27.

$$H_l = l \cdot \left\{(1-e) \cdot v_a^2 / \varphi \cdot e^3 \cdot g\right\} \cdot \Sigma(fx/d') \qquad [5.27]$$

Rose used dimensional analysis to develop an equation for the head loss in filters, as illustrated in Equation 5.28.

$$H_l = \left(1.067 \cdot C_D \cdot V^2 \cdot l\right) / \left(g \cdot d \cdot \varphi \cdot e^4\right) \qquad [5.28]$$

where: H_1 = head loss in the filter, m
C_D = Newton's drag coefficient

$$= (24/R_e) + \{3/(R_e)^{0.5}\} + 0.34 \qquad [5.29]$$

v = filtration velocity, m/s
l = filter depth, m
g = gravitational acceleration, m/s²
d = particle diameter, m
φ = particle shape factor, dimensionless
e = bed porosity

The clogging of the filter bed by impurities during the filtration process increases the head loss through the filter. A method of estimating the head loss through the filter makes use of this principle. The head loss can be predicted[13] by using Equation 5.30:

$$H_1 = H_0 + \sum_{i=1}^{n} (H_{li})_t \qquad [5.30]$$

where: H_1 = total head loss through the filter, m
H_0 = head loss of the clear water at time t = 0, m
$(H_{li})_t$ = head loss in the ith layer of media in the filter at time t, m

The head loss within an individual layer is associated with the amount of solids that are captured by that layer:

$$(H_{li})_t = a \cdot (acc_i)_t^b \qquad [5.31]$$

where: $(H_{li})_t$ = head loss in the ith layer of media in the filter at time t, m
$(acc_i)_t$ = amount of solids collected in the ith layer at time t, mg/cm³
a,b = experimental constants.

A filter run is defined as the time needed for a filter to function before it requires cleaning. The end of the filter run is determined by excessive head loss, or by excessive turbidity of the water to be filtered. Once the end of the filter run has been reached, the filter should be taken out of service for cleaning.[13]

Example 5.5

A dual media filter with the following characteristics is used in a water treatment plant:

Characteristic	Anthracite layer	Sand layer
Depth (cm)	60	50
Average particle size (mm)	1.7	0.85
Specific gravity	1.7	2.65
Particle shape factor	0.80	0.95
Porosity of bed (%)	40	40

For a filtration rate of 10 m³/m²/hr, compute the head loss by using Rose's equation. (Take the coefficient of kinematic viscosity to be $1.14 \cdot 10^{-6}$ m²/s).

Solution

1. Given for anthracite: l = 0.6 m, d = $1.7 \cdot 10^{-3}$, φ = 0.8, e = 0.4. For sand: l = 0.5 m, d = $0.85 \cdot 10^{-3}$, φ = 0.95, e = 0.4, v_f = 10/(60·60) = $2.78 \cdot 10^{-3}$ m/s.
2. Use Rose's equation (Equation 5.28) to compute the head loss for each media as: $H_1 = 1.067 \cdot C_D \cdot v^2 \cdot l/g \cdot d \cdot \varphi \cdot e^4$.

3. For the anthracite media:
   - Determine Reynold's number: $R_e = v_f \cdot d/v = 2.78 \cdot 10^{-3} \cdot 1.7 \cdot 10^{-3}/1.14 \cdot 10^{-6} = 4.15$
   - From Equation 5.29 find Newton's drag coefficient as: $C_D = (24/R_e) + [3/(R_e)^{0.5}] + 0.34 = (24/4.15) + [3/(4.15)^{0.5}] + 0.34 = 7.6$.
   - Compute the headloss from Rose's equation as presented in 2 above as: $H_1 = [1.067 \cdot 7.6 \cdot (2.78 \cdot 10^{-3})^2 \cdot 0.6]/[9.81 \cdot 1.7 \cdot 10^{-3} \cdot 0.8 \cdot [0.4]^4] = 0.11$ m.

4. For the sand media:
   - Determine Reynold's number: $R_e = v_f \cdot d/v = 2.78 \cdot 10^{-3} \cdot 0.85 \cdot 10^{-3}/1.14 \cdot 10^{-6} = 2.07$
   - From Equation 5.29 find Newton's drag coefficient as: $C_D = (24/R_e) + [3/(R_e)^{0.5}] + 0.34 = (24/2.07) + [3/(2.07)^{0.5}] + 0.34 = 14.0$.
   - Compute the headloss from Rose's equation as presented in 2 above as: $H_1 = [1.067 \cdot 14 \cdot (2.78 \cdot 10^{-3})^2 \cdot 0.5]/[9.81 \cdot 0.85 \cdot 10^{-3} \cdot 0.95 \cdot [0.4]^4] = 0.285$ m.

5. Determine the total head loss: headloss through the anthracite layer + headloss through the sand layer = 0.11 + 0.285 = 0.395 m.

5. Filter Clogging

The aforementioned equations provide the head loss for a clean filter bed. When the filter is operated for the removal of suspended impurities the porosity of the bed tends to change due to the collection of particles in the void spaces (see Figure 5.8). Usually, the rate of removal of particles or the decrease in impurities is assumed to be proportional to their concentration as shown in the following equation:[29]

$$\partial C/\partial y = -\lambda \cdot C \quad [5.32]$$

where: ∂C = rate of change of concentration of suspended solids, mg/L
∂y = rate of change of depth from inlet surface, m
λ = constant characteristic of the bed
   = coefficient of filtration, /m

Integrating Equation 5.32 for $y = 0$, and $C = C_o$ then,

$$C = C_o \cdot e^{-\lambda_o y} \quad [5.33]$$

where C_o = concentration of solid substances at the inlet surface, mg/L.

According to Equation 5.33 the concentration of impurities found in the liquid passing through the filter decreases logarithmically with the filter depth. The upper part of the filter bed is managing most of the filter operation as compared with the lower part.

The quality of the effluent liquid is indicted by the following equation:

$$C_e = C_o \cdot e^{-\lambda_o l} \quad [5.34]$$

where: C_e = concentration of impurities of the effluent, mg/L
C_o = concentration of impurities at the inlet surface, mg/L
λ_o = filtration coefficient, /m
l = filter depth, m

The impurities extracted from the water during the process of filtration are delivered to the filter bed, where they accumulate on and between the filter sand grains.

The rate of deposition of impurities can be determined by considering an element of the filter bed with a thickness of dy, at a depth of y below the top of the filter (see Figure 5.9). Impurities from the raw water accumulate within this thickness in a concentration of δ after the elapse of time t. During the following incremental period dt, the concentration of impurities increases by an amount given by: $d\delta = (\partial \delta/\partial t)dt$. The continuity equation may then be used to relate the rate of deposition to concentration of impurities entering and leaving the filter bed. Thus:

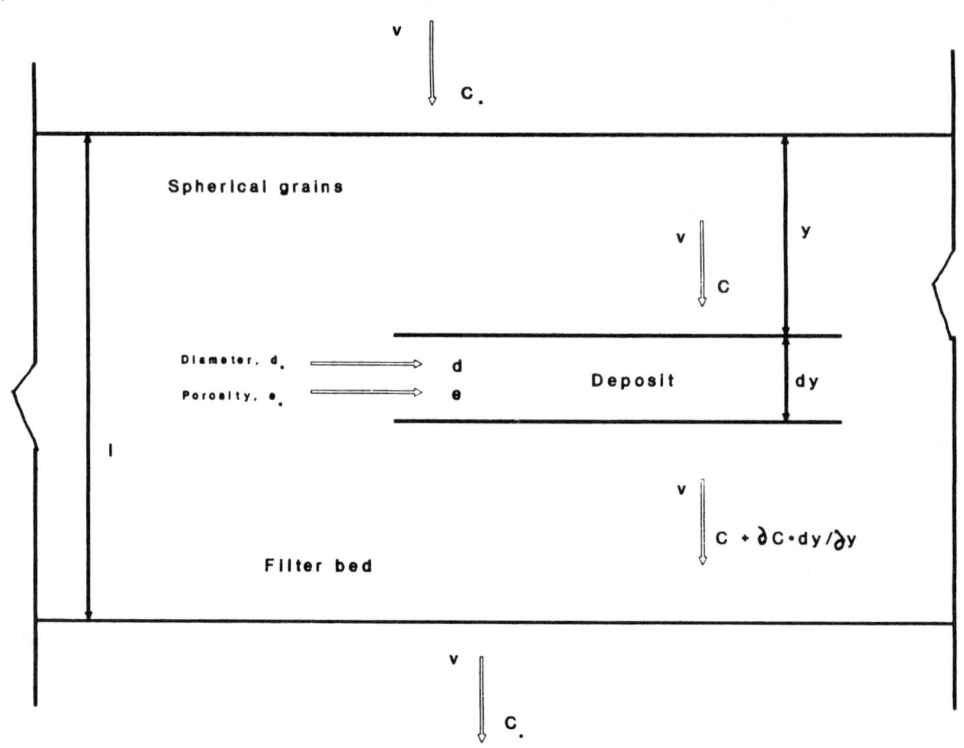

Figure 5.8 Clogging in a sand filter.[16,19]

$$\text{Deposition} = \text{Inflow} - \text{Outflow}$$

$$(\partial \delta / \partial t) dt \cdot dy = v \cdot C \cdot dt - v(C + [\partial C / \partial y] dy) dt$$

$$\text{or, } -(\partial C / \partial y) = (1/v)(\partial \delta / \partial t) \tag{5.35}$$

Taking the filtration coefficient to be equal to a constant value of λ_o, and for a concentration of impurities of $C = C_o \cdot e^{-\lambda_o y}$, then:

$$\partial C / \partial y = -\lambda_0 \cdot C_0 \cdot e^{-\lambda_0 y} \tag{5.36}$$

This equation yields the following equation:

$$\partial \gamma / \partial t = v \cdot \lambda_0 \cdot C_0 \cdot e^{-\lambda_0 y} \tag{5.37}$$

Integrating Equation 5.37 between the initial time of $t = 0$ when the rate of deposition is $\delta = 0$, then:

$$\delta = v \cdot \lambda_0 \cdot C_0 \cdot e^{-\lambda_0 y} \cdot t \tag{5.38}$$

where: δ = the rate of deposition
 v = filtration rate, m/s
 λ_o = filtration coefficient, /m
 C_o = concentration at inlet surface, mg/L
 y = depth of deposition of impurities from the inlet surface, m
 t = time taken for deposition of solids within the filter bed, s

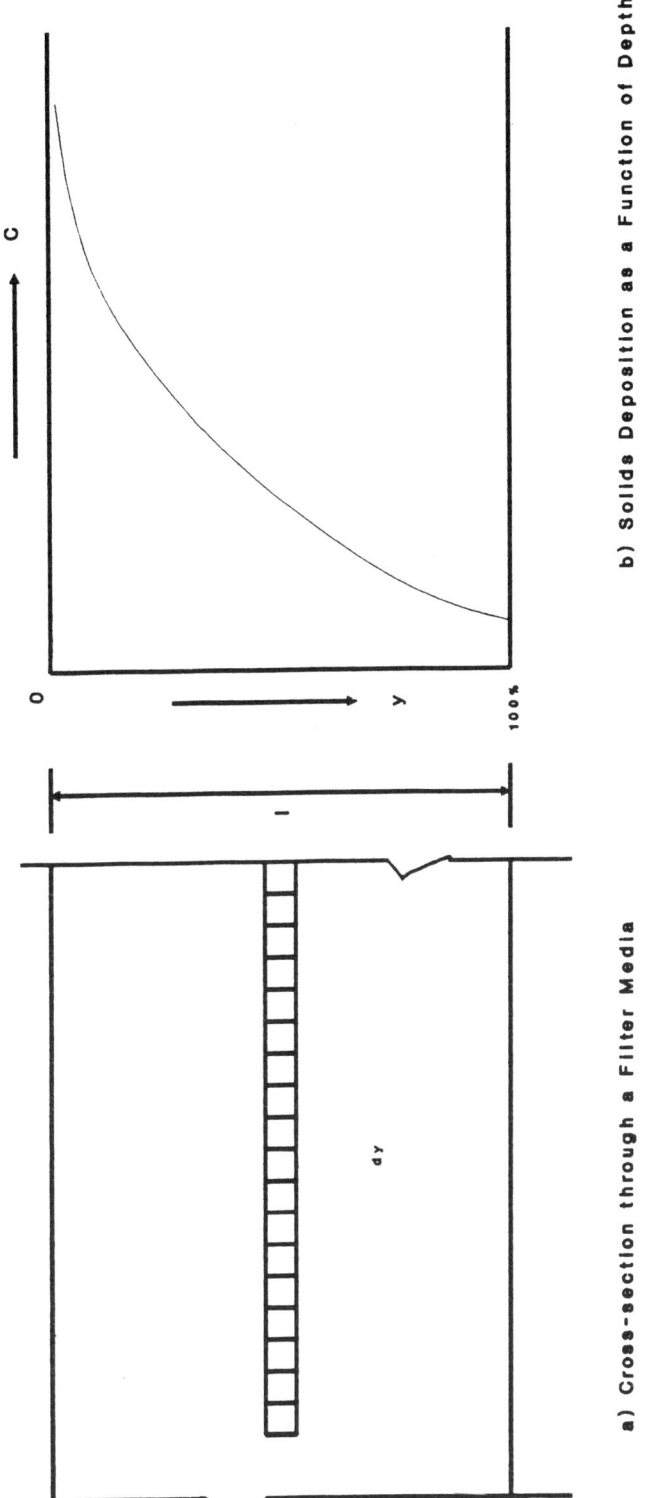

Figure 5.9 Solids deposition within a filter.[29,35]

Equation 5.38 signifies a logarithmic reduction of the concentration of suspended material with the filter depth and a linear increase with the elapsed time.[29]

Example 5.6

A filter is to treat water having a concentration of impurities of 18,000 mg/m³. If the bed thickness is 80 cm and the coefficient of filtration (the constant characteristic of the bed) is assumed to be 6 m⁻¹:

i. Compute the effluent concentration of impurities.
ii. Determine the percentage reduction of impurities achieved in the filter.
iii. Compute the concentration of impurities at half depth if the filtration rate is 0.12 m³/m²/min.
iv. Find the rate of deposition of impurities per unit time (take viscosity coefficient to be equal to $1.31 \cdot 10^{-6}$ m²/s).

Solution

1. Given: C_o = 18 mg/L, l = 0.8 m, λ_o = 6/m, and v_f = 0.12 m/min = $2 \cdot 10^{-3}$ m/s.
2. Use Equation 5.34 to determine the effluent concentration of impurities as: $C_e = C_o \cdot e^{-\lambda_o l} = 18 \cdot e^{-6 \cdot 0.8} = 0.15$ mg/L.
3. Determine the percentage reduction of impurities achieved in the filter as = $(100 \cdot (C_o - C_e))/C_o = 100 \cdot (18 - 0.15)/18 = 99\%$.
4. Use Equation 5.34 to determine the effluent concentration of impurities at half filter depth by taking y = 1/2 = 0.4 m. Thus, $C_e = C_o \cdot e^{-\lambda_o y} = 18 \cdot e^{-6 \cdot 0.4} = 1.63$ mg/L.
5. Use Equation 5.38 to determine the rate of deposition of impurities per unit time as: $\delta/t = v \cdot \lambda_o \cdot C_o \cdot e^{-\lambda_o y} = 2 \cdot 10^{-3} \cdot 6 \cdot 18 \cdot e^{-6 \cdot 0.8} = 1.8 \cdot 10^{-3}$.

6. Rapid Filtration

a. Introduction

Usually, the sand used as the filter medium in rapid sand filters has an effective grain size in the range of 0.4 to 1.2 mm. Due to the coarse sand used, the pores of the filter bed will be relatively large and the impurities contained in the raw water will penetrate further into the filter bed. Thus, ability of the filter bed to reserve the accumulated impurities is much more efficiently utilized. This ability permits the treatment of very turbid water.

Cleaning of a rapid filter is conducted by the method of backwashing (Figure 5.10). In backwashing, a high rate of water flow is directed from the bottom of the filter bed to expand and scour the sand. The backwash water carries the precipitated impurities out of the filter. Backwashing may be done quickly; it does not take more than half an hour. Filter bed cleaning may be carried out as frequently as needed.

b. Application of Rapid Sand Filters

For the treatment of various water sources, rapid sand filters can be used in many ways:

1. *Filters used as the only treatment.* This is for removal of iron and manganese from deep groundwater sources. In the case of a rather turbid surface water, coagulation followed by rapid sand filtration and chlorination can be employed.
2. *Filters used as preliminary treatment units.* This is to remove the major portion of the suspended particles in a raw water. In such a system slow sand filters may be incorporated after rapid filters. The action of the rapid filters (sometimes called roughing filters) is to lessen the load on the following slow sand filters.
3. *Filters used as final treatment units.* This is the final treatment to remove the remainder of the contaminants.

Rapid sand filters are generally open with the water flowing through the filter under gravity action. For certain operating conditions, other types of filters may prove more appropriate, such as:

- Pressure filters (Figure 5.11). In pressure filters the filter bed together with the filter bottom is enclosed in a watertight steel pressure vessel or similar. The driving force for the filtration process is the water pressure applied on the filter bed. This pressure may be so high that almost any needed length of filter run may be acquired. Pressure filters are not easy to install, run, and maintain.
- Up-flow filters (Figure 5.12). Up-flow filters supply a graded filtration scheme from coarse to fine filter media. The coarse bottom layer of the bed gets rid of the greater portion of the suspended impurities, with no great augmentation in the filter bed resistance.

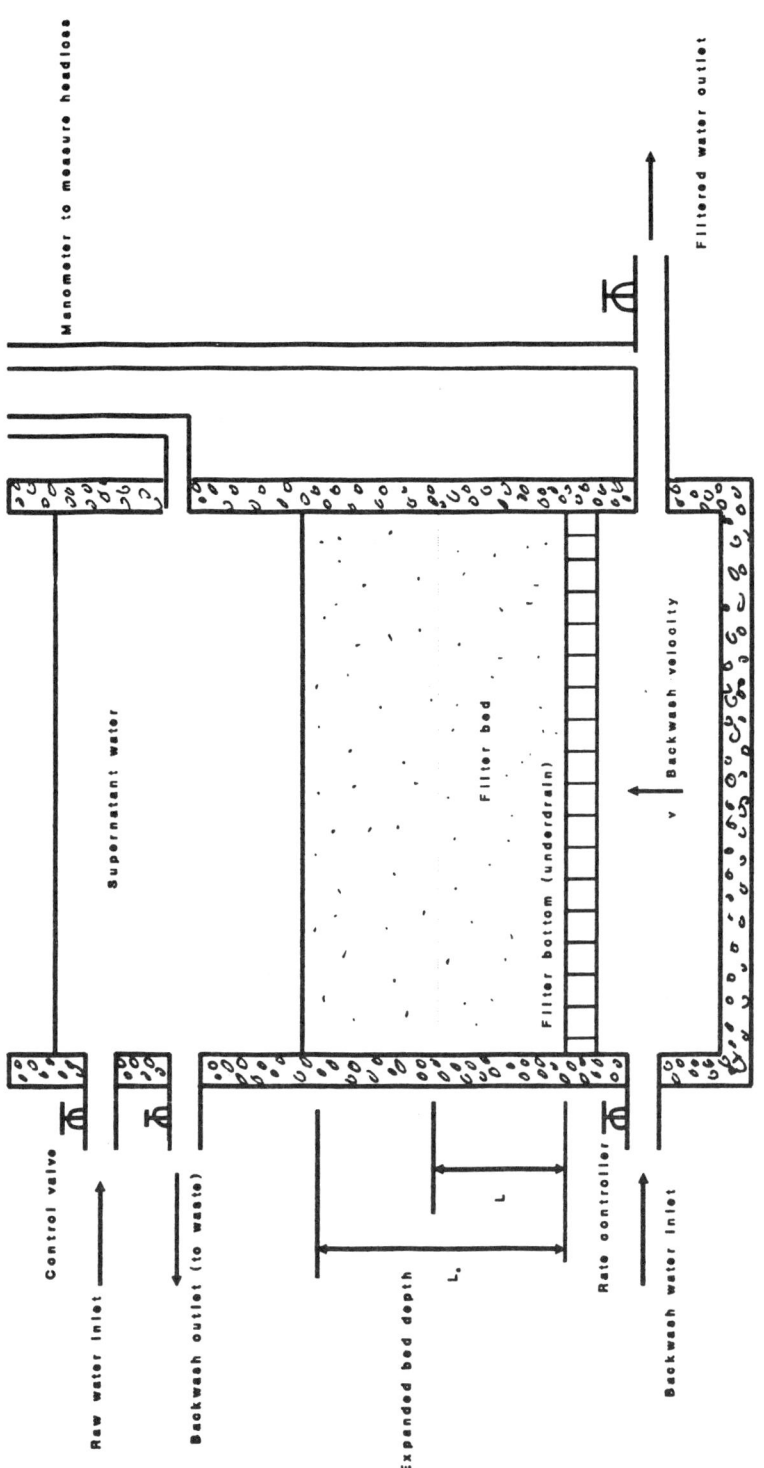

Figure 5.10 Schematic diagram: backwashing a rapid sand filter.

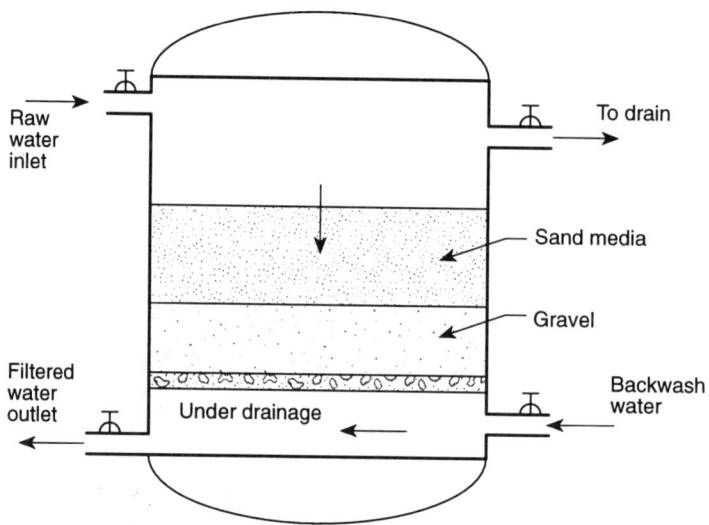

Figure 5.11 Schematic diagram of a pressure filter.[32,35]

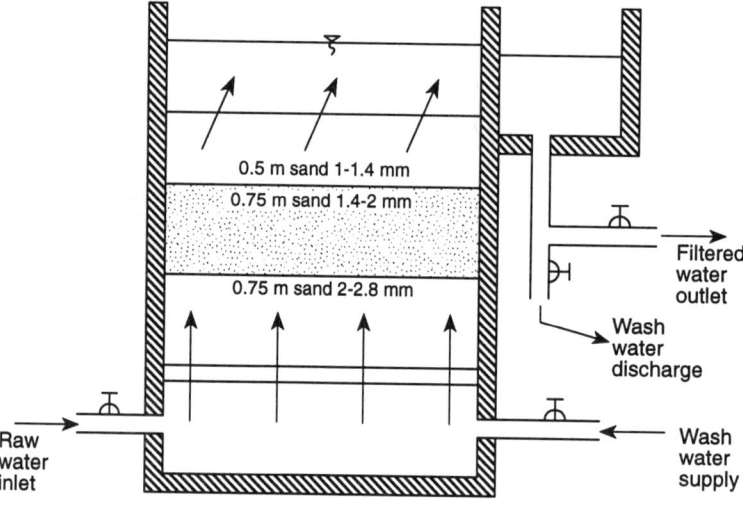

Figure 5.12 Schematic diagram of an upflow filter.[29,30,35]

- Multiple media filters (Figure 5.13). Multimedia filters are gravity, down-flow filters with the bed composed of different materials which are placed in coarse to fine layers along the direction of flow.

c. Operation of a Filter

The following steps outline the general filter operation aspects (see Figure 5.14):

1. Water enters the filter through valve A. The flow continues in vertical direction down through the filter bed, the underdrainage system (filter bottom) and out of the filter through valve B.
2. Due to the gradual clogging of the pores, the filter bed's resistance against the downward water movement gradually increases. This results in a reduction in the filtration rate unless counteracted by allowing more water to flow through the system.
3. When the filter is operated for some time, the filter's rate controller will be completely open. As such, a further clogging of the filter bed cannot be further counteracted and the filtration rate decreases. Then the filter needs to be taken out of service for cleaning.

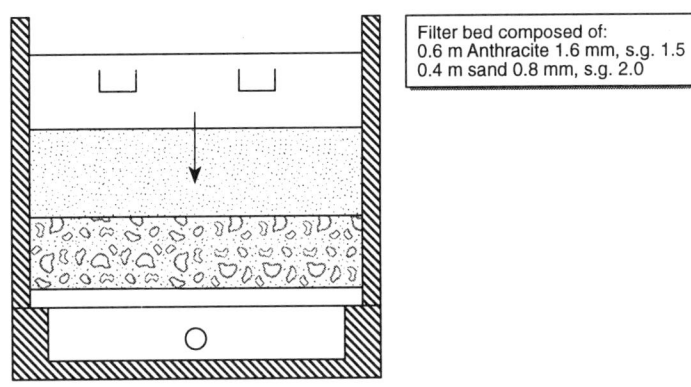

Figure 5.13 Dual media filter bed.[29,30,35]

4. Cleaning is carried about by backwashing. Here, valves A and B are closed, and valve C is opened to drain the remaining raw water out of the filter.
5. After some time, valve E is opened in order to introduce the wash water. The backwash rate must be high enough to expand the filter bed in order to scour the filter sand grains. The accumulated impurities are then removed with the wash water.
6. The wash water is collected in a wastewater trough and then discharged through valve D.
7. When backwashing is completed, valves E and D are closed and valves A and B reopened to permit the raw water to commence a new filter cycle.

d. Filter Control Systems

Filter control systems are numerous. Figure 5.15 illustrates one type filtration rate controller. Generally, rate controllers may be classified as:

- Inlet rate control devices which offer equal distribution or flow splitting.
- Outlet rate control devices which include level-operated valves, overflow weirs, and siphons.

Basically, filter rate control arrangements can be divided into:[29,30,35]

1. *Individual controller.* This type of controllers is employed in order to maintain filtered water production at a constant rate.
2. *Raw or filtered water controller.* The total flow of water through the filter works is managed by a raw water intake controller, or alternatively by a filtered water outlet controller.
3. *Declining rate controller.* Same as (2) above, but the filter units function through the application of individual declining-rate controllers.

Usually, filters are designed to function with a constant water level. This condition demands furnishing the filter with a rate controller either in the influent or along the effluent line. These controllers are able to regulate the bed resistance to water flow. They open gradually and automatically offset any increase in the filter bed resistance. In this manner the controllers maintain constant filter operating conditions.

e. Design Considerations in Filters

The parameters that need to be considered in the design of filters include:

- Filter media grain size. Filter bed problems are met when the grain size falls below the value of 0.8 mm. As such, coarse filter media grain sizes are suggested.
- Filter bed depth. A large bed thickness is advisable. Huisman[29,30] recommended a depth of 0.6 to 0.8 m for final treatment following coagulation and sedimentation, 0.8 to 1.2 m for pretreatment preceding slow sand filtration, and 1.5 to 3 m for removing ferrous iron from groundwater.
- Rate of filtration. For rapid filters a value of $1.4 \cdot 10^{-3}$ m/s (5 m/hr) may be taken for the filtration rate.
- Depth of supernatant water.

Two distinct filter constructions may be found for the same raw and filtered water measures. The constructions available include the filters that operate with an excess pressure and the filters that operate with a reduced pressure.

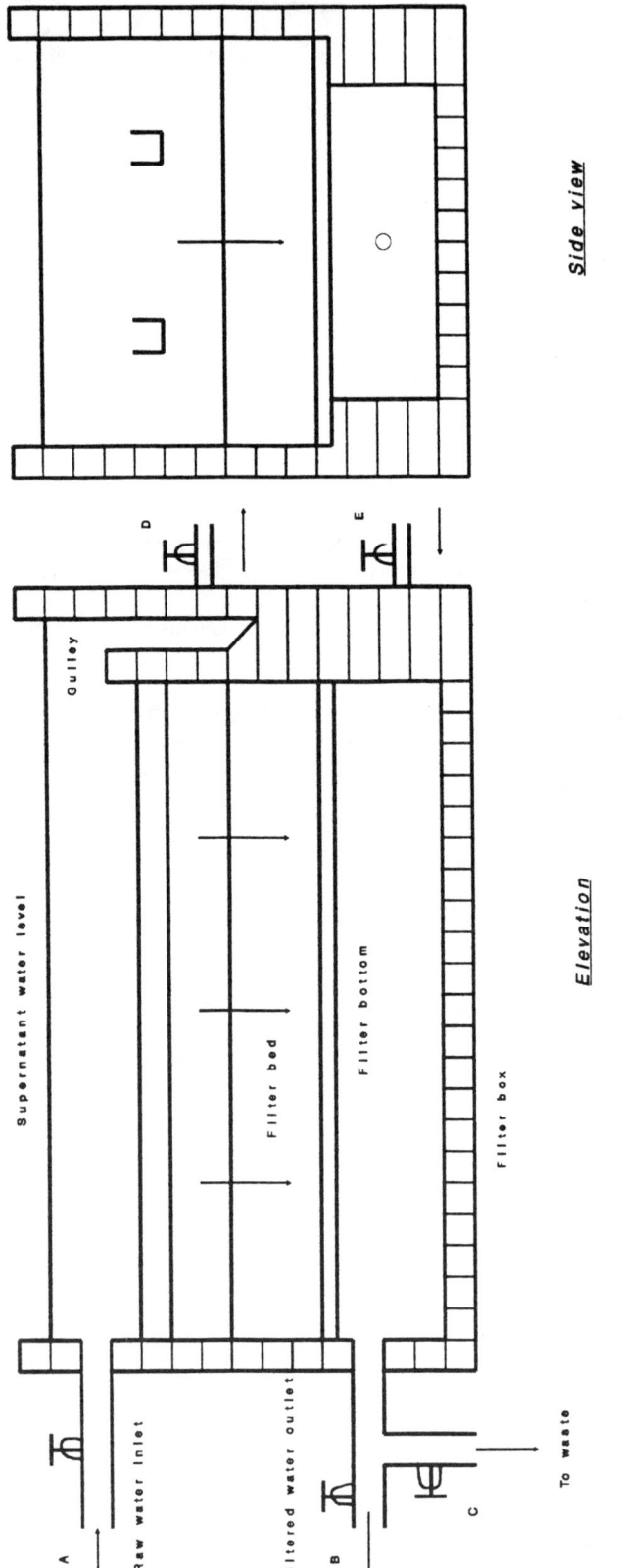

Figure 5.14 Rapid gravity sand filter.[29,30,35]

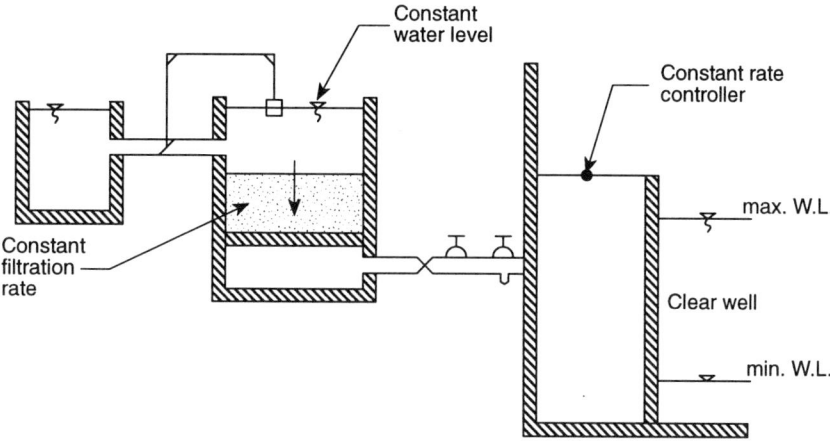

Figure 5.15 Filtration rate control device.[35]

Filters that are constructed to function under excess pressure possess a large depth of water, 1 to 1.5 m, on top of the filter bed. Filters that are constructed to operate under reduced pressure have a supernatant water depth of 0.25 to 0.4 m. The case of a filter constructed to operate under a reduced pressure permits a smaller depth of the filter box, thus decreasing the cost of construction. Nevertheless, reduced pressure filters introduce the danger of air binding in the filter.

The occurrence of negative heads may be prevented by providing a large depth of supernatant water when the maximum allowable filter resistance is reasonable. When the supernatant water depth is small, negative heads may occur immediately after commencement of filtration. Therefore, a reduction in the solubility of gases enclosed in the filter occurs. Air binding causes problems of deteriorating effluent quality, shortened filter runs, and promoting loss of filter material during backwashing.

f. Cleaning a Filter

Cleaning of the filter is essential in order to renew its capacity and/or improve the quality of the filtered water. Cleaning can be carried out manually or mechanically.

The manual cleaning of the filter may be performed by using jetting tubes. Jetting tubes are clumsy and costly in terms of labor when using short filter runs. Mechanical cleaning is done by changing the direction of flow, or introducing wash water at the bottom of the filter bed at a rate many times greater than the rate of filtration.

g. Hydraulics of Backwashing a Filter

The upward water flow intended for cleaning the filter expands the filter bed. This expansion produces a situation in which the accumulated impurities are scoured off the particles. As the bed expands the rate of increase of head loss decreases. At the point where the entire bed is about to get suspended the head loss is a constant value. When this stage is reached the upward backwash force balances the downward gravitational force of the particles in water. Any additional increase in the flow of backwashing increases the expansion but not the head loss.

Extreme expansion is not advised in order to avoid unwanted conditions such as the separation of particles, reductions in the scouring action, and an increase in water consumption for backwashing the filter.

In Figure 5.10 the resulting expansion may be evaluated as: $(l_e - l)/l$. Usually, this expansion is in the region of 5 to 25%.

At maximum frictional resistance developed in the bed the upward water force = (expanded depth · net unit weight of the medium · medium volume)/total volume:

$$\rho \cdot g \cdot h = l_e \cdot (\rho_s - \rho) \cdot g \cdot (1 - e_e) \qquad [5.39]$$

where: ρ = density of water, kg/m^3
e_e = porosity of expanded bed, dimensionless
l_e = expanded depth of filter, m

ρ_s = density of solid particle, kg/m³
g = gravitational acceleration, m/s²
h = increase in elevation during backwash, m

$$h/l_e = (\rho_s - \rho) \cdot (1 - e_e)/\rho$$
$$h/l_e = (s.g. - 1) \cdot (1 - e_e) \quad [5.40]$$

The rising water drag force acting on the particles tend to keep them in suspension. Thus, from settling theory:

$$(C_D \cdot A \cdot \rho \cdot v^2/2) \cdot f(e_e) = (\rho_s - \rho) \cdot g \cdot V \quad [5.41]$$

where: C_D = drag coefficient
A = surface area of the particle, m²
ρ = density of the fluid surrounding the particle, kg/m³
v = face velocity of the backwash water, m/s
$f(e_e)$ = a function of the porosity of the expanded bed
e_e = porosity of expanded bed, dimensionless
ρ_s = density of the particle, kg/m³
g = gravitational acceleration, m/s²
V = volume of particle, m³

It has been found experimentally that:[29]

$$f(e_e) = (v_s/v)^2 = (1/e_e)^9 \quad [5.42]$$

thus, $e_e = (v/v_s)^{0.22}$ or

$$v = v_s \cdot (e_e)^{4.5} \quad [5.43]$$

where: v = face velocity of the backwash water (rise rate), m/s
v_s = particle settling velocity, m/s
e_e = porosity of expanded bed, dimensionless

Considering the static and fluidized conditions, then:

$$(1 - e) \cdot 1 = (1 - e_e) l_e \quad [5.44]$$

$$l_e/l = (1 - e)/(1 - e_e) = (1 - e)/\left[1 - (v/v_s)^{0.22}\right] \quad [5.45]$$

where: l_e = expanded bed depth, m
l = bed depth, m
e = porosity of bed, dimensionless
e_e = expanded bed porosity, dimensionless
v = face velocity of the backwash water (rise rate), m/s
v_s = particle settling velocity, m/s

Water needed for backwashing may be supplied by the water distribution system, by special wash water pumps drawing water from a clear well, or by an elevated wash water reservoir or tank.

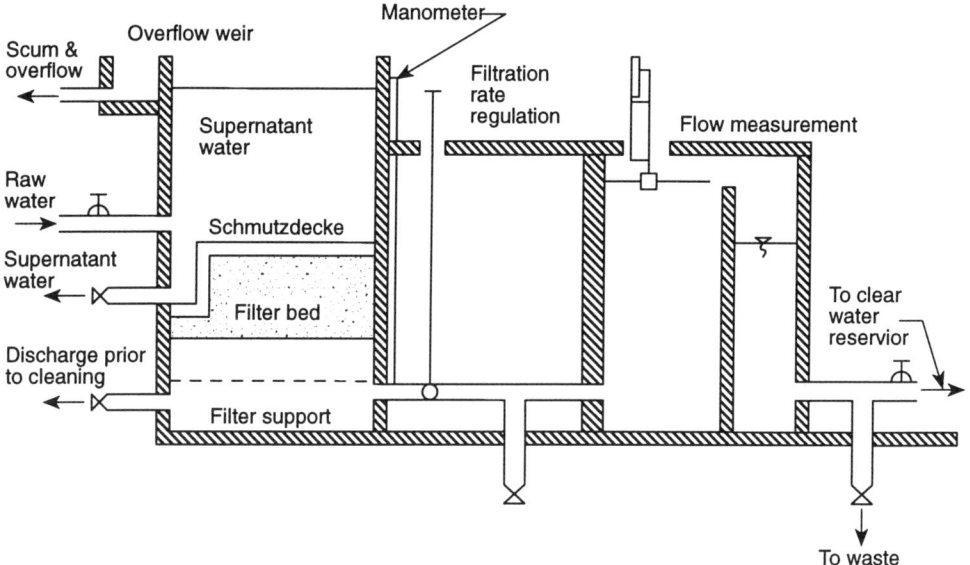

Figure 5.16 Basic elements of a slow sand filter.[36]

Example 5.7

A sand filter bed consists of unisized particles with a porosity of 35% and depth of 0.8 m. The filtration rate is kept at 250 L/m²/min. The measured settling velocity of the sand particles approached 7 m/min. Find the depth of bed expansion when the filter is backwashed at the rate of 0.3 m³/m²/min.

Solution

1. Given: $l = 0.8$ m, $e = 0.35$, $v_f = 250$ L/m²/min, $v = 0.3$ m/min, and $v_s = 7$ m/min.
2. Use Equation 5.45 to find depth of bed expansion as:

$$l_e/l = (1-e)/(1-e_e) = (1-e)/\left[1-(v/v_s)^{0.22}\right]$$

3. Substitute given values in the equation presented in 2 above as: $l_e/0.8 = (1 - 0.35)/[1 - (0.3/7)^{0.22}]$. This yields a value of $l_e = 1.04$ m.

7. Slow Sand Filtration

a. Introduction

A slow sand filter (Figure 5.16) consists of a bed of fine sand through which water is allowed to percolate. The majority of the suspended impurities in the raw water are held in the upper 0.5 to 2 cm layer of the bed media. This permits the filter to be cleaned by scraping away the top layer of sand. As low rates of filtration are used, the interval between successive cleanings will be fairly long, usually several months.

Slow sand filters have many advantages such as:

- Production of a high-quality water free from suspended impurities, and hygienically safe since it removes pathogenic organisms.
- Filters can be built using local materials and local skills and labor.
- Complex mechanical and electrical equipment, required for most other water treatment processes, can be avoided in the case of slow sand filtration.
- The simple technique of slow sand filters also implies cheap capital and running costs.

b. Principles of Operation of a Slow Sand Filter

A slow sand filter incorporates a tank that is open at its top. This tank is suitable to accommodate the bed of sand. Depth of the tank is about 3 m and its area can differ from a few tens to many hundreds of square meters.

At the bottom of the tank an underdrain system is positioned to support the filter bed and to ease the uniform egress of effluent. The filter is provided with a number of influent and effluent lines fitted with valves and control devices in order to maintain a steady filtration rate and a constant water level.

Algae play an important role in water purification by filtration. As autotrophic organisms, algae require light for photosynthetic processes. This is the reason for presence of algae in open filters. Algae are able to build up cell material from simple materials such as water, carbon dioxide, nitrates, phosphates, etc., with the help of solar energy. As algal cells grow and multiply, their volume increases. This condition impedes the downward water flow through the filter and demands the periodic removal of the algae by cleaning the filter. When filamentous species of algae are present they form a gelatinous mat on the surface of the filter. This mat removes suspended matter and bacteria by straining and adsorption processes.

Water bacteria will multiply in the filter, building a zooglea or schmutzdecke of bacterial slime. The schmutzdecke (dirty layer) is an adhesive medium for plankton and diatoms.

The efficiencies of straining and adsorption of the schmutzdecke are thus enhanced and less impurities reach the filter. This situation prolongs filter runs and decreases blockage.

An average filter run of about 2 months is often used. When filter runs are much longer, filtration rates can be raised allowing a greater plant production. If filter runs are shorter than expected, additional units will have to be added.

The length of filter run also depends on the maximum allowable head loss, which in turn increases with an increase in the depth of the supernatant water.

Under all circumstances negative heads (water pressures below atmospheric) ought to be avoided with slow sand filters. This is due to the fact that negative heads may cause the liberation of dissolved gases. The formed air bubbles accumulate in the filter bed and they increase the resistance against the downward water flow (see Figure 5.17). Likewise, rising air bubbles, of larger size, tend to make holes in the filter bed through which water passes without being adequately treated.

B. ADSORPTION PROCESS
1. Introduction

Adsorption applications are those operations in which a solute material is transferred from a fluid to the surface of a solid (adsorbent) upon contact with the latter. Adsorption is a surface phenomena. Good adsorbents must have both a high surface-to-volume ratio and an "active" or activated surface. This indicates that adsorbents are highly porous materials that are filled with fine capillaries.[41,43]

The adsorption process may be physical or chemical in nature. Physical adsorption, or van der Waals adsorption, is the result of intermolecular forces of attraction between molecules of the solid and the substance adsorbed. Physical adsorption is a readily reversible phenomenon. The adsorbed substances do not dissolve in the solid but remain entirely upon the surface.

Chemosorption, or activated adsorption, is the result of a chemical interaction between the solid and the adsorbed substance. The force holding adsorbent and adsorbate may alter significantly, but the adhesive force is generally much greater than that encountered in physical adsorption processes. The process is frequently irreversible, and original material may not be recovered upon desorption.

When an adsorbent solid is immersed in a pure liquid, heat evolves when adsorption of the liquid has occurred. When an adsorbent is mixed with a binary solution, adsorption of both solute and solvents occur. Since the total adsorption cannot be determined, the relative or apparent adsorption of the solute is evaluated instead.

The apparent adsorption of a given solute depends upon the concentration of solute, the temperature, and the type of adsorbent. The extent of adsorption of a given solute practically always decreases with an increase in temperature, and usually it is reversible, so that the same isotherm results whether the solute is desorbed or adsorbed.

2. Adsorption Isotherms

Adsorption is a surface phenomenon often used to remove trace levels of contaminants from either a liquid or a gas stream by contact with a solid surface referred to as the adsorbent (e.g., activated carbon, molecular sieves, silica gel, natural soils, etc.). This adsorption phenomenon is influenced not only by the type of adsorbent used but also by the molecular size and polarity of the contaminant, the nature of the solution or gas stream in which the contaminate is dispersed, and the contacting system employed. It is important to be able to relate the amount of contaminant adsorbed from the gas or water stream to the

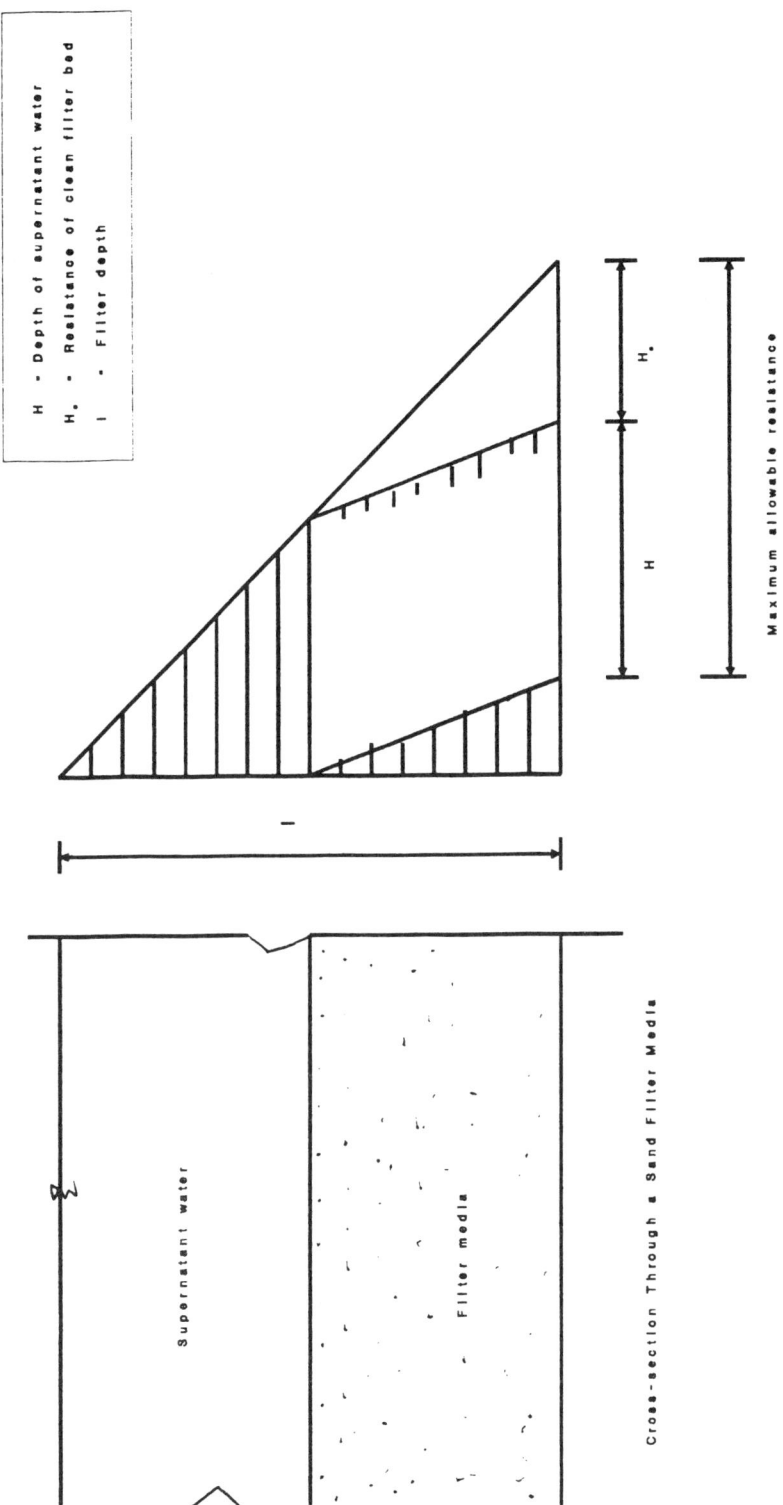

Figure 5.17 Pressure distribution within a slow sand filter.[29,30,35]

amount of adsorbent needed to reduce the contaminant to an acceptable level. A purely empirical equation often used to describe this adsorption or sorption phenomenon is the Freundlich isotherm.[1,5]

$$X/M = KC^{1/n} \qquad [5.46]$$

where: X/M = the mass (X) of the element or contaminant adsorbed from solution per unit mass of adsorbent (M)
K, n = constants fitted from the experimental data. The K intercept can be graphically determined at the point where the log of one of the residual concentrations of contaminant is located
$1/n$ = slope (rise over run)
C = concentration of the metal ion or contaminant in the solution phase at equilibrium

By taking the logarithm of both sides, Equation 5.46 is converted to a linear form[1,5] as presented in Equation 5.47.

$$\log(X/M) = \log K + (1/n) \cdot \log C \qquad [5.47]$$

If the experimental data fits the Freundlich adsorption isotherm a plot of log (X/M) vs. log C gives a straight line, as shown in Figure 5.18.

If a vertical line is erected from a point on the horizontal scale corresponding to the initial contaminant concentration (C_0) and the isotherm extrapolated to intersect that line, the X/M value at this point of intersection can then be read from the vertical scale. The value of $(X/M)_{C_0}$ represents the amount of contaminant adsorbed per unit weight of adsorbent when that adsorbent is in equilibrium with the initial contaminant concentration. This represents the ultimate sorption capacity of the adsorbent for that contaminant.

The Freundlich equation is most useful for dilute solutions over small concentration ranges. The 1/n value represents the slope or change in rate of effectiveness in uptake with varying amounts of adsorbent, and K (the ordinate intercept) the fundamental effectiveness of the adsorbent. High K and high 1/n values indicate high adsorption capacities.

Both graphical and computer analysis can be made for the Freundlich adsorption isotherm. A computer program for analysis of Freundlich adsorption isotherm is included in the Appendix (Table 13).

Example 5.8

The following is adsorption or sorption data for zinc by Dirab (Saudi Arabia) soil after 14 days of contact, using a continuously agitated slurry technique.

Determine the K and n constants, the $(X/M)_{C_0}$ value and the ultimate capacity of the Dirab soil for the sorption of zinc at an initial concentration of 1.65 mg/L by both graphically and by computer methods.

Dirab Soil Data for Freundlich Isotherm for Sorption of Zinc (14 days of contact)

Dirab soil concentration (mg/L) M	Residual zinc[a] concentration (mg/L) C	Zinc[a] adsorbed (mg/L) X	X/M (mg/mg)	X/M (mg/g)
Blank 0	1.65	0	0	0
250	0.90	0.75	0.003	3
500	0.47	1.18	0.00236	2.36
1000	0.30	1.35	0.00135	1.35
2000	0.15	1.50	0.00075	0.75
4000	0.12	1.53	0.00038	0.38

[a] Determined by atomic adsorption spectrophotometry.

From the graphical presentation the constants K and n were found to be 0.0034 for K at a zinc concentration of 1 mg/L, and 1.2 for n. From the computer program presented in the Appendix, which

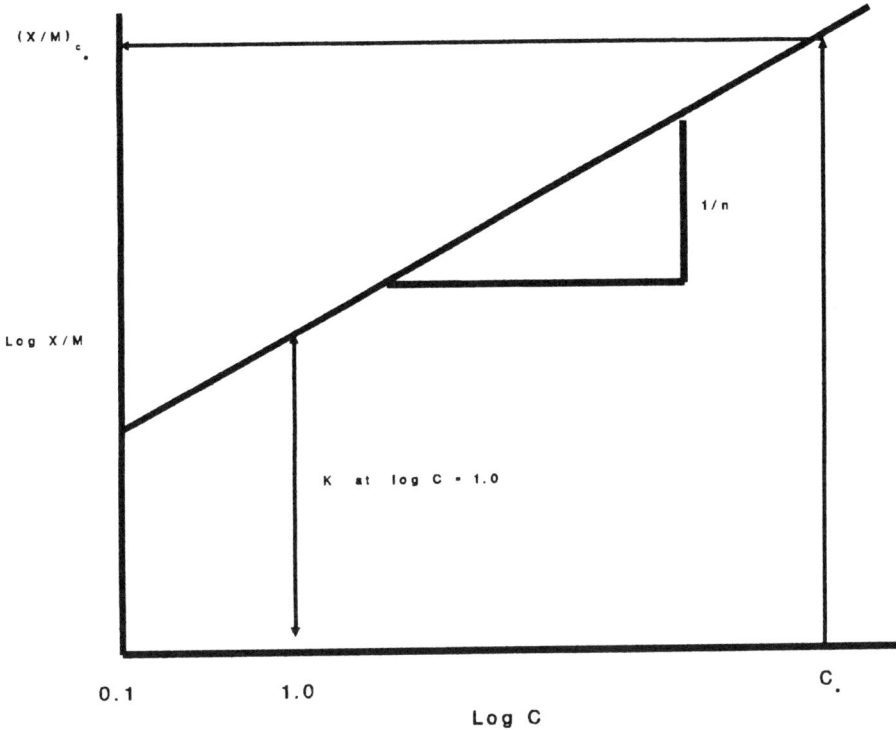

Figure 5.18 Determination of constants for the Freundlich adsorption isotherm (on log-log axis).

also includes a least squares regression analysis, the K value was found to be 0.0040 and the n value 1.02. The graphical $(X/M)_{C_8}$ value was found to be 0.0052 while the computer program gave a value of 0.0066. The $(X/M)_{C_0}$ value gives the ultimate capacity of the Dirab soil for sorption of zinc at the initial zinc concentration, which in this case was 1.65 mg/L. Using the computer program value, the ultimate capacity of the Dirab soil for the sorption of zinc at an initial concentration of 1.65 mg/L would be (0.0066·100 mg/g) 6.6 mg of zinc per gram of soil.

Should the laboratory data not fit the Freundlich isotherm, then other isotherms such as the Langmuir and Brunauer, Emmett, and Teller (BET) isotherms should be tried.

C. DESALINATION
1. Introduction

Desalination signifies the production of water suitable for human consumption from saline waters such as sea water and brackish water. The dissolved solids concentration of various types of water are presented in Table 5.4.

Desalination also refers to a process in which the energy required to separate water and salts, which are found together in raw sea water, is supplied in a predetermined and well-designed form.

Fundamentally, desalination of sea water can be accomplished by many different processes (see Figure 5.19).

1. *Distillation.* Distillation denotes the evaporation of water from a solution and its consequent condensation.
2. *Freezing.* Freezing of a solution of saline water enables the formation of ice crystals of pure water. The crystals are further separated from the mixture, and remelted.
3. *Reverse osmosis.* Reverse osmosis makes use of special membranes. By application of an appropriate pressure, pure water molecules may be made to flow through the membranes while dissolved substances are retained.

Figure 5.19 illustrates a comprehensive summary of various desalination methods.

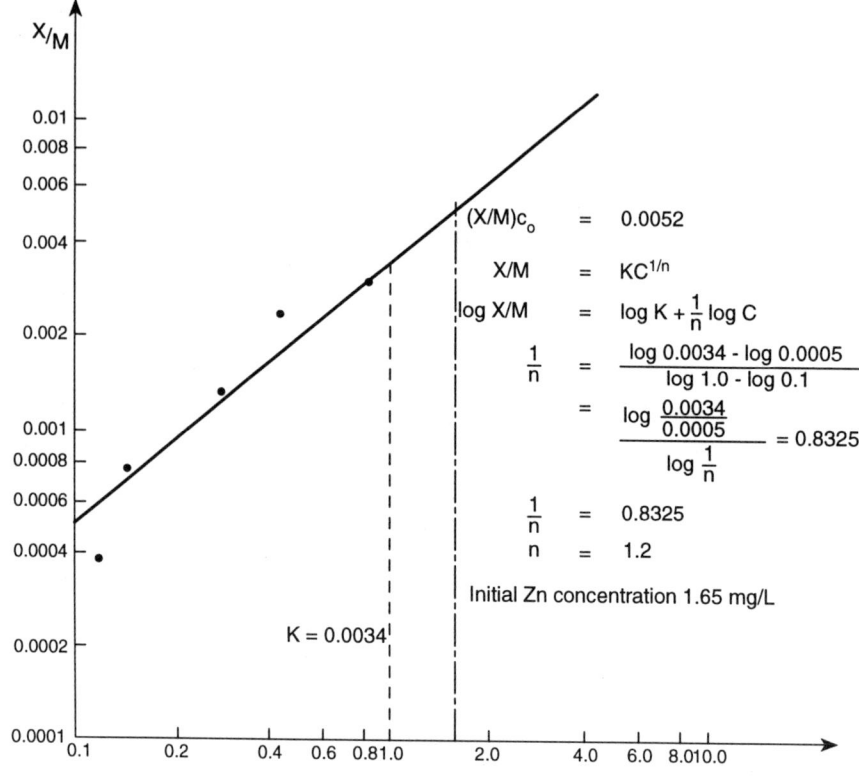

Example 5.21

Table 5.4 Dissolved Solids Content for Various Types of Water

Type of Water	Total dissolved solids (TDS) (mg/L)
Brackish water	1,500–12,000
Sea water (Middle East)	50,000
Sea water (North Sea)	35,000

Compiled from References 44 and 45.

2. Classes of Desalination Processes

The major classes of desalination processes may be grouped in two broad categories: namely, the thermal and power processes.

The thermal processes take their main input energy in the form of heat such as the case in multistage flash distillation (MSF), and multiple effect boiling distillation (MEB).

The power processes take their main input energy in the form of work. Examples of these processes are vapor compression distillation (VC), reverse osmosis (RO), electrodialysis (E), and freezing.

Every one of these processes is regulated by the same fundamental thermodynamic laws, for instance, boiling or freezing, such as the application of boiling point elevation in distillation. Other examples include the introduction of the freezing point depression concept in the freezing process, or the application of osmotic pressure value to the osmosis mechanism.

3. Distillation

a. Introduction

Distillation is a method of desalinization whereby saline water is boiled or evaporated in an appropriate container to produce two separate streams. One stream, with a low concentration of dissolved salts, is

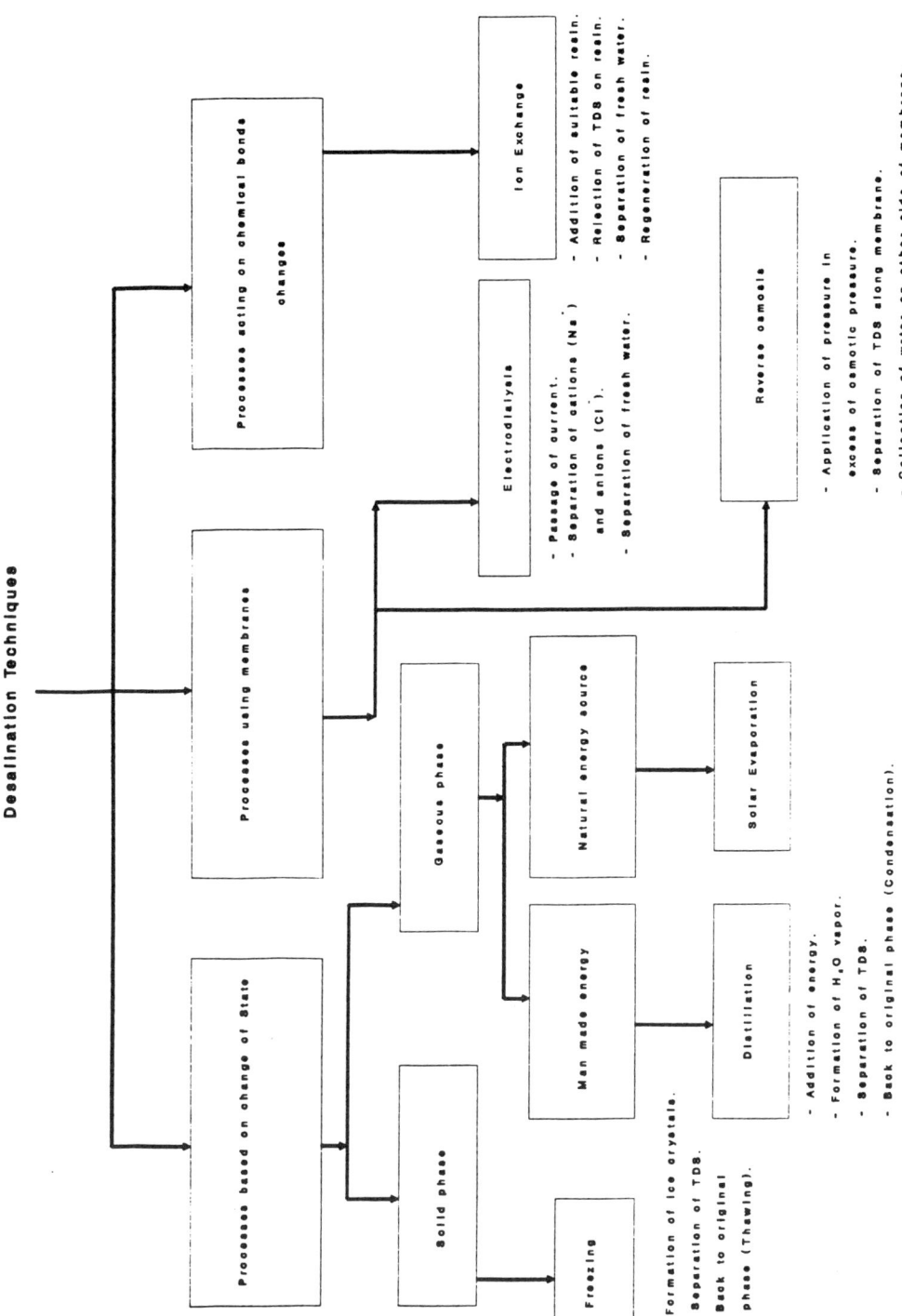

Figure 5.19 Desalination techniques.

denoted as the fresh water stream. The second stream contains the remaining dissolved salts, and is called the concentrate or brine stream. The vapors or stream are condensed to provide pure water.

The advantages of the distillation method include:

- The removal of organisms contained in the feed water, such as bacteria, viruses, etc.
- The removal of nonvolatile matter that can be present in the feed, such as dissolved gases like carbon dioxide (CO_2) and ammonia (NH_3).

Over 60% of the freshwater distilled from sea water is produced with heat.[44] From an economical and practical point of view, the boiling point of the water is governed by the atmospheric pressure of the water being boiled.

The reduction in boiling point is of paramount importance for both multiple boiling effect and for scale control.

Scale formation limits the top temperature which can be used in the distillation process. Chemicals that promote the formation of scale include gypsum ($CaSO_4$), carbonates (CO_3^{2-}), and hydroxides (OH^-). Gypsum causes major problems such as loss of efficiency due to scale formation on heat exchanger surfaces, difficulty in removal of scales, and closure of the plants to allow scale removal.

The concepts of pressure regulation and the multiple boiling technique create conditions for the successful implication of the process in different parts of the world.

b. Principle of the Distillation Process

Essentially, the process pertains to a change in phase. For water distillation two heat exchangers are required. One heat exchanger is for evaporating the raw water to steam while the other aids vapor condensation. Figure 5.20 shows the classic single-effect still.

The solids condense on the heat transfer surfaces and deposit a scale. Scale has been widely[2,44–46] defined as:

- Hard crystalline deposits which stick to heat transfer surfaces. This type of scale may be removed by physical methods, e.g., chipping or drilling.
- A deposit formed by precipitation from the solution of a material whose solubility decreases with an increase in the temperature.
- A dense crystalline deposit that is well bonded to the metal surface.

Types of scales encountered include:

1. *Alkaline scales.* Alkaline scales include essentially calcium carbonate ($CaCO_3$) and magnesium hydroxide ($Mg(OH)_2$), due to the alkalinity present in sea water. Scale formation limits the top temperature that can be used in a distillation process.
2. *Nonalkaline scales.* Nonalkaline scales include calcium sulfate ($CaSO_4$) and calcium phosphates and silicates. Calcium sulfate can precipitate from aqueous solutions in three discrete crystalline forms, namely: anhydrite ($CaSO_4$), hemihydrate ($CaSO_4 \cdot \frac{1}{2}H_2O$), and dihydrate ($CaSO_4 \cdot 2H_2O$). The development of calcium sulfate scale in plants that desalinize sea water can present significant difficulties such as loss of efficiency due to the insulating properties of the scale formed on surfaces of heat exchangers, difficulty in extraction of calcium sulfate scale which is insoluble in mineral acids, and shutting down of desalination plants to allow mechanical removal of formed scale. The practical way for detaching the calcium sulfate scale relies on the maintenance of an operating temperature below 120°C and not permitting the concentration of solids to become excessively high.

The scale control methods incorporate:

- Addition of an acid. Acids are added to remove the bicarbonate ion from the water before it is admitted to the desalination unit. Acids used include sulfuric acid (H_2SO_4) and hydrochloric acid (HCl). In multistage flash systems (MSF) the carbon dioxide (CO_2) formed during the bicarbonate decomposition is withdrawn from the feed water, for example, by passing it through a degassing tower.
- Use of chemicals. Chemicals are introduced to prevent or retard the formation of hard scale. Examples of the chemicals used include organic materials like starches, tannins, plant extracts; and polyphosphate-based additives such as sodium hexametaphosphate and polymaleic acid. Usually a small quantity of the chemical is added.
- Sponge ball cleaning (also referred to as Taprogge by the manufacturer). Flexible sponge balls of slightly bigger diameter than the tubes of the evaporators are forced through to scour any deposits from the surface of the tube. The process may be aided by adding an abrasive agent to the balls.

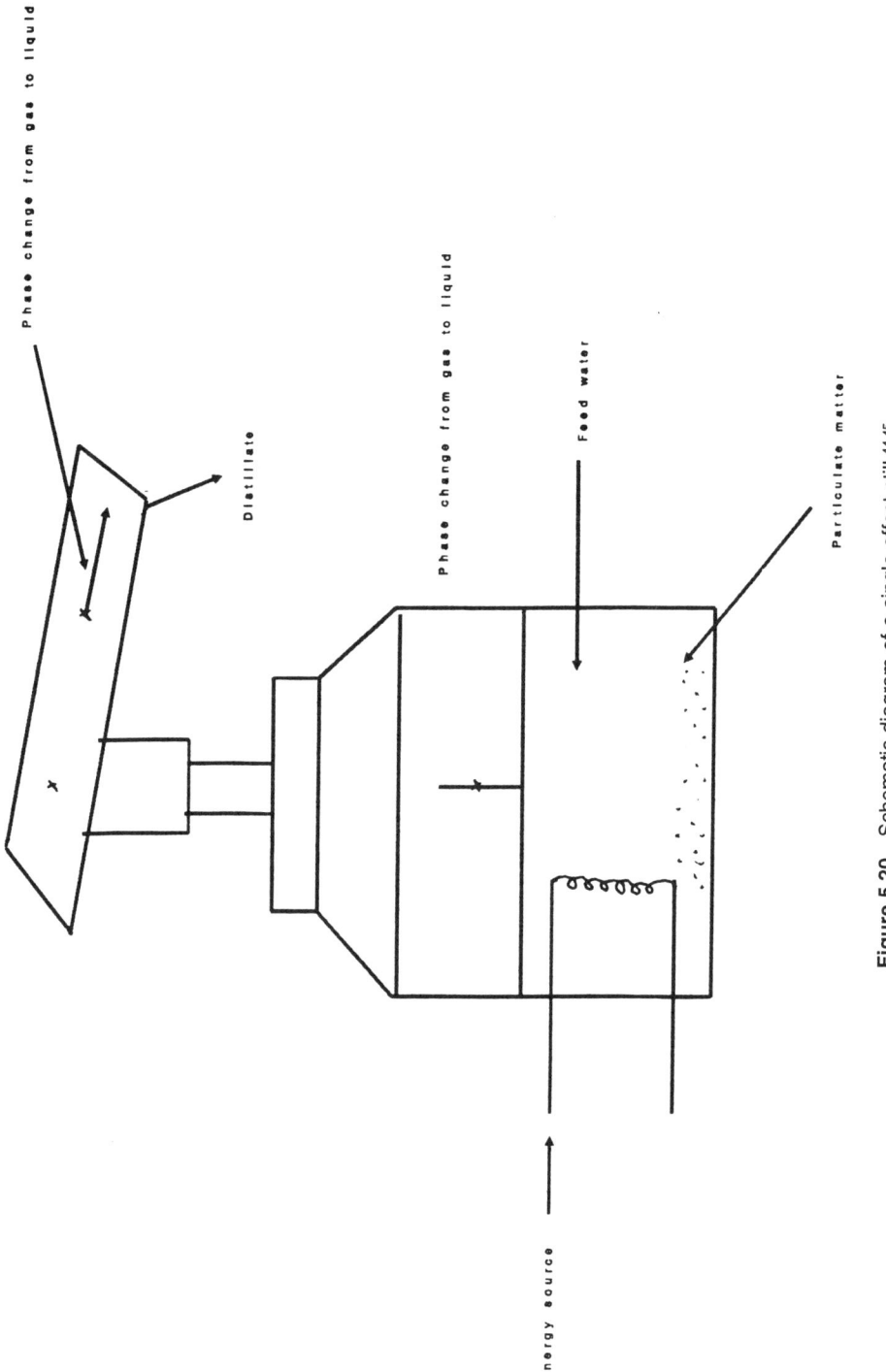

Figure 5.20 Schematic diagram of a single-effect still.[44,45]

- Removal of scale-forming species. Scale-forming species that are removed include the calcium ion (Ca^{2+}), magnesium ion (Mg^{2+}), bicarbonates (HCO_3^-), and sulfates (SO_4^{2-}) ions. Removability of these ions is effectuated by:
  - Acid addition: to remove the bicarbonates (HCO_3^-).
  - Cation-exchange resins: to get rid of the calcium ions (Ca^{2+}).
  - Lime-magnesium carbonate ($Ca(OH)_2 \cdot MgCO_3$) precipitation process: to remove the ions of calcium (Ca^{2+}) and the bicarbonates (HCO_3^-).
  - Use of ion-selective membranes: to permit the passage of monovalent ions through the membrane and reject divalent ions such as calcium (Ca^{2+}), magnesium (Mg^{2+}), and the sulfates (SO_4^{2-}).
- Mechanical/physical techniques of scale avoidance. In this method finely divided substances are added to a supersaturated solution to provide a surface that promotes the growth of crystals. Examples of these substances include calcium carbonate ($CaCO_3$), calcium sulfate ($CaSO_4$), barium sulfate ($BaSO_4$), magnesium hydroxide ($Mg(OH)_2$), glass particles, etc.

The fundamental mechanism in distillation implies a change of phase. In order to distil water, only two heat exchangers are required. One to evaporate water (by steam) and the other to condense the vapor (by cool water).

c. Double Distillation Systems

In a double distillation plant two stills of comparable capacity are commonly joined. The distillate from the first still feeds the boiler of the second one. Water quality is improved through the separation of impurities. Nevertheless, there is a carry-over of certain metal ions like silica and also there is some leakage from the surfaces of heating components.

d. Multistage Flash Evaporators

In a multi-stage flash (MSF) evaporator the system design includes:

- Establishment of the needed plant output.
- Finding the amount of available energy required to furnish the plant output.

An MSF distiller encompasses a number of phases or stages in series.
Examples of antifoams used in sea water distillation plants include:

- Ethoxylated fatty alcohols such as cetyl alcohol.
- Fatty acid esters of polyalkalene glycol, e.g., polyethylene glycol monooleate.
- Silicone compounds.

e. Multiple-Effect Distillation

In multiple-effect (ME) distillation the water in the first phase is heated to the boiling point under high pressure, whereas the second (and subsequent) phases operate at atmospheric pressure. Figure 5.21 illustrates the pressure scheme within an ME distillation unit.

The water vapor emerging from the first phase condenses in the heating tubes of the second stage, while the heat released by the latent heat of condensation and by the lowering of the temperature is used by the liquid water in the second stage. This condition aids the rate of evaporation. Cooling water is used in the last stage to finalize the condensation. Figure 5.22 outlines the system layout of a distillation unit.

The relationship between the heat exchanged in any phase can be represented as in Equation 5.48, whereby the heat exchanged in any operation is given as:

$$Q_i = u_i \cdot A_i \cdot \Delta T_i \qquad [5.48]$$

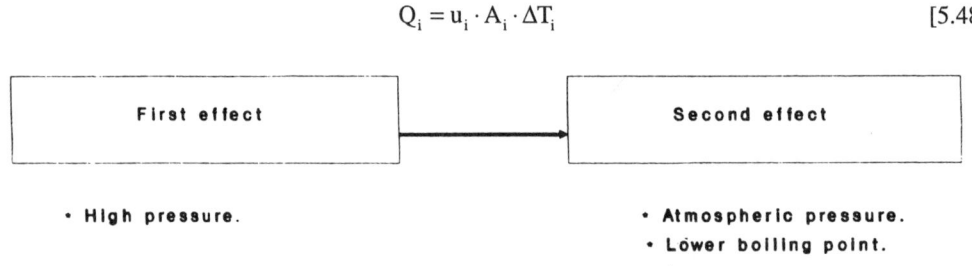

Figure 5.21 Pressure regulation within a multi-effect distillation unit.

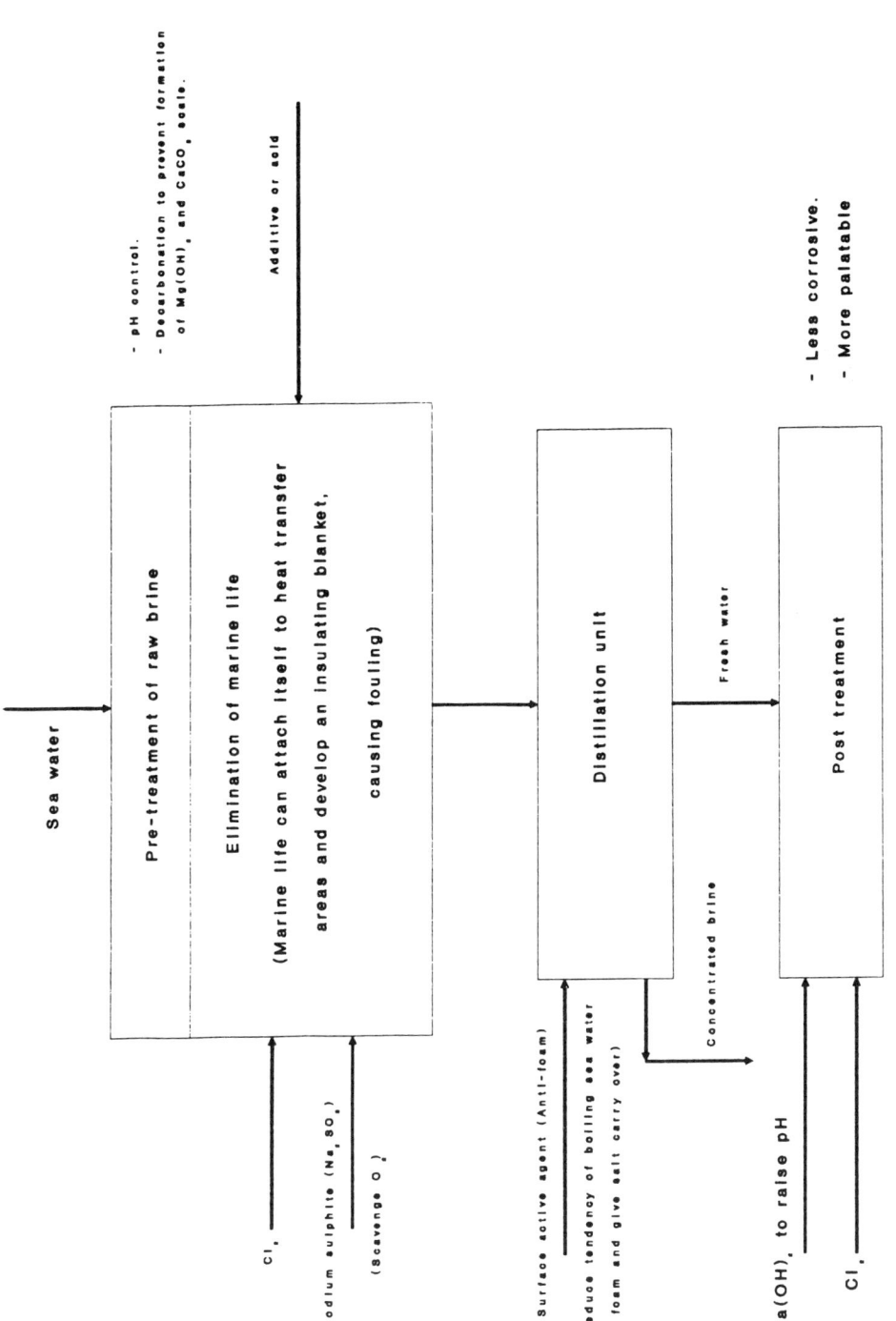

Figure 5.22 Distillation unit layout.

where: Q_i = heat exchanged in effect number i
u_i = heat transfer coefficient for heat exchanger number i
A_i = area of heat exchanger number i
ΔT_i = temperature difference between water in the effect and steam entering the heat exchanger

$$\Delta T_1 = T_0 - T_1 \quad [5.49]$$

T_0 = temperature of heating steam supplied to the first phase
T_1 = boiling temperature in the first phase

$$\Delta T_2 = T_1 - T_2 \quad [5.50]$$

T_2 = boiling temperature in second phase, etc.

The heat transformed into latent heat of evaporation in the first phase yields a certain amount of distillate. This amount of distillate is then utilized as steam in the second phase. In this way a similar amount of heat to the one given in the first stage develops in the second one.

Assuming identical areas of heat exchangers and equivalent amounts of heat transferred in each phase, then the relationship between the heat developed in each phase and the temperature difference can be formulated as presented in Equations 5.51 and 5.52.

$$Q_1 = Q_2 = Q_3 \ldots \quad \text{and} \quad A_1 = A_2 = A_3 \ldots \quad [5.51]$$

$$u_1 \cdot \Delta T_1 = u_2 \cdot \Delta T_2 = u_3 \cdot \Delta T_3 \quad [5.52]$$

Equations 5.51 and 5.52 indicate that the temperature drop in each phase is inversely proportional to the heat transfer coefficient.

Example 5.9

In a desalination unit a triple-phase still is used for water treatment. Dry steam is introduced to the first phase at a temperature of 120°C. Through a decrease in the pressure in the last phase the steam temperature is reduced to 50°C. The three stills have equal areas of heat exchange and have heat transfer coefficients in the ratios of 5:4:2, respectively. Determine the temperature difference in each phase.

Solution

1. Given: $u_1:u_2:u_3$ = 5:4:2, T_1 = 120, T_3 = 50.
2. Use Equation 5.52 to determine the temperature differences for the different phases of the still as:

$$u_1 \cdot \Delta T_1 = u_2 \cdot \Delta T_2 = u_3 \cdot \Delta T_3$$

3. Determine the total temperature drop in the three phases as equal to $\Delta T_1 + \Delta T_2 + \Delta T_3 = \Delta T = 120 - 50 = 70$.
4. Let $a = u_1 \cdot \Delta T_1 = u_2 \cdot \Delta T_2 = u_3 \cdot \Delta T_3$, then:

$$\Delta T_1/\Delta T = (a/u_1)/(a/u_1 + a/u_2 + a/u_3)$$

Multiply by u_1/a, then:

$$\Delta T_1/\Delta T = (1)/(1 + u_1/u_2 + u_1/u_3)$$

Similarly:

$$\Delta T_2/\Delta T = (1)/(u_2/u_1 + 1 + u_2/u_3), \text{ and } \Delta T_3/\Delta T = (1)/(u_3/u_1 + u_3/u_2 + 1).$$

5. Substitute values in Steps 1 and 3 in the equations formed in Step 4 above to find temperature differences as:

$$\Delta T_1/70 = (1)/(1 + 1.25 + 2.5).$$

This yields ΔT_1 = 14.7°C. Similarly: ΔT_2 = 18.4°C, and ΔT_3 = 36.8°C.

f. Solar Distillation

i. Introduction

Most of the conventional distillation units consume energy which is normally derived from electric power and fossil fuel for their operation. Nonetheless, solar energy may successfully be utilized in distillation plants despite the fact that it is a lower-grade energy. The techniques employed in distillation plants using solar energy have the following advantages:

- They are relatively simple.
- A very large component of the labor and materials used in solar stills can be indigenous.
- Most maintenance can be carried out by unskilled personnel.

ii. Basic Principles of Solar Distillation

A conventional solar still is simply an airtight rectangular basin, commonly manufactured by using galvanized iron sheets as the raw material (see Figure 5.23).

The most important parameter affecting the output of a solar still is the intensity of the solar radiation incident upon it.

The daily production of distilled water is defined as the ratio between the energy used in vaporizing the water from the still as related to the latent heat of vaporization of the water, or in a mathematical format, it may be as presented in Equation 5.53:

$$Q_D = E_{vp}/L_v \quad [5.53]$$

where: Q_D = daily output of distilled water, kg/m³/d
E_{vp} = energy used in vaporizing the water from the still, J/m²/d
L_v = latent heat of vaporization of the water, J/kg

The efficiency of the still may be determined from Equation 5.54:

$$\text{Eff} = E_{vp}/E_t \quad [5.54]$$

where: Eff = efficiency of the still in distilling the water, % (Usually, the efficiency does not exceed a value of 35% for a typical basin-type solar still)
E_{vp} = energy utilized in vaporizing the water from the still, J/m²/d
E_t = amount of the solar energy incident on the glass cover of the still unit, J/m²/d

Solar energy is a boundless, continuous, and renewable energy source. Nevertheless, this process is expensive in the sense that the capital cost of the collecting equipment is high. Major drawbacks of the solar distillation process include the following:

- Solar energy is a diffused and intermittent form of energy.
- Solar energy is only obtainable during the daytime.
- Solar energy depends on climatic factors and weather conditions.
- The intensity of solar energy changes seasonally.

Incident energy onto a solar still is a direct or an indirect form of radiation. In direct radiation the solar flux incoming at the collector is considered not to have suffered any scattering by the surrounding atmosphere. The indirect or scattered radiation refers to radiation that is scattered by elements of the atmosphere, or by the clouds and sometimes the surrounding earth surfaces.

The emerging energy from the still is comprised of any or a combination of convection, radiation, and reflection to the atmosphere; ground and edge conduction and convection; vapor and brine leakage from the basin; and the sensible heat of condensate and overflow.

The variables that affect the performance of the still include:

- The ambient temperature.
- The insolation intensity.
- The tightness of the unit in stopping the escape of vapors.
- The amount of heat losses through the base of the still.

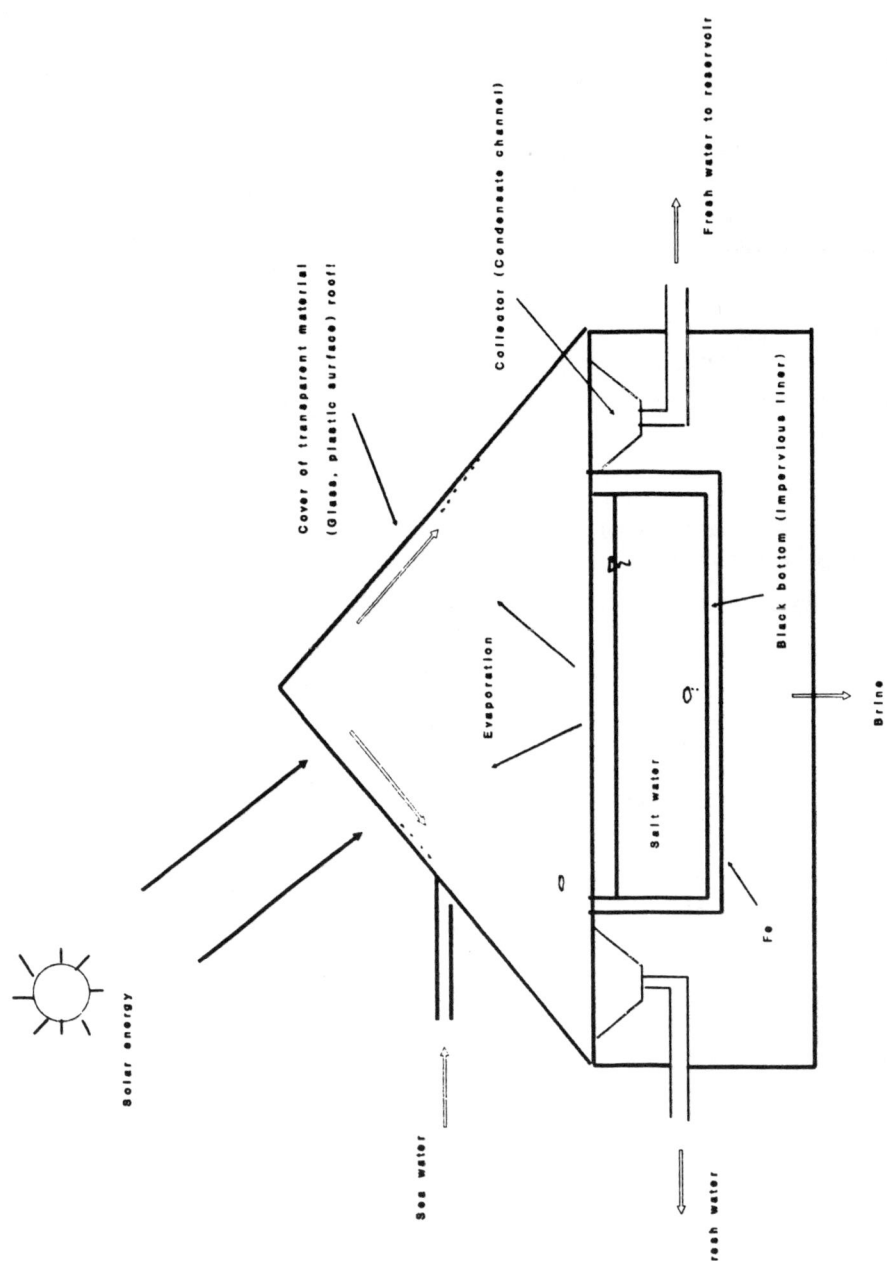

Figure 5.23 Schematic diagram of a conventional solar still.[44-46]

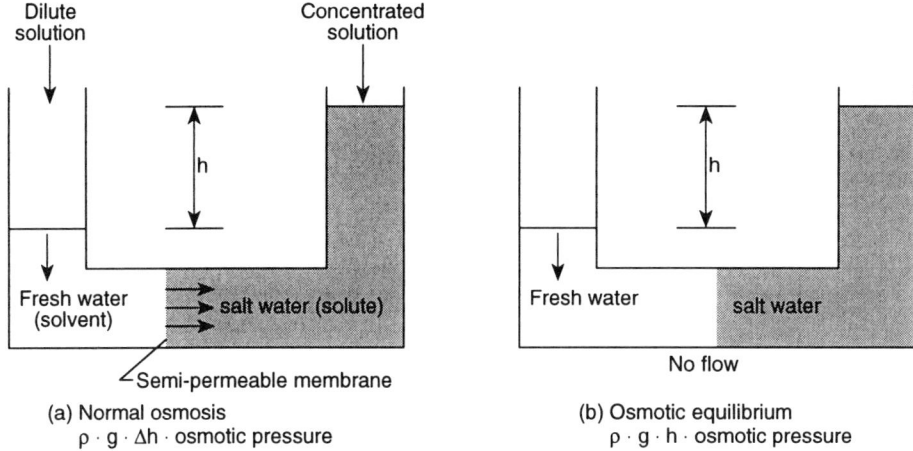

Figure 5.24 Normal osmosis and osmosis equilibrium concepts.

- The depth of brine: it is to be noted that the shallower the layer of brine in the still, the higher is the efficiency of production.
- The slope of the still cover.

4. Osmosis

a. Introduction

Osmosis is a word derived from the Greek word "osmos" which means impulse. Osmosis is the transfer of a solvent via an impermeable membrane to a solute. The flow is directed from the more diluted to the more concentrated solution. Figure 5.24 illustrates the concepts of normal and reverse osmosis processes.

The flow through the membrane can be prevented by increasing the pressure on the side where there is a higher solids concentration. The pressure needed to stop the flow is called osmotic pressure.

Osmotic pressure is a measure of the forces that brings together the solvent molecules. This action causes molecules of pure solvent to pass through the membrane into the solution. Pure solvent molecules replace other molecules that have been restricted by interactions with the solute. As such, the osmotic pressure depends upon the number, and not the kind, of the particles of solute in solution.

The flow of solvent via a membrane results in the establishment of a driving force. This driving force can be estimated by the difference in vapor pressure of the solvent on either side of the membrane. The transfer of solvent through the membrane from the dilute to the concentrated solution proceeds until the phase of hydrostatic pressure exceeds the driving force of the vapor pressure differential. For an incompressible solvent, the osmotic pressure at equilibrium may be determined as:

$$P_{osm} = \left(R_g \cdot T/V\right) \cdot \text{Ln}\left[P_o/P\right] \qquad [5.55]$$

where: P_{osm} = osmotic pressure, atmospheres
R_g = universal gas constant for all ages.

The universal gas constant (R_g) value can be found from the relationship:

$$R_g = P \cdot V/n \cdot T \qquad [5.56]$$

Since 1 mol of an ideal gas at atmospheric pressure (1 atm) occupies a volume of 22.414 L at 273 K; then:

$$R_g = (1 \cdot 22.414)/(1 \cdot 273) = 0.082 \text{ L} \cdot \text{atm/mol/K}$$
$$= 8.314 \text{ J/K/mol}$$

where: T = temperature, K
P_o = vapor pressure of solvent in dilute solution
P = vapor pressure of solvent in concentrated solution
V = volume per mole of solvent = 0.018 L for water
n = number of moles

Generally, the occurrence of a nonvolatile solute in a liquid decreases the vapor pressure of the solution. This phenomenon is attributed to a physical blocking phase at the surface of the liquid when there are particles, ions, or molecules of the solute.

Raoult's law assumes that for dilute solutions the reduction in vapor pressure of a solvent is directly proportional to the concentration of particles in solution. The law states that the extent of the physical blocking effect or depression of the vapor pressure is directly proportional to the concentration of the particles in solution.

The effect is directly related to the molal concentration for solutes that do not ionize. (A molal solution consists of one gram molecular weight dissolved in one liter of water, the resulting solution having a volume slightly in excess of one liter).

For solutes that do ionize, the effect is proportional to the molal concentration times the number of ions formed per molecule of solute modified by the degree of ionization. Raoult's law is only valid for dilute solutions.[47] Therefore, Equation 5.55 can be rearranged to relate osmotic pressure to the molar concentration of particles in the concentrated solution. (A molar solution consists of one gram molecular weight dissolved in enough water to make a one liter of solution).

$$P_{osm} = C \cdot R_g \cdot T \qquad [5.57]$$

where: P_{osm} = osmotic pressure, atmospheres
C = molar concentration of particles
R_g = universal gas constant for all gases
T = temperature, K.

Equation 5.57 is strictly valid for dilute solutions in which Raoult's law holds true.[47]

Example 5.10

A sample of water was analyzed for major ions. The concentration of the tested ions, in mg ion/L, are as follows:

Cations: Na^+ = 0.5, Mg^{2+} = 0.1, Ca^{2+} = 0.5, K^+ = 0.04.
Anions: Cl^- = 0.9, HCO_3^- = 0.06, NO_3^- = 0.1, SO_4^{2-} = 1.1.

The water sample was placed on one side of a semipermeable membrane while distilled water was positioned on the other side. Find the difference in osmotic pressure across the semipermeable membrane (assume the temperature is 20°C).

Solution

1. Determine the molar concentration of particles in the water sample as: molar ion concentration = concentration of ion (mg/L)/MW, where MW = molecular weight. Thus, Na^+ = 0.5/23 = 0.022; and similarly, Mg^{2+} = 0.004, Ca^{2+} = 0.013, K^+ = 0.001, Cl^- = 0.025, HCO_3^- = 0.001, NO_3^- = 0.002, SO_4^{2-} = 0.011.

2. Find the molar ion concentration of the sample of water as:

$$C = 0.022 + 0.004 + 0.013 + 0.001 + 0.025 + 0.001 + 0.002 + 0.011 = 0.079 \ M.$$

3. Determine the osmotic pressure by using Equation 5.57 as:

$$\begin{aligned} P_{osm} &= C \cdot R_g \cdot T \\ &= 0.079 \ (mol/L) \times 0.082 \ (L \cdot atm/K/mol) \ (273.16 + 20)K \\ &= 1.9 \ atm. \end{aligned}$$

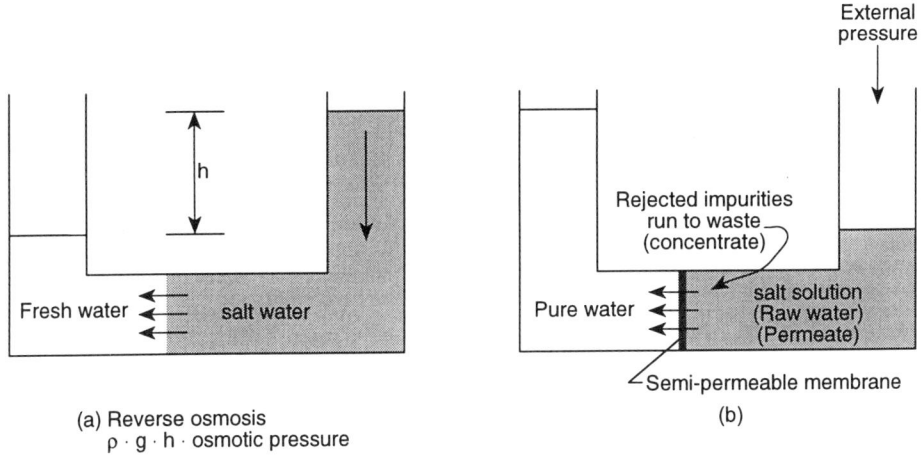

Figure 5.25 Reverse osmosis process.

b. Reverse Osmosis

Reverse osmosis is a physical process by which the dissolved substances in a solvent are separated with the aid of a semipermeable membrane.

By employing a pressure in excess of the natural osmotic pressure to the raw water, the membrane selectively permits the solvent (i.e., water) molecules to pass through while a high proportion of the electrolytes and dissolved organic materials are rejected (see Figure 5.25).

c. Advantages of the Reverse Osmosis Process

The advantages of the reverse osmosis process include:[2,44–46]

- The process is used for the purification of a raw water of a solids concentration ranging from 700 mg/L upwards, or it can be used for the treatment of brackish water, saline well water, or sea water.
- The reverse osmosis (Ro) process reduces the total dissolved solids of raw water. Reductions of up to 99% have been reported.
- RO removes biological and colloidal matter from water. Reductions of up to 98% have been reported.
- RO rejects microbial cells such as bacteria, viruses, and pathogens. The rejections achieved reached, in some cases, 100%.
- The process removes most dissolved organics — reductions of up to 97% have been recorded.

Figure 5.26 illustrates the layout for a RO process for desalinization of water. The figure also includes a summary of pretreatment objectives.

d. Reverse Osmosis Flow Process

The flow process in a reverse osmosis plant is based on the terms of the thermodynamics of an irreversible operation. The flow equation may be presented as follows:

$$Q_w = -\text{Diff} \cdot C_w \cdot V_w \cdot (\Delta P - \Delta P_{osm})/(R_g \cdot T \cdot B) \qquad [5.58]$$

where: Q_w = water flux
Diff = diffusion coefficient
C_w = concentration of water in membrane
V_w = partial molar volume of water in membrane
ΔP = applied pressure differential across membrane
ΔP_{osm} = osmotic pressure differential across membrane
R_g = universal gas constant
T = temperature
B = membrane thickness

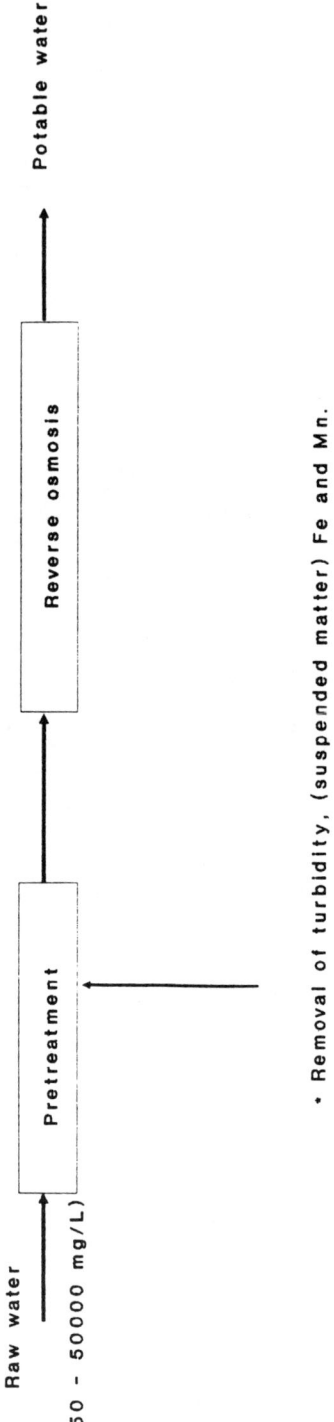

Figure 5.26 Pretreatment for the RO process.

Equation 5.58 can be simplified as:

$$Q_w = K_w \cdot A \cdot (\Delta P - \Delta P_{osm})/t \qquad [5.59]$$

where: Q_w = water flux
K_w = membrane permeability coefficient for water
A = membrane area
t = time.

The salt flow, Q_s, through the membrane is given by:

$$Q_s = -D_s \cdot (dC_s/dk) \qquad [5.60]$$

or

$$Q_s = k_s \cdot A \cdot \Delta c_s/t \qquad [5.61]$$

where: Q_s = salt flow
k_s = membrane permeability for salt
A = membrane area
ΔC_s = concentration differential across membrane
t = time.

The aforementioned equations illustrate that for an increase in the operating pressure there is an increase in water flow, while the salt flow remains constant.

e. Selection of Membranes

In membrane selection the main requirements include:

- Membrane needs to have a high salt rejection quality.
- There must be an adequate water flux.
- Membrane needs to be stabile.
- Membrane needs to be easily installed in a practical reverse osmosis device.
- A large area of very thin membrane must be supported so that it can withstand high pressure. Pressures applied are usually in the range of 3 to 7 mPa.
- Membrane must provide reliable performance in a commercial permeator.
- Membrane needs to have a high mechanical strength.
- Service life of membrane must be adequate.
- Membrane must possess a wide operating range in terms of ion content of the water source, pressure, temperature, resistance to chemical and biological attack, and the versatility to fit different requirements.
- The cost of the membrane must be low.
- Tendency for fouling of the membrane needs to be low, likewise, the membrane must be easy to clean.

Some of polymers selected for membranes include:

- Cellulose-based polymers (e.g., cellulose acetate, cellulose triacetate, and cellulose acetate butyrate).
- Commercial polymers (e.g., 66-nylon, polyethylene terephthalate, polyvinyl alcohol, and polyacrylonitrile).
- Graft copolymers (e.g., 66-nylon graft polyacrylic acid).
- Miscellaneous experimental polymers (e.g., acrylonitrile-N-[2-hydroxyethyl] acrylamide, polyaminopivalic acid, ethylene-N-[sulfoethyl]-methacrylamide).
- Nitrogen-linked polymers (e.g., aliphatic polyamides, aromatic polyamides, aliphatic/aromatic polyomides, polyimides, etc.).
- Polymer blends (e.g., polyvinyl alcohol — polyvinyl pyrrolidone — polyacrylonitrile — polyvinyl tetrazole).

f. Pretreatment of the Feedwater to an Osmosis Plant

Pretreatment of the feedwater before its introduction to a reverse osmosis plant is needed for different reasons, such as:

- Prevention of the hydrolysis of cellulose acetate-based membranes — this may be solved by an appropriate pH adjustment.
- Safeguard of the polyamide membranes (a problem that may be cured through dechlorination!).
- Prevention of membrane fouling.

Fouling involves trapping of some type of material within the RO device or on the surface of the membrane. Potential problems include membrane scaling, metal oxide precipitation, device plugging, colloidal fouling, and biological fouling.

Membrane scaling is due to deposition of salts that are dissolved in the feed water. The salts may include calcium carbonate, calcium sulfate, silica, strontium sulfate, barium sulfate, and calcium fluoride. Methods used for the control of scaling include:

- Conversion control to avoid exceeding the solubility of the salts.
- Elimination of one of the ions responsible for a scale-forming material (methods used to accomplish this include softening, adjusting pH by addition of sulfuric acid or hydrochloric acid to precipitate calcium carbonate and reduce scaling).
- Inhibiting the crystal growth of a scale-forming substance.

In metal oxide precipitation the most important elements are iron and manganese. Iron fouling involves the oxidation of the divalent ion (Fe^{2+}) to the trivalent ion (Fe^{3+}) ion and the settling of ferric hydroxide, $Fe(OH)_3$. Iron fouling can be avoided by using the following methods:

- Removal of the ion from feedwater by employing sedimentation and/or filtration, aeration and filtration, sodium softening resin, etc.
- Prevention of oxidation from the ferrous to the ferric state. This may be practiced by using, for example, citric acid adjusted to pH 4 with NH_4OH.

Device plugging is produced by mechanical filtration in which large particles are entrapped in the system. The device plugging problem is not of practical significance.

Colloidal fouling is brought about by the capture of colloids on the surface of the membrane. This is mainly due to coagulation of colloids. The colloids usually found pertain to the class of aluminum silicate (clays). These colloids are hydrophobic in nature, with a diameter ranging between 0.3 to 10 µm. The rate of colloidal fouling is managed by the concentration and stability of colloids. Generally, pretreatment includes:

- Filtration through sand, carbon, or diatomaceous earth.
- Coagulation using alum and iron salts, or using sedimentation followed by slow filtration.
- Addition of nonionic or anionic polyelectrolyte besides application of pressure filtration units.
- Ion exchange softening.

Cleaning methods employed to free the colloidally fouled membranes include hydrochloric acid at a pH of around 2.5, and sodium hydroxide at a pH of about 11, or citric acid.

Biological fouling can be caused by the growth of microorganisms in a reverse osmosis apparatus.

The problem of biological degradation of a membrane can be solved through the usage of an aromatic polyamide membrane. This membrane is resistant to bacterial attack. Thus, there is no need for disinfection. Since chlorine reacts with the membrane material, it is necessary to dechlorinate water supplies containing chlorine before introducing them to the reverse osmosis system. Biological growth can also be prevented by treatment with a formaldehyde solution.

The usual scalant found in sea water is calcium carbonate ($CaCO_3$) and it can be prevented by adjusting the pH from 6 to 6.5.

The economical life of a membrane is expected to be in the range of five years for water purification and treatment of sea water.

g. Reverse Osmosis Plants

Reverse osmosis plants can be classified according to their method of application as:

- Plants used for purification of fresh and potable water — product water from such facilities is intended for nuclear and fossil fuel boilers as coolant and process water, or to be used in industry as process and rinse water, or it may enter into the electronic and pharmaceutical production as rinse and dilution water.
- Plants for desalination of saline groundwater wells with a salinity of 1000 to 2500 mg/L, or for the treatment of brackish waters with a salinity of 2000 up to 9000 mg/L, or for the desalination of sea water up to 50,000 mg/L.

Figure 5.27 gives a schematic diagram of a reverse osmosis system.

Figure 5.28 illustrates the methodologies of pretreatment and posttreatment as applied in reverse osmosis plants.

Example 5.11

A salt solution at a temperature of 25°C has a vapor pressure of 3.1 kPa. The vapor pressure of pure water at this temperature is 3.17 kPa. Determine the osmotic pressure per unit volume.

Solution

1. Given: T = 25°C,
   P_0 = vapor pressure of solvent in dilute solution = 3.17 kPa
   P = vapor pressure of solvent in concentrated solution = 3.1 kPa
2. Use R_g = 8.314 J/K/mol and find temperature as:
   T = 273.16 + 25 = 298.16 K
3. Substitute in Equation 5.55 ($P_{osm} = (R_g \cdot T/V) \, Ln \, [P_0/P]$) to compute the osmotic pressure as:
   P_{osm} = (8.314 × 298.16) Ln (3.17/3.1)
   = 55.4 Pa/unit volume.

5. Electrodialysis

a. Introduction

Electrodialysis is a process in which ions are conveyed through ion-selective membranes from one solution to another under the action of a direct current electrical potential.

b. Electrodialysis Technology

Usually, the electrodialysis device is an array of alternating anion-selective and cation-selective membranes across which an electric potential is applied. The membranes are separated from each other by gaskets which form compartments through which fluids can flow (see Figure 5.29).

The general principles governing the operation of an electrodialysis plant are

- Membranes used can be constructed in such a way as to ease the passage of selective ions.
- Ions originate from salts dissolved in the water. Dissociation of the ions produces cations and anions.
- Dissociated ions are attracted to the oppositely charged electrodes.

The number of pairs to be analyzed in a single stack depends on:

- Electrodialysis capacity required.
- Uniformity of flow distribution among the compartments.
- Maximum voltage that can be accepted.

Typical membrane thickness is in the range of 0.15 to 0.6 mm, while the typical thickness of compartments between membranes[45] has dimensions of 0.5 to 2 mm.

The thickness of a cell pair is 1.3 to 5.2 mm (3.2 mm typical).

The effective area of a cell pair for ion transport is in the range of between 0.2 to 2 m^2.

Energy consumption when a concentrated electrolyte is demineralized is given by Equation 5.62:

$$E = i^2 \cdot R \qquad [5.62]$$

where: E = energy consumed, Joules
  i = electric current applied to the stack, amperes
  R = resistance, ohms

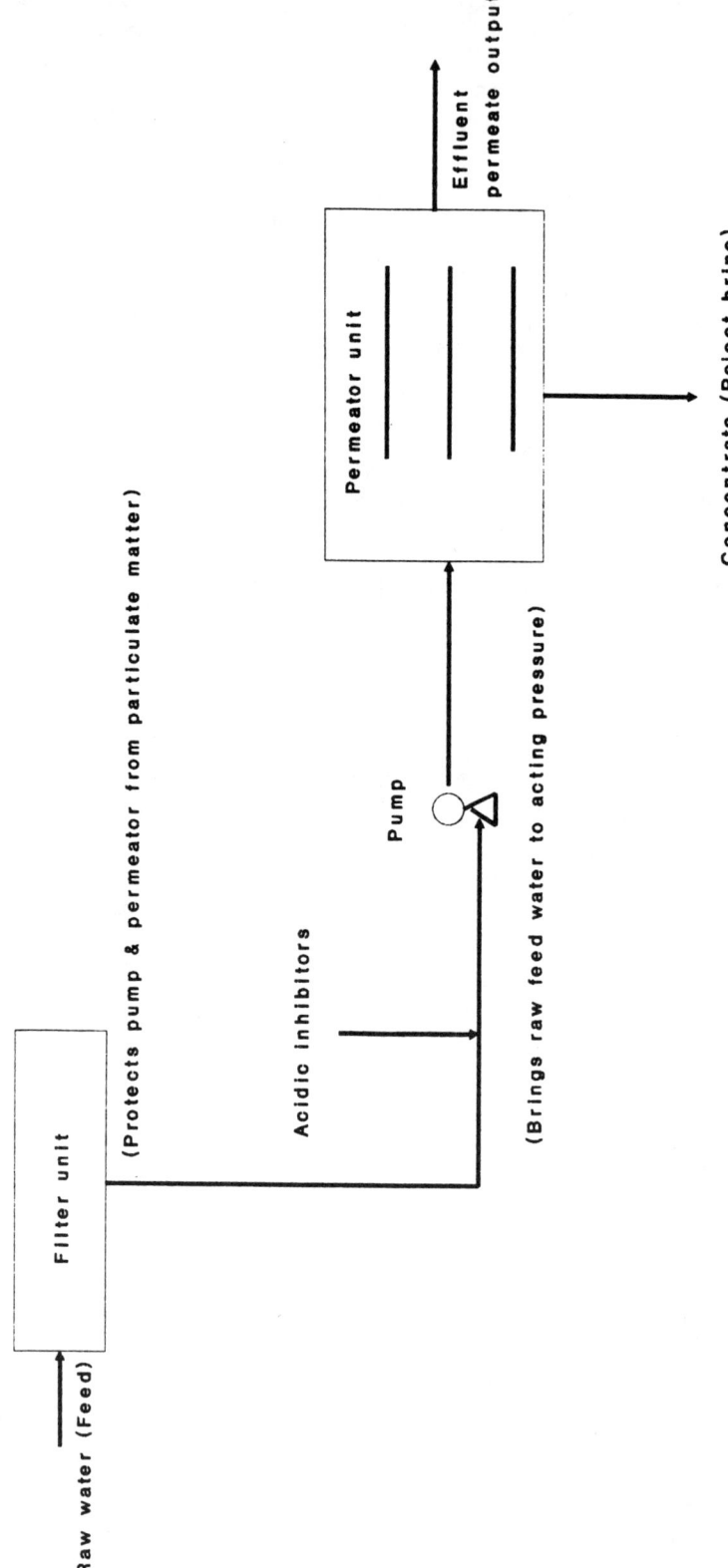

Figure 5.27 Schematic layout of a reverse osmosis plant.

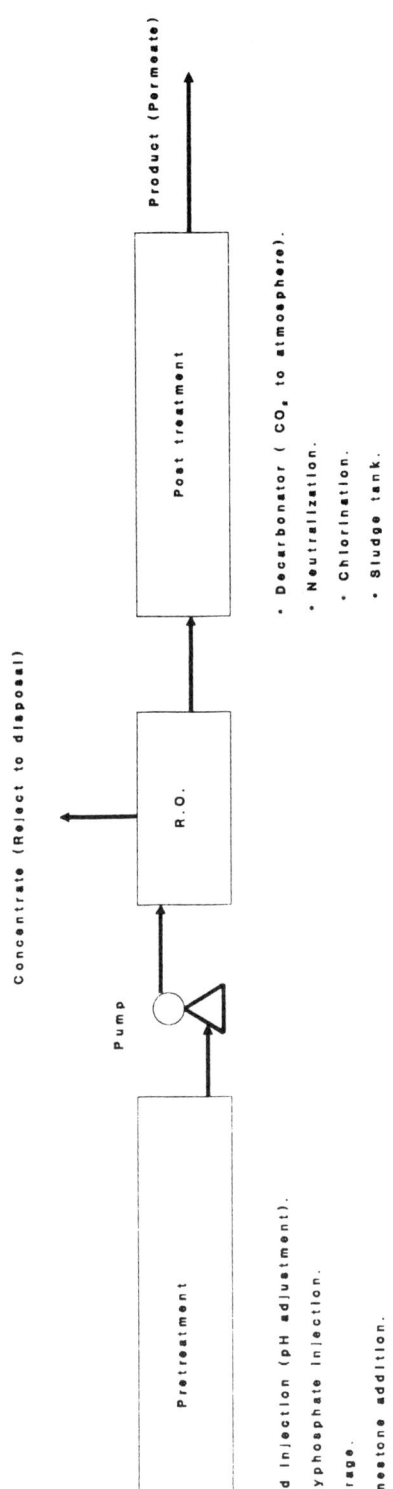

Figure 5.28 Methodologies for pretreatment and posttreatment in reverse osmosis plants.

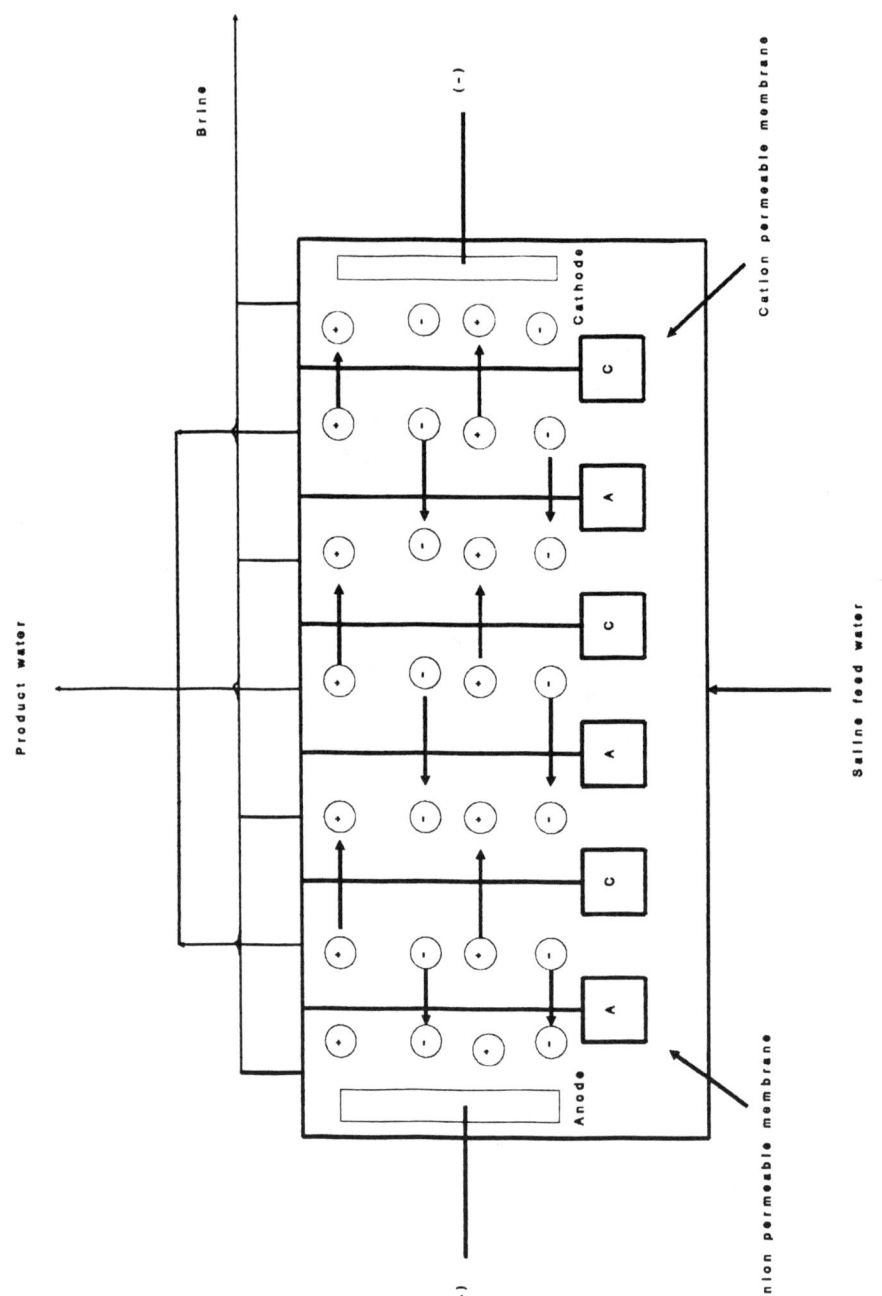

Figure 5.29 Schematic representation of the electrodialysis process.[10,48]

The efficiency of the current to transport counter-ions is between 85 to 95%. In a solution of sodium chloride (NaCl) 60% of the current is carried by Cl⁻ ions and 40% by Na⁺ ions. Thus, 25 to 35% of Cl⁻ ions passing through an anion-selective membrane must be transported to the membrane-solution interface by diffusion and convection. Interface diffusion plays a major role.[45]

Therefore, the deficiency in the quantity of electrolyte brought to the interface by electrical conductivity equals the amount of electrolyte brought to the interface by diffusion. Mathematically,[45] this may be formulated as indicated in Equation 5.63.

$$i \cdot (\text{Eff}' - \text{Eff})/(100 \cdot \text{Far}) = \text{Diff} \cdot (C_o - C)/B \qquad [5.63]$$

where
- i = current density, ampere/cm²
- Eff' = current efficiency for transport of counter-ions through the membrane, %
- Eff = current efficiency for transport of the same ions in the solution in contact with membrane, %
- Far = Faraday's constant (quantity of electricity needed to transport one gram equivalent of counter-ion through the membrane when its efficiency is 100%, that is, 26.8 Amp·hr).
- Diff = diffusion constant for the electrolyte at the temperature of the electrodialysis
- C_0 = concentration of the electrolyte in the interior of electrolyte in the depletion compartment
- C = concentration of electrolyte at the membrane-solution interface
- B = thickness of diffusion layer at the interface

The maximum value[45] of Equation 5.63 is obtained when C = 0. This is as presented in Equation 5.64:

$$i = i_{max} = (\text{Far} \cdot \text{Diff} \cdot C_o)/[100 \cdot B \cdot (\text{Eff}' - \text{Eff})] \qquad [5.64]$$

When $i = i_{max}$ the membrane is said to be concentration polarized.

c. Electrodialysis Process Disadvantages

Disadvantages of the electrodialysis process include:

- The process is uneconomical for sea water treatment. Nonetheless, the process may be used for the transformation of a brackish water with a total solids concentration of about 500 to 2000 mg/L to that of potable water.
- An expensive pretreatment is needed since the method is sensitive to organic macromolecules and ions.
- High-sulfate waters can be troublesome with the electrodialysis method (pass Cl⁻ more readily than SO_4^{2-}).
- The process removes only common mineral ions. As such, the use of this method alone cannot remove colloids and organics.
- The process is a sophisticated one since it requires skilled maintenance and supervision.

Example 5.12

An electrodialysis device is used to desalinize a brackish water of 2500 mg/L total dissolved solids content (TDS). The device has the following characteristics:

- Current efficiency for transport of counter-ions through the membrane = 85%.
- Current efficiency for transport of same ions in solution in contact with membrane = 64%.
- Thickness of diffusion layer at interface = 5·10⁻³ cm.
- Diffusion constant for electrolyte at the temperature of electrolysis = 1.5·10⁻⁵ cm²/s.

Take Faraday's constant = 26.8 Am·hr, and determine the value of the maximum current density.

Solution

1. Given: Far = 26.8 Amp·hr, Diff = 1.5·10⁻⁵ cm²/s, C_o = 2500 mg/L, B = 5·10⁻³ cm, Eff = 64%, Eff' = 85%.
2. Use Equation 5.64 to find the maximum current density as:

$$i_{max} = Far \cdot Diff \cdot C_o / (100 \cdot B \cdot (Eff' - Eff))$$

$$= (26.8 \cdot 1.5 \cdot 10^{-5} \cdot 2500) / (100 \cdot 5 \cdot 10^{-3} \cdot (85-64))$$

$$= 0.1 \, Amp/cm^2$$

D. ION EXCHANGE RESINS

Ion exchange is a unit process used for the displacement of ions of a certain species from an insoluble exchange material by ions of a different species in a solution.[1,10,43]

Some aluminosilicates (e.g., natural or synthetic zeolites and some clays) have the property of exchanging ions of their structure for ions in the solution. This reversible interchange is called ion exchange. Ion exchange resins are insoluble polymers which have active groups covalently bonded to the polymer.[42]

The ion exchange process may be a batch process or a continuous one. In the batch process the ion exchange resin is mixed and stirred with the solution to be treated. The resin employed in the process is then extracted in a clarifier to be regenerated for reuse. In the continuous process of ion exchange the resin is placed in a bed or a packed column, and the solution to be treated is permitted to flow through the column or bed.

Ion exchange resins may be of a natural or a synthetic organic source. The latter group possess greater stability, higher capacity, and ability to control their synthesis.[42] Generally, ion exchange resins may be grouped as:

1. *Cationic exchangers.* Cationic exchangers exchange cations. They are further classified as:
   - Weak acid resins: they contain the carboxylic (–COOH) groups as the exchange sites.
   - Strong acid resins: they contain sulfonic acid groups (–SO$_3$H) as the exchange sites.

2. *Anionic exchangers.* Anionic exchangers are capable of exchanging anions. They are further divided in two groups, namely:
   - Weak basic resins: they contain tertiary (–NR$_2$), secondary (–NHR), or primary (–NH$_2$) amino groups.
   - Strong basic resins: they possess the quaternary ammonium (–NR$_3$OH) group.

3. *Specific ion exchangers, amphoteric ion exchangers, selective chelating groups.* Amphoteric exchangers exchange both cations and anions. They posses sufficient selectivity to capture particular ions in preference to others.

The chemistry of the ion exchange process can be represented as follows:

$$Res^-H^+ + Na^+ \rightleftharpoons Res^-Na^+ + H^+$$

$$2Res^-Na^+ + Ca^{2+} \rightleftharpoons Res_2^{2-}Ca^{2+} + 2Na^+$$

Regeneration is the reverse of the above process and a concentrated NaCl solution can be used to renew this resin. The factors that govern the process include degree of crosslinking in the resin; valency of ions; concentration of ions in solution; type, nature and chemical composition of the exchange groups present on the matrix; characteristic of solution, resin life, and factors affecting the rate of diffusion of ions. Ion exchange is an appropriate technique that can have different applications, such as:

- Separation and concentration of ionic substances from liquids.
- Production, from a potable supply, of clean water for industrial usage.
- Removal of hardness ions, e.g., Ca^{2+}, Mg^{2+}.
- Production of deionized (demineralized) water.
- Recovery of heavy metals from wastewater, e.g., nickel, copper, zinc, mercury, and selenium.[42]
- Removal of alkalinity, toxic ions, or valuable ions.
- Recovery of mineral acids and their associated salts.
- Desalination of brackish water.
- Recovery of organic compounds such as phenol, protein, etc.

- Recovery and recycle of water and chemicals from wastes.
- Removal of ammonium ions from sewage effluents.

The employment of ion exchange resins for wastewater treatment subjects them to a more difficult environment than that in water treatment. The resins are more exposed to chemical attack by oxidation, thermal decomposition, and physical degradation. This is due to osmotic shock or attrition. Also, wastewater often contains large organic molecules that are sorbed irreversibly within the resin beads; high turbidity levels are another possible problem.[42]

Example 5.13

A sodium cation exchange resin is to be used to treat 6000 m³ of water per day.
The water has the following composition:

Ion	Concentration (meq/L)
Na^+	2.0 as Na^+
Ca^{2+}	4.0 as Ca^{2+}
Mg^{2+}	1.3 as Mg^{2+}
CO_2	1.0 as CO_2
HCO_3^-	2.0 as HCO_3^-
SO_4^{2-}	3.2 as SO_4^{2-}
Cl^-	2.1 as Cl^-

The finished water is to have a hardness of 2 meq/L (as $CaCO_3$). There is to be enough resin present to permit continuous operations for 7 days before regeneration is necessary. The exchange capacity of the resin is 46 kg/m³ (as $CaCO_3$). Determine the volume of resin required and the amount of sodium chloride (NaCl) needed for regeneration if 0.2 kg of NaCl is required for every 0.07 kg of hardness removed. What percentage of the flow can be permitted to by-pass the plant?

Solution:

The resin, for all practical purposes, removes 100% of the hardness, therefore a portion of the water to be treated can be by-passed and then recombined with the treated water.

In this case the original hardness is caused by the divalent metallic cations of Ca^{2+} and Mg^{2+}.

Ca^{2+} = 4.0 meq/L as Ca^{2+}
   1 meq/L as $Ca^{2+}CO_3^{2-}$ = (40 + 12 + 16·3)/2 = 50 mg/L
   4.0 meq/L as Ca^{2+} = 4·50 = 200 mg/L as $CaCO_3$
Mg^{2+} = 1.3 meq/L as Mg^{2+}
   1.3 meq/L as Mg^{2+} = 1.3·50 = 65 mg/L as $CaCO_3$

The total original hardness (Ca^{2+} + Mg^{2+})
   = 200 mg/L + 65 mg/L = 265 mg/L as $CaCO_3$
The finished water is to have a hardness of 2.0 meq/L as $CaCO_3$
   = 2.0·50 = 100 mg/L as $CaCO_3$.
Therefore the percentage of water that can by-pass the resin
   = 100·(100)/265 = 37.7%
or only 62.3% of the flow will require treatment.
The hardness to be removed
   = 265 mg/L − 100 mg/L = 165 mg/L as $CaCO_3$
The volume of water requiring treatment (taking into account the water to be by-passed) is 6000 m³ water/day · 7 days · 0.623
   = 26,166 m³.
The total hardness to be removed during the 7-day period is
   26,166 m³ · 1000 L/m³ · 165 mg/L · 1/10⁶ (mg/kg) = 4,317 kg.
The exchange capacity of the resin is 46 kg/m³.
Therefore the volume of the resin required
   = 4,317 kg/46 (kg/m³) = 93.8 m³.
For design purposes use a safety factor of 1.5. Then the required resin volume = 93.8 m³ · 1.5 = 140.7 m³.

The amount of NaCl required for regeneration would be 0.2 kg of NaCl · 4317 kg (hardness to be removed)/0.07 kg = 12,334 kg.

E. DISINFECTION PROCESS
1. Introduction
Disinfection addresses the destruction of pathogenic organisms and thus the prevention of waterborne diseases. It differs from sterilization, which is the destruction of all microscopic life.

Due to the small size of bacteria it is not feasible to guarantee their complete elimination from water by physical and chemical means alone. Nevertheless, for potable supplies it is necessary to make sure that all harmful microorganisms are killed in order to safeguard community health.

Disinfection methods can generally be grouped in two types:[2,5,6,10,49–53]

a. Physical Methods
The physical methods of disinfection include:

1. *Heat treatment.* Heat treatment is used at 100°C for a period of 15 to 20 min. This is different from pasteurization. Pasteurization is practiced in the food industry and it implies the application of heat treatment at 80°C for 10 min to kill vegetative cells.
2. *Ultraviolet rays (UV).* The water to be treated is allowed to flow through a chamber where it is exposed to UV light at a wavelength of 200 to 310 nm. The UV lamps need to be kept clean to allow exposure of the microorganisms to the maximum intensity. The advantages to be gained by using the UV method include:

    - The process is easy to install and operate.
    - No change in the characteristics of the water.
    - UV does not react with other compounds in the water.
    - The required contact period is short.
    - No taste or odors are created.

    The disadvantages in using UV include:

    - No residual is formed.
    - Pretreatment is required to avoid shielding of microbial cells by solids.
    - The method is costly.

3. *Metal ions.* Metal ions used in disinfection include silver (Ag^+) and copper (Cu^{2+}). The advantages of this method of disinfection include:

    - The use of low concentration of the metal ion.
    - No production of toxic compounds.
    - The method provides a residual.
    - The action of the ion is independent of the concentration of organisms.
    - Ions have effects on varying microbial species.

    The disadvantages in using ions include:

    - Requirement of a reasonable pretreatment.
    - Disinfection action is affected by temperature and pH variations.
    - The method is costly.
    - Chemical water characteristics affect disinfection.

b. Chemical Methods
Chemical methods of disinfection incorporate the application of oxidants (compounds that can accept an electron) to disinfect the water. Examples of chemical disinfectants include chlorine gas (Cl_2), ozone (O_3), chlorinated compounds such as chlorine dioxide (ClO_2), iodine (I_2), potassium permanganate ($KMnO_4$), etc.

2. Requirements of a Disinfectant
A good chemical disinfectant must possess the following important characteristics:[2,4–6,10–13,20,33,47,52,53]

- It needs to be effective and fast acting in order to kill pathogenic microorganisms that are present in the water.
- It must be somewhat soluble in water in the concentrations required for disinfection.
- It needs to be capable of providing a residual.

- It ought not to impart taste, odor, or color to the water.
- The disinfectant must not be toxic to human and animal life.
- It ought to be easy to handle, transport, and control.
- It must be easy to detect and measure in water.
- The chemical needs to be readily available at moderate cost.

The disinfection process is dependent on:

- Type and concentration of organisms.
- Type and concentration of disinfectant used.
- Presence of oxidants consuming compounds.
- Temperature.
- Disinfectant dose and contact time.
- pH.

3. Chlorination

a. Introduction

Poor sanitation that results in the fecal pollution of water sources presents a great threat to the public health. Effective chlorination of water supplies has in many cases achieved a substantial reduction in the enteric diseases that are primarily water related.[49]

General characteristics of chlorine include:

- It is a greenish-yellow toxic gas.
- It is found in nature in a combined state, chiefly with sodium as common salt.
- It has a penetrating and irritating odor.
- It is heavier than air.
- It vaporizes under normal atmospheric temperature and pressure.
- It is manufactured by the electrolysis of brine (the solution of sodium chloride, NaCl) with byproducts of caustic soda (NaOH) and hydrogen:
  $2Na^+ + 2Cl^- + 2H_2O$ electrolysis $\rightleftharpoons H_2 + Cl_2 + 2Na^+ + 2OH^-$
- It is slightly soluble in water, approximately 1% by weight at 10°C.

The likely chemical reactions that may result when chlorine is added to water are

1. *Reaction with water:*

    $Cl_2 + H_2O \rightleftharpoons HCl + HOCl$
    $HCl \rightleftharpoons H^+ + Cl^-$
    $HOCl \rightleftharpoons H^+ + ClO^-$

    Hypochlorous acid (HOCl) is the more effective disinfectant; it is referred to as available chlorine. The most effective disinfection occurs at acidic pH levels. Table 5.5 gives an outline of the recommended minimum chlorine residuals needed for bactericidal disinfection of water.

2. *Reaction with ammonia.* Ammonia reacts with chlorine in water to form chloramines:

    $NH_4^+ + HOCl \rightleftharpoons NH_2Cl + H_2O + H^+$ (monochloramine)
    $NH_2Cl + HOCl \rightleftharpoons NHCl_2 + H_2O$ (dichloramine)
    $NHCl_2 + HOCl \rightleftharpoons NCl_3 + H_2O$ (trichloramine, or nitrogen trichloride)

    The final stoichiometric reaction is

    $2NH_4^+ + 3HOCl \rightleftharpoons N_2 + 2H^+ + 3HCl + 3H_2O$

    From the last chemical reaction equation the amount of chlorine needed to oxidize 1 g of ammonia may be computed as $(Cl_2)/(NH_4^+) = 5.9$.

3. *Oxidation of inorganic substances.* Chlorine acts as an oxidant reducing inorganic chemicals such as hydrogen sulfide (H_2S), iron (Fe^{2+}), and manganese (Mn^{2+}).

4. *Oxidation of organic compounds.* Chlorine combines with organic substances to form trihalomethanes (THMs) and other chlorinated organics. THMs are produced within the water treatment system as byproducts of chlorination. THM precursors contain a wide variety of organics: dissolved, colloidal particulate forms, and humic substances. Humic substances are amorphous, acidic, predominantly aromatic (these are ring compounds that have special properties; in most cases the ring is made up of carbons, with a particular linkage of alternating single and double bonds) hydrophilic, chemically

Table 5.5 Public Health Service Recommended Minimum Chlorine Residuals for Bactericidal Disinfection of Water at 20 to 20°C

pH	Minimum free available Cl_2 residual after 10 min contact (mg/L)	Minimum combined available Cl_2 residual after 60 min contact[a] (mg/L)
6.0	0.2	1.0
7.0	0.2	1.5
8.0	0.4	1.8
9.0	0.8	1.8
10.0	0.8	Not applicable
>10.0	>1.0	Not applicable

[a] Not applicable to all waters, particularly those that are turbid, and does not apply at temperatures below 20°C.

From *Water Chlorination Principles and Practices,* American Water Works Association, M20, Denver, CO, 1973, 26–28. With permission.

complex macromolecules consisting of a complex mixture of naturally occurring organics, and are responsible for the natural color imparted to waters.[23]

THMs include chloroform ($CHCl_3$), bromodichloromethane ($CHBrCl_2$), dibromochloromethane ($CHBr_2Cl$), and bromoform ($CHBr_3$). Chloroform has been shown to be an animal carcinogen, and is suspected of being a human carcinogen.[23]

b. Breakpoint Chlorination

Ammonia in water can come about by the hydrolysis of urea: $(NH_2)_2CO + H_2O \rightleftharpoons 2NH_3 + CO_2$, or through the decomposition of an organic substance such as protein. Oxidation of ammonia by chlorine leads to the breakpoint concept.

The concentration of chlorine needed to produce the lowest of the total chlorine residual is termed the breakpoint dose. At higher chlorination applications the excess chlorine residual beyond the breakpoint stays available and the curve develops in a manner parallel to the zero demand line. At this level the available chlorine consists of free chlorine with a very low concentration of nitrogen trichloride when the pH is less than 7. Figure 5.30 illustrates the products of the reaction between ammonia and chlorine in water.

In Figure 5.30, in the region A to B the chlorine added reacts swiftly with the reducing substances found in the water. This reduces chlorine to the chloride ion (Cl^-), which is not a disinfectant. The residual chlorine is low and, for a small chlorine dosage, the disinfecting characteristics are almost absent.

The addition of more chlorine is presented in region B to C. The chlorine has completely oxidized the reducing agents and has created monochloramine and dichloramine through the reaction with ammonia. Monochloramine and dichloramine are called combined available chlorine.

In this region, B to C, the addition of chlorine affords relative increase in the combined chlorine residual. In the region C to D, the addition of further chlorine decreases the available chlorine. This reduction is due to the production of trichloramine, nitrogen, etc., which are not disinfectants. Therefore, the capability of the mixture to kill pathogens is reduced. Upon addition of more chlorine these reactions are complete and the ammonia is entirely oxidized. Finalization of the oxidation stage is achieved at point D. Consequently, more addition of chlorine stays as free available chlorine, HOCl, and behaves as a potent chlorine residual. Point D is denoted as the breakpoint.

Usually, chlorination is conducted beyond the breakpoint to guarantee the availability of free chlorine residual. The chlorine dosage needed to accomplish the breakpoint is a multiple of the ammonia content. For an effluent with a high concentration of ammonia breakpoint chlorination is costly. Therefore, the chlorination in this case is carried out to the region B to C to acquire a combined chlorine residual.

Depending on the preferred amount of residual chlorine and the point of employment, chlorination systems may be grouped under two broad categories:

- Prechlorination. Prechlorination involves the application of chlorine prior to any treatment process. This practice is sometimes used for the control of algae, taste, and odor.
- Postchlorination. Postchlorination refers to the application of chlorine after the application of other treatment processes, specifically, after filtration.

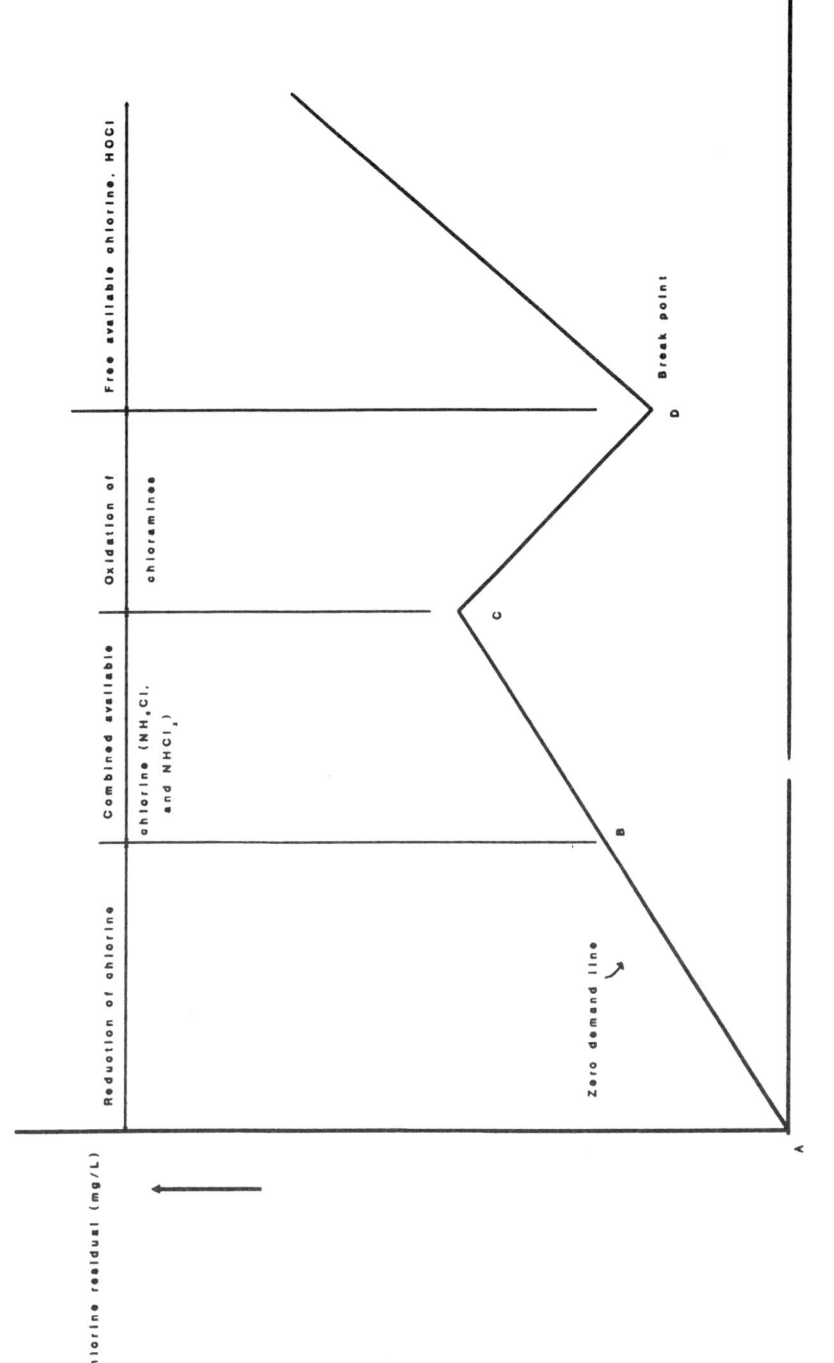

Figure 5.30 Reaction between ammonia and chlorine in water (breakpoint chlorination).[1,10,11,20,23,50,52,53]

c. Residual Chlorine

Many techniques are available to determine the concentration of residual chlorine. One of the simple methods used is the diethyl-para-phenylenediamine (DPD) method. In the DPD method the free available chlorine immediately reacts with N-diethyl-para-phenylenediamine, producing a red color if iodine is absent. Standard solutions of DPD (potassium permanganate) are employed to make different color intensities. Intensity of the red color is proportional to the chlorine residual. (See Section VI.B.3 of Chapter 8).

d. Kinetics of Chlorination

The ability of chlorine to kill pathogenic organisms is associated to the concentration of the disinfectant and the contact time between the pathogens and the disinfectant. This relationship follows Chick's law.

Chick's law provides a mathematical relationship of the diffusion of microorganisms and is presented in Equation 5.65:

$$dN/dt = -k \cdot N \quad [5.65]$$

where: N = number of viable microorganisms of one type at time t
t = time, d
k = constant, d^{-1}, or s^{-1}.

Integrating Equation 5.65 between limits of $N = N_o$ at $t = 0$ yields:

$$Ln(N/N_o) = -k \cdot t$$

or,

$$N/N_o = e^{-k \cdot t} \quad [5.66]$$

where: N = number of viable microorganisms of one type at time t
N_o = number of viable microorganisms of one type at time t = 0
t = time, d
k = constant, d^{-1}, or s^{-1}
e = Navier constant

The rate of kill is dependent on the number of microorganisms found initially. When the microorganisms have similar resistance, the kill tends to follow an exponential order. An entire kill is not achievable. The efficiency of disinfection is indicated as a percentage of the ratio of microorganisms killed to the original number of microorganisms.

Ckick's law is not valid for all disinfectants nor for all microorganisms. The kill depends upon such factors as ability to penetrate the cell wall, the time to penetrate vital centers within the cell, and the distribution of disinfectants and microorganisms, etc.

The chlorine concentration and the contact time may also be related by another equation:

$$C^n \cdot t = k \quad [5.67]$$

where: C = concentration of chlorine, mg/L
t = contact time, or time required for a given percent of microbial kill, min
n, k = experimental constants that are valid for a particular system

The effects of changes in the temperature may be modeled by an equation derived from the equation of Van't Hoff-Arrhenius[1,10] as shown in Equation 5.68.

$$Ln(t_1/t_2) = [E'(T_2 - T_1)]/R_g \quad [5.68]$$

where: t_1, t_2 = time required for the given kills, s
E' = activation energy (see Table 5.6), calories
T_1, T_2 = temperature corresponding to t_1 and t_2, Kelvin
R_g = universal gas constant
= 1 cal/K/mol = 0.082 L·atm/K/mol = 8.314 J/K/mol.

Table 5.6 Activation Energies for Aqueous Chlorine

pH	E' (calorie)
7.0	8,200
8.5	6,400
9.8	12,000
10.7	15,000

From Fair, G. M., et al., *J. Am. Water Works Assoc.*, 40: 1051, 1948. With permission.

Table 5.7 Ranges of Chlorine Dosage Wastewater Generally Required for Disinfection of Various Types of

Type of wastewater	Range (mg/L)
Raw	6–40
Primary effluent	5–24
Activated sludge effluent	2–9
Filtered effluent	1–6

From *Design of Municipal Wastewater Treatment Plants*, Vol. II, Water Environment Federation and American Society of Civil Engineers, 1992, 880. With permission.

The advantages to be gained through the use of chlorine in disinfection are that chlorine is effective, reliable, and offers a residual in the system.

The disadvantages when chlorine is used to treat water include:

- The hazards incurred during handling, transportation, and storage of the chemical.
- The provision of an appropriate and a safe site for the storage of the chemical.
- The reactions of chlorine with organic compounds present in solution and the formation of chlorinated organic compounds (THMs) that are injurious to human health.

In the case of chlorination of wastewater larger doses of the disinfectant are required. This is due to the occurrence of different materials that exert a chlorine demand such as ammonia. Ranges of the concentrations of chlorine demanded to produce a free chlorine residue in various wastewaters are as outlined in Table 5.7. Contact times of around half an hour at average flow, with a minimum of a quarter of an hour for peak flow, are common.[10]

Example 5.14

Find the contact time required for a disinfectant to achieve a 99.9% kill for a microbial system that has a rate constant (to base 10) of $4 \cdot 10^{-2}$ s^{-1}.

Solution

1. Given: rate of kill = 99.9%, k = $4 \cdot 10^{-2}$/s.
2. Determine the contact time from Chick's law (Equation 5.66) as:
   $t = -(1/k) \cdot \text{Log}(N/N_o)$.
3. Substitute given values in 1 in Chick's equation to find the contact time as:
   $t = -(1/4 \cdot 10^{-2}) \cdot \text{Log}[(100 - 99.9)/100] = 75$ s.

Another disinfectant used in wastewater treatment is ozone. This will be discussed in Chapter 6.

V. SLUDGE TREATMENT AND DISPOSAL

A. SLUDGE DIGESTION

Sludge digestion signifies the controlled degradation of organic substances of sludges, normally occurring under anaerobic conditions. Sludge digestion serves to decrease the volume of thickened sludge and changes the remaining solid substances to inert compounds that are probably free from pathogenic organisms.

The factors that influence the anaerobic sludge digestion process include pH, temperature, nutrients concentration and feed, toxic substances such as heavy metals, volatile acids, ammonia, type and characteristics of decomposed materials, shock loads, and mixing conditions. Aerobic sludge digestion is not as sensitive to environmental factors as the anaerobic digestion process, nevertheless, it consumes energy.

An aerobically digested sludge is brown to dark brown in color and has a flocculent appearance. It has a musty inoffensive odor and is not easily dewatered.

An anaerobic digester is composed of a closed vessel with an airtight floating cover. The sludge is withdrawn from the conical bottom of the vessel while the floatable cover permits volume changes due to sludge additions and removals. Sludge is intermittently fed to the digester and the supernatant is removed and returned to the secondary treatment unit. The sludge solids are acted upon by the acid-forming facultative and anaerobic bacteria. These organisms convert the sludge solids to soluble organic acids and alcohols. This action results in a reduction in the pH of the system and can result in the termination of the process by these species of bacteria. Strictly anaerobic bacteria, termed methane formers, takeover. The methane formers convert the formed acids and alcohols to carbon dioxide and methane gas, with traces of other gases such as H_2S. Methane formers are anaerobic bacteria that function within a narrow pH range of from 6.5 to 7.5.

An anaerobically digested sludge is dark brown to black in color. Its faint color resembles that of hot tar, burnt rubber, or sealing wax. Anaerobically digested sludge contains a large amount of gas.

The volumetric gas production or the specific yield can be determined from the relationship:

$$V_g = \left[Y_t \cdot VS \cdot \left(1 - \left\{ k_n / \left(t \cdot u_{smax} - 1 + k_n \right) \right\} \right) \right] / t \qquad [5.69]$$

where: V_g = volumetric gas production rate, or the specific yield, m^3 gas/m^3 digester/d
Y_t = ultimate gas yield, m^3 gas/kg VS added
VS = concentration of influent volatile solids, kg/m^3
k_n = kinetic coefficient, dimensionless
t = hydraulic detention time, days
u_{smax} = maximum specific growth rate of microorganisms, per day

Table 5.8 gives a general design information and criteria for a conventional anaerobic digester.

B. SLUDGE DEWATERING
1. Introduction
Sludge dewatering is a physical (mechanical) unit operation used to reduce the moisture content of sludge for one or more of the following reasons:

- To reduce the volume of sludge that requires disposal; this lowers transportation costs.
- To ease sludge handling process.
- To retard, if needed, biological decomposition.
- To increase calorific value by removal of excess moisture, when the sludge is required to be incinerated.
- To render sludge odorless.
- To reduce leachate production at the landfill site.

One of the greatest problems in wastewater engineering is the dewatering of the sludges produced in the various purification stages. The cost of sludge treatment and disposal can represent more than 50% of the total cost of construction and operation of a sewage treatment plant.[55]

Sludge dewatering is not sludge disposal, but a desirable operation in order to reduce the volume of the sludge by removing water.

The disposal of sewage sludges is not easily carried out. At present it can be injected into farm land at approved locations, turned into a saleable fertilizer, used for landscape fertilizer, or incinerated.

2. Sludge Dewatering Methods
The methods that are normally adopted for sludge dewatering include land disposal (drying beds), vacuum filters, pressure filters, centrifuges, and compaction (see Figure 5.31). These methods are of physical rather than chemical nature. A brief discussion of some of these processes is given below:

Table 5.8 Design Information for the Conventional Anaerobic Digester

Parameter	Value
Volatile solids loading, kg/m^3/d	0.3–2
Volatile solids destruction, %	40–50
Gas production, m^3 gas/kg VS	0.2–1.5
Influent sludge solids, kg/m^3/d	2–5
Total solids decomposition, %	30–40
pH	6.5–7.4
Alkalinity concentration, mg/L	2000–3500
Solids retention time, day	30–90
Digester capacity, m^3/capita	0.1–0.17
Gas composition, %	
Methane	65–70
Carbon dioxide	32–35
Hydrogen sulfide	Trace
Temperature, °C	30–35

Compiled from References 17, 21, and 54.

a. Drying Beds

Dewatering on land was an early method used to reduce the sludge moisture content prior to ultimate disposal (see Figures 5.32 and 5.33).

In drying beds the sludge is spread on the beds to a depth of 125 to 250 mm. The reductions that may be acquired in the moisture content is partly due to evaporation, but chiefly through drainage. Regional climatic conditions greatly influence the sludge dewatering on sand drying beds. The cake of the drying bed cracks as it dries. This cracking permits additional evaporation and thus the removal of the moisture.

In good climatic situations, the sludge may dry to 25% solids after a few weeks, but normally it takes 2 to 3 months. The drying time is shorter in regions of greatest sunshine, low rainfall, and low humidity. Localities with longer summer periods or places with low humidity are regarded as more favorable for sludge drying beds. Also the prevalence of wind and its velocity are factors that affect evaporation from sludge drying beds.[56]

Other factors that affect drying include the chemical and physical nature of solids, internal structure of the particles and size distribution, bed dimensions (especially the depth), and shape of the drying surface in relation to air flow.

The drains that are laid under the gravel in the drying beds are usually unjointed pipes made of cement, or of earthenware when the sludge is corrosive. There must be enough pipes, with the correct slope, to drain the whole mass of the sludge evenly. The area required for drying sludge is generally about 4 persons per square meter; which implies a large land area requirement. The dried sludge is most often removed by hand. This is a slow process that requires repeated handling and which can be avoided by mechanizing the beds.[20]

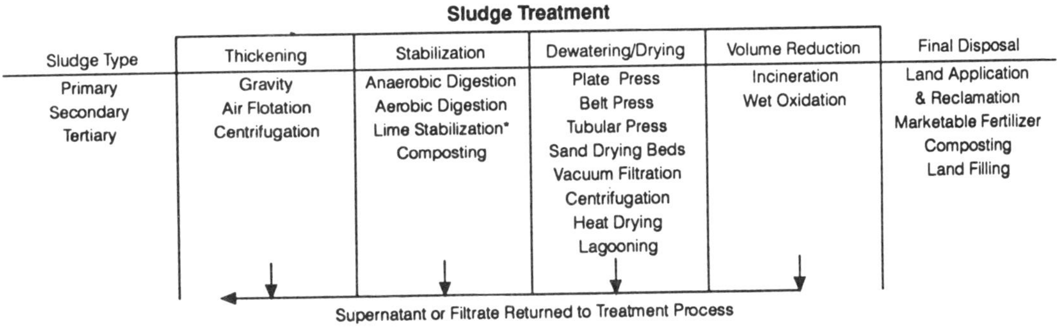

Sludge Treatment

Sludge Type	Thickening	Stabilization	Dewatering/Drying	Volume Reduction	Final Disposal
Primary Secondary Tertiary	Gravity Air Flotation Centrifugation	Anaerobic Digestion Aerobic Digestion Lime Stabilization* Composting	Plate Press Belt Press Tubular Press Sand Drying Beds Vacuum Filtration Centrifugation Heat Drying Lagooning	Incineration Wet Oxidation	Land Application & Reclamation Marketable Fertilizer Composting Land Filling

Supernatant or Filtrate Returned to Treatment Process

Figure 5.31 Alternative processes or options for treating and final disposal of sludges (biosolids). *Sludges can also be conditioned by adding inorganic chemicals such as ferric chloride or organic chemicals such as polymers.

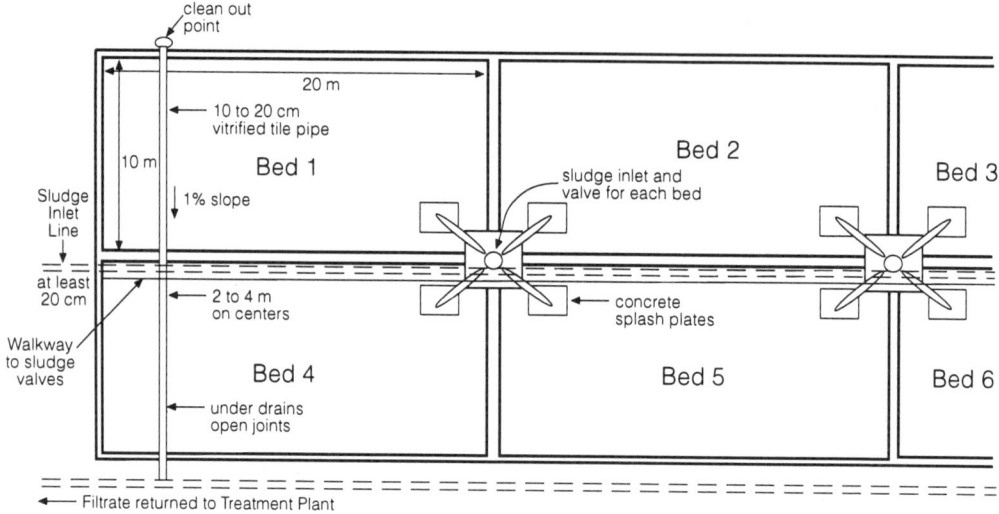

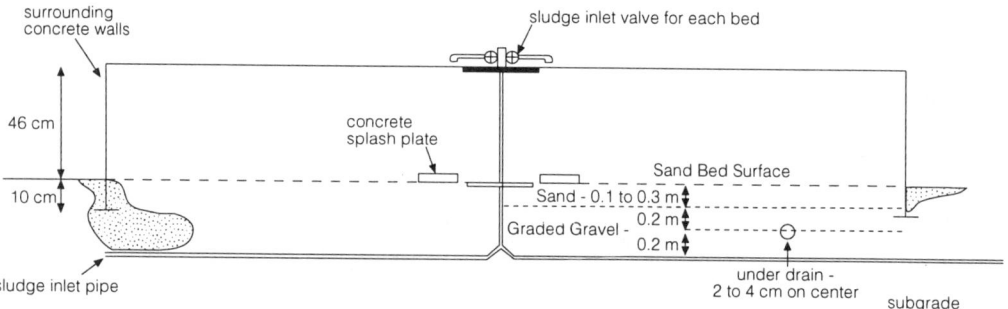

Permanent vehicle treadways should be provided at 6 m centers for truck or front end loader removal of dried sludge.

Figure 5.32 A typical sludge drying bed.

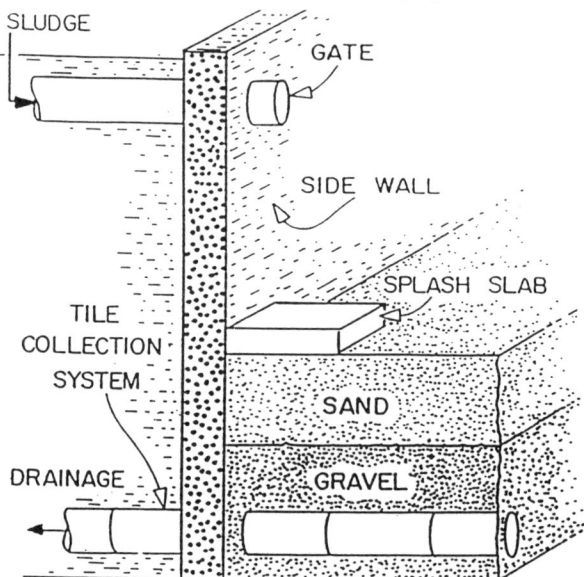

Figure 5.33 Construction details of a sludge drying bed.[8]

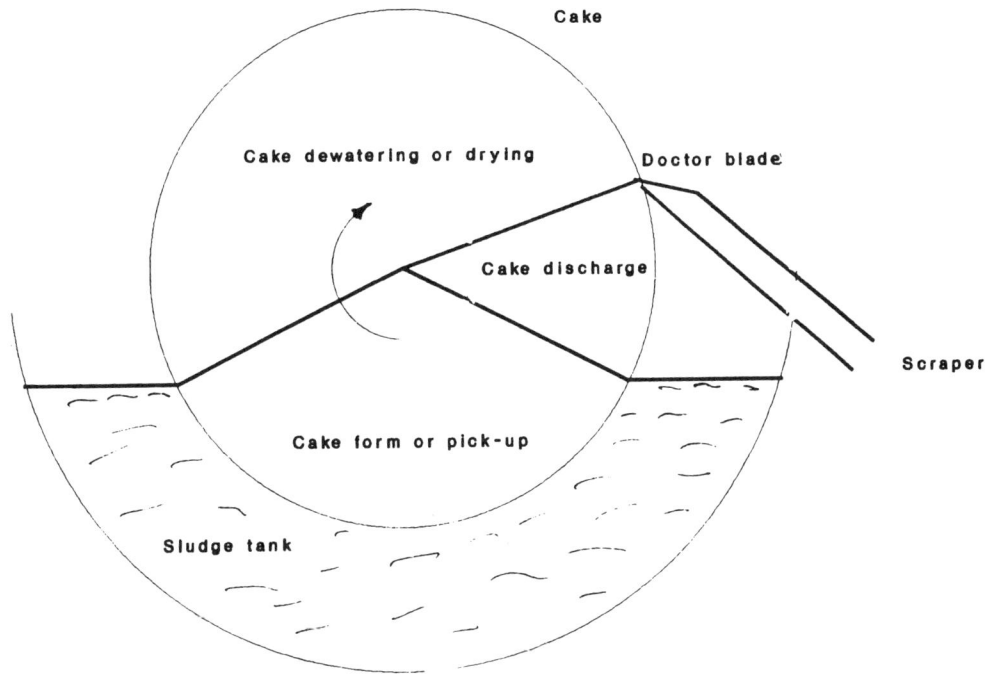

Figure 5.34 Schematic diagram of a vacuum filter.[1,4,58]

b. Vacuum Filtration

Vacuum filtration of sewage sludge deals with the filtration of sludges carried out by using suction or vacuum on a continuous basis to yield a cake of a reasonably low water content (see Figure 5.34).

Vacuum filtration is a continuous operation that is generally accomplished by using cylindrical drum filters. The drum is covered with a filter cloth which may be made of natural or synthetic fibers, wire mesh, fabric, etc. The drum rotates with the vacuum employed. Theoretically the maximum pressure that can be exercised is atmospheric. Nevertheless, in practice, operating pressures are around 60 to 90 kN/m^2, i.e., around 70% of the atmospheric pressure. Water is withdrawn through the media, and the sludge cake is scraped off. Compressed air may be introduced through the media to facilitate the liberation of the cake from the discharge point; the filtrate is recycled to the plant.

Sludge conditioning is crucial before vacuum filtration. With this method a solids content of around 25% can be secured. The moisture content of the cake is generally higher than the one provided by a filter press.

The factors that influence the yield of cake from vacuum filters include the nature of the sludge, characteristics of the sludge, type and concentration of filter aids, degree of agitation received by the sludge, type of filter cloth and its condition, and the operational parameters of the filter itself.

c. Pressure Filtration

Pressure filtration or the application of a positive pressure is a batch process as opposed to the continuous operation of a vacuum filter (see Figure 5.35).

Filter pressing has a long operational past that established its success for withdrawing water from sewage sludges. In a pressure filter the conditioned sludge is pumped gradually under an increasing pressure into the filter plates supporting cloth. The filter cloth is usually made of nylon, terylene, or polypropylene which holds the solids. Hydraulic rams retain the plates together and the liquid is forced through the filter cloth to the plate outlet ports. The filtrate is recycled to the plant. The cake capacity for filter presses vary from 0.014 to 17 m^3 per cycle with pressures up to 1,550 kN/m^2 being used.

This method of moisture removal may produce a solids content of 25 to 50%. The major advantages of the filter press include production of a sludge cake of a low moisture content, and production of a filtrate of low suspended solids, with a low capital cost investment.

d. Centrifugation

Dewatering by centrifuges may be defined as a sedimentation process under the influence of forces greater than the ordinary forces of gravity. Centrifugation methods are applicable when the particles are

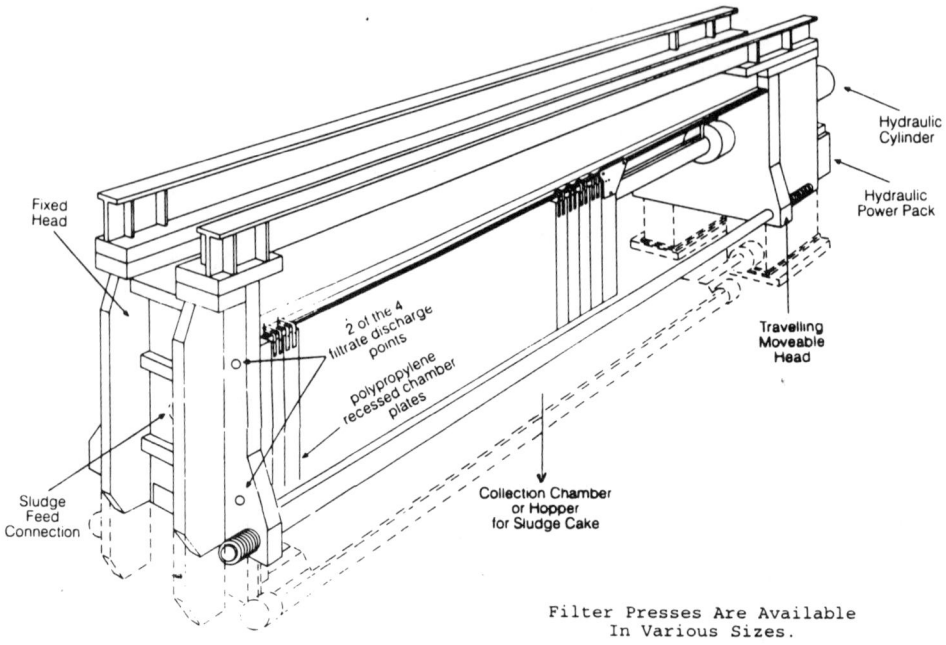

Figure 5.35 A filter press. (From Ditz, R. B., Netzsch, Inc., Filter Press Division, 1994. With permission.)

too small for sedimentation to develop but not small enough to make it essential to consider their diffusion.

The forces that act on the particles are the centrifugal and Stoke's frictional forces. Centrifuges fundamentally are settling vessels. In this kind of machine, centrifugal force acts on the suspended particle in the sludge, causing it to settle through the liquid portion. By rotating this settling vessel at a high speed, the settling forces acting on the solid particles can be increased several times.

Sedimentation in a centrifugal domain is subjected to the same parameters as sedimentation in a gravitational one. Influencing factors include the concentration of suspended phase, particle size and shape, solvation, electrostatic factors, viscosity effect, and the differential density between solid particles and liquid.

e. Elutriation

Elutriation is the washing of sewage sludge with settled or biologically treated wastewater. This is practiced to alter the constituents of the liquid phase in contact with the sludge particles.

During the mixing of the elutriant and the sludge, substances such as carbonates and phosphates are removed from the sludge together with products of decomposition and nonsettleable fine materials. The result is a more porous sludge and a reduced need for a sludge conditioner.[56] The main problem with the elutriation process is the quantity of solids recirculated to the treatment process.

The main advantages of elutriation include the removal of alkalinity, the reduction in the concentration of fine particles, and the removal of gas bubbles. The latter condition leads to more efficient settling.

Sludge can be taken through a single-stage or a multistage operation of elutriations. Elutriation also can be carried out in two steps, in which the recovered wash water from the second step is returned for reuse in the first step. This elutriation process is referred to as countercurrent elutriation.[56]

The overflow from the elutriator generally contains a fair amount of fine and colloidal matter which, after recirculation to the head of the plant, normally has to be readsorbed by the flocculated sludge.[20]

3. Factors Influencing Sludge Dewaterability

The main factors that affect the sludge dewatering include the presence and concentration of fine particles, sludge solids content and size, shearing strength, protein content, pH and particle charge, moisture content, and filter aids and conditioners used.

a. Presence of Fine Particles

The fine particles of the sludge have an important effect on its dewatering properties. The filtration properties become more difficult as the particle sizes decrease.[57] Likewise, the compressibility coefficient is decreased with a decrease in particle size. This means that the smaller the particle the more resistant it is to compression. Abdel-Magid[58] indicated that if the fine particles are flocculent and compressible, the greater is the resistance to filtration due to the closure or collapse of pores — a phenomenon that results from the particles being unable to create a support skeleton. Gale[59,60] stated that the presence of fines in a bed substantially increased the specific resistance, since not only is the specific surface increased but in certain circumstances the void ratio is also reduced — the smaller particles being lodged in the interstices between the larger particles.

Generally, all the methods used for conditioning the sludge remove or agglomerate the colloidal and fine particles.[58] Coackley[55,57,61] showed that the smallest particle size range has the highest specific resistance.

Abdel-Magid,[58] in his work on digested sludge, showed that the original specific resistance of the sludge is greater than the specific resistance of the separated particles; this suggests that a small proportion of the very small particles have a disproportionate effect on the total specific resistance of the sludge. Abdel-Magid[58] also indicated that the smaller particles are less deformable than the larger particles. The compressibility coefficient of the settleable solids was found to be greater than that of the original sludge, indicating that the solids are more deformable than the sludge solids as a whole. Abdel-Magid[58] also demonstrated the role of the supracolloidal particles on the filterability of digested sewage sludge, by showing that the supracolloidal fraction increased the specific resistance of the sludge drastically. This indicates that the removal of the colloids from digested sewage is crucial for efficient and better dewatering.

Karr and Keinath[62] showed that particle size plays a significant role in dewatering biological sludges. Experimentally, they fractionated sludges into prescribed particle size ranges and determined the relative concentration of solids present within each size range. Following this procedure, the specific resistance was determined. Their results revealed that the particle size range from $(1 \text{ to } 100) \cdot 10^{-6}$ m had a significant role in determining the rate of water loss of sludges by vacuum filtration.

b. Solids Content

The specific resistance of sewage sludge is theoretically independent of the concentration of sludge solids, even though several investigators have reported conflicting data.[55,62] Coackley[55] reported that the specific resistance decreased with a decrease in the solids content. The increase in filter yield with an increase in sludge solids concentration is understandable, since the higher the solids concentration the less the water that must be removed per unit weight of filter cake. Karr and Keinath[62] found that the specific resistance test is superficially influenced by the total solids concentration of a sludge when fines are present to the extent that they cause blinding of the sludge and filter medium.

c. Protein Content

Coackley[63] showed that proteinaceous material constitutes a large part of the sludges and that the sludge particles, even if they do not consist of pure protein, will be covered by a layer of protein that will govern the chemical and physical behavior of the particles. It was found that the protein content ranges between 15 to 30% of the organic matter present for digested sludge, and 28 to 50% in the case of activated sludge. This layer of protein can account for the hydrophilic properties of the sludge.

d. pH and Particle Charge

Measurement of the zeta potential of the sludge indicates that the charge on sludge particles is the same as that of protein isolated from the sludge.[63] Most of the colloids found in wastewater are negatively charged. The particles found in raw water supplies do not all possess the same zeta potential; values of zeta potential for individual particles often range from 10 to 20 mV. A common value for the medium zeta potential for particles in raw water supplies is –10 to –15 mV. The achievement of a zero zeta potential does not necessarily ensure a satisfactory flow of filtrate.

e. Water Content

The water in sludges is found in various forms. In many studies of the dewatering process, the concept of the various forms of water need to be kept in mind when considering the degree of ease with which the water can be removed from the sludge.

f. Shearing Strength

Mixing generally leads to deterioration of the dewatering characteristics of sludges.[58,60,64] Swanwick and Davidson[65] carried out small and full-scale experiments on sludges subjected to shearing forces of the magnitude associated with pump impellers. They found that there was an increase both in the specific resistance and the concentration of fines in proportion to the stirring speed. Pumping of digested sewage sludges also hampers the dewatering properties. Gale[60] indicated that mixing leads to an increase in the quantity of fine particles in the sludge which results in deterioration of its dewatering characteristics.

g. Anaerobic Digestion

Anaerobic digestion reduces the total solids content of the sludge. It also influences the way in which sludge cracks during subsequent dewatering on drying beds. Both of these effects are expected to reduce the time needed for dewatering sludges on drying beds.[65] Coackley[55] found that the specific resistance of the sludge decreased as digestion proceeded to the point where gas production ceased. The specific resistance then remained constant. The drainability increased between the 7th and the 21st days, when the sludge contains the most gas.[66]

The conversion of biomass material to methane for use as an energy source has attracted attention of many researchers throughout the world. This conversion is accomplished by anaerobic digestion. Other benefits of digestion include reduction or elimination of pathogens, depending upon temperature, and production of a stable slurry or sludge that can be used as a fertilizer and soil conditioner. Environmental factors that affect biological reactions include pH, temperature, nutrients, and toxic compounds which are amenable to external control in the anaerobic digestion process.[54]

h. Filter Aids

Although the rate and extent to which unconditioned wastewater sludge can be dewatered is usually low, the rate can be enhanced dramatically by addition of filter aids. The filtration of many sludges can be improved by the use of an appropriate filter aid.

Filter aids are finely divided materials which may be added to either the filter septum before the addition of the sludge, or to the sludge as a whole. The former addition is referred to as precoat while the latter addition is denoted a body aid. Both types of addition are expected to improve the filtrate flow. Precoat aids can be added to protect the filter septum against blockage by fine particles. The main function of a filter aid is to improve the characteristics of the filter cake, with particular emphasis on the rigidity and porosity of the cake structure. Some examples of physical types of filter aids are wood, flour, diatomaceous earth, fly ash, expanded perlite, and acacia seeds.

4. Filtration of Sludge

a. Introduction

Filtration is the main method used for the separation of a solid and a liquid from a suspension of the former from the latter. Filtration entails the retention of the solid on a screen or membrane while allowing solid-free liquid (filtrate) to percolate through.

Thus, the removal of a solid in the initial stages of filtration is due mainly to the filter medium (a membrane of paper, sand, cloth, synthetic substances, or similar porous material). Nevertheless, as the filter cake builds up, the cake itself becomes the filter. It follows, then, that the rate of filtration in the initial stages is controlled mainly by the characteristics of the filter medium, and in the latter stages by the characteristics of the filter cake.

b. Filterability Determination

One of the methods formerly used for finding the dewaterability of a sludge was by performing the cracking-time test. This test involves the filtration of a given volume of sludge under vacuum until a cake is formed which eventually cracks, with a resultant drop in pressure. The time required to reach this stage is frequently taken as a measure of the dewaterability of sewage sludges.[58]

Later, the dewatering efficiency of sewage sludges was found by one of two methods: the specific resistance concept, and the capillary suction time method (see Figure 5.36).

c. Specific Resistance Concept

For the determination of the basic specific resistance equation, Coackley[61,63] started with a combination of Poiseuille's and Darcy's laws for flow through a porous media as presented in Equation 5.70.

$$dV/dt = P \cdot A / \mu R \qquad [5.70]$$

where: V = volume of filtrate, m³
t = time of filtration, s
P = pressure applied, N/m²
A = area of filtration, m²
μ = viscosity of filtrate, N·s/m²
R = total resistance to filtration of the cake and filter medium
$R = R_c + R_m$ [5.71]
$R_c = r_s \cdot C \cdot V/A$ [5.72]
r_s = specific resistance of sludge cake, m/kg
C = solids content, kg/m³

From the above equations the rate of filtration, i.e., the ease of dewatering, is given by Carman and Coackley[55,57,61,67–71] in the following equation:

$$dV/dt = (P \cdot A^2) / \{\mu(r_s \cdot C \cdot V + R_m \cdot A)\} \qquad [5.73]$$

where: V = volume of filtrate, m³
t = time, s
P = pressure applied, N/m²
A = Area of filtration, m²
μ = viscosity of filtrate, N·s/m²
r_s = specific resistance of sludge cake, m/kg
C = solids content, kg/m³
R_m = resistance of filter medium, /m

For constant pressure, integration of Equation 5.73 yields:

$$t/V = (\mu \cdot r_s \cdot C / 2 \cdot P \cdot A^2) \cdot V + (\mu \cdot R_m / P \cdot A) \qquad [5.74]$$

Equation 5.74 may be put in the form of Equation 5.75:

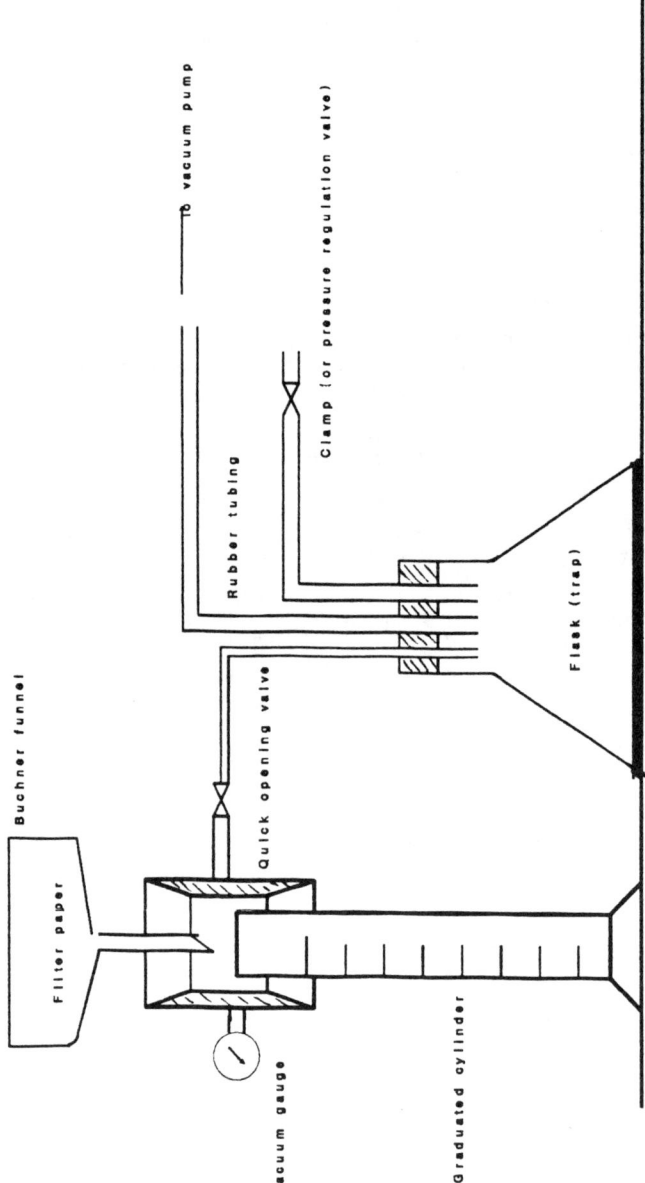

Figure 5.36 Schematic diagram of a specific resistance apparatus.[13,14]

Table 5.9 Sludge Characteristics as Related to Specific Resistance

Specific resistance value (m/kg)	Sludge characteristics
10^{11}–10^{12}	Easily filtered sludge
10^{14}–10^{15}	Poorly filtered sludge

Compiled from References 21 and 58.

$$t/V = b \cdot V + a \quad [5.75]$$

where:

$$b = (\mu \cdot r_s \cdot C)/(2 \cdot P \cdot A^2) \quad [5.76]$$

and

$$a = \mu \cdot R_m / P \cdot A \quad [5.77]$$

By plotting t/V against V a straight line of gradient b is obtained and r_s is found as indicated in Equation 5.78:

$$r_s = (2 \cdot b \cdot P \cdot A^2)/(\mu \cdot C) \quad [5.78]$$

where: r_s = specific resistance of sludge cake, m/kg
b = slope of the straight line of t/V vs. V, s/m^6
P = pressure applied, N/m^2
A = area of filtration, m^2
μ = viscosity of filtrate, N·s/m^2
C = solids content, kg/m^3

Therefore, the specific resistance may be defined as the resistance to filtrate flow caused by a cake of unit weight of dry solids per unit filter area.

Table 5.9 gives a general outline of the values of the specific resistance that provides an estimation of the degree of sewage sludge dewaterability.

The value of r_s for most wastewater sludges changes with pressure according to the relationship:

$$r_s = r_s' \cdot P^a \quad [5.79]$$

where: r_s = specific resistance to filtration at applied pressure P, m/kg
r_s' = a constant
a = coefficient of compressibility, (varies between 0 and 1).

Equation 5.79 can be put in the form:

$$\text{Log}(r_s) = a \cdot \text{Log}(P) + \text{Log}(r_s') \quad [5.80]$$

If "Log (r_s)" is plotted as a function of "Log (P)" then a straight line of slope "a" is obtained. The coefficient "a" is a measure of the compressibility of the cake to deformation. The greater the value of a, the more compressible is the sludge. When "a" = 0 the cake is incompressible.

d. Specific Resistance Test

The specific resistance can be found by conducting the filtration of a sludge sample on a Buchner funnel as follows:

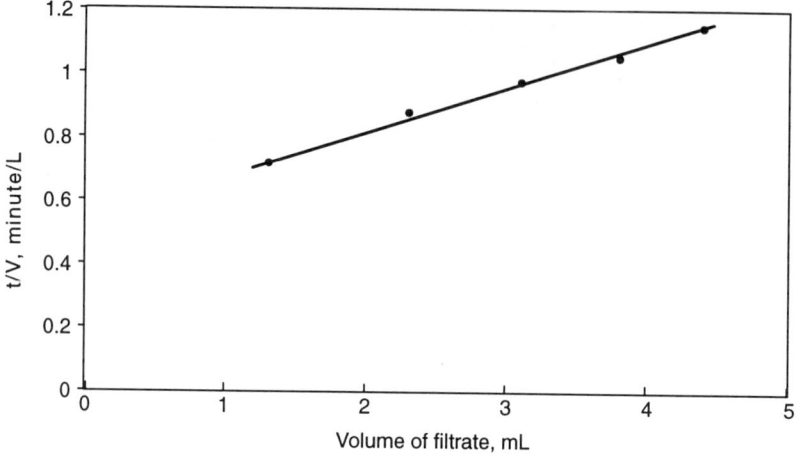

Example 5.15

1. The solids concentration of the sludge is determined.
2. A Whatman filter paper number 1, of known characteristics, is wetted with distilled water and placed in the funnel.
3. A vacuum is applied to remove excess water from the paper.
4. The sample is poured carefully on top of the filter paper.
5. Vacuum is then applied and the time is recorded.
6. Filtrate is collected in a graduated cylinder.
7. A record is taken of the volume collected vs. time to enable computations of the specific resistance to filtration.
8. Temperature of the filtrate is recorded to enable obtaining viscosity.

Figure 5.36 shows the layout of the experimental apparatus.

Example 5.15

In a specific resistance test using the Buchner funnel technique on a sludge sample, the following values were recorded:

Time, minutes				
1	2	3	4	5
Volume of filtrate collected, mL				
1.4	2.3	3.1	3.8	4.4

1. Find the specific resistance of the sludge given:
   - Vacuum applied = 69 kPa,
   - Filtrate viscosity = $1.1 \cdot 10^{-3}$ N·s/m²,
   - Volume of filtrate used = 50 mL,
   - Solids concentration = 0.075 g/mL,
   - Area of Whatman number 1 filter paper = 44.2 cm².
2. Is this sludge amenable to dewatering by vacuum filtration?

Solution

1. Given: $P = 69 \cdot 10^3$ N/m², $A = 44.2 \cdot 10^{-4}$ m², $\mu = 1.1 \cdot 10^{-3}$ N·s/m², $C = 0.075$ g/mL, variation of volume of filtrate collected with time of filtration.
2. Find the ratio of time to filtrate volume, t/V as:

Time, minutes				
1	2	3	4	5
Volume of filtrate collected, mL				
1.4	2.3	3.1	3.8	4.4
t/V, minute/mL				
0.714	0.870	0.968	1.05	1.14

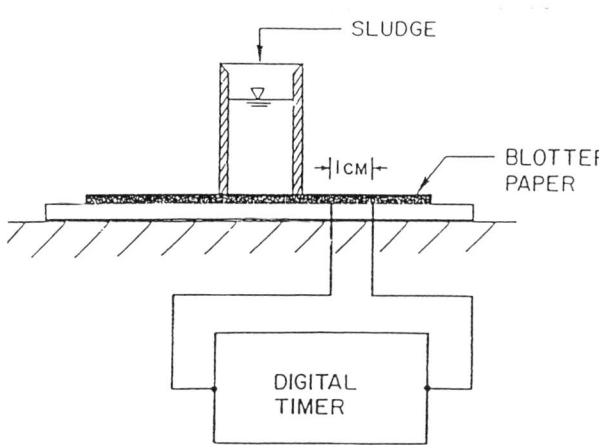

Figure 5.37 A capillary suction time device. (From Hammer, M. J., *Water and Wastewater Technology*, John Wiley & Sons, New York, 1977. With permission.)

3. Plot graph of t/V vs. V.
4. From graph find the slope of the straight line obtained as: $b = 8.25 \cdot 10^{12}$ s/m^6.
5. Find the sludge solids concentration as: $C = 0.075 \cdot 10^{-3}/10^{-6} = 75$ kg/m^3.
6. Use Equation 5.78 to compute the sludge specific resistance as:
   $r_s = 2 \cdot b \cdot P \cdot A^2/\mu \cdot C$
   $= (2 \cdot 8.25 \cdot 10^{12} \cdot 69 \cdot 10^3 \cdot (44.2 \cdot 10^{-4})^2)/(1.1 \cdot 10^{-3} \cdot 75)$
   $= 2.7 \cdot 10^{14}$ m/kg.
7. Since the sludge will not filter well if the specific resistance exceeds $1 \cdot 10^{12}$ m/kg, then this sludge sample (with $r_s = 2.7 \cdot 10^{14}$ m/kg) will not filter well in a vacuum filter.

e. Capillary Suction Time (CST)

Capillary suction time (CST) is the time needed by water to drain from a sludge and onto a blotter. Withdrawal of water by this method can be regarded to be a filtration process. The shorter the CST, the higher the filterability of the sewage sludge. Long CST values usually suggest problems in the filtration of the sewage sludge investigated.

Figure 5.37 illustrates the Capillary Suction Time Apparatus[13].

Table 5.10 offers an idea about the values of the typical capillary suction times that are found for certain sewage sludges.

f. Filter Leaf Test

The filter leaf test[13] simulates the principle of operation of a vacuum filter (see Figure 5.38). The procedure of conducting the test using the filter leaf method is as outlined below:

1. A filter cloth is wrapped around a hollowed-out disk.
2. The disk or the filter leaf is inserted into the sludge for about 20 s. This period of time is considered the length of time a typical vacuum filter might be in contact with the sludge.
3. Disc is raised out of the sludge for about 40 s and then the vacuum is released. This time resembles the drying cycle period.
4. The filter yield is estimated (in terms of kg of dry solids per m^2 of filter area per hr).

Table 5.10 CST[a] Values for Some Sludges

Sludge type	CST (s)
Primary sludge	70
Digested sludge	50
Conditioned sludge	20
Activated sludge	10

[a] CST = capillary suction time.

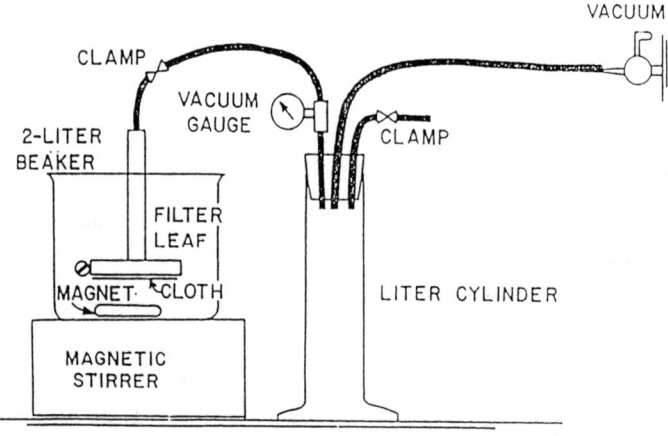

Figure 5.38 Filter leaf apparatus for vacuum filter design. (From Hammer, M. J., *Water and Wastewater Technology*, John Wiley & Sons, New York, 1977. With permission.)

5. Centrifugation

The operational variables accompanying the centrifugation process include the characteristics of the sludge (such as its water-holding capacity), rotational speed, hydraulic loading rate, depth of the liquid pool in the bowl, and use of polyelectrolytes to improve performance.[1]

The performance of a centrifuge is usually evaluated by its percent capture.[1] The percent capture is defined as:

$$\text{Percent capture} = \left[1 - \left\{\left(C_r(C_c - C_f)\right)/C_f(C_c - C_r)\right\}\right] \cdot 100 \quad [5.81]$$

where: C_r = concentration of solids in reject wastewater (centrate), mg/L, %
C_c = concentration of solids in the cake, mg/L, %
C_f = concentration of solids in sludge feed, mg/L, %.

The percent capture, for a constant feed content, increases as the concentration of solids in the centrate decreases.

A centrifuge needs to settle the solids and then reject them. Thus, two parameters must be addressed while modeling the scale-up between two geometrically similar centrifuges. The first parameter concerns settling of particles as measured by the Sigma equation. The second parameter concerns the rejection of solids by the centrifuge as advocated by the Beta equation.[13]

In the Sigma equation it is assumed that if two centrifuges (centrifuge number 1 and centrifuge number 2) are to have the same settling effects within the bowl, the relationship of Equation 5.82 must prevail.[13]

$$Q_1/\Sigma_1 = Q_2/\Sigma_2 \quad [5.82]$$

where: Q_1 = liquid flow rate into the first centrifuge, m³/s
  Σ_1 = a parameter related to the characteristics of the first centrifuge; this parameter does not depend on the sludge characteristics
  Q_2 = liquid flow rate into the second centrifuge, m³/s
  Σ_2 = a parameter related to the characteristics of the second centrifuge

For a solid-bowl centrifuge, the Sigma value is determined as:

$$\Sigma = \left([v_r]^2 \cdot V\right) \Big/ \left(g \cdot \operatorname{Ln}(r_2/r_1)\right) \qquad [5.83]$$

where: Σ = a parameter related to the characteristics of a centrifuge[13]
  v_r = rotational velocity of the bowl, rad/s
  V = liquid volume in the pool, m³
  g = gravitational acceleration, m/s²
  r_1 = radius from centerline to the surface of the sludge, m
  r_2 = radius from centerline to inside bowl wall, m

The aforementioned equations indicate that if the first centrifuge offers a satisfactory and a reasonable performance for its characteristics and flow rate, then a second centrifuge of similar geometry can be relied upon to produce equal results at a different flow rate.

The arguments presented so far ignore the movement and rejection of solids out of the bowl. Solids withdrawal is a parameter of considerable importance in centrifugation systems. Solids movement may be estimated through the Beta equation[13] as outlined in Equation 5.84:

$$W_1/\beta_1 = W_2/\beta_2 \qquad [5.84]$$

where: W_1 = solids loading rate for the first centrifuge, kg/hr
  β_1 = beta function for the first centrifuge
  W_2 = solids loading rate for the second centrifuge, kg/hr
  β_2 = beta function for the second centrifuge.

Beta function may be computed by using Equation 5.85:

$$\beta = v_w \cdot d_p \cdot n \cdot \pi \cdot Z \cdot D \qquad [5.85]$$

where: β = beta function for a centrifuge
  v_w = the difference in the rotational velocity between the bowl and the conveyor, rad/s
  d_p = the distance between blades or the scroll pitch, m
  n = number of leads
  Z = depth of sludge in the bowl, m
  D = bowl diameter, m

The scale-up procedure enables the measurement of the liquid flow rate as well as the solids rejection from the larger centrifuge which is of similar geometry. The lowest value of the two parameters governs the centrifuge capacity and dewatering performance.[13]

Example 5.16

A sewage sludge of 2% solids concentration is dewatered by centrifugation. The sludge is introduced to the solid bowl centrifuge at the daily rate of 15 m³. Due to an increase in sewage sludge that needs dewatering the old centrifuge needs to be scaled up to another geometrically similar and larger one. The characteristics of both centrifuges are as tabulated below. Find the flow rate at which the new centrifuge will perform as well as the old one.

Characteristic	Old centrifuge	New centrifuge
Bowl length, cm	25	60
Bowl diameter, cm	15	30
Bowl speed, rpm	5000	4200
Bowl depth, cm	2.5	5.0
Scroll pitch, cm	5	10
Number of leads	1	1
Conveyor velocity, rpm	4950	4150

Solution

1. Given: $Q = 15$ m³/d, $C = 2\%$, characteristics of the two centrifuges.
2. Use given data to form the following table:

Parameter	Old centrifuge	New centrifuge
l, cm	25	60
D, cm	15	30
r_2, cm	15/2 = 7.5	30/2 = 15
v_r, rad/s	$(2\pi/60)\cdot 5000 = 524$	$(2\pi/60)\cdot 4200 = 440$
Z, cm	2.5	5.0
$r_1 = r_2 - Z$, cm	7.5 − 2.5 = 5	15 − 5 = 10
d_p, cm	5	10
n, dimensionless	1	1
v_w, rad/s	5000 − 4950 = 50	4200 − 4150 = 50

3. Following is a schematic diagram of the old centrifuge illustrating essential dimensions and terms.
4. Use the Sigma Equation 5.82 to measure the solids settling characteristics by each centrifuge as: $Q_1/\Sigma_1 = Q_2/\Sigma_2$.
   - The volume, V, may be estimated as:
     $V = 2\pi([r_1 + r_2]/2)\cdot(r_2 - r_1)\cdot l$.
   - The parameter Sigma is to be computed by using Equation 5.83 as:
     $\Sigma = ([v_r]^2 \cdot V)/(g\cdot \mathrm{Ln}(r_2/r_1))$.

Parameter	Old centrifuge	New centrifuge
V, cm³	2,454	23,562
v_r, rad/s	524	440
$\mathrm{Ln}(r_2/r_1)$	Ln(7.5/5) = 0.40547	Ln(15/10) = 0.40547
g, cm²/s	981	981
Σ	1,693,984	11,468,055
Q, m³/d	15	?

5. Determine the liquid flow rate into the new centrifuge as:
   $Q_2 = (Q_1/\Sigma_1)\cdot \Sigma_2 = (15/1693984)\cdot 11468055 = 102$ m³/d.

   This signifies that the new centrifuge, as far as settling of solids is concerned, will attain equal performance for dewatering the sludge at a flow rate of 102 m³/d for the given conditions.

6. The movement of solids out of the centrifuge is considered through the Beta equation as presented in Equation 5.84 as:
   $W_1/\beta_1 = W_2/\beta_2$
   - Determine the solids loading rate for the two centrifuges as: W = flow rate (m³/d)·solids concentration·density (assume density of water = 1000 kg/m³).
   - Use Equation 5.85 to determine the Beta value for each centrifuge as:
     $\beta = v_w \cdot d_p \cdot n \cdot \pi \cdot Z \cdot D$.

Parameter	Old centrifuge	New centrifuge
W, kg/d	300	?
v_w, rad/s	50	50
d_p, cm	5	10
n, dimensionless	1	1
Z, cm	2.5	5.0
D, cm	15	30
β	29,452	235,619

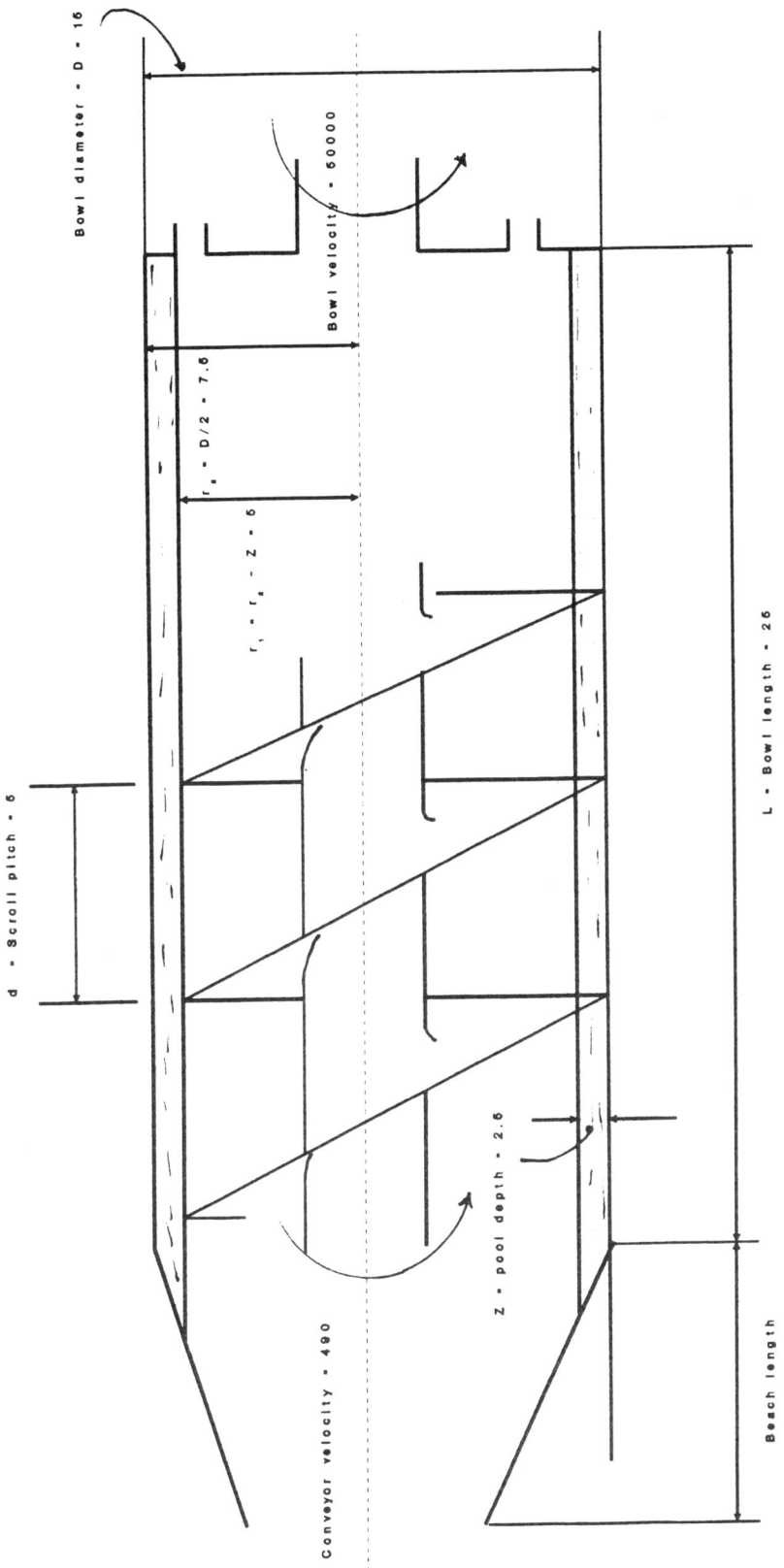

7. Use Equation 5.84 to evaluate the solids movement of the new centrifuge as:
   $W_2 = (W_1/\beta_1) \cdot \beta_2 = (300/29452) \cdot 235619 = 2400$ kg/d.
   This is equivalent to a value of $2400/(0.02 \cdot 1000) = 120$ m³/d (assuming the same solids concentration of 2% and the same density).
8. The Sigma equation anticipated settling characteristics for the new centrifuge yielding the same dewatering performance as the old one when the liquid flow rate is equal to 102 m³/d. The Beta equation indicates the movement of solids out of the new centrifuge to yield similar dewatering performance as the old centrifuge when the solids loading rate is equal to 120 m³/d.

The scale-up procedure involves the calculation of a liquid flow rate as well as a solids rejection rate and the lower value governs the centrifuge capacity. Therefore, in the case under study the liquid flow rate governs. Thus the new centrifuge is not to be operated at a liquid flow rate of more than 102 m³/d.

C. DILUTION

While dilution is clearly not the solution to pollution, nevertheless, discharges of small volumes of relatively dilute wastewater and discharges of treated wastewater do take place. These discharges have physical, chemical, and biological impacts on the receiving body of water. Mathematical models have been developed to predict the effect these discharges have on rivers, lakes, streams, estuaries, and oceans. The natural purification processes available in each of these waters depends on many related factors such as the volume of the body of the water course and its movement, the water body characteristics, ability of the water body to reoxygenate itself, the water use downstream of the disposal point, characteristics of the waste discharged, quantity of waste, etc.

The dilution capacity of the receiving body of water can be calculated by using the principle of mass balance. Figure 5.39 illustrates a schematic diagram of a treatment plant disposing of its final effluent into the neighboring water course.

Let the effluent from the plant, at point A, have a concentration of pollutant as C_w, while it is flowing at the rate of discharge of Q_w. Let the concentration of the same pollutant in the stream, upstream point B, have a value of C_{st}, and the stream water flows at the rate of Q_{st}. Let the concentration of the pollutant and the flow of the stream downstream of point C be C_m and Q_m, respectively. Then the mass balance yields the dilution law as symbolized in Equation 5.86.

$$C_w \cdot Q_w + C_{st} \cdot Q_{st} = C_m \cdot Q_m \qquad [5.86]$$

Example 5.17

The concentration of a certain pollutant upstream of a sewage treatment plant is 1 mg/L and the stream flow is 30 m³/s. The discharge of effluent from the sewage treatment plant is at the rate of 5 m³/s. The pollutant concentration downstream of the sewage treatment plant is 3 mg/L. Compute the pollutant concentration of the waste effluent stream.

Solution

1. Given: $C_{st} = 1$ mg/L, $Q_{st} = 30$ m³/s, $Q_w = 5$ m³/s, $C_m = 3$ mg/L (see figure on page 244).
2. Use the dilution law Equation 5.86 to compute the pollutant concentration at the wastewater effluent stream as:
   $C_w = (C_m \cdot Q_m - C_{st} \cdot Q_{st})/Q_w$.
3. Find the discharge of the mixture of stream water flow and the treatment plant effluent as:
   $Q_m = Q_w + Q_{st} = 5 + 30 = 35$ m³/s.
4. Substitute the values given in 1 above and the value of the flow of the mixture as computed in 3 above to find the concentration of the pollutant at the waste effluent stream as:
   $C_w = (3 \cdot 35 - 1 \cdot 30)/5 = 15$ mg/L.

D. DISPOSAL INTO NATURAL WATERS
1. Introduction

The disposal of treated wastewater into natural bodies of water may be appropriate, provided all governmental regulations for such discharges are considered and properly followed.

Pretreatment of the waste prior to discharge is essential in order to reduce concentrations of noxious and toxic elements that may hamper life or alternatively change the natural purification capacity of the receiving body of water.

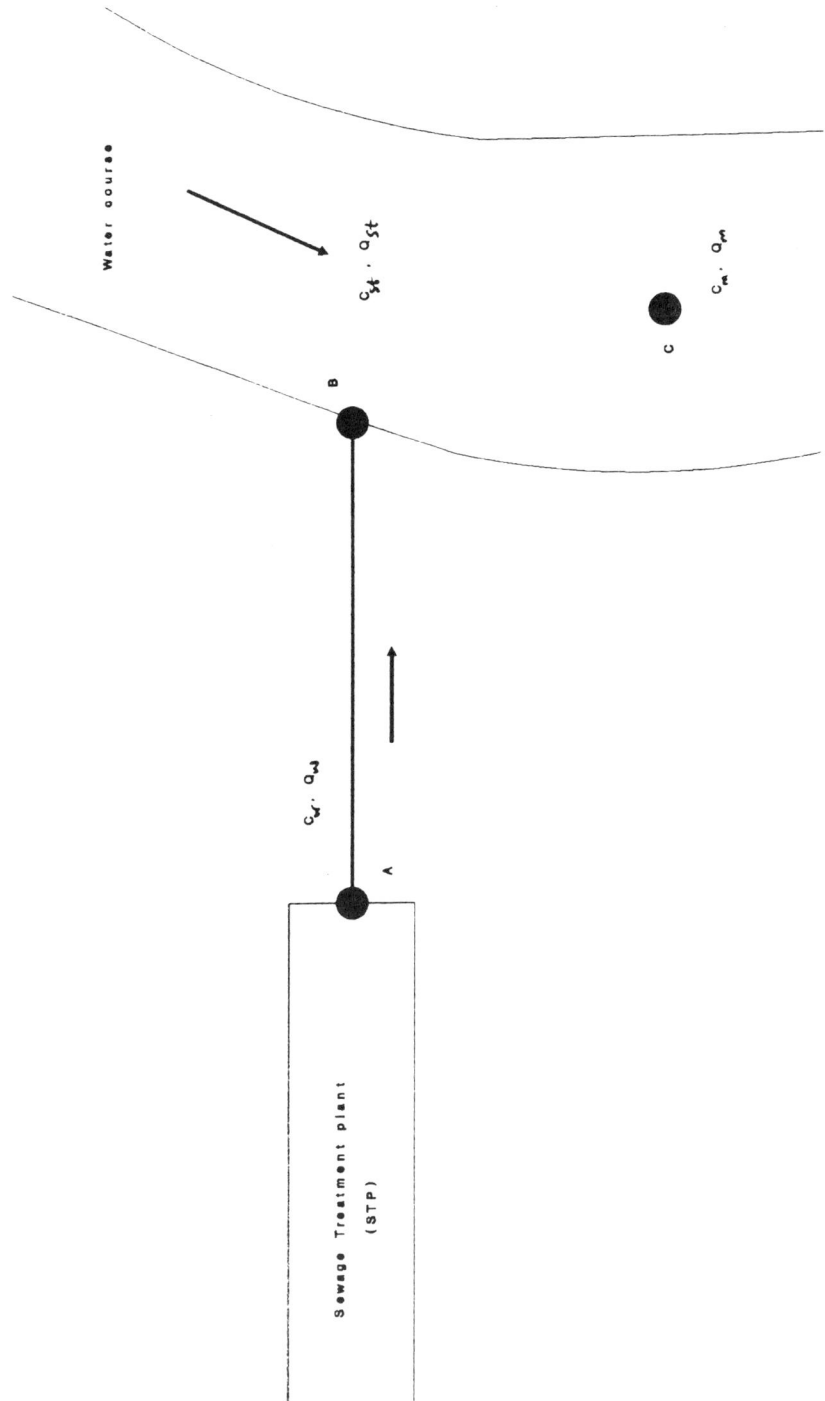

Figure 5.39 Schematic diagram of sewage treatment plant effluent discharged into a water course.

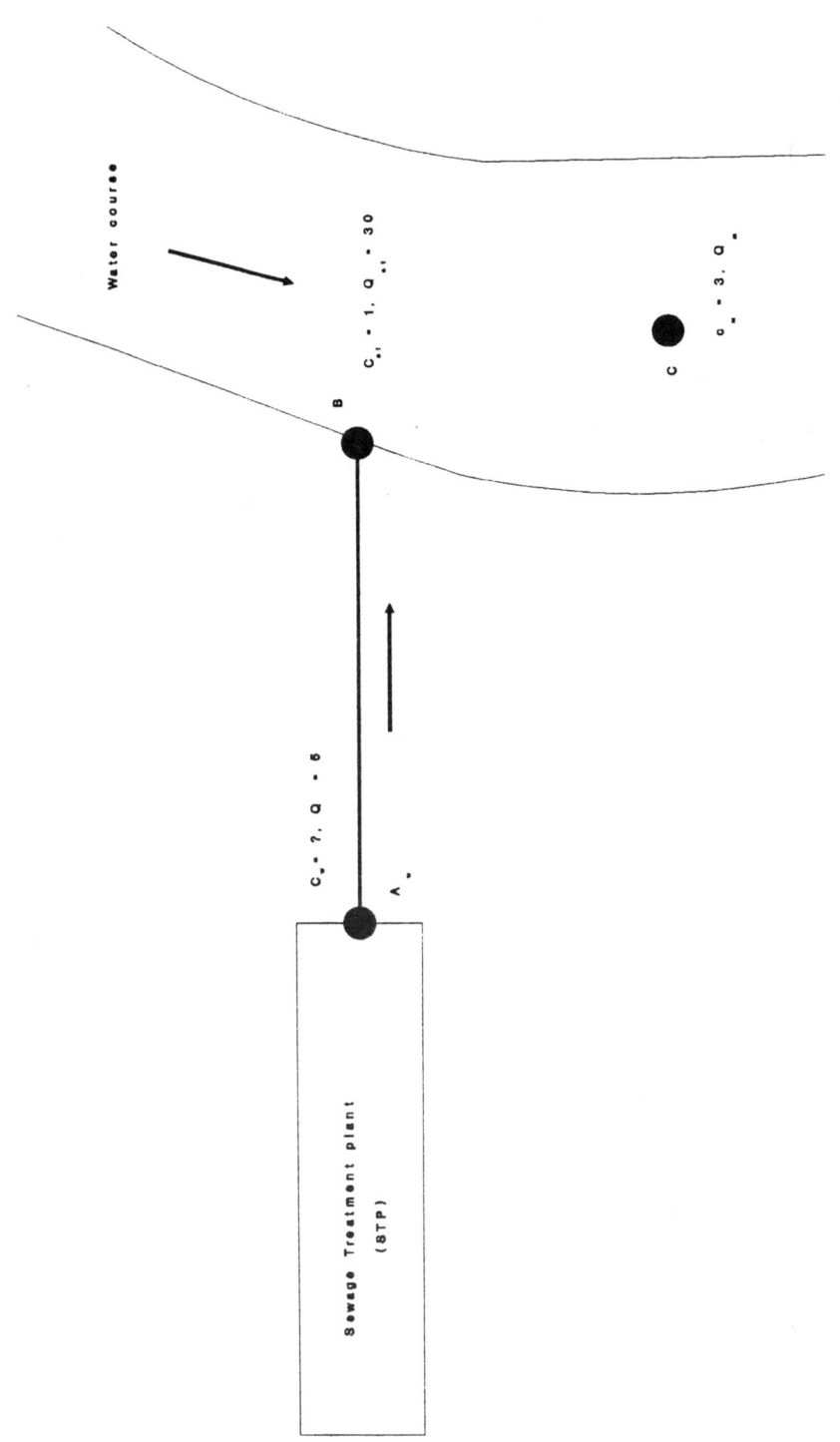

The effects of improperly treated wastewater discharges to streams or rivers include:

- The wastewater may render the stream unfit for ordinary beneficial uses.
- Discharges may change the quality of the stream water, which in turn affects the appearance and the biology and biochemistry of the stream.
- Improperly treated wastewater may render the water in the stream or river incapable of supporting fish life. This happens when the steam becomes devoid of dissolved oxygen. This dissolved oxygen is vital to sustain aquatic life, for example a Salmonide river needs to have a dissolved oxygen concentration of 7 mg/L while a Cyprinidae river requires a dissolved oxygen content of 5 mg/L.
- Improperly treated wastewater may create malodors and nuisances. These conditions may develop when the dissolved oxygen is absent. The stream or river may go anaerobic with the microorganisms using the combined oxygen in sulfates and nitrates, with the consequent result of the generation of odors, etc.
- Improperly treated wastewater may produce taste and odor problems downstream due, for instance, to phenolic compounds.
- Improperly treated wastewater may cause an increase in the concentration of toxic materials that are resistant to biodegradation in the river or stream water. Examples of these substances include sophisticated organic materials, chlorinated hydrocarbons, derivatives of pesticides, germicides, weed killers, etc.
- Improperly treated wastewater may create public epidemiological health hazards in form of classical diseases such as cholera, typhoid, and dysentery, or chronic poisoning by materials such as persistent organic synthetic compounds if consumed over long periods of time.
- The discharge of improperly treated wastewater may delay the mechanisms of self-purification in the water course.

Table 5.11 gives a general classification of the water quality for streams in terms of the 5-day BOD, the suspended solids, and the dissolved oxygen content.

2. Oxygen Renewal in Streams (Reaeration)

Dissolved oxygen is one of the most important constituents of natural bodies of water as it is essential for both chemical and biological reactions.

The sources of oxygen renewal, or reoxygenation, in a stream are twofold:

- Reaeration from the atmosphere.
- Photosynthesis of aquatic plants and algae.

When the concentration of dissolved oxygen falls below its equilibrium concentration, the net movement of the gas is initiated from the surrounding air phase to the water phase. The difference between the oxygen equilibrium concentration and the actual concentration is referred to as the oxygen deficit. The amount of reaeration within the stream is proportional to the dissolved oxygen deficiency.

Oxygen concentration as furnished by photosynthesis depends on the size of the algal population and the amount of sunlight captured by the algal cells.

When the sun is shining there is more incident radiation than when it is dark. This signifies a sinusoidal rate of photosynthesis. When large populations of algae are present, a diurnal variation in the concentration of dissolved oxygen develops.

Table 5.11 Stream Classification Patterns

Classification scheme	BOD_5^{20} (mg/L)	SS (mg/L)	DO as a % of saturation value
Very clean	≤1	≤4	
Clean	2	10	≥90
Fairly clean	3	15	75–90
Doubtful	5	21	50–75
Poor	7.5	30	<50
Bad	10	35	
Very bad	≥20	≥40	

Compiled from References 4 and 21.

The rate of reaeration is given by Equation 5.87 as:

$$r_r = k''(C_s - C) \quad [5.87]$$

where: r_r = rate of reaeration
C_s = dissolved oxygen saturation concentration, mg/L
C = dissolved oxygen concentration, mg/L
k'' = reaeration constant, /d, to base e.

The k'' can be estimated by finding the characteristics of the stream using the empirical formula proposed by O'Conner and Dobbins[1,31] for a natural stream as:

$$k'' \left(294(\text{Diff}_T \cdot v)^{0.5}\right) / (h)^{1.5} \quad [5.88]$$

where: v = mean stream velocity, m/s
h = average depth of flow, m
Diff_T = molecular diffusion coefficient for oxygen, m³/d.

The molecular diffusion coefficient is temperature dependent. As such, its change with temperature can be predicted by Equation 5.89 as:

$$(\text{Diff})_T = (\text{Diff}_c) \cdot (T_c)^{(T-20)} \quad [5.89]$$

where: $(\text{Diff})_T$ = molecular diffusion coefficient for oxygen at temperature of T °C, m²/d
Diff_c = molecular diffusion coefficient for oxygen at temperature of 20°C, m²/d; this may be taken to be equal to $1.76 \cdot 10^{-4}$
T_c = temperature correction coefficient, may be taken to be equal to 1.037
T = temperature, °C

Table 5.12 gives a layout of the reaeration constants for different water bodies.

3. Oxygen Depletion in Streams

Oxygen depletion, or deoxygenation, in a river or a stream is primarily due to:

- The continuation of the metabolic processes of microorganisms and their biodegradation of suspended and dissolved organic solids that are discharged into the streams by both natural processes or manmade sources.
- The required concentration of oxygen needed by the sludge and benthic deposits at the bottom of the stream.

a. Biodegradation of Organic Matter

Deoxygenation is brought about by oxidation of the organic matter. The amount of oxygen needed for the conversion of organic matter is related to the 5-day BOD.[20] The rate of dissolved oxygen consumption depends on the concentration of organics, the rate of biodegradation, and the dilution capacity of the stream. Thus, the rate of deoxygenation could be described by Equation 5.90 as:

$$r_D = -k' \cdot L \quad [5.90]$$

where: r_D = rate of deoxygenation
k' = first order reaction rate constant, /d.
L = ultimate carbonaceous BOD at point in question, mg/L
   $= L_0 \cdot e^{-k't} \quad [5.91]$
L_0 = ultimate carbonaceous BOD at point of discharge, mg/L
t = time, d.

Table 5.12 Reaeration Constants

Water body	Range of k″ at 20°C (to base e)[a]
Small ponds and back waters	0.11–0.23
Sluggish stream and large lakes	0.23–0.34
Large streams of low velocity	0.34–0.46
Large streams of normal velocity	0.46–0.69
Swift streams	0.69–1.15
Rapids and waterfalls	>1.15

[a] For other temperatures use:

$$(k'')_T = (k'')_{20} \cdot (1.024)^{(T-20)}$$

where: $(k'')_T$ = reaeration constant at a temperature of T°C.
$(k'')_{20}$ = reaeration constant at a temperature of 20°C.

From Fair, G. M. and Geyer, J. C., *Water Supply and Waste-Water Disposal*, John Wiley & Sons, New York, 1954, 845, 846. With permission.

Substituting Equation 5.91 in Equation 5.90, then:

$$r_D = -k' \cdot L_o \cdot e^{-k't} \qquad [5.92]$$

where: r_D = rate of deoxygenation
k' = first order reaction rate constant, /d
L_0 = ultimate BOD at point of discharge, mg/L
t = time, d.

b. Oxygen Depletion Due to the Oxidation of Sludge Deposits

Upon introduction of an organic load to a body of water such as a stream, the heavy materials settle to form a sludge deposit or a benthic layer on the bottom of the stream. The accumulated deposits exercise a significant oxygen demand on the overlying water layer, especially for slowly moving streams. The bulk of the deposited sludge is liable to be anaerobically biodegraded at the bottom of the stream, and aerobically oxidized at the interface between the sludge and the flowing water of the stream.

The amounts of settlement and removal of organic particles change depending upon the influence of certain factors such as the velocity of water movement in the stream, the degree of turbulence within the stream, and the characteristics and amount of settled material within the reach.

The effect of accumulating organic mud and sediments may be evaluated by the empirical equation developed by Fair et al.[72] as presented in Equation 5.93.

$$L_m = 3.14\left(10^{-2} \cdot L_o\right) \cdot T_c \cdot VS \cdot \left\{(5 + 160 \cdot VS)/(1 + 160 \cdot VS)\right\} \cdot (t)^{0.5} \qquad [5.93]$$

where: L_m = maximum daily benthal oxygen demand, g/m²
L_0 = BOD_5^{20} of benthal deposit, g/kg volatile matter
T_c = temperature correction factor

$$= BOD_5^T/BOD_5^{20} = \left(1 - e^{-k'_t}\right)/\left(1 - e^{-5k'_{20}}\right) \qquad [5.94]$$

VS = daily rate of volatile solids deposition, kg/m²
t = time during which settling takes place, d

4. Sag Curves in Streams

A simple stream oxygenation model has been developed by Streeter and Phelps; following is their mathematical model:

$$d(OX)/dt = k' \cdot L - k'' \cdot OX \qquad [5.95]$$

where: OX = oxygen deficit, mg/L
t = time, d
k′ = first order reaction rate constant, /d
L = ultimate BOD at point in question, mg/L
k″ = reaeration constant, /d

The Streeter and Phelps model demonstrates that the oxygen deficit increases with the increase in the pollutional load, and decreases with an increase of oxygen concentration through reaeration.

The model suffers from the following limitations:[1,4,10,12,14,15]

- It neglects the effects of oxygen production by algal photosynthesis action.
- It does not take in account oxygen depletion by the benthic deposits.
- The model assumes one pollution source or a point source.
- The equation does not anticipate factors affecting the organic load apart from the BOD.
- The model assumes steady-state conditions along each river reach.

Nevertheless, the model is reasonable when and if cautiously used.

Integrating Equation 5.95 and assuming that the initial oxygen deficit at the point of waste discharge, at time t = 0, is OX_0, mg/L, then:

$$OX_t = \{k' \cdot L_o [e^{-k't} - e^{-k''t}]/[k'' - k']\} + OX_o \cdot e^{-k''t} \qquad [5.96]$$

where: OX_t = oxygen deficit at time t, mg/L
k′ = first order reaction rate constant, /d
L_0 = ultimate BOD at point of discharge, mg/L
k″ = reaeration constant, /d
t = time of travel in the stream from the point of discharge, d
OX_0 = initial oxygen deficit at the point of waste discharge, at time t = 0

Effective biodegradation of the organic matter by microbial action starts immediately after its discharge. This decomposition consumes oxygen. Thus the rate of atmospheric reaeration increases with an increase in oxygen utilization for biodegradation of the waste. Eventually, a point is reached where the rate of utilization of oxygen for biodegradation of the waste equals the rate of atmospheric reaeration. At this point along the reach of the stream there is a balance. Downstream of this point the rate of reaeration is greater than the rate of oxygen depletion. Therefore, the amount of dissolved oxygen starts to increase. Eventually, the stream recovers unless another pollutional discharge takes place. This signifies the phenomenon of self-purification. The self-purification of the stream may thus be defined as the ability of a natural body of water to purify itself by decomposition of organic matter and other pollutants or contaminants.

A very important point in the sag curve is the point of lowest dissolved oxygen concentration. This point represents the maximum impact on the dissolved oxygen deficit due to organic waste disposal. This point is the critical oxygen deficit.

The critical oxygen deficit, OX_c, at the critical point, distance, x_c, can be determined by letting the rate of change of deficit, d(OX)/dt, equal zero which is the maximum deficit, then

$$OX = Ox_c \text{ and } L = L_o \cdot e^{-k't} \qquad [5.97]$$

thus, Equation 5.95 becomes:

$$OX_c = (k' \cdot L_o \cdot e^{-k' t_c})/k'' \qquad [5.98]$$

where: OX_c = critical oxygen deficit, mg/L
k′ = first order reaction rate constant, /d
L_0 = ultimate carbonaceous BOD at point of discharge, mg/L
t_c = critical time required to reach the critical distance. This time can be found by differentiating Equation 5.96 with respect to time t and putting d(OX)/dt = 0.

Thus:

$$t_c = (1/[k'' - k']) \cdot \text{Ln}\left\{[k''/k']\left(1 - \left([OX_o/L_o] \cdot ([k'' - k']/k')\right)\right)\right\} \qquad [5.99]$$

where: t_c = critical time, d
k'' = reaeration constant, /d
k' = first order reaction rate constant, /d
OX_0 = initial oxygen deficit at the point of waste discharge, at time t = 0
L_0 = ultimate BOD at point of discharge, mg/L

The critical distance may be found as in Equation 5.100 as:

$$x_c = t_c \cdot v \qquad [5.100]$$

where: x_c = critical distance, m
t_c = critical time, d
v = velocity of flow in the stream, m/d

The dissolved oxygen concentration varies along the reaches of the stream and are a function of the following:

- BOD exertion by the organic load, $k' \cdot L$
- Reaeration from the atmosphere, $k''(C_s - C)$
- Algal cells photosynthesis, AG
- Sludge deposits biological oxidation, SL
- Algal respiration rates, ALR.

Thus a steady state equation may be developed for all of the aforementioned factors affecting the dissolved oxygen deficit, as presented in Equation 5.101:

$$OX_t = \left\{k' \cdot L_o\left[e^{-k't} - e^{-k''t}\right]/[k'' - k']\right\} + OX_o \cdot e^{-k''t} + (SL + ALR - AG) \cdot \left(1 - e^{-k''t}\right)/k'' \qquad [5.101]$$

Equation 5.101 implies the need for evaluation of the many interacting parameters, including detailed testing and analysis of the stream under review. The mathematical findings for the various river reaches needs to be verified by actual field tests and measurements.

Example 5.18

A city has a population of 70,000 with each person, on the average, producing 0.5 m³ of wastewater per day. The wastewater treatment plant for this city has an effluent with a carbonaceous biochemical oxygen demand (CBOD$_5$) of 30 mg/L. The temperature of the wastewater is 27°C and there is 0.5 mg/L dissolved oxygen in the plant's effluent. The stream into which the effluent is discharged has a flow of 864,000 m³/day at 0.3 m/s with an average depth of 2 m. The temperature of the stream water is 22°C before the wastewater is added to the stream. The stream is 80% saturated with dissolved oxygen (DO), has a chloride concentration of 60 mg/L, and CBOD$_5$ of 3.0 mg/L. The deoxygenation coefficient K′ for the wastewater is 0.39/day at 20°C (base e). Determine the following:

1. The wastewater flow in m³/day.
2. The temperature, DO, and CBOD$_5$ of the mixture of the stream water and the wastewater.
3. The initial dissolved oxygen deficit of the stream just below the wastewater treatment plant's outfall.
4. The distance downstream to the critical DO deficit.
5. The minimum DO in the stream below the wastewater treatment facility.

Solution:

1. The wastewater flow = 70,000 persons·0.5 m³/person/day = 35,000 m³/day.
2a. The temperature of the mixture of the stream water and the wastewater:
   = [864,000 m³/d·22°C + 35,000 m³/d·27°C]/[864,000 m³/d + 35,000 m³/d]
   = 22.2°C.

2b. The dissolved oxygen of the mixture of the stream and the wastewater: the DO of the stream at 22°C, 80% saturation with DO and a 60 mg/L chloride concentration (see DO Table, DO vs. temperature) = 8.8 mg/L · 0.8 = 7.04 mg/L.
The 60 mg/L chloride concentration would reduce the DO by 60/100 · 0.008 = 0.0048 mg/L. (Also see the table for saturation values for DO at various temperatures.) This chloride concentration has a negligible effect on the DO concentration. The DO of the mixture
   DO = [864,000 · 7.04 mg/L + 35,000 · 0.5 mg/L]/899,000
      = 6.78 mg/L.
2c. The $CBOD_5$ of the mixture:
   L = [864,000 · 3 mg/L + 35,000 · 30 mg/L]/899,000
     = 4.05 mg/L.
3. The saturated DO of the stream at 22.2°C just below the outfall would be (see DO saturation vs temperature table):
   = 8.8 (at 22°C) – (0.2/1) · 0.1 (correction to 22.2°C)
   = 8.78 mg/L.
   Therefore, the initial DO deficit just below the wastewater outfall = (8.78 mg/L) – (6.78 mg/L) = 2.0 mg/L.
4. Use Equation 5.99 to determine time and distance to the critical DO point:
   $t_c = [1/(K'' - K')]\ln\{(K''/K')(1 - OX_0/L_0)[(K'' - K')/K']\}$.
   To apply Equation 5.99, K″, K′, OX, and L_0 must be determined. K′ (deoxygenation coefficient) of the wastewater is 0.39/d at 20°C (base e).
   Determine K′ at 22.2°C (temperature of the mixture of the wastewater and the stream) using the relationship:
   K′ at 22.2°C = 0.39 · $(1.047^{22.2-20})$ = 0.43/day.
   Use Equations 5.87 and 5.88 to determine K″ (reaeration coefficient). First use Equation 5.89 to determine the molecular diffusion coefficient $(Diff)_T$ at 22.2°C:

$$(Diff)_T = (Diff_c) \cdot (T_c)^{(T-20)}$$

$$(Diff)_c = 1.76 \cdot 10^{-4} \text{ at } 20°C$$

$$(Diff)_{22.2°C} = (1.76 \cdot 10^{-4}) \cdot (1.037)^{22.2-20}$$

$$= 1.91 \cdot 10^{-4} \text{ m}^2/\text{d}. \quad [5.89]$$

Substitute $(Diff)_T$, stream velocity, and stream depth in Equation 5.88.

$$K'' = 294 \cdot (Diff_T \cdot v)^{0.5}/h^{1.5}$$

$$= 294 \cdot (1.91 \cdot 10^{-4} \cdot 0.3)^{0.5}/2^{1.5}$$

$$= 0.79/\text{d (base e) at } 22.2°C. \quad [5.88]$$

The ultimate carbonaceous $CBOD_5$ L_0 can be determined from Equation 1.44: $L = L_0(1 - e^{-K't})$ (K′ based on natural ln e), where L = 5-day carbonaceous BOD, mg/L, and L_0 = ultimate carbonaceous BOD, mg/L

First, determine K′ at 22.2°C:
   $K'_{22.2}$ = $K'_{20} \cdot (1.047)^{22.2-20}$
   $K'_{22.2}$ = (0.39/d)·1.106 = 0.431/d at 22.2°C
   L   = 4.05 mg/L
   4.05 mg/L = L_0 (1 – $e^{-(0.431/\text{d} \cdot 5 \text{ d})}$)
   L_0 = 4.05/(1 – $e^{-2.155}$) = 4.58 mg/L

Substituting values in Equation 5.99:

$t_c = (1/[0.79 - 0.431]) \cdot \ln\{(0.79/0.431) \cdot [1 - ((2/4.58) \cdot (0.79 - 0.431)/0.431)]\}$

$t_c = (2.786) \cdot \ln[(1.833) \cdot (1 - 0.364)] = 0.43$ days.

The distance downstream to the critical point equals the time of travel multiplied by the stream velocity. In this case the distance is

$X_c = 0.43$ days $\cdot$ 0.3(m/s)$\cdot 86,400$(s/day)$/1000$(m/km)

$= 11.15$ km.

5. The minimum DO in the stream below the wastewater treatment facility can be determined by using Equation 5.96:

$OX_t = \{K' \cdot L_0[(e^{-K't} - e^{-K''t})/(K'' - K')]\} + OX_0 \, e^{-K''t}$

$= \{0.431 \cdot 4.58 \cdot [(e^{-0.431 \cdot 0.43} - e^{-0.79 \cdot 0.43})/(0.79 - 0.431)] + 2.0 \cdot e^{-0.79 \cdot 0.43} = 2.07$ mg/L.

Therefore the minimum DO is $8.78 - 2.07 = 6.71$ mg/L. This oxygen level in the stream would support not only Cyprinidae fish but also Salmonide fish if they were present.

VI. HOMEWORK PROBLEMS

A. DISCUSSION PROBLEMS

1. What are the major sources of wastewater in your area?
2. Outline major factors that influence the quantity and quality of the wastewater in your community.
3. Differentiate between unit operations and unit processes in wastewater treatment facilities. Give examples of each.
4. What are the reasons for the treatment of industrial wastes?
5. What are the major wastewater reuse fields?
6. What are the problems that must be considered if highly persistent detergents, pesticides, and other toxic compounds are present in the wastewater?
7. "The lethal concentration of some common toxic substances to fish is not a fixed quantity and it depends upon different conditions." Name five of these conditions.
8. Recommend a suitable method that can be adopted for disposal of acid wastes generated in petroleum refineries and chemical plants? State your reasons.
9. Explain how the filtration process improves water quality.
10. Sketch a plan and a cross section of an individual rapid sand filter.
11. Why is there such a significant difference in biological activity of a slow sand filter in comparison to a rapid sand filter?
12. Explain the terms loss of head and negative head as applied to filtration.
13. Describe briefly the method of cleaning a rapid sand filter.
14. Indicate how the Rose equation: $h_f = 1.067 \cdot C_D \cdot v^2 \cdot L/(g \cdot \phi \cdot d \cdot e^4)$ can be used to determine the pressure drop through a rapid gravity filter.
15. "To reduce building costs, rapid filters have sometimes been constructed with a shallow depth of water on top of the sand bed." Discuss this statement and point out the disadvantages of this construction.
16. Differentiate between "slow sand filters" and "rapid sand filters".
17. Write a brief note on the advantages and disadvantages of slow sand filters.
18. Briefly outline "desalination methods". Which method would you recommend for a town in the Middle East.
19. Indicate advantages and disadvantages of the reverse osmosis process.
20. "Solar distillation can be ideal for distillation plants." Discuss this statement indicating advantages and limitations of this method.
21. Briefly explain the following:
    1. Electrodialysis process and its application in an arid developing country.
    2. Problems involved in the reverse osmosis process.
    3. Solar distillation and future application.
    4. Freezing as a method of purifying sea water.
22. Briefly explain "disinfection". Elucidate your answer with appropriate sketches and diagrams.
23. Define the terms: chlorine demand, free available chlorine, combined available chlorine, and breakpoint chlorination.

24. Name three commonly used disinfectants and discuss the advantages and disadvantages posed by each.
25. Outline possible parallel competitive reactions occurring when chlorine is added to water.
26. Give examples of dechlorinating agents.
27. Outline the major methods that are used for the dewatering of sewage sludges. Which method would you recommend for your local town?
28. What is the significance of sewage sludge dewatering as a process preceding final sludge disposal?
29. Why are wastewater sludges often difficult to dewater?
30. What are the major factors that influence filterability of sewage sludges?
31. What are the environmental problems that may result by unauthorized discharges of industrial wastewaters that contain acids or alkalies?
32. Briefly outline the effects of discharging raw domestic wastewater into natural bodies of water.
33. What parameters need to be addressed in order to permit an industry to discharge its wastewater into a stream or river?

B. TRUE AND FALSE STATEMENT AND MULTIPLE CHOICE QUESTIONS (MCQS)
True or False Statements

Indicate whether the following statements are true (T), or false (F):
1. Untreated wastewater may contain pathogens.
2. Adsorption is associated with the surface properties of the liquid.
3. Slow sand filters are always backwashed before the minimum head loss of 0.6 to 0.7 m is reached.
4. Rapid sand filters are always backwashed before the maximum allowable turbidity is reached.
5. Disinfection implies the destruction of pathogens and all microscopic life.
6. The higher the temperature, the more effective the action of a disinfectant.

Multiple Choice Questions (MCQs)

Circle the most appropriate answer:
1. Given that a raw water contains high concentrations of pesticides (which are very soluble in water and are broken down very slowly by bacteriological and chemical oxidation), then the method of purification to be selected would be:
    a] Chemical oxidation.   b] Activated carbon.
    c] Biological breakdown. d] Chlorination.
    e] None of these.
2. Which one of the following treatment units are capable of producing a safe bacteriological water suitable for drinking purposes?
    a] Rapid sand filters.   b] Slow sand filters.
    c] Pressure filters.   d] Chemical coagulation.
    e] None of these.
3. The analysis of a raw water sample revealed that the majority of particles encountered are inorganic in nature and have a particle size in the range of 1 nm. Which one of the following physical treatments would you choose?
    a] Screening.   b] Sedimentation and flotation.
    c] Filtration.   d] Ion exchange.
    e] Adsorption.
4. The content of manganese in a groundwater that contains oxygen is mostly:
    a] High.   b] Low.
    c] Absent.   d] None of the above.
5. Rapid sand filters should always be backwashed before the maximum allowable
    a] detention time is reached.   b] filter resistance is reached.
    c] filtration rate is reached. d] turbidity is reached.
6. Efficient digestion (methane fermentation) requires:
    a] presence of light.   b] anaerobic conditions.
    c] low pH.   d] temperature lower than 20°C.
    e] absence of bacteria other than methane bacteria.

C. SPECIFIC MATHEMATICAL PROBLEMS

1. Determine the different per capita wastewater flows (i.e., maximum daily, average daily, and minimum daily flow) for a population of 30,000 using the following data: annual average water consumption of 150 L/capita/d.; DWF amounting to 85% of water consumption; and minimum flow equals 45% of average flow.

2. A sand filter of depth 1 m and porosity of 40% is composed of uniform sand grains of 0.3 mm in diameter with a shape factor of 0.92 and a specific gravity of 2.6. Water at a temperature of 20°C is passed through the bed at a filtration rate of 50 cm/hr. Using both Carman-Kozeny and Rose's equations, determine the bed head loss.

3. A rapid sand filter is used for treating a water flow of 7500 m^3/day. The filter dimensions are 3.5 m × 8 m with a depth of 0.8 m. An analysis of the sand showed that it has a porosity of 0.55 with an average grain size diameter of 0.6 mm and shape factor of 0.86. The supernatant water depth is 1.5 m while the water has a viscosity of $1.1 \cdot 10^{-3}$ N·s/m^2. The rate of filter backwashing is 25 m^3/m^2/hr for a period of 5 min. Compute:
   a] the filtration rate,
   b] the amount of water used for backwashing the filter, and
   c] the initial head loss through the filter (use both Rose and Carman-Kozeny equations).

4. A filter is to treat water having a concentration of impurities of 15,000 mg/m^3. If the bed thickness is 80 cm and the coefficient of filtration is assumed to be 6 m^{-1}, compute the effluent concentration of impurities. Find also the percentage reduction of impurities as provided by this filter.

5. How many rapid filters are needed to treat 40 m^3/min of water?

6. A model sand filter of 1 m length consists of a tube of 1.5 cm diameter. The bed is composed of uniform spherical sand grains of 0.5 mm diameter, with a porosity of 40%. The filter is used to treat water at a flow rate of 6 m^3/m^2/hr. Find the head loss across this filter by using Rose's equation for water at a temperature of 20°C.

7. A filter bed is formed with a 50-cm layer of uniform anthracite coal with an average size of 1.6 mm and a 50 cm layer of uniform sand with an average size of 0.5 mm. Water is to be filtered at the rate of 3 L/m^2/s. Assuming that the filter bed porosity throughout is 40%, and that the operating temperature in it is 20°C, determine the clear water head loss. Use both Rose and Carman-Kozeny equations for computing the head loss.

8. Surface water from a stream contains suspended solids of 130 mg/L. The village people in the vicinity have a water demand of 20 m^3/min. To purify this water the district sanitary engineer proposed screening, sedimentation, rapid gravity filtration, and chlorination. The filters to be used are composed of unsized sand grains with an effective diameter of 0.08 cm. The filter bed depth amounts to 110 cm while the supernatant water level is 140 cm above the filter bed. Filtration is kept constant at a rate of 0.1 cm/s.
   a] How many filters will the sanitary engineer use?
   b] What unit filter bed area is recommended?

9. A filter bed with a 75-cm layer of sand is composed of spherical unsized sand particles with a porosity of 45%. For a filtration velocity of 6250 L/m^2/day and a maximum bed head loss (as determined by Rose's equation) to be 12 cm, compute the needed average grain diameter. Is this a suitable size? (Assume the kinematic viscosity is equal to $1.3 \cdot 10^{-6}$ m^2/s.)

10. A rapid sand filter bed has a media depth of 1.5 m and is composed of uniform spherical sand grains with a diameter of 1.2 mm and porosity of 40%. The supernatant water depth is 1.2 m. The filter is operated at a constant filtration rate, "v", for which the filter bed resistance as computed by the Carman-Kozeny equation measures 0.3 m.
    a] What is the filtration rate "v"?
    b] Comment on your computations.
    (Take the viscosity coefficient to be equal to $1.31 \cdot 10^{-6}$ m^2/s).

11. a. Derive expressions for the rate of removal of solids in a filter and the rate of deposition of solids assuming a constant coefficient of filtration. Comment about your assumptions.
    b. The water to be filtered has a concentration of impurities of 20 mg/L. The constant characteristic of the bed is 5 m^{-1}. The filter has a depth of 1.2 m and a constant filtration rate of 2 mm/s.
       i] Compute the concentration of impurities at half depth, and
       ii] Determine the rate of deposition of impurities per unit time. (Take the viscosity coefficient to be equal to $1.31 \cdot 10^{-6}$ m^2/s).

12. In a treatment works slow sand filters are used to treat a flow of 1 m³/s at a filtration rate of 0.08 mm/s. Compute the number of filters to be used, and the unit area of each filter.
13. If, in Problem 12 above, rapid sand filters are preferred, determine the number of filters needed.
14. A filter bed thickness of 1.5 m is composed of a layer of anthracite coal with a porosity of 35% and a grain size of 1.1 mm. Compute the porosity of the bed after backwashing for a filter bed expansion of 20%.
15. A dual media filter bed has the following characteristics:

Item	Anthracite layer	Sand layer
Depth of bed (cm)	60	40
Average particle size (mm)	1.6	0.8
Specific gravity	1.6	2.65
Porosity (%)	40	40
Shape factor	0.8	0.9

The clear water head loss in the filter bed is desired to be 64 cm with an operating temperature of 15°C.
   a. Compute the allowable rate of filtration by using Rose's and Carman-Kozeny equation.
   b. Comment on the practicability of your answer.
   c. Why is the value of the specific gravity for the sand layer higher than that of anthracite coal in the filter?
16. A sand filter with a depth 0.85 m is composed of uniform sand particles with a diameter of 0.6 mm, a specific gravity of 2.65, and a shape factor of 0.73. The porosity of the packed bed is 0.5.
   i] Plot a curve for the head loss vs. the filtration rate over the filter velocity range of 1.0 to 12.0 m/hr at a water temperature of 15°C (use the Carman-Kozeny equation).
   ii] What conclusions can you draw from your graph?
17. The following design details have been furnished for a slow sand filter plant:

Item	Value
Daily demand	800 m³
Rate of filtration	0.15 m/hr
Area of filter required	240 m²
Size of each filter bed	15 m·10 m
Number of filter beds provided	2

As an environmental engineer you are asked to review and comment on this design. How would your report read?

18. The sorption data for zinc by a Dariyah soil sample after 3 days of contact using the continuously agitated slurry technique is presented in the following table.
    Determine the K and n constants for the Freundlich adsorption isotherm by both the graphical procedure and the computer program. Also evaluate the ultimate capacity of the Dariyah soil for the sorption of zinc at an initial concentration of 1.37 mg/L.

Table A Dariyah Soil Data For Freundlich Isotherm for Sorption of Zinc (3 days contact time)

Dariyah soil concentration (mg) M	Residual zinc concentration (mg/L) C	Zinc[a] adsorbed (mg/L) X	X/M
Blank 0	1.37	0	
250	0.78	0.59	0.00236
500	0.76	0.61	0.00122
1000	0.50	0.87	0.00087
2000	0.40	0.97	0.000485
4000	0.20	1.17	0.000292

[a] Determined by atomic absorption spectrophotometry.

19. A two-stage still has equal heat exchanger areas. The still phases have heat transfer coefficients in the ratios 5.6 to 4. Dry steam at a temperature of 120°C is fed to the first stage of the multistage still. Find the drop in temperature in the second stage when the temperature is 60°C. Comment on your answer.
20. A vertical tube evaporation unit consists of four stages. The temperature from the first to the last is 105, 88, 65, and 55°C, respectively. Dry steam is supplied to the first stage at a temperature of 250°F. Find:
    i] the temperature difference in each tube.
    ii] the ratios of heat transfer coefficients in the heat exchangers assuming equal heat exchanger areas.
21. A semipermeable membrane separates a mineral-free water from a water that contains the following impurities:

Impurity	Concentration (mg/L)
Calcium chloride, $CaCl_2$	2
Magnesium chloride, $MgCl_2$	3
Sodium sulfate, Na_2SO_4	4

    Determine the osmotic pressure that would be required across this semipermeable membrane at a temperature of 20°C.
22. A brackish water contains ions whose molar concentrations are as indicated in the table shown below:

Cations	Anions
Sodium, Na^+ 0.025	Chloride, Cl^- Missing
Potassium, K^+ 0.001	Bicarbonates, HCO_3^- 0.001
Magnesium, Mg^{2+} 0.01	Nitrates, NO_3^- 0.003
Calcium, Ca^{2+} 0.015	Sulfates, SO_4^{2-} 0.015

    The reverse osmosis process was used to desalinate the water. The osmotic pressure difference across a semipermeable membrane that had brackish water on one side and distilled water on the other side is 2.2 atm. Assuming a temperature of 22°C, compute the missing chloride ion concentration for the conditions outlined.
23. An electrodialysis apparatus treating water with a total dissolved solids concentration (TDS) of 2000 mg/L records the following data:
    - Current efficiency for the transport of some ions in solution in contact with the membrane — 65%.
    - Current efficiency for the transport of counterions through the membrane — 90%.
    - Diffusion constant for the electrolyte = 1.5×10^{-5} cm²/s.
    - Thickness of the diffusion layer at the interface = 0.05 mm.

    Taking Faraday's constant as 26.8 Ampere·hour, find the value of the maximum current density.
24. Chloroform ($CHCl_3$) may be produced experimentally from the reaction between chlorine and methane. The reaction products are chloroform and hydrochloric acid. Determine the amount of chlorine needed to yield 129 g of chloroform.
25. It is required to treat 5000 m³/d of water with 0.4 mg/L of chlorine. If the disinfectant is in the form of bleaching powder that contains 35% available chlorine, how many kilograms of bleach are needed to treat the daily water flow?
26. In a treatment plant a flow of 25,000 m³/d is disinfected by adding 0.5 kg/hr chlorine. Find the chlorine dosage and the chlorine demand if the chlorine residual after 10 min contact time is to be 0.3 mg/L.
27. Determine the contact time required to achieve a 99.99% kill for a disinfectant-microorganism system having a rate constant of 0.05/s.
28. A flow of water contains 5 mg/L of Cl_2. Determine the amount of sulfur dioxide (SO_2) needed to remove the chlorine (Cl_2) from the water?
29. Determine the percentage kill for a disinfectant-microorganism system with a rate constant (to base 10) of $5 \cdot 10^{-2}$/s. Take the required contact time for disinfection to be 28 s.
30. A wastewater sludge was tested for filterability using a Buchner funnel apparatus. The filter diameter used was 7.5 cm, and vacuum drawn on the sample amounted to 60 kPa. For 50 mL of filtrate, the dried solids collected on the filter was 4.25 g. The slope of the t/V vs. V data plotted

on graph paper was $0.72 \cdot 10^{12}$ s/m³. Taking a filtrate temperature of 20°C, calculate the specific resistance of the wastewater sludge.

31. In a specific resistance test the following results were obtained:
    Vacuum pressure = 60 kN/m²
    Filtrate temperature = 20°C
    Solids content = 24.8 kg/m³
    Area of filtration = $44.18 \cdot 10^{-4}$ m²
    Volume of sludge used = $100 \cdot 10^{-6}$ m³

Time (min)	Volume of filtrate (mL)
2	6.1
3	7.5
4	8.6
5	9.6
6	10.5
7	11.4
13	15.9
18	18.8
20	19.8

    Graphically plot the values of t/V against V and hence obtain the slope and calculate the specific resistance.

32. The Buchner funnel technique was used to evaluate the specific resistance on a sample of activated sludge. The test was repeated three times for the sample, and the results are recorded in the table below:

Time, t (s)	Volume of filtrate, V (mL)		
	Test (1)	Test (2)	Test (3)
60	1.3	1.4	1.5
120	2.4	2.5	2.6
240	4.1	4.2	4.3
480	6.8	6.9	7.0
900	10.3	10.4	10.5

    Find the degree of filterability of the sample given:
    Vacuum utilized = 68.95 kPa
    Volume of filtrate used = $50 \cdot 10^{-6}$ m³
    Solids content = 5.5%
    Filtrate temperature = 20°C
    Diameter of Whatman number 1 filter paper = 7 cm.

33. Determine the specific resistance of an activated sludge sample from the following data that were obtained by using a small laboratory test filter with a filter diameter of 7 cm.

Time (s)	Volume of filtrate (mL)
100	3.6
200	5.5
400	8.2
600	10.3
800	12.1
1000	13.6

    Vacuum = 60 kN/m², solids content = 5%, filtrate viscosity = $1.01 \cdot 10^{-3}$ N·s/m²

34. The value of the specific resistance of sludge (A) at a pressure of 90 kPa is $1.6 \cdot 10^{13}$ m/kg. A previously reported value in the literature for another sludge (B) indicated $2.8 \cdot 10^9$ s²/g at a pressure of 196 kPa. If both sludges (A) and (B) have the same compressibility coefficient of 0.65, compare the dewaterability of the two sludges.

35. The specific resistance of a sludge is $2 \cdot 10^{11}$ m/kg at an applied pressure of 49 kPa. Compute the specific resistance of the same sludge at a pressure of 90 kPa if the compressibility coefficient of the sludge is 0.7.

36. Find the specific resistance of a sample of sludge given the following data obtained from a small laboratory Buchner funnel test:

Time (min)	1	2	3	4	5
t/V (min/mL)	0.77	0.88	0.96	1.03	1.09

Vacuum applied   69 kPa
Solids concentration   80 mg/mL
Filtrate viscosity $1.1 \cdot 10^{-3}$ N·s/m²
Area of filtration $3.8 \cdot 10^{-3}$ m²

Comment on the results you obtained.

37. A resin with a cation exchange capacity of 3000 equivalents per m³ is used to treat 1200 m³ of wastewater per day. This wastewater is planned for reuse in agriculture irrigation on a long-term basis. The wastewater contains the following metals:

Cd^{2+} = 0.04 mg/L
Cu^{2+} = 0.8 mg/L
Ni^{2+} = 0.8 mg/L
Pb^{2+} = 20 mg/L
Zn^{2+} = 8 mg/L

The finished wastewater is required to meet the recommended limits presented in Table 2.10.

   a. What volume of resin would be needed based on a continuous operation for 5 days before regeneration was required?
   b. What percentage of the flow could be bypassed assuming that resin removes 100% of the metallic ions?

38. A total of 6000 m³ of wastewater are discharged each day into a clean stream (BOD_5 = 1 mg/L) with a minimum flow of 1 m³/s. The 20°C, 5-day BOD of the wastewater is 200 mg/L. The wastewater has a DO of 5.0 mg/L. The temperature of the stream is 24°C and is 90% saturated with oxygen. The stream velocity is 0.45 m/s. The temperature of the wastewater is 26°C. The K' and K'' values at 20°C are 0.25 and 0.56 per day, respectively (base e). Determine the time and distance in the stream to the critical oxygen deficit as well as the amount of dissolved oxygen at this point.

39. Suppose the secondary treatment processes in Example 5.18 were shut down due to a power outage and that the BOD_5 of the effluent discharged to the stream at this time was 165 mg/L. What effect would this have on the stream as it relates to T_c, X_c, and the critical oxygen deficit?

D. SUGGESTED EXERCISES

Make a field visit to a local wastewater treatment works and write a computer program to evaluate the performance and the design of an existing wastewater treatment plant that incorporates the established treatment units.

The computer program may be written for individual treatment sectors such as sedimentation, activated sludge, etc., or it may be in an integrated form of a computer package that includes the different treatment units of the works.

Data to be used in the evaluation procedure must come from the collected data at the plant site, and from the existing monitoring program. The design and performance must be compared to the original design estimates such as the design population; design average, peak, and present flows; design calculations for the individual treatment processes and units (preliminary, primary, secondary, and advanced or tertiary treatment); design influent and effluent for the BOD, and suspended solids.

The computer program must be capable of drawing and dimensioning the different treatment units. The preferred language for writing the program is the BASIC language due to its simplicity and ease.

The report to be submitted must include:

1. The written computer program.
2. Plan, elevation, and cross-sectional maps and a flow diagram of the works indicating the several processes involved.
3. Data collected at the treatment site showing the various tests conducted at the works on the incoming and outgoing effluent and the concentrations for each item.
4. Problems and limitations of the adopted design.
5. Suggestions for improvement as an outcome of the evaluation program carried out by you.
6. Discussion and conclusions.

REFERENCES

1. Metcalf and Eddy, Inc., *Wastewater Engineering Treatment Disposal Reuse,* McGraw-Hill, New Delhi, 1979.
2. **Barnes, D., Bliss, P. J., Gould, B. W. and Vallentine, H. R.,** *Water and Wastewater Engineering System,* Pitman Inter., Bath, 1981.
3. **Callely, A. G., Forster, C. F. F., and Stafford, D. A.,** *Treatment of Industrial Effluents,* Hodder and Stoughton, London, 1977.
4. **ElHassan, B. M. and Abdel-Magid, I. M.,** *Environment and Industry: Treatment of Industrial Wastes,* Institute of Environmental Studies, Khartoum University, Khartoum, 1986.
5. **Fair, G. M. and Geyer, J. C.,** *Water Supply and Waste-water Disposal,* John Wiley & Sons, New York, 1954, 845, 846.
6. **Hammer, M. J.,** *Water and Wastewater Technology,* John Wiley & Sons, New York, 1977.
7. **Mara, D.,** *Sewage Treatment in Hot Climates,* John Wiley & Sons, Chichester, 1980.
8. **Nathanson, J. A.,** *Basic Environmental Technology: Water Supply, Waste Disposal and Pollution Control,* Prentice-Hall, Englewood Cliffs, NJ, 1986.
9. **Negulescu, M.,** *Municipal Wastewater Treatment, Developments in Water Science,* Elsevier, Amsterdam, 1986, 23.
10. **Peavy, H. S., Rowe, D. R., and Tchobanoglous, G.,** *Environmental Engineering,* McGraw-Hill, New York, 1985.
11. **Salvato, J. A.,** *Environmental Engineering and Sanitation,* Wiley-Interscience, New York, 1983.
12. **Ditz, R. B.,** Netzsch, Inc., Filter Press Division, personal communication, 1994.
13. **Vesilind, P. A. and Peirce, J. J.,** *Environmental Engineering,* Ann Arbor Science, Collingwood, MI, 1982.
14. **Water Chlorination Principles and Practices,** American Water Works Association, M20, Denver, CO, 1973, 26–28.
15. **Masters, G. M.,** *Introduction to Environmental Engineering and Science,* Prentice-Hall, Englewood Cliffs, NJ, 1991.
16. **Tebbutt, T. H. Y.,** *Principles of Water Quality Control,* 2nd ed., Pergamon Press, Oxford, 1977.
17. **Wilson, F.,** *Design Calculations in Wastewater Treatment,* E. and F. N. Spon, London, 1981.
18. ASCE/AWWA, *Water Treatment Plant Design,* 2nd ed., McGraw-Hill, New York, 1990.
19. **Mullick, M. A.,** *Wastewater Treatment Processes in the Middle East,* The Book Guild, Sussex, 1987.
20. Degremont, *Water Treatment Handbook,* Degremont, Rueil-Malmaison, France. Cedex, France, 6th ed., Vols. I, II, 1991.
21. **Abdel-Magid, I. M.,** *Selected Problems in Wastewater Engineering,* Khartoum University Press, National Research Council, Khartoum, 1986.
22. **Ganczarczyk, J. J.,** *Activated Sludge Process: Theory and Practices, Pollution Engineering and Technology,* Vol. 23, Marcel Dekker, New York, 1983.
23. **Berger, B. B.,** Ed., *Control of Organic Substances in Water and Wastewater,* Noyes Data Co., Park Ridge, NJ, 1987.
24. **Frederick, S.,** Ed., *Standard Handbook for Civil Engineers,* McGraw-Hill, 1976.
25. **James, A.,** *An Introduction to Water Quality Modelling,* Wiley-Interscience, Chichester, U.K., 1984.
26. **Vernick, A. S. and Walker, E. C.,** *Handbook of Wastewater Treatment Processes, Pollution Engineering and Technology,* Vol. 19, Marcel Dekker, New York, 1981.
27. **Tebbutt, T. H. Y.,** *Basic Water and Wastewater Treatment,* Butterworths, London, 1990.
28. **Lee, M. D. and Visscher, J. T.,** Water Harvesting in Five African Countries, Occasional Paper Number 14, IRC, The Hague, The Netherlands, 1990.
29. **Huisman, L.,** Sedimentation and Flotation: Sedimentation and Flotation, — Mechanical Filtration, — Slow Sand Filtration, — Rapid Sand Filtration, Delft University of Technology, Herdruk, 1977.
30. **Huisman, L.,** *Slow Sand Filtration,* World Health Organization, Geneva, 1974.
31. **O'Conner, D. and Dobbins, W.,** The Mechanism of Reaeration in Natural Streams, *J. Sanit. Eng.,* SA6, 1956.
32. **Punmia, B. C.,** *Environmental Engineering,* Vol. 1, Water Supply, Standard Book House, Naisarak, Delhi, 1979, 6.
33. AWWA, *Water Quality and Treatment — A Handbook of Public Water Supplies,* McGraw-Hill, New York, 1971.
34. **Black, J. A.,** *Water Pollution Technology,* Reston Pub. Co., Virginia, 1977.
35. **Huisman, L., Sundaresan, B. B., Netto, J. M. D., and Lanoix, J. N.,** *Small Community Water Supplies,* Hofkes, E. H., Ed., John Wiley & Sons, New York, 1986.
36. **Pescod, M. B., Abouzaid, H., and Sundaresan, B. B.,** Slow Sand Filtration: A Low Cost Treatment for Water Supplies in Developing Countries, published for the WHO Regional Office for Europe by the Water Research Center, U.K. in collaboration with the IRC, The Netherlands, Stevenage, Hertfordshire, U.K..
37. IRC Report, International Appraisal Meeting held in Nagpur, India, Sept. 15–19, 1980, Slow Sand Filtration for Community Water Supply in Developing Countries, BS 16, IRC, The Hague, The Netherlands, 1981.
38. IRC, Research and Demonstration Project on Slow Sand Filtration, Guidelines for Operation and Maintenance of Slow Sand Filtration Plants in Rural Areas of Developing Countries, IRC, Rijswijk, The Netherlands, 1983.
39. IRC Report, National Workshop held in New Delhi, India, Jan. 19–21, 1987, Design and Construction of Slow Sand Filters, C. S. I. R. (India), IRC.
40. **Van Dijk, J. C. and Oomen, J. H. C. M.,** Slow Sand Filtration for Community Water Supply in Developing Countries — A Design and Construction Manual, Technical Paper Number 11, IRC, The Hague, 1982.
41. **Rich, L. G.,** *Unit Operations of Sanitary Engineering,* John Wiley & Sons, New York, 1974.

42. **Bolto, B. A. and Pawlowski, L.,** *Wastewater Treatment by Ion Exchange,* E. & F. N. Spon, New York, 1987.
43. **Perry, R. H.,** Ed., *Perry's Chemical Engineers' Handbook,* 6th ed., McGraw-Hill, New York, 1984.
44. **Buros, O. K.,** *The Desalting ABC's,* International Desalination Association, Topsfield, MA, 1990.
45. **Porteous, A.,** *Desalination Technology: Developments and Practice,* Applied Science Publishers, London, 1983.
46. **Malik, M. A. S., Tiwari, G. N., Kumar, A., and Sodha, M. S.,** *Solar Distillation,* Pergamon Press, Oxford, 1982.
47. **Sawyer, C. N. and McCarty, P. L.,** *Chemistry for Environmental Engineering,* 3rd ed., McGraw-Hill, New York, 1978.
48. **Lacy, R. E.,** Membrane Separation Process, *Chem. Eng.,* 4:56, 1972.
49. **McJunkin, F. E.,** Water, Engineers, Development and Disease in the Tropics, Agency for International Development, Department of State, Washington, D.C., 1975.
50. **Fair, G. M., Morris, F. C., Chang, S. L., Weil, I, and Burden, R. A.,** The Behavior of Chlorine as a Water Disinfectant, *J. Am. Water Works Assoc.,* 40: 1051, 1948.
51. Design of Municipal Wastewater Treatment Plants, Vol. II, Water Environment Federation and American Society of Civil Engineers, 1992, 880.
52. AWWA Committee Report, *Disinfection,* American Water Works Association, Denver, CO, 1978, 70(4), 219.
53. **Dyer-Smith, P., Brown, Beveri and Co.,** Water Disinfection Status and Trends, *J. Water Sewage Treat.,* 2(4), 1983, 13.
54. **Gunnerson, C. G. and Stuckey, D. C.,** Anaerobic Digestion Principles and Practice for Biogas Systems, World Bank Technical Paper Number 49, World Bank, Washington, D.C., 1986.
55. **Coackley, P.,** Development in our Knowledge of Sludge Dewatering Behavior, 8th Public Health Engineering Conference held in the Dept. of Civil Engineering, Loughborough University of Technology, 1975, 5.
56. WPCF, *Sludge Dewatering, Manual of Practice No. 20,* Water Pollution Control Federation, Washington, 1969.
57. **Coackley, P.,** Sludge Dewatering Treatment, *Process Biochem.,* 2(3), 1967, 17.
58. **Abdel-Magid, I. M.,** The Role of Filter Aids in Sludge Dewatering, Ph.D. Thesis, University of Strathclyde, Glasgow, 1982.
59. **Gale, R. S.,** *Recent Research on Sludge Dewatering,* Water Pollution Research Lab., Stevenage, Herts, 1971, 531–810.
60. **Gale, R. S.,** Filtration theory with Special Reference to Sewage Sludges, *J. Water Pollut. Control,* 1967, 622–631.
61. **Coackley, P.,** The Theory and Practice of Sludge Dewatering, *J. Inst. Public Health Eng.,* 64(1), 1965, 34.
62. **Karr, P. R. and Keinath, T. M.,** Influence of Particle Size on Sludge Dewaterability, *J. Water Pollut. Control Fed.,* 1978, 1911–1930.
63. **Coackley, P.,** Research on Sewage Sludge Carried out in the Civil Engineering Department of The University College London, *J. Proc. Inst. Sewage Purif.,* 1, 1955, 59–72.
64. **Karr, P. R. and Keinath, T. M.,** Limitations of the Specific Resistance and CST Tests for Sludge Dewatering, *Filtration and Separation J.,* 1978, 543–544.
65. **Swanwick, J. D. and Davidson, M. F.,** Determination of Specific Resistance to Filtration, *Water Wastewater Treat. J.,* 8(8), 1961, 386–389.
66. **Rudolfs, W. and Heukelekian, H.,** Relation Between Drainability of Sewage and Degree of Digestion, *Sewage Works J.,* 6, 1934, 1073–1081.
67. **Coackley, P.,** The Dewatering Treatment, Ph.D. thesis, London University, 1953.
68. **Coackley, P. and Allos, R.** The Drying Characteristics of Some Sewage Sludges, *J. Proc. Instit. Sewage Purif.,* 6, 1962, 557.
69. **Nebiker, J. H.,** Dewatering of Sewage Sludge on Granular Materials, Environmental Engineering Report Number EVE-8-68-3, 1968.
70. **Newitt, D. M., Oliver, T. R., and Pearse, J. F.,** The Mechanism of the Drying of Solids, *Trans. Inst. Chem. Eng.,* 27, 1949, 1.
71. **Carman, P. C.,** Fundamental Principles of Industrial Filtration, *Trans. Inst. Chem. Eng.,* 1938, 16, 168–188.
72. **Fair, G. M., Moore, E. W., and Thomas, H. A.,** The Natural Purification of River Muds and Pollutional Sediments, *J. Sewage Works,* 13, 270, 756, 1941.
73. **Bartlett, R.,** Ed., *The Public Health Engineering Data Book 1983/84,* IPHE, Sterling Publishers, London, 1984.
74. **Pultock, S. J., Fane, A. G., Fell, C. J. D., Robins, R. G., and Wainright, M. S.,** Vacuum Filtration and Dewatering of Alumina Trihydrate. The Role of Cake Porosity and Interfacial Phenomena, *Int. J. Miner. Process.,* 17, 1986, 204–205.

Chapter 6

Risk Assessment in Wastewater Reclamation and Reuse

CONTENTS

- I. Basic Concepts for Environmental or Ecological Risk Assessment 262
  - A. Definitions 262
  - B. Background or Scoping Activities 262
  - C. Data Collection 263
  - D. Data Evaluation 263
- II. Exposure Assessment 264
  - A. Characterization of Exposure Setting (Step 1) 264
  - B. Identification of Exposure Pathways (Step 2) 265
  - C. Quantifying Exposure (Step 3) 265
- III. Toxicity Assessment 267
  - A. Hazard Identification 267
  - B. Dose-Response Assessment 268
    - 1. Dose Response For Noncarcinogens 269
    - 2. Dose Response For Carcinogens 274
      - a. Weight-of-Evidence Categories 274
      - b. Generating Slope or Potency Factors 274
- IV. Risk Characterization 279
  - A. Risk Communication 280
  - B. Risk Assessment/Wastewater Reclamation and Reuse 280
    - 1. Introduction 280
    - 2. Viruses 281
    - 3. Bacteria 281
    - 4. Protozoa 282
    - 5. Helminths 282
    - 6. Survival of Waterborne Pathogens in Wastewater, Soil, and on Crops 283
  - C. Risk Based on Modeling 284
    - 1. Introduction 284
    - 2. Mathematical Models Available For Extrapolation For Chemical Exposures 284
    - 3. Mathematical Models Available For Extrapolation For Infectious Agents 287
- V. Removal of Pathogenic Microorganisms By Various Wastewater Treatment Processes 288
  - A. Chlorination 288
  - B. Ozonation 291
- VI. Public Opinion and Wastewater Reuse 291
- VII. Summary and Conclusions 293
- VIII. Homework Problems 296
- References 296

I. BASIC CONCEPTS FOR ENVIRONMENTAL OR ECOLOGICAL RISK ASSESSMENT

A. DEFINITIONS

To try and understand this relatively new emerging discipline in the environmental and ecological fields it is first necessary to become familiar with the terminology currently in use.

Following are some of the most important risk assessment terms and working definitions for these terms.[1-5]

- *Risk.* The probability of injury, disease, or death under specific circumstances. In quantitative terms, risk is expressed in values ranging from zero (representing the certainty that harm will not occur) to one (representing the certainty that harm will occur).
- *Risk assessment/analysis.* The use of the factual base to define the potential health effects of exposure of individuals or populations or ecosystems to hazardous materials and situations. May contain some or all of the following four steps:
    1. Hazard identification — a qualitative and cursory determination of whether a threat to human beings or the environment exists.
    2. Dose-response assessment — analysis of what degree of harm results from differing "doses" of the threat (exposure duration and intensity).
    3. Exposure assessment — measurement or prediction of duration and intensity of exposure.
    4. Risk characterization — prediction of occurrence of expected adverse effects (usually probabilistic, e.g., a one in one million [10^{-6}] excess chance of lung cancer).
- *Risk characterization.* The description of the nature and often the magnitude of human risk, including attendant uncertainty.
- *Risk management.* The decision-making process that uses the results of risk assessment to produce a decision about environmental action. Risk management includes consideration of technical, social, economic, and political information.
- *Risk communication.* Is the exchange of information about risk.

The approach generally used by EPA to do risk assessment/analysis is based on a four-step process described by the National Academy of Sciences. As indicated under the risk assessment/analysis definition, the process may use some or all of the four steps. The sequence for the risk assessment/risk management process is presented in Figure 6.1, which includes not only the four steps but additional categories such as gathering background information, data collection and evaluation, as well as the risk management step itself.[6]

Risk management differs from risk assessment in that management of risk usually considers political, economic, and social issues in the decision-making process, while risk assessment deals with gathering factual information used to estimate the probability of incurring some risk (e.g., cancer) due to exposure to some harmful physical, chemical, biological, or radioactive operation or substance in the environment. Risk assessment and risk management also provide a basis for setting regulatory priorities and for making decisions that include the land, water, and air environments.[1]

Drinking water standards, air pollution standards, and many other environmental standards will be based in part on the use of risk assessment/analysis and risk management techniques.

Risk assessment/analysis has been used most often in establishing regulations that relate to carcinogens. Risk assessment has provided a means by which a contaminant in air, water, or soil can be evaluated as to its potential for causing cancer.[5]

While cost-benefit analysis are sometimes a part of the risk management program, risk management goes far beyond just cost-benefit analysis.

B. BACKGROUND OR SCOPING ACTIVITIES

Before starting the sequence involved in the risk assessment process, it is generally necessary to do an initial planning phase. The objectives and scope of the activities must first be clearly spelled out. These are the collection and analysis of contaminants to be considered, their toxicity, exposure pathways, transport modeling programs, characteristics of potentially exposed populations, enforcement considerations, as well as the quantity and quality of monitoring equipment available.[6]

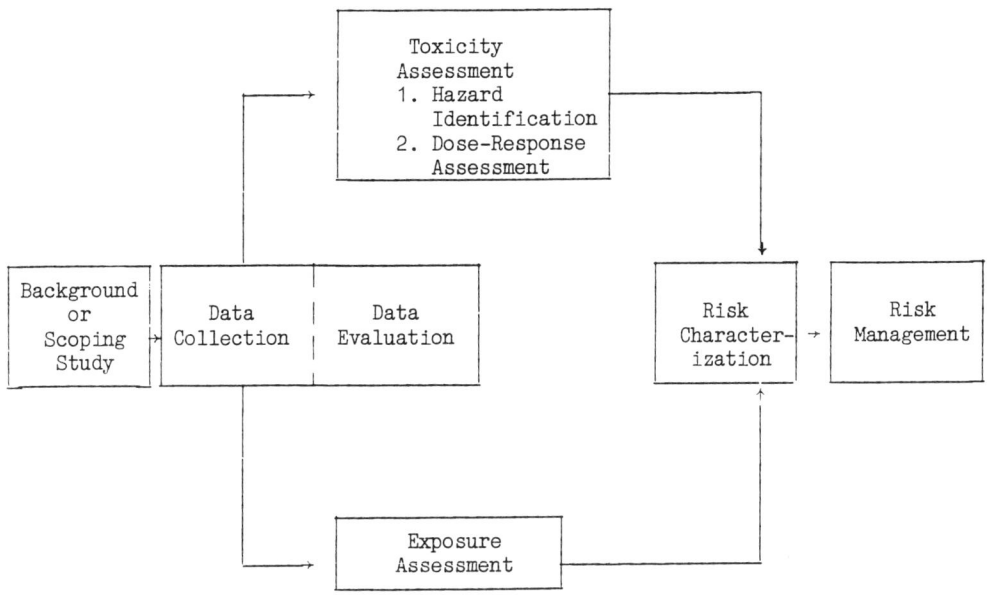

Figure 6.1 Sequence involved in risk assessment-risk management.[6]

C. DATA COLLECTION

Following the background or scoping activities, data collection and evaluation are needed. This involves:[6]

- Contaminant identification.
- Contaminant concentrations from key sources and its transport by air, water, or soil.
- Characteristics of the source.
- Characteristics of the environmental setting that affects the fate, transport, and persistence of the contaminants.

The sampling of background levels of a contaminant are important in order to separate naturally occurring site-related contaminant levels to non-site-related levels.

Table 6.1 is a list by media (air, water, soil, biota) of important parameters that should be considered in the risk assessment sampling process. These parameters will vary from site to site and may have to be expanded depending upon the problems encountered.

In carrying out the sampling program attention as to sampling locations, frequency, sampling equipment and analysis, and procedures must be carefully considered.

Sampling procedures should consider:[6]

- Each medium of concern.
- Background concentrations.
- All potential exposure points within each medium.
- Migration to potential exposure points, including data for models.
- Potential exposures based on possible future land uses.
- Sufficient data to satisfy concerns about distributions of sampling data and statistics.
- Number and location of samples.

All sampling data should be reviewed as soon as it is available to determine if the information is in accordance with the original plan and to decide what changes, if any, are necessary.

D. DATA EVALUATION

A great deal of analytical data will be available after the sampling analysis activities are completed. The following steps should be considered in organizing and presenting the data in a meaningful way.[6]

1. Gather all data available from the site investigation and sort by medium.
2. Evaluate the analytical methods used.

Table 6.1 Important Parameters that May Need to Be Obtained during Site Sampling Investigations

Type of modeling	Modeling parameters
Source characteristics	Geometry, physical/chemical conditions, emission rate, emission strength, geography
Soil	Particle size, dry weight, pH, redox potential, mineral class, organic carbon and clay content, bulk density, soil porosity
Groundwater	Head measurements, hydraulic conductivity (pump and slug test results), saturated thickness of aquifer, hydraulic gradient, pH, redox potential, soil-water partitioning
Air	Prevailing wind direction, wind speeds, stability class, topography, depth of waste, contaminant concentration in soil and soil gas, fraction organic content of soils, silt content of soils, percent vegetation, bulk density of soil, soil porosity
Surface water	Hardness, pH, redox potential, dissolved oxygen, salinity, temperature, conductivity, total suspended solids, flow rates and depths for rivers/streams, estuary and embayment parameters such as tidal cycle, saltwater incursion extent, depth and area, lake parameters such as area, volume, depth, depth to thermocline
Sediment	Particle size distribution, organic content, pH, benthic oxygen conditions, water content
Biota	Dry weight, whole body, specific organ, and/or edible portion chemical concentrations, percent moisture, lipid content, size/age, life history stage

Compiled from Reference 6.

3. Evaluate the quality of data with respect to sample quantitation limits.
4. Evaluate the quality of data with respect to qualifiers and codes.
5. Evaluate the quality of data with respect to blanks.
6. Evaluate tentatively identified compounds.
7. Compare potential site-related contamination with background.
8. Develop a set of data for use in the risk assessment.
9. If appropriate, further limit the number of chemicals to be carried through the risk assessment.

The report developed in the data evaluation process should contain historical and current data as well as uncertainties as to sampling and analysis. The report for the data evaluation should include and discuss levels of contaminants found and their relationship to detection limits. Location of significant levels of a contaminant and its concentration in air, water, soil, and biota should be pointed out.

Tables should be included in the text of the data evaluation report indicating chemicals detected, frequency of detection, range of concentration, detection limits, background levels, and concentrations found in each media.

II. EXPOSURE ASSESSMENT

Exposure assessment is the determination or estimate, qualitatively and quantitatively, of the magnitude, frequency, and route of exposure by which humans are potentially exposed to various contaminants. Conducting an exposure assessment involves analyzing contaminant releases; identifying exposed populations; identifying all potential pathways of exposure; estimating exposure point concentrations for specific pathways, based both on environmental monitoring data and predictive chemical modeling results; and estimating contaminant intakes for specific pathways.[6]

A human exposure assessment consists of a three-step or three-part process that can be summarized as follows (see Figure 6.2).

A. CHARACTERIZATION OF EXPOSURE SETTING (STEP 1)

1. *Characterize physical setting*.[6] Characterize the exposure setting with respect to the general physical characteristics of the site. Important site characteristics include the following:

   - Climate (e.g., temperature, precipitation).
   - Meteorology (e.g., wind speed and direction).
   - Geologic setting (e.g., location and characterization of underlying strata).
   - Vegetation (e.g., unvegetated, forested, grassy).
   - Soil type (e.g., sandy, organic, acid, basic).
   - Groundwater hydrology (e.g., depth, direction, and type of flow).
   - Location and description of surface water (e.g., type, flow rates, salinity).

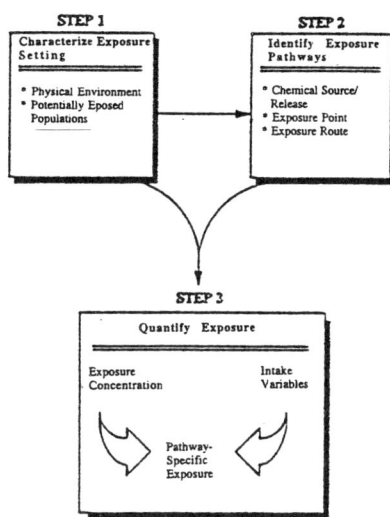

Figure 6.2 Three-step exposure assessment process.[6]

2. *Characterize exposed population.* Characterize the population with respect to:[6,7]
   - Proximity to the site.
   - Land use — residential, commercial, industrial, or recreational.
   - Potential future land use.
   - Societal community interests (subpopulation groups, such as an influx of construction workers).

B. IDENTIFICATION OF EXPOSURE PATHWAYS (STEP 2)

Figure 6.3 illustrates the potential human exposure pathways that a contaminant or toxic agent may use to enter the environment. An exposure pathway generally consists of four elements: (1) source; (2) transport medium (air, water, soil, biota); (3) point of potential human contact; and (4) an exposure route at point of contact (inhalation, ingestion, skin or eye contact).[6,7]

1. *Chemical source/release.* One of the key aspects of environmental risk assessment is associated with determining the source, transport, and fate of chemicals in air, surface water, ground water, soil, sediment, and biota.

   After a contaminant or toxic agent is released to the environment it may be[6,7]

   - Transported (e.g., convected downstream in water or on suspended sediment or through the atmosphere).
   - Physically transformed (e.g., volatilization, precipitation).
   - Chemically transformed (e.g., photolysis, hydrolysis, oxidation, reduction, etc.).
   - Biologically transformed (e.g., biodegradation).
   - Accumulated in one or more media (including the receiving medium).

   This information is important in developing information relating human exposure pathways for various contaminants or toxic agents.

   Table 6.2 presents some typical release sources, release mechanisms, and receiving medium.[6,7]

2. *Exposure points and exposure routes.* The next element in a human exposure pathway evaluation is identifying exposure points. Any point of potential human contact with a contaminant is an exposure point.

   After determining the exposure points, identifying exposure, routes such as ingestion, inhalation, skin, or eye contact can be made. Table 6.3 presents a population/exposure route matrix that can help determine potential exposure routes while Figure 6.4 presents contaminant routes within the body.[8,9]

C. QUANTIFYING EXPOSURE (STEP 3)

A summary of the estimated exposure concentrations for the various contaminants and the exposure pathways such as water, air, soil, skin, or food is prepared. Table 6.4 presents generic Equation 6.1 for calculating chemical intake while Table 6.5 presents specific Equation 6.2 for exposure to contaminants in drinking water. Similar equations can be used for air, soil, dermal, or food exposures to various contaminants.[6]

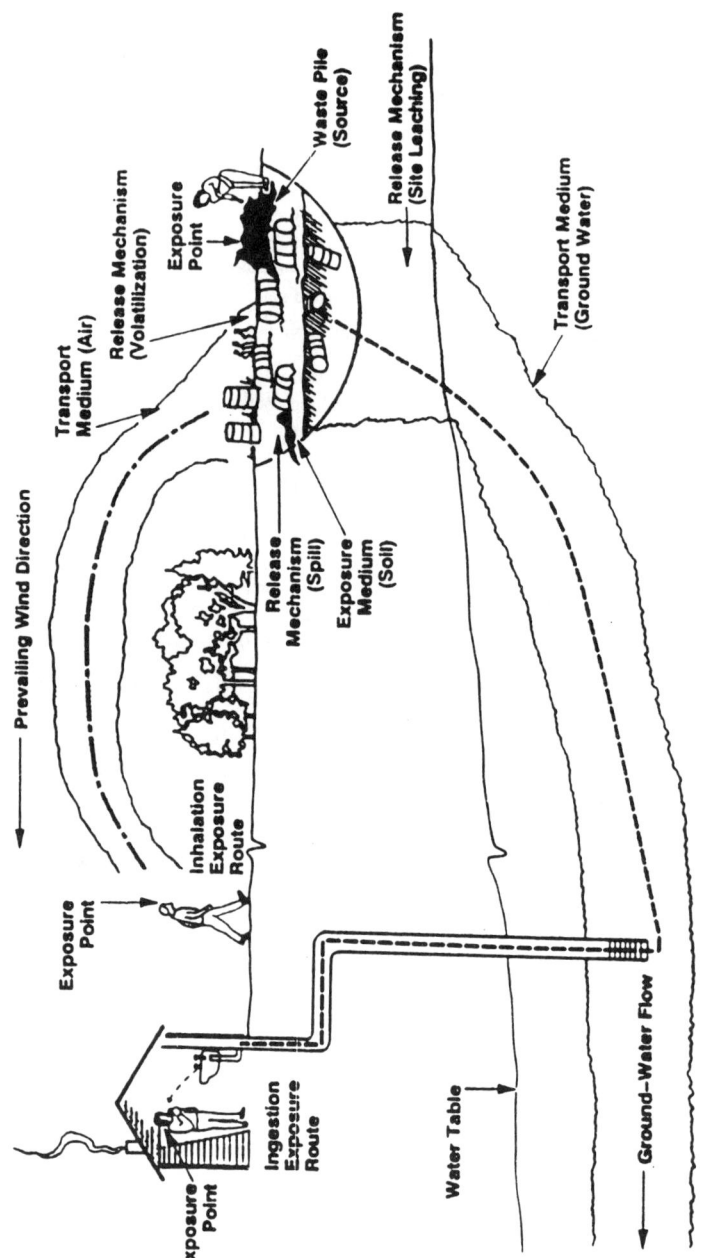

Figure 6.3 Illustration of exposure pathways.[6-8]

Table 6.2 Common Chemical Release Sources, Release Mechanisms, and Receiving Medium

Receiving medium	Release mechanism	Release source
Air	Volatilization	Surface wastes — lagoons, ponds, pits, spills
		Contaminated surface water
		Contaminated surface soil
		Contaminated wetlands
		Leaking drums
	Fugitive dust generation	Contaminated surface soil
		Waste piles
Surface water	Surface runoff	Contaminated surface soil
	Episodic overland flow	Lagoon overflow
		Spills, leaking containers
	Groundwater seepage	Contaminated ground water
Groundwater	Leaching	Surface or buried wastes
		Contaminated soil
Soil	Leaching	Surface or buried wastes
	Surface runoff	Contaminated surface soil
	Episodic overland flow	Lagoon overflow
		Spills, leaking containers
	Fugitive dust generation/deposition	Contaminated surface soil
		Waste piles
	Tracking	Contaminated surface soil
Sediment	Surface runoff,	Surface wastes — lagoons, ponds, pits, spills
	Episodic overland flow	Contaminated surface soil
	Groundwater seepage	Contaminated groundwater
	Leaching	Surface or buried wastes
		Contaminated soil
Biota	Uptake (direct contact, ingestion, inhalation)	Contaminated soil, surface water, sediment, groundwater or air
		Other biota

Compiled from Reference 6.

Example 6.1

Compute the lifetime individual chronic daily intake (CDI) for carbon tetrachloride (CCl_4) in water if the maximum contaminant level for the chemical is 5 parts per billion (5 µg/L, 0.005 mg/L).

Solution

1. Given: CW = 0.005 mg/L.
2. Use Equation 6.2 (see Table 6.5) to determine the lifetime individual chronic daily intake as: CDI = CW·IR·EF·ED/W·AT and assume IR = 2, EF = 365, ED = 70, and W = 70.
3. Substitute given and assumed values of steps 1 and 2 in Equation 6.2. Thus, CDI = 0.005·2·365·70/ 70·70·365 = 0.000143 mg/kg day.

As indicated in Table 6.5, this CDI is based on drinking 2 L of water per day for 70 years for a person with an average body weight of 70 kg.

III. TOXICITY ASSESSMENT

Toxicity assessment is generally accomplished in two steps, hazard identification and dose-response assessment.

A. HAZARD IDENTIFICATION

Hazard identification is the process of determining whether exposure to a contaminant can have adverse health effects (e.g., cancer, birth defects) for an exposed human population. As part of the hazard identification, evidence is gathered regarding the potential for a substance to cause adverse health effects (e.g., carcinogenic or noncarcinogenic) in humans. The major sources for this information include controlled epidemiologic investigations, clinical studies, and experimental animal studies.

Well-conducted epidemiologic studies showing a positive relationship between a contaminant and disease in humans is considered the most valid information, however, only limited data of this nature are

Table 6.3 Matrix of Potential Exposure Routes

Exposure medium/ exposure route	Residential population	Commercial/industrial population	Recreational population
Groundwater			
Ingestion	L	A	—
Dermal contact	L	A	—
Surface Water			
Ingestion	L	A	L, C
Dermal contact	L	A	L, C
Sediment			
Incidental ingestion	C	A	C
Dermal contact	C	A	L, C
Air			
Inhalation of vapor			
Phase chemicals			
Indoors	L	A	—
Outdoors	L	A	L
Inhalation of			
particulates			
Indoors	L	A	—
Outdoors	L	A	L
Soil/dust			
Incidental ingestion	L, C	A	L, C
Dermal contact	L, C	A	L, C
Food			
Ingestion			
Fish and shellfish	L	—	L
Meat and game	L	—	L
Dairy	L, C	—	L
Eggs	L	—	L
Vegetables	L	—	L

Notes: L = Lifetime exposure.
      C = Exposure in children may be significantly greater than in adults.
      A = Exposure to adults (highest exposure is likely to occur during occupational activities).
      — = Exposure of this population via this route is not likely to occur.

Compiled from Reference 6.

generally available and only for a few chemicals. When information relating a toxic agent's adverse human health effects are not available then effects from experiments with test animals such as rats, mice, rabbits, guinea pigs, hamsters, dogs, and monkeys are extrapolated to the anticipated potential effects in humans.

Studies using pharmacokinetics, which is a branch of pharmacology and deals with the absorption, distribution, and elimination of drugs by the body, may be used to help determine if a contaminant can have an adverse health effect in humans.

Studies using cultures of microorganisms such as the Ames mutagenicity test can be used as a short-term test in the evaluation of a chemical's potential mutagenicity. It is thought that chemicals that are mutagenic may also be carcinogenic.

Another way to predict a compound's potential for adverse health effects in humans is an analysis of the chemical structure. A compound with a structure similar to another compound whose effects are known can help estimate its potential in creating adverse health effects in humans.

B. DOSE-RESPONSE ASSESSMENT

Toxicology is defined as the study of the nature of the effects of poisons and their antidotes. Experimental toxicology is essentially a biological assay with the fundamental concept of a dose-response relationship. For biologically harmful agents it is useful to relate the dose administered to the response or damage produced. In plotting these two functions a dose-response curve is produced.[10]

The dose is plotted as the abscissa, usually in milligrams per kilogram per day (mg/kg/day). For human adults the dose is averaged over 70 years with an average body weight of 70 kg. The response is plotted

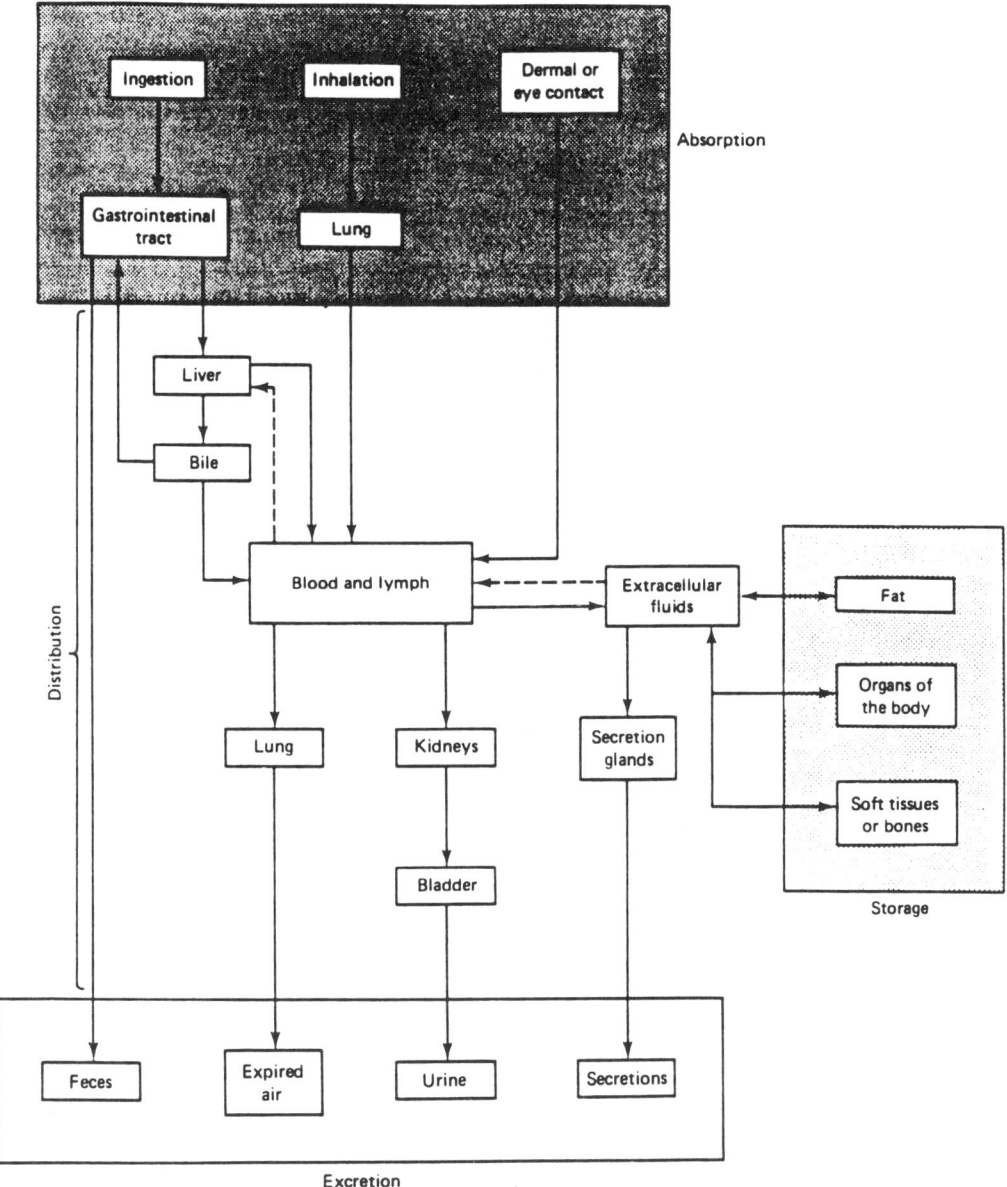

Figure 6.4 Fate of chemical toxicants in the body.[8,9]

on the ordinate and indicates the adverse health effects. Figure 6.5 is an example of a generalized dose-response curve.

The threshold dose is the point on the curve where it intersects the abscissa and is the dose below which no damage or response occurs.

For noncarcinogenic responses it is usually assumed there is a threshold dose; however, for carcinogenic agents it is conservatively assumed that no threshold dose exists and therefore the dose-response curve goes through the origin (see Figure 6.6).

Carcinogens are referred to as "nonthreshold" as it is believed that any level of exposure to certain chemicals can generate a carcinogenic response. That is, no dose is thought to be risk-free.[11]

1. Dose-Response For Noncarcinogens

As indicated in the previous section, noncarcinogenic agents are assumed to have a threshold dose. A reference dose (RfD) is the toxicity value used most often in evaluating noncarcinogenic effects from

Table 6.4 Generic Equation for Calculating Chemical Intakes

Equation:	$I = C \times \dfrac{CR \times EFD}{W} \times \dfrac{1}{AT}$	[6.1]

Where:
- I = intake; the amount of chemical at the exchange boundary (mg/kg body weight-day)

Chemical-related variable
- C = chemical concentration; the average concentration contacted over the exposure period (e.g., mg/L water)

Variables that describe the exposed population
- CR = contact rate; the amount of contaminated medium contacted per unit time or event (e.g., L/day)
- EFD = exposure frequency and duration; describes how long and how often exposure occurs. Often calculated using two terms (EF and ED):
- EF = exposure frequency (days/year)
- ED = exposure duration (years)
- W = body weight; the average body weight over the exposure period (kg)

Assessment-determined variable
- AT = averaging time; period over which exposure is averaged (days)

Compiled from Reference 6.

Table 6.5 Residential Exposure: Ingestion of Chemicals in Drinking Water and Beverages Made from the Water

Equation:	$(CDI)^a$ Intake (mg/kg-day) $= \dfrac{CW \times IR \times EF \times ED}{W \times AT}$	[6.2]

Where:
- CW = Chemical concentration in water (mg/L)
- IR = Ingestion rate (L/day)
- EF = Exposure frequency (days/year)
- ED = Exposure duration (years)
- BW = Body weight (kg)
- AT = Averaging time (period over which exposure is averaged — days)

Variable Values:
- CW: Site-specific measured or modeled value
- IR: 2 L/day (adult, 90th percentile; EPA 1989d)
  1.4 L/day (adult, average; EPA 1989d)
  Age-specific values (EPA 1989d)
- EF: Pathway-specific value (for residents, usually daily — 365 days/year)
- ED: 70 years (lifetime; by convention)
  30 years (national upper-bound time (90th percentile) at one residence; EPA 1989d)
  9 years (national median time (50th percentile) at one residence; EPA 1989d)
- W: 70 kg (adult, average; EPA 1989d)
  Age-specific values (EPA 1985a, 1989d)
- AT: Pathway-specific period of exposure for noncarcinogenic effects (i.e., ED × 365 days/year), and 70-year lifetime for carcinogenic effects (i.e., 70 years × 365 days/year).

[a] CDI = chronic daily intake.

Compiled from Reference 6.

exposure to toxic agents. A chronic reference dose (RfD) (formerly used as an acceptable daily intake, ADI) suggests a dose at which there is no noticeable response. The RfD is defined as an estimate of a daily exposure level for a human population, including sensitive subpopulations, that is likely to be without appreciable risk of deleterious effects during a lifetime. Chronic RfDs are specifically developed

Figure 6.5 Dose-response curve illustrating a threshold dose.

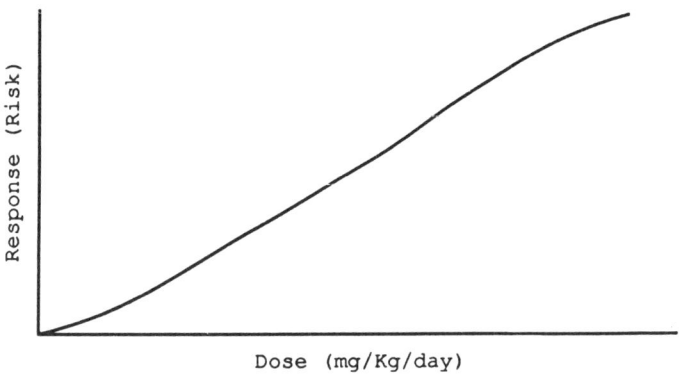

Figure 6.6 Dose-response curve illustrating no threshold dose.

to be protective for long-term exposure to a compound.[6,12] The RfDs are given in mg/kg/day and can be for either the oral or inhalation exposure route, with an exposure time being chronic, subchronic, or a single event. A chronic exposure is considered to be for a lifetime (70 years), while subchronic RfDs are used to characterize potential noncarcinogenic effects associated with short-term exposures of 2 weeks to 7 years. Single exposure events are referred to as developmental RfDs.

The EPA maintains a data base for RfDs through both the Integrated Risk Information System (IRIS) and the Health Effects Assessment Summary Tables (HEAST).[12,13] The Biomedical and Environmental Information Analysis Center at Oak Ridge National Laboratory at Oak Ridge, Tennessee also provides information regarding the toxic effects of various chemicals.[14]

The IRIS data base, developed by the U.S. Environmental Protection Agency (EPA), contains summary information related to human health risk assessment. IRIS, which is updated monthly, is the Agency's primary vehicle for communication of chronic health hazard information representing EPA consensus positions following comprehensive review by intra-Agency work groups. It is a useful information resource tool that points the user to the underlying human and/or animal data used to support the Agency's opinion.

The core of the system is a collection of computer files that contain hazard identification and dose-response risk information for approximately 500 chemicals. Combined with specific exposure information, the data in IRIS can be used for characterization of the public health risks of a given situation, which can then serve as input for a risk management decision designed to protect public health.[13]

Table 6.6 presents a few of the noncarcinogenic chemicals and their chronic RfDs. Oral route or oral exposure relates to administration of liquids through a stomach tube, as in the forced feeding of test animals (gavage).

Table 6.6 Chronic Noncarcinogenic RfDs for Selected Chemicals

Chemical	Year	Oral route	UF inhalation	MF	CL	Exposure pathway Oral route (mg/kg/day)	Inhalation (mg/kg/day)
Boron	1990	100	NA	1	M	9×10^{-2}	NA
Bromoform	1991	1,000	NA	1	M	2×10^{-2}	ND
Cadmium	1990	10	NA	1	H	1×10^{-3} (food)	
						5×10^{-4} (water)	UR
Carbon tetrachloride	1991	1,000	NA	1	M	7×10^{-4}	ND
Chloroform	1992	1,000	NA	1	M	1×10^{-2}	P
Chromium (VI)	1992	500	300	1	L	5×10^{-3}	2×10^{-6}
Cyanide	1990	500	NA	5	M	2×10^{-2}	NA
Fluorine	1990	1	NA	1	H	6×10^{-2}	ND
Malathion	1992	10	NA	1	M	2×10^{-2}	P
Mercury (inorganic)	1990	1,000	30	1	M	3×10^{-4}	3×10^{-4}
Monochloramine	1992	—	—	—	—	P	ND
Nickel	1992	300	NA	1	M	2×10^{-2}	UR
Nitrobenzene	1990	10,000	3,000	1	M	5×10^{-4}	2×10^{-3}
Phenol	1990	100	NA	1	M	6×10^{-1}	ND
Styrene	1990	1,000	NA	1	M	2×10^{-1}	P
Tin	1987	100	NA	1	M	6×10^{-1}	ND

Note: RfD — reference dose
UF — uncertainty factor
MF — modifying factor
CL — confidence level
L — low
M — medium
H — high
UR — under review
NA — not applicable or not available
ND — not determined
P — pending

An oral RfD value can be converted to a corresponding concentration in drinking water, assuming a human body weight of 70 kg and water consumption of 2 L/day, as follows:

$$\text{mg/L in water} = \frac{\text{oral RfD (in mg/kg/day)} \times 70 \text{ kg}}{2 \text{ L/day}} \quad [6.3]$$

Compiled from References 12–14.

Many new terms are now used in the evaluation of RfDs and the dose-response curves involved with these noncarcinogenic chemicals. Following is a list of some of these terms, their acronyms, and an appropriate definition.[6,12]

1. *Lowest-observed-effect level (LOEL).* The lowest dose administered that results in a response.
2. *No-observed-effect-level (NOEL).* In dose-response experiments, an exposure level at which there are no statistically or biologically significant increases in the frequency or severity of *any* effect between the exposed population and its appropriate control.
3. *Lowest-observed-adverse-effect-level (LOAEL).* In dose-response experiments, the lowest exposure level at which there are statistically or biologically significant increases in frequency or severity of adverse effects between the exposed population and its appropriate control group.
4. *No-observed-adverse-effect-level (NOAEL).* In dose-response experiments, an exposure level at which there are no statistically or biologically significant increases in the frequency or severity of adverse effects between the exposed population and its appropriate control; some effects may be produced at this level, but they are not considered to be adverse, nor precursors to specific adverse effects. In an experiment with more than one NOAEL, the regulatory focus is primarily on the highest one, leading to the common usage of the term NOAEL to mean the *highest* exposure level without adverse effect.

Figure 6.7 presents a graphical interpretation of these terms as they relate to the dose-response curve.[8]

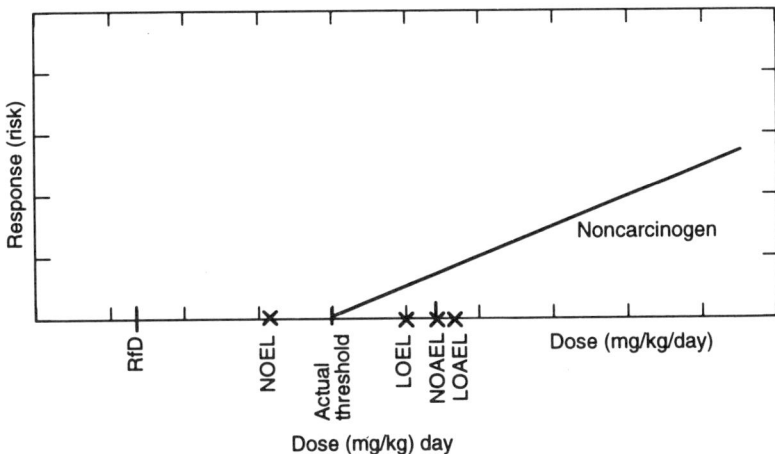

Figure 6.7 Noncarcinogenic RfD, LOEL, NOEL, LOAEL, and NOAEL as they relate to a threshold dose. (From Masters, G. M., *Introduction to Environmental Engineering and Science*, Prentice-Hall, Englewood Cliffs, NJ, 1991, 211. With permission.)

The RfD is obtained by dividing the NOAEL (or LOAEL) by an uncertainty factor (UF) and a modifying factor (MF):

$$\text{RfD} = \frac{\text{NOAEL or LOAEL}}{\text{UF} \times \text{MF}} \qquad [6.4]$$

The uncertainty factor generally consists of multiples of 10. The bases for application of different uncertainty factors and the modification factor are as follows:[6]

- A UF of 10 is used to account for variation in the general population and is intended to protect sensitive subpopulations (e.g., elderly, children).
- A UF of 10 is used when extrapolating from animals to humans. This factor is intended to account for the interspecies variability between humans and other mammals.
- A UF of 10 is used when a NOAEL derived from a subchronic instead of a chronic study is used as the basis for a chronic RfD.
- A UF of 10 is used when a LOAEL is used instead of a NOAEL. This factor is intended to account for the uncertainty associated with extrapolating from LOAELs to NOAELs.
- An MF ranging from 0 to 10 is included to reflect a qualitative professional assessment of additional uncertainties in the critical study and in the entire data base for the chemical not explicitly addressed by the preceding uncertainty factors. The default value for the MF is 1.

As indicated, the appropriate NOAEL or LOAEL is divided by the appropriate UFs and MF number to determine the RfD. If more than one UF is involved, all the applicable UFs are used as follows:

$$\text{RfD} = \text{NOAEL or LOAEL} / \text{UF}_1 \cdot \text{UF}_2 \cdot \ldots \text{UF}_n \times \text{MF} \qquad [6.5]$$

The potential for noncarcinogenic effects is evaluated by comparing an exposure level (intake) over a specified period of time with a reference dose (RfD). This ratio of exposure to toxicity is called a hazard quotient.

$$\text{Noncancer Hazard Quotient} = \frac{\text{Intake (CDI)}}{\text{RfD}} \qquad [6.6]$$

The greater the Intake/RfD above unity the greater the level of concern. However, Intake/RfD values are not statistical probabilities. For instance, a ratio of 0.002 does *not* mean that there is 2 in 1000 chances of an effect occurring.

Example 6.2

If a drinking water supply routinely contained 80 µg/L of chloroform (trichloromethane, $CHCl_3$, one of the trihalomethanes, THMs) what would be the noncancer hazard quotient associated with drinking this water?

Solution

1. Determine the CDI using Equation 6.2

$$CDI = \frac{CW \cdot IR \cdot EF \cdot ED}{W \cdot AT}$$

$$CDI = \frac{80 \times 10^{-3} \text{ mg/L} \times 2 \text{ L/day} \times 365 \text{ days/yr} \times 70 \text{ yr}}{70 \text{ kg} \times 365 \text{ days/yr} \times 70 \text{ yr}}$$

$$= 0.0023 \text{ mg/kg/day}.$$

2. From Table 6.6 the RfD for chloroform (oral route) is 1×10^{-2} mg/kg/day.
3. Using Equation 6.6:

$$\text{Noncancer Hazard Quotient} = \frac{\text{Intake (CDI)}}{\text{RfD}}$$

$$= \frac{0.0023 \text{ mg/kg/d}}{1 \times 10^{-2} \text{ mg/kg/d}}$$

$$= 0.23.$$

The hazard quotient is well below unity and thus would not pose a concern for $CHCl_3$ in the drinking water and would not be a potential problem for noncarcinogenic effects.

To assess the overall potential for noncarcinogenic effects by more than one chemical, a hazard index (HI) is used as follows:

$$HI = \frac{\text{Intake}_1}{\text{RfD}_1} + \frac{\text{Intake}_2}{\text{RfD}_2} \ldots + \frac{\text{Intake}_i}{\text{RfD}_i} \quad [6.7]$$

When HI exceeds 1, concern for potential health effects must be considered.

2. Dose Response For Carcinogens

As indicated earlier, carcinogens are assumed to have no threshold dose and that exposure to any amount or even a few molecules of a toxic agent can create uncontrolled cellular growth. Therefore to evaluate cancer risks, the threshold concept cannot be used. EPA uses a two-part evaluation system to calculate potential carcinogenic effects. This two-part system consists of a weight-of-evidence classification and a slope factor or potency factor determination.

a. Weight-of-Evidence Categories

An evaluation is first made to determine the potential that a toxic agent has to cause human or animal cancers. This evidence is characterized separately from human and animal studies; the parameters used are sufficient data, limited or inadequate data, and no data or no evidence. This information is combined and the agent or chemical is given a provisional weight-of-evidence classification. Five categories have been established and these categories are presented in Table 6.7.[6,12]

b. Generating Slope or Potency Factors

A toxicity value that quantitatively defines the relationship between dose and response is called a slope factor or a potency factor. The slope factor is used in risk assessment to estimate a lifetime probability of an individual developing cancer as a result of exposure to a particular level of a potential carcinogen. Slope or potency factors should always be accompanied by the weight-of-evidence factor to indicate the strength of evidence that an agent is a human carcinogen. At low doses of an agent it is difficult to measure responses, and therefore extrapolation of high-dose responses are used to predict low-dose responses.

Table 6.7 EPA Weight-of-Evidence Classification System for Carcinogenicity

Group	Description
A	Human carcinogen
B1 or B2	Probable human carcinogen
	B1 indicates that limited human data are available
	B2 indicates sufficient evidence in animals and inadequate or no evidence in humans
C	Possible human carcinogen
D	Not classifiable as to human carcinogenicity
E	Evidence of noncarcinogenicity for humans

Compiled from References 6 and 12.

This entails applying mathematical models. A number of mathematical models and procedures have been developed to extrapolate high-dose responses to low-dose responses. Some of these models include linearized multistage, Weibull, probit, logit, one-hit, and gamma multihit models. Various models and their use will be discussed in more detail later in this chapter. EPA recommends that the linear multistage model be employed to develop slope factors in the absence of adequate information to the contrary. The linear multistage model is considered to be one of the most conservative of the various models.

EPA indicates that, in general, after the data are fit to the appropriate model, the upper 95th percent confidence limit of the slope of the resulting dose-response curve is calculated. This value is known as the slope factor and represents an upper 95th percent confidence limit on the probability of a response per unit intake of a chemical over a lifetime (i.e., there is only a 5% chance that the probability of a response could be greater than the estimated value on the basis of the experimental data and model used).[6,12] Slope or potency factors are only linear at low doses.

Toxicity or lifetime risk for carcinogenic effects can be calculated in terms of the intake or the average daily dose (mg/kg/day) multiplied by the slope or potency factor (mg/kg/day)$^{-1}$.[6]

$$\text{Risk} = \text{CDI} \cdot \text{SF} \quad [6.8]$$

where: Risk = a unitless probability (e.g., 2×10^{-5}) of an individual developing cancer
CDI = chronic daily intake averaged over 70 years (mg/kg/day)
SF = slope factor, expressed in (mg/kg/day)$^{-1}$.

To calculate the average daily dose or chronic daily intake (CDI), a 70 year lifetime and a 70 kg human is used with an average inhalation rate of 20 m³/day and an average water consumption rate of 2 L/day. Toxicity values for carcinogenic effects also can be expressed in terms of risk per unit concentration of the substance in the medium where human contact occurs. Unit risks associated with unit concentrations in air or water can be calculated as follows:[6,12]

$$\text{air unit risk} = \text{risk per } \mu g/m^3$$
$$= \text{slope factor} \times 1/70 \text{ kg} \times 20 \text{ m}^3/\text{day} \times 10^{-3} \quad [6.9]$$

$$\text{water unit risk} = \text{risk per } \mu g/L$$
$$= \text{slope factor} \times 1/70 \text{ kg} \times 2 \text{ L/day} \times 10^{-3} \quad [6.10]$$

Multiplication by 10^{-3} is necessary to convert from mg (the slope factor, given in [mg/kg/day]$^{-1}$) to µg (the unit risk is given in [µg/m³]$^{-1}$ or [µg/L]$^{-1}$).

The sources presented earlier for information associated with RfDs are the same sources that can be used to supply slope factors and unit risk estimates for various chemicals.[12–14] Table 6.8 presents a few slope factors for inhalation and oral exposure pathways as well as the weight-of-evidence for these chemicals.

An understanding of the degree of uncertainty associated with the slope factor values is important when interpreting the data. All EPA-verified slope factors are accompanied by a weight-of-evidence classification that indicates the likelihood that an agent is a human carcinogen. A weight-of-evidence A category indicates that slope factors are less likely to change as the values were consistent in different species, sexes, study design, or dose-response relationships. A category A classification is associated with

Table 6.8 Carcinogenic Slope Factors for a Few Selected Chemicals

| Chemical | Year | Weight-of-evidence | | Exposure pathway | |
		Oral route	Inhalation	Oral route (mg/kg/day)$^{-1}$	Inhalation (mg/kg/day)$^{-1}$
Benzene	1990	A	A	2.9×10^{-2}	2.9×10^{-2}
Benzo (a) pyrene	1990	B2	B2	11.5	6.1
Bromoform	1991	B2	B2	7.9×10^{-3}	3.9×10^{-3}
Cadmium	1990	ND	B1	6.1	ND
Carbon tetrachloride	1991	B2	B2	1.3×10^{-1}	1.3×10^{-1}
Chlordane	1990	B2	B2	1.3	1.3
Chloroform	1990	B2	B2	6.1×10^{-3}	8.1×10^{-2}
Chromium (VI)	1990	ND	A	ND	41
Dieldrin	1990	B2	B2	16	16
Polychlorinated biphenyls 1260	1990	B2	B2	7.7	ND
Styrene	1990	B2	B2	3×10^{-2}	2×10^{-3}
Tetrachloro-ethylene (perchloro-ethylene)	1989	B2	B2	5.1×10^{-2}	5.2×10^{-7}
Toxaphene	1990	B2	B2	1.1	1.1
Trichloro-ethylene (TCE)	1989	B2	B2	1.1×10^{-2}	1.3×10^{-2}
Vinyl chloride	1989	A	A	2.3	0.295

Note: ND - not determined.
Compiled from References 12-14.

values based on human data for the chemical of concern. A weight-of-evidence C category indicates that toxicity values might change if additional information becomes available.[6]

Following are two examples utilizing the data presented in Table 6.8, one for an oral and one for an inhalation exposure route.

Example 6.3

The maximum contaminant level (MCL) for carbon tetrachloride has been set at 5 ppb (5 µg/L) in drinking water.

At this concentration of CCl_4 what would be the lifetime individual cancer risk using standard values for daily intake?

If all 250×10^6 Americans drank water with this CCl_4 concentration, what maximum number of extra cancers per year would be estimated?

Solution

1. From Equation 6.8

    Risk = CDI · SF

2. From Equation 6.2

$$CDI = \frac{CW \times IR \times EF \times ED}{W \times AT}$$

$$= \frac{5 \times 10^{-3} \text{ mg/L} \times 2 \text{ L/day} \times 365 \text{ days/yr} \times 70 \text{ yr}}{70 \text{ kg} \times 365 \text{ days/yr} \times 70 \text{ yr}}$$

$$= 0.0001428 \text{ mg/kg/day}$$

CW $= 5 \times 10^{-3}$ mg/L, IR = 2 L/day, EF = 365 day/yr,

ED = 70 yr, W = 70 kg, AT = 70 yr · 365 days/yr

3. From Table 6.8 the slope factor for carbon tetrachloride

   SF $= 1.3 \times 10^{-1}$ (mg/kg/day)$^{-1}$

4. Substitute CDI and SF in Equation 6.8

   Risk $= 0.0001428$ mg/kg/day $\times 1.3 \times 10^{-1}$ (mg/kg/day)$^{-1}$
   $= 18.6 \times 10^{-6}$

This suggests that over a 70-year period the individual probability of cancer from drinking this water would be 18.6 per million. If there were 18.6 cancers per million over a 70-year period, then in 1 year in a population of 250 million, the number of cancers caused by CCl_4 would be

$$\text{Cancers/yr} = 250 \times 10^6 \times \frac{18.6 \text{ cancers}}{10^6} \times \frac{1}{70 \text{ yr}}$$

$$= 66$$

If the normal cancer death rate expected in a population is 193 deaths/yr per 100,000, the total cancer deaths/yr in a 250×10^6 population would be

$$250 \times 10^6 \times \frac{193}{100,000} = 482,500$$

The estimated 66 deaths for drinking water containing 5 µg/L CCl_4 would only be

$$\frac{66}{482,500} \times 100 = 0.014\%$$

This would appear to be an insignificant contribution to the overall cancer deaths in this population.

Example 6.4

A foam sheeting operation discharges 100 tons of styrene per year from a stack 150 m tall. On a clear day in September (stability condition B) when the wind is 5 m/s from the southwest, how far downwind would the maximum concentration occur, and what would the estimated concentration of styrene be? If 100,000 people were exposed to this maximum styrene concentration for 70 years for 24 hours per day what would the estimated cancer risk be for a 70-kg person?

1. Use the following dispersion model to calculate the pollutant concentration downwind.[15,16]

$$C_{x,0} = \frac{Q}{\pi u \sigma_z \sigma_y} \exp\left[-\frac{1}{2}\left(\frac{H}{\sigma_z}\right)^2\right] \qquad [6.11]$$

where: C = pollutant concentration, g/m³
         Q = pollutant emission rate, g/s
         π = pi, 3.14159
         u = mean wind speed, m/s
         σ_y = standard deviation of horizontal plume concentration, evaluated in terms of downwind distance x, m
         σ_z = standard deviation of vertical plume concentration evaluated in terms of downwind distance x, m
         exp = base of natural logs, 2.71828183
         H = effective stack height, m
         x = downwind distance along plume mean centerline from point source, m
         y = crosswind distance from the centerline of the plume, m

The maximum ground-level concentration occurs where

$$\sigma_z = 0.707 \cdot H \qquad [6.12]$$

This is based on σ_z/σ_y being constant downwind.[16]

2. Calculate Q the pollutant emission rate in g/s.

   $Q = 100$ tons/yr
   $= 100$ tons/yr $\times 9.071.8 \times 10^2$ kg/ton $\times 1000$ g/kg
   $= 90.72 \times 10^6$ g

   $$Q = \frac{90.72 \times 10^6}{365 \text{ days/yr} \times 8.64 \times 10^4 \text{ s/day}}$$

   $= 2.88$ g/s

3. Determine σ_z using Equation 6.12:

   $\sigma_z = 0.707 \cdot H$
   $\sigma_z = 0.707 (150 \text{ m})$
   $= 106$ m.

4. Read the distance downwind to the point of maximum contaminant concentration using the vertical diffusion coefficient σ_z figure in Reference 15 or 16. With a $\sigma_z = 106$ m and B stability conditions, the distance downwind from the source to the point of maximum ground-level concentration is 1000 m. Then, using the lateral diffusion coefficient σ_y figure and a 1000-m distance from the source (stability B conditions) $\sigma_y = 160$ m.

5. Calculate the maximum downwind concentration using Equation 6.11:

   $$C_{x,0} = \frac{2.88 \text{ g/s}}{3.14159 (5 \text{ m/s})(106 \text{ m})(160 \text{ m})} \exp\left[-\frac{1}{2}\left(\frac{150\text{m}}{106\text{m}}\right)^2\right]$$

   $= 10.81 \times 10^{-6} \exp\left[-\frac{1}{2}(2)\right]$
   $= 10.81 \times 10^{-6} (0.3679)$
   $= 3.98 \times 10^{-6}$ g/m³ × μg/g
   $= 3.98$ μg/m³

6. Calculate the total styrene intake for a human over a 70-year period at a breathing rate of 20 m³/day.

   Total intake $= 20$ m³/day $\times 365$ days/yr $\times 70$ yr $\times 3.98$ μg/m³ $\times \dfrac{1}{1000 \text{ μg/mg}}$

   $= 2034$ mg

7. Calculate the chronic daily intake (CDI).
   (The absorption in this example is considered to be 100%.)

   Chronic daily intake $= \dfrac{2034 \text{ mg}}{70 \text{ kg} \times 70 \text{ yr} \times 365 \text{ days/yr}}$

   $= 0.00114$ mg/kg/day

8. Calculate the risk for cancer using the slope factor for styrene for the 100,000 city population.

   The slope factor from Table 6.8 for styrene is 2×10^{-3} (mg/kg/day)⁻¹

   Risk = CDI · SF

   $= 0.00114$ mg/kg/day $\times 2 \times 10^{-3}$ (mg/kg/day)⁻¹

   $= 2.3 \times 10^{-6}$

   This calculation suggests that the individual probability of cancer over a 70-year period under the stipulated conditions would be 2.3 per million or $\dfrac{2.3}{1,000,000} \times 100,000 \times \dfrac{1}{70 \text{ yr}} = 0.003$ cancers/yr for the 100,000 population. The annual cancer deaths that would normally be expected is 93 deaths/year per 100,000.[8]

   It would seem under these most adverse and unlikely conditions that the additional cancers per year is almost nondetectable.

The stability categories used in the Gaussian Model vary from A to F as follows:

A — extremely unstable
B — moderately unstable
C — slightly unstable
D — neutral
E — slightly stable
F — moderately stable

See References 15 or 16 for details.

Other air quality models are available from EPA through the Office of Air Quality Planning and Standards at Research Triangle Park, North Carolina. Such models as ISCLT or ISCST are much more sophisticated than the model used here, yet the same principles apply.

IV. RISK CHARACTERIZATION

Risk characterization summarizes and brings together the information assembled in the data collection, hazard identification, dose response, and exposure assessment operations. Both noncarcinogenic and carcinogenic risks are presented quantitatively and qualitatively.

Risk characterization provides a connection between risk assessment and risk management, so that decisions can be made not only on health risks but can also include the economics, technical feasibility, and regulatory aspects of the problem.

The toxicity information, which includes the hazard identification and dose-response steps needed for the risk characterization process include:[6]

- Slope factors for all carcinogenic chemicals.
- Discussion of weight-of-evidence and classifications for all carcinogenic chemicals.
- Type of cancer for Class A carcinogens.
- Chronic and subchronic RfDs and shorter-term toxicity values (if appropriate) for all chemicals (including carcinogens and developmental toxicants).
- Critical effect associated with each RfD.
- Discussion of uncertainties, uncertainty factors, and modifying factor used in deriving each RfD and "degree of confidence" in RfD (i.e., high, medium, low).
- Whether the toxicity values are expressed as absorbed or administered doses.
- Pharmacokinetic data that may affect the extrapolation from animals to humans for both the RfD and slope factor.
- Uncertainties in any route-to-route extrapolations.

The necessary exposure assessment information needed to carry out this risk characterization step include:[6]

- Estimated intakes (chronic, subchronic, and shorter-term, as appropriate) for chemicals.
- Important exposure modeling assumptions, including:
  - Chemical concentration at the exposure points
  - Frequency and duration of exposure
  - Absorption assumptions
  - Characterization of uncertainties
- List of which exposure pathways can reasonably contribute to the exposure of the same individuals over the same time period.

To quantify risks for both carcinogenic and noncarcinogenic effects, exposure pathways must be considered. The pathways for humans can be ingestion, inhalation, or dermal.

To calculate the carcinogenic risk EPA uses the linear multistage model. Equation 6.8 is used to make the necessary calculations and includes the chronic daily intake (CDI) and the slope factor (SF). The linear multistage model is used only at low risk levels (below estimated risks of 0.01). For risk levels above 0.01 the one-hit equation is often used and is as follows:[6]

$$\text{Risk} = 1 - \exp(-\text{CDI} \times \text{SF}) \qquad [6.13]$$

Table 6.9 Helpful Hints for Risk Communication

Try to:
• Use a mix of well written text, illustrative graphics, and summary tables
• Explain the major steps and the results of the risk assessment in terms easily understood by the general public (and especially by members of exposed or potentially exposed populations)
• Define highly technical terms early (e.g., in a glossary)
• Use a standard quantitative system — preferably the metric system — throughout, and units that are the same where possible (e.g., µg/L for all water concentrations)
Avoid:
• The use of large blocks of text unbroken by any headings, graphics, tables, lists, or other "visual dividers"
• The presentation of much quantitative information within the text (rather than in tables)
• The drawing of "risk management" conclusions (e.g., stating that the total or largest risk is insignificant)

Compiled from Reference 6.

where: Risk = a unitless probability (e.g., 2×10^{-5}) of an individual developing cancer
exp = the exponential
CDI = chronic daily intake averaged over 70 years (mg/kg/day)
SF = slope factor, in $(mg/kg/day)^{-1}$

The total estimated cancer risk for multiple chemicals and pathways is determined by combining the various risk factors in a simple additive manner.

This summation of risk of developing cancer is expressed as one probability number and assumes that the intake of each individual substance was small and that each substance acts independently of the other. As indicated earlier, the potential for noncarcinogenic effects to occur in an individual is not expressed in the same probability way used for potential carcinogens. Equation 6.6 is used to make the noncarcinogenic health effects determination and Equation 6.7 to assess noncarcinogenic potential health effects for more than one chemical.

A. RISK COMMUNICATION

Risk communication is considered to be part of the environmental management process. Risk communication means presenting information to the public in a clear, concise relevant manner. The essential information presented includes the environmental and health risks for the situations under consideration, specific decisions already made and also decisions being considered for this particular situation, and the plans for the final course of action to solve or control a particular problem. Risk communication is a tool for educating the public on the nature of risks they are faced with and why and how risk management decisions are being made. It is also a means by which governmental agencies can get feedback from the public on issues of concern to them. Risk communication can also help to educate the public as to actions they as individuals can take to help control risks in their own environment.[4]

Putting risks in an understandable context is very important, and of course can also be very difficult.

Risk communication must be at the local, state, and federal level. The public gets most of its information on environmental issues through the mass media, newspapers, radio, and television.[4,14]

Table 6.9 presents a summary of helpful hints for the risk communication operation.

B. RISK ASSESSMENT/WASTEWATER RECLAMATION AND REUSE
1. Introduction

The reuse of reclaimed wastewater has been increasing over the past decade, especially in arid and semiarid regions where reclaimed wastewater is used to conserve valuable potable water supplies. Continued population growth also increases pressure on cities, counties, and utilities to protect freshwater supplies and conserve potable supplies by using reclaimed wastewater for nonpotable purposes. While nonpotable reclaimed wastewater generally contains chemical contaminants at low levels (see Table 9.5), there is still a need to control pathogenic bacteria, protozoa, viruses, and helminths that the public could be exposed to. For instance, the exposure could occur by consuming raw food crops irrigated with inadequately treated wastewater, or through aerosols from spray irrigation or cooling towers, or through accidental ingestion of reclaimed wastewater used for recreational purposes. Another potential problem is a cross connection between a potable and a nonpotable water supply system.

Table 6.10 Infective Dose for a 50% Attack Rate for Virus in Humans

Organism	50% Infective dose	Mode of administration
Poliovirus 3 (Fox)[a]	4 ($TCID_{50}$)	Stomach (gavaged)
Poliovirus 1 (Sabin)[a]	72 ($TCID_{50}$)	Oral
Echovirus 12	35 (PFU)	Oral
Coxsackievirus A21	28 ($TCID_{50}$)[b]	Nasal
Adenovirus 4	1 ($TCID_{50}$)	Nasal

Note: TCID - tissue culture infective dose.
PFU - plaque-forming unit.

[a] Study conducted in healthy infants.
[b] Infective dose was found to approximate the illness dose.

From Sorber, C. A., *Municipal Wastewater Reuse News,* 57, 5, 1982.

Domestic wastewater is often one of the components in the reclaimed wastewater and contains all the microorganisms and parasites found in human excreta. The pathogens found in wastewater and the diseases associated with these pathogens are presented in the following sections. As indicated earlier, one step in the toxicity assessment process is hazard identification, and the identification of pathogenic organisms present in the reclaimed wastewater would be a first step.

2. Viruses

Viruses are the smallest of all microorganisms. They are small parasites or pathogens of plants, animals, and bacteria. All are so small they can be seen only by the magnification obtained with an electron microscope. Viruses can only be cultivated in living cells and do not multiply outside a suitable host cell. Viruses can pass through filters that retain bacteria. It has also been demonstrated that viruses are more resistant to disinfection by chlorination than are bacteria.[17,18]

Following are some of the most important viruses found in sewage or wastewater.[16,19-21]

- Polio virus (Types 1, 2, 3). Causes not only paralytic poliomyelitis, but also aseptic meningitis, fever, and influenza-like illnesses. Just recently it was reported that the polio disease has been brought under control even on a global basis.
- Enteric cytopathogenic human orphan (Echoviruses) (34 types). Causes aseptic meningitis, diarrhea, respiratory diseases, and epidemic exanthem.
- Coxsackie viruses (Group A — 26 types, Group B — 6 types). Causes aseptic meningitis, gastroenteritis, and may cause heart and respiratory diseases.
- Rotaviruses. Causes diarrhea, especially in infants and young children.
- Revovirus. Causes gastroenteritis.
- Adenovirus (32 types). Causes acute respiratory infection.
- Hepatitis-A virus. Causes infectious hepatitis.
- Norwalk agent. Causes gastroenteritis.

An important question is the dose-response aspects of exposure to an infective agent, that is, the number of organisms required to cause an infection or illness. It can be seen from Table 6.10 that inhaling even one adenovirus may be sufficient to induce infection in 50% of the subjects.[4]

3. Bacteria

Bacteria are unicellular prokaryotic organisms, or simple associations of similar cells. Cell multiplication is usually by binary fission. Prokaryotic organisms include blue-green algae and bacteria, and do not have a distinct nucleus compared to eukaryotic organisms which do have distinct nuclei and include plants and animals.[17]

Following are some of the major pathogenic bacteria found in wastewater and the diseases associated with them.[16,17,20,21]

- *Salmonella typhi:* causes typhoid fever, diarrhea.
- *Salmonella paratyphi:* causes paratyphoid fever, enteric infections.
- *Salmonella typhiniurium:* causes food poisoning, salmonellosis.
- *Shigella sonnei:* causes shigellosis (bacillary dysentery).

Table 6.11 Infective Dose for Selected Bacteria

Organism	Infective Dose (No. of organisms)	Attack rate	Minimum dosage[a] (No. of organisms)
Salmonella spp.	10^5–10^8	50%	10^5
Shigella spp.	10^1–10^2	50%	10^1
Escherichia coli	10^6–10^{10}	25–75%	10^6
Vibrio cholerae	10^3–10^{11}	25–75%	10^3
Streptococcus faecalis	>10^{10}	25–75%	10^{10}

[a] Minimum number of organisms required to produce disease in any of the subjects tested.

Compiled from References 20–22.

- *Mycobacterium tuberculosis:* causes tuberculosis.
- Enteropathogenic *Escherichia coli:* causes diarrhea or gastroenteritis.
- *Vibrio cholerae:* causes cholera, diarrhea, and dehydration.
- *Yersina enterocolitica:* causes diarrhea or gastroenteritis.
- *Francisella tularensis:* causes tularemia.
- *Leptospira interrogans:* causes leptospirosis, jaundice (Weil's disease).

Table 6.11 presents the dose-response information for selected pathogenic bacteria often found in wastewaters. For some organisms such as *Shigella,* an ingestion of only 10 organisms can result in a 50% attack rate.

The identification of pathogenic organisms is not only difficult but also time consuming. The usual practice is to determine the total coliform, fecal coliform, or fecal streptococci present in the wastewater and use this information as indicators of the presence of pathogenic organisms.

The organisms most commonly used as indicators of fecal pollution are members of the coliform group, particularly *Escherichia coli.* The presence of commensal bacteria of intestinal origin such as those of the coliform group and *Streptococcus fecalis* do not, in general, constitute a health hazard. However, their existence in certain numbers indicates that the water is likely to contain other organisms that could cause serious infections.[23] Serious questions are now being raised as to the appropriateness of using the coliform organisms as indicators of the presence of pathogenic organisms in water. It has been pointed out that research work in this area is urgently needed.[21]

4. Protozoa

Protozoa are single-celled eukaryotic protists. As indicated previously, eukaryotic means that the cells have a distinct nucleus, while protists include organisms which do not fall naturally into the plant or animal kingdom but often share characteristics common to both plants and animals. The protists generally include bacteria, algae, fungi, and protozoa.[17]

Protozoa are complete, self-contained organisms that can be parasitic, pathogenic, or nonpathogenic. Only a few aquatic protozoa are pathogenic, with only a few species infecting the intestinal tract of man and animals.[16,17] Protozoa are excreted in the feces in the form of cysts and are difficult to deactivate even with fairly high concentrations of chlorine.

Following are the major pathogenic protozoa found in wastewater and the diseases associated with them.[19–21]

- *Entamoeba histolytica:* causes amoebic dysentery, liver abscess, and colonic ulcerations.
- *Giardia lamblia:* causes anorexia, diarrhea, and infections of the small intestine.
- *Blantidium coli:* causes diarrhea, colonic ulceration, and dysentery.
- *Cryptosporidium:* causes diarrhea.

The typical survival time for cysts in wastewater is around 20 days.[19] Table 6.12 indicates the dose-response for a 50% attack rate for two pathogenic protozoans.

5. Helminths

Helminths, or parasitic worms, often involve two or more animal hosts, one of which can be human. Helminths may infect the intestine of man, and their ova or eggs are passed in the excreta which can then contaminate a water supply. Contamination may also be via aquatic species of other hosts such as snails or insects. Modern wastewater treatment processes are generally very effective at removing these

Table 6.12 Infective Dose for a 50% Attack Rate for Protozoans in Humans

Organism	Infective oral dose (No. of organisms)	Minimum dosage[a] (No. of organisms)
Entamoeba hystolytica	10–100	1
Giardia lamblia	1–10	1

[a] Minimum number of organisms required to produce disease.

Compiled from References 20 and 22.

Table 6.13 Infective Dose for a 76 to 100% Attack Rate for Pathogenic Helminths

Organism	Disease	Number for infective dosage
Enterobius vermicularis	Pinworm — enterobiasis	10^2
Ascaris lumbricoides	Roundworm — ascariasis	10^2
Necator americanus	Hookworm — necatoriasis	10^2
Taenia saginata	Beef tapeworm — taeniasis	10^2

From Shahalam, A. B. M., *J. Environ. Sci.,* Sept./Oct. 1989, 35. With permission.

organisms. Following are the major helminths found in wastewater and the diseases associated with them.[19–21]

- *Ascaris lumbricoides* (round worm): causes ascariasis, digestive disorders, bowel obstruction, obstruction of bile and pancreatic ducts.
- *Trichuris trichiura* (human whip worm): causes anemia, diarrhea, shifting of the rectum.
- *Taenia saginata* (beef tape worm): causes anorexia, loss of weight, nervousness, insomnia, abdominal pain.
- *Schistosoma* spp. (flat worms): causes damage to bladder, kidneys and liver, decreased ability to work.
- *Necator americanus* (human hook worm): causes anemia, and disability.
- *Echinococcus granulosus* (tape worm): causes damage to liver, lungs, kidney, and spleen.
- *Taenia solium* (pork tape worm): can cause fatal cysticercosis, especially if the cysts localizes in the eyes or heart.
- *Enterobius vermicularis* (pinworm): causes anal itching, disturbed sleep, and irritability.
- *Hymenolepidosis nana* (dwarf tape worm): causes diarrhea, abdominal pain, weight loss, and weakness.

Table 6.13 presents the dose-response information for a 76 to 100% attack rate for some pathogenic helminths.[21]

A recent report sponsored by the World Bank and WHO indicated that complete removal of the eggs of intestinal nematodes (worms) to a geometric mean of <1 viable nematode egg per liter is desirable for unrestricted use of wastewater for agricultural irrigation.[24]

Despite the presence of pathogenic organisms in wastewater there are no known outbreaks of communicable diseases related to the utilization of properly treated wastewater for irrigation.[25] However, there have been outbreaks of cholera, typhoid fever, and paratyphoid fever when raw sewage was illegally used for growing vegetables.[26] Even effluents from secondary treatment plants cannot be considered safe, and for this reason additional treatment is needed depending upon how the reclaimed water is to be used.

6. Survival of Waterborne Pathogens in Wastewater, Soil, and on Crops

The survival of pathogenic organisms in wastewater, soil, and crops depends on many factors including:[20,21,27]

- Indoor/outdoor environment
- Soil moisture content
- Methods of wastewater disposal
- Crop type
- Distance of crop parts from soil surface

- Seasons and temperature
- Wastewater and soil pH
- Time pathogen remained in wastewater before its use on soil and crops
- Treatment level of wastewater
- Method of soil cultivation or soil disturbance
- Depth of soil
- Condition of crop surface
- Sunlight
- Organic matter in soil
- Antagonistic soil microorganisms

There is a great variability in the survival of pathogenic microorganisms in the environment. *Ascaris lumbricoides* appears to survive the longest, and under favorable conditions can remain viable in soil for years.[27]

Table 6.14 summarizes survival time in soil for a number of microorganisms discussed earlier.[21,27,28]

Most pathogenic bacteria die off in wastewater, soil, and on crops in about 1 month, except for the tubercle bacilli which have a protective waxy coating. Viruses survive from less than 3 days up to 10 weeks (Hepatitis A). Pathogenic protozoans survive only for up to 7 or 8 days; however, the ova of the helminth, *Ascaris* can survive up to 7 years under proper soil conditions. Proper soil conditions include high moisture content, low temperature, high alkalinity, limited sunlight, high organic content, and low soil microflora.[21]

The preceding information has dealt with the risk assessment involved in the health effects caused by exposure to pathogenic microorganisms. The two aspects included here were the hazard identification of pathogenic bacteria, virus, protozoa, and helminths, as well as the limited available dose-response data. It is difficult to extrapolate high experimental doses to typically low doses for pathogenic microorganisms and it also is difficult to extrapolate animal data to human situations. The next section deals with the mathematical modeling or extrapolation from high to low dose-responses for both toxic chemical and biological agents.

C. RISK BASED ON MODELING
1. Introduction

An agent that is known or has potential for causing cancer in humans, or a pathogenic organism that can cause infectious diseases in humans, can be defined by a toxicity value that relates the dose to the response and is referred to as a slope factor or potency factor.

A slope factor is a plausible upper-bound (a number that is greater or equal to any number in a set) estimate of the probability of a response per unit intake of a toxic agent over a lifetime.

It is very difficult to measure the response of humans or test animals at low exposure levels to a toxic agent, thus extrapolation from relatively high doses to experimental animals (or the dose-responses noted in epidemiological studies) are used to develop slope factors at low exposure levels.

The dose-response curve in Figure 6.8 indicates the area where extrapolation in the low dose area is needed.[22]

This analysis ignores the possibility of a threshold or minimum infectious dose in that the dose-response curve is extrapolated to pass through zero.

2. Mathematical Models Available For Extrapolation For Chemical Exposures

Many mathematical models are available to carry out this extrapolation process for chemical, biological, and radiological exposures.

Among the models available for extrapolation of chemical dose-response curves are linear multistage, Weibull, probit, logit, one-hit, and gamma multihit models as well as various time-to-tumor models.[6]

EPA guidelines recommend the linearized multistage models, which is considered to be one of the more conservative models that overemphasizes risk.

Cothern and associates estimated the individual cancer risk rates for drinking water containing trichloroethylene (TCE). Their report included a total population of over 250×10^6 exposed to various concentrations of TCE ranging for 0.25 to 100 µg/L.[29]

The bioassay data were fit to four analytical models: logit, multistage, probit, and Weibull (see Figure 6.9). The dose-response curves for the models were fit to equivalent human dose response data by converting the relative surface area of the test animals to that of humans. Figure 6.9 shows the curves for

Table 6.14 Survival of Waterborne Pathogens in Wastewater, Soil, and Crops

Organism	Medium	Type of application	Survival time
Bacteria:			
Bacillus typhosa	Wastewater	—	<5 minutes
	Soil	AC[a]	29–70 days
	Vegetables	AC	31 days
Cholera vibrios	Wastewater	—	<39 days
	Spinach, lettuce	AC	22–29 days
	Nonacid vegetables	AC	2 days
Coliform	Grass	Sewage	14 days
	Tomatoes	Sewage	35 days
Fecal coliform	Wastewater	—	<1 day
	Sandy soil	—	8–55 days
	Tomatoes	—	>30 days
Leptospira	Soil	AC	15–43 days
Salmonella typhi	Radishes	Infected feces	53 days
	Soil	Infected feces	74 days
Shigella	Tomatoes	AC	2–7 days
Tubercle bacilli	Soil	AC	6 months
Typhoid bacilli	Soil	AC	7–40 days
Viruses:			
Polio virus	Wastewater	—	23 days
	Sandy soil	—	<11 days
	Lettuce	—	<36 days
	Radishes	—	<3 days
Coxsackieviruses	Wastewater	—	<41 days
New enteroviruses	Sandy soil	—	<41 to 110 days
	Vegetables	—	<60 days
	Root crop	—	<60 days
Hepatitis A	Wastewater	—	>10 weeks
Protozoa:			
Entamoeba histolytica	Vegetables	AC	3 days
	Soil	AC	8 days
Giardia lambia	Wastewater	—	7–8 days
Helminths:			
Ascaris ova	Soil	—	Up to 7 years
	Vegetables	AC	27–35 days
Necator americanus (hook worm)	Wastewater	—	<18 days
	Sandy soil	—	<10 days
	Soil	Infected feces	6 weeks
Taenia saginata	Wastewater	—	>16 days
	Sandy soil	—	<210 days, winter (only few days in summer)

[a] AC — artificial contamination.

Compiled from References 21, 27, and 28.

the four models extrapolated to the lower exposure levels. The log-log plot in Figure 6.9 shows the two non-zero data points in the upper right-hand corner for TCE. The solid lines are the point estimates and the error bars show the upper 95% confidence limits. The probit model decreases extremely quickly relative to dose while the Weibull model decreases very slowly.

Although the dose-response curves projected by each model in Figure 6.9 start at the same points, they diverge significantly at lower dose levels. At a concentration of 50 µg/L TCE in drinking water, the Weibull model provides a risk estimate approximating 1×10^{-2} at the upper 95% confidence level, whereas the probit model gives an estimate of 10^{-10}. As the authors indicate, this estimate provides a range of uncertainty equivalent to not knowing whether one has enough money to buy a cup of coffee or to pay off the national debt.

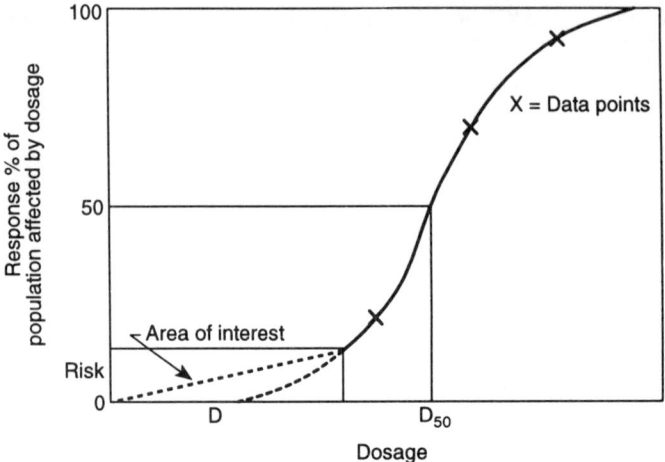

Figure 6.8 Linear extrapolation from a known point on the curve to zero. (From Hutzler, N. J. and Boyle, W. C., *Water Reuse,* Ann Arbor Science, MI, 1982, 293. With permission.)

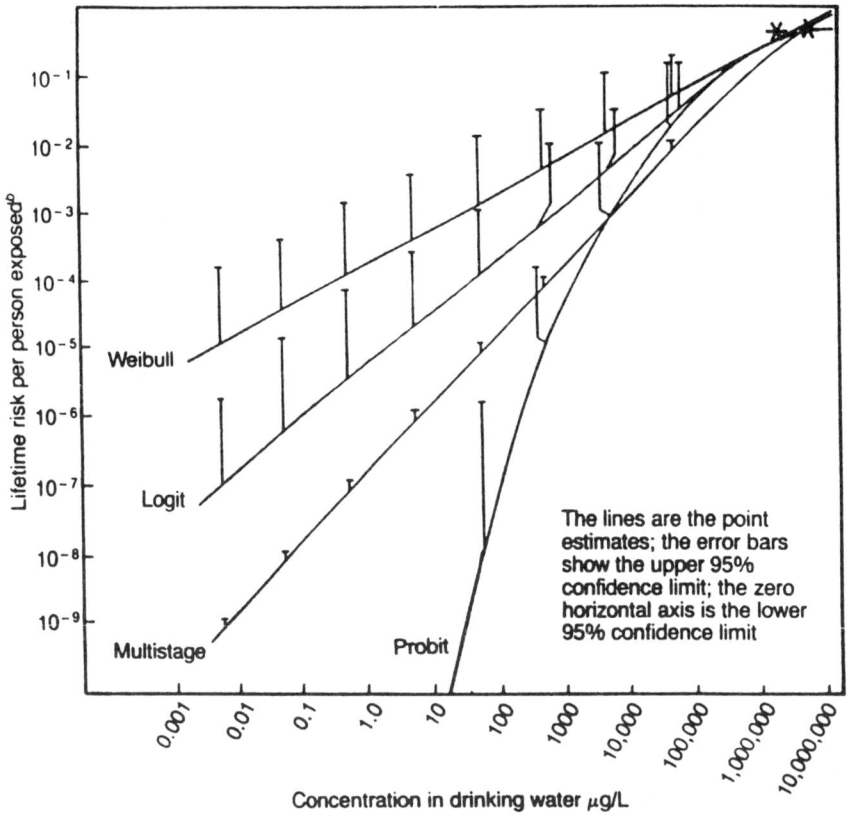

Figure 6.9 Bioassay data and model extrapolation for exposure to TCE by ingestion. a. Bioassay data are the starred points in the upper right-hand corner. The lifetime risk per person exposed was calculated assuming that the risk due to inhalation and dermal exposure is equal to that due to ingestion exposure. The error introduced by this simplifying assumption is less than the widths of the lines shown. b. Lifetime exposure through ingestion, inhalation, and dermal contact, assuming that ingestion exposure equals inhalation plus dermal exposure. (From Cothern, C. R., Coniglio, W. A., and Marcus, W. L., *Environ. Sci. Technol.,* 20, 2, 1986, 111. With permission.)

In this particular case, for TCE the linear multistage model did not give the most conservative risk estimate.

The various extrapolation methods may provide a reasonable fit to the observed data but they may still lead to large differences in slope factors for low doses.

3. Mathematical Models Available For Extrapolation For Infectious Agents

Asano and Sakaji tested three infectious agent models to analyze the risks associated with the use of reclaimed municipal wastewater.[30] These models were

1. The single-hit exponential model
2. The log-normal model
3. The beta-distributed infectivity probability model[31,32]

The following equation presents the single-hit exponential model:

$$P^* = 1 - \exp(-rN) \qquad [6.14]$$

where: P^* = the probability that a single individual exposed to a dose of N pathogens will become infected
N = the average number of pathogens (concentration of pathogens times volume of water consumed)
r = model parameter

The log-normal model assumes the pathogens are log-normally distributed. The probability, P^*, that a single individual will become infected after exposure to a dose of N organisms, is given by Equation 6.15:

$$P^* = \frac{1}{2^{0.5}} \int_{-\infty}^{Z} \exp\left(-\frac{z^2}{2}\right) dz \qquad [6.15]$$

$$Z = \frac{\ln(N) - \mu}{\sigma} \qquad [6.16]$$

where μ, σ are the model parameters.

The beta-distributed infectivity model is a stochastic model in which the pathogen-host interactions may not be characterized by a discrete value, but by a distribution of values. To estimate the probability of response in a human population, see Equation 6.17:

$$P^* = 1 - 1 + \frac{N^{-\alpha}}{\beta} \qquad [6.17]$$

where: P^* = probability of infection due to the ingestion of N pathogens
N = total number of pathogens in a known volume of reclaimed wastewater consumed
α, β = model parameters characterized by the dose-response data from References 31 and 32.

Asano and Sakaji indicated that the single-hit model was the least viable model in which a single organism was assumed to survive and infect the host cell organism. In subsequent work they used the beta-distributed model because it best represented the frequency distribution for pathogens of concern.[30-33]

Using the beta-distributed infectivity model, Asano et al. concluded that the annual risk of infection from exposure to chlorinated tertiary effluent containing 1 viral unit/100 L in recreational activities such as swimming or golfing was in the range of 10^{-2} to 10^{-7}, while exposures resulting from food-crop irrigation or groundwater recharge with reclaimed municipal wastewater was in the range of 10^{-6} to 10^{-11}.[33,34]

The recreational risk from exposure to tertiary treated wastewater ranged from 1 in 100 to 1 in 10 million, a wide order of magnitude. However, the risk from exposure to food crops irrigated or groundwater recharged with tertiary-treated effluent was 10^{-6} to 10^{-11}, and this is in the same risk magnitude as drinking heavily chlorinated water (e.g., Miami) for 1 year, which is estimated at 10^{-6}.[2]

Table 6.15 Expected Removal of Excreted Microorganisms by Various Wastewater Systems

Treatment process	Removal ($\log_{10}$ units) of			
	Bacteria	Helminths	Viruses	Cysts
Primary sedimentation				
Plain	0–1	0–2	0–1	0–1
Chemically assisted[a]	1–2	1–3[g]	0–1	0–1
Activated sludge[b]	0–2	0–2	0–1	0–1
Biofiltration[c]	0–2	0–2	0–1	0–1
Aerated lagoon[c]	1–2	1–3[g]	1–2	0–1
Oxidation ditch[b]	1–2	0–2	1–2	0–1
Disinfection[d]	2–6[g]	0–1	0–4	0–3
Waste stabilization ponds[e]	1–6[g]	1–3[g]	1–4	1–4
Effluent storage reservoirs[f]	1–6[g]	1–3[g]	1–4	1–4

[a] Further research is needed to confirm performance.
[b] Including secondary sedimentation.
[c] Including settling pond.
[d] Chlorination or ozonation.
[e] Performance depends on number of ponds in series and other environmental factors.
[f] Performance depends on retention time, which varies with demand.
[g] With good design and proper operation the recommended guidelines are achievable.

From Feigin, A., Ravina, I., and Shalhevet, J., *Irrigation With Treated Sewage Effluent,* Springer-Verlag, New York, 1991, 32. With permission.

V. REMOVAL OF PATHOGENIC MICROORGANISMS BY VARIOUS WASTEWATER TREATMENT PROCESSES

One of the major concerns in the utilization of treated wastewater is the possible transmission of disease. The number of pathogenic organisms in the treated wastewater depends on the treatment processes employed. While conventional wastewater treatment processes can provide a dramatic reduction in the number of viruses, bacteria, and parasites, it is still not adequate to provide the required margin of safety. Disinfection is presently the main method used to reduce the levels of microorganisms in the treated wastewater.

Disinfection is defined as the killing or destruction of all pathogenic organisms. Disinfection is often carried out using chlorine or a chlorine compound, however, ozonation can also be used for disinfection purposes.

Table 6.15 presents the removal in log units for excreted microorganisms by various wastewater systems.[35] Table 6.16 presents a summary of the removal for specific pathogenic organisms by various domestic wastewater treatment processes but does not include chlorination or ozonation.

The log removal is the negative log of the fraction of microorganisms remaining (similar to pH, which is the negative log of the [H] ion concentration). For instance, if plain primary sedimentation removes 0 to 1 log units this would mean from $10^0 = 1$, or 99% removal, to $10^{-1} = 0.10$ or 99.9% removal.

An examination of Table 6.16 suggests that microorganism removal by the various processes is highly variable and that an effective process for one organism may be completely ineffective for another.

A. CHLORINATION

Chlorine is the cheapest disinfectant and therefore chlorination is usually the method most often used for the disinfection of wastewater effluents. Chlorine, especially in the free available forms (Cl_2, HOCL, OCl^-), have excellent bactericidal properties; for instance a 99% kill of *E. coli* requires a residual of only 0.04 mg/L of chlorine as HOCL and a contact time of only 3.8 min at a temperature of from 0 to 6°C (see Figure 6.10).[10,36]

The equation for calculating the above 99% kill of *E. coli* is as follows:

$$c^{0.86} t_p = 0.24 \qquad [6.18]$$

where: c = chlorine concentration as HOCL (mg/L)
            t_p = contact time for 99% kill (minutes)

Table 6.16 Summary of the Percent of Removal of Selected Pathogenic Microorganisms by Various Wastewater Treatment Processes

Organism	Primary treatment	Activated sludge	Trickling filters	Stabilization pond
Bacteria coliform	48	91–98	97	99.96
Fecal coliform	—	97–99.9	95–97	99.6
Salmonella spp.	15	85–99	70–99	99.99
Salmonella typhi	99.5	74–99	99.5–99.9	99.5
Strep. fecalis	<50	85–95	—	—
Virus	<30	30–40	76	50–80
Polio virus	0–10	75–99	~85	—
Coxsackie	<50	0–50	~95	—
Amoebic cysts	50	Not affected	83–99	100
E. histolytica	nil	0–99	10–99.9	—
Helminths ova	90	—	18–26	100
Ascaris ova	66	93–99.2	77–99.8	100

Compiled from References 19, 21, 22, and 35.

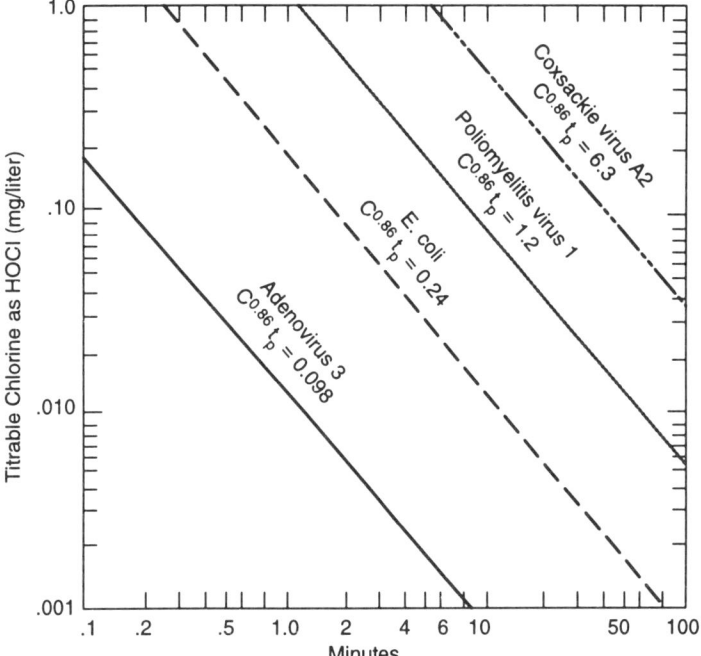

Figure 6.10 Relationship between concentration and time for 99% destruction of *E. coli* and three viruses by hypochlorous acid (HOCl) at 0 to 6°C. (From Berg, G., *J. N. Engl. Water Works Assoc.*, 78, 79, 1964. With permission.)

Therefore: $0.04^{0.86} \, t_p = 0.24$

$$t_p = \frac{0.24}{0.0627}$$

$$= 3.8 \text{ min.}$$

Figure 6.10 also illustrates the relative resistance of three viruses to hypochlorous acid (HOCl) at 0 to 6°C. While the polio and Coxsackie viruses are considerably more resistant than *E. coli* to HOCl, the adenovirus 3 virus is apparently the most sensitive: 0.1 mg/L of chlorine as HOCl destroyed 99% of *E. coli* in about 104 s, while the adenovirus 3 virus was destroyed in about one-tenth of that time. At this HOCl concentration (0.1 mg/L) the poliovirus required about 9 min and the Coxsackie virus required over 45 min for a 99% kill.

It has been concluded that, in general, with the exception of adenovirus 3, from 3 to 100 times more chlorine is required to kill viruses than to kill most bacteria.

Another equation used to predict reduction in coliform organisms is as follows:[36]

$$\frac{N_t}{N_0} = (1 + 0.23 C_t t)^{-3} \qquad [6.19]$$

where: N_t = number of coliform organisms at time t
N_0 = number of coliform organisms at time t_0
C_t = total amperometric chlorine residual at time t, mg/L (includes both free and combined chlorine residuals)
t = residence time, min

For example, secondarily (biologically) treated effluent has a typical coliform count of 1×10^7 colonies per 100 mL. The suggested water quality for long-term landscape irrigation is 2.2 total coliform per 100 mL. Thus the time required to reduce the coliform organisms from 10 million to 2.2 per 100 mL would be as follows:

$$\frac{N_t}{N_0}^{-1/3} = 1 + 0.23 C_t t$$

$$t = \frac{\left(\frac{N_t}{N_0}\right)^{-1/3} - 1}{0.23 C_t}$$

$$t = \frac{\left(\frac{N_0}{N_t}\right)^{1/3} - 1}{0.23 C_t}$$

For a 0.5 mg/L total amperometric chlorine residual (TACR)

$$t = \frac{\left(\frac{10^7}{2.2}\right)^{1/3} - 1}{0.23 \,(0.5)}$$

$$= \frac{165.64 - 1}{0.23 \,(0.5)} = 1432 \text{ min} = 23.9 \text{ hr}$$

For 1.0 mg/L TACR

$$t = \frac{165.64 - 1}{0.23 \,(1.0)} = 716 \text{ min} = 11.9 \text{ hr}$$

For 3.0 mg/L TACR

$$t = \frac{165.64 - 1}{0.23 \,(3)} = 238 \text{ min} = 4 \text{ hr}$$

Based on this equation, it would take a contact time of 4.0 hr for a 3.0 mg/L TACR to reduce the coliform count from $10^7/100$ mL to 2.2/100 mL.

The killing of pathogens by chlorination depends not only on the type of chlorine residual present, contact time, and temperature, but also on the pH, degree of mixing, turbidity, and the presence of interfering substances.

The longer the contact time and the higher the chlorine residual the greater the pathogenic kill, however, in general, the higher the pH the lower the kill.

Generally the free available forms of chlorine (Cl_2, HOCl, OCl$^-$) are more effective disinfectants than the chloramines (NH_2Cl, $NHCl_2$, NCl_3). Also, it is generally agreed that bacteria are more easily killed than viruses and viruses are more easily killed than cysts, while most helminths' ova are not affected by chlorination.[19]

Following are suggested conditions to be followed for good chlorination results:[19]

- BOD should be low (less than 20 mg/L, preferably less than 10 mg/L).
- COD should be low (preferably less than 40 mg/L).
- Suspended solids concentration should be low (preferably less than 10 mg/L). This can be achieved by tertiary treatment employing rapid sand filtration.
- Ammonia content should be as low as possible (preferably 1 mg/L) achieved by efficient operation of secondary treatment processes (nitrification).
- pH value should be low, preferably between 6 and 7. At pH 8 or more, the efficiency of disinfection is greatly reduced.
- An efficient system for rapid mixing of the concentrated chlorine solution with effluent at the point of addition improves disinfection of effluents with high chlorine demand.
- Chlorine dose must be high (usually 10 to 20 mg/L chlorine) and the actual requirement must be assessed regularly.
- Contact time should be at least 1 hr.
- The chlorination process must be carefully controlled, with continuous adjustment of the chlorine dosage according to chlorine requirements and effluent flow.

If these conditions are observed, it is technically feasible to produce an effluent containing not more than 100 coliform organisms per 100 mL, an acceptable standard for many wastewater reuse activities.

B. OZONATION

Ozone is a very effective disinfectant and is even more effective than chlorine at destroying virus. It reacts readily with organic matter and, unlike chlorine, does not react with ammonia. Although ozonation is more expensive than chlorination its adoptions for wastewater disinfection is growing.[19] Studies using ozone for viral destruction have shown that ozone has a virtually 100% efficiency after a 5-min contact time. This result was observed when the dose was about 15 mg/L and the residual was 0.015 mg/L. Another study yielded similar results although the multistage ozone contact chamber had a total detention time of 18 min. The virus inactivation results are presented in Table 6.17. The variation in dose rate had little effect on the removals; however, the preapplication concentration did seem to influence the removal rate.[28]

Although ozone has strong germicidal properties there is some reluctance to accept it as a disinfectant since it leaves no significant residual. A solution to this problem is to use ozone in conjunction with chlorine. Ozone could serve as the primary disinfecting agent, with chlorine applied at a low dose to provide residual disinfection.[28]

VI. PUBLIC OPINION AND WASTEWATER REUSE

It has been stated that there are four publics: (1) the general public, popularly known as the man in the street; (2) the organized public, whose citizen activities are channeled through organizations; (3) the representative public, comprising elected and appointed officials; and (4) the economically concerned public, those individuals and institutions whose interests may be affected, adversely or favorably, by water quality policies and decisions.[37]

However, the U.S. Army Corps of Engineers has provided a much more detailed listing of the public as they relate to the water resources area.[7]

While this list is much more specific, each item on the list could still be included in one of the broader groups mentioned above.

Table 6.17 Virus Inactivation by Ozonation

System	No. of experiments	Ozone dose, (mg/L)	Median total coliform (MPN/100 mL)	Geometric mean virus concentration, (PFU/100 gal[d]) Predisinfection	Final effluent	Average virus inactivation efficiency Log	Percent
A[a]	9	10	2	7.2×10^4	88	2.9	99.874
B[b]	10	10	27	5.6×10^6	169	4.5	99.997
	5	50	<2	7.0×10^5	66	4.0	99.990
C[c]	6	6	700	1.4×10^6	100	4.1	99.992

[a] Preozonation treatment include alum flocculation-coagulation and filtration.
[b] Preozonation treatment with direct filtration.
[c] Preozonation treatment with activated carbon adsorption.
[d] PFU/100 gal = plaque-forming unit per 100 gallons (One PFU is due to a virus manifesting itself after incubation by forming a clear transparent area on a specially prepared Petri dish or flat-sided flask.)

From Culp/Wessner/Williams, *Wastewater Reuse and Recycling Technology,* Noyes Data Corp., Park Ridge, NJ, 1980, 332. With permission.

1. Individual citizens, including the general public and key individuals who do not express their preferences through, or participate in, any groups or organizations.
2. Sporting groups.
3. Conservation/environmental groups.
4. Farm organizations.
5. Property owners and users, representing those persons who will be or might be displaced or affected by any alternative under study.
6. Business and industrial groups, including Chambers of Commerce and selected trade and industrial associations.
7. Professional groups and organizations, such as the American Institute of Planners, American Society of Civil Engineers, and others.
8. Educational institutions, including universities, high schools, and vocational schools. General participation is by a few key faculty members and students or student groups and organizations.
9. Service clubs and civic organizations, including service clubs in a community such as Rotary Club, Lions Club, League of Women Voters, and others.
10. Labor unions.
11. State and local governmental agencies, including planning commissions, councils of government, and individual agencies.
12. State and local elected officials.
13. Federal agencies.
14. Other groups and organizations, possibly including various urban groups, economic opportunity groups, political clubs and associations, minority groups, religious groups and organizations, and many others.
15. Media, including the staff of newspapers, radio, television, and various trade media.

Each of these public groups obviously has a right to express itself on all issues, and none should be excluded from public participation. However, it is the organized public which has demonstrated the most interest in the water resources issues and is determined to establish mechanisms through which it can effectively be heard.[37]

Public opinion is important in planning, constructing, and operating wastewater reclamation and reuse facilities, as it is the public who must pay for them as well as accept the direct utilization of the treated wastewater. Generally the public gets most of its information on environmental issues through the mass media — newspapers, radio, and television. There are many techniques that can be used to communicate with the public and these include:[38]

- Brochures
- Information packets
- Newsletters (perhaps one that is published regularly)
- Videotapes or slide shows
- Advertisements

- Fact sheets
- Press releases
- Open house and plant tours
- Educational and informational workshops
- Community meetings
- Community advisory groups
- Service group presentations
- Educational activities with schools
- The news media (e.g., radio and television interviews)
- A telephone hotline established by the appropriate organization
- Electronic devices and computers

The U.S. EPA uses the term risk communication to indicate informing the public regarding environmental or ecological risks. Risk communication has been divided into three components:[4]

1. The environmental and health risks involved in a particular situation.
2. Specific decisions EPA has made regarding such situations.
3. Reasons for final decisions and planned course of action.

The risk management decisions are also a means by which EPA can get feedback from the public on issues concerning them.

Bruvold has carried out extensive studies as to public opinion regarding wastewater reuse. Not only has he carried out his own extensive evaluations in this area, but has also evaluated results from other studies regarding public opinion on various reclaimed water reuse options.[39]

Table 6.18 summarizes the results from seven studies which Bruvold appraised, ranging from using reclaimed water for irrigation to full domestic use.[39]

The data in Table 6.18 point out that the respondents in these studies were more favorably disposed to water reuse as the degree of human contact diminished. It would appear that this information could help wastewater reuse planners set priorities as to which type of reuse project would have the greatest chance for public acceptance.

However, Bruvold also carried out another extensive public opinion survey in California using "salient" effluent use options. Salient options were defined as those limited to specific uses of effluent actually proposed for the respondents' community in the near future. In this research, information was presented to each respondent that covered, in lay language, the type of treatment planned, the type of use envisioned, and the environmental, health, and economic impacts of each option. Results indicated that factors such as water conservation, environmental and public health protection, and reduced treatment and distribution costs were more important than the degree of human contact for the salient use options.[39,40]

The salient options survey included 10 communities with 140 responds included in each community.

Table 6.19 shows weighted mean percentage of respondents opposed to 27 uses of reclaimed water for both the 7 studies presented in Table 6.18 and the 5 salient reuse options. The data in Table 6.19 suggest little or no relationship between the degree of contact and opinions toward reuse for the five salient options.

Bruvold (1988) developed two hypotheses from the two sets of results. First, in the general reuse surveys, degree of human contact was the more important determinant of public opinion on effluent use. Second, for the salient reuse options, the five factors of conservation, environment, health, treatment cost, and distribution cost were the more important determinants of public opinion data. One of the most important implications of this research was that it contradicts the commonly held notion that the degree of human contact provides clear guidelines for evaluating public acceptance of effluent use options, particularly the more innovative options.[40]

Those contradictions point out that if a wastewater reclamation and reuse project is planned in a community, then public opinion surveys should be made to try and make certain that the project is a success. It is also evident that more research is needed in the salient reuse options before general guidelines can be adopted for wastewater reuse and reclamation undertakings.

VII. SUMMARY AND CONCLUSIONS

The risk assessment/analysis process is undergoing expansion and refinement. The four-step process for risk assessment of hazard identification, dose-response assessment, exposure assessment, and risk characterization has been broadened to now include background studies, data collection and evaluations, risk

Table 6.18 Percentage of Respondents Opposed to 27 Uses of Reclaimed Water in the General Options Surveys

	Bruvold (N = 972)	Stone and Kahle (N = 1,000)	Kasperson et al. (N = 400)	Olson et al. (N = 244)	Bruvold (N = 140)	Milliken and Lohman (N = 399)	Lohman and Milliken (N = 403)
Food preparation in restaurants	56%			57%			
Drinking water	56	46	44	54	58	63	67
Cooking in the home	55	38	42	52		55	55
Preparation of canned vegetables	54	37		52			
Bathing in the home	37	22		37		40	38
Pumping down special wells	23			40			
Home laundry	23		15	19		24	30
Swimming	24	20	15	25			
Commercial laundry	22	16		18			
Spreading on sandy areas	13			27			
Irrigation of dairy pastures	14			15			
Irrigation of vegetable crops	14		16	15	21	7	9
Vineyard irrigation	13			15			
Orchard irrigation	10			10			
Pleasure boating	7	14	13	5			
Hay or alfalfa irrigation	8	9		8			
Commercial air conditioning	7			9			
Golf course hazard lakes	3	8		5	8		
Electronic plant process water	5	5	3	12			
Home toilet flushing	4	5		7		3	4
Residential lawn irrigation	3	6		6	5	1	3
Irrigation of recreation parks	3			5	4		
Golf course irrigation	2	5	2	3	4		
Irrigation of freeway greenbelts	1			5			
Road construction	1			4			
Stream or river discharge							
Bay or ocean discharge							

Note: N = Number of Respondents.

From Bruvold, W. H., *Water Pollut. Control Fed.*, 60, 1, 1988, 45. With permission.

management, and risk communication. The main objective of risk assessment is to provide a means of presenting and evaluating scientific information that is available to decide whether a hazard exists and what the magnitude of that hazard may be, whether it is chemical, biological, or radioactive in nature.

Table 6.19 Weighted Mean Percent Opposed to Uses of Effluent

Type of reuse	Degree of contact	General options surveys	Salient options surveys
Food preparation in restaurants	Very high	56%	
Drinking water	Very high	54	64%
Cooking in the home	Very high	48	
Preparation of canned vegetables	Very high	46	
Bathing in the home	High	33	
Pumping down special wells	High	27	41
Home laundry	High	23	
Swimming	High	21	66
Commercial laundry	High	19	
Spreading on sandy areas	High	16	65
Irrigation of dairy pasture	Moderate	14	
Irrigation of vegetable crops	Moderate	13	40
Vineyard irrigation	Moderate	13	
Orchard irrigation	Moderate	10	66
Pleasure boating	Low	10	24
Hay or alfalfa irrigation	Low	8	40
Commercial air conditioning	Low	7	
Golf course hazard lakes	Low	6	
Electronic plant process water	Low	5	34
Home toilet flushing	Low	4	
Residential lawn irrigation	Low	4	
Irrigation of residential parks	Low	3	26
Golf course irrigation	Low	3	
Irrigation of freeway greenbelts	Low	2	
Road construction	Low	2	
Stream of river discharge	Low		50
Bay or ocean discharge	Low		71

Compiled from References 39 and 40.

Setting of standards for drinking water, air pollution, and many other environmental areas is now being evaluated, in part, by using risk assessment/analysis and risk management techniques.

To quantify exposure to various agents that may be harmful to human health, dose-response curves for potential noncarcinogens and carcinogens are used. For the noncarcinogens, a threshold is assumed to exist, while for carcinogens no threshold is presumed to be present. A reference dose (RfD) is used to calculate risk for noncarcinogens, while a slope factor or potency factor is used for carcinogens. The RfDs and slope factors are available through EPA. The EPA Integrated Risk Information System (IRIS) has at present a computerized list of this information for approximately 500 chemicals. To develop the RfDs and slope factors, data from epidemiological studies, animal test studies, pharmacokinetic studies, cell or microorganisms studies, and the agent's own chemical structure are used to evaluate RfDs and slope factors. Under low dose-response conditions numerous mathematical models are available to extrapolate from high dose exposures to low dose exposures. Mathematical models are available for both chemical and biological agents. The model recommended by EPA is the linearized multistage model as it is one of the most conservative models and overemphasizes risk.

While the dose-response information for numerous chemical and radioactive nuclides is fairly well developed, the same cannot be said for infectious agents causing disease. Only limited dose-response information is available for the pathogenic bacteria, viruses, protozoa, and helminths.

Considerable work has been done regarding public opinion and wastewater reuse. The first studies indicated that the public was favorably disposed to wastewater reuse as the degree of human contact diminished, however, further research using five salient reuse options which included conservation, environment, health, treatment cost, and distribution cost found these factors to be more important to the public than the degree of human contact. It is evident that more research is needed in this area before general guidelines can be promulgated.

VIII. HOMEWORK PROBLEMS

1. What are the four steps described by the National Academy of Science involved in doing a risk assessment/analysis?
2. Explain the difference between risk assessment, risk management, and risk communication.
3. Characterize and describe the following acronyms:

    LOEL
    NOEL
    LOAEL
    NOAEL
    RfD
    IRIS
    CDI

4. What online EPA electronic database can provide details regarding risk information on specific chemical substances for use in decision-making and regulatory activities?
5. Define the following terms:

    Toxicology
    Threshold Dose
    Hazard Quotient
    Slope Factor
    Risk

6. What are the five EPA weight-of-evidence categories developed for potential carcinogens?
7. How would EPA classify vinyl chloride as it relates to carcinogenicity in humans?
8. What distinguishes a carcinogenic dose-response curve from a noncarcinogenic dose-response curve?
9. What is the difference between an RfD and an ADI?
10. Why are risk assessment and risk management important for environmental protection programs?
11. What are the three steps involved in carrying out an exposure assessment?
12. What are the four elements involved in the exposure pathways for a contaminant or toxic agent to enter the environment?
13. Estimate the cancer risk for a 70-kg worker exposed to a particular potential carcinogen under the following conditions. Exposure time is 40 hr a week for a period of 20 years (3 week vacation per year). The breathing rate is 20 m^3/day. The slope factor for the carcinogen is 1.16 (mg/kg/day)$^{-1}$. The absorption factor is estimated at 100% and the average concentration in the air is 60 µg/m^3.
14. A drinking water supply has an average carbon tetrachloride concentration of 30 µg/L. Estimate the maximum lifetime risk for an adult weighing 70 kg and drinking 2 L/day of this water. If a city of 45,000 is using this water supply, what number of extra cancers per year could be expected?
15. It is estimated that 3.5 kg/day of vinyl chloride is being released from a single stack with an average effective height of 60 m. On an overcast day (D class stability), the wind speed is 5 m/s. What is the distance to the maximum ground-level concentration downwind? What would the vinyl chloride concentration be at this point? Also, what would the cancer risk be for a person living at this point with an overall exposure time of 30%. Assume absorption of vinyl chloride to be 60%.
16. Bromoform, one of the members in the trihalomethane group (THMs), was detected in drinking water at a concentration of 4 µg/L. Estimate the maximum lifetime risk for an adult weighing 70 kg and drinking 2 L/day of this water.
17. Estimate the likely kill of *E. coli* in a treated wastewater effluent having an initial concentration of 1 × 10^6 coliform counts per 100 mL and a total residual chlorine concentration of 3.0 mg/L and a contact time of 30 min.

REFERENCES

1. EPA, Assessing Multiple Pollutant Multiple Source Cancer Risks From Urban Air Toxics, EPA-450/2–89–010, Office of Air Quality Planning and Standards, Appendix A, Research Triangle Park, NC, 1989.
2. **Conway, R. A., Ed.,** *Environmental Risk Analysis For Chemicals,* Van Nostrand Reinhold, New York, 1982, 4.
3. USEPA, Air Pollution Control Orientation Course Update, S1:422, Air Pollution Training Institute, U.S. Environmental Protection Agency, Research Triangle Park, NC, 1990.

4. USEPA, Environmental Decision-Making Today, An Interview With Lee, M., Thomas, EPA Journal, Office of Public Affairs, U.S. Environmental Protection Agency, Washington, D.C., 13, 9, 1987.
5. **North, W. and T. F. Yosie,** What it is: How it Works, EPA Journal, Office of Public Affairs, U.S. Environmental Protection Agency, Washington, D.C., 13, 9, 1987, 13.
6. USEPA, Risk Assessment Guidance For Superfund. Vol. 1. Human Health Evaluation Manual (Part A) Interim Final, Office of Emergency and Remedial Response, EPA/540/1–89/002, U.S. Environmental Protection Agency, Washington, D.C., December 1989.
7. **Canter, L. W. and R. C. Knox,** *Ground Water Pollution Control,* Lewis Publishers, Chelsea, MI, 1985, 308.
8. **Masters, G. M.,** *Introduction to Environmental Engineering and Science,* Prentice-Hall, Englewood Cliffs, NJ, 1991, 211.
9. Environ, *Elements of Toxicology and Chemical Risk Assessment,* Environ Corporation, Washington, D.C., 1988.
10. Culp/Wesner/Culp, *Handbook of Public Water Systems,* Van Nostrand Reinhold, New York, 1986, 109, 402.
11. **Davis, M. L. and D. A. Cornwell,** *Introduction to Environmental Engineering,* 2nd ed., McGraw Hill, New York, 1991.
12. USEPA, Health Effects Assessment Summary Tables, Annual FY-1991, Office of Research and Development, OERR 9200.6–303(91–1) NTIS No. PB 91–92 1199, U.S. Environmental Protection Agency, Washington, D.C., 1991.
13. Anon., IRIS User Support, Computer Science Corporation (MS-190), 26 W. Martin Luther King Drive, Cincinnati, Ohio, 1992.
14. TIRC, Toxicology Information Response Center, Oak Ridge National Laboratory, P.O. Box 2008, Oak Ridge, TN, 1992.
15. **Turner, D. B.,** *Workbook, of Atmospheric Dispersion Estimates,* U.S. Dept. of Health Education and Welfare, National Air Pollution Control Administration, Cincinnati, Ohio, 1969.
16. **Peavy, H. S., D. R. Rowe, and G. Tchobanoglous,** *Environmental Engineering,* 1st ed., McGraw-Hill, New York, 1985.
17. **Pelscar, Reid, and Chan,** *Microbiology,* 4th ed., McGraw Hill, New York, 1977, 7, 12.
18. **Engelbrecht, R. S., et al.,** *Virus Sensitivity to Chlorine Disinfection of Water Supplies,* EPA-600/2–78–123 Office of Research And Development, U.S. Environmental Protection Agency, Cincinnati, Ohio, 45268, 1978, 4.
19. **Schaefer, W. J.,** Health Aspects of Reuse of Treated Wastewater For Irrigation, World Health Organization, Intercountry Seminar on Wastewater Reuse, Manama, Bahrain, October 1984.
20. **Sorber, C. A.,** Public Health Aspects of Agricultural Reuse Applications of Wastewater, *Municipal Wastewater Reuse News,* AWWA Research Foundation, American Water Works Association Denver, CO, 57, 1982, p. 5, 10.
21. **Shahalam, A. B. M.,** Wastewater Effluent vs. Safety in its Reuse: State-of-the-Art, *J. Environ. Sci.,* Sept/Oct. 1989, 35.
22. **Hutzler, N. J. and W. C. Boyle,** Risk Assessment in Water Reuse, *Water Reuse,* E. Joe Middlebrooks, Ed., Ann Arbor Science, Ann Arbor, MI, 1982, 293.
23. **Rowe, D. R., K. Al-Dhowalia, and A. Whitehead,** *Reuse of Riyadh Treated Wastewater,* Project No. 18/1402, King Saud University, College of Engineering Research Center, Riyadh, Saudi Arabia, 1988.
24. WHO, The Engelberg Report, Health Aspects of Wastewater and Excreta Use in Agriculture And Aquaculture, Sponsored by The World Bank and World Health Organization, Geneva, June 1985, 2, 11.
25. **Shuval, H. I.,** *Water Renovation And Reuse,* Academic Press, New York, 1977, 6, 46.
26. **Searle, D.,** Health Fears Crop Irrigation Promise, *Construction Today,* December 1986, 41.
27. **Krenkel, P. A. and V., Novotny,** *Water Quality Management,* Academic Press, New York, 1980, 127.
28. Culp/Wesner/Williams, *Wastewater Reuse and Recycling Technology,* Noyes Data Corporation, Park Ridge, NJ, 1980, 332.
29. **Cothern, C. R., W. A. Coniglio, and W. L. Marcus,** Estimating Risk to Human Health, *Environ. Sci. Technol.,* 20, 2, 1986, 111.
30. **Asano, T., and R. H. Sakaji,** Virus Risk Analysis in Wastewater Reclamation and Reuse, *Chemical Water and Wastewater Treatment,* Springer-Verlag, New York, 1990.
31. **Haas, C. N.,** Estimation of Risk Due to Low Doses of Microorganisms: A Comparison of Alternative Methodologies, *Am. J. Epidemiol.,* 118, 4, 1983.
32. **Haas, C. N.,** Effects of Effluent Disinfection on Risks of Viral Disease Transmission Via Recreational Water Exposure, *J. Water Poll. Control Fed.,* 55, 8, 1983, 1111.
33. **Asano, T., et al.,** Evaluation of the California Wastewater Reclamation Criteria Using Enteric Virus Monitoring Data, *Water Science Technology,* Great Britain, 26, 7–8, 1992, 1513–1524.
34. **Asano, T., and G. Tchobanoglus,** Evolution of Tertiary Treatment Requirements in California, *Water Environment and Technology,* 4, 2, 1992, 36–41.
35. **Feigin, A., I. Ravina, and J. Shalhevet,** *Irrigation With Treated Sewage Effluent,* Springer-Verlag, New York, 1991, 32.
36. **Berg, G.,** The Virus Hazard in Water Supplies, *J. N. Engl. Water Works Assoc.,* 78, 79, 1964.
37. **Reilly, William K.,** *Toward Clean Water: A Guide to Citizen Action,* Introduction, The Conservation Foundation, Washington, D.C., August, 1976, 27.
38. **Santos, S. L.,** Developing A Risk Communication Strategy, *Am. Water Works Assoc. J,* 10, 1990, 45.
39. **Bruvold, W. H.,** Public Opinion on Water Reuse Options, *Water Pollut. Control Fed.,* 60, 1, 1988, 45.
40. **Lieuwen, A.,** *Effluent Use in the Phoenix and Tucson Metropolitan Areas,* Water Resources Research Center, University of Arizona, Phoenix, June, 1990, p. 32.

Chapter 7

Economics of Water and Wastewater Reclamation and Reuse Projects

CONTENTS

Forward 300
   I. Introduction 300
  II. Estimating Water Supply Systems Costs 302
     A. Estimating Pipe Installation Costs 303
     B. Estimating Treatment Costs 304
     C. Estimating Pump System Costs 304
     D. Estimating Easement Costs 306
     E. Estimating Storage Costs 307
     F. Estimated Overall Costs 307
 III. Estimating Reclaimed Wastewater and Reuse Systems Costs 308
     A. Estimating Wastewater Treatment Unit Costs 310
     B. Estimating Capital Costs For Force Mains 310
     C. Estimating Pumping Station Costs 311
        1. Estimating Capital Costs For Pumping Stations 311
        2. Estimating Operating Costs For Pumping Stations 312
     D. Estimating Distribution System Costs 312
     E. Reclaimed Wastewater Storage Costs 313
     F. Estimated Overall Reclaimed Wastewater Costs 314
 IV. Selection Criteria and Evaluation of the System 314
     A. Benefit-Cost Ratio (BCR) 315
     B. Positive Net-Present Value Criterion 316
     C. Cost Effectiveness Analysis 316
     D. Impact Analysis 316
  V. Finance and Cost Comparisons 317
     A. Financial and Economic Analysis Relationship 317
     B. Financing of Water and Reclaimed Wastewater Projects 317
     C. Cost Allocation Among Different Users 317
     D. Cost Comparison Over Time 318
 VI. Wastewater as an Economic Resource 320
VII. Homework Problems 320
     A. Discussion Problems 320
     B. Numerical Problems 321
References 322

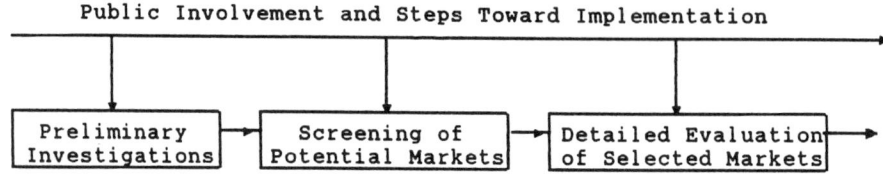

Figure 7.1 Planning phases for the reuse program.[2]

FORWARD

Engineering and economical factors are major variables that govern the selection of water supply and wastewater reclamation and reuse projects. This chapter deals with water and wastewater reclamation cost relationships. Also, this chapter presents procedures for the evaluation of water and reclaimed wastewater project alternatives. The information is presented in such a way as to give guidance and direction in evaluating the engineering and economical factors involved in water supply and wastewater reclamation and reuse projects.

I. INTRODUCTION

The planning process for a reclaimed wastewater reuse program requires three essential steps: preparation of preliminary investigations, screening of potential markets, and detailed evaluation of selected markets (see Figure 7.1).[1,2] Through all these stages, public involvement can provide guidance in the planning process, and public support is definitely needed if implementation of a reuse project proves feasible.[2] The preliminary phase investigations are intended to address information and data acquisition, project objectives, preliminary market assessment for potential users, reuse alternatives, public health factors, legal laws and regulations, economic considerations, environmental and social analysis, possible constraints for the reclaimed wastewater reuse scheme, and preparation of a report which includes an implementation plan. The screening phase for potential markets tries to formulate an economical comparison for the cost of freshwater compared to that of reclaimed wastewater. Users are categorized based on parameters such as the quantity of reclaimed water to be used, timing schedules, economic costs, water quality, and water quality demands. Governing factors relating to reclaimed wastewater reuse include:[1,2]

- Reclaimed wastewater quality
- Reclaimed wastewater quantity
- Reclaimed wastewater ability to meet fluctuations in demand
- Current and future rates and fees
- Aspects of reuse legislation
- Attitudes of agencies, organizations, and users toward reuse

In the final step of the screening phase a detailed evaluation is made in selecting the reclaimed wastewater users for the various alternatives being considered. In this phase a comprehensive analysis and evaluation of the reclaimed wastewater reuse project is conducted. During this phase funding options are compared, user costs evaluated, and a comparison made between unit costs of freshwater and that of reclaimed wastewater for each selected alternative. A detailed evaluation of the environmental, institutional, and social analysis as they relate to recommended alternatives is considered.[1,2]

The formulation of alternatives usually offers opportunities to take advantage of innovative processes rather than the selection of a traditional design. Alternatives must be feasible, that is, they must take into account what is environmentally, financially, economically, legally, technically, socially, politically, and institutionally feasible. The evaluation concept is illustrated in the Venn (or Euler) diagram shown in Figure 7.2, which assumes eight constraints. In Figure 7.2, alternatives that lie outside the intersection of these set of alternatives are not feasible and thus are not considered. The most feasible alternative is then chosen. All alternatives should be tested for feasibility periodically during the planning process, as additional data may shift them into or out of the region of feasibility.[3]

During planning of water and wastewater reclamation projects two types of evaluations need to be considered, namely subjective evaluation, and post-analysis evaluation (sometimes termed post-audit or ex-post evaluation). In the former case the planner compares the relative values of alternatives in order to select the plan. The latter case is done after a plan has been implemented, and predicted results are compared with actual results to determine how well the plan worked.[3]

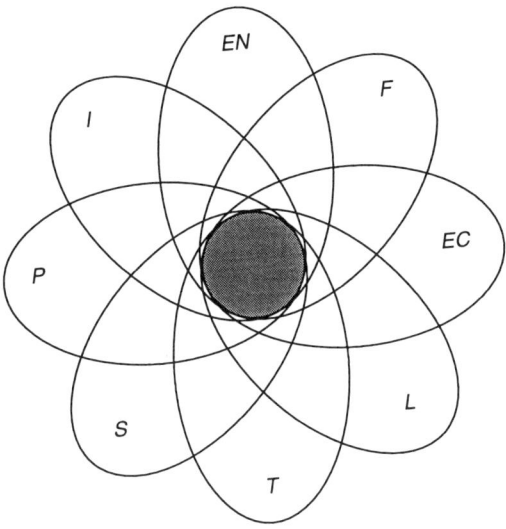

U = the universe of all possible alternatives
EN = the set of environmentally feasible alternatives
F = the set of financially feasible alternatives
EC = the set of economically feasible alternatives
L = the set of legally feasible alternatives
T = the set of technically feasible alternatives
S = the set of socially feasible alternatives
P = the set of politically feasible alternatives
I = the set of institutionally feasible alternatives
///// = $EN \cap F \cap EC \cap L \cap T \cap S \cap P \cap I$ = the set of feasible alternatives

Figure 7.2 Set of feasible alternatives. (From Helweg, O. J., *Water Resources Planning and Management*, John Wiley & Sons, New York, 1985. With permission.)

Water and wastewater reclamation projects inherently name a specific purpose which gives a similar benefit for all alternatives. A comparison in such cases may be reduced to the selection of the least expensive alternative.[4] In other cases comparison based on cost alone cannot be used due to some special overriding circumstances, these circumstances are called irreducibles or intangibles.[5] Evaluation of a good quality reclaimed wastewater depends basically on its convenience and on its cost in comparison to the cost of a freshwater supply. To make meaningful comparisons the components in each system must be clearly delineated and precise cost components presented.

Following is an outline of a three-step procedure that can be used in preparing estimates for cost comparisons for water and wastewater reclamation and reuse projects.[7]

1. *Order of magnitude estimates.* These estimates are used in the planning and initial evaluation stage of a project to screen alternatives. They typically provide accuracy in the range of ±30 to ±50% and are developed through semiformal means such as conferences, questionnaires, and generalized equations.
2. *Semidetailed or budget estimates.* These estimates are used in the design stages of a project. Their accuracy usually lies in the range of ±15%. These estimates are used to screen the different alternatives.
3. *Definite estimates.* These estimates are used in the detailed engineering/construction phase of a project. They are used as the basis for bids and their accuracy is about ±5%. Detailed estimates are gathered from specifications, drawings, site surveys, vendor quotations, federal agencies, professional journals, private organizations, and "in-house" historical records. These estimates often are prepared with the help of computer software packages.

Table 7.1 outlines a list of overhead and administrative cost estimates including direct costs, indirect costs, and other costs.[8]

Table 7.1 List of Cost Estimates for Water and Reclamation Wastewater Projects

Direct Costs

Labor
   Engineering, manufacturing, tooling, quality and reliability assurance, testing, planning, tool design, tool maintenance, packaging
Materials and subcontracts
   Raw materials, partially finished materials (forgings), (parts, sheet stock, fuels, and lubricants), tool materials, equipment and supplies
Other direct costs
   Travel costs, shipping, transportation, computer services, reproduction services, training costs

Indirect Costs

Labor burden
   Bonuses, health insurance, paid holidays, paid vacations, social security, supervisor's salary, pensions
Material burden
   Handling, inventory control, purchasing costs, storage
Overhead
   Amortization, claims, communications, custodial, depreciation, heating and cooling, industrial relations, insurance, lighting, maintenance, operating supplies, power, rental of buildings, waste disposal, and water

Other Costs

General and administrative (G&A) expenses:
   Administration, advanced design, bid and proposal costs, independent research and development, advertising, corporate expenses, executive salaries, finance, marketing, personnel department, research, training department, corporate taxes
Furnished equipment
   Fuels and lubricants, subassemblies, facilities, equipment, spare parts, experiments, selected services
Other
   Warrantee costs, hazardous pay, license fees
Fee (Profit or earnings)
   Return to stockholders, reinvestment, capital equipment and tooling

From Stewart, R. D. and Wyskida, R. M., *Cost Estimator's Reference Manual,* John Wiley & Sons, New York, 1987. With permission.

II. ESTIMATING WATER SUPPLY SYSTEMS COSTS

In evaluating a freshwater supply system, current and projected costs need to be considered. Current costs include both amortized capital and operational costs. Capital costs include both depreciation and interest, while operational costs include labor, maintenance, and materials.[1] Unit costs are generally higher for smaller projects than for larger ones.

A major contributor to the effective life span of a project is a well-developed and implemented maintenance program that keeps equipment running and buildings in a sound condition. Without proper maintenance, the usable life of any piece of equipment is much shorter than its design life, sometimes by as much as 30%.[9]

In water supply systems, revenues also depend on population growth and per capita demand; faulty forecasts can create a real problem for financial managers and planners.[10]

Estimating operational, maintenance, and replacement costs requires experience but is more sound than population forecasting. Major areas included in preparing cost estimates for a potable water supply project are[1,11,12]

- Acquisition: which incorporates the freshwater supply source, storage basin, and pumping stations.
- Treatment: which includes water treatment works. Many factors influence the construction costs of water treatment plants such as plant capacity, design criteria, treatment unit processes and operations, site conditions and land cost, climate, permit costs, competition among bidders and suppliers, and general local and national economic conditions.
- Distribution: which includes multiple storage tanks, distribution systems, and pumping stations.
- Other costs: which include administration and customer services.
- Operation and maintenance and replacement costs.
- Contingencies: which include differences between estimated prices and the actual prices, omissions of work not included in the estimate, unforeseen difficulties at the site, and changes in plans and other uncertainties.

- Willingness and ability to pay: which requires an evaluation of the people's willingness to pay for a water supply or wastewater treatment system. This can be evaluated by interviews, as well as a pricing approach that identifies what portion of a person's income they are willing to spend on their water supply or wastewater treatment operations.

Water and wastewater treatment costs can also be estimated by interviewing municipal and administrative personnel responsible for such systems, or through available planning and annual reports or an analysis of water and wastewater treatment rates and fees.[13–15]

The willingness-to-pay approach is very difficult to apply and is not widely used. It is particularly susceptible to consumer misunderstandings. Another difficulty is the practical problem of phrasing questions for the survey.[14]

Water supply and wastewater treatment projects are expensive, especially for small projects. The smaller the project, the higher the unit costs. Also, small systems are prone to problems related to:[1]

- Primitive existing treatment units.
- Inadequate maintenance of treatment units and equipment.
- Improper maintenance of the distribution systems.
- Inadequate funding.

Items that need to be considered in preparing a cost estimate for a water supply system include:[16]

- Cost of developing the water source.
- Cost of pumping (housing, and equipment).
- Cost of treatment works.
- Cost of trunk mains.
- Cost of storage facilities.
- Cost of distribution systems.
- Cost of other related parameters (such as overhead and administration work).
- Cost of operation and maintenance and replacement [O&M&R].

These estimated costs are generally based on past costs for similar operations or work. Current costs can be estimated by using appropriate construction or municipal cost indices.[17,18] Experience has shown that although water is sold to users by public utilities, the prices charged by the utilities do not always promote efficiency. In general, prices are low and the rate structures do not adequately reflect the costs of providing service to various types of customers. The result is excessive water use with little attention being paid to conservation.[19]

Capital or overhead costs for a water supply system can be broken down into various components; these components are presented in the following sections.

A. ESTIMATING PIPE INSTALLATION COSTS

Estimating pipeline costs begins with the selection of the pipe material. This involves a balance between cost (including installation) and life expectancy, with consideration being given to pressure requirements and safety factors for a particular situation.[20] This cost can be estimated by using a cost function (see Equation 7.1).[4]

$$CST_p = a \cdot D^b \cdot l \qquad [7.1]$$

where: CST_p = construction cost of water supply pipes, monetary units
D = pipe diameter, m
l = length of pipe, m
a, b = cost coefficients

Construction costs do not include special costs for site excavation work, equipment, general contractors' overheads or profit, engineering, land or legal costs, or fiscal or administrative and interest costs during construction. These costs are most often added following summation of the cost of the individual unit processes.[11,18]

Another equation used to estimate pipe installation costs is as follows:[3]

$$CST_p = 5.11 \cdot l \cdot D^{1.29} \qquad [7.2]$$

where: CST_p = installed pipe cost, monetary units
         l = length of the pipe, m
         D = diameter of the pipe, m (pipe diameter may be determined by using mathematical formulas derived from the continuity equation or from nomographs or tables)

Generally, the cost per pipeline capacity decreases markedly as the pipe size increases. Operational costs for water supply pipe systems may be estimated using Equation 7.3:[4]

$$CST_{op} = a \cdot CST_p \qquad [7.3]$$

where: CST_{op} = operation cost, monetary units
         a = cost coefficient
         CST_p = construction pipe cost, monetary units

Example 7.1

Determine the pipe installation costs, and the operational costs for a water supply system given the following design data:

Parameter	Value
Length of pipe (km)	1
Flow rate (m³/min)	60
Maximum velocity in pipes (m/min)	108
Cost coefficient	1.08

Solution

1. Given: Q = 60/60 = 1 m³/s, l = 1000 m,
   v = 108/60 = 1.8 m/s, a = 1.08.

2. Find the capital cost by using Equation 7.2 to determine the installed pipe cost as: $CST_p = 5.11 \cdot l \cdot D^{1.29}$
   Determine pipe diameter by using the continuity equation as:

   $Q = A \cdot V$ or, $D = [(4 \cdot Q)/(\pi \cdot v)^{0.5}]$
   $= (4 \cdot 1/\pi \cdot 1.8)^{0.5} = 0.84$ m.

   Find the pipe installation cost:

   $CST_p = 5.11 \cdot 1000 \cdot (0.84)^{1.29} = \$4081.00$

3. Determine the O&M cost by using Equation 7.3:

   $CST_{op} = a \cdot CST_p = 1.08 \cdot 4081.00 = \$4407.00$

B. ESTIMATING TREATMENT COSTS

The cost of water treatment processes, or a combination thereof, depends upon many factors such as construction costs which include earth work, concrete, steel, labor, pipes, valves, electrical installations, instrumentation, equipment, and housing. Operational and maintenance costs include items such as energy, maintenance material, chemicals, and labor. Costs can be determined from especially prepared cost curves for each unit process.[18] These curves have been developed to cover a wide range of flows. Two such ranges vary from 9.46 to 3785 m³/d (2500 gpd to 1 mgd) and from 3785 to 757,000 m³/d (1 to 200 mgd).[18] These ranges were developed as processes applicable in one range may not be applicable in another range. Costs involved in large flows often result in dramatic changes in the slope of the cost curves, commonly these dramatic changes are in the 3785 to 18,925 m³/d (1 to 5 mgd) range. Local variations in units costs can easily be incorporated into the cost calculations. Adjustment of cost curves may be necessary to reflect site-specific conditions, geographic or local conditions, as well as for standby power.[18]

C. ESTIMATING PUMP SYSTEMS COSTS

This cost is a function of the required energy. In the case of electric power the cost is a function of the amount of electricity used. This, in turn, is a function of the total head and discharge. The total head consists of the static and friction heads (see Equation 7.4).

$$H_T = H_S + H_f \qquad [7.4]$$

where: H_T = total head, m
H_S = static head, m
H_f = friction head, m

The friction head can be determined from Chezy-Manning's formula (see Equation 7.5).

$$H_f = \left(v^2 \cdot n^2 \cdot 1\right) / \left([R_h]^{4/3}\right) \qquad [7.5]$$

where: v = velocity of flow, m/s
n = Manning's roughness coefficient, s/m$^{1/3}$
l = length of pipe, m
R_h = hydraulic radius, m

The hydraulic radius (R_h) may be calculated from Equation 7.6:

$$R_h = A/w_p \qquad [7.6]$$

where A = Area, m^2, and w_p = wetted perimeter, m.
The required power can be computed by using Equation 7.7.

$$PR = (Q \cdot H_T \cdot \rho)/Eff \qquad [7.7]$$

where: PR = required power, kW
Q = flow rate or discharge, m^3/s
H_T = total head, m
ρ = specific weight of water = 9.79 kN/m^3 at 20°C
Eff = total efficiency of pump and motor

The pump system cost may be found by using Equation 7.8:

$$CST_{pu} = CST_i \cdot PR \qquad [7.8]$$

where: CST_{pu} = pump system cost, monetary units
CST_i = unit cost corresponding to installed power, monetary units/kW (may be found from agencies or from especially prepared cost curves or tables[18])
PR = required power, kW

The operation and maintenance energy costs can be estimated by using Equation 7.9:[3]

$$CST_E = a \cdot cst_E \cdot PR \cdot t \cdot 365 \qquad [7.9]$$

where: CST_E = annual energy cost, monetary units
a = cost coefficient (often taken as 1.08)[3]
cst_E = power price/kW.hr
PR = power, kW
t = daily hours of operation, hr/d
  = {required volume of water (m^3)}/{pumping rate (m^3/hr)}
  = V/Q $\qquad [7.10]$
365 = number of days in a year

Total O&M costs are based on energy requirements, maintenance, material specifications, and labor demands. The energy category includes operational energy requirements and building energy requirements. All energy components (such as diesel fuel, electricity, natural gas, nuclear energy, solar energy, etc.) are expressed in terms of kilowatt-hours (kW.hr) per year.[11]

Example 7.2

Estimate the required pump system costs (capital and operational) for the water supply system presented in Example 7.1 given the following additional information:

Parameter	Value
Static head (m)	30
Daily operating pumping hours	14
Pump system capital cost ($/kW)	800
Efficiency of pump and motor (%)	80
Cost of electricity ($/kW.hr)	0.06
Cost coefficient	1.08
Manning's roughness coefficient	0.013

Solution

1. Given: $H_S = 30$ m, t = 14 hr/d, $CST_i = \$800.00/kW$,

   a = 1.08, Eff = 0.8, $cst_E = 0.06$, n = 0.013.

2. For a pipe flowing full, the hydraulic radius can be determined using Equation 7.6:

   $R_h = A/w_p = (\pi D^2/4)/(\pi D) = D/4 = 0.84/4 = 0.21$ m

   The friction head may be determined from Chezy-Manning's formula as presented in Equation 7.5:

   $H_f = (v^2 \cdot n^2 \cdot l)/([R_h]^{4/3})$
   $= (1.8^2 \cdot 0.013^2 \cdot 1000)/([0.21]^{4/3}) = 4.39$ m

   Use Equation 7.4 find the total head loss:

   $H_T = H_S + H_f = 30 + 4.39 = 34.39$ m

   Find the required power by using Equation 7.7:

   $PR = Q \cdot H_T \cdot \rho/Eff = 1 \cdot 34.39 \cdot 9.79/0.8 = 421$ kW

   Determine the pump system cost from Equation 7.8:

   $CST_{pu} = CST_i \cdot PR = \$800.00 \cdot 421 = \$336,800.00$

3. Compute the O&M energy costs by using Equation 7.9:

   $CST_E = 1.08 \cdot cst_E \cdot PR \cdot t \cdot 365$
   $= 1.08 \cdot 0.06 \cdot 421 \cdot 14 \cdot 365 = \$139,405.00$

D. ESTIMATING EASEMENT COSTS

The easement costs can be calculated by using Equation 7.11:[3]

$$CST_{Es} = (l \cdot B \cdot CST_{la})/A \qquad [7.11]$$

where: CST_{Es} = easement cost, monetary units
l = length of pipe, m
B = easement width, m
CST_{la} = cost of land/hectare, monetary units/ha
A = area, m^2

Example 7.3

Estimate the easement costs for a water supply system with the following design data:

Item	Value
Length of pipe (km)	3
Easement width (m)	1.5
Cost of land ($/ha)	100,000

Solution

1. Given: l = 3000 m, B = 1.5 m, CST_{la} = $100,000/ha.
2. Find the easement cost, for an area of 1 ha [10,000 m²], by using Equation 7.11:
$CST_{Es} = (l \cdot B \cdot CST_{la})/A = 3,000 \cdot 1.5 \cdot 100,000/10,000 = \$45,000.00$

E. ESTIMATING STORAGE COSTS

The cost of water supply storage reservoirs may be estimated from Equation 7.12:[4]

$$CST_{res} = a_s \cdot V^b \quad [7.12]$$

where: CST_{res} = construction cost of the water supply storage reservoir, monetary units
a_s, b = cost coefficients, b≤1
V = reservoir volume, m³

Likewise, the operational costs of the water supply storage reservoir can be estimated from Equation 7.13:

$$CST_{ores} = a_{s1} \cdot V = a_{s2} \cdot CST_{res} \quad [7.13]$$

where: CST_{ores} = operation cost of water supply service reservoir, monetary units
a_{s1}, a_{s2} = cost coefficients
V = reservoir volume, m³

In addition to using the above equations to determine costs, tables are also available to estimate costs. Tables are particularly suitable for pipe cost; these tables separate the various pipes available by size and material. Tables 7.2 and 7.3 give further examples of key parameters that must be considered in estimating construction and operation and maintenance costs for various facilities of reclaimed wastewater and water supply systems.[4]

F. ESTIMATED OVERALL COSTS

The overall costs for the economic life of a project can be determined by the summation of various cost components. The total capital cost for a project can be calculated by using Equation 7.14:

$$CST_T = CST_p + CST_E + CST_{Es} + CST_{res} + CST_0 + CST_u \quad [7.14]$$

where: CST_T = total cost, monetary units
CST_p = installed pipe costs, monetary units
CST_E = energy costs, monetary units
CST_{Es} = easement costs, monetary units
CST_{res} = storage reservoir costs, monetary units
CST_0 = overall operation and maintenance costs, monetary units
CST_u = unforeseen costs (these costs are often added as a percentage of the total cost in order to take care of miscellaneous items[18])

Knowing the economic life of a project, the annualized cost of the project can be calculated.[3,21] For example, by multiplying the total capital cost by a capital recovery factor {commonly expressed as: $(CST_a/CST_T, ir, n)$} the annualized capital cost can be determined[3] (see Equation 7.15):

$$CST_a = \left(CST_T \cdot \left\{[ir/100] \cdot (1+[ir/100])^n\right\}\right) \Big/ \left((1+[ir/100])^n - 1\right) \quad [7.15]$$

where: CST_a = annualized capital cost, monetary units
CST_T = total present capital cost, monetary units (the present value is influenced by interest rates, inflation rates for investment and for the operational and planning periods, and for the anticipated lifetime of the system[4])
ir = discount rate, %
n = economic life of the project, years

Table 7.2 Examples of Parameters for Estimating Construction Costs of Water Supply and Reclaimed Wastewater Systems

Item	Main factors influencing cost	Key parameters of cost functions	
		Parameter	Symbol
Water pipes	Diameter, material, class of pipe	Diameter of pipe, length	$f(D,l)$
Ground storage reservoir	Storage capacity, construction materials, shape and structure of reservoir, soil conditions	Storage capacity	$f(V)$
Elevated storage reservoir	Storage capacity, height, construction materials, shape and structure, wind and earthquake loadings, soil conditions	Storage capacity and height of reservoir above ground level	$f(V,H)$
Pumping station	Pump capacity, pump head, no. and type of pumps used, construction material for station, class and material for pressure pipe	Pump capacity and pumping head	$f(Q,H)$
Water treatment	Plant capacity, type of process and treatment facilities, construction materials, topography of plant site, soil conditions, raw water intake	Plant capacity	$f(Q)$
Sewers	Diameter of sewer, depth of sewer, materials, shape of trench, type of soil, water table level, static and dynamic loading on sewer	Diameter of sewer, mean sewer invert depth (h_m), length	$f(D,h_m,l)$
Wastewater treatment	Type of treatment, construction material, area, plant capacity, influent BOD	Area, population equivalent, BOD	$f(A,PE)$ $f(BOD)$

From Orth, H. M., *Model-Based Design of Water Distribution and Sewage Systems,* John Wiley & Sons, New York, 1985. With permission.

Table 7.3 Examples of Parameters for Estimating Operation and Maintenance Costs of Water Supply and Reclaimed Wastewater Systems

Item	Main factors influencing cost	Key parameters of cost functions	
		Parameter	Symbol
Water pipes	Total length of pipes, material and quality of construction, topography of area, pressure in pipes	Total length of pipes, or percentage of construction cost	$f(l)$ or $f(CST)$
Sewers	Total length of pipes, materials and quality of construction, topography of area	Total length of sewers, or percentage of construction cost	$f(l)$ or $f(CST)$
Storage tank	Quality of construction, size of structure	Percentage of construction cost	$f(CST)$
Water treatment	Plant capacity, type of process and facilities, quality of raw water	Plant capacity	$f(Q)$
Wastewater treatment	Plant capacity, designed efficiency of plant type of process and facilities, quality of construction	Population equivalent, or BOD of influent wastewater	$f(PE)$ or $f(BOD)$
Pumping station	Pump capacity, pumping head, no. and type of pumps, pump efficiency, quality of construction, energy cost	Percentage of construction cost and pumping head (for Q = constant)	$f(CST, H)$

From Orth, H. M., *Model-Based Design of Water Distribution and Sewage Systems,* John Wiley & Sons, New York, 1985. With permission.

Different interest rates are used for economic analysis in private enterprise projects and in public works projects. These differences in economic analysis for private and public projects are accentuated by the existence of general property taxes and income taxes.[5] The total annual cost for a project is the sum of the annualized capital cost and the annual operating and maintenance costs.

III. ESTIMATING RECLAIMED WASTEWATER AND REUSE SYSTEMS COSTS

The estimated costs for reclaimed wastewater and reuse projects can be divided into two categories, capital and operations-and-maintenance (O&M) costs, and include the following:

- Costs of additional treatment required to meet reuse standards.
- Costs of conveyance and distribution of the treated wastewater.
- Costs of any storage, if needed.
- Costs of monitoring reclaimed wastewater quality.
- Costs of on-site hookup and use.
- Costs of operation and maintenance and replacement.
- Costs of customer billing and administration.
- Costs of salaries, overhead, and utility services.

Indirect costs for reuse projects depend upon the type of project and its effects on the environment. Examples of such indirect costs include damage to the environment such as to aquatic life, or the reduction of flow in a stream or river due to the project.

Other costs that have to be paid for by the users include:

- Costs of facilities needed for water quality monitoring.
- Costs of additional treatment that may be required by the users.
- Costs of re-piping for implementation of dual systems.
- Costs for insurance for workers and employees as they relate to accidents, disasters, and emergencies.
- Costs to try and change normal operations such as restricting access to irrigated lands, use of large volumes of water to control soil salinity, or compensation to an industry due to poor quality cooling water.
- Costs for purchase of equipment and devices related to the wastewater reuse project.

To promote the use of reclaimed wastewater the price must be reasonable and competitive. Some of the economic factors that are difficult to quantify in the preliminary review of a reclaimed wastewater project includes the following:[1]

- Costs for additional wastewater treatment to meet local disposal standards if the wastewater is not to be reclaimed.
- Costs for future demands for expanded wastewater treatment facilities.
- Costs for customers to retrofit for installation of a dual distribution system.
- Costs for flushing out minerals or trace metals in the reclaimed wastewater that is to be used for irrigational purposes.
- Costs for increased monitoring and administrative functions required for distributing two grades of water.
- Replacement costs of capital equipment over the life span of the project.

These factors need to be considered when evaluating the economic feasibility of reclaimed wastewater reuse alternatives. It is also difficult to quantify health, social, and environmental benefits in monetary terms. Nevertheless, these benefits can be ranked in order of priority. A project with a high economic return may not necessarily be adopted, especially if it has adverse social and/or environmental impacts.[22]

Examples of benefits that may be gained by the use of reclaimed wastewater may include the certainty of having a convenient, reliable, and adequate water supply (industry), or the nutrient value of the reclaimed water for irrigational purposes (agriculture).

If additional treatment is required, the estimated costs for such treatment can be based on specific treatment processes. Valuable information regarding wastewater treatment costs can be found in local or regional reclaimed wastewater works feasibility studies and in cost information records maintained at recently constructed water and wastewater treatment facilities in the area.[1] The cost of additional treatment does not include the cost of treatment that would be required to achieve discharge standards for local receiving waters, as this would be considered part of the necessary expense of disposal and would be incurred in any event, with or without reuse.[15] Advanced wastewater treatment to remove an additional 3.8 to 10% BOD_5, 5.2 to 13% suspended solids, and 61 to 68% phosphorus and ammonia-nitrogen has been found to increase capital costs from 42 to 99% and operational and maintenance costs from 37 to 55%.[17,23] This suggests that other more cost-effective alternatives should be explored if advanced wastewater treatment is required.[17] All costs must take into account the average and peak flows, BOD_5 removals, solids treatment, plant maintenance, collection systems, pumping stations, transportation and maintenance costs, billing and administrative costs, and infiltration and inflow conditions.[16]

The costs for distribution of the treated reclaimed wastewater affects the economics of reuse. The most economical distribution system for a reuse plan is determined by evaluating the trade-off between initial capital costs and operations-and-maintenance costs over the design period of the project. For the system serving one or more large users, a "reconnaissance-level" estimate of costs can be obtained for the conveyance system components by using the following steps.[1]

A. ESTIMATING WASTEWATER TREATMENT UNIT COSTS

As usual, the unit cost for biological wastewater treatment systems decreases with increasing plant capacity. The fundamental relationship between cost and plant capacity governs the scale of economics. Capital costs of wastewater treatment processes can usually be broken down into structural and equipment components.[1] Equation 7.17 presents a way to estimate the capital cost for a biological treatment unit:[4]

$$CST_W = a \cdot (PC)^b \qquad [7.17]$$

where: CST_W = cost of wastewater treatment, monetary units
PC = plant capacity (e.g., population equivalent, PE)
a, b = cost coefficients (0 < b < 1) (the actual values of the coefficient "a" may depend, for example, on the currency in consideration)

Similarly, the operational costs may be estimated by using Equation 7.18:

$$CST_{OW} = a_0 \cdot (PC)^c \qquad [7.18]$$

where: CST_{OW} = operational costs, monetary units
a_0, c = cost coefficients (c ≤ 1)
PC = plant capacity

B. ESTIMATING CAPITAL COSTS FOR FORCE MAINS

Studies have shown that the average velocity for cost-effective water mains is approximately 1.22 m/s (4 ft/s).[1] This velocity has been used in Equation 7.19 to express a suitable, though not necessarily optimum, pipe diameter in terms of flow for effluent distribution.

U.S. customary units[1]:

$$D = 0.566 \cdot Q^{0.5} \qquad [7.19a]$$

where D = pipe diameter, ft, and Q = rate of flow, ft³/s.
SI units (metric):

$$D = 1.025 \cdot Q^{0.5} \qquad [7.19b]$$

where D = pipe diameter, m, and Q = rate of flow, m³/s.

In each community a technically and economically optimum force main diameter may change depending on local operating expenses. Figure 7.3 depicts the relationship between capital and operating costs. The figure illustrates the trade-offs between initial capital costs and operations-and-maintenance costs over the life span of a project.[1] The selection of smaller pipe diameters results in larger friction heads, requiring larger pumps and more energy. An increase in pipe diameter gives smaller friction heads and a decrease in energy consumption and cost (see Curve I, Figure 7.3), but at a steadily increasing capital costs (see Curve II, Figure 7.3). Summing the capital and O&M costs over a range of increasing diameters yields total annual costs (see Curve III, Figure 7.3). The minimum point on Curve III signifies the lowest total annual costs, corresponding to the optimum force-main diameter. When local energy costs increase more rapidly than other costs, annual O&M costs will be higher for any given pipe diameter, displacing the total-cost curve to the right; in such situations, the optimum pipe diameter will be larger.

Having determined the suitable or optimum force-main diameter, the total capital costs of force mains per linear length may be estimated according to local market prices or from potential suppliers. The total capital costs can be estimated by determination of the distribution routes, the end-point elevations along the routes, and the potential users. Equipment costs and construction or installation costs are determined separately.

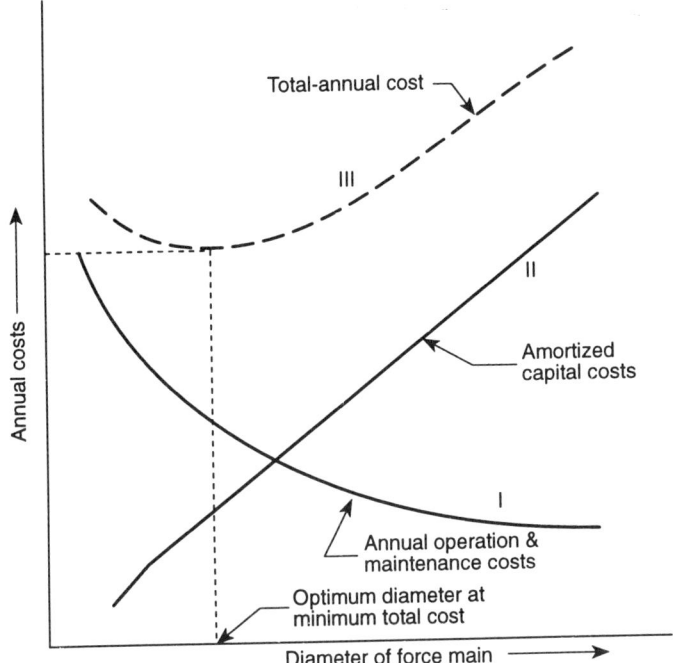

Figure 7.3 Relationship of the annual conveyance costs to the diameter of the force main.[1]

Example 7.4

For an average daily reclaimed wastewater flow of 25,920 m³, and a peaking factor of 1.5, select a suitable pipe size for a force main.

Solution

1. Given: $Q_3 = 25{,}920$ m³/d $= 25{,}920/24 \cdot 60 \cdot 60 = 0.3$ m³/s, and $P_f = 1.5$
2. Determine the peak flow rate as:

$$Q = Q_3 \cdot P_f = 0.3 \cdot 1.5 = 0.45 \text{ m}^3/\text{s}$$

3. Use Equation 7.19b to select the suitable size for the force main as:

$$D = 1.025 \cdot Q^{0.5} = 1.025 \cdot (0.45)^{0.5} = 0.69 \text{ m } (= 27 \text{ in.})$$

C. ESTIMATING PUMPING STATION COSTS
1. Estimating Capital Costs for Pumping Stations

Pumping stations for reclaimed wastewater are generally designed to handle the maximum daily flow. Peak flow rates may be adjusted accordingly when large storage units constitute a part of the distribution system. A smaller pumping station and distribution pipe size results when large storage units are introduced into the project. The determinations of the capital costs of pumping stations are set according to the average daily flows and design peaking factors.

Construction costs of pumping stations can be determined from a cost function such as the one indicated in Equation 7.20.[4]

$$\text{CST}_{pu} = a_p \cdot Q^b \qquad [7.20]$$

where: CST_{pu} = construction cost of pumping station, monetary units
 a_p, b = cost coefficients ($b \leq 1$)
 Q = design flow rate, m³/s

2. Estimating Operating Costs for Pumping Stations

Annual O&M costs may be classified into two major groups: power costs, and labor-and-materials costs. O&M costs are based on total capacity of the pumping station. Power costs can be estimated on the basis of power needed at an average daily flow, taking into account efficiencies of both pump and motor. For systems using electric energy, Equation 7.21 may be used to calculate the required horsepower (HP):[1]

$$PR_h = (Q \cdot H_T)/2800 \qquad [7.21]$$

where: PR_h = horse power required, HP
         Q = average daily flow, gallons per minute [gpm]
         H_T = total dynamic head, ft
         2800 = factor which takes into account a combined pump and motor efficiency of 70%.

The total dynamic head (TDH) signifies the sum of static and friction heads at average daily flows. Static head denotes the difference in elevation between the pumping station and suggested point of reclaimed wastewater reuse plus selected system pressure. Friction head losses for various pipes and flows may be found in tabular and graphical forms from hydraulic or fluid mechanics references.

Using the horsepower determined from Equation (7.21), the annual power costs may be estimated by using Equation (7.22):

$$CST_{PR} = PR \cdot t \cdot cst_{PR} \qquad [7.22]$$

where: CST_{PR} = annual power cost, monetary units
         PR = power, kW (Convert horsepower [HP] to kilowatt [kW] by multiplying by 0.7457)
         t = number of hours per year (it can be estimated according to the suggested use)
         cst_{PR} = unit cost of power, monetary units/kW.hr (the unit cost of power can be adjusted to local tariffs)

Otherwise, the operation costs can be determined as shown in Equation (7.23):

$$CST_{opu} = a_{pl} \cdot Q \cdot H_T \qquad [7.23]$$

where: CST_{opu} = operation cost of pumping stations, monetary units
         a_{pl} = cost coefficient
         Q = design flow rate, m³/s
         H_T = total dynamic head, m

Example 7.5

For the average flow rate of reclaimed wastewater given in Example 7.4, determine the annual O&M power costs if the unit cost is $0.05 per kW.hr (assume a total head loss of 25 ft).

Solution

1. Given: Q_a = 0.3 m³/s = 0.3/6.309·10⁻⁵
             = 4755.1 gpm (gallon per minute),
           H_T = 25 ft, cst_{PR} = $0.05/kW.hr.

2. Find the required horsepower from Equation 7.21:

   PR_h = Q·H_T/2800
         = 4755.1·25/2800 = 42.5 HP = 42.5·0.7457 = 31.7 kW.

3. Determine the annual power cost by using Equation 7.22:

   CST_{PR} = PR·t·cst_{PR} = 31.7·(365·24)·0.05 = $13,885.00.

D. ESTIMATING DISTRIBUTION SYSTEM COSTS

It has been found that the length of the water mains in a distribution system is a function of population density. As such, water-main lengths may be roughly estimated by using Equation 7.24:

$$l_m = 125 \cdot P_d^{-0.46} \quad [7.24]$$

where: l_m = length of main, miles/1000 population
P_d = population density, persons/square mile

Distribution system costs may be found from Equation 7.25:

$$CST_d = 1 \cdot cst_d \quad [7.25]$$

where: CST_d = cost of distribution system, monetary units
1 = length of mains, miles

$$L = l_m \cdot P/1000 \quad [7.26]$$

P = number of people served
cst_d = unit cost of force mains including installation, monetary units/linear foot

A distinction can be made between distribution and transmission systems. The former refers to the smaller diameter lines that feed individual reuse sites, while the latter refers to larger diameter water lines.[24]

Example 7.6
Water is to be delivered to a community with a population of 1500 living on one square mile of territory in an urban setting. Assume a cost of $30.00 per linear foot for the use of an average force-main diameter of 6 in., and determine the capital costs for this system.

Solution

1. Given: P = 1500, A = 1 mile², D = 6 in., cst_d = $30.00/ft.
2. The population density is

   P_d = P/A = 1500/l = 1500 capita/mile.²

3. Use Equation 7.24 to determine the length of water main as:

   l_m = 125·$P_d^{-0.46}$ = 125·1500$^{-0.46}$
   = 4.3 mile/1000 people = 4.3·(1500/1000) = 6.5 miles.

4. Estimate the capital costs of this system by using Equation 7.25 as: $CST_d = 1 \cdot cst_d$.

   Thus, CST_d = (6.5 miles·1760 yard/mile·3 ft/yard)·$30.00/ft
   = $1,029,600.

E. RECLAIMED WASTEWATER STORAGE COSTS

Storage systems may be required to balance diurnal or seasonal fluctuations in supply and demand or for operational purposes in order to equalize daily demand and supply, as well as to accommodate periods when treatment is inadequate to meet reuse standards.[25] Operational storage is typically provided to reduce pumping costs and distribution system line sizes, as well as improving system reliability.[24]

Properly designed and covered storage systems not only prevent deterioration in water quality but can also result in improvement in water quality through physical and biological processes. However, open storage units can result in contamination of the stored water.

The two general types of storage facilities are open earthen basins and covered storage reservoirs. Open earthen storage basins are appropriate for use in agricultural-irrigation and industrial reuse systems. These basins can be either lined or unlined, depending on local soil conditions and land topography. Covered concrete storage reservoirs are commonly used for potable water supply systems. Use of covered reservoirs reduces disinfectant losses (such as chlorine) and reduces the algal growth problems. The cost for steel storage tanks falls somewhere between those for lined earthen and concrete reservoirs.[1]

Storage and distribution systems for reclaimed water projects represent the principal costs for such projects.[24] Provision for storage of reclaimed wastewater at the point of reuse usually reduces the distribution costs. This is due to the fact that the diameter of the distribution pipes can be decreased. Storage costs depend on the availability and price of the land, the storage volume needed, geological structures at the selected site, and on the demand for additional treatment during or after the storage period. Storage costs include capital costs for construction as well as operational costs.[1]

F. ESTIMATED OVERALL RECLAIMED WASTEWATER COSTS

The overall reclaimed wastewater costs signify the summation of the individually computed components for the reuse project. Significant items for total costs estimations include: total fixed costs (such as labor, regular maintenance) and total variable costs (such as pumping costs). All capital costs are expressed in terms of annual or amortized costs so that costs can be budgeted over time. These capital annual costs can then be added to the projected annual operational and maintenance costs to yield a total annual cost, which can be used to compare reuse options. A fairly detailed cost breakdown of a project is necessary for all expected costs during the construction period as well as for the entire period of operation. Cost breakdowns have two significant advantages:

1. It gives an overview of relevant cost items and therefore it constitutes a basis for a reliable cost estimate.
2. It reduces the influence of random errors by calculating a total cost from the sum of individual cost components.[4]

Usually, the planning period for economic comparison is taken to be 20 years.[1] Therefore, equipment with a service life of 10 years is replaced at the end of 10 years. Similarly, if a structure has a service life of 30 years, it would have a salvage value of one-third its original value at the end of 20 years.[1] This salvage value is based on a straight-line depreciation relationship.

The economic life of a project refers to the length of time before the project becomes obsolete or uneconomical to operate. Several factors can affect this period; to mention a few: speed of obsolescence, environmental changes, and wearing out of equipment.[3] The expected average life span for different types of assets differs greatly. A conservative life span estimate to use in an economic study usually is shorter than the full expected life span of the asset under consideration.[5]

Annual capital recovery costs are typically two to four times the operation and maintenance costs. Total financial costs of sanitation or sewerage projects are estimated to be around 1.2 to 9 times the cost of water supply, depending on the water service level. The higher values reflect the hydraulic costs for collection and treatment.[26]

IV. SELECTION CRITERIA AND EVALUATION OF SYSTEMS

Managers and policy makers require information that enables them to make decisions to implement an efficient management or policy program. The costs of potable freshwater may be compared to the estimated costs of a reclaimed wastewater system. This comparison can demonstrate whether or not reclaimed wastewater is competitive, or complementary, with that of a potable water supply. Likewise, this comparison enables better decision making for a particular project. Nevertheless, this comparison neither illustrates conclusive evidence of the economics of wastewater reuse, nor does it constitute the sole basis for subsequent decision making.[1,2] Reclaimed wastewater reuse cost comparisons can be made with or without project alternatives. Estimated unit costs for reclaimed wastewater often do not relate very closely to the user's cost for utilization of the reclaimed wastewater. The steps to be followed in planning a reclaimed wastewater project are illustrated in Figure 7.4. This figure illustrates the major phases for preliminary investigations, market assessment, reuse alternatives, analysis of data, and the ultimate implementation plan. Usually, the planning process involves techniques such as cost-benefit and cost-effectiveness analyses; these provide a systematic approach to resource allocation. Cost-benefit analyses examine the relationship between costs and the realization of public projects, usually expressed in monetary terms.[27]

Generally, total benefits are equal to the area under the market demand curve from the origin to the allocation of interest.[19] A benefit, from a reclaimed wastewater point of view, may be defined as the present or future cost of not supplying fresh water plus the revenue from the sale of reclaimed wastewater.[24] The direct and indirect costs for a project can result in a change of plans.[27] Similarly, a cost may be regarded as the incremental cost of treatment, distribution, and on-site hookup and use of the reclaimed wastewater.[24]

The basic purpose of a cost-benefit analysis is to identify, and quantify the value in monetary terms of all quantifiable benefits and costs occurring in each year throughout the life of the project. Once these estimates are made the benefits and costs in the future years are reduced to a common denominator.[28]

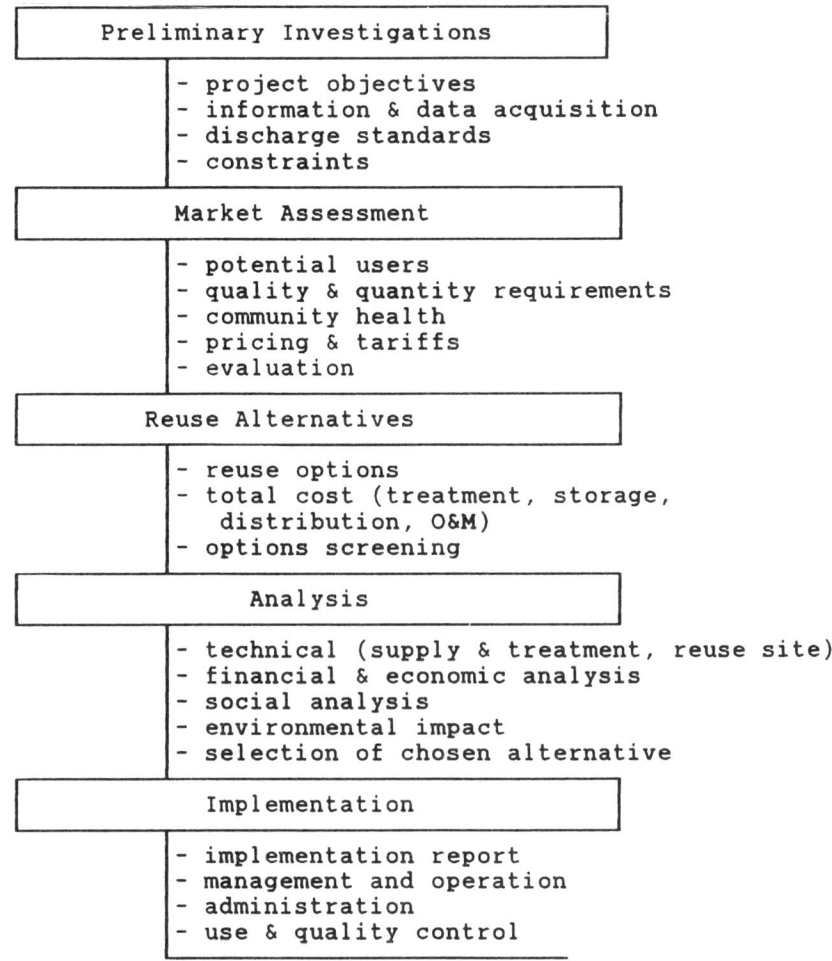

Figure 7.4 The reclaimed wastewater planning process.[27]

Cost information and benefits are related through different concepts that are discussed in the following sections.

A. BENEFIT-COST RATIO (BCR)

The BCR criteria implies that an activity should be undertaken if the ratio of the present value of benefits to the present value of costs exceeds 1.0. The benefit-cost technique makes the most precise statement about which chosen policies are most efficient, and it also imposes the greatest demand for precise information in order to produce such statements.[19] Nonmonetary issues are not factored into these calculations.[2] The BCR may be determined by using Equation 7.27:[29]

$$\text{BCR} = \frac{(\text{present value or worth of net benefits})}{(\text{present value of capital investments, costs})}$$

$$\text{BCR} = (\text{CST}_B - \text{CST}_D - \text{CST}_C)/\text{CST}_I \qquad [7.27]$$

where: BCR = benefit-cost ratio, dimensionless (as long as the ratio is equal to or greater than unity [BCR ≥ 1.0] the project is acceptable)

CST_B = present value of all the benefits the project generates for the public (or the users) (benefits are widely understood to mean cost savings, such as accident reductions and intangible service, to the users of the facility),[30] monetary units

CST_D = present value of the disbenefits associated with the project, monetary units

CST_C = present value of the annual costs of operation and maintenance of the project over its useful life, monetary units

CST_I = present value of the capital investment in the project, monetary units

Alternatively, the BCR may be computed by annualizing net benefits and initial capital investment, as shown in Equation 7.28:

$$BCR = CST_b / CST_c \quad [7.28]$$

where: CST_b = annual equivalent net benefits, monetary units
CST_c = annual equivalent capital investment, monetary units

The annual equivalent net benefits may be determined from benefit-cost data by direct computational techniques or by using prepared economical or statistical computer programs or software such as spreadsheets (examples of which may include Quattro Pro from Borland International; Excel from Microsoft Corporation; Lotus 1–2-3 from Lotus Systems, Inc.; Multiplan from Microsoft Corporation; Super-Calc 3 from Sorcim Corporation; The Financial Planner from Ashton Tate; Microplan from Digital Research; etc). These computer programs are a great help when cost components are categorized in order to evaluate a project.

Several methods are available to determine the BCRs. As a result, considerable engineering judgment is required in originating BCR studies, as well as in evaluating BCR studies done by others.[30]

The BCRs have been subjected to considerable criticism based on oversimplification of complex inputs, susceptibility to misinterpretation, potential for misuse, and difficulties in classifying and defining project impacts.[30–32]

B. POSITIVE NET-PRESENT VALUE CRITERION

This criterion implies that an activity should be undertaken whenever the present value of the net benefits is greater than zero.

Both the BCR criterion and the present-net value criterion do not guarantee efficiency. However, the Maximum Net-Present Value Criterion suggests that resources should be committed to those uses maximizing the present value of net benefits received. If accomplished properly, the latter analysis will correctly identify an efficient allocation.[19]

C. COST-EFFECTIVENESS ANALYSIS

Cost-effectiveness analysis frequently involves an optimization procedure whereby a systematic method for finding the lowest-cost means of accomplishing the objective is followed.

This technique requires the establishment of a single base criterion for the evaluation of projects, such as an annual water production program with a specific quality expressed as an increase in supply or decrease in demand.[2]

D. IMPACT ANALYSIS

This procedure attempts to identify and quantify the consequences of various actions and their impacts. This technique does not optimize or even compare the information gathered.[19]

When benefits are difficult or impossible to quantify, the cost-effectiveness analysis and impact analysis yield reasonable estimates. For water projects, the preferences are linked to social goals such as economic development, job creation, environmental preservation, and hazard reduction. The benefits are measured in terms of the achievement of these social goals, and much of water resources planning is linked to analyses of these benefits on a multi-objective basis.[10]

Example 7.7

Calculate the BCR for a project with total annual benefits of $22.70, total annual disbenefits of $0.20, total annual operating costs of $3.00, and total annual equivalent cost of capital investment of $15.00 (all costs in million dollars).

Solution:

1. Given: $CST_B = 22.7$, $CST_D = 0.2$, $CST_C = 3$, and $CST_I = 15$ (all units in millions of dollars).
2. Use Equation 7.27 to determine the BCR as:
   $BCR = (CST_B - CST_D - CST_C)/CST_I$
3. The BCR = $(22.7 - 0.2 - 3)/15 = 1.3$.
   Since the computed BCR is greater than 1, the project is feasible. The BCR ratio, in effect, indicates that for each dollar of capital invested, a return of $1.3 in revenue each year will occur for the economic life of the project.

V. FINANCE AND COST COMPARISONS

A. FINANCIAL AND ECONOMIC ANALYSIS RELATIONSHIP

A financial analysis refers to costs and benefits for a specific user or participant. The production of a reclaimed wastewater with acceptable characteristics will promote its own use if it is sold at a reasonable price and, in fact, will help in improving the treatment and operation of the production facility. Likewise, the confidence of the consumers in the wastewater treatment facility will be improved. This condition will result in attracting and retaining other users of the treated wastewater such as public, commercial, and industrial institutions.

The economic evaluation of reclaimed wastewater projects seeks to determine the net benefit to society of the proposed investment. Therefore, it considers all costs and benefits which will result in real resource terms (opportunity costs) to reflect the true value of the project.[33]

Economic costs are based on the physical conditions of the community (for example, abundance or scarcity of labor, availability of water, etc.) and are therefore quite objective; however, financial costs are subject to interest rates, loan maturities, and availability of central government subsidies.[34]

B. FINANCING OF WATER AND RECLAIMED WASTEWATER PROJECTS

Financing of water or reclaimed wastewater reuse units must be carefully evaluated. This is to ensure that the amount of money needed for proper operation and monitoring of a project is available.

For a water supply or a reclaimed wastewater reuse system, financial assistance can often be obtained from governmental agencies, from the public sector as from bonds (revenue bonds, general obligation bonds, industrial development bonds, etc.), service charges, connection fees, system development charges, and public revenues (taxes, assessments, and fees).[16] Other financing alternatives include use of bond banks, water banks, capital reserve, state corporations, lease-purchase agreements, cost sharing with the developer or user, private activity bonds, and federal and state government loans and grants.[17] Financial entities participating in a project may include investor-owned utilities, private firms, individual entrepreneurs, and users of project outputs of goods and services.[12] This helps in the planning, design, and construction of pollution-control facilities.[1] Under some circumstances, such grants can be applied to the implementation of reuse programs. Nonetheless, long-term financing should not be dependent on subsidies.

Most large-scale water projects are financed in part by national or international agencies.[3] Privatization (provision of water to the public by the private sector) may be introduced to help solve investment problems in water and wastewater facilities.[10] Privatization may be considered when legal taxing and borrowing limitations prohibit governments from funding and providing a needed service or facility. In such instances it may be possible to contract with a private organization, pay a fee, and transfer the responsibility — subject to conditions that will protect the public interest.[17]

The creation of a totally self-supporting reuse program financed wholly by reclaimed wastewater user fees is difficult. To satisfy the capital requirements for implementation of a reuse program, the majority of the construction and related capital costs are generally financed through long-term water and wastewater revenue bonds,[2] or in developing countries via loans, grants from external support agencies, developers, or industrial contributors.

C. COST ALLOCATION AMONG DIFFERENT USERS

Financial guidelines should be developed to assist the utility manager in ensuring financial stability of a project.[16] It is common for large engineering projects to serve more than one principal purpose (multipurpose projects). Thus, in their economic evaluation special procedures need to be followed. Some of the most important methods for allocating costs are briefly discussed below:[35]

1. *The separable costs-remaining benefits method.* In this method the amounts allocated to each purpose are in proportion to the remaining benefits for each purpose after subtracting from each its separable cost. Each benefit in this context is the single-purpose benefit or, if lower, the alternative cost of providing that benefit independently.
2. *Marketability method.* This method involves apportionment of common costs in ratio to revenues that might be received for the services provided if sold at market prices. This method is often difficult to apply in practice, since such services are not readily marketable. This method may also be used when agreement on the method of separable cost-remaining benefits is either inappropriate or unattainable.
3. *Benefit method.* This method involves apportionment of common costs in the ratio of estimated economic values of different benefits to be derived from the proposed development.
4. *Use of facilities method.* This method involves the allocation of common costs in proportion to the use made of a facility which is jointly shared by various participants. This method assumes that a unit of measurement can be found which can be applied to all users.
5. *Special cost method.* This method involves apportionment of common costs in proportion to special costs incurred on behalf of each of the participants.
6. *Alternative single-purpose expenditure method.* This method involves the division of common costs in proportion to the hypothetical alternative expenditures which would be incurred if each of the participants were separately to construct a facility to provide benefits which they would receive from this joint project.

Irrespective of the method chosen it is difficult to justify an allocation of common costs.[35] When cost comparisons are made for different countries, a single currency with the official or parallel exchange rates must be stated; this is especially true for developing countries.[6] The costs that are recovered from the users must also reflect actual costs over time if they are to be recovered in actual values.[6] In the case of developing countries, and where equipment is imported, the cost of the currency of the country of origin should be recorded and this cost should be converted to local currency using a shadow conversion obtained from concerned governmental agencies, central banks, or related institutions.[36] Shadow prices signify an attempt to correct distorted market prices and to evaluate the real cost to the economy. The official rate needs to be mentioned, whereas the parallel rate would merit consideration for special items in developing areas where the effect has much greater significance. If steep inflation rates, artificial exchange rates, or devaluation have occurred during the project, the method of accounting must be explained.[13] Likewise, it is also essential to indicate the year to which costs apply in order to aid in cost evaluations and comparisons.[6]

D. COST COMPARISON OVER TIME

The need to compare costs over time is particularly critical in situations where data are available only for certain years. This problem can be resolved by considering escalation and inflation situations.

Escalation may be defined as the time-oriented increase in costs. Escalation is caused by one or more of the following factors:

- Inaccuracy of original resource estimates.
- Incomplete design or changes in design, production quantity, or schedule at the time of estimation (bidding design changes).
- Unforeseen increases in labor burden (man-hours) to do a given amount of work.
- Increased skill level and the corresponding wage level to accomplish given tasks.
- Nonproductive labor encountered in providing work activities and work outputs.
- Unanticipated, unplanned, or excessive customer involvement.
- Increased government regulation and control.
- Inflexibility of the work force.
- Inability or failure to hire personnel at lower labor rates.
- Unanticipated failure of suppliers or subcontractors to deliver on time and within original cost estimates.[8]

Escalation results in the need for more labor hours or material to perform the same task (or the production of less output with the same amount of labor hours or materials[37]). When prices tend to increase over time, an escalation clause is sometimes included in contracts. This is to provide for any increase in costs that are not anticipated. Accurate cost estimating using the most advanced planning estimating methods and tools, subsequent aggressive management cost control of the project, and good planning and contracting practices can do a lot to reduce escalation problems.[37]

Inflation refers to a sustained increase in the general level of prices for goods and services in an economy and is brought about by the rising costs of materials, subcontracts, parts, supplies, goods, and services.[6,8,38] Inflation is caused principally by the injection of funds into the economy that unbalance the law of supply and demand for money.[8]

The rate of inflation refers to the percentage increase in the general price level, usually from one year to another. A positive rate of inflation in a given period denotes that:

1. The purchasing power of a currency during that period declines, or
2. The cost of an activity during that period rises annually at a rate equal to the rate of inflation.[35]

Generally, indexes are used to measure inflation and for adjusting prices and costs. The indexes that are most often used in economic analyses include the following:

1. *Consumer Price Index (CPI)*. The CPI is a weighted average of selected prices in the economy at any given time, divided by the prices of the same goods in a base year, or reference point.[6] Equation 7.29 gives a mathematical definition of the CPI:

$$\text{CPI} = (\text{CST}_n/\text{CST}_r) \cdot 100 \quad [7.29]$$

where: CPI = consumer price index in a given year
CST_n = cost of a market basket in a given year (nominal spending), monetary units
CST_r = cost of a market basket in the base year (real spending), monetary units

2. *Gross National Product Deflator (GNP deflator)*. The gross national product is the sum of the value of all final goods and services produced by the economy during a specified period of time, usually one year.[38] The GNP deflator is the economy's aggregate price index. The deflator is used to correct money values of the gross national product for price changes, so as to isolate the changes which have taken place in the physical output of goods and services.[39] The GNP deflator refers to the ratio of the nominal GNP to the real GNP. The GNP measures national income generated both within and outside the country. The GNP deflator may be calculated using Equation 7.30:

$$\text{GNP deflator} = (\text{GNP}_n/\text{GNP}_r) \cdot 100 \quad [7.30]$$

where: GNP_n = the nominal gross national product (determined by calculating all outputs at current prices)[38]
GNP_r = the real gross national product (determined by calculating all outputs at the prices that prevailed in some agreed-upon year)[38]

3. *World Bank's Manufacturers Unit Value (MUV) Index*. This is the cost, insurance, and freight (CIF) index of U.S. dollar prices of the industrial countries' manufactured exports to developing countries. It is frequently used to measure changes in the price of industrial exports to developing countries.[6]
4. *U.S. Gross Domestic Product (GDP) deflator*. The GDP is the market value of all goods and services currently produced within a country during a particular time interval. The GDP deflator may be a more appropriate method to use in circumstances where the U.S. inflation rate is a more relevant measure of changes in the cost of specific items than the local inflation rate.[6]

Example 7.8

The cost of a commodity in 1980 was $2650.00 and its price in 1994 was $3180.00. Find the Consumer Price Index in 1994.

Solutions

1. Given: CST_n = $3180.00, CST_r = $2650.00.
2. Find the CPI by using Equation 7.29 as:

   CPI in 1994 = $(\text{CST}_n/\text{CST}_r) \cdot 100$

   $= (3180/2650) \cdot 100 = 120$

Example 7.9

Compute the real GNP for the year 1994 given that the nominal GNP for the same year is $500,000.00 and the GNP deflator is 40%.

Solution

1. Given: $GNP_n = \$500,000.00$, GNP deflator = 40%.
2. Use Equation 7.30 to determine the real GNP as:
   GNP deflator = $(GNP_n/GNP_r) \cdot 100$.
3. Substitute values indicated in Step 1 in Equation 7.30. Thus, $(40/100) = 500,000/GNP_r$. This yields, $GNP_r = \$1,250,000.00$.

Generally in project evaluation and in comparison of costs the following steps can be followed:

1. Identification of project inputs (financial, equipment, staff, services, employee, community, etc.).
2. Outlining cost procedures and assumptions (annual costs, inflation, deflator, opportunity cost [the foregone value of the next best alternative that is not chosen], time schedule, etc.).
3. Identification of project outputs.
4. Evaluation of alternatives.

Opportunity costs are needed when free market conditions do not exist and distortions arise through such factors as direct controls, fiscal measures, and monopoly conditions. Among the items where cost adjustments are frequently required are land, unskilled labor, power, and imported goods.[33]

Some advocate that cost comparisons of potable water and a regional wastewater reuse system for irrigational purposes favor wastewater reuse when new development is taking place, as raw or potable water lines do not already exist, and a dual distribution system can be installed as construction proceeds.[40]

VI. WASTEWATER AS AN ECONOMIC RESOURCE

The benefits that can be gained from a wastewater reclamation and reuse project may be grouped as follows:[33]

1. *Primary benefits.* These are direct benefits resulting from the project and to which a monetary value can be allocated. These benefits may further be subdivided into:
   1. Direct primary benefits, such as a customer using reclaimed wastewater, and
   2. Indirect primary benefits, such as reduction in dust storms due to irrigation, or a decrease in the use of fertilizers due to the nitrogen and phosphorus present in the reclaimed wastewater.
2. *Secondary benefits or externalities.* These embrace the spillover effects resulting from, or induced by, the project, i.e., forward and backward linkages. These benefits which accrue to the economy are not taken into account in quantities or prices of inputs or outputs of the project itself. These effects are mainly of two kinds:
   1. Forward linkages or multiplier effects, which result from expenditure of incomes generated by the project, and
   2. Backward linkages, which are increases in income generated by additional activity occasioned by a project in industries which supply its inputs.
3. *Public benefits.* These benefits are outside the market and have no accepted monetary value and they include items such as enhanced environmental quality and aesthetic improvements (beautification).

Generally, the practical benefits to be considered during project evaluation would be the direct benefits. Other benefits which are difficult to quantify may be included, and a subjective assessment made of their importance.

VII. HOMEWORK PROBLEMS

A. DISCUSSION PROBLEMS

1. What are the essential steps that need to be followed in the planning process for a reclaimed wastewater reuse project in your locality?
2. What factors govern reclaimed wastewater reuse?
3. Give the general classifications for cost comparisons.
4. Why are costs liable to be higher for smaller water supply systems than for larger ones?
5. What are the most important cost components that are to be included in cost evaluation for:

1. a water supply system?
2. a wastewater disposal scheme?
3. a reclaimed wastewater reuse project?

6. What are the benefits of storage units in a wastewater reclamation project?
7. Indicate how to obtain an estimate of the costs for storage systems in a water reclamation and reuse project.
8. How are benefits related to costs in an economic evaluation process?
9. Differentiate between benefit-cost ratio criterion and cost-effectiveness analysis.
10. What are the limitations of the benefit-cost ratio (BCR)?
11. What merits lie behind the use of discount rates in a wastewater reclamation project?
12. Why does the unit cost of a biological wastewater treatment plant decreases with an increase in treatment plant capacity?
13. What is the significance of a shadow price in the estimation of a wastewater reclamation and reuse project?
14. Why is it important to break down costs in a project cost assessment?
15. What are the differences between distribution and transmission piping systems? Do these differences affect the cost estimates for each system?
16. Discuss different methods used to measure inflation. Review merits and drawbacks for each method.
17. Define the terms: "economic analysis" and "financial analysis", and outline their differences.
18. Indicate the different methods that can be practically used in a project economical analysis.
19. Consider a project of reclaimed wastewater reuse in an agricultural farm land. What factors besides construction requirements are to be taken into account during the evaluation of total costs for such a project?
20. Suggest a practical and fair way that may be applied by a municipality to gather water rates for a rural community?
21. What are the main causes of escalation?
22. What are the main factors that initiate inflation in an economy?
23. What is the significance of a positive rate of inflation, in a given period, in the economy?
24. Give a general classification of benefits that can be gained from a wastewater reclamation project.

B. NUMERICAL PROBLEMS

1. Determine the uniform amount at the end of each of 30 years at 8% interest that will be equivalent to $500,000 after 30 years.
2. Estimate the total annual capital and O&M costs of a water supply project with design parameters as presented in the following table.

Item	Value
Discharge (m^3/s)	1.8
Operating pumping hours (/d)	12
Efficiency of pump (%)	85
Efficiency of motor (%)	88
Length of pipe (km)	1.5
Manning's roughness coefficient	0.013
Static head (m)	25
Maximum velocity in pipes (m/s)	1.8
Easement width (m)	1.2
Cost of land ($/ha)	200,000
Pump system capital cost ($/kW)	900
Cost of electricity ($/kW.hr)	0.05
O&M cost coefficient	1.08
Discount rate (%)	8
Project economic life (years)	25

3. For an hourly reclaimed wastewater flow rate of 800 m^3 and a peaking factor of 1.25, select a suitable size for the force main for the preliminary project investigation.

4. Using the data given in Problem 3 and a total headloss of 20 m, determine the annual O&M power cost given a unit power cost of $0.04/kW.hr.

5. In a particular city, the municipality received suggestions for projects dealing with the following areas: recreational, transportation, water supply, hospital, and agricultural irrigation. Each agency responsible for these areas suggests that its project is justified based on a benefit-cost analysis. Using the data of net benefits and costs, as tabulated below, select the best projects if the capital budget of the city for next year is set at $26 million. Show your reasons for selection.

Project	Benefits (million $)	Costs (million $)
I	16	15.3
II	18	16
III	30	28
IV	40	30
V	60	48

6. A certain project has the following schedule of costs and benefits (values are given in thousands of dollars). Using a discount rate of 10 percent, find:
   1. Net benefits.
   2. The BCR.

Year	Costs	Benefits
1	50	0
2	20	40
3	20	30
4	20	50
5	20	30
6	20	35

7. Two mutually exclusive projects with equal life spans are to be evaluated. The present worth of costs and benefits of the two projects are as presented in the table below. Select the preferred project by using the BCR criterion.

Project	Cost	Benefit
A	$4000.	$7000.
B	$6000.	$9000.

8. Given that the nominal GNP for 1993 is $350,000.00 and a GNP deflator of 35%, compute the real GNP for the same year.

9. Fill in the blanks in the following table of GNP statistics.

Year	1991	1992	1993
Nominal GNP ($K)	?	4207.2	6209
Real GNP ($K)	4715	?	4889
GNP deflator (%)	115	120	?

REFERENCES

1. **Donovan, J. F. and Bates, J. E.,** Guidelines for Water Reuse, Municipal Environmental Research Laboratory, Office of Research and Development, U.S. Environmental Protection Agency, Cincinnati, OH, EPA-600/8-80-036, 1980.
2. USEPA, Manual Guidelines for Water Reuse, U.S. Agency for International Development, Washington, D.C., EPA/625/R-92/004, Sept. 1992.
3. **Helweg, O. J.,** *Water Resources Planning and Management,* John Wiley & Sons, New York, 1985.
4. **Orth, H. M.,** *Model-Based Design of Water Distribution and Sewage Systems,* John Wiley & Sons, Chichester, 1986.
5. **Linsley, R. K., Franzini, J. B., Freyberg, D. L., and Tchobanoglous, G.,** *Water-Resources Engineering,* McGraw-Hill, New York, 1992.
6. **Rassas, B.,** A Primer on Comparing and Using Cost Data in Water and Sanitation Reports, Technical Note Information for Action, Water and Sanitation for Health Project (WASH), Alexandria, VA, August 1992.
7. **DeGarmo, E. P., Sullivan, W. G., and Canada, J. R.,** *Engineering Economy,* Macmillan, New York, 1984.

8. **Stewart, R. D. and Wyskida, R. M.**, *Cost Estimator's Reference Manual*, John Wiley & Sons, New York, 1987.
9. **Renner, D. C.**, Establishing a Maintenance Program, *Water Eng. Manage. J.*, 140(2), February 1993, 20–35.
10. **Grigg, N. S.**, *Water Resources Planning*, McGraw-Hill, New York, 1985.
11. **Qasim, S. Y., Lim, S. W., Motley, E. M., and Heung, K. G.**, Estimating Costs for Treatment Plant Construction, *American Water Works Association J.*, 84(4), August 1992, 56.
12. **Goodman, A. S.**, *Principles of Water Resources Planning*, Prentice-Hall, Englewood Cliffs, NJ, 1984.
13. **Roark, P.**, Evaluation Guidelines for Community-Based Water and Sanitation Projects, WASH Technical Report No. 64, Office of Health, Bureau for Science and Technology, U.S. Agency for International Development, Washington, D.C., 1990.
14. **Thomas, M. V.**, Valuation of Conserved Potable Water: What is Water Worth?, Proceedings of the Conference on Urban and Agricultural Water Reuse, Orlando, FL, 28 June to 1 July 1992, Water Environment Federation, Alexandria, VA, 1992, 183.
15. **Houser, D., Izmir, G., and Sears, J.**, *The Economics of Water Reuse Environmental Benefits*, Proceedings of the 65th Annual Conference and Exposition of the Water Environment Federation, New Orleans, 20 to 24 September 1992, Water Environment Federation, Alexandria, VA, 1992, 213.
16. **WPCF, Task Force on Financing and Charges**, *Financing and Charges for Wastewater Systems*, a special publication, 2nd ed., American Public Works Association (APWA), American Society for Civil Engineers (ASCE), and Water Pollution Control Federation (WPCF), Washington, D.C., 1984.
17. **Salvato, J. A.**, *Environmental Engineering and Sanitation*, 4th ed., John Wiley & Sons, New York, 1992.
18. **Gumerman, R. C., Culp, R. L., and Hasen, S. P.**, *Estimating Water Treatment Costs*, Municipal Environmental Research Laboratory, Office of Research and Development, U.S. Environmental Protection Agency, Cincinnati, OH, EPA-600/2-79-162c, Vol. 1–4, 1979.
19. **Tietenberg, T.**, *Environmental and Natural Resources Economics*, Scott, Foresman and Company, Glenview, IL, 1988.
20. **ASCE, Committee on Pipeline Planning**, Pipeline Division, Report of the Task Committee on Engineering Practice in the Design of Pipelines, *Pipeline Design for Water and Wastewater*, American Society of Civil Engineers, New York, 1975.
21. **Lindeburg, M. R.**, *Engineering Economic Analysis: An Introduction*, Professional Publications, Belmont, 1993.
22. **UN**, Criteria for and Approaches to Water Quality Management in Developing Countries, Natural Resources Water Series No. 26, Department of Technical Co-operation for Development, United Nations, New York, 1991.
23. **WPCF**, Advanced Waste Treatment: Has the wave crested?, *Water Pollut. Control Fed. J.*, 50(7), July 1978, 1706.
24. **WPCF, Task Force on Water Reuse**, *Water Reuse Manual of Practice*, SM-3, Water Pollution Control Federation, Alexandria, VA, 1989.
25. **Dent, R.**, The Smart Money is Still on "I. Q. Water" for Reuse, Proceedings of the Conference on Urban and Agricultural Water Reuse, Orlando, FL, 28 June to 1 July, 1992, Water Environment Federation, Alexandria, VA, 1992, 385.
26. **Gunnerson, C. G. and Jones, D. C.** Costing and Cost Recovery for Waste Disposal and Recycling, Water Supply and Urban Development, Department Operations Policy Staff, The World Bank, Washington, D.C., Report No. UDD-37, January 1984.
27. **Rossi, P. H. and Freeman, H. E.**, *Evaluation of a Systematic Approach*, Sage Publications, London, 1982.
28. **WHO**, Expert Committee on Vector Biology and Control, Environmental Management for Vector Control, TRS 649, World Health Organization, Geneva, 1980.
29. **Cassimatis, P.**, *A Concise Introduction to Engineering Economics*, Boston Unwin Hyman, London, 1988.
30. **Collier, C. A. and Ledbetter, W. B.**, *Engineering Economic and Cost Analysis*, Harper and Row, New York, 1988.
31. **Riggs, J. L.**, *Engineering Economics*, McGraw-Hill, New York, 1982.
32. **Lund, J. R.**, Benefit-Cost Ratios: Failures and Alternatives, *Water Resour. Plan. Manage. J.*, 18(1), Jan./Feb., 1992, 94.
33. **Porter, R. B. and Fisher, F.**, *Sewage Effluent Reuse — Economic Aspects in Project Appraisal*, Reuse of Sewage Effluent, Proceedings of the International Symposium Institution of Civil Engineers, London 30–31 October 1984, Thomas Telford, London, 1985, 267.
34. **Kalbermatten, J. M., Julius, D. S., and Gunnerson, C. G.**, Appropriate Technology for Water Supply and Sanitation: A Summary of Technical and Economic Options, Transportation, Water and Telecommunication Department, The World Bank, Washington, D.C., December 1980.
35. **Institution of Civil Engineers**, *An Introduction to Engineering Economics*, William Clows and Sons, London, 1976.
36. **Cairncross, S., Carruthers, I., Curtis, D., Feachem, R., Bradley, D., and Baldwin, G.** Evaluation for Village Water Supply Planning, TPS 15, International Reference Center for Community Water Supply and Sanitation (IRC), The Hague, The Netherlands, February 1980.
37. **Stewart, R. D. and Stewart, A. L.**, *Microestimating for Civil Engineers*, McGraw-Hill, New York, 1986.
38. **Baumol, W. J. and Blinder, A. S.**, *Economics Principles and Policy*, Harcourt Brace Jovanovich, New York, 1991.
39. **Bannock, G., Baxter, R. E., and Rees, R.**, *The Penguin Dictionary of Economics*, Penguin Books, Harmondsworth, U.K., 1978.
40. **Gorder, P. J. and McNeil, R.**, Urban Reuse in the Denver Metropolitan Area: Strategies and Case Studies, Proceedings of the Conference on Urban and Agricultural Water Reuse, Orlando, FL, 28 June to 1 July 1992, Water Environment Federation, Alexandria, VA, 1992, 145.

Chapter 8

Reclaimed Wastewater Monitoring — Sampling and Analysis

CONTENTS

I. Introduction 326
II. Reclaimed Wastewater Sampling and Analysis 327
III. The Objective of Sampling 327
IV. Planning For Sampling 327
    A. Purpose of Sampling 328
    B. Locating Sampling Sites 328
    C. Types of Samples 328
    D. Time, Frequency, and Number of Samples Needed 328
    E. Sampling Equipment Required 329
    F. Volume of Sample Required 331
    G. Preservation of Samples 331
    H. Sample Identification 331
    I. Sampling Personnel 331
    J. Chain of Custody 332
    K. Special Instructions 333
V. Precision and Accuracy 333
VI. Procedures for Analysis of Important Constituents in Reclaimed Wastewater Planned for Reuse 333
    A. Physical Properties 334
        1. Alkalinity 334
        2. Color 334
        3. Solids (Residue) 335
        4. Specific Conductance or Electrical Conductivity 337
        5. Temperature 338
        6. Turbidity 338
    B. Analysis for Inorganics and Non-Metals 339
        1. Boron 340
        2. Chlorides (Cl^-) 340
        3. Chlorine (Cl_2) 342
        4. Dissolved Oxygen (DO) 344
        5. Nitrogen 345
        6. pH 347
        7. Phosphates 349
        8. Sulfates 349

  C. Organic Constituents 351
    1. Biochemical Oxygen Demand (BOD) 351
    2. Chemical Oxygen Demand (COD) 353
    3. Total Organic Carbon (TOC) 354
    4. Oil and Grease 354
    5. Pesticides 355
    6. Surfactants 355
  D. Metals 355
  E. Biological Characteristics 358
    1. Tests for the Coliform Bacterial Group (Indicator Organisms) 358
    2. Total and Fecal Coliform (Membrane Filter Technique) 360
  F. Radioactivity 363
    1. Introduction 363
    2. Regulatory Agencies 363
    3. Standards 363
    4. Radiation Detection and Measurement 365
VII. Flow Measurements 365
  A. Introduction 365
  B. Weirs365 365
  C. Parshall and Palmer-Bowlus Flumes 370
    1. Parshall Flumes 370
    2. Palmer-Bowlus Flumes 371
  D. Venturi Meters 371
  E. Electromagnetic Flowmeters 371
VIII. Reliability (Redundancy, Quality Control, Levels of Protection) 375
  A. Introduction 375
  B. Potential Problems 375
  C. Federal Reliability Guidelines 375
  D. Regulations Regarding Reliability 375
IX. Legal Aspects of Sampling and Monitoring 379
  A. Introduction 379
  B. Examples of State Legislation Regarding Sampling and Monitoring 379
  C. Federal Regulations Regarding Sampling and Monitoring 381
X. Homework Problems 383
Appendix: Statistical Principles of Normal or Gaussian Distribution 387
References 391

I. INTRODUCTION

The purpose of monitoring reclaimed wastewater that is planned for a beneficial reuse has many aspects that must be considered — the first and foremost being the protection of a community's health and welfare. Other important purposes and objectives of a monitoring, sampling, and analysis program dealing with wastewater reuse include:[1-5]

- Indicating the reliability of the wastewater treatment operation.
- Assisting in evaluating and providing information for the control and operation of the various treatment processes involved in the reclamation of the wastewater.
- Providing warnings for temporary operational failures.
- Determining compliance with local, state, and federal permits and regulations.
- Assessing the suitability of the treated wastewater for various reuse purposes, such as cooling water, landscape or agricultural irrigation, groundwater recharge, etc.
- Identifying potential problems that may be emerging so that appropriate control measures can be initiated.
- Assisting in reviewing the impact of the use of reclaimed wastewater on the environment (water, soil, and air).
- Establishing credibility with the public by providing complete and accurate information about the quality of the treated wastewater.

- Determining treated wastewater flows to evaluate detention times and calculate the unit costs for establishing fees.
- Determining peak flows and minimum flows.
- Providing information regarding long-term trends in order to plan for expansion or modification of a reuse facility.
- Providing information for judicial review.
- Scientific research.

II. RECLAIMED WASTEWATER SAMPLING AND ANALYSIS

Regular monitoring of the quality of the reclaimed wastewater is essential for both the users of this water and for the managers of reuse projects.

Monitoring programs include both sample collection and analysis and must of course follow well established guidelines, standards, and procedures to ensure that the results reported are accurate and reliable.

Some of the best established sources for the most acceptable procedures for sample collection and analysis include the following:

Standard Methods for the Examination of Water and Wastewater, 18th ed., 1992. APHA, AWWA, WEF, Washington, D.C.[6]

Federal Register, Vol. 38 and 41, Title 40, Chapter 1, Subchapter D, Part 136, pages 28758 and 52780, 1973, 1976.[7]

Annual Book of ASTM Standards, Part 31, Water 1985, American Society for Testing and Materials, Philadelphia, PA.[8]

Methods For Chemical Analysis of Water and Wastes, ERIC, Technology Transfer, U.S. Environmental Protection Agency, Cincinnati, OH, 45260, 1979.[9]

Methods For Collection And Analysis of Water Samples For Dissolved Minerals and Gases, U.S.G.S., Survey Techniques of Water-Resources Inventory, Book 5, Chapter A1, 1970, Department of the Interior, Washington, D.C.[10]

While the above references are considered some of the best available, it must also be kept in mind that as new technologies and procedures are developed they will be incorporated in the most recent publications. Therefore, in order to stay current in the area of sampling and analysis consultation with the most recent citations is necessary.

III. THE OBJECTIVE OF SAMPLING

The objective of sampling is to collect a representative portion of the water to be evaluated, that is, small enough in volume to be conveniently handled in the laboratory and yet representative of the water being examined. To accomplish this objective the sample must be collected in clean containers and preserved in such a way that nothing is added, lost, or changed between the time the sample is collected and the time it is analyzed.[1,2,5,6]

IV. PLANNING FOR SAMPLING

The following must be considered before sampling for a project starts:[1,2,5,6]

- Purpose of the sampling
- Locating the sampling sites
- Type of samples to be collected
  - Grab
  - Time-proportional composite
  - Flow-proportional composite
- Time, frequency, and number of samples needed
- Sampling equipment required
- Volume of sample needed
- Preservation of the sample
- Sample identification

- Sampling personnel
- Chain of custody
- Special instructions

A. PURPOSE OF SAMPLING

It is necessary to know if the sampling and analysis is for physical, chemical, biological, or radiological purposes, and how the information obtained will be used, such as in planning, research or design, process control, or for regulatory purposes.[5,6]

B. LOCATING SAMPLING SITES

Certain general principles apply in locating sampling sites, and following are a few factors that must be considered:[1,5,6]

- Accessibility to the sampling sites should be reasonably easy.
- Representative samples should be taken at points where the reclaimed wastewater is well mixed.
- Sampling should be at critical points in the water reuse system.
- Sampling should be at the center of a channel where the velocity is high and thus avoid settled solids.
- Sampling should be below the water surface in order to avoid scum and grease deposits.

C. TYPES OF SAMPLES

There are generally three types of samples.[2,5,11,12]

- Grab samples
- Time-proportional composite samples
- Flow-proportional composite

A grab (spot, catch, instantaneous, snap) sample is a single volume of water collected all at one time from a single place. For example, dipping a sample bottle into a tank of water, or opening a faucet or sample tap would result in a grab sample.

Grab samples are useful where the composition of the source is relatively constant or if the flow is not continuous. Grab samples are good for analysis where batch dumping of discharges are of short duration.

Composite samples are a series of grab samples taken at different times from the same sampling point, and mixed together. Thus a time-proportional composite sample is made up of several equal-volume grab samples.[11]

Flow-proportional samples are composite samples made up of grab samples whose volume is dependent upon water or wastewater flow at the time of sampling. Since composites are most commonly used for averaging, more statistical weight must be given to samples taken during high flow periods. This is accomplished by using larger sample volumes for high flows and smaller sample volumes for low flows.[2]

D. TIME, FREQUENCY, AND NUMBER OF SAMPLES NEEDED

The time and frequency of sampling is important; for example, grab samples taken for chlorine residual determinations (if not continuous) should be taken at times of peak flows. Samples used to determine dissolved oxygen (DO), pH, and fecal coliforms should be taken at minimum and maximum flows. Flow-proportional composite samples should be taken over a 24-hour period and probably as often as once an hour.[2]

Reclaimed wastewater monitoring varies from state to state and depends on the type of planned reuse. The *Manual — Guidelines For Water Reuse,*[12] gives examples of how varied sampling can be; following is an excerpt from this reference.[12]

> For unrestricted urban reuse, Arizona requires sampling for fecal coliform daily, while for agricultural reuse of non-food crops sampling for fecal coliform is only required once a month. Arizona also requires that turbidity be monitored on a continuous basis when a limit on turbidity is specified.
>
> California, Florida, and Washington also require the continuous on-line monitoring of turbidity. Oregon, on the other hand, requires that turbidity be monitored hourly for unrestricted urban and recreational reuse as well as agricultural reuse on food crops and sampling for total coliform be conducted either once a day or once a week, depending on the type of reuse application.

Washington requires continuous on-line turbidity monitoring for agricultural reuse on food crops, while California requires that total coliform samples be taken on a daily basis and turbidity be monitored on a continuous basis for unrestricted urban and recreational reuse, as well as agricultural reuse on food crops. For unrestricted and restricted urban reuse, as well as agricultural reuse on food crops, Florida requires the continuous on-line monitoring of turbidity and chlorine residual. Even though no limits on turbidity are specified in Florida, continuous monitoring serves as an on-line surrogate for SS. In addition, Florida requires that the TSS limit must be achieved prior to disinfection and that fecal coliform samples be taken daily for treatment facilities with capacities greater than 22L/s (0.5 mg/d). Florida also requires an annual analysis of primary and secondary drinking water standards for reclaimed water used in irrigation. Other states determine monitoring requirements on a case-by-case basis depending on the type of reuse.[12]

A single sample would be insufficient to give complete information for proper evaluation of a reclaimed wastewater planned for reuse. However, if an overall standard deviation for a given test is known, the following equation can be used to statistically assess the number of samples that should be taken in order to estimate a mean concentration value.[6]

$$N \geq \left(\frac{ts}{U}\right)^2 \qquad [8.1]$$

where: N = number of samples
t = student's t statistic for a given confidence level
s = overall standard deviation
U = acceptable level of uncertainty

Example 8.1.

If the standard deviation for a test is 30 mg/L and the acceptable level of uncertainty is ±6 mg/L, and a 95% confidence level is desired, how many samples should be taken?

Solution:

1. Given: s = 30 mg/L, U = 6 mg/L, p = 0.95.
2. Determine the $(1 - p)$ value as:

   $1 - p = 1 - 0.95 = 0.05$.

3. Find the student's t value for a given (1 confidence level) of 0.05 by interpolating from Table 8.1 as equal to 1.97.

4. Substitute given and computed values in Equation 8.1 to determine the number of samples:

$$N \geq (t \cdot s/U)^2 = \geq \left(\frac{1.97 \cdot 30}{6}\right)^2 = \geq 97$$

Therefore, at least 97 samples should be taken.

For information regarding basic statistical principles involving sampling, see the Appendix for this chapter.

A graphical solution for this problem can also be made. See Figure 8.1.

If $s/U = {}^{30}\!/_{\!6} = 5$, according to Figure 8.1 and using the 95% confidence level curve, the number of samples that should be taken is from 95 to 100 in order to estimate a mean concentration for the constituent being analyzed.[6]

E. SAMPLING EQUIPMENT REQUIRED

The sample containers used to transport the sample from the site to the laboratory can be glass, plastic, or hard rubber. The next consideration is size, and the third consideration is cleanliness of the containers.

For most purposes, wide-mouth plastic sample jars with plastic caps are acceptable. However, if sterilization is required or if tests such as organic analyses are to be made, glass may be required. If

Table 8.1 Student t Values for a Given Confidence Level

t	1 − p	t	1 − p
0.0	1.000	2.5	0.0124
0.1	0.920	2.6	0.0093
0.2	0.841	2.7	0.0069
0.3	0.764	2.8	0.0051
0.4	0.689	2.9	0.0037
0.5	0.617	3.0	0.00270
0.6	0.548	3.1	0.00194
0.7	0.483	3.2	0.00136
0.8	0.423	3.3	0.00096
0.9	0.368	3.4	0.00068
1.0	0.317	3.5	0.00046
1.1	0.272	3.6	0.00032
1.2	0.230	3.7	0.00022
1.3	0.194	3.8	0.00014
1.4	0.162	3.9	0.00010
1.5	0.134	4.0	0.0000634
1.6	0.110	4.1	0.0000414
1.7	0.090	4.2	0.0000266
1.8	0.072	4.3	0.0000170
1.9	0.060	4.4	0.0000108
2.0	0.046	4.5	0.0000068
2.1	0.036	4.6	0.0000042
2.2	0.028	4.7	0.0000026
2.3	0.022	4.8	0.0000016
2.4	0.016	4.9	0.0000010

Compiled from Reference 13.

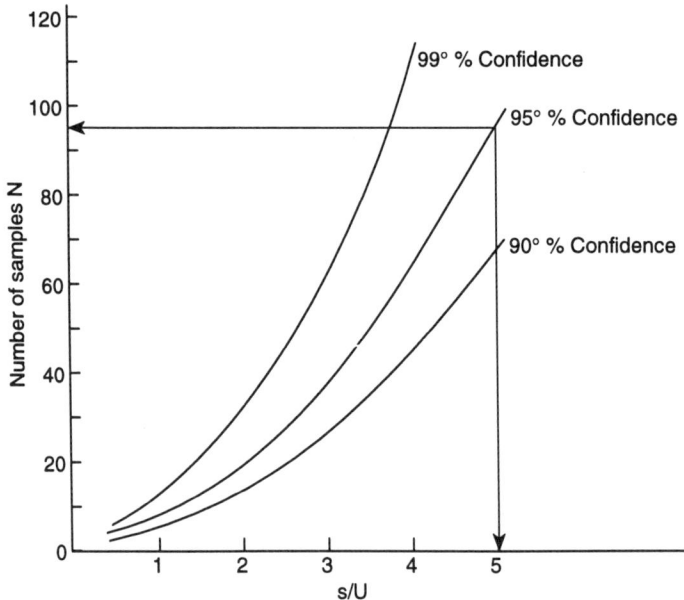

Figure 8.1 Approximate number of samples required in estimating a mean concentration. (From AWWA, *Standard Methods for the Examination of Water and Wastewater,* American Public Health Association, Washington, D.C., 1992. With permission).

bacteriological analyses are to be made, the sample containers must be sterilized; and if chlorine or its compounds are present in the water or wastewater, then a weak sodium thiosulfate solution must be added to the sample container before sterilization.

Cations of such metals as Cd, Co, or Zn can adsorb readily on the surface of some plastic or glass containers. A laboratory evaluation of this problem should be made for the sampling containers that are to be used on each project. See Table 8.2 for the recommended types of container to use for each constituent to be tested.[2,11]

The size of the sample required varies from test to test. Table 8.2 also presents the recommended volumes for each constituent to be tested.[11] See *Standard Methods* for an even more detailed and extensive list.[6]

A 2-L sample is generally adequate for most physical and chemical analyses, however, samples for chemical, bacteriological, or microscopic examination should be collected separately.[6]

The sample containers should be thoroughly washed and also distilled-water rinsed. Sample containers should not be cleaned with a phosphate detergent if the samples are to be used in a phosphorus determination, nor with a dichromate cleaning solution if the samples are to be used for metal analysis.[1]

F. VOLUME OF SAMPLE REQUIRED

A large enough sample must be collected to perform each analysis. As indicated previously, a couple of liters is generally adequate, but samples less than 100 mL are not recommended.[2]

Table 8.2 also presents information not only as to the recommended size of sample needed for each analysis, but also the type of container, the preservative to be used, and the recommended holding time.[11]

G. PRESERVATION OF SAMPLES

It is best to analyze the sample as soon as possible after collection, however, if immediate analysis is not possible then preservation measures are required.[1,6,14]

The primary reasons for the preservation of a sample are as follows:

- To retard biological action.
- To retard precipitation or the hydrolysis of chemical compounds and complexes.
- To reduce volatility of constituents.

The general methods used for sample preservation are pH control, chemical addition, refrigeration, and freezing.[14]

See Table 8.2 for recommendations for sample preservation; however, this information can also be found in the other standard references.[6-10]

H. SAMPLE IDENTIFICATION

Records should be kept of every sample that is collected. A sample identification label or tag should be made out at the time of collection. Each label or tag should include the following information:[1,5]

- Sample identity code
- Signature of sampler
- Description of sampling location, detailed enough to accommodate reproducible sampling (it may be more convenient to record the details in the field record book)
- Sampling equipment used
- Date of collection
- Time of collection
- Type of sample (grab or composite)
- Water temperature
- Sampling conditions such as weather, water level, flow rate of source, etc.
- Any preservative additions or techniques
- Record of any determinations done in the field
- Type of analyses to be done in the laboratory

I. SAMPLING PERSONNEL

Clear organizational responsibilities need to be established well before sampling starts. Sampling personnel should be aware of their responsibilities regarding picking up sampling equipment, acquiring keys or permission for obtaining access to the sampling sites, and delivering (or shipping) the samples to the laboratory.[15]

Table 8.2 Recommendation for Sampling and Preservation of Samples According to Measurement

Constituent	Volume Required (mL)	Container[a]	Preservative[b]	Holding Time[c]
Bacteriological				
Coliform	100	P,G	Cool, 4°C	Analyze as soon as possible
Physical Properties				
Color	500	P,G	Cool, 4°C	24 hours
Hardness	100	P,G	Cool, 4°C; HNO_3 to pH <2[d]	6 months
Odor	500	G only	Cool, 4°C	24 hours
pH	25	P,G	Determine on site	2 hours
Taste	500	G only	Cool, 4°C	24 hours
Total dissolved solids	100	P,G	Cool, 4°C	7 days
Temperature	1000	P,G	Determine on site	No holding
Turbidity	100	P,G	Cool, 4°C; store in dark	24 hours
Metals				
Total (dissolved and undissolved)	100 (for each metal)	P,G	HNO_3 to pH <2	6 months[e]
Total mercury	500	P,G	HNO_3 to pH <2	28 days (glass) 13 days (hard plastic)
Inorganics, non-metallics				
Acidity	100	P,G	Cool, 4°C	24 hours
Alkalinity	200	P,G	Cool, 4°C	24 hours
Chlorine	500	P,G	Determine on site	No holding
Fluoride	300	P,G	None required	28 days
Nitrate	100	P,G	Cool, 4°C; add H_2SO_4 to pH <2	48 hours
Carbon dioxide	100	P,G	Analyze immediately	No holding
Dissolved oxygen				
Electrode	300	G only	Determine on site	No holding
Winkler	300	G only	Fix on site	4–8 hours
Sulfate	50	P,G	Cool, 4°C	28 days
Organics[e]				
Pesticides and herbicides	2000	G	Cool, 4°C	7 days
Trihalomethanes	25–250	G	Cool, 4°C	Analyze as soon as possible

[a] Plastic (P) or glass (G). For metals, polyethylene with a polypropylene cap (no liner) is preferred.

[b] If the sample is stabilized by cooling, it should be warmed to 25°C for reading, or temperature correction made and results reported at 25°C.

[c] It should be pointed out that holding times listed above are recommended for properly preserved samples based on currently available data. It is recognized that for some sample types, extension of these times may be possible while for other types, these times may be too long. Where shipping regulations prevent the use of the proper preservation technique or the holding time is exceeded, such as the case of a 24-hour composite, the final reported data for these samples should indicate the specific variance.

[d] Where HNO_3 cannot be used because of shipping restrictions, the sample may be initially preserved by icing and immediately shipped to the laboratory. Upon receipt in the laboratory, the sample must be acidified to a pH <2 with HNO_3 (normally 3 mL 1:1 HNO_3/litre is sufficient). At the time of analysis, the sample container should be thoroughly rinsed with 1:1 HNO_3 and the washings added to the sample (volume correction may be required).

[e] Sample bottles should be sealed with bottle caps that are Teflon-lined.

Compiled from References 11 and 14.

J. CHAIN OF CUSTODY

The "Chain of Custody" has become very important, especially in these litigious times. "Chain of Custody" is a legal term and relates to a continuous record of a sample from the time of collection through analysis.

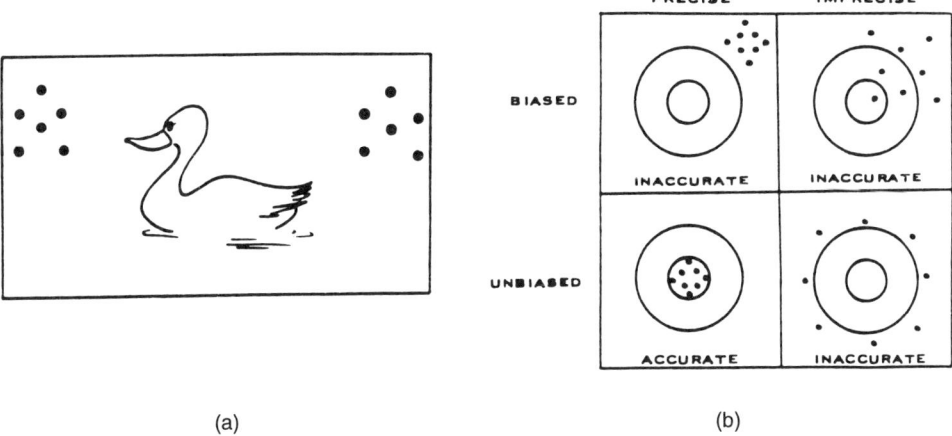

(a)   (b)

Figure 8.2 (a) On an average, the duck was dead! A hunter fired both barrels of a shotgun at a duck. The first hit 2 ft in front, the second hit 2 ft behind. The hunter had precision but not accuracy and no duck was on his table. (b) Diagram illustrating bias, precision, and accuracy. (From Chase, G. D. and Rabinowitz, J. L., *Principles of Radioisotope Methodology,* Burgess Publishing, Minneapolis, MN, 1965. With permission.)

The sampling procedures described in Section IV are followed and include sample identification, sampling equipment used, and sample preservation and transportation.

The sample's whereabouts from the time of collection through the time of analysis must be recorded.

The "Chain of Custody" record for each sample or groups of samples must include the sample number; signature of the collector; date, time, and address of collection; sample type, signatures of persons involved in the chain of possession, and inclusive dates of possession.[6]

K. SPECIAL INSTRUCTIONS

The samplers must have proper sampling instructions, especially if unusual situations take place in the sampling program. For instance, if a well on the sampling schedule has been shut down, then the well will need to be pumped to purge the stagnant water inside the casing. A useful technique is to visit the sampling locations ahead of time to determine if there are any particular access or sampling problems that should be noted.[15]

V. PRECISION AND ACCURACY

For a measurement to be accurate, it must be both precise and unbiased. Precision is a measure of the agreement among individual measurements. When a number of samples are analyzed in the laboratory, the results can be mathematically evaluated. If the values or readings or data are very close together then the results show good precision. However, precise data does not necessarily mean accurate data. Sometimes a set of precise data can be very different from the true or "absolute" value because of indeterminate errors inherent in the analytical procedures or due to the sample's characteristics. Precision can be expressed mathematically as standard deviation or relative standard deviation.[13,14]

Of course, nothing is "absolute".

Bias, on the other hand, is influenced primarily by determinate errors which introduce a constant error into the data. Bias may be measured in terms of the deviation of the sample mean from the true mean.

Accuracy is the measure of the departure of a measurement from the true or "absolute" value. Accuracy can be determined by adding a known quantity of a chemical standard (spiked sample) to a sample and then makes an analysis to determine the percent recovery. Accuracy can be expressed as a relative error.[14]

Figure 8.2(a) and (b) illustrates the terms precision, accuracy, and bias.[13]

VI. PROCEDURES FOR ANALYSIS OF IMPORTANT CONSTITUENTS IN RECLAIMED WASTEWATER PLANNED FOR REUSE

The procedures for analysis of reclaimed wastewater planned for reuse can be divided into six major categories:

- Physical properties
- Inorganic and non-metals
- Organics
- Metals
- Biological
- Radiological

A. PHYSICAL PROPERTIES

The physical properties of major concern in evaluating the quality of a water for reuse purposes include:

- Alkalinity
- Color
- Odor
- Solids (residue)
  - Total solids
  - Dissolved solids
  - Suspended solids
  - Fixed and volatile solids
  - Settleable solids
- Specific conductance
- Taste
- Temperature
- Turbidity

Alkalinity is included here under physical properties as *Standard Methods*[6] indicates it cannot be divorced entirely from the physical properties of a water sample.

Details for taste and odor are not included as this mainly applies to a potable water supply.

1. Alkalinity

The details regarding the analysis for alkalinity in a water or wastewater sample is presented in Chapter 3, Section V.D. The specific quantitative determination of alkalinity can be made using either indicator solutions or a pH meter (potentiometrically). Using a calibrated pH meter gives a greater sense of security as the colorimetric end points are sometimes difficult to discern. The combination of a pH meter and the indicator solutions is a good way to become familiar with color changes at the titration end points.

Following is an abbreviated procedure for the determination of alkalinity.

Procedure

1. Calibrate pH meter.
2. Pour a 100 mL sample into a clean dry 150-mL beaker using a 100-mL graduated cylinder; turn on the magnetic stirrer and record initial temperature.
3. Set the temperature compensation on the pH meter, rinse and dry the pH electrodes, immerse in sample, and record initial pH.
4. If the pH is greater than 8.3, add a few drops of phenolphthalein indicator solution, then add 0.02 N H_2SO_4 dropwise to the pH 8.3 end point and record volume of titrant; this is the "phenolphthalein end point". The phenolphthalein color change is from red to colorless.
5. Add a couple of drops of either bromocresol green indicator solution or methyl orange.
6. Titrate to pH 4.5 without undue delay; as the endpoint is approached make smaller additions of acid and be sure equilibrium is reached before adding more titrant.
7. Record volume of 0.02 N H_2SO_4 titrant used; this is the total or "methyl orange" endpoint. Color change for the bromocresol green indicator is from blue to yellow, for methyl orange it is from yellow-orange to pink or red.

The calculations for total alkalinity, hydroxide alkalinity, carbonate alkalinity, and bicarbonate alkalinity are presented in Chapter 3, Section V.D. and Figure 3.5.

2. Color

Color in water can be caused by minerals containing such elements as iron and manganese, or by humus and peat materials, aquatic life (plankton), and industrial or municipal wastes.[6]

Color in water is classified as either true color or apparent color. True color is due to the colloidal organic compounds in the water sample. True color is determined after the water sample has been filtered or centrifuged. Apparent color is the true color plus any other color-producing substances found in suspension, and is determined on the original sample without filtration or centrifugation.[11]

Color is measured by comparing the color of a sample with the color of a standard. The standard is based on the color produced by dissolving 1.246 g of potassium chloroplatinate (K_2PtCl_6) and 1.000 g of crystallized cobaltous chloride ($CoCl_2 \cdot 6H_2O$) in 1 L of water. This stock standard has a color of 500 Color Units. Other standards can be prepared by dilution.[6]

The color of a sample may be determined by visual comparison or by instrumental methods (spectrophotometric, tristimulus filter method).[6]

The proposed ADMI (American Dye Manufacturers Institute) tristimulus filter method presented in *Standard Methods*[6] for determination of color is an extension of the tristimulus method and has the advantage that the readings are independent of the hue of the sample. At the present time many permits and regulations indicate this method for color determination.

The drinking water color standard is related to the consumer's acceptance of the water rather than its safety. In the U.S. the secondary water quality (MCL) for color for drinking water is 15 CU,[21] while the WHO guidelines for color in drinking water is 15 true color units (TCU).[23]

In the wastewater reuse fields such as in the pulp and paper industry and the chemical and textile industries the recommended water quality requirements for color are 10 CU for the bleached pulp from the paper industry, 20 CU for the chemical industry, and 5 CU for the textile industry.[12]

3. Solids (Residue)

The total amount of solid material in a wastewater can be divided into five categories:

- Total solids
- Dissolved solids
- Suspended solids
- Fixed and volatile solids
- Settleable solids

Following are the apparatus, procedures, and calculations for the solids or residue determination in water or wastewater.

TOTAL SOLIDS
(TOTAL RESIDUE DRIED AT 103 TO 105°C)

Determination of total residual matter, suspended and dissolved, in water or wastewater.

Apparatus

- Porcelain evaporating dishes
- Muffle furnace, for operation at 550 ± 50°C
- Drying oven, for operation at 103 to 105°C
- Desiccator
- Analytical balance, 200 g capacity, capable of weighing to 0.1 mg

Procedure

1. Ignite clean evaporating dish at 550 ± 50°C for 1 hr in a muffle furnace.
2. Cool dish in air for 2 min, desiccate until cool, and weigh.
3. Transfer a measured volume of well-mixed sample to preweighed dish.
4. Evaporate sample to dryness in oven at 98°C overnight.
5. Dry evaporated sample for 1 hr at 103 to 105°C.
6. Cool dish in desiccator and weigh.

Calculation

$$\text{mg/L total residue} = \frac{(A - B) \times 1000}{\text{sample volume (mL)}} \qquad [8.2]$$

where: A = weight of dried sample + dish (mg)
B = weight of dish (mg)

TOTAL DISSOLVED SOLIDS
(TOTAL FILTRABLE RESIDUE DRIED AT 103 TO 105°C)

Filtrable residue is material that passes through a standard glass-fiber filter and remains after evaporation and drying to constant weight at 103 to 105°C.

Apparatus
All apparatus listed for total residue, and in addition:

- Glass-fiber filters, circular, without organic binder
- Filtration apparatus; Gooch crucible
- Suction flask

Procedure

1. Place filter on the Gooch crucible. Apply vacuum and wash filter with three successive 20-mL volumes of distilled water. Continue suction to remove all traces of water. Discard washings.
2. Ignite clean evaporating dish at 550 ± 50°C for 1 hr in a muffle furnace. Cool and store in desiccator until needed. Weigh immediately before use.
3. Filter a measured volume of well-mixed sample through prewashed filter, wash with three successive 10-mL volumes of distilled water, and continue suction for about 3 min after filtration is complete.
4. Transfer filtrate to preweighed dish and evaporate to dryness at 98°C overnight.
5. Dry evaporated sample for 1 hr at 103 to 105°C.
6. Cool dish in desiccator and weigh.

Calculation

$$\text{mg/L total filtrable residue} = \frac{(A-B) \times 1000}{\text{sample volume (mL)}} \qquad [8.3]$$

where: A = weight of dried residue + dish (mg)
B = weight of dish (mg)

TOTAL SUSPENDED SOLIDS
(TOTAL NONFILTRABLE RESIDUE DRIED AT 103 TO 105°C)

Total nonfiltrable residue is the retained material on a standard glass-fiber filter after filtration of a well-mixed sample.

Apparatus
All apparatus listed for total filtrable residue is required.

Procedure

1. Place filter on Gooch crucible. Apply vacuum and wash filter with three successive 20-mL volumes of distilled water. Continue suction to remove all traces of water, and discard washings.
2. Remove crucible and filter combination and dry in oven at 103 to 105°C for 1 hr. Cool in desiccator and weigh immediately before use.
3. Filter a measured volume of well-mixed sample through a preweighed filter-crucible apparatus. Wash filter with three successive 10-mL volumes of distilled water.
4. Carefully remove crucible and filter combination and dry for 1 hr (to constant weight) at 103 to 105°C. Cool in a desiccator and weigh.

Calculation

$$\text{mg/L total nonfiltrable residue} = \frac{(A-B) \times 1000}{\text{sample volume (mL)}} \qquad [8.4]$$

where: A = weight of filter + residue (mg)
B = weight of filter (mg)

FIXED AND VOLATILE SOLIDS
(FIXED AND VOLATILE SOLIDS IGNITED AT 500°C)

The residue from the total solids, total dissolved solids, or total suspended solids are ignited to a constant weight at 500 ± 50°C. The remaining solids represent the fixed total, fixed dissolved, or fixed suspended solids, while the weight lost on ignition is the volatile solids.[6]

Calculation

$$\text{mg volatile solids/L} = \frac{(A - B) \times 1000}{\text{sample volume, mL}} \qquad [8.5]$$

$$\text{mg fixed solids/L} = \frac{(B - C) \times 1000}{\text{sample volume, mL}} \qquad [8.6]$$

where: A = weight of residue + dish or filter before ignition, mg
B = weight of residue + dish or filter after ignition, mg
C = weight of dish or filter, mg

SETTLEABLE SOLIDS

Settleable solids represent that portion of the suspended solids that are of sufficient size and weight to settle in a given period of time, usually one hour. Two or more glass 1-L Imhoff cones graduated in milliliters (mL) at the lower end of the cone are used. These cones are generally held in a wooden rack.

The testing procedure is as follows:[3,14]

1. A measured quantity, usually 1 L, of a well-mixed sample is gently poured into a well-cleaned cone (use soap and a brush). Wetting the inside cone surface before use can help to prevent solids from sticking to the sloping glass sides. The sample is allowed to settle or stand for a period of one hour.
2. After the sample has stood 45 min, gently rotate the cone between the hands so as to loosen the solids that adhere to the sides.
3. Allow to settle 15 min longer.
4. Read from the graduations the volume of solid material deposited in the cone, making allowances for any unfilled portions of the cone below the level of the settled solids.

Results are expressed as milliliters of solids per liter which settled in 1 hr:

$$\text{mL of solids} \times \frac{1000}{\text{mL of sample}} = \text{mL of settleable solids per liter (mL/L)} \qquad [8.7]$$

While the results can be reported volumetrically they can also be determined gravimetrically.[2] The gravimetric analysis is made by first determining the total suspended solids of the sample, as described previously.

After the sample has settled for 1 hr in the cone, siphon 250 mL from the center of the cone at mid-depth and determine the suspended solids.

The calculations are as follows:[3,14] mg/L settleable solids = mg/L, total initial suspended solids − mg/L of suspended solids of sample at mid-depth after 1 hr settling.

For untreated wastewater the settleable solids are approximately 75% organic and 25% inorganic.[3]

4. Specific Conductance or Electrical Conductivity

Specific conductance or electrical conductivity of a solution is a measure of its ability to carry an electrical current. This electrical conductivity is due to the anions and cations in the solution. The conductivity or conductance of a solution is the reciprocal of its resistance and is given in units of μmhos, mhos, or Seimens (all a reciprocal of ohms).

1 milliSiemens per meter (mS/m) = 10 μmohs/cm
1 microSiemen per centimeter (μS/cm) = 1 μmohs/cm

To report conductivity in the International System of Units (SI) of mS/m divide µmhos/cm by 10.[5]

The specific conductance, κ, of a solution is the conductance of 1 cm^3 of solution between electrodes of 1 cm^2 area which are 1 cm apart, and has units mho/cm. Conductivity measurements are made in a conductivity cell using an AC Wheatstone bridge (in reality, resistance measurements are made and converted to conductance on the instrument read-out). Since the cell used for measuring conductivity does not have electrodes exactly 1 cm^2 in area and 1 cm apart, the reading must be corrected to standard conditions by using a cell constant, q.[16,17]

Commercial conductivity meters are readily available, and these meters measure not only specific conductance but also temperature and salinity. The readings can also be corrected to 25°C when the temperature dial on the meter is set to the sample temperature. Specific conductance for such instruments are reproducible to within ±0.4%.[18]

Following is a brief description of how to measure conductivity using a small conductivity meter.

Equipment and Supplies

Conductivity cell, 32 to 120°F, temperature compensated, beakers, plastic wash bottle. Standard KCl (0.01 N) solution having conductance of 1411 µmohs/cm, distilled water.

Calibration

1. Rinse cell cup three times with distilled water.
2. Rinse the cell cup once with standard KCl.
3. Fill the cell cup with standard solution KCl (0.01 N).
4. Select the range and push button to read directly, reading should be 1411 µmhos/cm, if reading is different set the needle by using calibration control screw located at the bottom of the instrument.

Sample Testing

1. Rinse the cell cup three times with the sample to be tested.
2. Fill the cell cup with sample.
3. Select the range and push button to read directly conductivity in µmohs/cm.

5. Temperature

Temperature is one of the most important measurements and can be easily made. Temperature has a profound effect on all chemical and biological processes. Temperature measurements can be used in the following ways:[6,19]

- Calculate percent saturation of dissolved oxygen for a water.
- Determine a water's stability with respect to calcium carbonate.
- Calculate salinity at various temperatures.
- Determine the source of a water supply, whether surface or groundwater.
- Determine infiltration to a wastewater collection system.
- Help locate hot-water discharges to a municipal treatment system or to a surface body of water.

Standard Methods[6] indicates that temperature measurements are normally made with any good mercury-filled Celsius thermometer. As a minimum, the thermometer should have a scale marked for every 0.1°C, with markings etched on the capillary glass. The thermometer should have a minimal thermal capacity to permit rapid equilibration. Periodically, check the thermometer against a precision thermometer certified by the National Institute of Standards and Technology (NIST, formerly the National Bureau of Standards) that is used with its certificate and correction chart. For field operations use a thermometer having a metal case to prevent breakage.[6]

6. Turbidity

Turbidity in water is caused by the presence of suspended matter such as clay, silt, colloidal organic particles, viruses, bacteria, algae, planktons, and other microscopic organisms.[20] Turbidity can be defined as the optical property of the water sample to cause light to be scattered or absorbed rather than transmitted through the sample.

Turbidity in water is known to:[21]

- Interfere with disinfection because it can form a shield around disease-causing organisms.
- Make it difficult to maintain a chlorine residual.

- Interfere with bacteriological testing of the water.
- Act as a food source for microorganisms.

Historically, turbidity in water or wastewater was based on the Jackson Candle Turbidimeter and reported in Jackson Turbidity Units (JTUs). This method had limited use as it can only measure turbidities greater than 25 JTUs. The present-day technique most often used to determine turbidity in water or wastewater is the nephelometric method.

The nephelometric method of measuring turbidity measures the light scattered at a 90° angle to a beam of light. As the amount of scattered light increases, the turbidity value increases. Turbidity determined by the nephelometric method is expressed as Nephelometric Turbidity Units (NTUs).[20]

While it is not recommended to correlate turbidity and suspended solids concentration, there are cases where a relationship, if carefully monitored, can be developed.

After the extensive testing of a specific reclaimed wastewater, a relationship between turbidity and the removal of pathogenic organisms including viruses has been established.[22]

Several nephelometric turbidimeters are available and approved by EPA. Manufacturers include Hach Chemical Co. (Model 2000, 18900, or equivalent); HF Instruments (Fisher Scientific Co.); and Turner Designs. Sample tubes and operational instruction are generally provided with each instrument.[20]

Following is a summary of the overall operational procedure for turbidity measurements.[20]

Procedure

- Collect sample
- Warm up instrument
- Calibrate instrument
- Clean sample cell
- Mix sample
- Fill sample cell
- Wipe outside of cell
- Place in instrument and cover
- Adjust scale range
- Dilute (if necessary)
- Read and record NTU
- Clean sample cell
- Calculate (if necessary)

Calculations

$$\text{NTU} = \frac{A \times (B+C)}{C} \qquad [8.8]$$

where: A = NTU of dilute sample
B = mL of water added
C = mL of undiluted sample

Table 8.3 presents a guide as to the turbidity reading that might be anticipated depending on the source of the sample.[11]

The present U.S. National Interim Primary Drinking Water Standard for turbidity is 1 NTU based on a monthly average, and up to 5 NTU may be allowed for a monthly average if it can be demonstrated that no interference occurs with disinfection or microbiological determinations.[21] The World Health Organization *Guidelines For Drinking Water Quality* recommends a turbidity of below 1 NTU and always less than 5 NTU.[23]

The turbidity criteria, guidelines, or regulation requirements for reclaimed wastewater vary depending on the planned reuse.

For instance, a turbidity criteria limit of 5 NTUs is often quoted for agricultural reuse.[12]

B. ANALYSIS FOR INORGANICS AND NON-METALS

The results of the analyses for the detection of inorganics and non-metallic constituents can be used to assess not only the quality of the reclaimed wastewater planned for reuse, but also to evaluate the various treatment processes.

Table 8.3 Scale of Nephelometric Turbidity Measurements

1000 NTU	Highly turbid surface water
100 NTU	Average for Missouri River
10 NTU	Average for lake water
1 to 5 NTU	Coagulated and settled water (filter influent)
0.1 NTU	Finished-water turbidity (filter effluent)

From AWWA, *Introduction to Water Quality Analysis,* Vol. 4, Denver, CO, 1982. With permission.

The major inorganic and non-metal constituents of concern for reclaimed wastewater reuse include:

- Boron
- Chlorides
- Chlorine
- Dissolved oxygen
- Nitrogen
  - Ammonia nitrogen
  - Nitrate nitrogen
  - Nitrite nitrogen
  - Total nitrogen
- pH
- Phosphates
- Sulfates

1. Boron

As indicated in Chapter 2, Section V.A.3, a major toxicity problem for crops irrigated with reclaimed wastewater is boron. Boron concentrations in freshwater rarely go above 1.0 mg/L and are generally less than 0.1 mg/L. Typical wastewater influent boron concentration are in the 1.0 mg/L range, while typical effluent concentrations for plants providing secondary treatment and chlorination are around 0.75 mg/L.[24]

Several analytical methods are available to determine boron concentrations in water and wastewater. Following are some of these methods:[25]

- Colorimetric methods: guinalizarin, purpurin, alizarin 5, etc.
- Mannitol potentiometric method: involves boron, mannitol, and titration with a NaOH solution.
- Indirect method: involves boron fluoride ion (BF_4^-) decolorized with $FeCl_3$ and sulfosalicyclic acid solutions.
- Carmine method: involves color change from bright red to bluish red or blue.
- Curcumin method: involves formation of a red color (see Figure 8.3).
- Inductively coupled plasma (ICP) method.

The ICP method for boron has an estimated detection limit of 5 µg/L and an upper limit concentration of 50 mg/L. However, the ICP unit and the spectrometer that goes with it are very expensive and generally are only located at major laboratory facilities.

The curcumin method for boron is applicable in the 0.10 to 1.0 mg/L range, while the carmine method is suitable for the determination of boron concentrations in the 1 to 10 mg/L range. Of course, the ranges for these methods can be extended by dilution or concentration of the sample.[6] At present only the carmine, curcumin, and ICP methods are presented in the *Standard Methods For Examination Of Water And Wastewater,* 18th edition, 1992.[6]

The *1985 Annual Book of ASTM Standards,* Vol 11.02, presents the carmine, curcumin, and mannitol potentiometric methods.[7]

Detailed step-by-step directions for the determination of boron in water or wastewater can be found in either of these publications.[6,7]

2. Chlorides (Cl⁻)

Humans excrete about 6 g of chlorides per day, mainly in urine. Sodium chloride (common salt), a major flavoring compound in the diet, passes directly through the digestive system unchanged, and on the average increases the Cl⁻ concentration in wastewater about 20 to 75 mg/L above that found in the original water supply.[24,27]

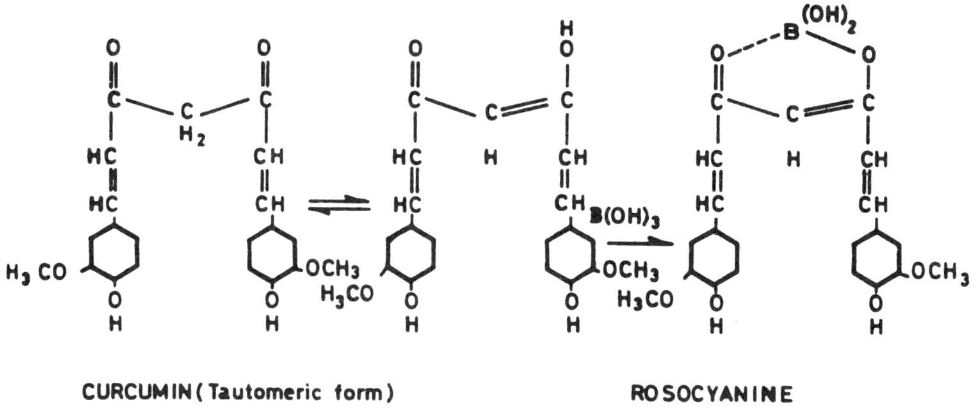

CURCUMIN (Tautomeric form)   **ROSOCYANINE**

Figure 8.3 Boron forms a red-colored product, rosocyanine, with curcumin.[25]

Chlorides can create taste problems in drinking water if over 250 mg/L, causing a salty taste when combined with sodium. However, not only can high chloride concentrations create problems in drinking water, but may also create problems for the reclaimed wastewater that is planned for reuse.

The problems associated with high chloride concentrations in reclaimed wastewater include corrosion, especially of steel and even some stainless steels, but also crop damage mainly in fruit trees and ornamental shrubs. Levels of chlorides in raw municipal sewage range from 10 to 650 mg/L, while chloride concentrations in secondarily treated wastewater range from 40 to 200 mg/L.[26]

Five methods are presented in *Standard Methods* for the determination of chlorides.[6] These include:

- Argentometric method (Mohr-silver nitrate)
- Mercuric nitrate method
- Potentiometric method
- Automated ferricyanide method
- Ion chromatographic method

The argentometric method, employing silver nitrate as the titrant and potassium chromate as the indicator, is often used. The mercuric nitrate method has some advantages, such as the distinct end point provided by the diphenylcarbazone indicator.[27] However, it must be kept in mind that extra precautions must be exercised in disposal of solutions containing mercury.

Three methods are currently acceptable for NPDES chloride monitoring. They are the silver nitrate, mercuric nitrate, and the automated colorimetric ferricyanide methods.[19]

Following is an abbreviated procedure for determination of chloride concentrations using the mercuric nitrate method. Standard solutions can be purchased from various chemical suppliers.

A 50 mL water sample is titrated with mercuric nitrate, forming soluble mercuric chloride. In the pH range 2.3 to 2.8, diphenylcarbazone indicates the end point by formation of a purple complex with excess mercuric ions.

Apparatus

- Erlenmeyer flasks, 250 mL
- Buret

Reagents

- Indicator-acidifier reagent
- Standard mercuric nitrate titrant, 0.0141 N
- Blank: distilled water + $NaHCO_3$

Procedure

1. Measure a 50-mL sample with graduated cylinder and pour into a 250 mL Erlenmeyer flask.
2. Add 1.0 mL of indicator-acidifier reagent; color of solution should be green-blue.

3. Titrate treated sample with 0.0141 N mercuric nitrate to a definite purple end point.
4. Determine blank by titrating 50 mL distilled water containing 5 mg $NaHCO_3$.

Calculation

$$\text{mg/L Cl}^- = \frac{(A-B) \times N \times 35,450}{\text{sample volume (mL)}} \quad [8.9]$$

where: A = mL titration for sample
B = mL titration for blank
N = normality of $Hg(NO_3)_2$

The 1985 *ASTM Standards*[8] presents four methods for determining chloride ion concentrations in water: the mercurimetric titration, silver nitrate titration, colorimetric, and ion-selective electrode methods. Following is a brief summary of the ion-selective electrode method.[8]

1. The chloride ion is measured potentiometrically using a chloride ion-selective electrode in conjunction with a double junction, sleeve-type reference electrode. Potentials are read using a pH meter having an expanded millivolt scale, or a selective-ion meter having a direct concentration scale for chloride.
2. The electrodes are calibrated in known chloride solutions, and the concentrations of unknowns are determined in solutions with the same background. Samples and standards should be at the same temperature.
3. Standards and samples are diluted with an ionic strength adjustor that also minimizes possible interferences such as ammonia, bromide, iodide, cyanide, or sulfide.

This method has been found to be simple, direct, dependable, and fast, and does not result in the creation of a hazardous waste that must be carefully disposed of. Of course, one must have the appropriate apparatus such as a pH meter or a selective-ion meter to use in conjunction with a chloride-ion selective electrode.[28]

3. Chlorine (Cl_2)

To control waterborne diseases, disinfection of drinking water and wastewater is widely practiced throughout the world. As indicated earlier, (see Chapter 5, Section IV.E. and Chapter 6, Section V) disinfection is the killing of pathogenic organisms while sterilization is the killing of all organisms. Chlorine is the most often used disinfectant for control of pathogens in water and wastewater.

Chlorine is a toxic, greenish-yellow gas about 2½ times heavier than air and only slightly soluble in water. One volume of liquid chlorine under pressure vaporizes to about 450 volumes of gas. Free chlorine is a strong oxidizing agent and therefore is extremely corrosive. It will react readily with reducing agents such as hydrogen sulfide (H_2S), iron (Fe^{2+}), manganese (Mn^{2+}), and nitrites (NO_2^-).

Figure 8.4 presents the principal characteristics of the free available chlorine (Cl_2, HOCl, OCl^-) and the combined chlorine (mono, di-, and trichloramines). The chloramines have less potent bactericidal power than the free chlorine.[29]

HOCl has been found to be a more effective disinfectant than OCl^-. The degree of breakdown or ionization or dissociation of HOCl is pH dependent. Table 8.4 presents the theoretical percentages of HOCl and OCl^- in pure water at various pH values, and is based on the following reactions:

$$HOCl \rightleftarrows H^+ + OCl^- \quad [8.10]$$

The ionization constant for this reaction is 2.9×10^{-8} at 25°C.

$$\frac{[H^+][OCl^-]}{[HOCl]} = 2.9 \times 10^{-8} \text{(at 25°C)}$$

$$\text{or,} \quad \frac{[OCl^-]}{[HOCl]} = \frac{2.9 \times 10^{-8}}{H^+} \quad [8.11]$$

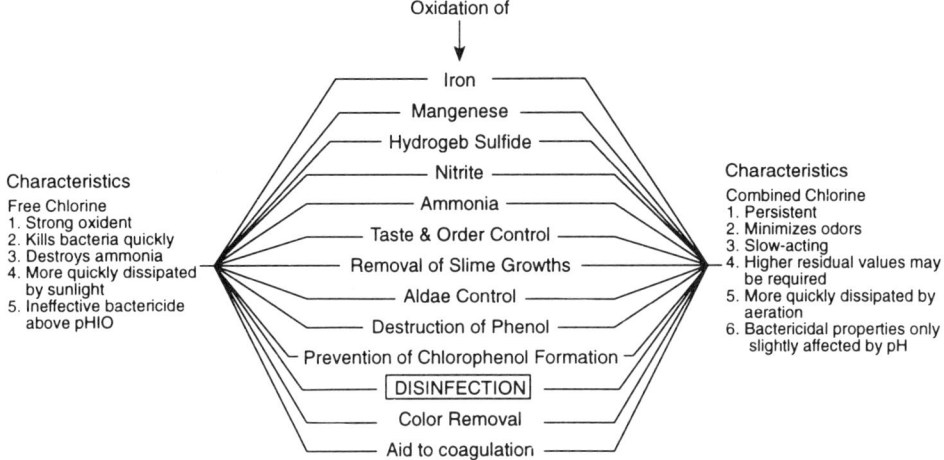

Figure 8.4 A guide to the properties of chlorine residuals. (From Anon., *Manual of Water Utility Operations*, Texas Water Utilities Association, Austin, 1975. With permission.)

For instance, at pH 7 [H⁺] = 10⁻⁷ and with a dissociation constant of 2.9×10^{-8} at 25°C the following percentages of HOCl can be calculated. The same calculations can be made for the various pH values.

$$\frac{[HOCl]}{[HOCl]+[OCl^-]} = \frac{1}{1+\frac{[OCl^-]}{[HOCl]}} = \frac{1}{1+\frac{2.9 \times 10^{-8}}{[H^+]}}$$

substituting [H⁺] = 10⁻⁷.

$$\text{The percentage HOCl} = \frac{1}{1+2.9 \times 10^{-1}}$$

$$= \frac{1}{1+0.29}$$

$$= 0.775$$

$$= 77.5\%$$

While chlorine may react with water to form both free available chlorine and combined chlorine, it can also react with organic compounds such as humic and fulvic acids which may be present in the water and produce trihalomethanes (THMs) such as chloroform, bromoform, bromodichloromethane, and other compounds, some of which can be carcinogenic for man.

Table 8.4 Theoretical Percentages of Hypochlorous Acid (HOCl) and Hypochlorite Ion (OCl⁻) at Various pHs at 25°C

pH	% HOCl	% OCl⁻
4	100.0	0.0
5	99.7	0.3
6	97.2	2.8
7	77.5	22.5
8	25.6	74.4
9	3.3	96.7
10	0.3	99.7
11	0.03	99.97

Other problems associated with chlorine and its compounds include taste and odor problems, especially when phenols are present. Chlorine residuals in wastewater discharges can also kill fish in surface waters such as streams, rivers, lakes, reservoirs, or estuaries.

Chlorine concentrations in water and wastewaters are analyzed basically in three ways: chemical titration, colorimetrically, and by electrochemical instrumentation.[6] The chemical titration methods include the following:

- Iodometric method I.
- Iodometric method II.
- DPD ferrous titrimetic method.

The colorimetric methods include:

- DPD — colorimetric method.
- Syringaldazine (FACTS) method.

The electrochemical instrumentation methods include:

- Amperometric titration method.
- Low-level amperometric titration method.
- Iodometric electrode technique.

In wastewater, the separation of the concentrations of free and combined chlorine is generally not made as the chlorination process is seldom carried past the breakpoint.

The most accurate and complete test for determining free available and combined chlorine concentrations is the amperometric method. This is due to the fact that the amperometric method is not subject to interference from color, turbidity, iron, manganese, or nitrite nitrogen.[6] *Standard Methods*[6] indicates that for total chlorine determinations containing significant amounts of organic matter, the DPD ferrous titrimetric method, the DPD colorimetric method, the amperometric, or the iodometric back titration method be used.[6]

The DPD method and amperometric method can be used not only to determine the chlorine residuals in polluted waters but also to allow for an estimation of the monochloramines and dichloramines present.[6]

4. Dissolved Oxygen (DO)

Dissolved oxygen (DO) is a very necessary constituent in water and reclaimed wastewater in order to prevent anaerobic conditions from developing. A lack of DO can result in anaerobic conditions which produce undesirable gases such as H_2S, CH_4, and NH_3.

Two methods most often used to determine DO concentrations are the Winkler method (iodometric method) and the sodium azide modification of the Winkler method, as well as the membrane electrode method.[6]

Extreme care must be exercised in collecting the samples in the standard 300 mL BOD bottles in order to exclude the small air bubbles that collect around the neck of the bottle. If the Winkler method is used, the DO determination must be made at the time of collection or "fixed" at the site. If the membrane electrode method is used the DO concentration must be determined at the site.

The choice of a method for DO analysis depends on the presence of interfering compounds, the accuracy desired, the number of samples to be analyzed, and the flexibility of the laboratory.

Following is an abbreviated laboratory procedure for the determination of DO by the modified Winkler method:

Apparatus

- Buret, graduated at 0.1 mL increments with a 50-mL capacity
- BOD bottles, 300 mL capacity
- Erlenmeyer flask, 250 mL
- A 10 mL measuring pipet

Reagents

- Manganous sulfate solution
- Alkaline iodide-sodium azide solution
- Sulfuric acid (concentrated)
- 0.025 N sodium thiosulfate solution or 0.025 N phenylarsine oxide (PAO) solution
- Starch solution

Standardized solutions can be purchased from chemical suppliers, or see *Standard Methods* for reagent preparation.

Procedure

1. Completely fill a 300-mL BOD bottle with the sample to be analyzed by siphoning the sample slowly into the bottle and allowing it to overflow for a period to displace the volume of the bottle two or three times.
2. By holding the tip of the pipet below the surface of the liquid, add 2 mL manganous sulfate solution and 2 mL alkaline iodide-sodium azide solution. (Sample is fixed.)
3. Replace stopper, avoiding trapping air bubbles, and mix well by gentle inversion. Repeat mixing after floc has settled halfway. Allow floc to settle again.
4. Remove stopper and add 2 mL of concentrated sulfuric acid. Hold pipet above the surface of the liquid. Mix until no floc is visible but not in direct sunlight.
5. Withdraw 203 mL of the solution into the Erlenmeyer flask and titrate with sodium thiosulfate solution or PAD solution until the yellow color almost disappears.
6. Add 1 mL starch solution and continue titration until the blue color just disappears.
7. Record the mL of the thiosulfate or PAD used. Disregard any return of the blue color.

Calculations

The dissolved oxygen present is expressed in mg/L and is equal to the number of mL of sodium thiosulfate or PAD solution used in the titration.

If the membrane electrode method is used, then follow the manufacturer's instructions. The instrument probe must be calibrated frequently and the membrane changed regularly. The membrane must be kept moist and carefully cleaned before storing otherwise it will dry out. Reactive gases and sulfur compounds can also damage the membrane.[30]

5. Nitrogen

Organic matter is mainly composed of carbon, oxygen, hydrogen, nitrogen, sulfur, and phosphorus (C, O, H, N, S, and P). The nitrogen in the organic matter plays an important role in the life functions of all plants and animals. Figure 8.5 presents both the aerobic and anaerobic processes involved in the nitrogen cycle.[31] In the areas of water and wastewater the forms of nitrogen of most interest are ammonia-nitrogen (NH_3-N), nitrite-nitrogen (NO_2^--N), nitrate-nitrogen (NO_3^--N), and organic nitrogen.

Ammonia-nitrogen in raw domestic wastewater generally ranges between 10 and 40 mg/L, while in the secondarily treated effluent NH_3-N concentration varys from 5 to 20 mg/L.[24]

Standard Methods[6] presents seven procedures that may be followed to determine NH_3-N concentrations in water or wastewater. These are as follows:

- Kjeldahl method (distillation)
- Nesslerization method (colorimetric)
- Phenate method (colorimetric)
- Titrimetric method
- Selective-electrode method
- Selective-electrode method using known addition
- Automated phenate method

For high NH_3-N concentrations, distillation and titration techniques are recommended.

Nitrite-nitrogen (NO_2^--N) is a very unstable form of nitrogen and is a temporary step in the nitrogen cycle. NO_2^--N is generally at the nondetectable (ND) level in untreated wastewater and from ND to 0.05 mg/L in secondary effluents.

Standard Methods[6] presents two methods for NO_2^--N determination and these are as follows:

- Colorimetric method (sulfanilamide naphthylamine hydrochloride method) (detection range is 10 to 1000 mg/L NO_2^--N)
- Ion chromatographic method (minimum detection limit is 0.1 mg/L NO_2^--N)

Nitrate-nitrogen (NO_3^--N) is the final product in the aerobic decomposition of organic matter containing nitrogen. Under anaerobic conditions, the oxygen combined in the nitrates (NO_3^-) is utilized and nitrites (NO_2^-) and nitrogen gas (N_2) are produced (see Figure 8.5).

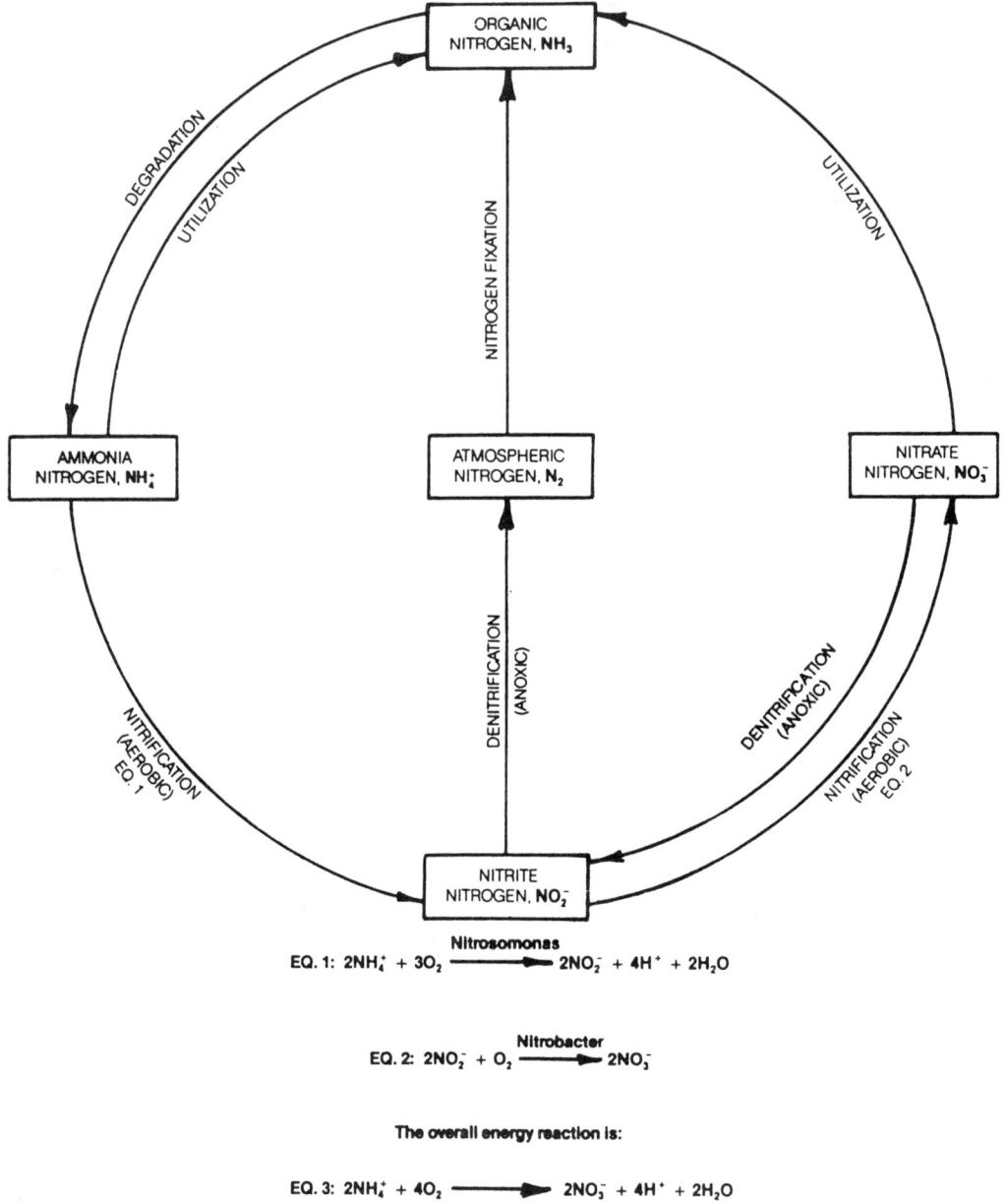

Figure 8.5 Wastewater nitrogen cycle. (From WPCF, *Operation of Municipal Wastewater Treatment Plants*, Alexandria, VA, 1991. With permission.)

NO_3^--N concentrations in untreated wastewater are generally at nondetectable levels, while in secondary effluent NO_3^--N concentrations can range from nondetectable to 50 mg/L.[19]

Standard Methods[6] presents seven methods for NO_3^--N determination, two of which are proposed methods. Following are the first five methods:[6]

- Ultraviolet screening method (screen a sample then select a method)
- Ion chromatographic method (minimum detection limit is 0.1 mg/L NO_3^--N)
- Nitrate electrode method (detection range is 0.14 to 1400 mg/L NO_3^--N)
- Cadmium reduction method (detection range is 0.01 to 1 mg/L NO_3^--N)
- Automated cadmium reduction method (detection range is 0.5 to 10 mg/L for either NO_2^- or NO_3^--N)

Table 8.5 Relation between pH, H⁺, and OH⁻ at 25°C

Activity of H⁺ moles/L	Activity of OH⁻ moles/L	pH
1.	0.000 000 000 000 01	0
0.1	0.000 000 000 000 1	1
0.01	0.000 000 000 001	2
0.001	0.000 000 000 01	3
0.000 1	0.000 000 000 1	4
0.000 01	0.000 000 001	5
0.000 001	0.000 000 01	6
0.000 000 1	0.000 000 1	7
0.000 000 01	0.000 001	8
0.000 000 001	0.000 01	9
0.000 000 000 1	0.000 1	10
0.000 000 000 01	0.001	11
0.000 000 000 001	0.01	12
0.000 000 000 000 1	0.1	13
0.000 000 000 000 01	1.	14

The cadmium reduction method measures both nitrite and nitrate nitrogen present in a sample by reducing the NO_3^- to NO_2^- (Cu-Cd column) and then colorimetrically determining the NO_2^- concentration.

The determination of the organic nitrogen concentration indicates the amount of undigested nitrogenous organic matter present in both treated and untreated wastewater.

The concentration of organic nitrogen in untreated domestic wastewater ranges from 8 to 35 mg/L (averaging around 13 mg/L), while in secondary effluents it may range from 5 to 30 mg/L.[19,26]

Standard Methods[6] presents the following two procedures for determination of organic nitrogen:

- Macro-Kjeldahl method.
- Semi-micro Kjeldahl method.

The selection of either method depends upon the organic nitrogen concentration in the sample. The macro-Kjeldahl method can be used for both high or low organic nitrogen concentrations while the semi-micro method should be used if the sample contains 0.2 to 2 mg of organic nitrogen.

The above two methods can be used to give the organic nitrogen concentration in a sample and also the NH_3-N concentration.

6. pH

pH is defined as the negative log of the hydrogen ion concentration. Water (H_2O) is slightly ionized to hydrogen (H^+) and hydroxide (OH^-) ions. A hydrogen (H^+) ion concentration of 0.000001 mol/L (for hydrogen, 1 mol/L = 1 mg/L) would be a pH of 6. The pH scale ranges from 0 to 14, with 7 being neutral. pH values less than 7 are considered acidic while those values above 7 are considered basic.

$$\left(-\log[H^+]\right) + \left(-\log[OH^-]\right) = 14 \qquad [8.12]$$

$$pH + pOH = pKw = 14 \qquad [8.13]$$

Table 8.5 presents the relationship between pH (H^+) and (OH^-) ion concentrations at 25°C.[3,30]

For instance, if the hydroxide (OH^-) ion concentration is 0.000001 mol/L (10^{-6} [OH^-]) then the pH of the solution would be 8. See Table 8.5 or make the following calculations:

$$-\text{Log}\left(10^{-14}/10^{-6}\right) = \log 10^{-8} = 8 = pH$$

For pH values that fall between whole numbers such as, say, a pH of 5.64, by definition, the (H^+) ion concentration may be calculated as follows:

$$\log(H^+) = -5.64 \text{ or } (0.36 - 6 = -5.64)$$

$$(H^+) = 10^{0.36} \cdot 10^{-6}$$

$$= 2.3 \times 10^{-6} \text{ mol/L } (H^+)$$

pH, just like temperature, is one of the most basic and important tests that can be made on water or wastewater. All chemical and biological reactions or activity are related to the pH of the processes involved.

There are several ways to make a good estimate of the pH as well as ways to make accurate and precise pH determinations.

A good estimate of the pH of a water sample can be made by using Hydrion® pH paper. This simple test depends upon the color change in the paper. Another simple pH test can be made with a hand-held electrochemical cell, however, it too gives only a good estimate. There are colorimetric kits available on the market for pH determinations, but this method is subject to interference due to color, turbidity, and dissolved solids present in the sample. Both *Standard Methods*[6] and *ASTM Standards*[8] present as the recommended procedure for pH determination the electrometric measurement using a glass electrode or electrodes as the sensor.[6,8]

The pH (hydrogen ion concentration) is measured as the difference in potential across a glass membrane in contact with the sample and with a reference solution. The sensor apparatus might be combined into one probe or divided into an indicating electrode (for the sample) and a reference electrode (for the reference solution). Before using, the meter must be calibrated with a solution of known pH (a buffer) and then checked for proper operation with a buffer of a different pH value.[1]

Following is a procedure for measuring pH using an electrometric glass electrode instrument.

Equipment

- pH meter with temperature compensation adjustment, range 0–14, 0.1 pH readability.
- Combined electrode which contains both glass and reference electrodes.
- Filling solution for electrode.
- Beakers: 150 mL or 250 mL size
- Cleaning tissues
- Thermometer, 1°C subdivision
- Wash bottle, plastic

Reagents

- pH 4.00 buffer solution
- pH 7.00 buffer solution
- pH 9.20 buffer solution
- Distilled water

Procedure

1. *Calibration*
    1. Store meter on "standby" with electrode immersed in distilled water.
    2. Pour buffer solutions into separate beakers.
    3. Find temperature of pH 7 buffer solution and adjust temperature compensation knob.
    4. Remove electrodes from water, rinse thoroughly with distilled water, dry with soft tissue, and immerse in pH 7.00 buffer solution.
    5. Set slope at 1.00.
    6. Press meter knob to "pH" and use calibration knob to adjust needle on meter to read 7.00.
    7. Press meter knob to "standby", remove electrodes from solution, rinse thoroughly with distilled water, dry with soft tissue, and immerse in pH 4.00 solution.
    8. Press meter knob to "pH", wait till needle reaches steady position and note the pH reading; the needle should be within 0.1 unit of 4.00. If not, start again at Step 4.
    9. Repeat Steps 6 and 7 using pH 9.20 buffer solution.
    10. Remove electrodes from solution, rinse thoroughly with distilled water, and immerse in beaker of distilled water.
    11. Discard buffer solutions in beakers.

2. pH Measurement of Sample

1. Pour sample into beaker.
2. Find temperature of sample and adjust temperature compensation knob.
3. Remove electrode from water, rinse thoroughly with distilled water, dry with soft tissue, and immerse in sample.
4. Press meter knob to "pH" and agitate the sample by stirring.
5. Record pH and temperature when needle reaches a steady position.
6. Remove electrode from sample, rinse thoroughly with distilled water, and immerse in beaker of distilled water.

7. Phosphates

Phosphorus is a nonmetallic element required by all forms of life. In nature it is chiefly available to plants and animals as orthophosphates.

The common compounds of phosphorus are orthophosphates ($H_2PO_4^-$, HPO_4^{2-}, PO_4^{3-}), condensed phosphates (pyro-, meta-, and other polyphosphates such as $Na_3(PO_3)_6$ used in synthetic detergent formulations), and organic phosphorus. All condensed phosphates gradually hydrolyze in water to the stable ortho form. Orthophosphates, in turn, are synthesized back into living animal or plant tissue.

Typical total phosphate concentrations in untreated wastewater range from 2 to 20 mg/L, including 1 to 5 mg/L of organic phosphorus and 1 to 15 mg/L of inorganic phosphorus.[31]

Typical total phosphate concentrations in treated wastewater range from 2 to 10 mg/L.[19]

While phosphorus is absolutely necessary for plant growth, it can also cause problems in the disposal of wastewater by stimulating excessive growth of algae and aquatic plants in the receiving waters. The excessive growths of algae blooms cause esthetic problems and also cause wide swings in the dissolved oxygen content that may kill fish. Decomposition of dead aquatic plants (organic matter) can result in the depletion of the dissolved oxygen in the water and create anaerobic conditions. At the same time taste and odor problems can result.

While phosphates can cause eutrophication problems in surface waters, they are not a problem in reclaimed wastewater when it is to be used for agricultural or landscape irrigation. In fact, it is a beneficial nutrient and can promote plant growth, thus reducing the amount of fertilizer needed to produce a good crop.

The analysis for phosphorus combines two steps. The first is the conversion of the phosphorus forms (condensed or organic) to dissolved orthophosphate, and then the colorimetric determination of the dissolved orthophosphate.

Standard Methods[6] presents three procedures for orthophosphate determinations: the vanadomolybdo-phosphoric acid method (suitable in the range of 1 to 20 mg P/L), the stannous chloride method (suitable down to 7 µg P/L), and the ascorbic acid method (suitable in the range of 0.001 to 6 mg/L).[6]

Procedures for the determination of the various forms of phosphorus in reclaimed wastewater involve colorimetric methods.

The orthophosphates can be determined by one of the three methods presented in *Standard Methods*.[6] The polyphosphates can be determined by analyzing for orthophosphate before and after acid hydrolysis:

ortho (after hydrolysis) − ortho (before hydrolysis) = polyphosphate (as PO_4)

The total phosphates can be determined by wet-digesting the sample to convert all forms to the orthophosphates, for which an analysis is then made.[32]

8. Sulfates (SO_4^{2-})

Typical sulfate concentrations in untreated wastewater range from 20 to 50 mg/L and, in general, increase about 15 to 30 mg/L due to domestic use.[26,33]

As can be seen from Figure 8.6, sulfates play a major role in the production of H_2S under anaerobic conditions. H_2S can cause problems in the environment, especially with regard to odor and corrosion.

However, sulfates are important in the growth of plants, and S is an essential nutrient. Thus the presence of SO_4^{2-} in reclaimed wastewater can be helpful, particularly for soils deficient in sulfur.[26]

Standard Methods[6] presents five procedures for determination of SO_4^{2-}. The ion chromatographic method is suitable for SO_4^{2-} determination above 0.1 mg/L. There are two gravimetric methods for SO_4^{2-} concentrations above 10 mg/L. The others are the automated methylthymol blue method and the turbidimetric method, which are applicable in the 1 to 40 mg/L range.

Following is an abbreviated procedure using the turbidimetric method for sulfate (SO_4^{2-}) concentration determinations.

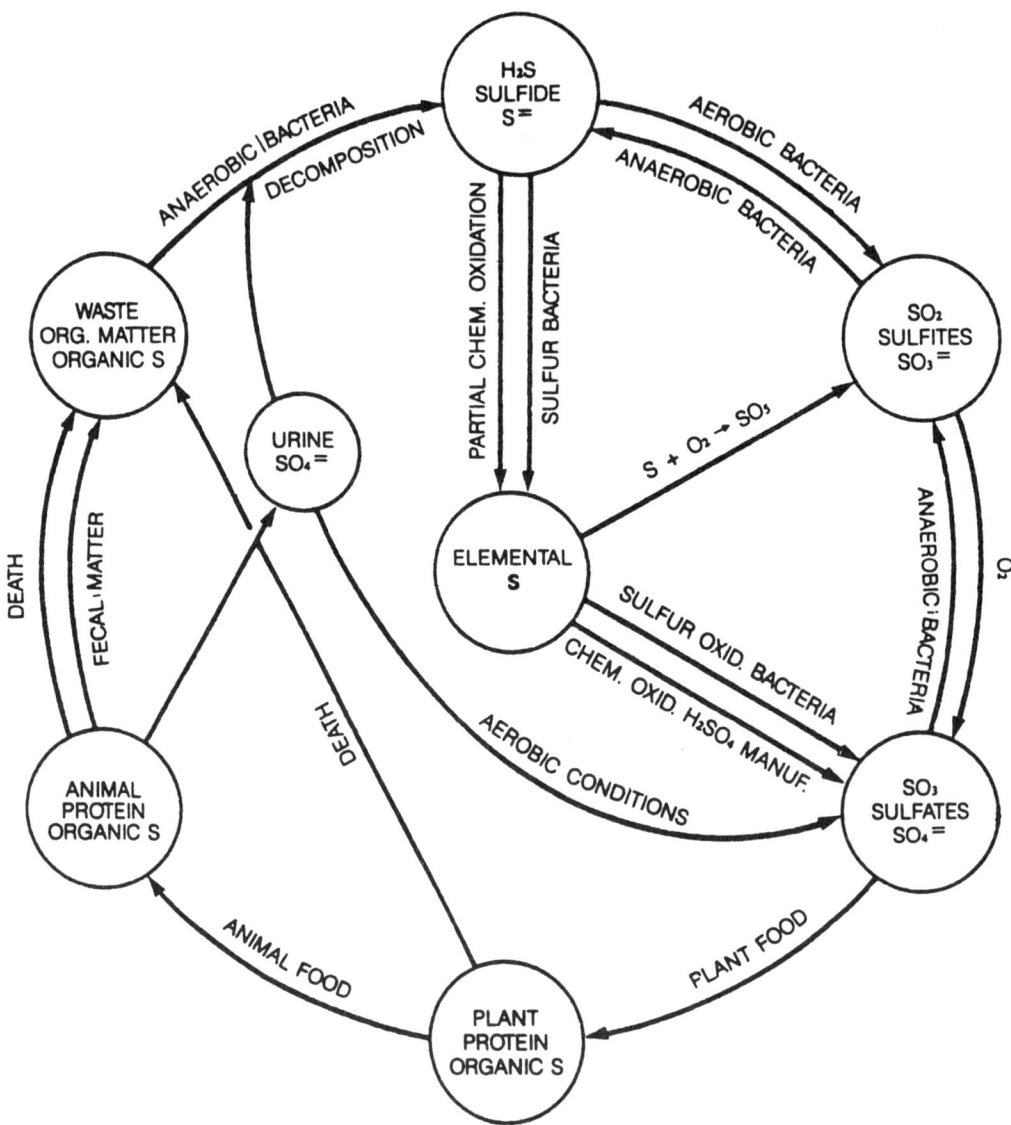

Figure 8.6 The sulfur cycle. (From WPCF, *Operation of Municipal Wastewater Treatment Plants,* Alexandria, VA, 1991. With permission.)

Sulfate ion is precipitated in a hydrochloric acid (HCl) medium with barium chloride ($BaCl_2$) so as to form barium sulfate ($BaSO_4$) crystals of uniform size. Light absorbance of the $BaSO_4$ suspension is measured by a nephelometer or transmission photometer and the sulfate ion concentration is determined by comparison of the reading with a standard curve.

Apparatus

- Magnetic stirrer
- Spectrophotometer, for use at 420 nm
- Measuring spoon
- Erlenmeyer flasks, 250 mL

Reagents

- Conditioning reagent
- Barium chloride crystals, $BaCl_2$
- Standard sulfate solution

Standard solutions can be either prepared in the lab or purchased from chemical suppliers.

Procedure

1. Measure a 100 mL sample, or a suitable portion made up to 100 mL, into a 250 mL Erlenmeyer flask.
2. Add 5 mL conditioning reagent and mix in stirring apparatus. While stirring, add a spoonful of $BaCl_2$ crystals and begin timing immediately. Stir for 1 min at constant speed.
3. Immediately after the stirring period has ended, pour the solution into the absorption cell of the photometer and measure turbidity (% transmittance) at 30-s intervals for 4 min. Because maximum turbidity usually occurs within 2 min and the readings remain constant thereafter for 3 to 10 min, consider turbidity to be the maximum reading obtained in the 4-min interval.
4. Estimate sulfate concentration in the sample by comparing the turbidity reading with a calibration curve prepared by carrying sulfate standards through the entire procedure.

C. ORGANIC CONSTITUENTS

The organic constituents most often used to evaluate the quality of reclaimed wastewater planned for reuse include the following:

- Biochemical oxygen demand (BOD)
- Chemical oxygen demand (COD)
- Total organic carbon (TOC)
- Oil and grease
- Pesticides
- Surfactants (detergent residues or foaming agents)

1. Biochemical Oxygen Demand (BOD)

The BOD test is an empirical bioassay test which measures the rate (and extent) of the aerobic degradation of organic matter in terms of the amount of oxygen consumed in its degradation. In a very oversimplified way, the complex reactions involved can be summarized as follows:[3,6,16]

$$\text{Microorganism} + \text{Organic Matter} + O_2 \rightarrow$$
$$\text{More Microorganisms} + CO_2 + H_2O + \text{Residual Organic Matter} \qquad [8.14]$$

The BOD test determines the relative amount of oxygen necessary for biological oxidation of wastewaters, effluents, and polluted water. It is the only test available to determine the amount of oxygen required by bacteria while stabilizing decomposable organic matter. Complete stabilization requires too long an incubation period for practical purposes; therefore, the 5-day period has been accepted as a standard. Samples are incubated in the dark at $20 \pm 1°C$. Dissolved oxygen levels are measured initially and at the end of the 5-day period using the methods presented in Section VI.B.4. The 5-day incubated sample must deplete more than 2 mg/L dissolved oxygen and have more than 0.5 mg/L dissolved oxygen left.[14]

To ensure the presence of excess dissolved oxygen (DO) throughout the test, it is stipulated that a DO depletion of no more than 70% be allowed for a valid test. Since water at 20°C, the common temperature of the BOD_5 test, contains only about 9 mg/L DO, it is common that samples for the BOD_5 measurement must be diluted. Dilution is carried out with a water that is supplemented with all the inorganic minerals and buffered at a pH value necessary for microbial growth. Oxygen is supplied by saturating the sample or dilution water with air. Microorganisms are supplied by seeding the dilution water with an appropriate inoculum (usually sewage, treated sewage, or in some cases microorganisms acclimated to the particular substrate of interest).[16]

The kinetics for the BOD_5 test were presented in Chapter 3, Section V.S.1. Detailed instructions for running the BOD_5 test are presented in *Standard Methods*.[6] However, following is a brief outline of this procedure that will also provide information that can be used to determine the reaction rate constant K and the ultimate BOD_L.

The apparatus and regents needed to run the BOD_5 test are the same as those required for the DO test (see Section VI.B.4). However, it will also be necessary to have a thermostatically controlled 20°C incubator available.

Procedure

1. First, the percent dilution to be used must be determined. Raw sewage or wastewater usually contains 100 to 300 mg/L BOD_5 so that 1 or 2% dilution can be used. That would be from 3 to 6 mL of raw wastewater in a 300-mL BOD bottle. Trickling filter and activated sludge effluents generally need a 5 to 10% dilution.
2. Fill four 300-mL BOD bottles about half full with dilution water. Then using a large-tipped pipet, measure the predetermined amount of sample into the four 300-mL BOD bottles (3.0 mL if 1% dilution is desired). Fill each bottle with dilution water and insert stoppers. See that all air bubbles are excluded.
3. Fill two bottles with straight dilution water and insert stoppers the same way (blanks).
4. Incubate one dilution water bottle with three diluted sample bottles at 20°C.
5. Run a dissolved oxygen determination on both remaining bottles and record the initial DO content.
6. Run a dissolved oxygen determination on the first sample bottle after 1 day, on the second sample bottle after 3 days, and on the third sample bottle and the dilution water bottle after 5 days. Note that in the dilution water bottle there should not be a change in DO concentration of more than ±0.2 mg/L in the 5-day period.

Calculations

$$\frac{\text{Initial DO of diluted sample} - \text{DO of sample after t days}}{\text{Percent of sample added}} \times 100 = BOD_t \ (mg/L) \qquad [8.15]$$

The BOD_5 test generally determines only the carbonaceous BOD ($CBOD_5$) in raw wastewater; this is due to the fact that untreated wastewater ordinarily does not contain many nitrifying bacteria. However, effluents from facilities providing biological secondary treatment often contain sufficient nitrifying organisms to exert a significant nitrogenous oxygen demand (NOD or BOD_N) in the 5-day BOD test.[34]

Figure 8.7(a) presents the conceptualized BOD curve showing the influence of the carbonaceous and nitrifying organisms.[34] Figure 8.7(b) shows the typical oxygen demand curve for a biologically treated effluent that contains sufficient nitrifying organisms to have an influence on the 5-day BOD test.[34] To determine the $CBOD_5$ nitrification, inhibitors are used. A compound, 2-chloro-6-(trichloromethyl) pyridine (TCMP) can be used for this purpose; water soluble formulations of this compound are available ($C_6H_3Cl_4N$). See *Standard Methods*[6] for procedures using this compound in determining the $CBOD_5$.

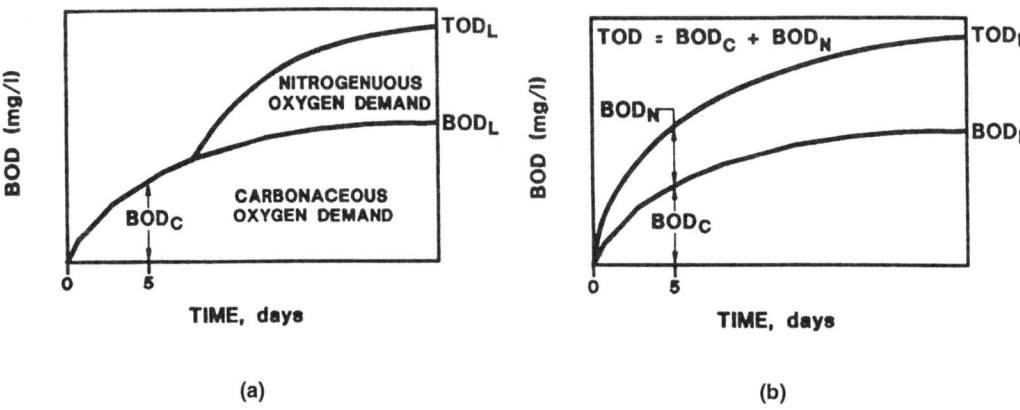

Figure 8.7 a. Conceptualized BOD curve showing carbonaceous and nitrogenous demand. b. The oxygen demand curve for a biologically treated effluent.[34]

BOD — biochemical oxygen demand
BOD_c — carbonaceous oxygen demand
BOD_L — ultimate biochemical oxygen demand
BOD_N — nitrogenous oxygen demand
TOD — total oxygen demand
TOD_L — ultimate total oxygen demand

2. Chemical Oxygen Demand (COD)

COD is defined as a measure of the amount of oxygen required for the chemical oxidation of organic matter in a sample, under acidic conditions, to carbon dioxide and water. The strong chemical oxidant most often used is potassium dichromate. All organic compounds, however, are not oxidized in this process: neither benzene, pyridine, or toluene are oxidized. Nor is ammonia oxidized, which of course by definition is not an organic compound.

The COD test has several advantages over the BOD_5 test in that the results are available in a couple of hours, and the test results are not affected if toxic compounds are present. The COD test also does not require as much manipulation as the BOD test and the procedures are more easily standardized. However, the COD result does not really indicate how readily the organic matter in the sample will be biologically degraded. In general, CODs are higher than BOD_5, and for a specific wastewater there can often be established a correlation for BOD_5/COD measurements.

The common range for COD in raw wastewater is from 200 to 400 mg/L, while for biologically treated effluents the range is from 10 to 80 mg/L.[19]

Standard Methods[6] presents three methods for COD determinations:

- Open reflux method.
- Closed reflux, titrimetric method.
- Closed reflux, colorimetric method.

Following is an abbreviated procedure for the closed reflux, titrimetric method.[6,28]

Apparatus

- Digestion vessels (culture tubes and caps)
- Heating block
- Oven (150 ± 2°C)

Reagents

- 0.0167 *M* standard potassium dichromate digestion solution
- Sulfuric acid-silver reagent
- Ferroin indicator solution
- 0.10 *N* ferrous ammonium sulfate solution "FAS" [$Fe(NH_4)_2SO_4 \cdot 6H_2O$]

Standardization of the FAS solution must be made daily against the $K_2Cr_2O_7$ digestion solution, and is carried out as follows:

- Add 6.0 mL of digestion solution, 14.0 mL of silver-sulfuric reagent, and 10.0 mL of deionized-distilled water to a 25 × 150 mm culture tube.
- Add 2 drops of ferroin indicator and titrate with standard FAS solution to a sharp color change from blue-green to reddish-brown.

Perform the standardization titration in duplicate.

- If the range between the volume of FAS used in the two titrations is greater than 0.20 mL, then perform another standardization titration.
- When the range between the volume of FAS used in titrations is less than 0.20 mL, calculate the mean volume of FAS used.

Calculate the FAS concentration:

$$\text{FAS molarity} = \frac{6.0 \text{ mL } K_2Cr_2O_7}{\text{mL FAS}} \times 0.10 \qquad [8.16]$$

Procedure

CAUTION: Wear face shield and protect hands from heat produced when contents of vessels are mixed. Mix thoroughly before applying heat to prevent local heating of vessel bottom and possible explosive reaction.[28]

1. To prevent damage from strong acid spillage place wet paper towels under tube rack and all other glassware.
2. Before use, wash culture tubes and caps with 20% H_2SO_4.
3. Place 10.00 mL of sample in culture tube with a volumetric pipet; use a smaller amount for high COD wastes and dilute to 10 mL with deionized water (graduated cylinder).
4. Add 6.0 mL of digestion solution.
5. Carefully run 14.0 mL of silver-sulfuric reagent down the inside wall of the vessel so that an acid layer is formed under the sample-digestion solution layer.
6. Tightly cap tubes and invert several times to mix thoroughly.
7. Prepare a blank as above, using 10.0 mL of deionized water.
8. Place tubes in an oven preheated to 150°C and reflux for 2 hr.
9. Cool tubes to room temperature and place them in a test tube rack.
10. Remove caps and add small magnetic stirring bars.
11. Add 2 drops of ferroin indicator to the blank tube and stir rapidly on a magnetic stirrer while titrating with FAS solution to the sharp color change from blue-green to reddish-brown at the end point, although the blue-green color may reappear within minutes.
12. Titrate the samples in the same manner; they should use less FAS titrant than the blank.

Calculation:

$$\text{mg/L COD} = \frac{(a-b)M \times 8000}{\text{mL sample}} \qquad [8.17]$$

where: a = mL FAS used for the blank
b = mL FAS used for the sample
M = molarity of FAS

Other EPA approved COD analytical methods are commercially available. One of these approved analytical methods involves a COD digester, reagent vials, COD reactor, and a calibrated colorimeter.[35]

3. Total Organic Carbon (TOC)

The total organic carbon (TOC) in a wastewater is a more convenient and direct expression of the total organic content of a sample than either the BOD_5 or COD. However, the TOC test does not provide the same kind of information.[6] If a relationship can be established between BOD_5, COD, and TOC for a specific wastewater, then the TOC can be used to not only estimate the BOD_5 and CODs in the sample, but also can be used to evaluate a wastewater treatment plant's performance.[36]

TOC can be measured manually or automatically. However the most recent *Standard Methods*[6] presents only procedures utilizing the automatic total organic analyzer. The total organic analyzer provides a rapid and simple means of determining organic carbon levels in aqueous samples. The organic carbon determination is free of the many variables inherent in the BOD_5 and COD tests.

The carbon analyzer basically involves the complete oxidation of a sample through the use of a catalytic combustion tube. A carrier gas conveys the resulting carbon dioxide and steam from the combustion tube, through a condenser where the steam is removed, and into an infrared analyzer sensitized for carbon dioxide detection. As the amount of carbon dioxide is proportional to the initial sample carbon concentration, the response can be compared to a calibration curve and the total carbon determined. The organic carbon is evaluated either by acidifying and purging the sample of all inorganic carbon prior to analysis or by providing a dual combustion tube for total carbon analysis and a low-temperature combustion tube for inorganic carbon analysis, the difference is then taken as total organic carbon.[37]

4. Oil and Grease

Included in the organic solids in wastewater are fats, oils, and grease (FOG). In this group of compounds are fats, fatty acids, soaps, waxes, and oils of many different types. Oil and grease are natural constituents of wastewater. It is reported that fecal matter contains more than 25% fats and grease. Other domestic sources of fats and grease include household kitchens, while commercial and industrial sources include slaughter houses, food processors and restaurants, automobile service stations, and petroleum refineries and storage depots.[38]

Oil and grease from animal and vegetable sources have been reported to be more easily biodegradable than those of petroleum origin.[39] "Oil and grease" are defined as any material extracted by a solvent from an acidified sample and not volatilized during the test.[6] The solvents most often used are hexane, petroleum ether, and freon. In the past, freon has been specifically recommended, but with the environmental problems (ozone layer destruction) associated with the chlorofluorocarbons, alternative solvents are now being used.[6] It would appear only a matter of time before the use of freon is discontinued. The common range for oil and grease in the influent and effluent at a wastewater treatment facility ranges from less than 5 to 50 mg/L.[19]

Standard Methods[6] presents the following three procedures for oil and grease determination.

- Partition — gravimetric method
- Partition — infrared method
- Soxhlet extraction method

For a general discussion, apparatus reagents, procedure, calculations and precision see *Standard Methods*.[6]

5. Pesticides

Insecticides, pesticides, and herbicides are manufactured from chlorinated hydrocarbons and from organophosphorus compounds. A hydrocarbon in which one (or more) hydrogen atom is replaced by a chlorine atom can be classified as a chlorinated hydrocarbon. The organophosphorus compounds are aromatic compounds and many contain sulfur, nitrogen and, of course, phosphorus in their structure.[27,38]

Chlorinated hydrocarbons are known for their persistence in the environment and can be especially dangerous as they can accumulate in the tissue of higher forms of life.[38]

Standard Methods[6] presents two procedures to determine organochlorine pesticide concentration in water samples. The first procedure involves liquid-liquid extraction, followed by gas chromatographic (GC) analysis. The second procedure presented utilizes liquid-liquid extraction followed by gas chromatographic/and mass spectrometric (GC/MS) analysis.

These methods can also be used to determine polychlorinated biphenyl (PCB) concentrations.[6]

6. Surfactants

Synthetic detergents are a group of surface active agents that can be used as substitutes for soap. The synthetic detergents are of three major types: anionic, nonionic, and cationic.[27] The anionic detergent, linear alkylate sulfonate (LAS), is the most widely used of the surfactants. The major domestic sources of LAS are from laundry detergents and cleaning compounds. The industrial and commercial sources include wool scouring plants, dyeing and rubber processing, as well as commercial laundries.[38]

The current detergent levels found in raw domestic wastewater range from 1 to 20 mg/L while levels found in secondarily treated effluents range from 0.1 to 1.5 mg/L.[6,39] The test for LAS and other anionic surfactants consists of adding methylene blue dye to a water or wastewater sample, which forms a blue-colored salt. The salt is soluble in chloroform and the color intensity produced is proportional to the concentration of the anionic surfactant. Organic sulfates, sulfonates, carboxylates, phosphates, and phenols, and inorganic cyanates, chlorides, nitrates, and thiocyanates can cause interferences.[14] See *Standard Methods*[6] and *ASTM Standards*[8] for the details regarding apparatus, reagents, procedures, precision, and bias for surfactant determination.

D. METALS

Some of the metals and elements found in treated wastewater are essential for life, others present a relatively small hazard, while still others are a potentially serious hazard.

The metals of major concern in the utilization of reclaimed wastewater include:

- Cadmium (Cd)
- Calcium (Ca)
- Chromium (Cr)
- Copper (Cu)
- Lead (Pb)
- Magnesium (Mg)
- Mercury (Hg)
- Molybdenum (Mb)

- Nickel (Ni)
- Potassium (K)
- Sodium (Na)
- Zinc (Zn)

Some trace metals such as Cd, Cu, Mb, Ni, and Hg can accumulate in the environment and be toxic to plants and animals. The presence of such metals may limit the suitability of the reclaimed wastewater for irrigation or other uses.

While most of the above listed metals will be found only at trace levels, others such as the alkali metals, sodium and potassium, and alkaline earths calcium and magnesium are clearly not at trace levels and are significant components as salts in most waters.[38] These metals are almost always present in all wastewater as chlorides, sulfates, or bicarbonate salts. The Na, Ca, and Mg concentrations in reclaimed wastewater can be used to calculate the sodium adsorption ratio (SAR) and the adjusted sodium adsorption ratio (Adj. SAR).

The concentrations of Na, Ca, and K, can be determined by the use of flame emission photometry (electrothermal technique). Flame emission photometry involves measuring the amount of light given off by atoms drawn into a flame. At certain temperatures, the flame raises the electrons in the atoms to a higher energy level. When the electrons fall back to a lower energy level, the atoms lose (emit) radiant energy which can then be detected and measured.[1]

Standards for each element to be analyzed must be prepared and tested in order to prepare calibration curves. The transmission value for the sample can then be located on the calibration curve and the concentration determined. Many atomic absorption instruments can also be used for flame emission photometric determinations.[1]

However, for many elements atomic absorption spectrophotometry is more sensitive as there are a greater number of unexcited atoms in the flame which are available to absorb the radiant energy.[1]

Atomic absorption spectrophotometry provides a rapid and easily performed technique for the analysis of metals in wastewater. Nearly all of the more than 30 elements that can be analyzed by atomic absorption can be analyzed by standard colorimetric methods. However, the colorimetric methods can be tedious and time consuming, requiring detailed sample pretreatment procedures. Atomic absorption methods provide for metal analyses with minimum sample preparation and, in many instances, analysis can be determined in the micrograms per liter ($\mu g/L$) range, or at the parts per billion level (ppb). This level of detection is generally not possible using colorimetric methods.[38]

Basically, atomic absorption spectrophotometry utilizes the principle that metallic elements in a sample introduced into a flame, such as an air-acetylene flame (2100°C), are dissociated and then converted to their atomic ground state. These atoms then absorb light energy from a hollow-cathode lamp. The amount of absorption is proportional to the concentration of the metallic element being analyzed in the sample.

Figure 8.8 indicates the major components in an atomic absorption spectrophotometer.[1] This unit consists of a suitable light source which emits a stable and intense light of a particular wavelength. Each element has characteristic wavelengths which it will readily absorb. A light source with a wavelength readily absorbed by the element to be determined is directed through the flame and a measure of its intensity is made without the sample, and then with the sample introduced into the flame. The decrease in intensity observed with the sample is a measure of the concentration of the element. A different light source must be used for each element. Fortunately, sources such as the hollow-cathode lamps are available for most naturally occurring elements.[40]

The sensitivity of flame atomic absorption spectrometry is defined as the metal concentration that produces an absorption of 1% (an absorbance of approximately 0.0044). The detection limit is defined as the concentration that produces absorption equivalent to twice the magnitude of the background fluctuations. Sensitivity and detection limits vary with the instrument, the element determined, and the technique selected. The optimum concentration range usually starts from the concentration of several times the sensitivity and extends to the concentration at which the calibration curve is no longer linear. To achieve best results, concentrations of samples and standards within the optimum concentration range of the spectrometer should be used.[40]

Table 8.6 presents a comparison of the sensitivity for flame emission and atomic absorption spectrophotometric methods for most of the elements and trace metal of major significance in the reuse of reclaimed wastewater.[1] The sensitivity for the various metals for atomic absorption in Table 8.6 and 8.7 are not in complete agreement, however, they do come from the same references.

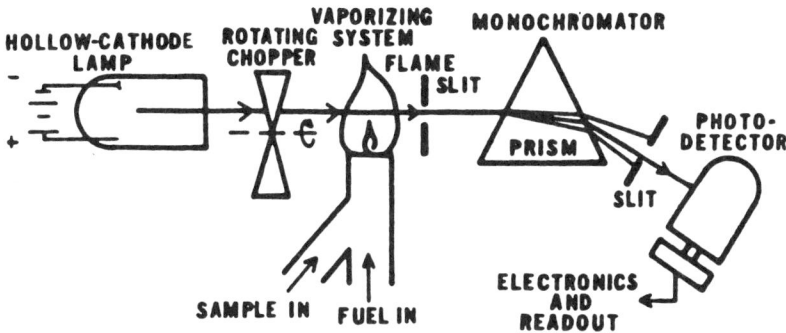

Figure 8.8 Schematic diagram of an atomic absorption spectrophotometer.[1]

Table 8.7 presents detection limits, sensitivity, and optimum concentration detection ranges for atomic absorption for the same elements and metals.[1]

The general arrangement of an atomic absorption spectrophotometer is no different than that of an emission flame photometer except for the addition of a light source (hollow-cathode lamp).

The advantages of atomic absorption spectroscopy as compared to emission spectroscopy is that atomic absorption is quite specific for many elements. Absorption depends upon the presence of free unexcited atoms in the flame, and these are in much greater abundance than excited atoms. Elements such as lead, mercury, and zinc are not easily excited in the flame and give poor sensitivities with the flame photometer but are readily detected by atomic absorption procedures. Also, the atomic absorption method is less subject to interference from other elements present in the water sample.

The elements and metals listed in Table 8.6 and 7 can also be detected by using the inductively coupled plasma (ICP) technique. ICP techniques are applicable over a broad linear range and are quite sensitive for some elements and metals. In general, detection limits for ICP methods are higher than those for the atomic absorption methods.[6] However, each element or metal must be considered separately depending upon the sensitivity, detection limit, and optimum concentration range for that element or metal. The ICP technique can be used to determine trace elements in a solution simultaneously, or sequentially for multielement samples.[41] The principles involved in the ICP technique are presented in the following excerpt from *Standards Methods*.[6]

> An ICP source consists of a flowing stream of argon gas ionized by an applied radio frequency field typically oscillating at 27.1 MHz. This field is inductively coupled to the ionized gas by a water-cooled coil surrounding a quartz "torch" that supports and confines the plasma. A sample aerosol is generated in an appropriate nebulizer and spray chamber and is carried into

Table 8.6 Comparison of Sensitivity for Flame Emission and Atomic Absorption (AA) Methods

Metals or Elements	Sensitivity (µg/L)	
	Flame	AA
Cadmium	2,000	10
Calcium	3	10
Chromium	100	10
Copper	10	5
Lead	2,000	150
Magnesium	100	3
Mercury	10,000	500
Molybdenum	—	200
Nickel	—	50
Potassium	1	5
Sodium	2	5
Zinc	200,000	5

Compiled from Reference 1.

Table 8.7 Atomic Absorption Concentration Ranges

Metals or Elements	Detection limit (µg/L)	Sensitivity (µg/L)	Optimum concentration range (µg/L)
Cadmium	5	25	50–2,000
Calcium	10	80	200–7,000
Chromium	50	250	500–10,000
Copper	20	100	200–5,000
Lead	100	500	1,000–20,000
Magnesium	1	7	20–500
Mercury[a]	0.2	—	0.2–10
Molybdenum	100	400	1,000–40,000
Nickel	40	150	300–5,000
Potassium	10	40	100–2,000
Sodium	0.2	15	30–1,000
Zinc	5	20	50–1,000

[a] Cold vapor technique.
Compiled from Reference 1.

the plasma through an injector tube located within the torch. The sample aerosol is injected directly into the ICP, subjecting the constituent atoms to temperatures of about 6000 to 8000 K. Because this results in almost complete dissociation of molecules, significant reduction in chemical interferences is achieved. The high temperature of the plasma excites atomic emission efficiently. Ionization of a high percentage of atoms produces ionic emission spectra. The ICP provides an optically "thin" source that is not subject to self-absorption except at very high concentrations. Thus linear dynamic ranges of four to six orders of magnitude are observed for many elements.

The efficient excitation provided by the ICP results in low detection limits for many elements. This, coupled with the extended dynamic range, permits effective multielement determination of metals. The light emitted from the ICP is focused onto the entrance slit of either a monochromator or a polychromator that effects dispersion. A precisely aligned exit slit is used to isolate a portion of the emission spectrum for intensity measurement using a photomultiplier tube. The monochromator uses a single exit slit/photomultiplier and may use a computer-controlled scanning mechanism to examine emission wavelengths sequentially. The polychromator uses multiple fixed exit slits and corresponding photomultiplier tubes; it simultaneously monitors all configured wavelengths using a computer-controlled readout system. The sequential approach provides greater wavelength selection while the simultaneous approach can provide greater sample throughput.[6]

Figure 8.9 is a schematic of the Plasma Quad$_c$ ICP-MS developed by V G Elemental, Cheshire, U.K.[42] The detection limits for this unit in most cases were even lower than those presented in Table 8.8.

Table 8.8 presents the estimated detection limit and upper limit concentration for the ICP method for the elements and metals of most interest in the wastewater reclamation and reuse field.[6,42]

E. BIOLOGICAL CHARACTERISTICS

The microbial agents of concern in treated wastewater planned for reuse can be placed in the following groups:

- Bacteria
- Fungi
- Protozoa (cysts)
- Helminths (nematodes) (worms, eggs)
- Viruses

1. Tests For The Coliform Bacterial Group (Indicator Organisms)

The microorganisms of prime concern are the pathogens (disease causing organisms). Testing for all the conceivable pathogens is impossible; the accepted procedure is to test for what are called indicator

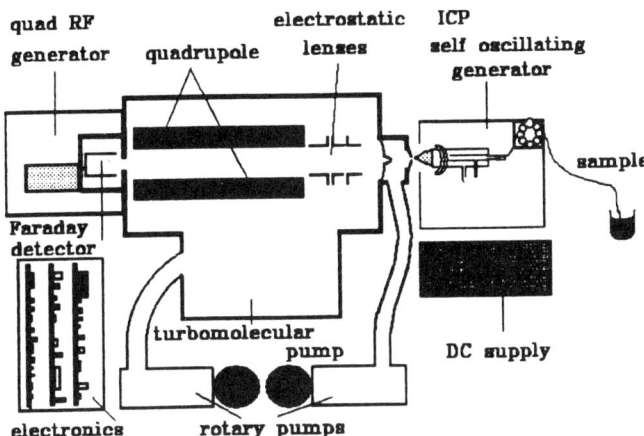

Figure 8.9 Plasma Quad$_e$ (PQ$_e$) schematic. (From Tye, C. T. and Hitchen, P., *American Environmental Laboratory,* International Scientific Communications, Inc., Shelton, CT, 1992. With permission.)

organisms. The most commonly used total and fecal coliform indicator organisms are *E. coli,* enterococci, and in some cases fecal streptococcus.[31]

Both total and fecal coliform bacteria are found in large numbers throughout the environment, especially in untreated wastewater. The indicator organisms are present in much greater numbers than the pathogens and therefore more readily detected. A low coliform count suggests few pathogens and vice versa. There are four EPA approved methods for the detection of total coliform.[43]

- Multiple-tube fermentation technique (MTF)
- Membrane filter technique (MF)
- Presence-absence coliform test (P-A)
- Minimal media ONPG-MUG test (auto analysis Colilert® system) (MMO-MUG)

The approved procedures for the first two procedures, MTF and MF, will be found in *Standard Methods,*[6] or the EPA publication *Microbiological Methods for Monitoring, Water and Wastes, U.S.A.*[44] For the MTF technique 10 fermentation tubes must be used. The third approved procedure, P-A, is presented in *Standard Methods.*[6] The fourth approved procedure for the minimal medium ONPG-MUG (MMO-MUG) test will be found in an article titled "National Field Evaluation of a Defined Substrate Method for the Simultaneous Detection of Total Coliforms and *Escherichia coli* From Drinking Water: Comparison with Presence-Absence Techniques."[45]

Table 8.8 Inductively Coupled Plasma (ICP) Method — Estimated Detection Limit and Upper Limit Concentration

Metal or Element	Estimated detection limit (µg/L)	Upper limit concentration (µg/L)
Cadmium	4	50
Calcium	10	100
Chromium	7	50
Copper	6	50
Lead	40	100
Magnesium	30	100
Molybdenum	8	100
Nickel	15	50
Potassium	100[a]	100
Sodium	30[a]	100
Zinc	2	100

[a] Sensitive to operating conditions.

Compiled from References 6 and 41.

The testing for fecal coliforms/*Escherichia coli* must be made if any routine or repeat sample is positive for total coliform. *E. coli* may be tested for in lieu of fecal coliform. For details on these procedures see the *Federal Register*.[43]

Total and fecal coliform bacteria are not generally considered pathogenic; however, strains of enteropathogenic *Escherichia coli* have been implicated in gastrointestinal disease outbreaks, and even causing deaths (Jack-In-The-Box hamburgers).

If the MF or MTF methods are used for the detection of total or fecal coliform they are reported in units of colonies per 100 mL for the MF method and as MPN per 100 mL for the MTF method.[31]

The common ranges for total coliform and fecal coliform in raw wastewater and secondarily treated wastewater are as follows:[12,19,33]

	Total coliform	Fecal coliform
	(number/100 mL)	
Raw wastewater	5×10^4 to 5×10^9	10^4 to 10^9
Secondarily treated wastewater	50 to 1000	<2 to 200

It is not possible to include here all the details for the various testing procedures for coliform; however, following are abbreviated methods using the membrane filter technique (MF) to determine total coliform and fecal coliform.

1. *Outline of the MF method for total coliform* — The sample is obtained in a sterile container and is filtered through a sterile 0.45 μm membrane filter. The filter is then placed on a sterile pad saturated with liquid media and is incubated at $35 \pm 0.5°C$ for 22 to 24 hr. Green metallic sheen colonies are counted under low magnification. The enrichment procedure is suggested for assessing drinking water.

The total coliform group includes all of the aerobic and facultative anaerobic, Gram negative, nonspore-forming, rod-shaped bacteria that ferment lactose in 24 to 48 hr at 35°C. This includes the following genera: *Escherichia, Citrobacter, Enterobacter,* and *Klebsiella*. Bacteriological samples cannot be preserved. Analyze samples preferably within 6 hr of collection. Otherwise, analyze within 24 hr of collection.[14]

2. *Outline of the MF method for fecal coliform* — The sample is obtained in a sterile container and is filtered through a sterile 0.45 μm membrane filter. The filter is then placed on a sterile pad saturated with liquid media and is incubated at $44.5 \pm 0.2°C$ for 22 to 24 hr. Blue colonies are counted under low magnification.

Fecal coliforms are part of the total coliform group. Fecal coliforms are defined as Gram negative, nonspore-forming, rod-shaped bacteria that ferment lactose in 24 ± 2 hr at $44.5 \pm 0.2°C$. The major species is *Escherichia coli*. Bacteriological samples cannot be preserved. Analyze samples preferably within 6 hr of collection. Otherwise analyze within 24 hr of collection.[14]

2. Total and Fecal Coliform (Membrane Filter Technique)

Apparatus:

- Filtration units, with membrane filters (adequately sterilized)
- Culture dishes (adequately sterilized)
- Incubators
- Colony counter

Reagents and Materials:

- Sterile buffered distilled water
- M-Endo medium, for total coliform
- M-FC broth, for fecal coliform
- Absorbent pads

Procedures:

1. Disinfect working area with chlorine or iodine solution.
2. Select sample volumes to be filtered and set out required number of sterile petri dishes; label dishes with a proper code, water source, and sample volume filtered.

3. Place sterile absorbent pad in each petri dish with forceps sterilized by immersing in ethanol and passing through flame; cover dishes immediately.
4. Using sterile pipet, transfer 2 mL of medium to each petri dish and re-cover dish (use M-Endo medium for total coliform and M-FC broth for fecal coliform).
5. Remove sterile funnel units from wrapping and assemble funnel base on filtration flask; invert funnel top on wrapping paper.
6. Place sterile membrane filter on funnel base with flamed forceps, grid side up; replace funnel top.
7. Deliver measured volume of well-mixed sample into funnel using sterile pipet or sterile graduated cylinder; if less than 10 mL volume is used, add small amount sterile buffered distilled water to funnel prior to sample delivery.
8. Apply vacuum after sample has been completely delivered; rinse funnel three separate times with about 20 mL of sterile buffered distilled water for each rinse.
9. Remove cover from prepared petri dish, pour out excess medium, and transfer membrane filter with flamed forceps, touching only the outer 1/8 in. of the filter; roll the filter onto the pad with the grid side up.
10. Replace the cover on the petri dish and invert the dish; place the inverted total coliform plates in a 35°C incubator. Place the inverted fecal coliform dishes in zip-loc bags (2/bag) and immerse completely in a 44.5°C water bath incubator.
11. Allow to incubate 24 ± 2.0 hrs and count the coliform colonies using a 10-power binocular microscope with fluorescent illumination.
12. Typical total coliform colonies will have a pink to dark-red color with a metallic surface sheen; the desired filter count is 20 to 80 coliform colonies and less than 200 total bacterial colonies per membrane.
13. Typical fecal coliform colonies are blue; the desired filter count is 20 to 60 fecal coliform colonies per membrane.

Calculation:

$$\text{Total coliform colonies/100 mL} = \frac{\text{Total coliform colonies counted} \times 100}{\text{mL sample filtered}} \quad [8.18]$$

$$\text{Fecal coliform colonies/100 mL} = \frac{\text{Fecal coliform colonies counted} \times 100}{\text{mL sample filtered}} \quad [8.19]$$

The Colilert® test systems for coliform and *E. coli* are now commercially available and some are USEPA approved. The new methods are based on enzymatic assays. The assay method is extremely simple and involves only a few steps. A reagent is added to the sample to adjust the pH. The sample is then incubated for 24 hr and the results read.[46]

Another especially good source for EPA approved methods for detection of coliform is the *Manual for the Certification of Laboratories Analyzing Drinking Water,* Criteria and Procedures, Quality Assurance, 3rd ed., EPA 570/9–90–008 A, Office of Drinking Water (WH-550 D), Washington, D.C., 1991.

The World Health Organization (WHO) has been evaluating the strict microbial standards required in the U.S. for reuse of reclaimed wastewater, especially with regard to agricultural irrigation. For instance, the California State Department of Health permits a level of 2.2 coliforms per 100 mL, depending on the crop being irrigated and on the irrigation method being used.[47] The WHO experts, including epidemiologists and public health officials, reviewed the extensive epidemiological studies and reports available in this area and concluded that the actual risk associated with irrigation using treated wastewater is much lower than previously estimated, and that the early microbial standards and guidelines for effluents to be used for unrestricted irrigation of vegetables and salad crops normally consumed uncooked were unjustifiably restrictive, particularly in respect of bacterial pathogens.[47]

On the basis of this new evidence, the Engelberg Report[48] recommended new guidelines containing less stringent standards for fecal coliforms than those previously suggested. However, they were stricter than previous standards in respect of numbers of helminth eggs, which were recognized to be the main actual public health risk associated with wastewater irrigation in those areas where helminthic diseases are endemic.[47,48] This group of experts not only recommended guidelines for fecal coliform but also intestinal nematodes for unrestricted irrigation of crops. See Chapter 2, Table 2.32 for details. These helminth egg guidelines are intended as a design goal for wastewater treatment systems and not as a standard requiring routine testing of effluent quality.[47] Table 8.9 presents two tentative methods for the quantitative determination of helminth eggs in wastewater.[47]

Table 8.9 Tentative Methods for Quantitative Determination of Helminth Eggs in Wastewater

Sedimentation Method

This method relies on a sedimentation procedure selected from more than 20 different procedures tested under field conditions in several countries by Professor J. Schwartzbrod (Faculty of Pharmaceutical Sciences and Biology, University of Nancy, France). It consists of the following steps:

a. A grab sample of wastewater (i.e., a single sample taken neither at a set time nor at a set flow) is taken and transported unpreserved to the laboratory at ambient temperature within a few days. It is then shaken and a measured volume (S) of 1 L removed and left to settle for at least 8 hours.
b. The supernatant is carefully removed and discarded, without disturbing the sediment.
c. The sediment (100–200 mL) is recovered from the container. The container walls are then washed with 25–50 mL of distilled water, which is added to the sediment. All the recovered material is centrifuged at 1000 g for 15 minutes.
d. The supernatant is removed and discarded. Aceto-acetic buffer (pH 4.5)[a] is added at a volume equal to that of the pellet and the mixture is stirred.
e. Ether is then added at a volume equal to twice that of the buffer and the mixture is stirred for 10 minutes.
f. The mixture is centrifuged at 1000 g for 6 minutes.
g. The supernatant is discarded and the pellet resuspended with about 5 mL (about 5 times the pellet volume is generally needed for resuspension) of saturated zinc sulfate solution (33%, relative density 1.18). The volume (V) of the product is measured.
h. A portion (P) of the product is transferred to a microscopic counting cell[b] using a Pasteur pipette, and the eggs are counted at 100 × magnification.
i. The total number of eggs per liter (N) present in the original sample of wastewater is determined from the formula:

$$N = \frac{X}{P} \cdot \frac{V}{S}$$

where: X = number of eggs counted
P = volume of product in the counting cell (mL)
V = total volume of product (mL)
S = volume of wastewater sample (L)

In the example described, the wastewater sample volume (S) is 1 L. In order to evaluate the helminth concentration relative to the goal of ≤1 egg per liter, the sensitivity of this method may be improved by increasing the sample volume to ≥2 L and/or increasing the size of the portion of the product read under the microscope.

[a] This contains 15 g of sodium acetate with 3.6 mL of acetic acid made up to 1 L with distilled water.
[b] Microscope counting cells that hold 0.3–1.0 mL are acceptable and are commercially available, e.g., McMaster cell (0.3 mL), Sedgwick-Rafter cell (1.0 mL).

Centrifugation Flotation Method

This method is based on the centrifugation flotation procedure as described by Ockert and Teichmann, and consists of the following steps:

a. A grab sample of wastewater is taken and transported unpreserved to the laboratory at ambient temperature within a few days. It is then shaken and a measured 1-L volume is removed and left to settle for at least 8 hours.
b. The supernatant is carefully removed and discarded without disturbing the sediment.
c. The sediment is transferred to 20-mL centrifuge tubes (maximum 3 mL per tube). The walls of the sedimentation beaker should be cleaned thoroughly using a spray bottle, and the rinsing water added to the sediments in the centrifuge tubes. They are then centrifuged for 10 minutes at 700 × g and the supernatants are discarded.
d. A volume of 3 mL of sodium nitrate solution (500 g/L relative density 1.3) is added to the sediment in each tube.
e. The solutions are then centrifuged for 3 minutes at 1000 × g.
f. The supernatant (now containing the helminth eggs) is removed and kept in a 1500-mL flask (preferably conical) containing 1 L of distilled water.
g. A volume of 3 mL of sodium nitrate solution is again added to the sediment in each tube, and the mixtures are centrifuged at 1000 × g for 3 minutes. The supernatant is carefully removed and added to the 1500-mL flask containing the first supernatants.
h. The procedure in point (g) is repeated (the sediment is thus centrifuged with sodium nitrate a total of three times).
i. The beaker containing all the supernatants diluted in water is left for several hours, to allow all the helminth eggs to settle to the bottom.
j. The supernatant from this beaker is carefully removed and discarded, and the sediment is transferred to centrifuge tubes. The walls of the sedimentation beaker should be thoroughly cleaned and the rinsing water added to the sediment in the centrifuge tubes. The contents of the tubes are then centrifuged for 4 minutes at 1000 × g.
k. After centrifugation, the lower 1 mL of fluid is carefully removed from each tube with a Pasteur pipette, placed in a counting cell, and examined under the microscope at 100× magnification.
l. The number of eggs counted will be the total number recovered from the 1-L wastewater sample.

From WHO, *Health Guidelines for the Use of Wastewater for Agriculture and Aquaculture*, M9/675073-001, 1989. With permission.

F. RADIOACTIVITY
1. Introduction

Radioactivity in our environment has both natural and anthropogenic (man-made) sources. Since the end of World War II the anthropogenic sources in our environment have increased due to nuclear testing (now underground), nuclear power plant accidents (e.g., Chernobyl, Three Mile Island), spent reactor fuel (e.g., Tomsk), accidental sinking of nuclear submarines (e.g., Komsomlets), radioactive waste dumps (National Priorities List — Superfund Sites), mining and reprocessing of nuclear material (uranium mill tailings), medical use of radioisotopes, as well as industrial and research uses of radionuclides.

The radionuclides of most concern in our environment, especially in our water and wastewater, can be found in the following publications:

- *International Commission on Radiological Protection.* Recommendations of the International Commission on Radiological Protection. *Annals of the ICRP,* 1(3): 1–53 (1977) (ICRP Publication 26.)[49]
- *International Commission on Radiological Protection.* Limits for intakes of radionuclides by workers. *Annals of the ICRP,* 2–8, (1979 to 1982) (ICRP Publication 30 and supplements).[50]
- *Federal Radiation Council.* Background Material for the Development of Radiation Protection Standards, Rep. No. 2 (Sept.), U.S. Government Printing Office, Washington, D.C., 1961.[51]
- *National Committee on Radiation Protection and Measurements (NCRP).* Maximum Permissible Body Burdens and Maximum Permissible Concentrations of Radionuclides in Air and Water for Occupational Exposure, NBS Handbook No. 69, 1959.[52]
- *National Council on Radiation Protection and Measurements (NCRP). A Handbook of Radioactivity Measurement Procedures.* NCRP Report No. 58, 1985a.[53]
- *U.S. Environmental Protection Agency.* Water Pollution Control; Radionuclides; Advance Notice of Proposed Rulemaking, 40CFR, Part 141, 34836; *Federal Register,* 51, No. 189, 1986.[64]

2. Regulatory Agencies

The primary agencies in the U.S. that have regulatory authority in the radiation field are the EPA, the Nuclear Regulatory Commission (NRC), the Department of Energy (DOE), Department of Transportation (DOT), Department of Defense (DOD), and state agencies.

There are also national and international scientific advisory organizations such as The National Council on Radiation Protection and Measurements (NCRP) and the International Commission on Radiological Protection (ICRP). Each of these organizations provides recommendations regarding human radiation protection and radioactive waste management, however, these organizations have no regulatory authority. The NCRP is chartered by the U.S. Congress to collect, analyze, develop, and disseminate information and recommendations about radiation protection and its measurement. The ICRPs functions are much the same, but on an international basis.[55]

3. Standards

The U.S. Safe Drinking Water Act (PL 93–523) includes radiological standards for drinking water. Table 8.10 presents these standards entitled, "The National Primary Drinking Water Standards For Radionuclides."[21,56]

Table 8.11 presents the World Health Organization guidelines for radioactive constituents.[57] To better understand the information included in Table 8.10 and Table 8.11, the following explanations of the various terms and units used are presented.[58]

Radiation is generally reported in units of curies (Ci), or Becquerels (Bq), roentgens (R), Grays (Gy), or rads, Sieverts (Sv) or rems.

A curie of radioactivity is defined as 3.70×10^{10} disintegrations per second. It also refers to that quantity of a nuclide containing 3.70×10^{10} disintegrations per second.

A common fraction of the curie is a picocurie (pCi), which is 10^{-12} of a curie, or 2.22 disintegration per minute (dpm). A more current unit now used to describe the strength of radioactive source is the Becquerel (Bq). A becquerel is equal to one disintegration per second; therefore[58]

$$1 \text{ Ci} = 3.7 \times 10^{10} \text{ Bq} \qquad [8.20]$$

Other terms associated with ionizing radiation are as follows. The Roentgen (R) is used to measure the exposure dose from ionizing radiation in the form of gamma rays and X-rays. One R is defined as the collection of one electrostatic unit (esu) of charge in one cubic centimeter (cm^3) of air (0.001293 g) at standard temperature and pressure.

Table 8.10 The U.S. National Interim Primary Drinking Water Standards for Radionuclides

Radionuclides	Unit
Alpha emitters	
Radium-226	5[a]
Radium-228	5
Gross alpha activity (excluding radon and uranium)	15[b]
Beta and photon emitters[c]	
Tritium	20,000
Strontium-90	8
Beta particle and photon radioactivity	4 mrem[d] (annual dose equivalent)

[a] pCi/L = ~0.19 Bq/L.
[b] 15 pCi/L = ~0.56 Bq/L.
[c] Based on a water intake of 2 L/d. If gross beta particle activity exceeds 50 pCi/L, other nuclides should be identified and quantified on the basis of a 2-L/d intake.
[d] 4 mrem = 4 µSv.

Compiled from References 21 and 56.

Table 8.11 World Health Organization Guidelines for Radioactive Constituents

Constituent	Guideline value (Bq/L)	Remarks
Gross alpha activity	0.1	If the levels are exceeded, more detailed radionuclide analysis may be necessary.
Gross beta activity	1	Higher levels do not necessarily imply that the water is unsuitable for human consumption.

Compiled from Reference 57.

$$R = 1 \text{ esu}/0.001293 \text{ g of air} \quad [8.21]$$

While the roentgen measures exposure in a volume of air, it is the ionizing radiation absorbed in the tissue which produces harmful biological effects. A unit used to measure absorbed dose is the radiation dose called the rad. One rad is defined as the deposition of 100 ergs of energy per gram of tissue (or other material).[58]

$$1 \text{ rad} = 100 \text{ ergs}/\text{g} \quad [8.22]$$

A more current term that is replacing the rad is the "Gray" (Gy). By definition:

$$1 \text{ Gy} = 100 \text{ rad} \quad [8.23]$$

Some types of ionizing radiation are more effective than others, per unit of energy absorbed, in producing biological damage. For example, alpha particles, fast neutrons, and protons are approximately 10 times as damaging to tissue per unit of energy absorbed as X-rays, gamma rays, and most beta particles. The unit which takes this biological effect into account is the roentgen equivalent man called the rem. The rem is equal to the absorbed dose in rads times the relative biological effect (RBE). The RBE is the ratio of the absorbed dose necessary for a given biological effect from X-rays or gamma rays to the absorbed dose for the same effect produced by any other form of radiation. Therefore,[58]

$$\text{rems} = \text{rad} \times \text{RBE} \quad [8.24]$$

A mrem would therefore be 10^{-6} of a rem. A more current term used to express absorbed dose is the "Sievert" (Sv)

$$1 \text{ Sv} = 10^2 \text{ dose equivalents (rems)} \quad [8.25]$$

4. Radiation Detection And Measurement

The detection and measurement of radioactivity requires both special sample preparation and instrumentation. The following section presents both field and laboratory instrumentation that can be used for detection and measurement of radioactivity, however, for sample preparation refer to Reference 49 to 54, as well as *Standard Methods*.[6]

Field and laboratory methods are available to identify and quantify concentrations of radionuclides in the environment.

All of these methods for detection of radioactivity require two basic elements. A sensing element and an indicating element. Sensing elements convert the energy of the radiation into electrical energy. The electrical energy may then pass directly to the indicating element — a scaler, meter, or recorder — or it may first pass through certain intermediate electronic circuits for the purpose of amplification or analysis.

Sensing elements for the detection of radiation depend upon the formation of ions for their operation. They can be divided into two general categories: (1) those which depend upon the collection of these ions; and, (2) those which do not depend upon ion collection. Ionization chambers, proportional counters, and Geiger counters require the collection and measurement of the ions produced, while photographic methods, cloud chambers, and scintillation counters do not.[13]

Selection of a radiometric method depends upon the number of radionuclides of interest, their activity, and type of activity emitted, as well as the sensitivity required and the sample size available.

Field instruments measure the radioactivity level at the site. Table 8.12 presents field-type radiation detection instruments.[55] These portable field instruments are relatively rugged and relatively insensitive to changes in temperature and humidity, but sensitive enough to discriminate between natural background radiation and excess radiation from anthropogenic sources.[55]

Laboratory methods for detection and measurement of radioactivity involve both chemical and instrumental techniques to quantify low-level radioactivity in sample media. For alpha counting, the samples, after appropriate preparation, are loaded into gas proportional counters, scintillation detectors, or alpha spectrometry systems for measurement.

Beta-emitting samples are counted in gas proportional counters, or in scintillation detectors.

For laboratory gamma radiation detection, crystals such as sodium iodide containing a trace of thallium iodide [NaI(TI)] or germanium diodes [Ge(Li)] are used. These crystals are connected to a photocell known as a photomultiplier tube. The crystals and photomultiplier are referred to as a scintillation detector. The basic scintillation system for detecting and identifying gamma radiation consists of a high-voltage supply, a preamplifier, and an N scaler. However, only a NaI and Ge(Li) detector coupled with a multichannel analyzer can provide energy spectra of the gamma rays and therefore verify the identity of a specific radionuclide.[13,55]

Table 8.13 presents the various types of laboratory radiation detection instruments often used for alpha, beta, and gamma radioactivity measurements.

VII. FLOW MEASUREMENTS

A. INTRODUCTION

Flow measurements of reclaimed wastewater provide a very important piece of information in the operation of a reuse project. Flow measurements can be used to indicate flow rates, volumes of water used, identify problems such as leakage, evaluate operational costs in order to set charges, as well as supply needed information for planning and conservation programs.[3,4,21,59]

There are many devices available to provide flow measurement and all have three factors in common: area, velocity, and a coefficient characteristic.

Devices that are most commonly used in the wastewater fields include:[3,4,59]

- Weirs
- Parshall and Palmer-Bowlus flumes
- Venturi meters
- Electromagnetic flow meters

B. WEIRS

A weir is essentially a dam or bulkhead placed across an open channel such as a river, stream, canal, or flume. The entire flow passes through a notch in the weir, or over the entire width of the weir as in the case of a suppressed weir. The water depth above the weir rises, and this rise is directly proportional to

Table 8.12 Types of Field Radiation Detection Instruments

Instruments	Range of counting rate and other characteristics	Typical uses	Remarks
Beta-Gamma Surface Monitors[a]			
Portable Count Rate Meter (Thin Walled or Thin Window G-M Counter)	0–1,000; 0–10,000; 0–100,000 count/min	Surfaces, hands, clothing	Simple, reliable, battery powered
Alpha Surface Monitors			
Portable Air Proportional Counter with Probe	0–100,000 count/min over 100 cm^2	Surfaces, hands, clothing	Not accurate in high humidity; battery powered; fragile window
Portable Gas Flow Counter with Probe	0–100,000 count/min over 100 cm^2	Surfaces, hands, clothing	Not affected by the humidity; battery powered; fragile window
Portable Scintillation Counter with Probe	0–100,000 count/min over 100 cm^2	Surfaces, hands, clothing	Not affected by the humidity; battery powered; fragile window
Tritium Monitors			
Flow ionization chambers	0.10 pCl/m^3/min	Continuous monitoring	May be sensitive to other sources of ionization

[a] None of these surface monitors is suitable for tritium detection.

From *Instrumentation and Monitoring Methods for Radiation Protection*, National Council on Radiation Protection and Measurements, Bethesda, MD, 1978. With permission.

Table 8.13 Types of Laboratory Radiation Detection Instruments

Type of instrument	Typical activity range (mCi)	Typical sample form	Data acquisition and display
Gas Proportional Counters	10^{-7} to 10^{-3}	Film disc mount, gas	Ratemeter or scaler
Liquid-Scintillation Counters	10^{-7} to 10^{-3}	Up to 20 ml of liquid gel	Accessories for background subtraction, quench correction, internal standard, sample comparison
NaI (TI) Cylindrical or Well Crystals	10^{-4} to 10^{-3}	Liquid, solid, or contained gas, <4 ml	Ratemeter
			Discriminators for measuring various energy regions
			Multichannel analyzer, or computer plus analog-to-digital converter
			Computational accessories for full-energy-peak identification, quantification, and spectrum stripping
Ionization Chambers	10^{-2} to 10^3	Liquid, solid, or contained gas (can be large in size)	Ionization-current measurement; digital (mCi) readout, as in dose calibrators
Solid-state Detectors	10^{-2} to 10	Various	Multichannel analyzer or computer with various readout options

From *A Handbook of Radioactivity Measurement Procedures*, National Council on Radiation Protection and Measurements, Bethesda, MD, 1985. With permission.

the flow over or through the weir and can be used to determine the discharge.[3,60] All level measurements are made relative to the crest of the weir (see Figure 8.10).[3,21,59,60]

The weir openings can be v-notch — with various angles (see Figure 8.11), rectangular-contracted (see Figure 8.12), rectangular-suppressed (see Figure 8.13), or trapezoidal-Cippoletti (see Figure 8.14). The flow over or through the weir is expressed as an empirical equation based on the weir dimensions and the upstream head (H). The equations for calculation of the flow for each type of weir is given along with Figures 8.11 to 8.14.

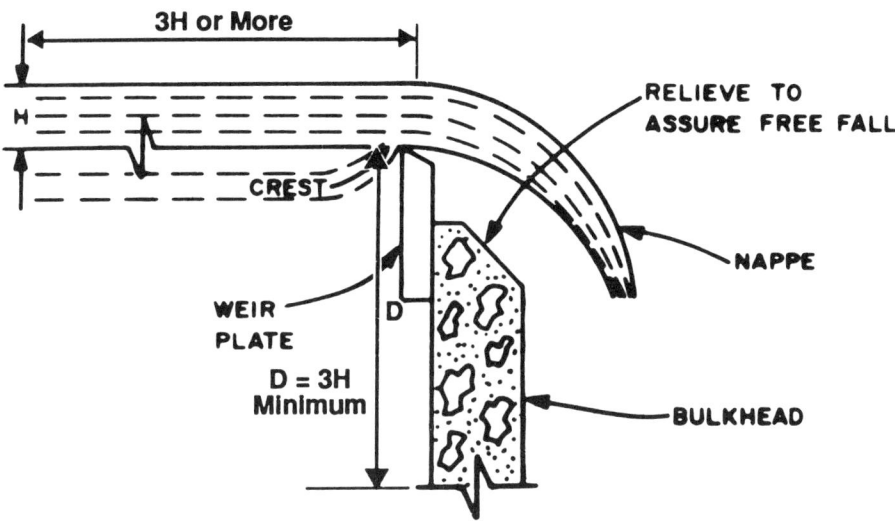

Figure 8.10 Head on weir (H).[60]

Example 8.2

A 90° V-notch weir was installed in a manhole to measure the water flow. If the maximum height of flow was 12.7 cm measured 60 cm upstream from the weir, what was the maximum flow in m³/d?

Solution:

1. Use Equation 8.26A (see Figure 8.11)

$$Q = 1.40\ H^{2.5}$$

2. Given H = 0.127 m
3. Substitute H and calculate Q in Equation 8.26A:

$$Q = 1.40\ (0.127)^{2.5} = 0.00805\ m^3/s$$

$$= 695.3\ m^3/d$$

Example 8.3

A rectangular contracted weir, 1.2 m in height, is to be installed in a channel 2.4 m wide. The maximum flow over the weir is 1.7 m³/s when the total depth of water measured in a stilling well upstream from the weir is 2.1 m. What length of weir should be installed?

Solution:

1. Use Equation 8.27 (see Figure 8.12)

$$Q = 1.84\ (L - 0.2\ H)^{3/2}$$

2. Determine H, head on weir

$$H = 2.1\ m - 1.2\ m = 0.9\ m$$

3. Given Q = 1.7 m³/s
4. Substitute in Equation 8.27 and determine L (the weir length)

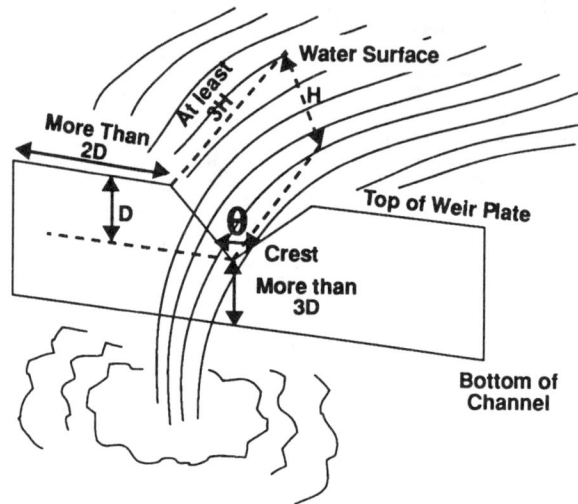

Figure 8.11 V-notch weir for measuring flows.[21]

V-Notch Angle (θ)	Flow (Q)	
90°	$1.4\ H^{2.5}$	[8.26A]
60°	$0.79\ H^{2.5}$	[8.26B]
45°	$0.57\ H^{2.5}$	[8.26C]
30°	$0.37\ H^{2.5}$	[8.26D]
22½°	$0.27\ H^{2.5}$	[8.26E]

where: Q = discharge, m³/s
        H = head on weir in, m
        K = constant dependent on the units of flow

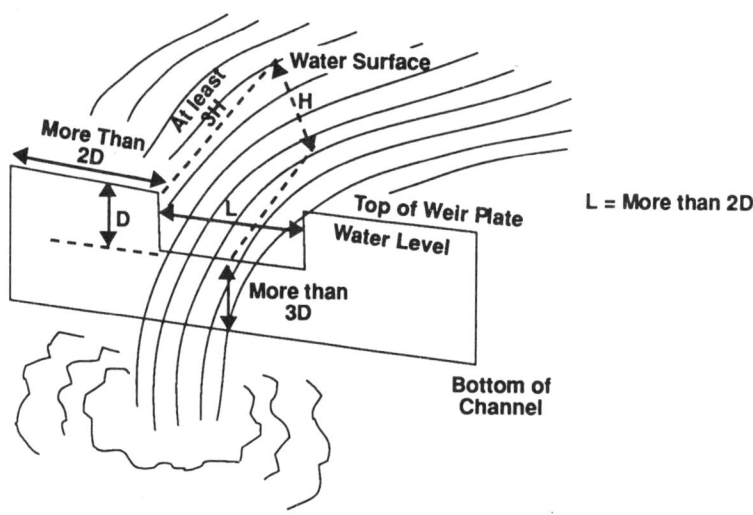

Figure 8.12 Rectangular contracted weir.[3]

$$Q = 1.84\ (L - 0.2\ H)\ H^{3/2} \qquad [8.27]$$

where: Q = discharge, m³/s
        L = length of weir, m
        H = head on weir, m

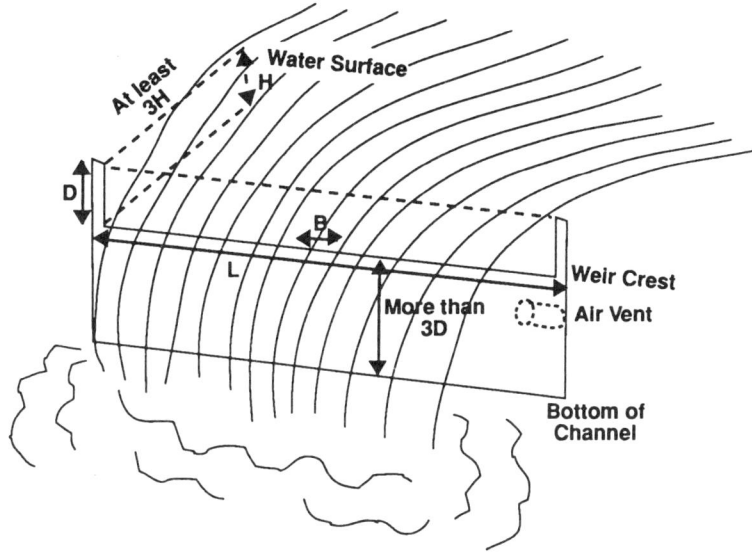

Figure 8.13 Rectangular suppressed weir.[3]

$$Q = 1.84\, L\, H^{3/2} \qquad [8.28]$$

where: Q = discharge, m³/s
        L = length of weir, m
        H = head on weir crest, m

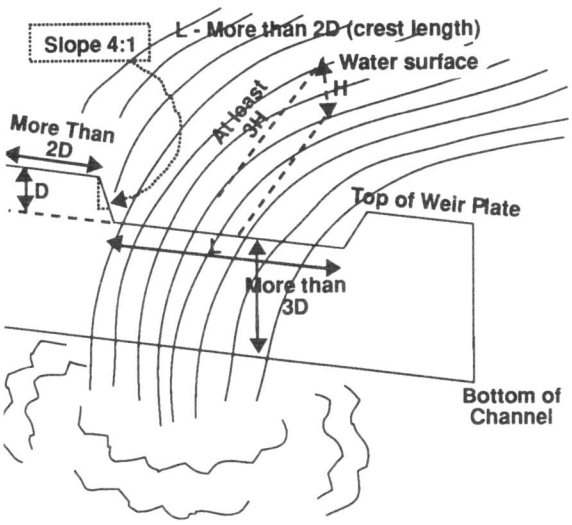

Figure 8.14 Cipolletti (trapezoidal) weir.[21]

$$Q = 1.858\, L\, H^{3/2} \qquad [8.29]$$

where: Q = discharge, m³/s
        L = length on weir, m
        H = head on weir, m

$$1.7 \text{ m}^3/\text{s} = 1.84 \left[L - 0.2(0.9) \right] 0.9^{3/2}$$

$$L = 1.26 \text{ m}$$

C. PARSHALL AND PALMER-BOWLUS FLUMES

Parshall and Palmer-Bowlus flumes can be used for flow measurements of either raw or treated wastewater.

1. Parshall Flumes

The Parshall flume operates on the principle that when the flow in an open channel passes through a constriction a hydraulic head is developed upstream from the constriction that is proportional to the flow. The Parshall Flume has several desirable characteristics for flow measurements including that there are no obstructions in the channel that can collect debris and interfere with the flow.

Parshall flumes also have a small loss of head and are self cleaning. Figure 8.15 presents a plan and elevation view of a free-flowing Parshall flume.[60] Parshall flumes are rated depending upon the throat width (W). Small Parshall flumes can be purchased prefabricated, while large flumes are constructed on the site. Dimensions can be found in Reference 61.

With free-flow conditions as depicted in Figure 8.15, the flow rate is related to the depth measured at a specified location by an equation of the form

$$Q = CH_a^n \quad [8.30]$$

where Q is the flow rate, H_a is the depth measured as indicated in Figure 8.15, and C and n are empirical constants which vary with flume size.

The flume can also operate in a "submerged flow" mode, which occurs when the downstream depth becomes so high that the break in the floor slope of the flume can no longer be a complete control point for the flow. In submerged flow, two depths must be measured in order to determine the flow rate. Therefore, it is desirable that flume installations be designed for free-flow conditions.[61]

Following is a more specific formula that can be used to calculate flows for Parshall flumes with widths (W) from 300 mm to 3 m.[62]

$$Q = 2.23 \times 10^{-2} \, W \left(\frac{H_a}{304.8} \right)^{1.522(W/304.8)^{0.026}} \quad [8.31]$$

where: Q = the flow in m³/min
   W = the width of the throat, mm
   H_a = the upstream depth, mm

If a flume is to measure the flow accurately there must be no disturbance of the normal flow pattern immediately up or downstream. A straight uniform channel should be maintained for a reasonable distance in both directions. Approach zones equal to at least 10 to 20 times the throat width are desirable. Downstream changes in direction must be sufficiently distant that their backwater curves do not reach the flume.[62]

Example 8.4

Compute the upper head (H_a) on a Parshall flume with a throat width of 0.457 m for a flow of 0.157 m³/s, (9.42 m³/min).

Solution:

1. Use Equation 8.31:

$$Q = 2.23 \times 10^{-2} \, W \left(\frac{Ha}{304.8} \right)^{1.522 \, (W/304.8)^{0.026}}$$

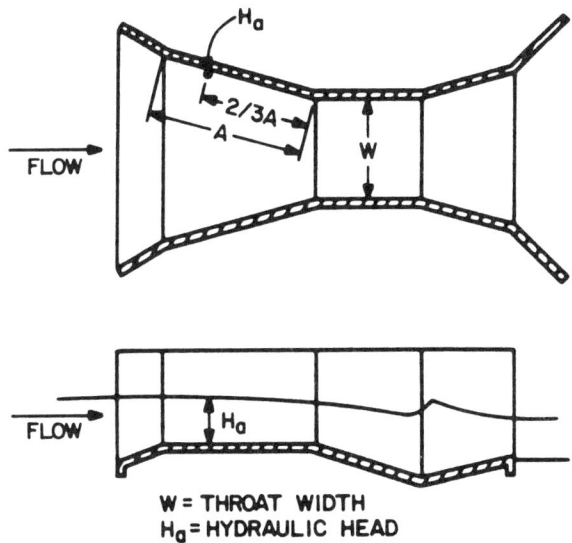

Figure 8.15 Free-flowing Parshall flume — head/width parameters.[60]

2. Given W = 0.457 m, Q = 0.157 m³/s (9.42 m³/min)
3. Substitute   W = 457 mm and Q = 9.42 m³/min

$$942 \text{ m}^3/\text{mm} = 2.23 \times 10^{-2}(457)\left(\frac{Ha}{304.8}\right)^{1.522\left(\frac{457}{308.8}\right)^{0.026}}$$

Ha = 290 mm = 29 cm

2. Palmer-Bowlus Flumes

Palmer-Bowlus flumes are long-throated flumes that can be theoretically analyzed and are normally installed in circular sewers at manholes.

The throat for a Palmer-Bowlus flume can be either rectangular or trapezoidal in shape.

Figure 8.16 presents an example of a Palmer-Bowlus flume with a trapezoidal throat. Equations for both types of throats are available for calculating flows.

Fairly complex theoretical equations, some requiring trial and error solutions, are needed in order to calibrate and operate a Palmer-Bowlus flume. EPA publication 600/2–84–186 provides recommendations regarding specifications, installation, calibration, and performance monitoring for Palmer-Bowlus flumes.[61]

D. VENTURI METERS

The Venturi meter operates on the principle that a fluid flow through a convergence or constriction section of known shape and area causes a pressure drop at the constriction. It has been demonstrated that the difference in pressure drop is proportional to the square of the flow.

Figure 8.17 shows a cross section of a typical Venturi meter with the hourglass shape indicating high and low pressure taps for either manual or automatic sensing.[60] Venturi meters are used to measure wastewater flows and are equipped with cleaning devices for the pressure taps.[59]

The following equation can be used to calculate the flow for a horizontal Venturi meter if the diameter and pressure reading at the low and high pressure taps are known as well as the temperature of the water and the discharge coefficient.[63]

$$Q = K\,A_2\sqrt{2g\left(\frac{P_1 - P_2}{\gamma}\right)} \qquad [8.32]$$

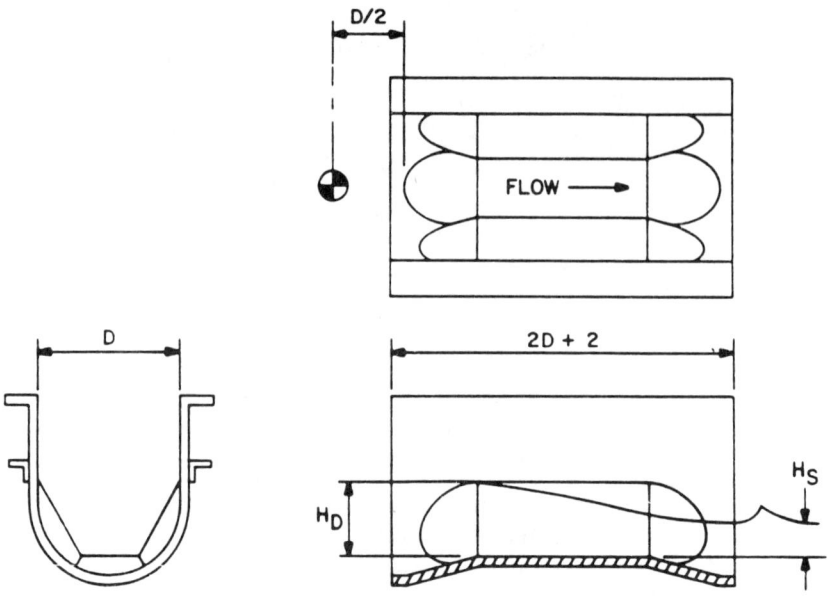

Figure 8.16 Palmer-Bowlus flume free-flow/depth relation.[60]

where: Q = flow, m³/s
K = discharge coefficient (common range is from 0.98 to 1.02)
A_2 = area of throat, m²
P_1 = high pressure reading, KPa (KPa = KN/m²)
P_2 = low pressure reading, KPa
g = acceleration due to gravity, 9.81 m/s²
γ = specific weight of water, KN/m³

Example 8.5

The Venturi meter illustrated in Figure 8.17 was used to measure a flow. The high pressure reading was 82 KPa and the low pressure reading was 39 KPa. The diameter of the Venturi at the high pressure tap was 0.57 m and at the low pressure tap it was 0.23 m. What is the flow in m³/s if the discharge coefficient for the Venturi is 0.98 and the wastewater temperature is 20°C?

Solution:

1. Calculate the area of the Venturi Meter throat

$$A_2 = \frac{\pi d^2}{4} = \frac{\pi (0.23 \text{ m})^2}{4} = 0.0415 \text{ m}^2$$

2. At 20°C the specific weight of water γ is 9.789 kN/m³
3. Given K = 0.98, g = 9.81 m/s², P_1 = 82 KPa

$$P_2 = 39 \text{ KPa}, \gamma = 9.789 \text{ KN/m}^3$$

4. Substitute in Equation 8.32

$$Q = 0.98 \left(0.0415 \text{ m}^2\right) \sqrt{2\left(9.81 \text{ m/s}^2\right)\left(\frac{82 \text{ KPa} - 39 \text{ KPa}}{9.789 \text{ KN/m}^3}\right)}$$

$$Q = 0.38 \text{ m}^3/\text{s}$$

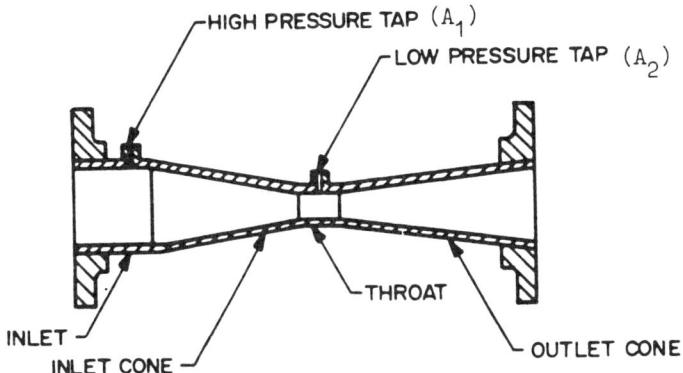

Figure 8.17 Classic Venturi meter.[60]

E. ELECTROMAGNETIC FLOWMETERS

The magnetic flowmeter operates on the principle that any fixed-dimension conductor, whether it is a bar of steel or a column of conductive liquid passing through the lines of force of a fixed magnetic field, will generate an electromotive force directly proportional to the rate at which the conductor is moving through the field. The electromotive force will be AC if the magnetic field is energized by an AC voltage. If a fixed magnetic field with a DC voltage is used, the electromotive force will be DC.[64]

For instance, if water, which is a conductive fluid, flows through a pipe and is normal to a magnetic field, an electromotive force will be induced across the fluid in a direction normal to both the magnetic field and the flow. The induced voltage is measured by placing insulated electrodes across the pipe so that a line connecting them is perpendicular to the magnetic field (see Figure 8.18).[60]

The electrical signal that is sent to the converter/transmitter is summed, referenced, and converted from a magnetically induced voltage to the appropriate scaled output. The magnetic flowmeter output is linearly proportional to the flow.[60]

There are several methods available for calibration of a electromagnetic flowmeter and these include:[64]

- The volumetric method
- Comparison with a reference flowmeter
- The dilution method
- The salt velocity method
- The velocity traverse method

From Equation 8.34, with the area-average velocity and the pipe diameter known, the volumetric flow Q can be readily calculated:

$$Q = VA = \frac{\pi D^2}{4} \times A \qquad [8.34]$$

where: Q = flow, m³/s
A = area, m² = $\frac{\pi D^2}{4}$
D = pipe diameter, m
V = velocity, m/s

The pipe diameter D is often the same value as the distance between the electrodes, but this is not always the case.

Magnetic flowmeters (mag meters) are suitable for application under the following general conditions:[60]

- Minimum head loss is desired
- The process fluid has a conductivity greater than 5 µmhos/cm
- The liquid is corrosive or abrasive
- The liquid has a solids concentration less than 10% by weight
- The pipe always flows full

Mag meters are not recommended for the following applications:

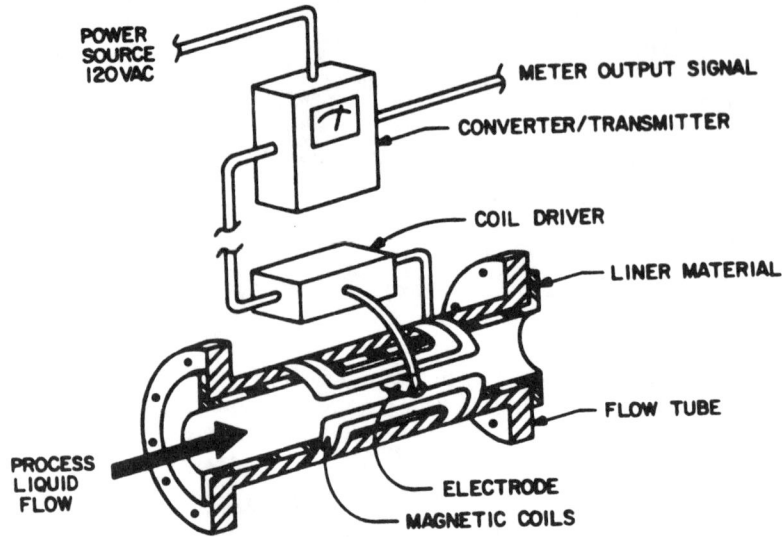

Magnetic flow meter construction.

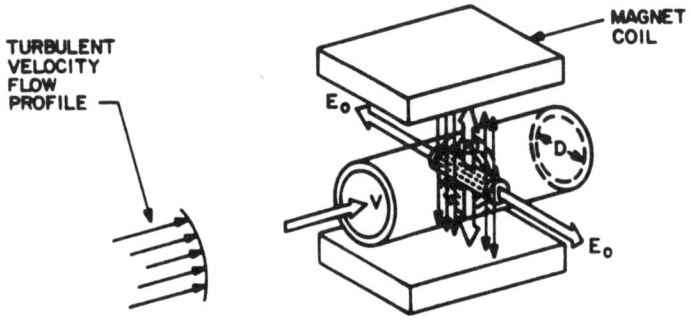

OUTPUT VOLTAGE = E_o = KBDV × 10^{-4} VOLTS
WHERE K = COEFFICIENT TO ACCOUNT FOR NON-IDEALITY
B = MAGNETIC FLUX DENSITY, TELSA
D = PIPE DIA. (DISTANCE BETWEEN ELECTRODES), METERS
V = AREA - AVERAGE VELOCITY OF FLOW, M/S

Mag meter induced voltage.

Figure 8.18 Electromagnetic flowmeter.[60]

$$\text{Output voltage} = E_o = KBDV \times 10^{-4} \text{ volts} \qquad [8.33]$$

where: K = coefficient to account for non-ideality
B = magnetic flux density, Telsa
D = pipe diameter
(distance between electrodes), m
V = area – average velocity of flow, m/s

- Nonconducting liquid process streams
- Gas streams
- Streams with powdered or granular dry chemicals
- Liquid streams with a solids concentration greater than 10% by weight

One problem with electromagnetic flowmeters is that they are subject to extraneous electrical effects that can interfere with proper operation of the unit. Servicing and trouble-shooting electromagnetic flowmeters requires substantial equipment and skilled technicians.[64]

VIII. RELIABILITY (REDUNDANCY, QUALITY CONTROL, LEVELS OF PROTECTION)

A. INTRODUCTION

For a reclaimed wastewater project to be successful, there must be assurance that the systems involved are reliable and that their failure could not pose a threat to public health or welfare.

The various processes involved in treating the reclaimed wastewater must be reliable and have the capability of producing the water quality required for the planned reuse project. The design of the treatment system must incorporate essential elements in the treatment processes in order to prevent the use of inadequately treated reclaimed wastewater. Not only is it important to incorporate reliability and redundancy into the design of the treatment facility, but also careful attention must be paid to the plant's construction and operation.

B. POTENTIAL PROBLEMS

Areas where problems can develop in a wastewater reclamation project include the power supply, individual treatment units (preliminary, primary, secondary, or advanced), mechanical equipment (especially disinfection), distribution systems, maintenance operations, as well as operating personnel problems. No system is perfect or infallible — the principle that "if something can go wrong it will go wrong" needs to be considered (Murphy's Law). One way to deal with such a situation is to provide backup systems.

C. FEDERAL RELIABILITY GUIDELINES

The *Manual — Guidelines for Water Reuse*[12] indicates that the federal guidelines for reliability for a reuse project must, in general, consider eight design principles and four other significant factors:[12]

- Duplicate power sources
- Standby power
- Multiple units and equipment
- Emergency storage
- Piping and pumping flexibility
- Dual chlorination
- Automatic residual control
- Automatic alarms

Other significant factors include:

- Engineering report
- Qualified personnel
- Effective monitoring program
- Effective maintenance and process control program

For a wastewater reuse project to be permitted by a state, the treatment facility must meet EPA Class I reliability guidelines or the equivalents.

Table 8.14 presents a summary of the Class I equipment EPA guidelines for reliability requirements.[12]

D. REGULATIONS REGARDING RELIABILITY

Following are excerpts from tentative, preliminary regulations for the State of Oregon dealing with reliability for reclaimed wastewater reuse.[65] While these preliminary draft regulations do not follow the sequence presented in Section VIII.C, they do, however, cover these items. Also included is a brief review of the Florida Department of Environmental Regulation (DER) directives for *Reuse of Reclaimed Water And Land Application.*[66]

1. No by-passing of inadequately treated sewage is allowed.

    This section is fairly straight forward. The Department is determined to assure the public that the use of reclaimed water is safe. It is, therefore, essential to prevent the inappropriate use of inadequately treated reclaimed water by not allowing by-passing of any essential component

Table 8.14 Summary of Class I Reliability Requirements

Unit	Class I requirement
Mechanically-cleaned bar screen	A backup bar screen shall be provided (may be manually cleaned).
Pumps	A backup pump shall be provided for each set of pumps which performs the same function. Design flow will be maintained with any one pump out of service.
Comminution facilities	If comminution is provided, an overflow bypass with bar screen shall be provided.
Primary sedimentation basins	There shall be sufficient capacity such that a design flow capacity of 50% the total capacity will be maintained with the largest unit out of service.
Filters	There shall be a sufficient number of units of a size such that a design capacity of at least 75% of the total flow will be maintained with one unit out of service.
Aeration basins	At least two basins of equal volume will be provided.
Mechanical aerator	At least two mechanical aerators shall be provided. Design oxygen transfer will be maintained with one unit out of service.
Chemical flash mixer	At least two basins or a backup means of mixing chemicals separate from the basins shall be provided.
Final sedimentation basins	There shall be a sufficient number of units of a size such that 75% of the design capacity will be maintained with the largest unit out of service.
Flocculation basins	At least two basins shall be provided.
Disinfectant contact basins	There shall be sufficient number of units of a size such that the capacity of 50% of the total design flow may be treated with the largest unit out of service.

Compiled from Reference 12.

of the wastewater treatment system. While the Department has no authority over the harvesting or sale of food crops, the Department would notify appropriate authorities (Oregon Department of Agriculture and the Oregon Health Division) if unsuitable reclaimed water were used in a manner not consistent with these rules.

2. Alarms.

Alarm devices required for various unit processes should be installed to provide warning of:

- Loss of power from the normal power supply.
- Failure of a biological treatment process.
- Failure of a disinfection process.
- Failure of a coagulation process.
- Failure of a filtration process.
- Any other specific process failure which would or could lead to noncompliance with these rules.

All required alarm devices should be independent of the normal power supply of the sewage treatment facility.

The person to be warned should be the plant operator, superintendent, or any other responsible person designated by the management of the sewage treatment facility and capable of taking prompt corrective action.

Individual alarm devices may be connected to a master alarm to sound at a location where it can be conveniently observed by the attendant. In case the sewage treatment facility is not attended full time, the alarm(s) should be connected to sound at a police station, fire station or other full-time service unit with which arrangements have been made to alert the person in charge at times that the sewage treatment facility is unattended.

3. Standby power.

The purpose behind this section is to assure that inadequately treated reclaimed water is not released due to a power failure. This can be accomplished by providing an auxiliary power generating system or by two separate, independent power feed systems to the plant. It can also be met with facilities for emergency storage or disposal:

- Where short-term retention or disposal provisions are used as a reliability feature, these should consist of facilities reserved for the purpose of storing or disposal of untreated or partially treated wastewater for at least a 24-hour period. The facilities should include all the necessary diversion devices, provisions for odor control, conduits, and pumping and pump back equipment. All of the equipment other than the pump back equipment should be either independent of the normal power supply or provided with a standby power sources.

- Where long-term storage or disposal provisions are used as reliability feature, these should consist of ponds, reservoirs, downstream sewers leading to other treatment or disposal facilities or any other facilities reserved for the purpose of emergency storage or disposal of untreated or partially treated wastewater. These facilities should be sufficient capacity to provide disposal or storage of wastewater for at least 20 days, and should include all the necessary diversion works, provisions for odor and nuisance control, conduits, and pumping and pump back equipment. All of the equipment other than the pump back equipment should be either independent of the normal power supply or provided with a standby power source.
- Diversion to a less demanding reuse is an acceptable alternative to emergency disposal of partially treated wastewater provided that the quality of the partially treated wastewater is suitable for the less demanding reuse.
- Subject to prior approval by the Department, diversion to a discharge point which requires lesser quality of wastewater is an acceptable alternative to emergency disposal of partially treated wastewater.
- Automatically actuated short-term retention or disposal provisions and automatically actuated long-term storage or disposal provisions should include, in addition to the above provisions of this section, all the necessary sensors, instruments, valves, and other devices to enable fully automatic diversion or untreated or partially treated wastewater to approved emergency storage or disposal in the event of failure of a treatment process, and a manual reset to prevent automatic restart until the failure is corrected.

4. Redundant facilities.

In order to provide assurances that reclaimed water will be consistently well-treated and safe for use, there must be confidence that failure of equipment will not result in poorly treated effluent being offered for reuse. Particularly in the case of reclaimed water, a single incident of inadequately treated water will nullify all the previous days and even years of well-treated water. It is, therefore, prudent to provide sufficient redundant equipment so that inevitable breakdowns can be accommodated without sacrificing effluent quality. The guidance presented here is from California's requirements:

Primary Treatment. All primary treatment unit processors shall be provided with one of the following reliability features.

- Multiple primary treatment units capable of producing primary effluent with one unit not in operation.
- Standby; primary treatment unit process.
- Long-term storage or disposal provisions.

Biological Treatment. All biological treatment unit processes shall be provided with one of the following reliability features:

- Alarm and multiple biological treatment units capable of producing oxidized wastewater with one unit not in operation.
- Alarm, short-term retention or disposal provisions, and standby replacement equipment.
- Alarm and long-term storage or disposal provisions.
- Automatically actuated long-term storage or disposal provisions.

Secondary Sedimentation. All secondary sedimentation unit processes shall be provided with one of the following reliability features:

- Multiple sedimentation unit capable of treating the entire flow with one unit not in operation.
- Standby sedimentation unit process.
- Long-term storage or disposal provisions.

Coagulation. All coagulation unit processes shall be provided with the following mandatory features for uninterrupted coagulant feed:

- Standby feeders,
- Adequate chemical storage and conveyance facilities,
- Adequate reserve chemical supply, and
- Automatic dosage control.

All coagulation unit processes shall also be provided with one of the following reliability features:

- Alarm and multiple coagulation units capable of treating the entire flow with one unit not in operation;
- Alarm, short-term retention or disposal provisions, and standby replacement equipment;
- Alarm and long-term storage or disposal provisions;
- Automatically actuated long-term storage or disposal provisions, or
- Alarm and standby coagulation process.

Filtration. All filtration unit processes shall be provided with one of the following reliability features:

- Alarm and multiple filter units capable of treatment the entire flow with one unit not in operation.
- Alarm, short-term retention or disposal provisions and standby replacement equipment.
- Alarm and long-term storage or disposal provisions.
- Automatically actuated long-term storage or disposal provisions.
- Alarm and standby filtration unit process.

Disinfection. All disinfection unit processes where chlorine is used as the disinfectant shall be provided with the following features for uninterrupted chlorine feed:

- Standby chlorine supply,
- Manifold systems to connect chlorine cylinders,
- Chlorine scales, and
- Automatic devices for switching to full chlorine cylinders.

Automatic residual control of chlorine dosage, automatic measuring and recording of chlorine residual, and hydraulic performance studies may also be required.

All disinfection unit processes where chlorine is used as the disinfectant shall be provided with one of the following reliability features;

- Alarm and standby chlorinator;
- Alarm, short-term retention or disposal provisions, and standby replacement equipment;
- Alarm and long-term storage or disposal provisions;
- Automatically actuated long-term storage or disposal provisions; or
- Alarm and multiple point chlorination, each with independent power source, separate chlorinator, and separate chlorine supply.

Other alternatives to reliability requirements. Other alternatives to reliability and redundancy may be accepted if the applicant demonstrates that the proposed alternative will assure an equal degree of reliability.

5. Cross connection prevention.

This section of the rules requires that piping and valves in the distribution system for reclaimed water be marked in a manner so that they can be differentiated from pipes and valves that are carrying domestic water. The concern, of course, is to alert plumbers or others that the reclaimed water piping is not part of a domestic water supply so that it will not be cross-connected into a domestic drinking water system. The rule references marking and construction standards in section 2, 3, 4, and 5 of the Final Draft of the "Guidelines for Distribution of Nonpotable Water" of the California-Nevada Section of the American Water Works Association, as revised September 14, 1983. A copy is available from the Department of Environmental Quality upon request. The sections referenced in this document discuss more issues than just marking and construction to prevent cross-connection. Those aspects not directly related to cross-connection should be ignored.

The Department recognizes that cross-connection may not be a significant issue in rural areas where there is a great separation distance between the reclaimed water system and any domestic water supply, either community or individual. The rule does allow the Department to grant exceptions if the reclaimed water system will be more than 100 feet away from a domestic water system. If the reclaimed water system is not an existing system, but must be constructed, it is likely that the Department will require that it be marked regardless of its separation distance from a domestic water supply. This is because marking is a cheap addition relative to the installation cost of the system itself.

6. Annual report required.

One of chief reasons for this section of the rules is to assure that the owner of the wastewater treatment plant will review the past year's use of reclaimed water and verify that compliance with these rules has been maintained. The report should cover the following major areas:

- Is the wastewater essentially the same as was originally represented in the original permit application or reclaimed water use plan? The main concern here is if new industrial or commercial users have come on line and caused a change in the character of the wastewater being treated at the treatment plant.
- How effectively has the treatment plant operated and met the effluent limitations for reclaimed water as specified in the permit? The Department is not interested in the actual discharge data; this would have been submitted on monthly discharge monitoring report forms. The Department is more interested in the general performance characteristics of the major components of the wastewater treatment plant and, whether due to age, design problems, wear, or other reason, a specific unit process or component of the process should be repaired or replaced.
- Was the reclaimed water provided to users and applied in a manner consistent with the reclaimed water reuse plan and the rules for use of reclaimed water? This part of the report can be covered by a summary of the amount of water provided to individual users, the types of crops grown, and any problems that arose in the distribution and/or use of the reclaimed water.
- Are there any additions or modifications to the treatment or the reclaimed water distribution system contemplated for the up-coming year?
- In the case of use of reclaimed water on food crops, have there been any problems demonstrated by reduced crop yields, reduced product quality, or other indication that could be attributed to the use of reclaimed water?

The Florida Department of Environmental Regulations has extensive, detailed regulations covering *Reuse Of Reclaimed Water And Land Application*.[66] This 70-page document covers all items mentioned in Section VIII.C and much, much more.

Section 17–610.462 of the regulations deals with reliability, and specifically with Class I reliability requirements, staffing, and the engineering report for the reuse project.[66]

IX. LEGAL ASPECTS OF SAMPLING AND MONITORING

A. INTRODUCTION

As of 1990, 13 states had specific criteria for water quality reuse; another 13 had no specific criteria, that is, no published criteria; and 24 did not permit reuse.[67] The states with the greatest number of reclaimed wastewater reuse projects include Arizona, California, Florida, Georgia, Kansas, South Carolina, and Texas.

B. EXAMPLES OF STATE LEGISLATION REGARDING SAMPLING AND MONITORING

The states that have passed reclaimed wastewater legislation and have specific regulations, as well as those states with no specific water reuse criteria, issue permits for the operation of water reclamation reuse projects.

Following are a few selected provisions in such a permit regarding sampling and monitoring for the City of Boca Raton, FL.[68]

This permit was issued on July 23, 1992, ID Number 5050M00565 by the Florida Department of Environmental Regulations. This permit consists of nine pages of condensed typing.[68] Under Item 7 — General Guidelines — on page 4, is the following provision regarding sampling and monitoring.

The permittee, by accepting this permit, specifically agrees to allow authorized Department personnel, upon presentation of credentials or other documents as may be required by law and at reasonable times, access to the premises where the permitted activity is located or conducted to:

- Have access to and copy any records that must be kept under conditions of the permit;
- Inspect the facility, equipment, practices, or operations regulated or required under this permit; and
- Sample or monitor any substances or parameters at any location reasonably necessary to assure compliance with this permit or Department rules.

Reasonable time may depend on the nature of the concern being investigated.

Also under General Conditions, Item 14 (C), on page 5, the following provision is made regarding monitoring records.

Records of monitoring information shall include:

- the date, exact place, and time of sampling or measurements;
- the person responsible for performing the sampling or measurements;
- the dates analyses were performed;
- the person responsible for performing the analyses;
- the analytical techniques or methods used;
- the results of such analyses.

Under the Permit Section, Specific Conditions, Item 2, page 6 of 9, the following parameters are specified regarding sampling, monitoring, and reporting.[68]

Sampling, reporting and effluent limitations for reclaimed water that is to be discharge for this Wastewater Treatment Plant (WWTP) for the period allowed to operate under this permit shall be in accordance with Florida Administrative Code (FAC) Chapters 17–601 and 17–600 and are as follows:

Parameter	Effluent Limit	Minimum Frequency	Sample Type	Sample Location
Flow	(b) MGD	Continuous (b)	Venturi meter with recorder and totalizer (b)	Upstream and downstream of filters (b)
$CBOD_5$	(a) mg/L	daily, 7/wk (a)	24 hour Composite	WWTP Influent and Reclaimed Effluent
TSS	(a) mg/L	daily, 7/wk (a)	Grab (e)	WWTP Influent and Reclaimed Effluent prior to disinfection
pH units	6.0 to 8.5	Continuous (h)	On-line pH meter (h)	Reclaimed Effluent
Chlorine Residual	(c) Min. 1.0 mg/L	Continuous	On-line chlorine analyzer	Reclaimed Effluent
Fecal Coliforms	Below detectable limits (d)	daily, 7/wk (d)	Grab (e)	Reclaimed Effluent
Turbidity	(f)(g)	continuous	On-line turbidity meter	Reclaimed Effluent prior to disinfection

(a) Limits, Maximum (mg/l).

	annual	monthly	weekly	one time grab
$CBOD_5$	20	30	45	60
TSS	5.0	5.0	5.0	5.0

Influent samples are for monitoring and reporting purposes. Pursuant to F.A.C. Rule 17–601.500(4)(a), influent samples shall be collected so that they do not contain digester supernatant or return activated sludge, or any other plant process recycled waters. WWTP influent $CBOD_5$ sampling may be used for the influent reuse $CBOD_5$ monitoring.

During the interim period, prior to the construction of the 9 MGD Reuse Treatment System, when the city provides reclaimed water to the F.A.U. Campus and for on-site irrigation and process water, the minimum frequency for $CBOD_5$ is weekly and TSS is daily, 7 days per week, and the sample type for $CBOD_5$ is 16 hour flow proportioned composite.

Annual average daily flow shall not exceed 9 MGD upstream of filters. During the interim period mentioned in 2(a); the annual average daily flow shall not exceed 1.8 MGD; minimum frequency is continuous; sample type is magnetic meter with recorder and totalizer; and sample location is the effluent subsequent to the hydropneumatic tank.

The minimum acceptable contact time shall be 15 minutes at the peak hourly flow. Higher residuals or longer contact times may be needed to meet the design and operational criteria for high-level disinfection as described in Rules 17–600.400(5)(c) and 17–600.440(5)(f), F.A.C.

Over a thirty (30) day period, 75 percent of the fecal coliform values shall be below the detection limits. Any one sample shall not exceed 25 fecal coliform values per 100 ml of sample pursuant to F.A.C. Rule 17–600.440(5). During the interim period mentioned in 2(a), the minimum frequency for fecal coliform sampling is daily, 7/week.

Grab samples will be collected during periods of peak hydraulic and/or organic loading.

The permittee shall determine a maximum turbidity limit based on a correlation between turbidity data and TSS data.

The permittee shall develop, and the department shall approve, an operating protocol designed to ensure that the high level disinfection criteria will be met before the reclaimed water is released to the system storage or to the reclaimed water reuse system. The operating protocol shall be reviewed and updated and shall be subject to department review and approval at least annually. Reclaimed water produced at the treatment facility that fails to meet the criteria established in the operating protocol shall not be discharged into system storage or to the reuse system. Such substandard reclaimed water (reject water) shall be either stored for subsequent additional treatment or shall be discharged to another permitted reuse system requiring lower levels of pretreatment or to a permitted effluent disposal system. The protocol shall include a correlation between TSS and Turbidity determined from infield sampling. F.A.C. Rule 17–610.463.

Hourly measurements during the period of required operator attendance may be substituted for continuous measurement.

C. FEDERAL REGULATIONS REGARDING SAMPLING AND MONITORING

The U.S. federal government does not issue permits for the operation of wastewater reclamation and reuse projects.

However, under the *Federal Water Pollution Control Act Amendments* of 1972 (PL 92–500) provision was made for the legal inspection, sampling, and monitoring of a wastewater treatment facility to determine if the conditions of their NPDES permit was being met.

The administration of the NPDES program is carried out by the EPA unless a state has established primacy and thus have their own permitting program — 37 states are currently administering NPDES permit programs under agreement with EPA.[69,70]

PL 92–500 provides for two types of monitoring:[69]

- Self-monitoring, where the facility must monitor itself; and
- Compliance monitoring, which may consist of checking and/or inspecting the self-monitoring program or monitoring conducted by the regulatory agency.

According to Section 308 of the Act, EPA may request and collect information by various means including the conduct of an inspection wherever there is an existing NPDES permit, or where a discharge exists or is likely to exist and no permit has been issued.

Much of the compliance monitoring in the NPDES program takes place at the state level. The *Clean Water Act* provides for the transfer of federal program authority to the states to conduct NPDES permit compliance monitoring. EPA Regional Administrators and some state water pollution control agencies have signed formal cooperative agreements that ensure timely, accurate monitoring of compliance with permit conditions.[69]

The permits issued under the NPDES program impose precise and detailed pollution control requirements on direct dischargers. Permits are written so that they:

- Limit discharges of effluents based on national technology-based guidelines and, where necessary, on water quality standards;
- Require permittees to *monitor their discharges and report results* and any violations to the permitting agency; and
- Where necessary, impose compliance schedules that the permittee must adhere to in abating pollution and in complying with specified effluent limitations.

Following is a part of the permit issued by the Kentucky Pollutant Discharge Elimination System (KPDES) for the City of Bowling Green, dealing with effluent limitations, monitoring, and testing requirements as of September 1991.[71]

EFFLUENT LIMITATIONS AND MONITORING REQUIREMENTS[71]

During the period beginning on the effective date of this permit and lasting through the term of this permit, the permittee is authorized to discharge from outfall(s) serial number(s): 001, Municipal Discharge.

Such discharges shall be limited and monitored by the permittee as specified below:

Effluent Characteristics	Discharge Limitations				Monitoring Requirements		
	lbs/day		Other Units (Specify)				
	Monthly Avg.	Weekly Avg.	Monthly Avg.	Weekly Avg.	Measurement Frequency	Sample Type	Sampling Location
Flow, Design (10.6 MGD)	—	—	—	—	Continuous	—	Influent or Effluent
Biochemical Oxygen Demand (5 day), Carbonaceous	2210	3320	25 mg/l	37.5 mg/l	1/Weekday	Composite	Influent & Effluent
Total Suspended Solids	2650	3980	30 mg/l	45 mg/l	1/Weekday	Composite	Influent & Effluent
Fecal Coliform Bacteria, N/100	—	—	200	400	1/Weekday	Grab	Effluent
Ammonia (as N)	884	1330	10 mg/l	15 mg/l	1/Weekday	Composite	Effluent
Dissolved Oxygen shall not be less than 7 mg/l					1/Weekday	Grab	Effluent
Total Residual Chlorine (TRC)			See PART I, Page I-2		1/Weekday	Grab	Effluent
Biomonitoring shall not exceed 3.3 chronic toxicity unit(s)			See PART IV, Pages IV-1 and IV-2				Effluent

In addition to the specific limits, the monthly average effluent $CBOD_5$, and suspended solids concentration shall not exceed 15 percent of the respective monthly average influent concentration (85% removal).

The pH of the effluent shall not be less than 6.0 standard units nor greater than 9.0 standard units and shall be monitored once per weekday by grab sample.

There shall be no discharge of floating solids or visible foam in other than trace amounts.

The effluent shall not cause a visible sheen on the receiving water.

The total residual chlorine daily maximum final effluent concentration shall not exceed 0.019 mg/l. Testing for this parameter shall be conducted according to either the amperometric titration method or the DPD colorimetric method as specified in Sections 408(C) or (E), *Standard Methods for the Examination of Water and Wastewater,* 16th Edition. Using one of these methods, if chlorine is not detected, the permittee shall report chlorine as being present at less than 0.01 mg/l (the detection level used) on its discharge monitoring report and shall be considered to be in compliance with the above permit chlorine limitation. Monitoring frequency shall be once per weekday by grab sample. Monitoring for total residual chlorine shall begin upon the effective date of this permit.

Compliance with this requirement shall be achieved within 12 months of the effective date of this permit condition. Quarterly reports of progress beginning from the effective date of this permit condition shall be submitted to the Permit Issuing Authority.

X. HOMEWORK PROBLEMS

1. Define the following: pH, turbidity, BOD_5, COD, TOC, BOD_c, BOD_L, BOD_N and coliform.
2. What are at least three of the best references to use for recommended procedures for water and wastewater sample collection and analysis?
3. What are at least three procedures that must be considered in order to accomplish the sampling objectives?
4. A wastewater treatment plant has an average daily flow of 45,420 m^3/d (12 mgd). A flow-proportional composite sample is to be taken with a total volume of 2 L collected 12 times during the day. The flow at one of the times of sampling was 18,925 m^3/d (5 mgd). What volume of sample should be collected at that time?
5. What are the three major types of samples?
6. What are the primary functions of the preservation of a water sample after collection?
7. What are 12 pieces of information that should appear on a water sample identification label?
8. What are two basic methods used to preserve samples?
9. What tests are commonly taken at the same time as the bacteriological samples are collected?
10. Explain the difference between accuracy and precision.
11. The evaluation of a reclaimed wastewater for reuse can be determined by four broad basic types of analyses. List all these analyses.
12. The standard deviation for a test has been reported to be 16 mg/L. If an acceptable level of uncertainty is ±7 mg/L and a 95% confidence level is desired, how many samples should be taken?
13. Distinguish between total solids and dissolved solids, also distinguish between volatile solids and fixed solids.
14. Determine the weights in mg/L of total solids in a wastewater sample using the following laboratory data:

    Volume of sample = 100 mL
    Weight of crucible = 130.2806 g
    Weight of crucible + sample = 130.3644 g after drying at 103°C for 24 hr.

15. Calculate the efficiency of a wastewater treatment plant in removing suspended solids, using the following laboratory results:

	Influent	Effluent
Volume of sample	250 mL	500 mL
Weight of Gooch crucible and solids after drying	15.4772 g	16.6199 g
Weight of Gooch crucible	15.4217 g	16.6089 g

16. Determine the percent of volatile solids in a wastewater sample given the following data:

    Volume of sample = 100 mL
    Crucible weight = 19.9712 g
    Crucible plus dry solids = 20.0623 g
    Crucible plus ash = 20.0090 g

17. What is the difference between the true and apparent color of water?
18. Matching. Select the best answer and enter the number in the blank.

    ____ a. NH$_3$ test
    ____ b. pathogen
    ____ c. color change, pink to colorless at pH 8.2
    ____ d. Alkalinity
    ____ e. Kjeldahl
    ____ f. Chloramines
    ____ g. Orthophosphate test
    ____ h. Turbidity
    ____ i. Dilution water used for BOD$_5$ test after incubation
    ____ j. Boron

    1. Disease-causing microorganism
    2. NH$_3$-N determination
    3. Curcumin method
    4. DO drop should not be more than 0.2 mg/L
    5. Indicating boric acid solution
    6. Stannous chloride method
    7. Phenolphthalein
    8. 0.02 N H$_2$SO$_4$ titrant
    9. NH$_3$ + Cl$_2$
    10. NTU

19. Matching. Select the best answer and enter the number in the blank.

___ a. 20°C
___ b. color
___ c. 35°C
___ d. Gray
___ e. 44°C
___ f. Sodium thiosulfate
___ g. Settleable solids
___ h. Chloride test
___ i. Specific conductance
___ j. 1 currie

1. Eliminates Cl_2 interference
2. Imhoff cone
3. 0.0141 N mercuric nitrate titration solution
4. Potassium chloroplatinate and cobaltous chloride
5. Incubation temperature for total coliform test
6. 3.7×10^{10} Bq.
7. Incubation temperature for BOD_5 test
8. Incubation temperature for fecal coliform
9. 100 rads
10. μmohs/cm

20. Matching. Select the best answer and enter the number in the blank.

___ a. Ferrous ammonium sulfate (FAS)
___ b. Oil and grease
___ c. Dissolved oxygen test
___ d. Chlorine
___ e. Fecal coliform
___ f. Conductivity
___ g. Sulfate test
___ h. Nitrosomonas
___ i. Total coliform
___ j. Nitrobacter

1. Modified Winkler
2. Turbidimetric method
3. Ammonia → nitrite
4. Blue colonies
5. Titrant used in COD test
6. Pink to dark red metallic sheen colonies
7. Nitrite → nitrate
8. DPD-colorimetric method
9. Soxhlet extraction method
10. Calibrate with 0.01 N KCl

21. Determine the pH of the following wastewater samples:
    a. Wastewater containing 1.0 mg/L of hydrogen ion
    b. Wastewater containing 0.002 mg/L of hydrogen ion
    c. Wastewater containing $0.17 \cdot 10^{-6}$ mg/L of hydroxide (OH^-) ion

22. If the pH of a given water is 4.5, determine the concentration of the hydrogen ion and hydroxide ion in moles/liter and in mg/L.

23. a. What ions, compounds, constitute a free chlorine residual?
    b. What group of compounds constitute a combined chlorine residual?

24. Give the pH transition intervals for the following common acid-base indicators as well as the color changes:

    Phenolphthalein
    Bromocresol green
    Methyl orange

25. The ionization constant for hypochlorous acid (HOCl) is $K = 2.88 \times 10^{-8}$ mol/L at 25°C. Determine the percentage of HOCl and OCl^- present at pH value 7.5.

26. Sketch both the nitrogen cycle and the sulfur cycle under aerobic and anaerobic conditions.

27. Sketch the idealized BOD curve showing the carbonaceous and nitrogenous phases.

28. What is the BOD_5 of a sample of wastewater if a 2 mL sample in a 300 mL BOD bottle had an initial dissolved oxygen of 7.5 mg/L and a final dissolved oxygen of 3.9 mg/L?

29. The BOD_5 of a wastewater sample was found to be 40.0 mg/L. The initial oxygen concentration of the BOD dilution water was equal to 9.0 mg/L, the dissolved oxygen concentration measured after incubation was equal to 2.74 mg/L and the size of sample used was equal to 40 mL. If the BOD bottle used was equal to 300 mL volume, what was the initial dissolved oxygen concentration in the wastewater sample?

30. Three of the following reagents are used in the dissolved oxygen (DO) test. Indicate the three reagents used and the order in which they are added to the BOD bottle.

    a. Potassium dichromate
    b. Mercuric sulfate
    c. Alkali-iodide azide
    d. Manganese sulfate
    e. Silver nitrate
    f. H_2SO_4
    g. Ferroin

31. What is added to the BOD dilution water in order to provide a suitable environment for the microorganisms to grow?
32. In setting up a BOD test, the 300 mL bottles contained the following samples:

Sample	Initial DO	5-day DO
Blank 0	7.9 mg/L	7.9 mg/L
Raw 6 ml	7.8 mg/L	2.3 mg/L
Primary 10 mL	7.7 mg/L	2.8 mg/L
Final 50 mL	7.6 mg/L	3.2 mg/L

    Calculate the 5-day BOD of the samples. What is the percentage reduction between raw and primary? What is the overall plant removal efficiency?
33. a. What was the indicator used in the COD titration procedure?
    b. In the COD test, what catalyst was used to oxidize the straight chain aliphatic components?
    c. What compounds are *not* oxidized in the COD test?
    d. What oxidizing agent is used in the COD procedure?
34. A 100 mL sample is subjected to a COD analysis. After the reaction 14.0 mL of the ferrous ammonium sulfate reagent is required to titrate the oxidizing agent remaining in the sample. When the ferrous ammonium sulfate is used to titrate the bank, 25 mL is required. The normality of the ferrous ammonium sulfate is 0.1 N. What is the amount of COD in mg/L present in the original sample?
35. Multiple choice. Circle the best answer. In some cases, there are more than one correct answer.
    a. Bacteriological samples should be refrigerated after collection at: 1. 0°C, 2. 10°C, 3. 4°C, 4. 25°C, 5. 32°F
    b. What is the turbidity of a sample of 10 mL was diluted with 40 mL of turbidity-free water and the diluted sample read 16 NTUs? 1. 20 NTUs, 2. 40 NTUs, 3. 80 NTUs, 4. 100 NTUs
    c. A COD/BOD ratio has a value of 1.8. This means:
        1. There is probably more biochemically oxidizable material than chemically oxidizable material.
        2. There is probably more chemically oxidizable material than biochemically oxidizable material.
        3. Such ratio numbers refer to reagents used, not to data values.
        4. None of the above.
    d. The "saturation" values for dissolved oxygen in surface waters depend primarily upon the:
        1. Salinity of the waters
        2. Temperature of the waters
        3. Presence of oxidizing bacteria
        4. Thickness of the surface films
    e. The pHs of six grab samples were: 6.0, 8.0, 6.9, 5.5, 8.9, and 8.5. From this, we may determine that the pH of the composite sample was:
        1. 4.6
        2. 8.2
        3. 7.0
        4. Cannot be determined from the above data
    f. The amperometric titration method is used to measure:
        1. BOD
        2. pH
        3. Residual coliform
        4. Total coliform
    g. BOD dilution water should be:
        1. Tap water with nutrients and buffer added
        2. Boiled to kill all bacteria
        3. Aerated distilled water with nutrients and buffer added
        4. Treated with sodium thiosulfate
    h. The most reliable BOD samples use at least the following amount of DO:
        1. 0.5 mg/L
        2. 2.0 mg/L

3. 4.0 mg/L
   4. 7.0 mg/L
  i. Specific conductance provides an estimate of:
   1. pH
   2. Dissolved solids
   3. Hardness
   4. Trace elements
36. What are the four approved EPA methods for the detection of total coliform?
37. Indicate whether the following are true or false. Coliforms are:
   ____ a. aerobic and facultative microorganisms
   ____ b. Gram positive
   ____ c. spore formers
   ____ d. ferment lactose
   ____ e. cocci
38. Name the major forms of nitrogen which are important in reclaimed wastewater. Which forms predominate?
39. What are at least seven trace metals that are of major concern in the beneficial use of reclaimed wastewater?
40. Indicate the basic components of the following units:
   a. Atomic absorption spectrophotometer (AA)
   b. Inductively coupled plasma unit (ICP)
41. List the names and purposes of the old and new units used in radiation measurements, also explain the importance of each unit.
42. In the U.S. National Interim Primary Drinking Water Standard, what level of gross alpha activity is permitted (MCL) and how does this compare to the World Health Organization guidelines? Also, compare the U.S. drinking water standards (MCL) for gross beta activity and that of the WHO guidelines.
43. List the field instruments utilized to measure beta, gamma, and alpha radiation.
44. What are the components in a basic scintillation system for detecting gamma radiation?
45. A 90° V-notch weir is placed in a channel with the base of the notch raised from the channel bottom. Determine the discharge in m^3/min for a height of flow equal to 150 mm measured in a stilling well behind the weir.
46. A rectangular contracted weir, 1.2 m high, is to be installed in a channel 2.4 m wide. The maximum flow over the weir is 1.7 m^3/s when the total depth back of the weir is 2.1 m. What length of weir should be installed?
47. Compute the upstream depth (in mm) for a Parshall flume with a throat width of 0.6 m for a flow of 0.17 m^3/s.
48. Determine the throat diameter for a Venturi meter that is to be used in a 380 mm effluent discharge line from a wastewater treatment plant. The average flow through the plant is 8000 m^3/day and it is desired to have a minimum head differential of 32 cm at low flow. Minimum flow is 60% of the average daily flow and maximum flow is 150% of the average flow. The coefficient for the Venturi meter is 0.96 when the wastewater temperature is 20°C. What is the head differential at maximum flow for this Venturi meter?
49. Visit your local wastewater treatment plant and evaluate how they are meeting reliability regulations, especially with concern to the following:
   By-passing
   Alarms
   Standby power
   Redundant facilities
   Disinfection
   Cross connections
50. What are the NPDES monthly and weekly maximum daily limits for fecal coliform in a wastewater effluent?

APPENDIX FOR CHAPTER 8
STATISTICAL PRINCIPLES OF NORMAL OR GAUSSIAN DISTRIBUTION

1. INTRODUCTION

The application of statistical principles is not only necessary in the setting up of a water or wastewater sampling program, but also in the analysis of the data collected.

2. NORMAL OR GAUSSIAN DISTRIBUTION

When an infinite number of measurements are made the individual values will be distributed similar to that of the curve shown in Figure A. This curve demonstrates Normal or Gaussian Distribution, and is identified by its bell shape. Many types of distribution curves exist other than Normal or Gaussian. A few of these other distributions include, binomial, t, chi, f, and log-normal. However, Normal and Gaussian Distribution are the most important and most often used in the monitoring of water and wastewater.

The theoretical Normal or Gaussian curve is best defined by two terms, the mean, μ, and the standard deviation, σ. The theoretical average for an infinite number of measurements is called the actual or true mean, μ. We would really like to know this value, μ, but this is impossible as an infinite number of measurements cannot be made. The theoretical standard deviation, σ, for an infinite number of measurements is equal to the square root of the theoretical mean, $\sigma = \sqrt{\mu}$. As stated, it is impossible to make an infinite number of measurements, therefore the mean or average of a limited number of tests (10 or 20) is used to estimate the average or mean and is symbolized as $\bar{x}$. The value of $\bar{x}$ approaches the value of μ as the number of measurements becomes very large. The theoretical standard deviation, σ, is also estimated by using a limited number of test data and is symbolized by the letter s.

3. STATISTICAL TERMINOLOGY
3.1 Standard Deviation

The standard deviation, s, is a measure of the dispersion, and variability of individual data values from the mean. Following is the formula used to calculate or define the standard deviation.

$$s = +\sqrt{\frac{\Sigma(x_i - \bar{x})^2}{n-1}} \qquad [1]$$

where: s = the standard deviation (always positive)
 x_i = a data or measurement value
 $\bar{x}$ = the mean of the measurements or data
 n = number of observations

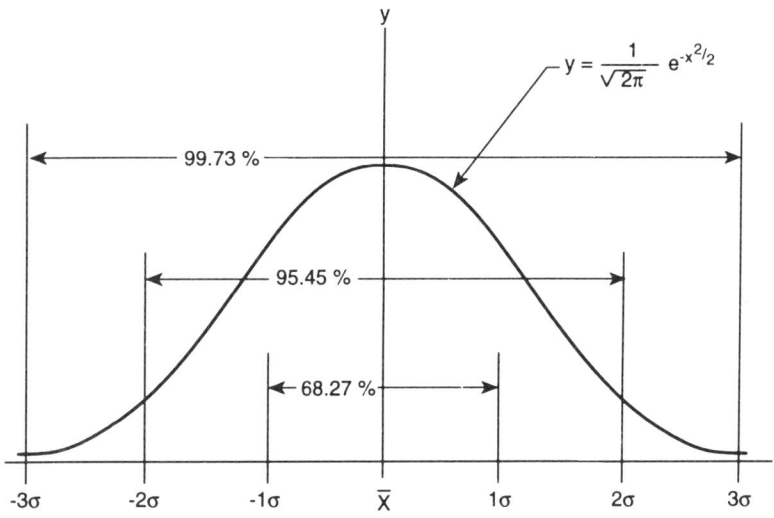

Figure A Normal or Gaussian distribution.

3.2 Variance

Another term that is used as an indicator of the dispersion and variability of data is the variance. The variance in the technical term for the square of the standard deviation, σ^2 = variance. If data were expressed in mg/L the variance would be in mg^2/L^2, therefore taking the square root of the variance provides the more familiar units mg/L.

3.3 Mean or Average, x

$$\bar{x} = \frac{\text{sum of all measurements}}{\text{number of measurements}} = \frac{\Sigma x}{n}$$

3.4 Median

The median of a set of data is the observation in the middle, that is, the number unit is located such that half of the observations are less than it and half are greater. To find the median of a set of observations, the data are arranged in order of magnitude. The median could, however, fall between two measurements.

3.5 Mode

The mode is the most frequent or common value (for the theoretical Normal or Gaussian distribution curve the mean, median, and mode are all equal).

3.6 Range

One indicator of the spread of a data set is the range, which is defined as the difference between the largest and the smallest values in the set.

$$\text{Range} = \text{largest x} - \text{smallest x} \qquad [2]$$

3.7 Standard Error of the Mean

A term often used in statistics is the standard error of the mean. The standard error of the mean is the standard deviation of a distribution of means or any other statistical measure computed from the samples.

The formula for the standard error of the mean is

$$\sigma_{\bar{x}} = \frac{\sigma}{\sqrt{n}} \qquad [3]$$

where: $\sigma_{\bar{x}}$ = standard error of the mean
σ = standard deviation of the sample
n = number of values

3.8 Coefficient of Variation or Relative Standard Deviation

The coefficient of variation indicates how tightly the measurements are grouped about the mean, and is computed as follows:

$$CV = \frac{s}{\bar{x}} \qquad [4]$$

where: CV = coefficient of variation
s = standard deviation
$\bar{x}$ = mean of average of the values, data, or measurements

The smaller the CV value the closer the data are grouped about the mean. The relative standard deviation is generally expressed as a percentage and is calculated by multiplying the CV by 100.

$$\text{Percent Relative Standard Deviation} = \frac{s}{\bar{x}} \cdot 100 \qquad [5]$$

where: s = standard deviation of a set of data
        $\bar{x}$ = mean or average of the data

For instance if s ± 1.5, x = 33, then the relative standard deviation is 4.5%.

The standard deviation and relative standard deviation are mathematical expressions of the precision of a set of data.

3.9 Relative Error
The relative error or the percentage relative error is a measure of the accuracy of a set of data, and reflects the closeness of a measured value to the true value. The percent recovery or relative error is determined by the addition of a known quantity of a chemical standard to a sample and a subsequent analysis determines the percent recovery of that known quantity.

3.10 Confidence, Limits, Intervals, and Levels
The confidence limits are the limits within which, at some specified level of probability, the true value lies. For example, when values are assigned a ±/σ, these are confidence limits. The confidence limit defines the boundary values for the confidence interval. For example, 20 ± 6 indicates confidence limits of 14 and 26, while values from 14 to 26 represent the confidence interval.

The confidence level is the assurance that a sample or result of a measurement will not be further from the true mean that some designated value.

Confidence levels can be expressed quantitatively in terms of the standard deviation as follows:

Number of standard deviations (plus or minus from the mean)	Percent of values included
± 1 σ	68.27
± 2 σ	95.45
± 3 σ	99.73

It is sufficiently accurate to round these numbers to 68, 95, and 99%.

See Figure A for a graphical presentation of this phenomenon. For instance, if the standard deviation for a certain set of tests is ±10 mg/L and the average or mean value for this test is 120 mg/L, then a 95% confidence level (2σ) indicates that if another test or measurement is made there would be a 95% chance or probability that the test result would be between 120 ± 20 mg/L. Also the standard deviations measured plus or minus from the mean gives the percentage of the area under the curve or the probability that another measurement will fall in this range.

4. PROBABILITY
Probability deals with the prediction of events. Probability, p, is the likelihood or chance that a particular phenomenon will occur and is represented by a scale of values from zero to unity. The probability of occurrence or an absolute certainty is represented by p = 1, and or an absolute impossibility by p = 0. If p is the probability of occurrence of an event, and q is the probability of an event not occurring, then

$$p + q = 1. \qquad [6]$$

Tables are available that can be used to estimate the probability or occurrence of an event. Table A presents the areas under the Normal or Gaussian probability curve, where the standard normal variable, t, is used to calculate the probabilities.

$$t = \frac{\chi - \mu}{\sigma} \qquad [7]$$

where: t = standard normal variable
        χ = particular numerical value
        μ = absolute or theoretical mean
        σ = standard deviation.

5. AREA DETERMINATION UNDER THE NORMAL OR GAUSSIAN PROBABILITY CURVE

The areas under the Normal or Gaussian distribution can be determined for various conditions and thus used to predict the occurrence of a particular event or measurement to happen. To calculate this area Equation 7 can be used. For a graphical presentation of the terms used in Equation 7 see Figure B.

The total area under the Normal or Gaussian curve is taken to be 1.0000. A simple demonstration of the application of Equation 7 and Table A is as follows: if χ is the distance of a particular value from the theoretical or absolute mean is 1, and $\mu = 0$, and the standard deviation is ± 1, then

$$t = \frac{1-0}{1} = 1$$

(see Figure B).

From Table A a value of $t = 1$ gives an area under the curve of 0.34134 between $\mu = 0$ and $\chi = 1$. Taking into account that $\sigma = \pm 1$, then the total area under the curve is $2 \times 0.34134 = 0.6828$, or 68%. Similar calculation can be made for $\pm 2\sigma$ (95.45%), and $\pm 3\sigma$ (99.73%).

Example

The effluent standard for chlorides for a wastewater treatment plant is to 90 mg/L. The mean concentration of chlorides from this plant has been found to be 70 mg/L, with a standard deviation of 15 mg/L.

Determine the probability of exceeding the 90 mg/L Cl⁻ effluent standard.

Solution.

1. Determine the t value $t = \frac{\chi - \mu}{\sigma} = \frac{90 - 70}{15} = 1.33$

2. From Table A t = 1.33 the area under the curve from $\mu = 70$ to $\chi = 90$ is 0.40824. Taking into account that the distribution curve has a total area of 1 and that half the area under the curve is 0.5, the probability of exceeding the 90 mg/L standard is $0.5 - 0.40824 = 0.09176$ or 9.2%.

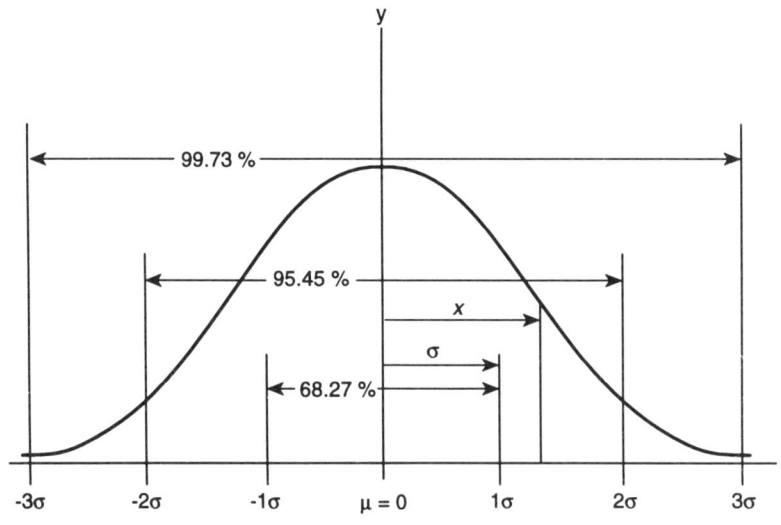

Figure B Graphical presentation of μ, χ, and σ.

Table A Areas Under the Normal or Gaussian Distribution Curve Between the Absolute Mean and the Calculated Value of t[a]

t	00	01	02	03	04	05	06	07	08	09
0.0	00000	00399	00798	01197	01595	01994	02392	02790	03188	03586
0.1	03983	04380	04776	05172	05567	05962	06356	06749	07142	07535
0.2	07926	08317	08706	09095	09483	09871	10257	10642	11026	11409
0.3	11791	12172	12552	12930	13307	13683	14058	14431	14803	15173
0.4	15542	15910	16276	16640	17003	17364	17724	18082	18439	18793
0.5	19146	19497	19847	20194	20540	20884	21226	21566	21904	22240
0.6	22575	22907	23237	23565	23891	24215	24537	24857	25175	25490
0.7	25804	26115	26424	26730	27035	27337	27637	27935	28230	28524
0.8	28814	29103	29389	29673	29955	30234	30511	30785	31057	31327
0.9	31594	31859	32121	32381	32639	32894	33147	33398	33646	33891
1.0	34134	34375	34614	34850	35083	35314	35543	35769	35993	36214
1.1	36433	36650	36864	37076	37286	37493	37698	37900	38100	38298
1.2	38493	38686	38877	39065	39251	39435	39617	39796	39973	40147
1.3	40320	40490	40658	40824	40988	41149	41309	41466	41621	41774
1.4	41924	42073	42220	42364	42507	42647	42786	42922	43056	43189
1.5	43319	43448	43574	43699	43822	43943	44062	44179	44295	44408
1.6	44520	44630	44738	44845	44950	45053	45154	45254	45352	45449
1.7	45543	45637	45728	45818	45907	45994	46080	46164	46246	46327
1.8	46407	46485	46562	46638	46712	46784	46856	46926	46995	47062
1.9	47128	47193	47257	47320	47381	47441	47500	47558	47615	47670
2.0	47725	47778	47831	47882	47932	47982	48030	48077	48124	48169
2.1	48214	48257	48300	48341	48382	48422	48461	48500	48537	48574
2.2	48610	48645	48679	48713	48745	48778	48809	48840	48870	48899
2.3	48928	48956	48983	49010	49036	49061	49086	49111	49134	49158
2.4	49180	49202	49224	49245	49266	49286	49305	49324	49343	49361
2.5	49377	49396	49413	49430	49446	49461	49477	49492	49506	49520
2.6	49534	49547	49560	49573	49585	49598	49609	49621	49632	49643
2.7	49653	49664	49674	49683	49693	49702	49711	49720	49728	49736
2.8	49744	49752	49760	49767	49774	49781	49788	49795	49801	49807
2.9	49813	49819	49825	49831	49836	49841	49846	49851	49856	49861
3.0	49865	49869	49874	49878	49882	49886	49889	49893	49897	49900
3.1	49903	49906	49910	49913	49916	49918	49921	49924	49926	49929
3.2	49931	49934	49936	49938	49940	49942	49944	49946	49948	49950
3.3	49952	49953	49955	49957	49958	49960	49961	49962	49964	49965
3.4	49966	49968	49969	49970	49971	49972	49973	49974	49975	49976
3.5	49977	49978	49978	49979	49980	49981	49981	49982	49983	49983
3.6	49984	49985	49985	49986	49986	49987	49987	49988	49988	49989
3.7	49989	49990	49990	49990	49991	49991	49992	49992	49992	49992
3.8	49993	49993	49993	49994	49994	49994	49994	49995	49995	49995
3.9	49995	49995	49996	49996	49996	49996	49996	49996	49997	49997
4.0	49997	49997	49997	49997	49997	49997	49998	49998	49998	49998

[a] If t = 1.27 then the value to use in the calculations is 0.39796.

REFERENCES

1. USEPA, Analytical Methods for Trace Metals. EPA 430/1–79–006, 107 T., Office of Water Program Operations, Cincinnati, OH, 1979.
2. Anon., *Operational Control Tests for Wastewater Treatment Facilities,* ERIC/CSMEE 154 T., EPA Grant #900953010, Linn-Benton Community College, Albany, OR, 1981.
3. NYDEC, Manual of Instruction for Wastewater Treatment Plant Operations, Vols. 1 and 2, Dept. of Environmental Conservation, Health Education Service, State of New York, Albany, 1988.
4. AWWA, *Introduction To Water Treatment,* Vol. 2, American Water Works Association, Denver, CO, 1984.
5. USEPA, Handbook for Sampling and Sample Preservation of Water and Wastewater. EPA-600/4–82–029, Office of Research and Development, U.S. Environmental Protection Agency, Cincinnati, OH, 1982.
6. AWWA, *Standard Methods for the Examination of Water and Wastewater,* 18th ed., APHA, AWWA, WEF, American Public Health Association, Washington, D.C., 1992.

7. *Federal Register,* Vol. 38 and 41, Title 40, Chapter 1, Subchapter D, Part 136, pages 28758 and 52780, 1973, 1976, U.S. Government Printing Office, Washington, D.C.
8. ASTM, *Annual Book Of ASTM Standards,* Part 31, Water, American Society for Testing and Materials, Philadelphia, 1985.
9. USEPA, Methods Chemical Analysis of Water and Wastes, ERIC, Technology Transfer, U.S. Environmental Protection Agency, Cincinnati, OH, 1979.
10. USGS, Methods for Collection and Analysis of Water Samples for Dissolved Minerals and Gases, U.S.G.S. Survey Techniques of Water Resources Inventory, Book 5, Chapter 41, Department of the Interior, Washington, D.C., 1970.
11. AWWA, *Introduction to Water Quality Analysis,* Vol. 4, American Water Works Association, Denver, CO, 1982.
12. USEPA, Manual — Guidelines For Water Reuse, Office of Water, EPA/625/R-92/004, U.S. Environmental Protection Agency, Washington, D.C., 1992.
13. **Chase, G. D., and J. L. Rabinowitz,** *Principles Of Radioisotope Methodology,* Burgess Publishing, Minneapolis, MN, 1965.
14. **Adams, V. D.,** *Water and Wastewater Examination Manual,* Lewis Publishers, Chelsea, MI, 1991.
15. AWWA, *New Dimensions In Safe Drinking Water,* Chapter 5, American Water Works Association, Denver, CO, 1988.
16. **Jenkins, D., et al.,** *AAPSE Water Chemistry Laboratory Manual,* American Association of Professors in Sanitary Engineering, EPA Grant No. 1P4-WP-309–01, University of California, Berkeley, 1972.
17. **Peavy, H. S., D. R. Rowe, and G. Tchobanoglous,** *Environmental Engineering,* McGraw-Hill, New York, 1985.
18. **Quinlan, J. F. and Donald R. Rowe,** *Hydrology and Water Quality in the Central Kentucky Karst,* Phase I, Water Resources Research Institute, Lexington, 1977, 1978.
19. **Kerri, K. D.,** Project Director, *Operation Of Wastewater Treatment Plants,* 3rd ed., California State University, School of Engineering, Sacramento, 1991.
20. Anon., *Turbidity, Operational Control Tests For Wastewater Treatment Facilities,* EW 008 452, ERIC, Ohio State University, Columbus, 1981.
21. *Handbook for Sampling and Sample Preservation of Water and Wastewater,* EPA 600/4-82-029, Office of Research and Development, U.S. EPA, Cincinnati, OH, 1982.
22. **Faller, J. A. and Ryder, R. A.,** Clarification and Filtration to Meet Low Turbidity Reclaimed Water Standards, *Water Environ. Technol.,* 68, January, 1991.
23. WHO, Guidelines For Drinking Water Quality, Vol. 2, World Health Organization, Geneva, 1984.
24. **Rowe, D. R., K. H. Al-Dhowalia, and A. Whitehead,** *Reuse of Riyadh Treated Wastewater,* College of Engineering, Research Center, King Saud University, Final Project Report No. 18/1402, November, 1987.
25. **Lishka, R. J.,** *Training Course Manual,* Training Program, Robert A. Taft Sanitary Engineering Center, Cincinnati, OH, 1963.
26. **Feigin, A., I. Ravina, J. Shalhevet,** *Irrigation With Treated Sewage Effluent,* Springer-Verlag, Berlin, 1991.
27. **Swayer, C. N. and P. L. McCarty,** *Chemistry for Environmental Engineering,* 3rd ed., McGraw-Hill, New York, 1978.
28. **Hullett, Rose,** Manager-Director, Ogden Environmental Laboratory, Western Kentucky University, Bowling Green, Personal Communication, April, 1993.
29. Anon., *Manual of Water Utility Operations,* Texas Water Utilities Association, Austin, 1975.
30. **Kerri, K. D.,** Project Director, *Water Treatment Plant Operation,* Vol. 2, California State University, School of Engineering, Sacramento, 1991.
31. WPCF, *Operation of Municipal Wastewater Treatment Plants,* MOP11, Vol. I and II, Water Pollution Control Federation, Alexandria, VA, 1991.
32. **Cohen, J. H.,** *Training Course Manual,* Training Program, Robert A. Taft Sanitary Engineering Center, Cincinnati, OH, 1963.
33. **Metcalf and Eddy,** revised by G. Tchobanoglous and F. L. Burton, *Wastewater Engineering Treatment Disposal And Reuse,* 3rd ed., McGraw-Hill, New York, 1991.
34. **Chansler, J. M. and D. H. Chaudhary,** Use of CBODs Testing for a More Accurate Treatment Plant Performance Evaluation, Proceedings of the 64th Annual Florida Water Resources Conference, Sponsored by Florida AWWA, FPCA, and FWPCOA, Orlando, FL, November 1990, 199.
35. Anon., *Oxygen Demand, Chemical (COD)* Hach Company, Loveland, CO, 42–25, 1992.
36. **Eckenfelder, W. W., Jr.,** *Principles Of Water Quality Management,* CBI Publishing, Boston, MA, 1980.
37. **Adams, C. E., et al.,** The Development of Design Criteria for Wastewater Treatment Processes, Proceedings of a Seminar, Vanderbilt University, Nashville, TN, 1974.
38. USEPA, Federal Guidelines — State And Local Pretreatment Programs, EPA-430/9–76–017a, Office of Water Program Operations, U.S. Environmental Protection Agency, Washington, D.C., 1977.
39. WPCF, *Water Reuse,* 2nd ed., Manual of Practice, MOP SM-3, Water Pollution Control Federation, Alexandria, VA, 1989.
40. Pye Unicam Ltd., *SP-9 Series Atomic Absorption Spectrophotometers Users Manual,* Publication No. 4013, 22997171, Cambridge, U.K., 1980.
41. USEPA, Inductively Coupled Plasma-Atomic Emission Spectrometric Method for Trace Element Analysis of Water And Wastes, EPA-600/4–82–055, Environmental Monitoring and Support Laboratory, U.S. Environmental Protection Agency, Cincinnati, OH, 1982.

42. **Tye, C. T. and P. Hitchen,** ICP-MS Instrumentation, *American Environmental Laboratory,* International Scientific Communications, Inc., Shelton, CT, February, 1992.
43. *Federal Register,* Vol. 54, No. 124, 40 CFR, Part 141, Subpart 141.21, p. 27565, U.S. Government Printing Office, Washington, D.C., 1989.
44. USEPA Microbiological Methods for Monitoring the Environment, Water and Wastes, U.S., Environmental Monitoring and Support Laboratory, EPA-600/8–78–017, U.S. Environmental Protection Agency, Cincinnati, OH, 1978.
45. **Edberg, C. S., M. J. Allen, and D. B. Smith,** National Field Evaluation Of A Defined Substrate Method for the Simultaneous Enumeration of Total Coliform and *Escherichia coli* from Drinking Water: Comparison with the Standard Multiple Tube Fermentation Method, *Appl. Environ. Microbiol.,* Vol. 54, p. 1595, 1988.
46. Anon. Hurry-Up Coliform Detection Method, *Waterworld News,* Pennwell Publishing, Tulsa, OK, Vol. 8, No. 6, 1992.
47. WHO, Health Guidelines for the Use of Wastewater for Agriculture and Aquaculture, M9/675073–001, World Health Organization, Geneva, 1989.
48. WHO, The Engelberg Report, Health Aspects of Wastewater and Excreta Use in Agriculture and Aquaculture, Sponsored by The World Bank and World Health Organization, Published by the International Reference Centre for Wastes Disposal (IRCWD), Dubendorf, Switzerland, 1985.
49. ICRP, International Commission on Radiological Protection, Recommendations of the International Commission on Radiological Protection, *Annals of the ICRP,* 1 (3):1–53 (1977).
50. ICRP, International Commission on Radiological Protection, Limits for intakes of radionuclides by workers, *Annals of the ICRP,* 2–8, (1979–1982).
51. Anon., Federal Radiation Council, Background Material for the Development of Radiation Protection Standards, Rep. No. 2, (Sept.), U.S. Government Printing Office, Washington, D.C., 1961.
52. National Committee on Radiation Protection and Measurements (NCRP), Maximum Permissible Body Burdens and Maximum Permissible Concentrations of Radionuclides in Air and Water for Occupational Exposure, Handbook No. 69, 1959.
53. National Council on Radiation Protection and Measurements (NCRP), A Handbook of Radioactivity Measurement Procedures, NCRP Report No. 58, 1985.
54. U.S. Environmental Protection Agency, Water Pollution Control; Radionuclides; Advance Notice of Proposed Rulemaking, 40 CFR, Part 141, 34836, *Federal Register* 51, No. 189, 1986.
55. USEPA, Risk Assessment Guidance For Superfund Volume 1 Human Health Evaluation Manual (Part A) Interim Final, EPA, Office of Emergency and Remedial Response, EPA/540/1–89–002, U.S. Environmental Protection Agency, Washington, D.C., 1989.
56. **Pontius, F. W.,** Technical Editor, *Water Quality and Treatment,* 4th ed., McGraw-Hill, New York, 1990.
57. WHO, Guidelines For Drinking Water Quality, Vol. I, World Health Organization, Geneva, 1984.
58. **Kelly, D. and J. Schienle,** *Radiological Health,* National Environmental Health Association, Denver, CO, 1982.
59. AWWA, *Introduction To Water Sources and Transmission,* Vol. I, American Water Works Association, Denver, CO, 1985.
60. **Skrentner, R. G.,** *Instrumentation Handbook for Water and Wastewater Treatment Plants,* Lewis Publishers, Chelsea, MI, 1990.
61. **Kulin, G.,** USEPA, Recommended Practice for the Use of Parshall Flumes and Palmer-Bowlus Flumes in Wastewater Treatment Plants, EPA 600/2–84–186, 292 N, Office of Research and Development, U.S. Environmental Protection Agency, Cincinnati, OH, 1984.
62. **Steel, E. W. and T. J. McGhee,** *Water Supply And Sewerage,* 6th ed., McGraw-Hill, New York, 1991.
63. Anon., *Professional Engineering Examinations,* Vol. I, The National Council of Engineering Examiners, Alexandria, VA, 1965–1971.
64. USEPA, Recommended Practice for the Use of Electromagnetic Flowmeters in Wastewater Treatment Plants, EPA 008530, EPA 600/2–84–187, U.S. Environmental Protection Agency, Columbus, OH, 1984.
65. **Hansen, F.,** Director, Oregon Department of Environmental Quality, Personal Communication, Preliminary Draft, "Guidance for Reclaimed Water Regulations," OAR 340–55, 1992.
66. State of Florida, Department of Environmental Regulations, Reuse of Reclaimed Water and Land Application, Tallahassee, Florida, 1990.
67. **Siemak, R. C., et al.,** National Guidelines For Water Reuse: Pros and Cons, in Proc. Conserv. 90 National Conf. and Exp. Offering Water Supply Solutions for the 1990s, Phoenix, Arizona, 1990, 51.
68. **Chansler, J. M.,** Permittee, Director of Public Utilities, City of Boca Raton, Boca Raton, FL. Letter from Florida Department of Environmental Regulation, Permit No. DC 50–199935, July 23, 1992.
69. USEPA, NPDES — Self Monitoring System — User Guide, Office of Water, Enforcement and Permits, U.S. Environmental Protection Agency, Washington, D.C., March, 1985.
70. **Goldfarb, W.,** *Water Law,* 2nd ed., Lewis Publishers, Chelsea, MI, 1991.
71. **Hopkins, S. O.,** Systems Manager, Water-Sewer Division, BGMU, Bowling Green, Ky., Letter from Natural Resources and Environmental Protection Cabinet, Dept. of Env. Protection, (NPDES No. KY0022403), Frankfort, Ky., September, 1991.

Chapter 9

Legal Aspects of Wastewater Reclamation and Reuse

CONTENTS

I. Introduction ... 395
II. Water Rights ... 396
III. Governmental Agencies Involved ... 397
    A. At the Federal Level ... 398
        1. 1899 Rivers and Harbors Act ... 399
        2. PL 80–845 Water Pollution Control Act 1948 ... 399
        3. PL 84–660 Federal Water Pollution Control Act of 1956 ... 399
        4. PL 87–88 Federal Water Pollution Control Act Amendments 1961 ... 399
        5. PL 89–234 Water Quality Act 1965 ... 400
        6. PL 91–224 National Environmental Policy Act (NEPA) 1969 ... 400
        7. PL 92–500 Federal Water Pollution Control Act 1972 ... 401
        8. PL 93–523 Safe Drinking Water Act 1974 ... 402
        9. PL 94–580 Resource Conservation and Recovery Act (RCRA) 1976 ... 403
        10. PL 95–217 The Clean Water Act (CWA) 1977 ... 403
        11. 42-U.S.C. 9601–9657 The Comprehensive Environmental Response, Compensation and Liability Act (CERCLA) "Superfund" 1980 ... 405
        12. PL 97–117 Clean Water Amendments 1981 ... 406
        13. PL 100–4 The Water Quality Act 1987 ... 407
        14. 1991–1992 Pending Clean Water Act Amendments ... 408
    B. At the State Level ... 408
    C. At the Local Level ... 411
IV. Water Quality Criteria and Water Quality Standards ... 413
V. Enforcement and Litigation ... 415
VI. Summary and Conclusions ... 420
VII. Homework Problems ... 422
    A. Discussion Questions ... 422
    B. Enrichment Questions ... 422
References ... 423

I. INTRODUCTION

The legal aspects of water rights are dealt with at five levels in the U.S.:[1]

1. U.S. Constitution
2. Statutes and ordinances
3. Administrative regulations

4. Executive orders
5. Common law court decisions

The U.S. Constitution does not include the word water; however, a broad interpretation of such words as commerce, transportation, property, and navigation gives the federal government the powers needed for the development of extensive water resources programs in the U.S.[1-3] The major sections of the U.S. Constitution interpreted by the U.S. Supreme Court that gives the federal government authority over the nation's water are

Article 1, Section 8, Number 1 and 3
Article IV, Section 3, Number 2
Article VI, Number 2.

Also, under the Fifth Amendment of the U.S. Constitution private property (water quantity rights) cannot be taken without just compensation.[1,2]

The U.S. Constitution gives the federal government authority as the supreme lawmaker in the U.S. in the area of water resources.

Bills passed by Congress or a state legislature become law (written law) after the consent of the president for federal statutes, or the governor of a state for state statutes. Statutes are formal documents or regulations that usually concern the rights of the general public. Statutes do not usually take away a citizen's right to sue.[4] At the county, municipal, or parish level of government, regulations or rulings passed by the local governing bodies are referred to as ordinances. Ordinances are generally less permanent and less constitutional in character than statutes.[2,5]

Congress and the state legislatures can also enact legislation (statutes) giving agencies the authority to exercise legislative, judicial, and executive powers. The extent of powers granted by legislation, creating such administrative bodies, and the implicit limitations on such power, form a part of what is now known as administrative law (administrative regulations). An example of an administrative body established in 1970 to combat air, land, and water pollution using administrative regulations is the Environmental Protection Agency (EPA). The administrative regulations that EPA use for environmental control form part of what we now call environmental law.[2,6]

Executive orders issued by presidents or governors regarding water resources have the force of law. In 1970 President Nixon, by executive order, created the EPA. President Carter in 1977 issued two executive orders: one entitled "Flood Plain Management" and the other "Protection of Wetlands". This constitutional power in the U.S. is concentrated in a single individual responsible only to the voting public. For example, President Abraham Lincoln, so it goes, made a decision on a matter that was opposed by all members of his cabinet. Lincoln announced the results as "One aye, seven noes — the ayes have it."[2,6]

The two leading legal systems in the Western World are civil law based on Roman law, and common law developed in England. Civil law is primarily legislative in origin while common law is primarily judicial in origin. The English colonists in America brought with them the principles of common law and the common-law judicial tradition. Common law is frequently referred to as judge-made law and is based on precedent. Common law has grown out of tradition and usage, and usually concerns private rights such as nuisances, trespass, negligence, violation of drainage rights, violation of riparian rights, or interference with the reasonable use of groundwater.[2,4]

II. WATER RIGHTS

The policies and the ways in which water rights are determined include the following:

1. Riparian rights are rights held by the owners of the land along the banks of bodies of water. The majority of states allow riparian owners domestic, agricultural, and some commercial use of the water so long as the quantity and quality is not substantially impaired.[6,7] There are no riparian rights to groundwater, diffused water surfaces, or water in artificial water bodies. The riparian water rights system is in effect in all states east of the Mississippi River except Mississippi, and in Arkansas, Iowa, and Missouri.[2] Most Western states where irrigation is important, follow the Spanish, rather than English law. They rely in whole or in part on the prior appropriation rule.[6]
2. The prior appropriation rule is that "the first in time shall be the first in right." This doctrine of water rights was developed in the western U.S., where the pioneers applied the same principles to water rights

as they did to minerals and land. Now appropriation rights are generally obtained by applying to a designated state agency. An important aspect of the appropriation doctrine is that the water is to be used for "beneficial use", generally meaning for economic development and ordinarily used in the areas of agriculture, industry, mining, or for municipal purposes. Commonly, priority is given in the following order: domestic, municipal, agriculture, industrial, and recreational use. The water appropriated in this manner must be used or the concept of "use it or lose it" is applied.[2,3,7]

3. The reasonable use rule permits use of water provided the use is reasonable. The riparian rights or "natural flow doctrine" has gradually been replaced by the "reasonable use rule" as the riparian rights rule tends to be antidevelopment in nature. Many factors must be considered in applying the reasonable use rule and decisions can be made on a case-by-case basis legislatively, bureaucratically, or by the courts.[2,7]

4. Absolute ownership rights are acquired through purchase of the land including the water rights conveyed in the deed. Legal water rights can be either private or public. No water right — private or public — is absolute. However, private water rights cannot be taken for public use without paying compensation.

5. Administrative bodies are given regulatory power by either the federal or state government and can thus exercise water rights to sources of water. Permits such as licenses and certificates of appropriation can be issued by most state governments and these states have statutes containing special procedures for adjudicating the water rights allocation.[2]

6. Pueblo water rights are those held by communities or cities established under Spanish law in former Mexican or Spanish territories in California and New Mexico. Pueblo water rights are superior to riparian or prior appropriative rights.[2,7]

7. Indian water rights were established by the Winters Doctrine in 1908 and expanded in 1963 by the U.S. Supreme Court decision to include federal lands such as national parks and national forests. The amount of water reserved appears to be interpreted as sufficient to satisfy agricultural and related needs. The 1963 U.S. Supreme Court decision indicated enough water be reserved to irrigate all the "practicably irrigable acreage." The amount of water needed to irrigate acreage practically has not yet been determined, however, this is not the only problem involved in Indian water rights. Other unresolved problems include groundwater, the transfer of water rights to non-Indians, and the eligibility of municipal and industrial users to water reserved for Indian reservations. Only the courts can decide these problems.[2,3,8]

III. GOVERNMENTAL AGENCIES INVOLVED

In the U.S. water resource development has evolved over the past some 180 years, starting in 1808 with the Gallatin Report dealing with canal and river improvements for navigation. The Corps of Engineers was established about 1824 and was the first major federal water regulatory agency. The next major water areas that the federal government became involved in were flood control, irrigation, and water power.[3] The first federal water pollution control legislation was the Rivers and Harbors Act of 1899 prohibiting the discharge of refuse into navigable waters. The key sections of this Act were 407, 411, and 413, and are often referred to as the 1899 Refuse Act.[9] However this legislation, nor any other legislation in the first half of the twentieth century was directed at promoting wastewater reuse.[8,10]

The first comprehensive federal water-quality legislation was the Federal Water Pollution Control Act (FWPCA) of 1948 which established criteria for all interstate and coastal waters, granted assistance for research and treatment facilities, and provided for enforcement of laws against pollution.[6] However, no federal funds were dedicated to carry out this legislation.

Table 9.1 presents some of the major federal legislation dealing with water and water pollution control in the U.S. As stated earlier, this legislation started with the Rivers and Harbors Act of 1899.[9] Since the FWPCA Act of 1948 many other laws have been enacted. These major legislative acts will be discussed in the next section.

The agencies involved in implementing this legislation has grown steadily, and following is a list of some of these federal agencies involved in the water, wastewater, and water reuse programs.[6,8,11]

  U.S. Environmental Protection Agency (EPA)
  U.S. Army Corps of Engineers (COE)
  U.S. Geological Survey (USGS)
  U.S. Soil Conservation Service (SCS)
  The Bureau of Land Management (BLM)

Department of Interior (USDI)
Council on Environmental Quality (CEQ)
U.S. Department of Defense (DOD)
U.S. Department of Transportation (DOT)
Fish and Wildlife Service (FWS)
National Marine Fisheries Service (NMFS)
Forest Service
Park Service
Coast Guard
U.S. Department of Commerce (DOC)
Bureau of Reclamation
National Oceanic and Atmospheric Administration (NOAA)

Historically, the state and local governments had the primary responsibility for water pollution control. However, with the passage of PL 92–500 in 1972, the federal government took major responsibility for the control of water pollution. Nevertheless, since 1972 the federal legislation appears to have come full circle, and starting with the Water Quality Act of 1987 state responsibilities for implementing the Clean Water Act have been reemphasized. Each state now has a state revolving fund loan program, as well as toxic pollutant and nonpoint source pollution control programs, and have regained to a certain degree control of project-level water-quality programs.[12] Most states have now established agencies similar to the federal EPA. These agencies are responsible to carry out the state and federal programs required or enacted by federal legislation. These agencies are very important in that they must often manage programs required by federal law. The law of water diversion and distribution is predominantly state law. States can make laws regarding water resources as long as the legislation is not in conflict with federal law.[1] At present, 13 states have specific criteria for water quality reuse; another 13 have no specific criteria, that is, no published criteria; and 24 do not permit reuse. The areas where water reuse is most often used is in arid regions, especially California, Texas, and Arizona.[13] The names of the state agencies that handle environmental programs vary from Conservation and Natural Resources to the Environmental Quality Department. Most enforcement of environmental laws is conducted at the state level once their pollution control programs have been federally approved.[11]

At the local government level many organizations are involved in water and wastewater operations. These organizations generally leave enforcement of environmental law to state and federal agencies.

A. AT THE FEDERAL LEVEL

Federal law is the supreme law of the land, and should states pass laws that are in conflict with federal statutes they may be ruled null and void by a federal court.[1] The first involvement of the federal government in the nation's waters was in 1874 with the establishment of a commission to study flooding

Table 9.1 Some Major Federal Water and Water Pollution Control Legislation

	River and Harbor Act of 1899
PL 80–845	Water Pollution Control Act of 1948
PL 84–660	Federal Water Pollution Control Act of 1956
PL 87–88	Federal Water Pollution Control Act Amendments of 1961
PL 89–234	Water Quality Act of 1965
PL 91–224	The National Environmental Policy Act (NEPA) 1969
PL 92–500	Federal Water Pollution Control Act of 1972
PL 93–523	The Safe Drinking Water Act (SDWA) 1974
PL 94–580	The Resource Conservation And Recovery Act (RCRA) 1976
PL 95–217	Clean Water Act of 1977
42-U.S.C. 9601–9657	The Comprehensive Environmental Response, Compensation And Liability Act (CERCLA) "Superfund," 1980
	Superfund Amendments and Reauthorization Act (SARA) 1986
PL 97–117	Clean Water Act Amendments 1981
PL 100–4	The Water Quality Act of 1987
	Pending Clean Water Act Amendments 1991–1992.

Note: PL = Public Law; U.S.C. = U.S. Code.

Compiled from References 10 and 11.

on the Mississippi River. This resulted in the creation of the Mississippi River Commission with the responsibility to prepare plans to improve navigation and control flooding on the Mississippi River.[3] The first involvement of the federal government in irrigation was in 1877 with passage of the Desert Land Act. This Act promoted the irrigation of arid land in four states and eight territories. Around 1879 the federal government started enacting legislation dealing with water power and the building of dams. As indicated earlier, the first federal water pollution control legislation was passed in 1899, the Rivers and Harbors Act. Up until 1948 the federal government took almost no responsibility for the control of water pollution, this being considered a state and local responsibility.[14]

Following are some of the most significant federal water and water pollution control legislation and some of the most significant provisions in that legislation (see Table 9.1).

1. 1899 Rivers and Harbor Act

As indicated earlier, Sections 407, 411, and 413 were used to enforce provisions of this Act and are generally referred to as the Refuse Act.[9] This Act forbid the dumping of refuse into any navigable water or tributary of any navigable waterway. It also forbid the depositing of material on the banks of a navigable waterway where it could be washed into the water and impede navigation. The Act prohibited all pollutional discharges except liquid waste from streets and sewers and activities carried out by the U.S. as public works.

The Act was interpreted to apply only to municipal sewage and not to industrial waste. Virtually any discharges into any water was a violation of this Act and could subject polluters to civil and criminal penalties. To avoid these sanctions a permit from the Secretary of the Army, acting through the Corps of Engineers, was to be obtained. However, from 1899 to May of 1971 only 415 permits were granted.[9] President Nixon, in December 1970, announced a program to include water pollution from industrial sources by requiring firms and individuals to file for Refuse Act permits. By June 1, 1972, 2,559 applications had been processed by the Corps, reviewed by EPA, and referred to appropriate state agencies for certification.[15] This Act has now been superseded by the Federal Water Pollution Control Act of 1972.

Environmental legislation for most of the first half of the twentieth century was put on hold due to two world wars and the great depression. The first comprehensive federal water pollution control legislation was the Water Pollution Control Act of 1948.

2. PL 80–845 Water Pollution Control Act 1948

This was the first temporary federal water pollution control legislation that formed the basis for a series of acts (amendments) forming the present body of laws dealing with controlling water pollution. The administration of this legislation was assigned to the Public Health Service (PHS) and was supposed to provide a trickle of funds to state and local governments to help solve their water pollution problems.[16]

The law had no specific water quality goals, and did not require the states to set standards. Provision for funding consisted of 1×10^6 for planning and design of municipal wastewater treatment facilities, 22.5×10^6 for construction of these plants, and 1×10^6 to study industrial pollution.[12] However, no funds were appropriated for these purposes.[3]

Federal enforcement was possible, but only when the pollution was interstate in nature, and only with the formal consent of the local agency where the pollution originated. Even then, exhaustive studies and conferences were required before enforcement began.[17]

Eight years later more permanent water pollution control legislation was enacted.

3. PL 84–660 Federal Water Pollution Control Act of 1956

This was the first federal legislation to provide funding for support of publicly owned sewage treatment plants. Support was limited to the treatment systems and did not include the cost of land.[18] This Act provided 5×10^9 mostly directed for construction of wastewater treatment facilities using conventional treatment technologies, and requiring stricter effluent standards.[8] The act provided a 30% incentive grant to subsidize construction of municipal wastewater treatment plants.[3] Funding was authorized for research, planning, and technical assistance for state programs.[16] However, no provisions for wastewater reclamation, reuse, or recycling were included in PL 84–660. Nevertheless, the research resulting from this legislation set the stage for developments in this area. The next amendments to this Act came in 1961.

4. PL 87–88 Federal Water Pollution Control Act Amendments 1961

This legislation was an expansion of the PL 84–660 Act and included renewal of funds for construction grants and also provided for enforcement, research, training, and long-range planning. Although the states were presumed to have the primary responsibility for water pollution control they, for the most part, provided only a small fraction of the necessary funds.[14] With each successive statute, federal assistance to municipal treatment agencies increased. The federal share went from 30 to 40 to 55 to 75% and even to 85% in some cases, until 1990 when the federal construction grants program was terminated.[10]

5. PL 89–234 Water Quality Act 1965

This was the first serious federal program aimed at the control of water pollution. The water pollution control program was shifted from the Public Health Service to the Federal Water Pollution Control Administration (FWPCA) in the Department of Health, Education and Welfare. However in 1966 this program was transferred to the Department of the Interior.[9] At present this program is administered by the EPA which was established in 1970. The most important provisions in the 1965 Act was the establishment of water quality standards and implementation plans for clean-up of all interstate and coastal waters. This Act also authorized larger appropriations and higher construction grants for building wastewater treatment facilities.[15]

The 1965 Act required the states to designate "beneficial uses" to which a particular reach of a waterway were to be put and to establish appropriate water quality standards consistent with that use. Enforcement of this legislation was difficult. In the 6 years following passage of the 1965 law, only 56 federal enforcement actions were taken, only 4 got past the conference level, and only 1 ever got to court.[17]

6. PL 91–224 The National Environmental Policy Act (NEPA) 1969

The main principle embodied in NEPA was the requirement of an environmental impact statement (EIS). An EIS was required for every federally funded project that might significantly affect the environment, which of course could include water reclamation and reuse projects. Projects that would require EISs were those that might affect water quality, fish and wildlife populations, noise levels, or air quality in our environment.

While each federal agency has its own requirements regarding what must be included in an EIS, the Council on Environmental Quality issued guidelines in the preparation and content of the statements. A brief outline of the content of an EIS is as follows.[19]

1. A description of the proposed action, a statement of its purposes, and a description of the environment affected, including information, summary technical data, and maps and diagrams where relevant, adequate to permit an assessment of potential environmental impact by commenting agencies and the public.
2. The relationship of the proposed action to land use plans, policies, and controls for the affected area. This requires a discussion of how the proposed action may conform or conflict with the objectives and specific terms of approved or proposed federal, state, and local land use plans, policies, and controls, if any, for the area affected, including those developed in response to the Clean Air Act or the Federal Water Pollution Control Act.
3. The probable impact of the proposed action on the environment, including environmental costs in the decision-making process.
4. Alternatives to the proposed action, including, where relevant, those not within the existing authority of the responsible agency.
5. Any probable adverse environmental effects which cannot be avoided (such as water or air pollution, undesirable land use patterns, and damage to life systems, urban congestion, threats to health, or other consequences adverse to the environment).
6. The relationship between local short-term uses of man's environment and the maintenance and enhancement of long-term productivity.
7. Any irreversible and irretrievable commitments of resources that would be involved in the proposed actions.
8. An indication of what other interests and considerations of federal policy are thought to offset the adverse environmental effects of the proposed action.

Over the past 20 years NEPA has contributed to the growing concern regarding environmental problems and has profoundly affected the federal government's actions regarding the environment. While NEPA targets no specific pollutional or human health problems, or prescribes any specific abatement techniques or remedial actions, or establishes a timetable to solve environmental problems, it does however seek to provoke thought and promote open and well-informed decision making.[11]

NEPA tells all federal agencies to take the following steps:[11]

- Think about the environment when planning any major action.
- Closely examine the need for and objective of projects, especially resource-intensive projects.
- Diligently search for alternatives that will not degrade the environment.
- Evaluate anticipated environmental effects to the extent necessary to delineate choices to decision makers and the public.
- Keep the federal decision-making process open, honest, and cooperative.

A number of states and local agencies have implemented a system of environmental impact assessments, modeled after NEPA. These statements are referred to as "little NEPAs." California, the District of Columbia, Puerto Rico, and the city of Chicago have their own versions of EISs. By 1989, 19 states had enacted "little NEPAs" by statute or executive order.[11]

7. PL 92–500 Federal Water Pollution Control Act 1972

This legislation was the most dramatic as well as the most comprehensive and assertive legislation ever enacted by the federal government to clean up the nation's waters. The objective of this Act was to restore and maintain the chemical, physical, and biological integrity of the nation's waters. The Act moved water pollution control from a water quality-based approach to a preventive approach by using three main procedures: first a construction grants program for reducing municipal waste discharges, secondly a National Pollutant Discharge Elimination System (NPDES) permit program for control of point source discharges, and thirdly a water quality management planning program for nonpoint source control.

The goals of the 1972 Clean Water Act (CWA) (shorthand for the Federal Water Pollution Act) are explicitly spelled out in Section 101 of the Act, "Declaration of Goals and Policy" section (a):

> The objective of this Act is to restore and maintain the chemical, physical, and biological integrity of the Nation's waters. In order to achieve this objective it is hereby declared that, consistent with the provisions of this Act — (1) it is the national goal that the discharge of pollutants into the navigable waters be eliminated by 1985; (2) it is the national goal that wherever attainable, an interim goal of water quality which provides for the protection and propagation of fish, shellfish, and wildlife, and provides for recreation in and on the water be achieved by July 1, 1983; (3) it is the national policy that the discharge of toxic pollutants in toxic amounts be prohibited; (4) it is the national policy that Federal financial assistance be provided to construct publicly owned waste treatment works; (5) it is the national policy that areawide waste treatment management planning processes be developed and implemented to assure adequate control of sources of pollutants in each State; (6) it is the national policy that a major research and demonstration effort be made to develop technology necessary to eliminate the discharge of pollutants into the navigable waters, waters of the contiguous zone, and the oceans; and (7) it is the national policy that programs for the control of nonpoint sources of pollution be developed and implemented in an expeditious manner so as to enable the goals of this Act to be met through the control of both point and nonpoint sources of pollution.[20]

The definition of the term navigable waters and pollution took on new meaning. Navigable waters was defined as "the waters of the United States" thus the Congress had extended the federal permitting authority to waterways to which it had never before been applied, particularly small tributary streams and wetlands.[17] While the definition of pollutant became so broad that virtually any substance discharged to a body of water was to be controlled.

> The term 'pollutants' means dredged spoil, solid waste, incinerator residue, sewage, garbage, sewage sludge, munitions, chemical wastes, biological materials, radioactive materials, heat, wrecked or discarded equipment, rock, sand, cellar dirt and industrial, municipal, and agricultural waste discharged into water.

The Act committed the federal government to provide 75% of the cost associated with the construction of wastewater treatment facilities and authorized $18 billion to try and meet the goals of this legislation. The $18 billion was even more money than that committed for the interstate highway system or the space program.

The NPDES permit system was and is the key to enforcing the effluent limitations and water quality standards of the Act. Every point source discharger must obtain a permit from either EPA or an authorized state agency. NPDES permits are also required on a phased-in basis for municipal and industrial storm water point sources. The permit specifies the allowable amount, composition of the effluent, and a specific schedule to achieve compliance. The permit is both a technology-based effluent standard and a water quality-based standard. For industrial waste effluents the discharge limitations were based on the application of the best practicable control technology currently available (BPCTCA) and this was to be provided by 1977. The acronym BPCTCA is often shortened to BPT. BPCTCA or BPT is defined as the level of treatment that has been proven to be successful for a specific industrial category and that is currently in full-scale operation.[21] EPA set these standards or limitations for each industrial category.

Publicly owned wastewater treatment works (POTW) were to meet secondary treatment standards as defined by EPA by July 1, 1977. These POTW plants were to apply secondary treatment to the effluent they discharge.

By 1983 industrial point dischargers were to provide the best available technology economically achievable (BATEA). The acronym BATEA was often shortened to BAT. BATEA or BAT is defined as the level of treatment beyond BPCTCA that has been proven feasible in laboratory and pilot studies and that is, in some cases, in full operation.[21] The BAT requirements were strongly disputed, in that the costs for reduction in contaminant levels was high compared to the benefits.

By 1983 the municipalities were to provide the best practicable waste treatment technology (BPWTT) over the life of the works. BPWTT was scarcely defined in the Act and EPA did not appear to view it as a technological standard.[17]

The principal control mechanism for water pollution control was the national technology-based effluent limitations which would move towards the goal of zero discharge. If there is an "available" and "economically achievable" pollution control technique that totally eliminates the discharge of pollutants, then BAT can be set at "no discharge" and that goal, for that particular industry, would have been met. In fact by 1976, 63 industrial subcategories had established BAT standards of zero discharge.[17]

The water quality standards involved in the NPDES permitting system are used to determine whether additional pollution reductions are necessary if a particular stretch of water on which a discharger is located is to be used for a designated purpose. If the water quality standard cannot be met on the basis of the effluent standard alone, then additional pollution reduction may be required.[19] See Chapter 8, Sections IX. B. and C. for examples of NPDES permits.

The water quality management planning program for nonpoint source control was dealt with in Sections 208, 303, and 304 of this Act. A study issued by the Council on Environmental Quality (CEQ) in 1972 showed that stormwater runoff in urban areas carries large volumes of pollutants into receiving waters. This report indicated that runoff problems in many urban areas needed to be controlled if water quality standards are to be met. Planning programs authorized by this Act focused attention on defining the specific nature of nonpoint sources on particular water reaches, identifying the sources of the runoff, and developing solutions which would be effective at a reasonable cost. EPA was also to support research to design better control methods for abating nonpoint pollution with initial attention to runoff from coal mining, agriculture, and construction activities.[19]

Throughout it all, this Act still made it clear that the congressional policy was to recognize that states had the primary responsibility to see that the goals of this legislation were met.

A brief review of the 1972 Act indicates its regulatory structure:[17]

- Identification of all point source dischargers.
- Establishment of near-term and long-term enforceable treatment requirements.
- Reliance on uniform source controls based on the capabilities of technology (effluent standards) rather than on receiving water conditions.
- Reliance on water quality standards to back up the effluent standards.
- Providing certain safety valves for undue economic hardships that might be imposed by long-term requirements.

8. PL 93–523 Safe Drinking Water Act 1974

While PL 92–500 was one of the most significant pieces of legislation aimed at control of water pollution, another very important piece of legislation that had an impact on our nation's water was the Safe Drinking Water Act (SDWA) of 1974. While this Act provides for regulations for drinking water, these regulations also indirectly affected the quality of the wastewater that may be discharged into streams and rivers that are a source for a public water supply.[22]

This Act required the administrator of EPA to develop national health standards for public water supply systems in order to protect the public health. State and local governments again had the primary responsibilities for implementing and enforcing these standards. Citizens are to be informed of any violation of standards and can file citizen suits in federal courts to secure compliance. Assurance of an adequate supply of chlorine for disinfection of drinking water during emergencies was also established, as well as regulation regarding deep-well injection of wastes.[19] This Act has been amended in 1977, 1986, and 1988. The EPA set primary standards to protect health from organic, inorganic, and microbiological contaminants and for turbidity in drinking water in 1977. These primary drinking water standards were finalized in 1986. The primary drinking water standards are based on the toxic effects for certain organic and inorganic compounds and are set at maximum contaminant levels (MCL). The MCL is defined as the maximum allowed level of a contaminant in water which is delivered to any user of a public water system.[23] However, in the case of turbidity the MCL is measured at the point of entry into the distribution system.

The water quality standards have been established in terms of Primary and Secondary Standards and are presented in terms of Primary MCLs and Secondary MCLs. These standards are presented later in Section IV of this chapter.[24]

9. PL 94–580 Resource Conservation and Recovery Act (RCRA) 1976

The Resource Conservation and Recovery Act of 1976 called for a national effort to deal with solid waste problems. Solid wastes are broadly defined in the Act to include waste solids, sludges, liquids, and contained gases; excluded are solid and dissolved materials in domestic sewage, return irrigation flows, and industrial discharges subject to permits under Section 402 of the Federal Water Pollution Control Act, and certain radioactive wastes covered by the Atomic Energy Act of 1954. The main goals are to control solid waste management practices that can endanger public health or the environment and to promote resource conservation and recovery. The key provisions of RCRA for achieving these goals are[25]

- Federal financial and technical assistance to state and local governments for development of comprehensive state solid waste management programs that include environmental controls on all land disposal of solid wastes.
- Regulation of hazardous wastes from the point of generation through disposal.
- Resource recovery and conservation activities.
- Regulatory control of hazardous wastes by EPA if states do not establish control programs that meet federal standards.
- A cabinet-level interagency study of resource conservation policies.
- Public participation in the development of all regulations, guidelines, and programs under the Act.
- Research, demonstrations, studies, and information activities related to a wide range of solid waste problems.

Indirectly, this Act gives EPA the authority to protect our water resources by controlling disposal of hazardous wastes. EPA is required to implement programs to protect groundwater and surface waters from possible contamination by improper disposal of hazardous wastes.[3]

EPA administrative enforcement under RCRA has varied from 189 actions in 1981 to 453 actions in 1989. RCRA civil referrals to the Department of Justice increased in 1989 to a record high of 169 cases; RCRA criminal referrals also increased.

During 1989 EPA and the states identified 664 facilities in significant violation of RCRA regulations. By the end of 1989, 78 facilities had been returned to compliance. Of the 88% still out of compliance, 300 were on compliance schedules, and 284 had enforcement actions pending against them. Included in the statistics are actions taken to regulate underground gas and oil storage tanks.[11]

RCRA penalties, like those of many environmental laws, are increasing. For example, the largest civil judicial penalty ever assessed for violation of RCRA was a 2.8×10^6 fine upheld by a federal appeals court. The court permanently prohibited operation of a privately operated landfill in northern Indiana and ordered the company to conduct corrective action.[26]

10. PL 95–217 The Clean Water Act (CWA) 1977

The CWA of 1977 endorsed the water pollution control goals set in PL 92–500, 1972. The act also provided special incentives for municipalities receiving federal construction grant funds to use Innovative and Alternative (I/A) technologies for wastewater treatment. I/A technologies are wastewater treatment

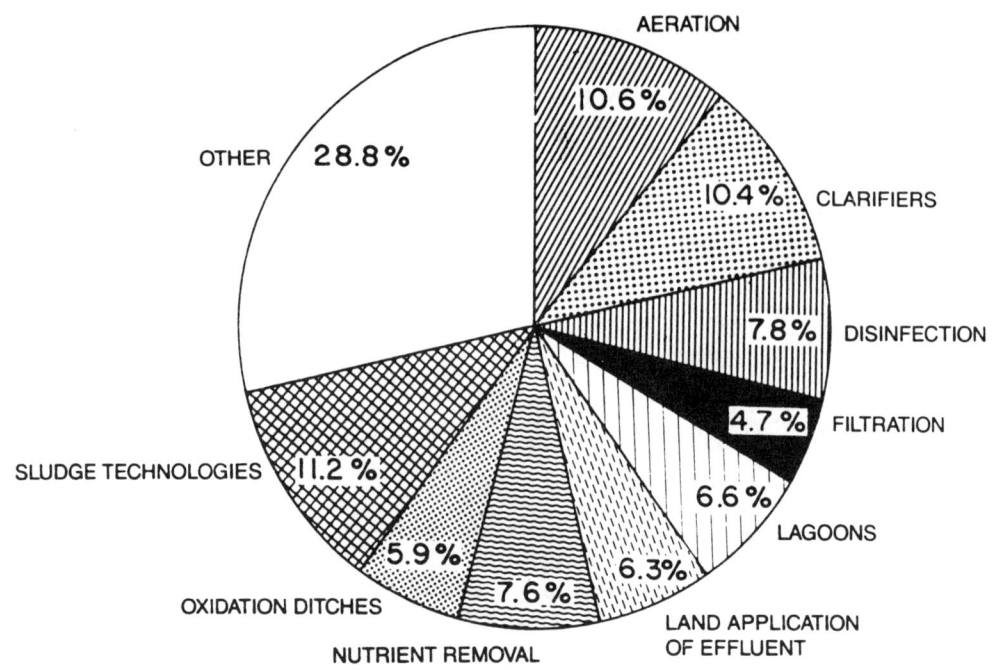

Figure 9.1 Innovative technologies funded by EPA up to 1986.[27]

processes or components that either reuse and recycle wastewater and sludge, reduce costs and energy compared to conventional treatment methods, or provide simple and economical treatment for small communities. Incentives for choosing an I/A technology in the past included a 20% increase in the federal grant share, the requirement for states to use a certain portion of construction grant funds for I/A technology projects, and the availability of 100% grants to modify or replace funded projects which failed (M/R grants).[27]

An innovative technology project is considered to be a new wastewater treatment process or component which has not been fully proven; but, based upon results from research and demonstration projects, appears promising. An innovative technology project provides a benefit, such as reduced costs or environmental benefits, along with an acceptable element of risk. A breakdown of innovative technologies funded by EPA is shown in Figure 9.1. The percentages are based on the number of awards up to 1986.[27]

The Act indicated that grants would not be made after September 30, 1978 for any project unless innovative and alternative (I/A) wastewater treatment processes had been fully studied. Up to 1986 the EPA-I/A technology program has awarded over 3500 grants to more than 1600 municipal wastewater treatment facilities, and at that time 400 of those facilities were operational.[27]

A problem with the 18×10^9 funding in PL 92–500 was that much of it had gone for growth such as collector sewers, interceptor sewers, and construction of treatment plants whose primary purpose was to meet growth needs rather than pollution control.

The 1977 Act reauthorized 4.5×10^9 for fiscal year 1978 and 5×10^9 annually for fiscal years 1979 to 82. During fiscal year 1977 EPA obligated 7.7×10^9 in new grants and spent $3.5 billion on current projects.[25] In October 1977 EPA announced a major new policy, emphasizing land treatment of municipal wastewater. EPA also released a manual to assist local planners and consultants in making maximum use of land treatment alternatives.

In 1972, 571 municipalities serving 6.6 million people were using land treatment compared to 304 municipalities serving 0.9 million people in 1940. Table 9.2 presents some of the successful land treatment applications in 1978.[25]

The 1977 CWA expanded the constituents regulated in wastewater discharges from the traditional BOD, suspended solids (SS), pH, and fecal coliform to include the more toxic pollutants. The Act required that effluent limitations be placed on 65 named classes of toxic chemical and that BAT be required by 1984 for treating these toxic chemicals.

Table 9.2 Selected Land Treatment Systems

Location	Average flow (million gallons per day)	Average annual application rate (feet per year)	Degree of preapplication treatment	Application technique	Number of years in operation
Slow rate					
Pleasanton, California	1.4	8.5	Secondary (plus aerated holding ponds)	Sprinkler (portable pipe)	20
Walla Walla, Washington					
Industrial	2.1	1.7	Aeration	Sprinkler (buried pipe)	5
Municipal	6.8	—	Secondary	Sprinkler and surface (ridge and furrow)	78
Bakersfield, California (existing system)	14.7	6.9	Primary	Surface (border strip and ridge and furrow)	38
San Angelo, Texas	5.8	10.3	Primary	Surface (border strip)	18
Muskegon, Michigan	28.5	6.0	Aerated lagoons	Springler (center pivot)	3
St. Charles, Maryland	0.6	10	Aerated lagoons	Sprinkler (surface pipe)	12
Rapid infiltration					
Phoenix, Arizona	13	364	Secondary	Surface (basin flooding)	3
Lake George, New York	0.7	140	Secondary	Surface (basin flooding)	38
Fort Devens, Massachusetts	1.3	94	Primary	Surface (basin flooding)	35
Overland flow					
Pauls Valley, Oklahoma	0.2	19–45	Raw (screened) and oxidation lagoon	Surface (bubbling orifice) and sprinkler (fixed and rotating nozzles)	2
Paris, Texas (industrial)	4.2	5.2	Raw (degreased and screened)	Sprinkler (buried pipe)	13

Note: 1 million gallons per day = 43.8 liters per second.
1 foot per year = 0.305 meters per year.

Compiled from Reference 25.

The Act required development of pretreatment standards for any pollutants that interfere or are incompatible with the treatment and operation of a publicly owned treatment works (POTW). This requirement improved the feasibility of reclamation, recycling, and reuse of municipal and industrial wastewaters.

All publicly owned wastewater treatment plants with total flows in excess of 18,900 m^3/d (5 mgd) or which received industrial wastewater were required to establish pretreatment programs. The 1977 Act changed the direction of water pollution control from treatment of wastewater and then discharge, to treatment and then reuse.

11. 42-U.S.C. 9601–9657 The Comprehensive Environmental Response, Compensation and Liability Act (CERCLA) "Superfund" 1980

While RCRA dealt with the regulatory control of hazardous wastes at facilities in operation or those that were to be put into operation, it did not deal with abandoned hazardous sites. Thousands of abandoned hazardous waste sites have been identified and often the responsible parties could not be located. Responding to this situation, Congress enacted CERCLA (pronounced sir-klah) which came to be known as the Superfund. CERCLA provides two basic types of response capabilities: an emergency response capability for handling major chemical spills and hazardous substance incidents, and a remedial response capability for undertaking the long-term cleanup of abandoned hazardous waste sites.[28]

CERCLA or the Superfund legislation was passed in 1980, amended in 1986, and reauthorized in 1990 for 3 years without change until September 30, 1994.[11,29] The 1986 amendments are referred to as Superfund Amendments and Reauthorization Act (SARA). CERCLA and SARA require parties responsible for sites contaminated by hazardous material to conduct cleanups under government supervision.

Hazardous material means hazardous wastes as defined by the Clean Water Act, Clean Air Act, Toxic Substances Control Act, Solid Waste Disposal Act, and as defined by the administrator of the enforcement agency. Oil and gas are not classified as hazardous materials.

CERCLA provided a $1.6 billion Super Trust Fund. CERCLA also provided the federal government with the authority to arrange for pollution removal and remedial action whenever a hazardous substance was released or even presented a substantial threat of release into the environment.

In 1986 SARA established an $8.5 billion fund for a 5-year period to finance Superfund response activities. The source of the revenue for the fund was a tax on domestic and imported petroleum products, a chemical feedstock tax, a corporate environmental tax, governmental general revenues, and fund interest and cost recoveries.[1]

Where hazardous waste releases had occurred, EPA was to use the Superfund to clean up the site and then take action against the party responsible. For reimbursement EPA could move directly against the polluter. In 1989 an EPA inventory identified 27,000 contaminated sites, and a National Priorities List (NPL) ranked the 1077 most dangerous sites. In 1984 EPA took administrative action against 137 violators, 224 in 1988, and 212 actions in 1989.[11] If a cleanup order is ignored, EPA can proceed to clean up the site with Superfund monies, but responsible parties then may be liable for punitive damages up to triple the cost of the federal remedial work.

In 1989 EPA referred 153 Superfund cases to the Department of Justice for civil action. Remedial-action consent decrees valued at $594 million were completed for 48 of these sites. Reimbursements for past work by EPA brought this total to over $800 million. For the first time since the program began, private parties undertook more remedial actions in 1989 than were begun by the government under the Superfund.[11]

In fiscal 1989 and 1990, the enforcement program collected twice as much money in injunctive relief and cost recovery than in the entire previous history of the program ($1 billion in 1989 alone). The program filed 252 Superfund cleanup suits — more than in the first 6 years of the program. In fiscal 1989 and 1990, the Department of Health and Human Services (DHHS) increased the number of Superfund sites receiving public health assessments to 1300.[26]

The Emergency Planning and Community Right-to-Know Act (EPCRA), Title III of SARA, requires specified facilities to report releases of any of 300 toxic chemicals. EPA maintains a computer database, called the Toxic Chemical Release Inventory (TRI), which aggregates the information. In 1987 EPA received 75,000 reports of releases from 18,000 facilities. The database, available to EPA and the public through computer telecommunications, is proving valuable to both public agencies and private citizens who seek to enforce environmental laws.

Under Section 311 of SARA a National Contingency Plan (NCP) was developed that presented preferrable cleanup methods, cost effectiveness criteria, cleanup priorities, and the government's role in the cleanup process.[1]

One problem in the Superfund program is that the Superfund laws are in conflict with the bankruptcy laws. The CERCLA Act operates on the basis that post cleanup is an appropriate time to consider who pays and how much, while the bankruptcy code presumes that the bankrupt person is entitled to a fresh start with no financial liabilities. The courts are now in the process of sorting this out.[29]

12. PL 97–117 Clean Water Amendments 1981

This Act dramatically reduced the federal share in the construction grants programs. The new law was also designed to cut down on the time it took for a construction grant to go through the grants program. Even in 1977 it took 2 to 3 years to move a project through planning and design into the construction phase.[30]

Litigation, environmental problems, escalating costs, administrative tieups, contractor and bidding problems, and the lack of municipal funding were some of the problems associated with the delays in getting construction underway.

Figure 9.2 indicates the months from the award of a construction contract vs. number of projects involved in 1977.[30] The median time was 219 months, meaning that 50% of the projects took less than 219 months to start construction and 50% took more than 219 to start.[30] While this seems like a long time, it was taking 7 to 9 years in 1981 to start construction.[10]

Even after many years of effort, many municipal discharges had yet to be cleaned up. More than half of the country's 20,000 municipal dischargers, including 106 of the larger urban areas, were still unable to comply with the July 1977 secondary treatment goals. Thousands of construction grants projects had been started, approximately $26.6 billion in federal funds obligated, but only 2223 projects worth

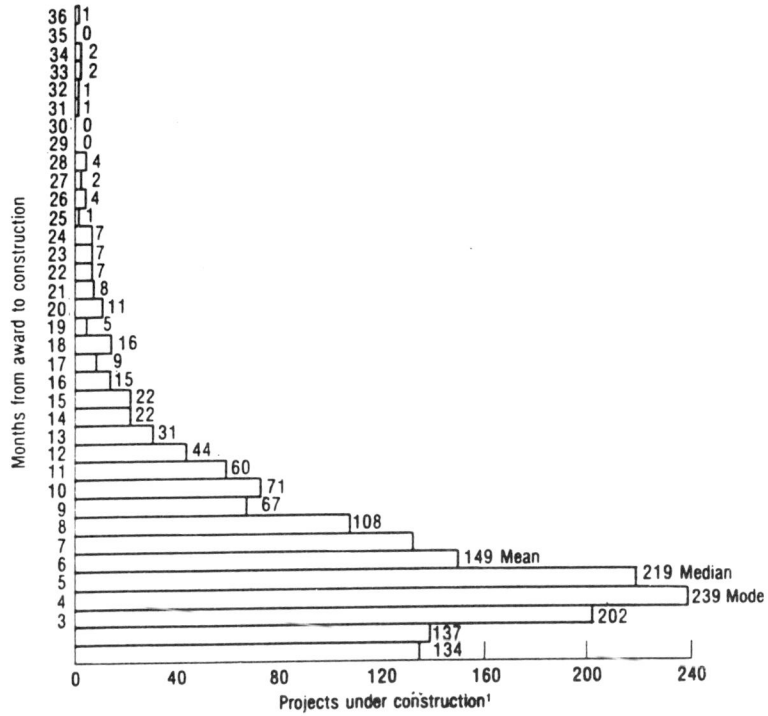

Figure 9.2 PL 92–500 Project construction delays.[30]

$2.8 billion had been completed. Most of the completed projects were small, relatively simple facilities; projects for the larger treatment authorities had to be divided into phases or segments requiring many years' worth of regular grants.[10]

Under this Act, enforcement activities for both municipal and industrial dischargers centered on revising national permits. On the municipal side, 3371 "major" POTWs (i.e., plants handling 1 million gallons or more a day) had NPDES permits by 1982. Of this total, 2956 were second-round permits and 775 were first-round permits. Some 2805 of the second-round permits contained final limits requiring secondary treatment or better; 553 of those contained limits on toxic pollutants. In addition, EPA and the states re-issued approximately 50% of the permits originally written for 12,208 "minor" POTWs.

The industrial side was judged in terms of meeting the 1977 BPT deadline. Approximately 81% of the major industrial dischargers met that deadline on time. In 1982, 96% of industrial dischargers had installed the necessary treatment equipment or were on compliance schedules to install it. These facilities reduce BOD by 69%, suspended solids by 80%, oil and grease by 71%, and dissolved solids by 52% of the amounts discharged in 1972. Monitoring data available showed that BOD reductions often yielded significant reductions of toxic organic pollutants as well. BPT treatment also resulted in the incidental removal of heavy metals.[10]

13. PL 100–4 The Water Quality Act 1987

Implicitly and explicitly, states are recognized as the level of government primarily responsible for making the federal government's programs succeed. The construction grant program was to be phased out in 1990 and a revolving loan fund program for the states was phased in 1989 and would continue to 1994 (see Table 9.3). The states must use the fund to make low-interest loans to communities in need of wastewater treatment facilities. A state fund "revolves" as the monies loaned out are returned with interest.[10] The state revolving fund may also be used to carry out the state-wide nonpoint source management plans.

In this Act EPA was to identify toxics in sewage sludge and establish numerical limits for maximum concentrations of these substances. The NPDES permits would now include sludge management requirements for POTWs. Separate NPDES permits would now be required for stormwater discharges, however, an entire city could be covered by one permit instead of permits for each individual outfall.

Table 9.3 Phasing Out Construction Grants and Phasing in the Revolving Loan Fund Program

Construction grants	Capitalization grants
1987 $2.4 billion	1987 -0-
1988 $2.4 billion	1988 -0-
1989 $1.2 billion	1989 $1.2 billion
1990 $1.2 billion	1990 $1.2 billion
1991 -0-	1991 $2.4 billion
1992 -0-	1992 $1.8 billion
1993 -0-	1993 $1.2 billion
1994 -0-	1994 $0.6 billion

From Kovalic, J. M., *The Clean Water Act of 1987,* Water Pollution Control Federation, Alexandria, VA, 1987. With permission.

This Act also gave industrial dischargers additional time to meet BAT and BCT effluent limitations. These limitations were not to extend beyond 3 years after March 31, 1989.

Administrative penalties under this Act were two-tiered. Tier I assesses a penalty of $10,000 per violation up to a maximum of $25,000. At this level there is an opportunity for an informal hearing. Tier II assesses a penalty of $10,000 per day of violation with a maximum of $125,000. There is also an opportunity for a formal hearing.

Civil and criminal judicial penalties include $2,500 to $25,000 per day of violation, 1 year of imprisonment, or both, for negligent acts. Where knowing violation of the Act occurred, the penalty was to be between $5,000 and $50,000 per day, 3 years of imprisonment, or both. For cases of known endangerment, the Act provides for a fine of $250,000 and imprisonment for a maximum of 15 years, or both.

This Act directed its attention at financing methods for construction of wastewater treatment facilities, non-point sources, stormwater NPDES permits, toxics in wastewater (priority pollutants), and enforcement and compliance regulations.

14. 1991–1992 Pending Clean Water Act Amendments

This potential legislation focuses on three important areas:

1. Water quality standards for toxic pollutants.
2. Storm water discharges.
3. Wetlands.[29]

A controversy between EPA and The Army Corps of Engineers over wetland regulations contributed in part to the delay of the Clean Water Act Reauthorization Legislation in the 1991 session of Congress.

States were slow in complying with 1987 CWA amendments, only four states and two territories met the statutory deadline of February 4, 1990 for adoption of numerical standards for toxics. EPA threatened to set standards in states failing to adopt the required numerical criteria. By November 1991 only 22 states and territories had not promulgated numeric water quality criteria for priority toxic pollutants.

The water quality for toxics will likely have far-reaching and immediate effects on NPDES permits and could reopen these permits for revisions.

EPA extended the deadlines for submittal of stormwater discharge permits applications, in part due to the difficulties experienced by industry and municipalities in gathering the information needed for these complex applications.

The actual definition of wetlands has been a contentious issue. Wetlands are generally considered to be an area having hydric (moist) soils, a periodic source of water, and vegetation especially suited for moist soils. The wetlands definition has both numerical standards and criteria for vegetation and hydrology. A conflict between developmental interests and wetland protection interests is presently underway and a clear, acceptable definition for regulatory agencies is yet to be determined.[29]

B. AT THE STATE LEVEL

The law of diversion and distribution of water is predominantly state law. These water rights are almost always included in state constitutions and statutes. Most water rights problems are assessed by state courts or administrative agencies, unless of course they are interstate in nature.

Where water diversion and distribution are concerned, state law generally reigns supreme unless the state violates a right embodied in the federal law. A state law in conflict with federal law is null and void.[1]

Historically, the states took an early lead in the water and water pollution control fields as it was apparent that control of a body of water used by a number of cities was not a matter that could be easily delegated to the cities themselves.

By the late nineteenth century, science and technology had developed methods for the treatment of water and wastewater. Local and state sanitary commissions, which in many cases evolved into departments of public health, were established with the main goal being to eliminate cholera and typhoid epidemics caused by contaminated drinking water. By World War I most states had established institutions to ensure safe drinking water, disposal of sewage, and control of water pollution.[31]

Although state and local governments still remain extremely important in the environmental protection area, over the past 40 years the nation has looked increasingly to the federal government for leadership.[31] From the late 1950s to the 1990s this leadership has been expressed in a number of far-reaching environmental statutes. However, in the 1980s this trend has been reversed; for example, the CWA of 1956 authorized $5 billion per year for construction of wastewater treatment facilities. This program is now being cut back and replaced by a federal capitalization grants program and the establishment of a State Revolving Loan Fund (SRF). The federal construction grants program ended in 1990 and the state program phased in with its termination in 1994. The 1972 CWA increased the federal share of POTW financing from 55 to 75% and, in some cases, 85% for I/A projects. Then in October 1984 the 55% was again reinstated, and now these grants are terminated.[1]

Where POTWs are concerned, states may make SRF grants at low interest rates or interest-free loans, for terms up to 20 years, with all payments of principal and interest credited to the SRF.

When it comes to the responsibility to enforce water and wastewater control regulations the federal government has generally left this up to the states.

The heart of the CWA is the NPDES permit system. NPDES permits are the key to enforcing the effluent limitations and water quality standards of the Act. Every point source must obtain a permit from EPA or an authorized state agency. The NPDES permits may be granted for up to a maximum of 5 years — 37 states currently administer NPDES permit programs under agreements with the EPA.

These NPDES permits contain three major parts: effluent limitations, compliance schedules, and monitoring and reporting requirements.

For EPA to delegate NPDES permitting authority to a state, the state must convince EPA that its program is consistent with the CWA and that the state possesses the resources and statutory authority to implement it.[1] Specifically, the state must show it has the necessary regulations, funding, and staff needed to conduct inspections, review reports, compile statistical information, and carry out enforcement of the NPDES requirements. If a state fails to adequately administer the NPDES program, EPA may cancel approval of the state's program and again take over enforcement responsibilities. A state can also lose its NPDES program if it fails to develop an acceptable pretreatment program.[1]

Following is the NPDES permitting process used by the Department of Health and Environmental Control (DHEC) in South Carolina.[32]

NPDES Permit

Introduction

The National Pollutant Discharge Elimination System (NPDES) permit program requires any person proposing to discharge waste into waters of the State must first apply and obtain a permit from DHEC. Note: EPA has delegated the Federal aspects of this program to the State (DHEC).

Authorizing Statute(s)

I. Federal

Federal Water Pollution Control Act of 1972 (P.L. 92–500) as amended by the Clean Water Act of 1977 (P.L. 95–217) as amended by the Water Quality Control Act of 1987 (P.L. 100–4).

II. State

South Carolina Pollution Control Act (S.C. Code of Laws, 1976, Title 48, Chapter 1)

Title of Regulation

South Carolina Regulation 61–9 NPDES permits (Amended April 27, 1990)

Agency Contact

Division of Domestic Wastewater or
Division of Industrial and Agricultural Wastewater
Bureau of Water Pollution Control
Department of Health and Environmental Control
2600 Bull Street
Columbia, S. C. 29201
(803) 734–5300

Summary of Regulation for Permit

I. Applicability

Any person wanting to discharge wastewater to surface waters of the State must first obtain a NPDES permit from DHEC prior to discharging. Public and private sanitary wastewater dischargers must submit applications to the Division of Domestic Wastewater. Industrial and all other dischargers must apply to the Division of Industrial and Agricultural Wastewater Division.

II. Summary Of Permit Process

A. Permit Application

Requires the proposed discharger to supply information to DHEC, such as discharge location, type manufacture, wastewater characterization, etc. Upon completion of required information, the applications are submitted to DHEC for processing.

B. Permit Draft

A permit is drafted based on any applicable guidelines/regulations and/or water quality considerations. Appropriate limitations are proposed in the permit to adequately protect water quality and human health. A copy of the draft permit is provided to the proposed discharger for review and comment.

C. Public Notice

Once a proposed discharger, EPA and the appropriate EQC District Office have had an opportunity to comment on the proposed permit limitations and conditions, a public notice is placed in a newspaper of general circulation in the area of the proposed site and notices are placed near the proposed site. Comments are accepted for 30 days after the public notice. A public hearing may be requested by any interested or affected party. A hearing notice must be given 30 days prior to the hearing.

D. Issuance/Effective Permit

Once DHEC issues the NPDES permit, the applicant or any interested affected party has 15 days to appeal the permit before it becomes effective.

Further appeal process of the permit is initially to the Board of Health and Environmental Control; thereafter to the Court of Common Pleas. Note, the appeal to DHEC's Board must follow the Procedure for Contested Cases, Regulation 61–72.

III. Permit Processing Time

The application must be submitted to DHEC at least 180 days prior to the intent to discharge. Depending on the complexity of the permit and public input more time to process the NPDES permit may be needed.

IV. Fees

A bill for an annual fee will be sent to the permittee by DHEC. The fee amount is in relation to the amount of wastewater to be discharged, a minimum of $200 to a maximum of $800.

V. Other Requirements/Issues

   A. If a wastewater treatment facility needs to be constructed, an operator of proper grade must be provided to operate and maintain the facility. The level of operator will be stated in the NPDES permit.

   B. The permit to construct will not be issued until the NPDES permit has become effective.

   C. Easements must be obtained from property owners, if wastewater crosses their property from permittee facility in reaching waters of the State.

   D. DHEC encourages any person intending to obtain an NPDES permit to meet at the earliest opportunity with DHEC to gain a comprehensive understanding of the regulatory requirements.

See Chapters 8 and 10 for examples of NPDES permits.

C. AT THE LOCAL LEVEL

While the federal and state governments pass legislation, develop regulations, and provide financial and technical assistance to local agencies responsible for water and wastewater control, the actual carrying out of these programs is at the local level. As President Truman said, "The buck stops here."

Local pollutional problems are generally best addressed at the local level because the local agencies are closest to the problem and are able to respond more quickly and effectively than federal or state agencies. However, federal and state involvement in devising strategies to effectively control intermunicipal and interstate water pollution problems is definitely required.

At the local level there are many organizations involved in the water resources area. Generally, these local organizations have three major responsibilities to the community: first to supply water, secondly to collect and treat the wastewater, and thirdly to control stormwater runoff. Federal and state governments do assist financially in the construction of water and wastewater control facilities, but the operational costs of such facilities are borne primarily by local agencies.

Anywhere from 70 to 130% of the water delivered to a metropolitan area is collected and treated before being discharged. A couple of reasons why the wastewater flows can be higher than the water flows is due to inflow and infiltration. Municipal wastewater treatment plants handle more than just domestic wastewater from homes and apartments. On a national basis about 45 to 55% comes from homes and apartments, 10 to 15% from inflow and infiltration, and the balance from commercial and industrial sources. Pollutants that can be found at high levels in municipal wastewater are oxygen-demanding materials (BOD, COD), suspended solids (both organic and inorganic), nutrients (N, P), flammable compounds, grease and oil, heavy metals, color and foaming compounds, as well as toxic substances. One source of the toxic substances is industrial discharges. These discharges can have an adverse effect on the operation of a publicly owned treatment works (POTW) as well as reduce the potential for water reclamation, recycling, and reuse.

The CWA, PL 92–500 of 1972, required the development of pretreatment guidelines and standards to provide a uniform approach to the control of industrial pollutants introduced into POTWs.[34]

Pretreatment is defined as the application of physical, chemical, and biological processes to reduce the amount of pollutants or alter the nature of the pollutant properties in a wastewater prior to discharging such wastewater into a publicly owned wastewater treatment system.[34]

Pretreatment requirements were again addressed in the CWA, PL 100–4, in 1987 Section 307, which required that effluent limitations be established for certain pollutants from various industrial categories. Pretreatment standards and effluent guidelines were set for over 50 industrial categories from the dairy products to the nonferrous metal forming and metal powders industries.[35] These standards placed restrictions on 129 priority pollutants in 65 classes of substances. A priority pollutant has been identified as having the greatest potential to harm human health or the environment. The harm or toxicity to human health is based on human epidemiological studies or testing of laboratory animals and the correlation with its potential for causing carcinogenicity, mutagenicity, teratogenicity, or loss of reproduction. The priority pollutants can be broken down into the following major categories:

- Halogenated compounds
- Pesticides, herbicides, and insecticides

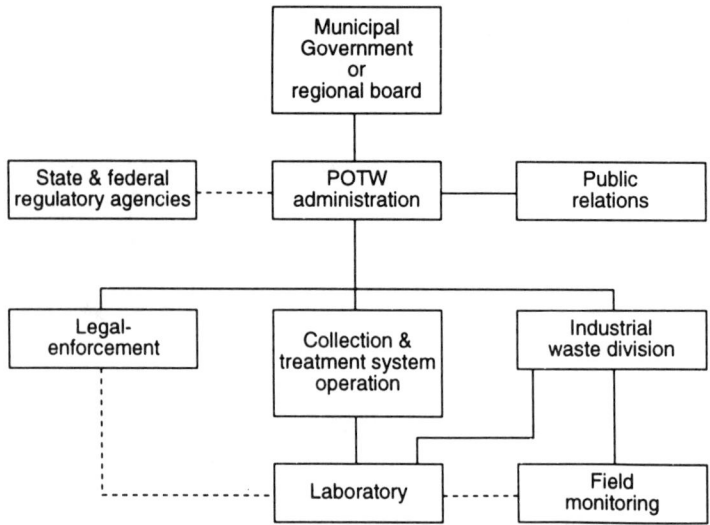

Figure 9.3 Conceptual organization of an industrial wastewater control program.[34]

- Organic compounds
- Metals and non-Metals
- Cyanides
- Asbestos

Section 307 of PL 100–4 also requires development of pretreatment standards for any pollutant that interferes with the performance or is not compatible with the operation of a POTW.

The objectives of these pretreatment regulations are to:

- Prevent introduction of pollutants that will interfere with plant operation, or with disposal of municipal sludges.
- Prevent introduction of pollutants that will pass through the treatment works or will be incompatible with them.
- Improve the feasibility of recycling and reclaiming municipal and industrial wastewaters and sludges.[36]

With the priority pollutant standards and effluent limitations, wastewater reclamation, recycling, and reuse become more favorable and the reuse of properly treated wastewater should increase.

As indicated earlier, while federal and state governments pass laws and set standards and regulations, the actual day-to-day operation of this system falls to local agencies.

Figure 9.3 presents a conceptual organizational structure required to carry out the industrial pretreatment program. As can be seen, the local or municipal government is at the heart of the program. Of course, a much more complex organizational system would be required in big metropolitan areas such as Los Angeles or New York.[34]

As of 1986, EPA had approved 22 states for administering their own pretreatment programs.[37]

The major steps involved in developing a pretreatment program at either federal, state, or local level are[38]

- Making an inventory of all industrial wastes discharged to a POTW.
- Evaluating the legal authority for control and enforcement.
- Determining the information required to develop an industrial waste ordinance.
- Estimating revenue sources to ensure adequate funding.
- Designing a monitoring and enforcement program.
- Determining pollutant removals by existing POTW.
- Establishing POTW tolerance to selected pollutants.

Following development of this body of information, EPA may approve a municipal program. Some of the steps required for a municipality to implement this program are[38]

- Adoption of an industrial pretreatment ordinance.
- Purchase of monitoring and laboratory equipment for use by the POTW if economically justified.
- Regular inspection of regulated industries to ensure compliance.

Local prosecutors have tended to leave the enforcement of water and wastewater laws to federal and state governments. However, this is changing, especially where a city has the trained expertise to carry out local environmental investigations. Recent examples of this local enforcement has taken place in Los Angeles and New York.[11]

IV. WATER QUALITY CRITERIA AND WATER QUALITY STANDARDS

Table 9.4 presents an abbreviated history of the development of drinking water standards, starting in Germany in 1853 up to the latest U.S. Safe Drinking Water Act (SDWA) of 1988.[39] The SDWA is to be reviewed every 3 years, indicating that the standards in this Act are not static but are revised regularly depending on technical, political, and social changes in society.
The key provisions of the U.S. SDWA are

- Generation of national primary and secondary drinking water standards.
- Definition of treatment techniques needed to meet these standards.
- Establishment of state management and regulatory programs for enforcement of the SDWA.
- Removal of suspended and colloidal matter from most water supply sources.
- Killing or inactivation of all pathogens in a water supply.
- Elimination of the contact of a water supply with lead pipes.
- Protection of groundwater sources.

In order to generate national primary and secondary drinking water quality standards it was first necessary to develop criteria. There is a similarity between the Air Quality Criteria Documents and Drinking Water Criteria Documents in that they both tell us what science has been able to measure thus far of the obvious as well as the insidious effects of contaminants on humans and the environment. These criteria documents take into account all of our previous experience in evaluating environmental hazards and provide information in making judgments as to air and water quality. The criteria do not have a legal basis nor do they imply an ideal condition. Air and water criteria reflect the latest scientific knowledge useful in indicating the kind and extent of all identifiable effects on health and welfare which may be expected from the presence of a pollutant in our air or water.[39,40]

The development of water quality standards involves an intensive technological evaluation that includes assessments of:[39]

- Occurrence in the environment
- Human exposure in specific and general populations
- Adverse health effects
- Risks to the population
- Methods of detection
- Chemical transformations of the contaminant in drinking water
- Treatment technologies and costs

In selecting contaminants for regulation, the most relevant criteria are (1) potential health risk, (2) ability to detect a contaminant in the drinking water, and (3) occurrence or potential occurrence in drinking water.

For each of the substances or contaminants the EPA selects, there are two methods for development of regulatory measures: either the EPA must establish a maximum contaminant level (MCL), or if it is not economically or technically feasible to monitor the contaminant level in the drinking water, a treatment method to remove the contaminant from the water supply must be specified.[39]

While criteria are without a legal basis, air and water quality standards do have a legal basis and spell out specific numbers that can be used by an established authority for administrative action and enforcement. However, the primary air quality standards are set to protect human health, regardless of whether the standard is economically or technically achievable, while the maximum contaminant level (MCL) for a drinking water standard requires a balance between public health benefits and what is technologically and economically feasible.

Table 9.4 An Abbreviated History of Drinking Water Standards

1853	F. Cohn in Germany used microscope and related water quality to algae and other microorganisms
1854	Dr. John Snow investigated cholera outbreak in London and determined it was related to contaminated drinking water
1885	Escherich discovered *Bacterium coli*
1891	Miquel established plate count standards for water quality
1893	Theobald Smith in New York used fermentation tubes for the first time
1893	Interstate quarantine regulations promulgated by the Secretary of Treasury
1903	Whipple introduced first standards for water quality using coliforms as indicators
1905	First *Standard Methods For The Examination of Water and Wastewater* published
1912	Common cup was banned on interstate carriers
1914	First U.S. Public Health Service (PHS) drinking water standards — *only bacteriological*
1925	First revision of PHS standards: Source protection Chemicals added Plate count dropped
1942	Second revision of PHS standards — the standards were separated into two parts: Standards with additional chemicals added Waterworks practice manual
1946	Third revision of PHS standards — waterworks practice manual published separately: membrane filter was allowed in 1957
1962	Fourth revision of PHS standards: Waterworks practice part dropped CCE, ABS-detergents, barium, cadmium, cyanide, nitrate, and silver added Fluorides — climate considerations Radioactivity — included for the first time Included rationale used for chemical standards
1974	Safe Drinking Water Act (SDWA) signed into law on December 16, 1974
1975	National Interim Primary Drinking Water Regulations, promulgated by the Environmental Protection Agency, became effective June 24, 1977
1977	Amendments to SDWA
1979	Trihalomethane regulations became effective: For population served: >75,000 — November 29, 1980 >10,000 — November 29, 1982
1986	Amendments to SDWA signed into law on June 19, 1986
1988	Amendments to SDWA

Note: ABS = alkyl benzene sulfonate; CCE = carbon chloroform extract.

From Calabrese, E. J., Gilbert, C. E., and Pastides, H., *Safe Drinking Water Act, Amendments Regulations and Standards,* Lewis Publishers, Chelsea, MI, 1989.

Air and water quality standards have both primary and secondary standards. The primary and secondary drinking water quality standards are presented in terms of MCLs (enforceable standards). An MCL is defined as the permissible level of a contaminant in water at the free-flowing outlet of the ultimate user of a public water system, except in the case of turbidity, where the maximum permissible level is measured at the point of entry to the distribution system.[41] In the process of setting MCLs, EPA creates maximum contaminant level goals (MCLGs) (nonenforceable health goals). The MCLGs are set at levels that present no known or anticipated health effects, including a margin of safety, regardless of technological feasibility or cost. The SWDA directs EPA to set MCLs as close to MCLGs as is feasible.[39]

Drinking water standards in the U.S. have been established in terms of Primary and Secondary standards. These standards are classified as being either chemical (organic, inorganic), physical, biological, or radiological. On an international basis the World Health Organization has developed *Guidelines For Drinking-Water Quality* Vol. 1, 2, and 3. These guidelines are intended to help countries develop drinking water standards that ensure an aesthetically pleasing water that does not result in any significant risk to the health of the consumer. These guidelines are intended to supersede both the European standards for drinking water and International standards for drinking water, the last editions of which were published in 1970 and 1971, respectively.[42]

As indicated earlier, the SDWA regulations and standards indirectly affect the quality of wastewater that may be discharged to a body of water and, thus, affect wastewater reclamation and reuse.

The factors that affect the quality of municipal wastewater effluent include[36]

- Type of treatment system.
- Quality of water supply (primarily the effects of TDS).
- Water usage.
- Impact of any industrial waste discharges.
- Infiltration into collection system.

Table 9.5 presents the U.S. maximum contaminant levels (MCLs), the WHO-recommended levels for drinking water, and the typical concentrations of constituents found in a municipal secondary effluent.[22,23,36,42]

For most of the constituents in the secondarily treated effluents the concentrations are below U.S. MCLs and WHO guidelines for drinking water.

Of course, color, odor, turbidity, and total coliforms are much higher in the secondarily treated wastewater than for the standards or guidelines. The organics, such as biochemical oxygen demand (BOD), chemical oxygen demand (COD), and total organic carbon (TOC) are as they should be — absent in the drinking water regulations and guidelines but clearly present in the treated effluent.

The standards and guidelines presented in Table 9.5 are the water quality requirements for the beneficial use of water for drinking purposes (domestic use). However, there are other beneficial uses for which water is needed such as agricultural irrigation, industrial, steam electric power, commercial, as well as many other uses.[22] Varying degrees of water quality are required depending upon its use. The two categories generally given the highest procurement rights are domestic use and agricultural irrigation. These two categories will be considered here.

While Table 9.5 presents standards and guidelines for drinking water, Table 9.6 presents criteria for not only a raw water supply, which generally requires conventional treatment before use, but also for groundwater recharge. Groundwater may only require withdrawal and then chlorination before distribution to the consumer. Table 9.6 also presents very specific criteria for many chemical parameters, especially metals, for water to be used for agricultural and landscape irrigation.[22,43]

The use of treated wastewater for agricultural irrigation and landscape irrigation is well established in arid and semiarid regions, and in some regions 70 to 85% of the treated wastewater is used for this purpose.[36] The required quality of this water varies in relationship to health, economic and environmental concerns, institutional barriers, and social and legal concerns. Some specific concerns are salinity, soil permeability, specific ion toxicity, nutrients present (nitrogen, phosphorus), bicarbonate (HCO_3), pH, and residual chlorine.[36] While Table 9.6 presented specific criteria for the water quality to be used for agricultural and landscape irrigation, Table 9.7 presents water quality guidelines that can be used to interpret water quality requirements for irrigational purposes. These guidelines are intended for estimating the potential hazards to crop production associated with long-term use.[36]

The criteria and guidelines presented in Tables 9.6 and 9.7 have mainly dealt with physical and chemical characteristics of water used for agricultural and landscape irrigation. While the criteria in Table 9.6 presented fecal coliform levels suggested in the U.S. for various water uses, WHO has also developed guidelines for the microbiological quality of treated wastewater intended for crop irrigation, and they included guidelines for intestinal nematode eggs.

WHO guidelines recommend that treated wastewater should contain:

- <1 Viable intestinal nematode egg per liter (on an arithmetic mean basis) for restricted or unrestricted irrigation; and
- <1000 Fecal coliform bacteria per 100 mL (on a geometric mean basis) for unrestricted irrigation.

Unrestricted irrigation refers to irrigation of trees, fodder, and industrial crops, fruit trees, and pasture; restricted irrigation to irrigation of edible crops, sports fields, and public parks.[44]

V. ENFORCEMENT AND LITIGATION

There are two types of penalties used for enforcement of the Clean Water Act: administrative and judicial. The majority of environmental violations are handled by administrative actions taken by federal, state, or local regulatory agencies. Administrative penalties are divided into two classes.

Class I, specifying a certain dollar amount and an informal hearing.
Class II, specifying a certain dollar amount and a formal hearing.[10]

Table 9.5 Comparison of U.S. and WHO Standards with Typical Secondary Effluent Quality

Parameters[a]	U.S. drinking water MCL	WHO guideline value	Municipal secondary effluent typical[f]
	(mg/L except as noted)		
Primary regulations:			
Arsenic	0.05	0.01	0.005
Barium	1	0.7	0.1
Cadmium	0.010	0.003	0.001
Chromium	0.05	0.05	0.05
Fluoride	4.0	1.5	0.7
Lead	0.015	0.01	0.02
Mercury	0.002	0.001	0.0002
Nitrate (as N)	10.0	10.0	3.0
Selenium	0.01	0.01	0.001
Silver	0.05	No guideline	0.005
Endrin	0.0002		ND[g]
Lindane	0.004	0.002	ND
Methoxychlor	0.1	0.020	ND
Toxaphene	0.005		ND
2,4-D	0.1	0.30	ND
2,4–5-TP Silvex	0.01		ND
Turbidity, TU	1[a]	No guideline	25.0
Total coliform bacteria, colonies/100 mL	[b]	Must not be detectable in any 100 mL sample	1×10^7
Benzene	0.005	0.010	
Carbon tetrachloride	0.005	0.002	
Para-dichlorobenzene	0.075	—	
1,2-Dichloroethane	0.005	0.030	
1,1-Dichloroethylene	0.007	0.030	
1,1,1-Trichloroethane	0.20	—	
Trichloroethylene	0.005	0.070	
Vinyl chloride	0.002	0.005	
Gross alpha, pCi/L	15	2.7 pCi/L (0.1 Bq/L)[c]	
Gross beta, pCi/L	50	27 pCi/L (1 Bq/L)[c]	
Radium 226&228, pCi/L	5	—	
Tritium, pCi/L	20,000	—	
Strontium 90, pCi/L	8	—	
Secondary regulations:			
Copper	1.0	2.0	0.1
Iron	0.3	No guideline	0.2
Manganese	0.05	0.5	0.05
Zinc	5.0	No guideline	0.5
Color, units	15	No guideline	60.0
Foaming agents (as MBAS)[h]	0.5	No guideline	1.5
Odor, TON	3	Inoffensive	40.0
Fluoride	1.4–2.4[d]	1.5	0.7
Chloride	250	No guideline	130
Corrosivity (non-corrosive pH, SU)	6.5–8.5	6.5–8.5	6.0–9.0
Sulfate	250	No guideline	100
Total dissolved solids	500	No guideline	730
Other:			
Total trihalomethanes	0.10		0.005
BOD	ND	ND	25.0
COD	ND	ND	100.0
TOC	ND	ND	40.0
NH_3-N	ND	ND	16.0

[a] Monthly average; [b] ≤40 Samples/month no more than 1 positive, ≥40 samples/month no more than 5% positive; [c] Bq/L = Becquerel per liter, Bq × 2.702703 × 10^{-11} = 1 curie, 1 curie = 3.7 × 10^{10} disintegrations per second (dps); [d] Varies with temperature; [e] See Table 2.19; [f] Unchlorinated; [g] Nondetectable = ND; [h] MBAS = Methylene blue active substances. Compiled from References 22, 23, 36, 42, 43, and 50.

Table 9.6 Water Quality Criteria for Raw Water Supply, Groundwater Recharge, and Agricultural and Landscape Irrigation

	Raw water supply	Groundwater recharge	Long-term agricultural irrigation	Long-term landscape irrigation
	(mg/L except as noted)			
Aluminum	—		5.0	5.0
Arsenic	0.05	0.05	0.1	0.1
Bacteria[a]	—		—	—
Fecal coliforms	2,000	23	1,000	2.2
Total coliforms	20,000	—	—	—
Barium	1	2.0	—	—
Berylium	—		0.1	0.2
BOD	—	10.0	—	20
Boron	0.75	0.02	0.75	0.75
Cadmium	0.01	0.01	0.01	0.01
Chloride	250		—	100
Chromium	0.05	0.15	0.1	0.1
Cobalt	—		0.05	0.05
COD	—	30.0	—	—
Color	75 cu		—	—
Copper	1.0	2.0	0.2	0.2
Cyanide	0.2	0.2	—	—
Fluoride	1.4–2.4		1.0	1.0
Iron	0.3	0.10	5.0	5.0
Lead	0.05	0.05	5.0	5.0
Lithium	—		2.5	2.5
Magnesium	—		—	—
Manganese	0.05	0.10	0.2	0.2
MBAS[b]	0.5	0.50	—	—
Mercury	0.002	0.01	—	—
Molybdenum	—		0.1	0.01
Nickel	—		0.2	0.2
Nitrogen				
Ammonia	0.05	5.0	—	—
Nitrate (as N)	10.0	10.0	—	—
Nitrite (as N)	1.0	0	—	—
Odor	V.F.[c]	V.F.	—	—
Oil	V.F.	V.F.	—	V.F.
Organics				
CCE	0.3		—	—
PCB	0.001		—	—
Oxygen	—	Aerobic	—	—
pH	—	5.0–9.0	6.0–9.0	6.0–9.0
Phenols	0.01	0.001	50	50
Selenium	0.01	0.01	0.02	0.02
Silver	0.05	0.10	—	—
Solids				
Settleable	—		—	—
Suspended	—	10.0	—	15
Total dissolved	—		500–5,000	1,200
Sulfate	250		200	200
Vanadium	—		0.1	0.1
Zinc	5.0	10.0	2.0	2.0

[a] Per 100 mL.
[b] MBAS — Methylene blue active substance (foaming agent).
[c] V.F. — virtually free.

Compiled from References 22 and 43.

Table 9.7 Guidelines for Interpretation of Water Quality for Irrigation

Potential irrigation problem	Units	Degree of restriction on use		
		None	Slight to moderate	Severe
Salinity (affects crop water availability)				
EC_W*	dS/m or mmho/cm	<0.7	0.7–3.0	>3.0
TDS	mg/L	<450	450–2000	>2000
Permeability (affects infiltration rate of water into the soil. Evaluate using EC_W and SAR together)[a]				
SAR = 0–3		and EC_W = >0.7	0.7–0.2	<0.2
= 3–6		= >1.2	1.2–0.3	<0.3
= 6–12		= >1.9	1.9–0.5	<0.5
= 12–20		= >2.9	2.9–1.3	<1.3
= 20–40		= >5.0	5.0–2.9	<2.9
Specific ion toxicity (affects sensitive crops)				
Sodium (Na)[b,c]				
Surface irrigation	SAR	<3	3–9	>9
Sprinkler irrigation	mg/L	<70	>70	
Chloride (Cl)[b,c]				
Surface irrigation	mg/L	<140	140–350	>350
Sprinkler irrigation	mg/L	<100	>100	
Boron (B)	mg/L	<0.7	0.7–3.0	>3.0
Miscellaneous effects (affects susceptible crops)				
Nitrogen (total-N)[d]	mg/L	<5	5–30	>30
Bicarbonate (HCO_3) (overhead sprinkling only)	mg/L	<90	90–500	>500
pH		Normal range 6.5–8.4		
Residual chlorine (overhead sprinkling only)	mg/L	<1.0	1.0–5.0	>5.0

[a] See Chapter 2, Section V.A.2. for calculation of sodium adsorption ratio (SAR).
[b] Most tree crops and woody ornamentals are sensitive to sodium and chloride.
[c] With overhead sprinkler irrigation and low humidity (<30%), sodium or chloride greater than 70 or 100 mg/L, respectively, has resulted in excessive leaf absorption and crop damage to sensitive crops.
[d] Total nitrogen should include nitrate-nitrogen, ammonia-nitrogen, and organic-nitrogen. Although forms of nitrogen in wastewater vary, the plant responds to the total nitrogen.
* "EC_W" refers to electrical conductivity of water.
Compiled from Reference 36.

Judicial penalties are divided into civil and criminal categories. Civil law is based on Roman law and deals with relationships between individuals, as opposed to criminal law which deals with offenses against the state.[10] Civil law judgments are based on codified principles rather than on precedents, and civil law courts do not generally employ trial by jury. Criminal law defines offenses against the state and regulates their prosecution and punishment. In the U.S., power to define crime rests with the states and with the federal government. U.S. criminal law is based on English common law, however, the trend now is toward states enacting penal or criminal codes.

The procedure in criminal cases is much the same throughout the U.S. A grand jury usually examines the evidence against a suspect and either dismisses the case or draws up an indictment. Trial is by jury or before a judge alone. The public prosecutor (usually called the district attorney) presents the government's case and counsel represents the accused. If the accused is found innocent, he or she is discharged, if guilty the judge pronounces sentence. If convicted, the defendant may appeal.[45]

About 10% of EPA cases are generally referred for civil judicial action, most of which are settled by court-approved consent decrees. Only slightly over 1% of EPA cases are referred for criminal action.[11]

Table 9.8 Litigation Initiated by the Department of Justice Under the Refuse Act, 1967–1973

Fiscal year	Actions initiated	Convictions	Fines paid
1967	56	*	*
1968	41	*	*
1969	46	42	*
1970	129	59	*
1971	191	127	174,800.00
1972	198	200	216,500.00
1973	106	132	336,450.00

Note: After PL 92–500 was enacted, a 3-year phase-in period took place and then EPA began referring Clean Water and Safe Drinking Water violations to the Department of Justice for civil actions.

Table 9.9 indicates the number of administrative actions initiated by CWA/SDWA and RCRA from fiscal year 1974 to 1989 as well as the number of EPA referrals to the Department of Justice, fiscal years 1974 to 1989, for civil water cases.

Compiled from Reference 11.

Table 9.9 EPA Administrative Enforcement Actions Initiated by CWA/SDWA and RCRA, Fiscal Years 1974–1989, and EPA Civil Referrals to the Department of Justice (Water Cases)

Fiscal year	CWA/SDWA	RCRA	EPA referrals to justice dept. civil (water)
1974	0	—	0
1975	738	—	20
1976	915	0	67
1977	1,128	0	93
1978	730	0	137
1979	506	0	81
1980	569	0	56
1981	562	159	37
1982	329	237	45
1983	781	436	56
1984	1,644	554	95
1985	1,031	327	93
1986	990	235	119
1987	1,214	243	92
1988	1,345	309	123
1989	2,146	453	94

Compiled from Reference 11.

Felony charges stand out as a major change in environmental enforcement. A felony in criminal law is an offense more serious than a misdemeanor. State laws vary, but in most U.S. jurisdictions (and in federal statutes) felonies are those crimes punished by death or imprisonment for more than 1 year. Felonies are usually tried by a jury.

A misdemeanor charge is a criminal offense less grave than a felony. By federal and state law it is usually punishable by a fine or by imprisonment for less than 1 year. From 1899 to 1970 only 415 actions were initiated under the Rivers and Harbors Act. A section of this law, called the Refuse Act, forbade the discharge of refuse matter into navigable waters. In 1972 federal prosecutors obtained 200 criminal convictions for discharges of pollutants into navigable waters. That same year, amendments to the Federal Water Pollution Control Act (PL 92–500) superseded the River and Harbors Act of 1899 (Sections 407,

Table 9.10 State Environmental Enforcement Actions for Water and RCRA Programs, Fiscal Years 1985–1989

	Water	RCRA
Administrative orders		
1985	2,936	459
1986	2,827	519
1987	1,663	613
1988	2,887	743
1989	3,100	1,189
Judicial referrals		
1985	137	82
1986	221	25
1987	286	86
1988	687	46
1989	489	129

Compiled from Reference 11.

411, and 413, called the Refuse Act) as the primary basis for judicial actions against industries. Table 9.8 presents litigation initiated by the Department of Justice under the Refuse Act from 1969 to 1973.[11]

The states have also been active in enforcement actions, using both Administrative Orders and Judicial Referrals. Table 9.10 presents the number of State Environmental Agency enforcement actions for fiscal years 1985 to 1989.[11]

Despite the number of referrals, relatively few environmental cases go to trial. Most end in a court-approved consent decree, in which a violator promises to correct a violation within a set time frame and may pay a penalty for past violations.[11]

EPA also has the responsibility under the CWA to bring publicly owned wastewater treatment facilities into compliance with national standards. In 1989 EPA took enforcement action against 61 municipalities. Of these, 19 civil court actions were taken. The highest CWA penalty obtained against a municipality was imposed in 1989, when a Denver sewage district was assessed a $1.1 million fine.[11]

In one of the largest oil spills in U.S. history, the Department of Justice obtained an indictment against Exxon Corporation and Exxon Shipping for a total (under criminal law) of ten felony and misdemeanor counts. The settlement was for $1 billion for federal and state civil damages and criminal restitution resulting from the Exxon Valdez (the name of the oil tanker) oil spill.[26]

Exxon's payments for settlement of the fines and damages will be spread over 10 years, with $100 million of it imposed as a fine for breaking the criminal law. Another $900 million, which is considered a payment for environmental damage, will go into a fund that will finance research and projects to restore the environment. The government has the right to demand an additional $100 million if it discovers further damage to the environment.[46]

As of November 13, 1991 the Exxon Corporation had paid Alaska and the federal government the first $125 million of the $1 billion settlement. It paid $50 million to the state, $75 million to the federal government, and a $25 million criminal fine to the U.S.[47]

Joseph J. Hazelwood, captain of the Exxon Valdez, was acquitted on March 23, 1990 of the most serious charges involving the oil spill but was convicted on a single charge of misdemeanor negligence. The jury rejected charges that the captain was drunk when the tanker ran aground on a reef and spilled nearly 11 million gallons of North Slope crude oil into Prince William Sound.

The spill wreaked devastation on birds, fish, and other wildlife, and coated hundreds of miles of coastline with oil.[48]

While the Exxon Valdez oil spill was the largest in U.S. history, it was much smaller than the 68-million-gallon oil spill off the Brittany coast of France. This case has taken 14 years to reach settlement. See Table 9.11 for a listing of oil spill settlements in U.S. Courts.[49]

VI. SUMMARY AND CONCLUSIONS

The first federal water pollution control legislation was the Rivers and Harbors Act of 1899, with key sections 407, 411, and 413 being referred to as the Refuse Act of 1899. These sections of the Act were used even up to 1973 for enforcement litigation and fines. Most of the first half of the twentieth century

Table 9.11 Oil Spill Settlements in U.S. Courts

Source of Spill	Location	Number of gallons	Year occurred	Year settled	Settlement amount
Amoco Cadiz tanker	Brittany Coast, France	68 million	1978	1992	$204 million
Exxon Valdez tanker	Prince William Sound, Alaska	11 million	1989	1991	1.125 billion
Ashland Oil storage tank	Monongahela Ohio Rivers, PA	700,000	1988	1989	4 million
Exxon Bayway pipeline	Arthur Kill, NY–NJ	567,000	1990	1991	15 million
Shell Oil Refinery	Carquinez Strait, CA	500,000	1989	1989	19 million

Note: Legal claims against damage caused by the Amoco Cadiz oil spill off France have been settled after 14 years. The spill was the largest ever.

From Grady, W., *Chicago Tribune,* Sec. 1, 10, Jan. 25, 1992.

concentrated on the two world wars and the great depression. Not until 1948 was any temporary federal water pollution control legislation initiated. The first real, permanent federal water pollution control legislation was passed in 1956. This was the first legislation to provide funds for construction of wastewater treatment facilities, planning and technical assistance grants for state programs, and funding for research. The 1956 Act was amended in 1965, 1966, and 1970. The Federal Water Pollution Control Act amendments of 1972 (PL 92–500) were dramatic, comprehensive, and assertive. This act replaced the previous language used in the CWAs and carried the message that water reuse and nutrient recycling deserved serious consideration as desirable wastewater treatment processes. An evaluation of the goal in the 1972 legislation, which was to restore and maintain the chemical, physical, and biological integrity of the nation's waters, revealed that wastewater reuse and recycling were not being incorporated to any great extent in the wastewater treatment processes.

This resulted in the CWA of 1977 which provided incentives for municipalities to develop innovative and alternative technologies (I/A). In some cases I/A technologies that included water reuse plans could receive up to 85% federal sharing rather than 75% for conventional treatment methods. The 1977 CWA expanded the conventional regulated constituents such as BOD, SS, pH, and fecal coliform, to now include toxic pollutants. Pretreatment regulations and standards were included which made wastewater reclamation, recycling, and reuse more attractive. Amendments to the CWA were made in 1981 and 1987. However, by now the legislation appears to have come full circle and in 1990 the federal construction grants program was terminated and a State Revolving Loan Fund established.

The heart of the federal water pollution control legislation is the pretreatment and the NPDES programs. These programs are now largely carried out at the state and local levels.

At the same time that the water pollution control legislation was being developed, other federal legislation that also impacted on the quality of our nation's water was passed, such as NEPA, SDWA, RCRA, CERCLA and SARA.

The development of water quality criteria and standards are not static but are updated regularly depending upon technical, political, and social changes. Criteria for various beneficial uses of water are moving ahead, yet much has to be done in this area. The U.S. Drinking Water Standards are reviewed and updated every 3 years while the WHO Guidelines For Drinking Water Quality are revised when additional scientific information becomes available.

The federal government has taken a large number of environmental enforcement and litigation cases, however, most environmental enforcement is conducted at the state level. Local enforcement of environmental law has generally been left to the state and federal agencies, however, as local expertise becomes available the local agencies are now more active in prosecution of environmental crimes.

In the past the federal government has passed legislation, established regulations with specific goals, set deadlines, and presented guidelines for application of technology to control water pollution. This command and control method has accomplished much but is often viewed as inadequate.

The new approach concerns economic development and global competitiveness, technological innovation, and a wider variety of environmental risks. This new approach is called pollution prevention, and is aimed at reducing or eliminating pollutants at their source rather than at the end-of-the-pipe treatment of wastes after they have come into existence.

Progress on effective control of our nation's waters depends on the application of management tools and technological systems. A systems approach to control water pollution indicates a holistic, life-cycle approach that defines problems, establishes objectives, generates options, analyzes and selects solutions, and then continually assesses results.[26]

VII. HOMEWORK PROBLEMS

A. DISCUSSION QUESTIONS

1. What are the five legal levels in the U.S. that establish and deal with water resources and water rights problems?
2. Define riparian rights, the prior appropriation rule, reasonable use, and absolute ownership as they relate to the legal water rights system in the U.S.
3. Matching (use the one most appropriate number)

   ____ a. Gallatin Report   1. Based on Spanish law
   ____ b. Common law   2. Dealt with land irrigation
   ____ c. Mississippi River Commission 3. Dealt with navigation
   ____ d. Pueblo Water Rights   4. Offended party
   ____ e. Defendant   5. Based on English law
   ____ f. Civil law   6. Accused party
   ____ g. Desert Land Act   7. Winters Doctrine
   ____ h. Written law   8. Dealt with flooding
   ____ i. Plaintiff   9. Based on Roman law
   ____ j. Indian Water Rights   10. Statutes

4. Describe the difference between common law, statute law, and ordinances.
5. Fill in the correct word(s):

   Primary drinking water standards are based on _____ and secondary drinking water standards are based on _____ .

6. According to the federal government, who has the primary responsibility for the prevention and control of water pollution and enforcement of regulations?
7. Briefly, what were the goals of the 1972 Clean Water Act?
8. Define the following acronyms: BPCT, BPCTCA, BPWTT, BATEA, BAT, and BPT.
9. What is the average length of time an NPDES permit is good for?
10. What are the major parts of an NPDES permit?
11. Will new NPDES permits be required now that toxic pollutants have become so important?
12. How many states have approved EPA programs to administer pretreatment programs and what are the major steps in developing an approved program?
13. What action may EPA take if a state does not adequately administer its NPDES program?
14. What are the major elements involved in a systems approach to control water pollution problems?
15. What are at least three important things needed to solve a complex water pollution problem?
16. Name two of the most important enforcement mechanisms available in the Clean Water Act.
17. What elements must a state show that it has for EPA to approve the state to administer its NPDES programs?
18. The Capitalization Grants were terminated in 1990. What will take the place of these grants?
19. List the steps involved in the water pollution control process used in the U.S.
20. What are the major factors needed to carry out an effective enforcement program?
21. Why did the federal government transfer control of the water pollution program to EPA in 1970?
22. Describe the difference in general terms between criteria and standards.
23. What is the difference between the U.S. and WHO biological water quality criteria for water used for agricultural irrigation?
24. Define a priority pollutant. What are some of the major categories priority pollutants can be divided into, and at present how many are there on the list?

B. ENRICHMENT QUESTIONS

1. Describe the NPDES and pretreatment program in operation in your municipal or local water pollution control agency.
2. Does your state have primacy for their NPDES and pretreatment programs?
3. What surface rights system is used in your state to allocate water, and how are these water rights administered?
4. Discuss I/A technologies, and is this program still relevant now that federal funding of wastewater treatment facilities has been terminated?

5. If you were to participate in developing a criteria document for lead in drinking water, what information would you collect (journals, books, proceedings, conferences, etc.), and how would you finally make a recommendation as to the appropriate lead level in drinking water?
6. How much wastewater reclamation and reuse is practiced in your community at both the governmental and industrial levels?
7. What are the principal local and state water agencies in your community?
8. What local ordinances concerning industrial wastewater discharges to the local sewerage system are in force in your community and how are these ordinances enforced?
9. What water pollution problems exist in your local area and what will be the situation in 5 years?
10. What type of wastewater treatment facility is operating in your community? Is tertiary or advanced treatment used at this facility, and if so, what kind?
11. What other sources than the State Revolving Fund would you investigate in order to secure funds to construct or expand a wastewater treatment facility in your community? Consider local, state, and federal levels.

REFERENCES

1. **Goldfarb, W.,** *Water Law,* 2nd ed., Lewis Publishers, Chelsea, MI, 1988.
2. U.S. Constitution, Constitutional Convention, 1787, Philadelphia, The Bill of Rights, 5th Amendment, 1791.
3. **Viessman, W., Jr. and C. Welty,** *Water Management Technology and Institutions,* Harper & Row, New York, 1985, 30.
4. USEPA, Air Pollution Training Institute, Air Pollution Law, in Air Pollution Control Orientation Course (SI: 422), U.S. Environmental Protection Agency, Research Triangle Park, NC, 1981.
5. *New "Standard" Dictionary,* Funk & Wagnalls, New York, 1961.
6. *The Encyclopedia Americana, Int. ed.,* Grolier Inc., Danbury, CT, 1986.
7. **Babbitt, H. E.,** *Water Supply Engineering,* McGraw-Hill, New York, 1967, 585.
8. **Lohman, L. C.,** *Water Reuse,* E. J. Middlebrooks, Ed., Ann Arbor Science, Ann Arbor, MI, 1982, chap. 2.
9. **Reitze, A. W., Jr.,** *Environmental Law,* Vol. 1, 2nd ed., North American Int., Washington, D.C., 1972.
10. **Kovalic, J. M.,** *The Clean Water Act of 1987,* Water Pollution Control Federation, Alexandria, VA, 1987.
11. Council on Environmental Quality, 20th Annual Report of the Council on Environmental Quality, U.S. Printing Office, Washington, D.C., 1989.
12. **Hickey, J. A.,** Great Lakes — Clean Waters Shifting Sands, *Water Environ. Technol.,* 4, 54, 1992.
13. **Siemak, R. C., et al.,** National Guidelines For Water Reuse: Pros and Cons, in Proc. Conserv. 90 National Conf. and Exp. Offering Water Supply Solutions for the 1990s, Phoenix, AZ, 1990, 51.
14. **Spihaus, A., et al.,** Waste Management And Control, National Academy of Sciences, National Research Council, Pub. 1400, Washington, D.C., 1966, 15.
15. Council on Environmental Quality, The Third Annual Report of the Council on Environmental Quality, U.S. Government Printing Office, Washington, D.C., 1972, 121.
16. Council on Environmental Quality, The First Annual Report of the Council on Environmental Quality, U.S. Government Printing Office, Washington, D.C., 1970, 43.
17. **Anon.,** *Toward Clean Water: A Guide to Citizen Action,* The Conservation Foundation, Washington, D.C., 1976, 16.
18. **Jewell, W. J., and B. L. Seabrook,** A History of Land Application as a Treatment Alternative, EPA-430/9–79–012, Office of Water Program Operations, U.S. Environmental Protection Agency, Washington, D.C., 1979, 33.
19. Council on Environmental Quality, The Fifth Annual Report of the Council on Environmental Quality, U.S. Government Printing Office, Washington, D.C., 1974, 139.
20. *Federal Water Pollution Control Act,* Section 101, U.S. Government Printing Office, Washington, D.C., 1972.
21. **Eckenfelder, W. W., Jr.,** *Principles of Water Quality Management,* CBI Pub., Boston, MA, 1980, 3.
22. **Culp, G. et al.,** *Wastewater Reuse And Recycling Technology,* Noyes Data Corp., Park Ridge, NJ, 1980.
23. *Safe Drinking Water Act,* Public Law 93–523, 1974, 1977, 1986, 1988.
24. **Culp, G. L., G. Wesner, and R. Culp,** *Handbook of Public Water Systems,* R. B. Williams and G. L. Culp, Ed., Van Nostrand Reinhold, New York, 1986, 10.
25. Council on Environmental Quality, The Ninth Annual Report of the Council on Environmental Quality, U.S. Government Printing Office, Washington, D.C., 1978, 112.
26. Council on Environmental Quality, The 21st Annual Report of the Council on Environmental Quality, U.S. Government Printing Office, Washington, D.C., 1990.
27. *Innovative and Alternative Technology Projects,* 1986 Progress Report, Office of Municipal Pollution Control (WH-595), U.S. Environmental Protection Agency, Washington, D.C., 1986.
28. Council on Environmental Quality, 13th Annual Report of the Council on Environmental Quality, U.S. Government Printing Office, Washington, D.C., 1982, 129.

29. *Federal Environmental Law,* Annual Report, 1992 ed., Government Institutes, Inc., Rockville, MD, 1992.
30. Council on Environmental Quality, The Eighth Annual Report of the Council on Environmental Quality, U.S. Government Printing Office, Washington, D.C., 1977, 29.
31. Council on Environmental Quality, The 16th Annual Report of the Council on Environmental Quality, U.S. Government Printing Office, Washington, D.C., 1985, 2.
32. Environmental Permitting in South Carolina, S.C. Department of Health and Environmental Control, Columbia, SC, 1991.
33. Kentucky Pollutant Discharge Elimination System, Department for Environmental Protection, Division of Water, Frankfort, KY, 1987.
34. Federal Guidelines — State and Local Pretreatment Programs, Vol. 1, EPA-430/9–76–017a, Office of Water Program Operations, U.S. Environmental Protection Agency, Washington, D.C., 1977, A-2, B-4.
35. **Corbitt, R.,** *Standard Handbook of Environmental Engineering,* McGraw-Hill, New York, 1990, 3–11.
36. *Water Reuse,* 2nd ed., Manual of Practice SM-3, Water Pollution Control Federation, Alexandria, VA, 1989, 17–20.
37. Municipal Compliance with the National Pollutant Discharge Elimination System, Report to EPA, EPA Construction Grants Program, EN 014–426 40SN, U.S. Environmental Protection Agency, Washington, D.C., 1986, 3.
38. *Pretreatment of Industrial Wastes,* Manual of Practice No. FD-3, Water Pollution Control Federation, Alexandria, VA, 1981, 11.
39. **Calabrese, E. J., C. E. Gilbert, and H. Pastides,** *Safe Drinking Water Act, Amendments, Regulations and Standards,* Lewis Publishers, Chelsea, MI, 1989.
40. *Air Quality Criteria for Carbon Monoxide,* U.S. Department of Health, Education and Welfare, AP-62, U.S. Government Printing Office, Washington, D.C., 1970.
41. *Glossary Water and Wastewater Control Engineering,* 3rd ed., American Public Health Association, Washington, D.C., 1981, 225.
42. Guidelines for Drinking Water Quality, Vol. 1, 2, and 3, World Health Organization Publication Center, Albany, NY, 1984.
43. **Williams, R. B.,** *Water Reuse,* Ed. E. J. Middlebrooks, chap. 5, H. Bouwer, chap. 6, and A. J. Englande, Jr. and R. S. Keimers, III, chap. 31, Ann Arbor Science Publishers, Ann Arbor, MI, 1982.
44. **Mara, D., and S. Cairncross,** Guidelines for the Safe Use of Wastewater and Excreta in Agriculture and Aquaculture, World Health Organization Publication Center, Albany, NY, 1989.
45. **J. S. Levy and A. Greenhaw, Ed.,** *The Concise Columbia Encyclopedia,* Columbia University Press, New York, 1983.
46. Exxon Coughs Up Over Oil Spill, *New Scientist,* March 23, 1991, 14.
47. Payment is Made in Exxon Case, *New York Times,* Nov. 14, 1991.
48. Valdez Captain Cleared of Felony but is Convicted on Minor Charge, *New York Times,* March 23, 1990.
49. **Grady, W.,** Amoco fine hiked in 'disaster waiting to happen', *Chicago Tribune,* Section 1, Jan. 25, 1992, 10.
50. Guidelines for Drinking Water Quality, Vol. 1, Recommendations, World Health Organization Publication Center, Albany, NY, 1993.

Chapter 10

Case Studies

CONTENTS

Forward 426
Case Study I — City of Boca Raton, Florida 426
   I. Introduction 426
  II. Wastewater Reuse in Florida 428
     A. Background and Development of the Boca Raton Project 429
     B. Existing Wastewater Treatment Facilities at Boca Raton 429
     C. Reclaimed Wastewater Production Facilities 431
     D. Monitoring the Reclaimed Wastewater Quality 433
     E. Cost of Construction and Operation of the Boca Raton Reuse Project 434
     F. Compliance With Regulations 437
 III. Summary and Conclusions 438
Case Study II — Logan Aluminum Plant, Kentucky 438
   I. Introduction 438
  II. Wastewater Treatment and Reuse in Kentucky 439
 III. Wetlands 439
     A. Wetland Classification 441
     B. Removal of Nutrients and Pollutants With Wetland Systems 442
 IV. Wetlands in Kentucky 444
     A. Natural Wetlands 444
     B. Constructed Wetlands 444
  V. Design Considerations For Constructed Wetlands 447
     A. Constructed Wetlands Classification 447
     B. Water Balance and Hydrologic Factors 449
     C. Organic Loading Rates 450
     D. Nutrient and Contaminant Removal 450
 VI. Background and Development of the Logan Aluminum Wetlands Project 455
     A. Plant Location 455
     B. The Logan Aluminum, Inc. Plant 456
     C. Logan Aluminum Wastewater Treatment 456
     D. Wetlands Construction 456
     E. Treatment Process 457
 VII. KPDES Wetland Permit 457
VIII. Cost of Construction and Operation of Wetlands 458
 IX. Summary 458

Case Study III — Clayton County, Georgia 459
  I. Introduction 459
  II. Weather and Meteorological Data 459
  III. Wastewater Reclamation and Reuse in Georgia 460
  IV. Land Treatment Design Considerations 462
      A. Basic Land Treatment Processes 462
      B. Types of Slow-Rate Land Treatment Systems 462
      C. Land Requirements 463
      D. Design Hydraulic Loading Rate — Type 1 Systems 464
          1. Monthly Hydraulic Loading Rate Based on Soil Permeability 464
          2. Hydraulic Loading Rate Based on Nitrogen Loading 464
          3. Storage Requirements 465
  V. Clayton County 466
      A. Clayton County Water Authority 466
      B. Background and Development of the Clayton County Wastewater Treatment Project 468
      C. Design Considerations 468
      D. Major Components in the Clayton County Land Treatment Facility 469
      E. Land Treatment Operation 469
      F. System Performance 470
  VI. Effluent Limitations and Monitoring Requirements 470
  VII. Manpower and Costs 472
  VIII. Conclusions 474

Overview of Wastewater Reclamation and Reuse in the Sultanate of Oman 474
  I. Introduction 474
  II. Weather and Meteorological Data for Oman 475
  III. Water Resources in Oman 475
  IV. Wastewater Treatment and Reuse in the Sultanate of Oman 475
  V. Wastewater Treatment and Reuse at the Sultan Qaboos University 476
  VI. Wastewater Quality and Reuse Standards in Oman 476
Homework Problems 478

References — Case Study I 479
References — Case Study II 480
References — Case Study III 481
References — Sultanate of Oman 481

FORWARD

Three current examples of wastewater reclamation and reuse projects are presented in this chapter, as well as an overview of the status of wastewater reclamation and reuse in the Sultanate of Oman.

The first case study deals with the urban reuse of treated wastewater in the City of Boca Raton, Florida. The second case study deals with reuse of treated wastewater at the Logan Aluminum plant at Russellville, Kentucky, and the third case study deals with the irrigation of forest land with treated wastewater in Clayton County, Georgia.

CASE STUDY I
IN-CITY RECLAMATION IRRIGATION SYSTEM CITY OF BOCA RATON PROJECT IRIS

I. INTRODUCTION

The present population of Florida is over 13 million, with around 40 million vacationers coming to that state each year.[1-3] Florida's population is growing at the phenomenal rate of 1000 persons per day with approximately 80% of these people living in the coastal areas. Figure 10.1 indicates the actual population of Florida up to 1980 as well as the projected population through the year 2020.[4]

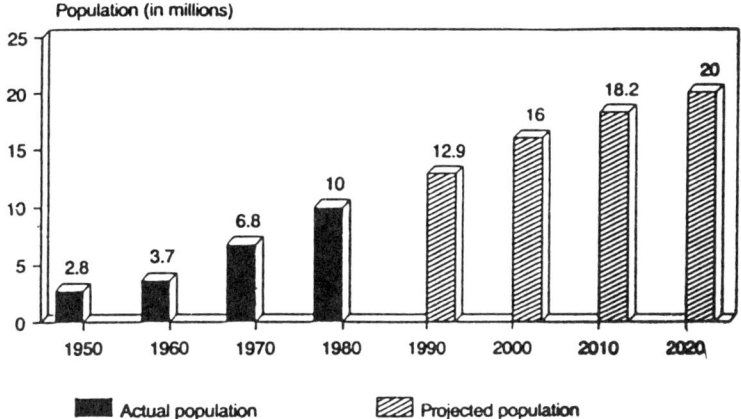

Figure 10.1 Florida's population growth — 1950 to 2020. (From Chansler, J. M., *Water Eng. Manage.*, May, 1991, 31. With permission.)

While Florida is a water-rich state with rainfall of between 130 to 150 centimeters (cm) [50 to 60 inches (in.)] per year and over 7000 freshwater lakes, based on projected water demands there are many areas in Florida that are at or near their total available consumption of freshwater supplies.[4] The present average domestic daily per capita water consumption in Florida is 0.68 cubic meters (m^3, [180 gallons (gal)]) which is almost double the national average.[3,4] Figure 10.2 presents the actual domestic water demands for Florida up to 1980 and the projected domestic water demands to the year 2020.[4]

A recent survey in Florida showed that 30% of the potable water was consumed by households (domestically), 20% by industry, and 50% went for irrigational purposes.[4]

The major source of this water is from shallow aquifers that are vulnerable to overdraft, contamination, and saltwater intrusion.[1] The other major water sources are from lakes and rivers, many of which are fed by water coming from the Everglades or wetlands. At present, more than half of Florida's wetlands have

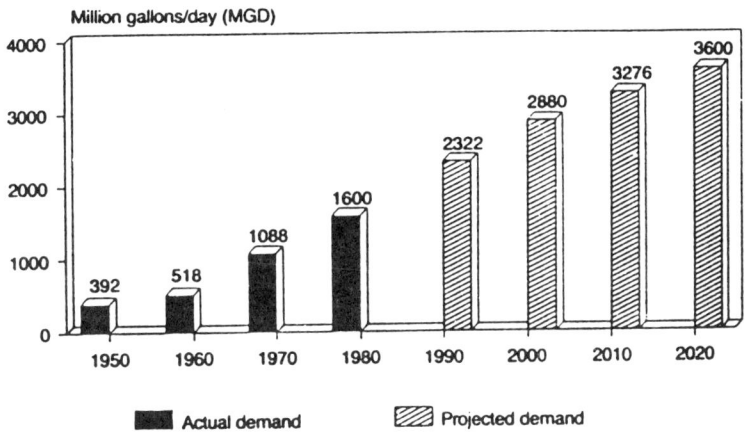

1950–60: 140 GPCD
1970–80: 160 GPCD
1990–2020: 180 GPCD

Figure 10.2 Domestic water demand in Florida — 1950 to 2020. (From Chansler, J. M., *Water Eng. Manage.*, May, 1991, 31. With permission.)

1 MGD = 1 million gallons per day
= 3785 cubic meters per day (m^3/d)
1 GPCD = 1 gallon per capita per day
= 0.003785 cubic meters per capita per day (m^3 pcd)

Table 10.1 Water Reuse in Florida

	1986	1990	Percent increase
Total number of facilities	118	214	81
Total reuse flow	0.78 Mm³/d (206 mgd)	1.21 Mm³/d (320 mgd)	55
Total reuse capacity	1.37 Mm³/d (362 mgd)	2.21 Mm³/d (585 mgd)	62
Total domestic wastewater flow[a]		3.785 Mm³/d (1000 mgd)	
Percentage reused			58[b]

[a] Estimated amount of domestic wastewater treated at central treatment facilities.
[b] This is actually the percent of capacity: 36% was reused in 1990.

Compiled from References 4 and 6.

been lost — 1.3 million hectares (ha) [3.2 million acres (ac)]. These wetlands have been cleared and/or drained to meet the unprecedented growth and development that has and is taking place in Florida.[5]

II. WASTEWATER REUSE IN FLORIDA

Recent droughts have made it necessary in many areas in Florida to not only impose restrictions on water use, but also to turn to the development of wastewater reclamation and reuse programs. The Florida Department of Environmental Regulations has set a goal to reuse 40% of the state's total wastewater flows. In the most densely populated areas of the state, the South Florida Water Management District has indicated the need for 100% wastewater reuse by the year 2010.[4]

Florida, along with California and Arizona, is one of the national leaders in the area of wastewater reuse.[2] Table 10.1 presents the number of wastewater reuse projects in Florida for 1986 and 1990 as well as the total volume of wastewater being reused.[4,6]

Following are some of Florida's nationally acclaimed wastewater reuse projects.[2]

St. Petersburg dual distribution system — Highly treated reclaimed water is made available in a separate piping system for landscape irrigation, including the irrigation of more than 6000 residential lawns. This system has been operating since 1977 and an average of more than 0.076 Mm³ (20 mil gal) of reclaimed water is reused each day.

Orlando wetlands — Orlando has developed an artificial wetlands using reclaimed water from its Iron Bridge advanced treatment facility. The wetlands system was planned as a multipurpose use area, featuring recreation such as hiking, jogging, and nature observations.

Naples — Nine golf courses are irrigated with about 0.02 Mm³ (5 mg) of reclaimed water per day.

Cocoa Beach — About 0.01 Mm³ (3 mil gal) per day of reclaimed water is used to irrigate more than 500 properties and a golf course.

Loxahatchee River environmental control district — More than 0.01 Mm³ (3 mg) per day of high-quality reclaimed water is used to irrigate seven golf courses.

Tallahassee spray irrigation system — Florida's capital city irrigates about 809 ha (2000 ac) using reclaimed water for growing corn, soybeans, and other fodder crops. In some parts of the year, cattle graze upon irrigated pasture. Tallahassee has been using reclaimed water for agricultural irrigation since 1966.

Project APRICOT — This project provides reclaimed water for irrigation of residential lawns and greenspace areas in Altamonte Springs. Reclaimed water also is used in a car wash.

CONSERV II — This award-winning project serves portions of both Orange County and Orlando. Highly treated reclaimed water is piped about 32 kilometers (km), [20 miles (mi)] west of Orlando and is used to irrigate about 2833 ha (7000 ac) of citrus groves and 4.0 ha (10 ac) of ferns. The capacity of the system is 0.17 Mm³ (44 mil gal) per day. This project received the 1989 Grand Conceptor Award for engineering excellence from the American Consulting Engineers Council.

Curtis Stanton Energy Center — This coal-fired power plant cools its boilers with 0.13 Mm³ (3.5 mil gal) per day of reclaimed water from the Orange County Eastern Service Area treatment plant.

Tampa's McKay Bay Refuse-to-Energy Facility — This facility uses reclaimed water from the Hookers Point treatment plant for noncontact cooling water.

Another important urban wastewater reuse project was dedicated on December 6, 1991 at Boca Raton, Florida. The official name for this $40 million project is IRIS, and stands for In-City Reclamation Irrigation System.[7]

A. BACKGROUND AND DEVELOPMENT OF THE BOCA RATON PROJECT

Southern Florida has received 10% below-average rainfall for the past 30 years.[4] However, the recent droughts from 1989 to 1991 were so severe that many counties in the Boca Raton region imposed restrictions on the use of potable water. Several counties required a 15% reduction in total potable water consumption and limited irrigation and other outside uses to 4-hour periods 3 days per week and in some areas to 2 days per week.[8]

Due to climatic and sandy soil conditions and the never-ending growing season, between 50 to 70% of the potable water consumption in an urban setting in Florida is used for landscape irrigation.[8] Prior to restrictions the daily per person water consumption in Boca Raton was well over 1.5 m^3 (400 gal) per day, with 70% of this water being used for outside purposes — mainly landscape irrigation. Since the water use restrictions were imposed the water consumption has decreased to 1.23 m^3 (325 gal) per capita per day (pcpd) or a 20% reduction.[9] Even with this decline in water consumption the City of Boca Raton administration realized that one of the most effective means of conserving the potable water supply was to treat and use the wastewater. This treated wastewater normally was pumped to an ocean outfall. The treated wastewater can now be used for nonpotable purposes such as irrigation of city parks and recreational areas, road medians, cemeteries, church properties, and on other green landscape areas.[8]

B. EXISTING WASTEWATER TREATMENT FACILITIES AT BOCA RATON

Before the reclaimed wastewater facilities were constructed the city provided preliminary, primary, and secondary treatment, as well as disinfection of the wastewater by chlorination, and tertiary treatment for some 0.01 Mm^3/d (3 mgd) of the 0.041 to 0.045 Mm^3/d (11 to 12 mgd) wastewater flows.[9] The preliminary treatment consists of fine screening and grit removal; primary treatment consists of primary clarifiers; and the secondary, biological treatment employs the activated sludge process which, of course, is used in combination with the secondary clarifiers. The secondarily treated effluent is then disinfected by chlorination. About 25 to 30% of the secondarily treated effluent is given advanced or tertiary treatment (two automatic backwash filters) and this water is then used for in-plant purposes as well as for landscape irrigation at Florida Atlantic University.[8] The secondarily treated effluent that has not been given advanced or tertiary treatment is then pumped, as indicated earlier, through a 8 km (5 mi) long outfall terminating about 1.5 km (0.95 mi) offshore in the Atlantic Ocean at a depth of 27 meters (m) (90 feet (ft)).

The sludge from the primary treatment process is dewatered prior to disposal. The waste activated sludge, which has settled in the bottom of the secondary clarifiers, is pumped to the rotary drum thickeners where 75% of the water in the sludge is removed. This thickened sludge is then pumped to two 24.4 m (80 ft) diameter primary digesters and one 24.4 m (80 ft) diameter secondary digester. Sludge that has settled in the bottom of the primary clarifiers is pumped directly to the digesters.

In the digesters, anaerobic bacteria are used to break down the organic solids. Methane gas and carbon dioxide are by-products of this process, and the methane gas is used to heat the digesters. The digested sludge is dewatered by three belt filter presses prior to landspreading of the sludge.[10]

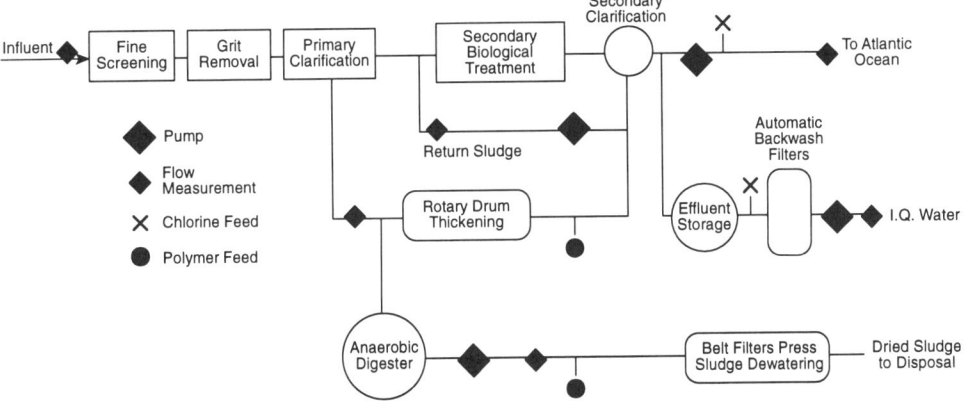

Figure 10.3 Schematic flow diagram: City of Boca Raton wastewater treatment facility. I.Q. = irrigation quality. (From Glades Road Wastewater Treatment Expansion, Boca Raton, FL, 1989. With permission.)

Table 10.2 Basic Design Specifications for the City of Boca Raton's Wastewater Treatment Facility

Key Treatment Units	Number of Units	Size and/or Ratings
Bar screens	3	20 mgd (each)
Grit collectors	2	40 mgd (each)
Primary clarifiers	2	100-ft diameter (each)
Aeration trains	3 Trains of 3 basins each	85-ft wide/255 ft long (each train)
Secondary clarifiers	5	2 @ 105-ft diameter
		3 @ 110-ft diameter
Effluent pump station	3 Low-capacity/transfer 3 high-capacity	26 mgd
Effluent storage tank	1	2.5 mgd
Automatic backwash filter	1	1.5 mgd
Rotary drum thickeners	4	2 @ 175 gpm
		2 @ 80 gpm
Anaerobic digesters	3	80 ft diameter/23½ ft deep
Belt filter presses	3	2-meter belt width (each)

From Glades Road Wastewater Treatment Expansion, Boca Raton, FL, 1989. With permission.

Table 10.3 Effluent Limitations and Monitoring Requirements — Part 1 — Final[11]

1. These limitations are to be achieved by the effective date of the permit, and shall remain in effect until permit expiration for outfall Serial Number 001, sanitary wastewater. (May 31, 1994)

   Such discharges shall be limited and monitored by the permittee as specified below:

	DISCHARGE LIMITATIONS		MONITORING REQUIREMENTS		
PARAMETERS	Monthly average	Weekly average	Measurement frequency	Sample type	Sampling point
Flow, m³/day (MGD)	Report	Report	Continuous	Recording flow meter and totalizer	Effluent
Biochemical oxygen demand (5 Day)	30 mg/L	45 mg/L	7/Week	24-hr composite	Influent and effluent
Total suspended solids	30 mg/L	45 mg/L	7/Week	24-hr composite	Influent and effluent
Fecal coliform					
Bacteria, geometric mean (No./100 mL)	See Item 9		7/Week	Grab	Effluent
pH	See Item 6		Continuous	Meter	Effluent
Total chlorine residual	Report	Report	7/Week	Grab	Effluent

From Vidaurreta, E. J., NPDES Permit No. 0026344, 1989. With permission.

Figure 10.3 presents a schematic flow diagram of the City of Boca Raton's wastewater treatment facility, and Table 10.2 gives the basic design specifications for this plant.[10]

This plant is operated in compliance with the National Pollution Discharge Elimination System (NPDES). See Table 10.3 for the effluent limitations and monitoring requirements in this permit.[11] The City of Boca Raton also has an Industrial Pretreatment Program in place and this program is in operation.[8]

NPDES No. FL002634

Part I (Continuation)

Effluent Limitations And Monitoring Requirements

2. Any bypass of the treatment facility which is not included in the effluent monitored above is to be monitored for flow and reported on a Discharge Monitoring report (DMR).

3. There shall be no discharge of floating solids or visible foam in other than trace amounts.
4. The effluent shall not cause a visible sheen on the receiving water.
5. In addition to the specified limits, the monthly average effluent BOD_5 and suspended solids concentrations shall not exceed 15% of the respective influent values (85% removal).
6. The pH of the effluent shall not be less than 6.0 standard units or greater than 8.5 standard units and shall be monitored continuously with a recorder. The values shall not deviate outside the required range more than 1% of the time (7 hours and 26 minutes) in any calendar month and no individual excursion shall exceed 60 minutes. An "excursion" is an unintentional and temporary incident in which the pH value of discharge wastewater exceeds the range set forth in this permit.
7. The effluent (100%) shall not be lethal to more than 50% of the appropriate test organisms. The testing for this requirement must conform with Part IV of this permit. Lethality to more than 50% of the test of 48 hours duration will constitute a violation of Florida Administrative Code Section 17–4.244(4) and the terms of this permit.
8. Samples taken in compliance with the monitoring requirements specified in this permit shall be taken at the nearest accessible point after final treatment but prior to the actual discharge or mixing with the receiving waters.
9. a. The arithmetic average of the monthly geometric average fecal coliform values collected during an annual period shall not exceed 200 per 100 mL of effluent sample.
    b. The geometric mean of the fecal coliform values collected during a month shall not exceed 200 per 100 mL of sample.
    c. No more than 10% of the samples collected (the 90th percentile value) during a period of 30 consecutive days shall exceed 400 fecal coliform values per 100 mL of sample. Note: to report the 90th percentile value, list the fecal coliform values obtained during that month in ascending order. Report the value of the sample that corresponds to the 90th percentile (multiply the number of samples by 0.9). For example: for 30 samples, report the corresponding fecal coliform for the 27th value of ascending order.
    d. Any one sample shall not exceed 800 fecal coliform values per 100 mL of sample.
10. For the parameter, *Fecal Coliform*, no violations for annual average will be deemed to have occurred until data have been compiled for the first 12 months after the effective date of the permit.
11. a. The discharge of pollutants from Outfall 001 shall not exceed the limiting permissible concentration for the liquid and suspended particulate phases of the waste material as described in Section 227.27(a)(2) and (3), Section 227.27(b), and Section 227.27(c) of the Ocean Dumping Criteria following dilution as measured at the edge of any regulatory mixing zone.

C. RECLAIMED WASTEWATER PRODUCTION FACILITIES

The wastewater reclamation facilities includes chemical feed and storage equipment, a flash mixer, two automatic backwash filters (ABW-filters), six Dynasand filters, chlorine storage and feed equipment, (high-level disinfection) two chlorine contact chambers, a 0.01 Mm^3 (3 mil gal) reuse storage tank, and transfer pumps.[8,9,12]

Figure 10.4 presents a schematic flow diagram for the Boca Raton wastewater reclamation facility.[12]

Phase I of this project will result in 0.034 Mm^3/d (6 mgd) coming from the Dynasand filters and another 0.011 Mm^3/d (3 mgd) from the existing travelling bridge filters (automatic backwash filters). A 0.056 Mm^3/d (15 mgd) distribution and land application system will also be included in this system. Following are the major components in the reclaimed wastewater treatment facility.[12,13]

1. *Low Lift Pump Station.* Above-ground reservoir with three (3) variable speed, vertical turbine centrifugal pumps. Each pump will have an intermediate capacity of 13.2 cubic meters per minute (m^3/min) (3500 gpm) with a 7.0-m (23-ft) total dynamic head (TDH) delivering 0.038 Mm^3/d (10.08 mgd) with one pump out of service.
2. *Polymer Feed System.* Consists of 9.5 m^3 (2500 gal) polymer storage tank with mix chamber and feed to the flash mixing tank. The flash mixing tank will have two (2) $5.5 \times 3.0 \times 3.0$ m ($18 \times 10 \times 10$ ft) mixing compartments in parallel. Each compartment will have one submersible low-speed propeller-type mixer for mixing polymer solution with secondary treatment effluent. Discharge is via a splitter box

to a 61 cm (24-in.) line to the existing traveling bridge filters and to a 91-cm (36-in.) line to the proposed upflow Dynasand filters.

3. *Dynasand Filters.* Six (6) concrete basin filter cells containing four (4) Dynasand filter modules in each cell, 4.645 m^2 (50 ft^2) per module, for a total of 111.5 m^2 (1200 ft^2) 0.14 m^3/min/m^2 (3.5 gpm/ft^2) at 0.22 Mm^3/d (6 mgd)]. This is to the chlorine contact basins. Air requirements are 1.8 m^3/min (62.4 cfm) for Phase I (24 filter modules, 0.07 m^3/min [2.6 cfm] of air per module) and will be provided by two (2) 2.2 m^3/min (78 cfm) air compressors (maximum air demand provided with the single largest unit out of service). Filter reject water will be returned to the headworks of the WWTP. Gravity discharge is to the chlorine contact basins. (These filters must have the capacity to consistently reduce the total suspended solids (TSS) in the filtered water to 5.0 mg/L or less. The Boca Raton filters are designed to reduce the TSS to less than 2 mg/L.[9])

4. *Traveling Bridge Filter (TBF) (ABW-filters) Effluent Pumps.* Replace existing pumps with three (3), constant speed vertical turbine centrifugal pumps. Each pump will have an intermediate capacity of 3.95 m^3/min (1045 gpm) @ 4.6 m (15 ft) TDH [0.01 Mm^3/d, 3 mgd] with one pump out of service). Influent is filtered secondary treated wastewater from the TBF and discharge is to the receiving channel of the chlorine contact tank. The existing traveling bridge filters are automatic backwash, medium depth [0.6 to 0.68 m (2 to 2.25 ft)], monomedia sand filters. The filter beds are 3.8 m (12.5 ft) wide and 14.0 m (46 ft) long, providing a total surface area per filter of 53.2 m^2 (575 ft^2). The rated capacity of each filter is 0.0057 Mm^3/d (1.5 mgd). The specified filter rate is 0.08 m^3/min/m^2 (2 gpm/ft^2) with a peak rate of 0.16 m^2/min/m^2 (4 gpm/ft^2).

5. *Chlorine Contact Basins/Reclaimed Water Transfer Pumps.* Two (2) concrete-reinforced chlorine contact basins in parallel with each basin having a design flow capacity of at least 50% of the total design flow. Each tank will have five (5) channels ("end around" baffling). The dimensions of each channel will be approximately 1.78 × 18.3 × 3 m side water depth (SWD) (5 × 97.8 m^3 = 489 m^3) (5 ft 10 in. × 60.0 ft × 10 ft SWD) (5 × 26,165 gal = 130,900 gal). The volume of both basins = 978 m^3 (261,800 gal) and are designed to provide a contact time of 25 min at an equalized peak flow of 0.057 Mm^3/d (15.0 mgd). The chlorine feed equipment has capacity to maintain a total chlorine residual of 1.0 or 2.0 mg/L, thereby providing a product number of 25 to 50 based upon a fecal coliform count of 1000 colonies/100 mL or less following filtration.

The final reclaimed water distribution facility at the Glades Road site has an average capacity of 0.057 Mm^3/d (15 mgd) and a peak capacity of 0.11 Mm^3/d (30 mgd) and includes two 0.01 Mm^3 (3 mil gal) ground storage tanks and a high-service pump building. Five 20 m^3 (5400 gal) per minute high-service

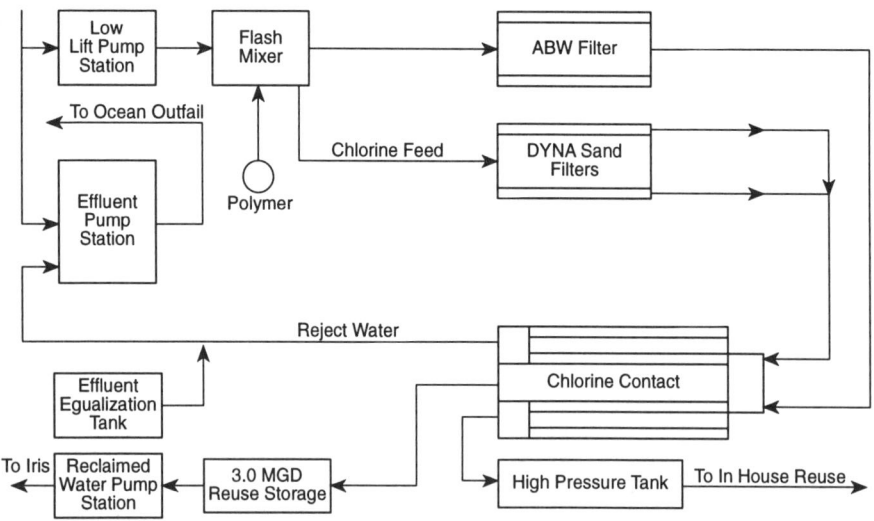

Figure 10.4 Schematic flow diagram: City of Boca Raton water reclamation facility.[12]

Table 10.4 Boca Raton Reuse Phase I Permit — Florida Department of Environmental Regulation

Sampling, reporting and effluent limitations for reclaimed water that is to be discharged from this wastewater treatment plant (WWTP) for the period allowed to operate under this permit from July 23, 1992 to July 23, 1994.

PARAMETER	EFFLUENT LIMIT	MINIMUM FREQUENCY	SAMPLE TYPE	LOCATION
Flow	(b) mgd	Continuous (b)	Venturi meter with recorder and totalizer (b)	Upstream and downstream of filters (b)
CBOD$_5$	mg/L[a]	daily, 7/wk[a]	24-hour composite	WWTP influent and reclaimed effluent
TSS	mg/L[a]	daily, 7/wk[a]	Grab (e)	WWTP influent and reclaimed effluent prior to disinfection
pH units	6.0 to 8.5	Continuous (h)	On-line pH meter (h)	Reclaimed effluent
Chlorine residue	(c) Min. 1.0 mg/L	Continuous	On-line chlorine analyzer	Reclaimed effluent
Fecal coliforms	Below detectable limits (d)	Daily, 7/wk (d)	Grab (e)	Reclaimed effluent
Turbidity	(f)(g)	Continuous	On-line turbidity meter	Reclaimed effluent prior to disinfection
	annual	monthly	weekly	one time grab
CBOD$_5$	20	30	45	60
TSS	5.0	5.0	5.0	5.0

[a] Limits, maximum (mg/L).
Compiled from Reference 13.

pumps [@ 54.2 m (178 ft) (TDH)] — four operating and one standby — will be used to satisfy the estimated peak demand for reclaimed water. The installation of these pumps will be phased in as well as the installation of a second ground storage tank, in accordance with the overall development of the system.[8,9]

This $40 million wastewater project will be constructed in four phases over a 10-year period. The first phase will be completed in April 1994, and have a capacity of 0.034 Mm^3/d (9 mgd). The final phase of this project will be completed in the year 2001, with an ultimate capacity of 0.057 Mm^3/d (15 mgd). The estimated total cost to complete phase I of this wastewater reclamation treatment facility is $10 million, while the total cost including the distribution facilities will be an additional estimated $30 million.

D. MONITORING THE RECLAIMED WASTEWATER QUALITY

The City of Boca Raton has been issued a Phase I operational permit for this 0.1034 Mm^3/d (9.0 mgd) Reuse/Land Application System. This permit provides for specific monitoring conditions, some details of which are presented in Table 10.4.[13]

SPECIFIC CONDITIONS OF PHASE I OPERATIONAL PERMIT

Influent samples are for monitoring and reporting purposes. Pursuant to F.A.C. Rule 17–601.500(4)(a), influent samples shall be collected so that they do not contain digester supernatant or return activated sludge, or any other plant process recycled waters.

During the interim period, prior to the construction of the 9 MGD Reuse Treatment System, when the city provides reclaimed water to the F.A.U. Campus and for on-site irrigation and process water, the minimum frequency for CBOD$_5$ is weekly and TSS is daily, 7 days per week, and the sample type for CBOD$_5$ is 16 hour flow proportioned composite.

(b) Annual average daily flow shall not exceed 9 MGD upstream of filters. During the interim period mentioned in 2(a); the annual average daily flow shall not exceed 1.8 MGD; minimum frequency is continuous; sample type is magnetic meter with recorder and totalizer; and sample location is the effluent subsequent to the hydropneumatic tank.

(c) The minimum acceptable contact time shall be 15 minutes at the peak hourly flow. Higher residuals or longer contact times may be needed to meet the design and operational criteria for high-level disinfection as described in Rules 17–600.400(5)(c) and 17–600.440(5)(f), F.A.C.

(d) Over a thirty (30) day period, 75 percent of the fecal coliform values shall be below the detection limits. Any one sample shall not exceed 25 fecal coliform values per 100 ml of sample pursuant to F.A.C. Rule 17–600.440(5). During the interim period mentioned in 2(a), the minimum frequency for fecal coliform sampling is daily, 7/week.

(e) Grab samples will be collected during periods of peak hydraulic and/or organic loading.

(f) The permittee shall determine a maximum turbidity limit based on a correlation between turbidity data and TSS data.

(g) The permittee shall develop, and the department shall approve, an operating protocol designed to ensure that the high level disinfection criteria will be met before the reclaimed water is released to the system storage or to the reclaimed water reuse system. The operating protocol shall be reviewed and updated and shall be subject to department review and approval at least annually. Reclaimed water produced at the treatment facility that fails to meet the criteria established in the operating protocol shall not be discharged into system storage or to the reuse system. Such substandard reclaimed water (reject water) shall be either stored for subsequent additional treatment or shall be discharged to another permitted reuse system requiring lower levels of pretreatment or to a permitted effluent disposal system. The protocol shall include a correlation between TSS and Turbidity determined from infield sampling. F.A.C. Rule 17–610.463.

(h) Hourly measurements during the period of required operator attendance may be substituted for continuous measurement.

E. COST OF CONSTRUCTION AND OPERATION OF THE BOCA RATON REUSE PROJECT

In Boca Raton two sites for the reclaimed water production and distribution facilities were evaluated: the Glades Road Public Utilities Complex site and the Second Avenue Water Treatment Plant site. After careful, detailed evaluations the Glades Road site was selected.[8]

Also, two alternatives were considered in locating the distribution mains within the street right-of-way: in the grassed area adjacent to the pavement or underneath the street pavement. Preliminary estimates indicated that placing the distribution mains under the street pavement could increase the total estimated cost of the project by as much as $5.3 million.[8]

It was therefore decided to locate the distribution mains in the grassed areas adjacent to the pavement.

The final estimated cost of the project that is now underway has been divided into three main categories:[8,12]

The reclaimed water production and distribution facilities	$10,000,000.
The reclaimed water transmission mains system	$15,000,000.
The reclaimed water distribution mains	$15,000,000.
Estimated total cost (in 1990 dollars)	$40,000,000.

As indicated earlier, the project is to be carried out in four phases. The estimated cost for phase 1 is $10 million and is to be completed by April 1994. The final phase is to be completed in the year 2001.[12] Included in the estimated $40 million cost for this project is $8 million for design and construction engineering fees, administrative costs, legal fees, and land right-of-way easements.[8]

An operating budget for the 1994 Boca Raton system includes the following line items:[12]

Electricity	$27,600.
Supplies	$28,000.
Automatic meter reading	$38,800.
Total	$95,200.

This allocation to the reuse program represents five percent (5%) of the overall wastewater system budget.

Table 10.5 User Charges for Reclaimed Water

Location	Year operation began	Reuse methods	User charge
Altamonte Springs	1989	Irrigation	Residential $8/month Commercial $0.50/1,000 gallons + avail. charge of $3/ERC
Cape Coral	1990	Irrigation	Residential $5-$6/month (planned)
Cocoa Beach	1985	Irrigation	Residential $6/month
Dunedin	1991	Irrigation	$0.50/1,000 gal
Encon (Jupiter)	1985	Irrigation	$0.27/1,000 gal
Largo	1987	Irrigation	$0.20/1,000 gal
Naples	1988	Irrigation	$0.06/1,000 gal
Orange County	1986	Cooling water, wetlands creation, citrus, golf course/parks irrigation	No user charge
Sanford	1990	Irrigation	$3.25/month base charge + $0.05/1,000 gal
St. Petersburg	1977	Irrigation	Residential $10.36/1 acre or less, $5.32 for each additional acre per month. Commercial $0.30/1,000 gal
Tarpon Springs	1989	Irrigation	$0.55/1,000 gal

From Riley, S. and Hungate, R., *Fla. Water Resources* J., August, 1991, 38. With permission.

Table 10.6 Summary of Unit Costs: Slow-Rate/Large-User Land Applications

	Total cost per 1,000 gallons (1990 dollars)		
Wastewater Treatment Plant	Treatment	Transmission	Treatment and transmission
Broward County	$0.42	$0.44	$0.86
Coral Springs, I.D.	0.48	0.33	0.81
Hollywood	0.47	0.36	0.83
Margate	0.46	0.22	0.68
Pembroke Pines	0.46	0.37	0.83
Plantation	0.49	0.30	0.79
Sunrise 1	0.46	0.17	0.63
Sunrise 2	0.46	0.27	0.73
Sunrise 3	0.49	0.22	0.71

From Davis, P. A., et al., *Proc. Urban Agri. Water Reuse*, WEA and FPCA, Orlando, FL, 1992, 407. With permission.

To help fund this project, a user charge system has been established which includes an availability charge and a consumption charge.[12]

The reclaimed water supply is metered. Following are the user charge rates:[12]

Flat rate availability (3/4 in. meter)	$3.00 monthly
Gallonage charge (0–25,000)	$0.35/1000 gallons
Gallonage charge (>25,000)	$0.55/1000 gallons

Current user charges for reclaimed water in Florida vary widely. Table 10.5 presents user charge information for some of the larger water reuse programs in Florida.[14] In Broward County, which is next

Table 10.7 City of Boca Raton Water Reclamation Facility

	ADF (mgd)	TSS (mg/L)	Cl_2 resid	Turb (NTUs)	Fecal #/100 (MAX)	Cl_2 DOSE (mg/L)	N-NO_2 (mg/L)	N-NO_3 (mg/L)	$CBOD_5$[a] (mg/L)	# of filters on line
					1992 ANNUAL REPORT[12,13] **REUSE DATA 1992**					
AVG	0.748	0.5	3.3	0.55	X	16	0.3	3.0	6.0	X
MIN	0.666	0.2	2.5	0.39	0	14	0.1	1.0	4.5	2
MAX	0.895	0.9	3.8	0.73	1	19	0.8	5.2	7.5	2
					1993 ANNUAL REPORT[12,13] **REUSE DATA 1993**					
JAN	0.756	0.4	3.3	.28	9	18	0.2	2.4	5.0	2
FEB	0.819	0.5	2.7	.65	0	19	0.2	2.8	4.5	2
MAR	0.919	0.3	2.5	.78	1	18	0.2	2.8	9.2	2
APR	1.051	0.2	2.8	.69	0	15	0.7	1.7	11.8	2
MAY	0.871	0.4	3.0	.54	0	17	0.1	2.2	7.8	2
JUN	0.807	0.5	2.9	.52	1	13	0.3	0.4	6.8	2
JUL	0.812	0.7	2.7	.50	1	16	0.0	1.3	5.2	2
AUG	0.670	0.7	3.1	.54	0	17	0.1	1.0	3.5	2

[a] For November and December only.

Table 10.8 Purgeable Organics in the Reclaimed Wastewater for the City of Boca Raton

EPA METHOD 624 — PURGEABLE ORGANICS			
CAS No.	PARAMETER	CONCENTRATION (μg/L)	MDL[a] (μg/L)
71–43–2	Benzene	0.0	(0.20)
75–27–4	Bromodichloromethane	5.7	(0.20)
75–25–2	Bromoform	0.0	(0.20)
74–83–9	Bromomethane	0.0	(0.50)
56–23–5	Carbon Tetrachloride	0.0	(0.50)
108–90–7	Chlorobenzene	0.0	(0.20)
75–00–3	Chloroethane	0.0	(0.50)
67–66–3	Chloroform	15.8	(0.20)
74–87–3	Chloromethane	0.0	(0.50)
124–48–1	Dibromochloromethane	1.1	(0.20)
95–50–1	o-Dichlorobenzene	0.0	(0.20)
541–73–1	m-Dichlorobenzene	0.0	(0.20)
106–46–7	p-Dichlorobenzene	0.0	(0.20)
75–34–3	1,1-Dichloroethane	0.0	(0.25)
107–06–2	1,2-Dichloroethane	0.0	(0.20)
75–35–4	1,1-Dichloroethene	0.0	(0.40)
156–60–5	trans, 1,2-Dichloroethene	0.0	(0.25)
78–87–5	1,2-Dichloropropane	0.0	(0.40)
10061–01–5	cis,1,3-Dichloropropene	0.0	(0.50)
10061–02–6	trans,1,3-Dichloropropene	0.0	(0.50)
100–41–1	Ethylbenzene	0.0	(0.20)
75–09–2	Dichloromethane	0.0	(0.50)
79–34–5	1,1,2,2-Tetrachloroethane	0.0	(0.24)
127–18–4	Tetrachloroethene	0.0	(0.14)
108–88–3	Toluene	0.0	(0.20)
71–55–6	1,1,1-Trichloroethane	0.0	(0.30)
79–00–5	1,1,2-Trichloroethane	0.0	(0.30)
79–01–6	Trichloroethene	0.0	(0.20)
75–01–4	Vinyl Chloride	0.0	(0.17)

MDL = Method Detection Limit = method detection limit × dilution factor. A value of 0.0 = BMDL (below method detection limit)

Compiled from References 12 and 13.

door to Boca Raton, a summary of total unit costs for slow-rate/large-user land applications of wastewater reuse has been prepared.[15] Slow-rate land application of reclaimed wastewater includes irrigation of golf courses, other green spaces, and single and multifamily residential sites.[15]

Table 10.6 presents the unit costs for several large-user land application systems in Broward County.[15]

From the information presented in Tables 10.5 and 10.6, it would appear that the user charge of around $0.45 per 1000 gallons represents slightly over 50% of the actual total cost to treat and transmit the reclaimed wastewater. Of course, as wastewater reclamation and reuse become more widespread, user charges will gradually catch up to the actual production costs.

F. COMPLIANCE WITH REGULATIONS

The Florida Department of Environmental Regulation issued the City of Boca Raton a permit to operate their wastewater reclamation and reuse facility.[13] This permit indicates the sampling, reporting, and effluent limitations for the various parameters (see Table 10.4).

In all cases, the City of Boca Raton complies with the parameters listed in their permit. See Table 10.7 for average, minimum, and maximum monthly values for the items listed in the permit for the year 1992, as well as the average monthly values up to August 1993.

Not only does the City comply with all the parameters in the permit but they also have checked for purgeable organics — all of which were at concentrations below the U.S. Drinking Water Standards maximum contaminant levels (MCLs) (see Table 10.8).[13]

The City also checked the reclaimed wastewater for at least 7 physical properties, 18 metals, 15 inorganic non-metallic constituents, 63 individual organic compounds, 10 pesticides, 3 herbicides, as well as total coliform and fecal coliform.[13] Only the color parameter exceeded the U.S. Drinking Water Standards.

III. SUMMARY AND CONCLUSIONS

Even though Florida is a water-rich state with an average rainfall of between 130 to 150 cm/yr (50 to 60 in./yr) and with numerous freshwater lakes and large aquifers, both shallow and deep, still many areas in the state are at or near using their total available freshwater supplies.

One of the major problems for Florida is the phenomenal population growth of 1000 persons a day plus the 40 million vacationers that visit the state every year. The domestic water consumption on a per capita basis is almost twice that of the national average, and a well-to-do city such as Boca Raton has, in the past, consumed domestically more than 1.5 m^{3}pcd (400 gpcd).

The recent two-year drought in Florida extending into 1991 resulted in water restrictions to reduce demand and conserve the available potable water supply. The administration for the City of Boca Raton decided it was time to reduce its potable water consumption and replace it with reclaimed wastewater that could be used for nonpotable purposes such as landscape irrigation. The city has now embarked on a $40 million 10-year program that is to be carried out in four phases. The first phase is to be completed by April 1994. This first phase of the project provides for a reclaimed water capacity of 0.034 Mm^3/d (9 mgd) and costs around $10 million dollars.

The City of Boca Raton's water reclamation facility, when completed, will consist of chemical feed and storage equipment, a flash mixer, two automatic backwash filters, six Dynasand filters, chlorine storage and feed equipment, two chlorine contact chambers, two reuse storage tanks, and transfer pumps. The final phase of this project will be completed in the year 2001 and have a capacity of 0.056 Mm^3/d (15 mgd).

The City of Boca Raton has compiled with all of the items in their wastewater reuse permit and has also evaluated at least another 118 water quality parameters. While this water is clearly not to be used for drinking purposes, it still meets all the U.S. Drinking Water Standards, except for color.

This urban wastewater reclamation and reuse project is beneficial in many ways, such as:[2,8,12]

- Reduces the demand for valuable groundwater which is suitable for drinking water purposes.
- Eliminates surface water discharges to the Atlantic Ocean.
- Saves approximately $7.7 to $8.7 million investment in development of water supply wells and expansion of the Boca Raton water treatment facility.
- Contributes to groundwater recharge.
- Allows for multiple use of the reclaimed water.
- Provides aesthetic value by contributing to keeping the landscape fresh and green.

CASE STUDY II
THE LOGAN ALUMINUM WETLANDS PROJECT

I. INTRODUCTION

Kentucky, like Florida, is a water-rich state with abundant water resources, however, in Kentucky, availability is variable with periodic episodes of drought as well as flooding.[1]

The water in Kentucky comes from some 143,200 km (89,000 mi) of rivers and streams, 159 public lakes, 19 reservoirs, and an estimated 75,000 public lakes larger than 0.10 hectares (ha) (0.25 acre, ac), with ground water supplying about 20% of Kentucky's drinking water needs.[1]

The rainfall in Kentucky varies greatly, however, climatic data for Russellville, located in Logan County, which is very close to the Logan Aluminum plant, indicated a precipitation of 134.4 cm (52.91 in.) for 1990 and 122.4 cm (48.2 in.) for 1991.

Kentucky's population of around 3.6 to 3.7 million people withdraw 15.14×10^6 m^3 (4 billion gallons, gal) of water every day. This indicates a total water consumption of 4.09 m^3 per person per day (1081 gallons per capita per day, gpcd) based on a 3.7-million population. About 10% of this water is used for drinking water purposes, 80% for agriculture, power generation, and oil and gas production, and the balance for commercial purposes.[1] Of the 15.14×10^6 m^3 (4 billion gal) of water withdrawn daily, nearly

95% comes from surface sources and 5% from groundwater sources. Much of the water withdrawn from streams and rivers is used by the state's 16 hydroelectric generation plants or for cooling purposes at the state's 58 coal-fired power plant units and then returned to the stream or river.[1] While Kentucky does have abundant water resources, periodically about 10% of the state's population is vulnerable to water supply shortages due to drought.[1]

In general, Kentucky's climate can be described as moderate with plentiful rainfall, mostly between March and June. Snowfall averages from 25 to 51 cm (10 to 20 in.) a year but seldom stays on the ground more than a few days. The growing season varies from 180 days in the northeast to 210 days in the lower Mississippi floodplain. The mean temperature for Louisville, the largest city in Kentucky, is 1°C (33°F) in January and 25°C (77°F) in July.[3]

II. WASTEWATER TREATMENT AND REUSE IN KENTUCKY

In 1991, there were an estimated 3969 municipal, industrial, and small package wastewater treatment plants operating in Kentucky. This included 258 municipal wastewater treatment plants (MWWTPs), 1936 package plants, 928 industrial treatment facilities, the remaining facilities being small plants providing a limited degree of treatment. Most of the wastewater in Kentucky is treated by 63 major and 173 minor municipal wastewater treatment plants.[1] The minor municipal wastewater treatment plants process less than 3785 m^3/d (1 mgd/d) while the 1936 package plants treat mainly domestic wastewater from small residential communities, and have discharges up to 37.85 m^3/d (0.01 mgd) for each plant.[1] Up to 1991, the Kentucky Department of Environmental Protection had approved 72 pretreatment programs for the municipal wastewater treatment plants. These programs required approximately 520 industries to pretreat their wastewater before discharging it to the municipal wastewater treatment facilities.[1]

Kentucky, with its abundant water resources, has not had to emphasize wastewater reclamation and reuse in order to conserve the potable water supplies. As of 1993, Kentucky had 112 small reuse wastewater treatment facilities. The total volume of the wastewater treated at these facilities for reuse was 10,750 m^3/d (2.84 mgd) — 105 of these reuse facilities were for slow-rate agricultural irrigation, 2 were for slow-rate forest irrigation, 4 were for slow-rate golf course irrigation, and 1 was for subsurface injection.[4]

Kentucky has no guidelines, criteria, or other regulations regarding wastewater reuse projects but permits them on a case-by-case basis.

In general, in Kentucky the wastewater reuse systems are small in size, and here the main driving force for wastewater reuse is due to environmental concerns and regulations.

III. WETLANDS

In order to better understand the Logan Aluminum Inc. case study dealing with the development of their wetlands system, it is first necessary to become familiar with the definitions, design, operation, and the expected performance for such natural treatment systems.

At present, there is no clear definition of wetlands. In 1989, four government agencies had jurisdiction over wetland development: the Fish and Wildlife Service (FWS), the Army Corps of Engineers (COE), the Environmental Protection Agency (EPA), and the Department of Agriculture (USDA). These agencies in 1989 issued a manual defining a wetland as any depression where water accumulates for 7 consecutive days during the growing season, where certain water-loving plants are found, and where the soil is saturated enough with water that anaerobic bacterial activity can take place.[5] After considerable controversy, the federal administration designated wetlands as areas having 15 consecutive days of inundation during a growing season or 21 days in which the soil is saturated with water up to the surface. The growing season for this definition was shorter and the variety of plants that qualified an area as a wetlands were reduced.[5]

Section 404 of the Clean Water Act (CWA) is still the principal vehicle used for regulating wetland use. Congress has not substantively amended this provision since 1977.[6]

At present, there are three major bills pending regarding the wetlands issue: two in the Congress and one in the Senate. One congressional bill is supported mainly by industry while the other bill is supported mainly by the environmentalists. The bill in the Senate attempts to strike a compromise between the industrial group and the environmentalists' position.[6]

At the federal level, the departments and independent agencies involved in policies and programs dealing with the wetlands include:[7]

- U.S. Department of the Interior (DOI)
  - Fish and Wildlife Service (FWS)
  - Bureau of Land Management (BLM)
  - National Park Service
- The Army Corps of Engineers (COE)
- The Environmental Protection Agency (EPA)
- U.S. Department of Agriculture (USDA)
  - National Forest Service
  - Soil Conservation Service
- U.S. Department of Commerce (DOC)
  - National Oceanic and Atmospheric Administration (NOAA)
- The Federal Coordinating Committee on Science Engineering and Technology (FCCSET)

The Forest Service in the USDA has published guidelines for restoring riparian areas and wetlands throughout the National Forest System. Not only is the Forest Service involved in such a program but so is the Bureau of Land Management (BLM) in the DOI. Figure 10.5 presents an explanation as to what constitutes a riparian area and the benefits that are involved.

The benefits provided by a healthy riparian area are also the same benefits that can be gained by having healthy wetlands.

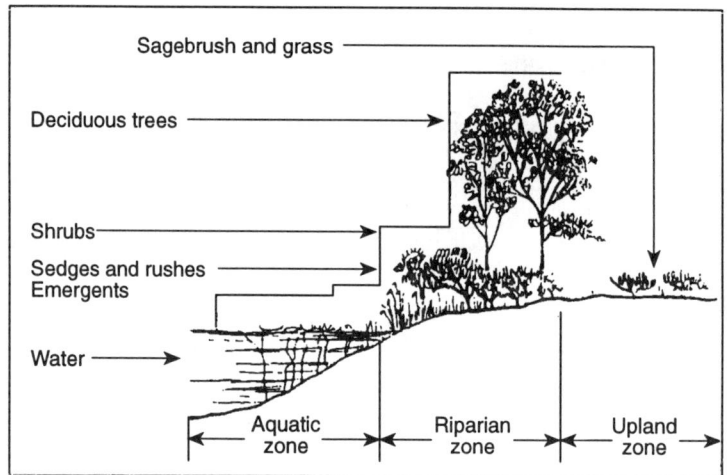

Figure 10.5 What is a riparian area?[7]

The green area immediately adjacent to streams, rivers, and lakes — known as a riparian area — is identified by the presence of vegetation that requires large amounts of free or unbound water.
Healthy riparian areas can provide the following benefits

- Produce more forage than uplands, resulting in higher livestock-weaning rates.
- Shelter livestock during weather extremes.
- Reduce flood velocities and bank erosion, minimizing property loss.
- Stabilize streambanks with dense vegetation that reduces damage from animal trampling, ice scouring, and erosive flood waters.
- Increase late summer streamflows for irrigation, stockwater, and fisheries by recharging underground aquifers and providing bank water storage.
- Filter sediment, protecting water quality, prolonging irrigation pump life, and reducing siltation of ponds and irrigation ditches.
- Improve wildlife habitat by providing food, water, and cover.
- Improve fisheries by providing food, cool water, and cover.
- Provide recreation sites for picnics and camping.

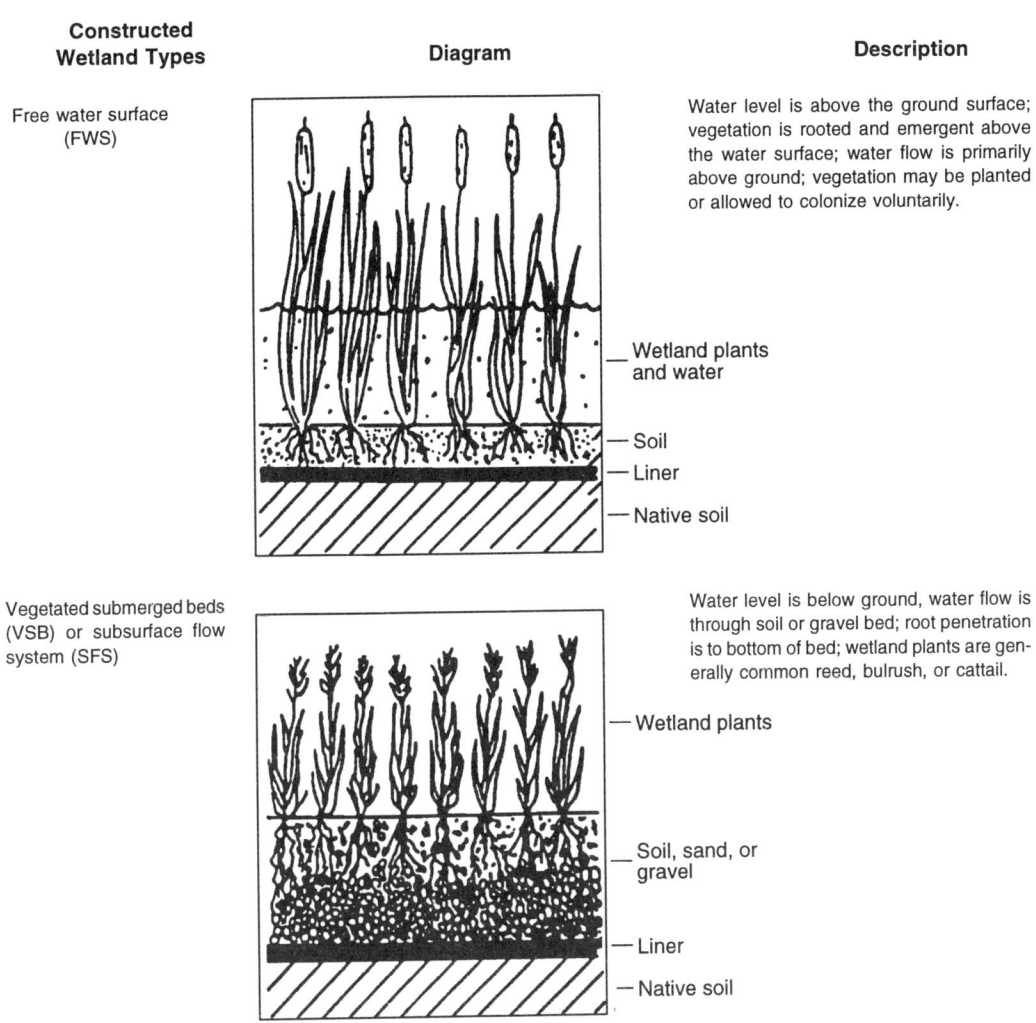

Figure 10.6 Classification of constructed wetland types commonly used for wastewater management.[8]

A. WETLAND CLASSIFICATION

There are two basic types of wetlands — natural and constructed. Natural wetlands occur throughout the landscape as transitional areas between aquatic ecosystems and uplands (savanna, pampas, champaign, or moors), while man-made (constructed wetlands) are planned, designed, and constructed.[8]

Natural wetlands include a variety of plants such as cattails (*Typha*), rushes (*Scirpus*), and reeds (*Phragmites*). They can also contain some floating plants such as water hyacinth (*Eichhornia crassipes*), duck weeds (*Lemna* and *Spirodela* spp.), pennywort (*Hydrocotyle umbellata*), and water ferns. The submerged plants found in natural wetlands include aquatic plants such as waterweed, water mil foil, and water cress. Willows, cypress, and ash trees tend to be found in swamps.[8–10]

Constructed wetlands can take many forms. Most constructed wetlands contain predominately herbaceous plants (seeded plants that wither away to ground level after each growing season, as distinguished from a tree or shrub). Constructed wetlands are generally divided into two categories: free water surface (FWS) wetlands, in which the majority of the water flows over the sediment and through the above-ground plant zone; and the vegetated submerged bed (VSB) wetlands, designed to conduct water through the bed of the system to make contact with the plant roots.[8] The VSB type of constructed wetland system is also often referred to as a subsurface flow system (SFS).[9]

Figure 10.6 illustrates the general difference between the two types of constructed wetlands.[3]

Table 10.9 Typical Removal Efficiency of Wetland Systems

	Removal (%)	
Parameter	Natural wetland (secondary effluent)	Constructed wetland (primary effluent)
BOD_5	70–96	50–90
SS	60–90	—
N	40–90	30–98
P	10–50	20–90

Compiled from References 9 and 12.

Table 10.10 Summary of Nutrient Removal from Natural Wetlands

			Percent Reduction			
Project	Flow, m³/d	Wetland type	TDP[a]	NH_3-N	NO_3-N	TN[b]
Brillion Marsh, WI	757	Marsh	13	—	51	—
Houghton Lake, MI	379	Peatland	95	71	99[c]	—
Wildwood, FL	946	Swamp/marsh	98	—	—	90
Concord, MA	2,309	Marsh	47	58	20	—
Bellaire, MI	1,136[d]	Peatland	88	—	—	84
Coots Paradise, Town of Dundas, Ontario, Canada	—	Marsh	80	—	—	60–70
Whitney Mobile Park, Home Park, FL	≈227	Cypress dome	91	—	—	89

[a] Total dissolved phosphorus.
[b] Total nitrogen.
[c] Nitrate and nitrite.
[d] May–November only.

Compiled from Reference 9.

Although natural and constructed wetlands differ in their physical, chemical, and biological characteristics, their function as to BOD_5, NH_3-N, fecal coliform, NO_3-N and phosphorus removals are quite similar.

B. REMOVAL OF NUTRIENTS AND POLLUTANTS WITH WETLAND SYSTEMS

Organic removal (BOD_5) has been indicated to occur as a result of aerobic, facultative, or anaerobic bacterial activity. Nitrogen removal is apparently by a combination of plant uptake and biochemical transformations. Phosphorus removal has been linked to precipitation and microbial immobilization mechanisms, while heavy metal removals have been indicated to occur as a result of precipitation-adsorption phenomena.[11]

Table 10.9 presents a summary of the percent removal of nutrients and pollutants in natural and constructed wetland systems.[9,12] There is a wide range in the percent removal of nutrients and pollutants and this would indicate large variations in climatic conditions such as temperature and sunlight, as well as large variations in physical conditions such as water surface area, water depth, and the type of vegetation present.[9,12]

Table 10.10 presents a summary of nutrient removals for several specific natural wetlands systems.[9]
Table 10.11 presents a summary of nutrient removals for specifically constructed wetland systems.[9]

The performance data for natural and constructed wetlands treating mainly secondarily treated effluents look good. However, a different picture emerges when an evaluation is made of wetlands receiving a raw, strong, highly polluted wastewater from an animal production facility. Table 10.12 presents Kentucky Pollutant Discharge Elimination System (KPDES) requirements, recorded constructed wetland performance (removal rates), loading rates that have been determined by actual field tests, as well as the expected quality of the effluent.

All of the parameters evaluated in Table 10.12 indicate that the quality of the expected effluent will be in violation of the KPDES requirements.[13]

Table 10.11 Summary of Nutrient Removal from Constructed Wetlands

Project	Flow, m³/d	Wetland type	BOD$_5$, mg/L Influent	BOD$_5$, mg/L Effluent	SS, mg/L Influent	SS, mg/L Effluent	Percent reduction BOD$_5$	Percent reduction SS	Hydraulic surface Loading rate, (m³/ha-d)
Listowel, Ontario (12)	17	FWS[a]	56	10	111	8	82	93	—
Santee, CA (10)	—	SFS[b]	118	30	57	5.5	75	90	—
Sidney, Australia (13)	240	SFS	33	4.6	57	4.5	86	92	—
Arcata, CA	11,350	FWS	36	13	43	31	64	28	907
Emmitsburg, MD	132	SFS	62	18	30	8.3	71	73	1,543
Gustine, CA	3,785	FWS	150	24	140	19	84	86	412

[a] Free water surface system.
[b] Subsurface flow system.
Compiled from Reference 9.

Table 10.12 Comparison of Kentucky Pollutant Discharge Elimination System (KPDES) Permit Values, Recorded Constructed Wetland Performance Values, Projected Agricultural Loading Rates, and Expected Performance

	Bowling Green Regional Office — August 1992			
Parameter	KPDES	Recorded wetlands performance (for Kentucky)	Loadings	Status (effluent)
BOD_5 mg/L	25 mg/L	49 to 96% (Av. Rer. 57%)[a]	2340 mg/L	1333 mg/L (NOV)[b]
TSS mg/L	30 mg/L	12 to 98% (Av. Rer. 77%)	1588 mg/L	365 mg/L (NOV)
NH_3-N mg/L	4–10 mg/L[c]	81 to 94% (Av. Inc. 33%)[d]	2000 mg/L	2660 mg/L (NOV)
Fecal coliform (#/100 mL)	200/100 mL[c]	82 to 100% (94% removal with chlorination)	$>1 \times 10^6$/100 mL	>60,000 /100 mL (NOV)
Metals[a] (site specific)		No reduction noted	9.8 mg/L copper (Pig manure)	9.8 mg/L copper (NOV)

[a] Average removal rate.
[b] NOV = notice of violation for KPDES.
[c] Seasonal summer/winter.
[d] Average increase.
[e] Metals may be required for certain animal metabolisms.

Compiled from Reference 13.

IV. WETLANDS IN KENTUCKY

A. NATURAL WETLANDS

In the 1780s, Kentucky had an estimated 635,400 ha (1.57 ac) of natural wetlands. By the 1980s, approximately 121,400 ha (300,000 ac) remained. This represents an 81% loss of natural wetlands.[7] The greatest losses occurred in western Kentucky where 52% of the state's bottomland forests were cleared between 1957 and 1974. The U.S. Fish and Wildlife Service (FWS) indicates that an estimated 1460 ha (3600 ac) of wetlands continue to be lost every year in Kentucky.[1]

Most of Kentucky's remaining wetlands are privately owned. Only 10,520 ha (26,000 ac) are owned and managed by public agencies. The majority of wetlands are located in the Jackson Purchase and Western Coalfield regions, although all of the state's 120 countries have wetlands located within their borders to some degree.

A statewide project to identify and map wetlands is currently underway. Identification of wetlands through aerial photography and interpretation, and a review of soil maps is being conducted under the National Wetland Inventory (NWI) by the U.S. Fish and Wildlife Service.[1]

B. CONSTRUCTED WETLANDS

The Tennessee Valley Authority (TVA), in cooperation with the National Wildlife Federation, the Kentucky Division of Water, and the U.S. Environmental Protection Agency, implemented demonstration projects to investigate and promote the feasibility and benefits of using constructed wetlands for treating domestic wastewater. Using full-scale wetland treatment systems at three small towns in western Kentucky, the marsh-pond-meadow, surface flow marsh, and gravel subsurface flow marsh design concepts were evaluated using various plant species, slopes, loading rates, and substrates.[14] The constructed wetlands demonstration projects were located in the western Kentucky cities of Benton, Hardin, and Pembroke.

The objectives for these demonstration projects were as follows:[14]

- Evaluate the relative advantages/disadvantages of three types of constructed wetlands.
- Determine ability to comply with permit parameters.
- Evaluate basic design and operation factors and develop criteria.
- Evaluate cost effectiveness.
- Promote technology and transfer to users and regulators.

Table 10.13 presents a summary of key design factors for all three of these projects, as well as the construction costs.[14]

Table 10.13 Summary of Project Site Design and Cost Factors

Factor	Benton	Hardin	Pembroke
Wetlands type	Gravel marsh and surface flow marsh	Gravel marsh	Marsh-pond-meadow
Application rate	92 ha/m³/s (10 acres/mgd)	146 ha/m³/s (16 acres/mgd)	379 ha/m³/s (41 acres/mgd)
Population served	4200	545	1100
Design flow	0.048 m³/s (1,100,000 gpd)	0.0044 m³/s (100,000 gpd)	0.0039 m³/s (90,000 gpd)
Influent	Primary lagoon effluent	Comminuted, aerated	Comminuted, aerated
Marsh vegetation	*Scirpus*, other	*Phragmites, Scirpus*	*Typha, Scirpus, Phragmites*
Slope	0.1%	0.2% Bottom 0.0% Top	0.1%
Substrate	Gravel, native soil	Limestone, river rock	Native soil, limestone
Liner	Native impermeable soil	Compacted clay	Native impermeable soil
Cost			
Construction[a]			
Wetlands	$250,000	$234,000	$149,000
Hardwood storage	—	$92,000	—
Land	—	$10,000	$35,000
Engineering	$10,000	$30,000	$28,000
Total capital	$260,000	$366,000	$212,000
Cost per GPD			
Construction[a]	$0.23	$2.34	$1.66
Total capital	$0.24	$3.66	$2.36

[a] Includes demonstration specific design requirements and monitoring equipment.

Compiled from Reference 14.

Table 10.14 Benton, KY Constructed Wetland Sewage Treatment System, Gravel and Surface Flow Marshes

Design Flow:	0.048 m³/s (1,000,000 gpd) average
Pretreatment:	Sedimentation and biochemical oxidation in 6.5 ha (16-acre) primary lagoon
Marsh:	Secondary lagoon converted to three parallel cells, 1, 2, and 3
Application rate:	292 ha/m³/s (10 acres/mgd)
Total surface area:	4.4 ha (10.8 acres), 3 cells, equal size
Length/width ratio:	7.6:1 each cell
Slope:	0.1%
Liner:	3.0 to 4.6 m (10–15 ft) *in situ* impermeable clay

	Cell 1	Cell 2	Cell 3
Vegetation:	Cattail, iris, sweetflag, and arrowhead	Woolgrass bulrush	Softstem bulrush
Water depth:	0 to 0.5 m (0 to 1.5 ft)	0 to 0.5 m (0 to 1.5 ft)	−0.6 to +0.5 m (−2.0 to +1.5 ft)
Substrate:	Native soil	Native soil	46 cm (18 in.) base of 1.9–2.5 cm (¾″–1″) limestone, and 15 cm (6 in.) cap of ⅜″–¾″ limestone

Posttreatment: Chlorination

Compiled from Reference 15.

An abbreviated description of the Benton site is included here, but for full details see References 14 and 15.

The Benton site has a constructed wetlands with a capacity of 3785 m³/d (1 mgd). The population served is about 5000. The constructed wetlands were designed to upgrade and polish the effluent from the existing lagoon. The original two-cell (series) 10.5 ha (26 ac.) lagoon system had frequently been hydraulically overloaded and the National Pollutant Discharge Elimination System (NPDES) permit

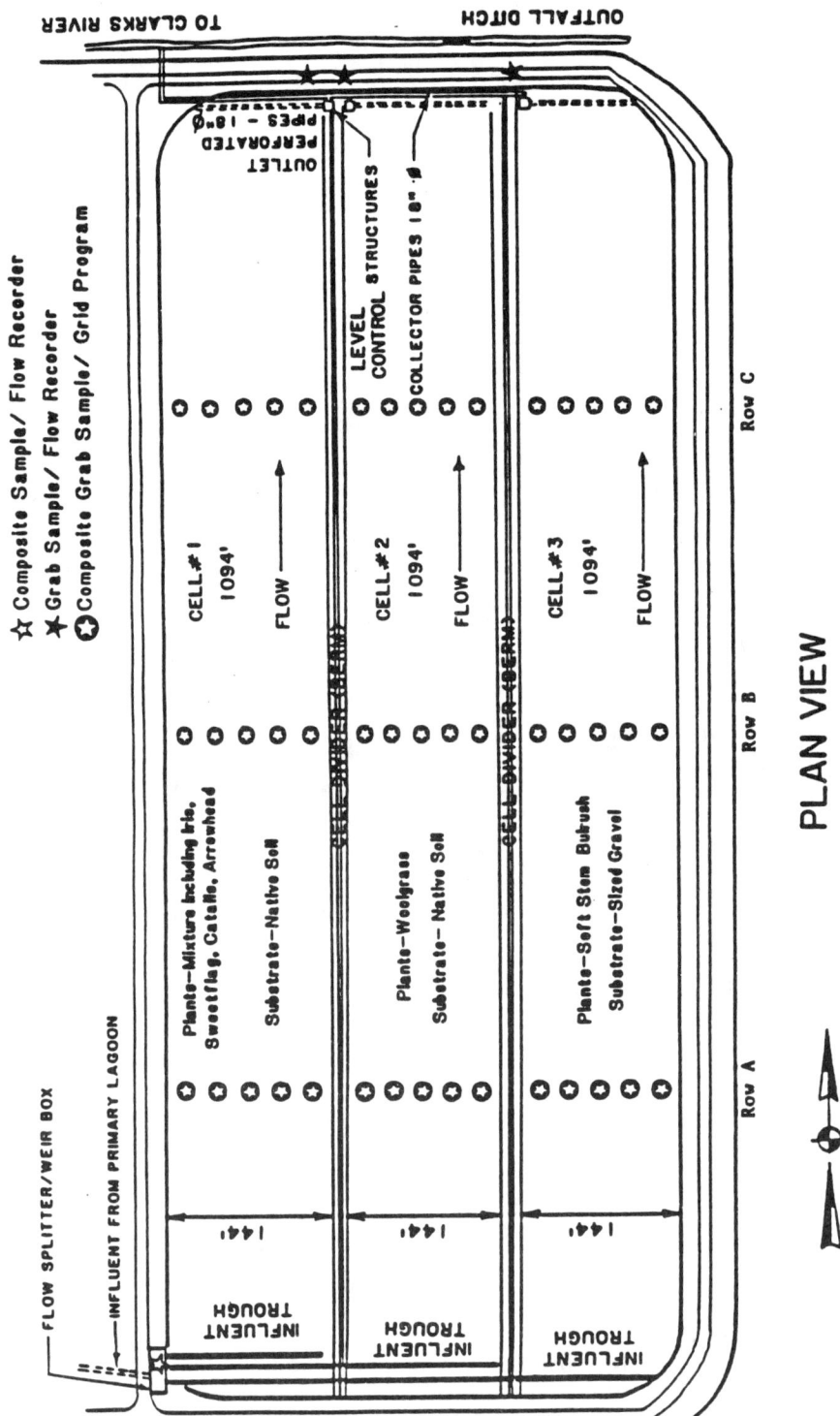

Figure 10.7　Benton, KY constructed wetlands treatment system, gravel and surface flow marshes.[14,15]

Table 10.15 Benton, KY Constructed Wetlands NPDES Effluent Samples, January–December 1987

Parameter	Minimum	Average[a]	Maximum	NPDES limits[a]
BOD, mg/L	1.0	5/10	14	25/25
TSS, mg/L	1.0	12/27	42	30/30
NH_3-N, mg/L	1.5	9/9	16	4/10
E. coli, No./100 mL	<1.0	25/20[b]	>5000	200/200[b]
DO, mg/L	2.5	6/6	9.0	7/7

[a] Summer/winter: May–October/November–April.
[b] Geometric mean.

Compiled from Reference 14.

limits for various parameters occasionally exceeded. The 4-ha (10 ac) secondary lagoon was modified into a three-cell constructed wetland, designed to receive effluent from the primary 6.5-ha (16 ac) lagoon. The design is summarized in Table 10.14 and illustrated in Figure 10.7. The three cells are equally sized, but cell 3 receives 50% of the total flow, and cells 1 and 2 each receive 25% of the flow.[14,15]

The wetlands treatment system began operating in January 1987, with all the flow from the lagoon directed to cell 3. The NPDES data for effluent samples collected by the city is summarized in Table 10.15.[14]

Cells 1 and 2 began operation in January 1988, with the demonstration monitoring program beginning in March 1988. The wetlands vegetation was planted in stages from the fall of 1986 to the spring of 1988.

Table 10.16 presents the median, average, maximum and minimum values for the various parameters monitored and the average percentage concentration reduction for cells 1, 2, and 3 for the period from March to December 1988.[15]

The following conclusions regarding the Benton constructed wetland operation are taken directly from Reference 15.

The initial demonstration monthly monitoring data and dye study results for the Benton constructed wetland resulted in the following conclusions:[15]

- The summer dye study revealed that the three cells experienced substantial short circuiting. Most short circuiting occurred in the first half of each cell in the portion of the cross section (west) containing the deepest water.
- All three cells effectively reduced BOD_5. The gravel cell (cell 3) was the most effective. Effluent concentrations never exceeded the NPDES permit limit of 25 mg/L (monthly average).
- Settling and filtering are principal removal mechanisms and the reductions occur relatively quickly within each cell.
- Microbial metabolism is not as prevalent in the wetland cells during the summer, probably because of the low concentrations of soluble BOD_5 in the lagoon effluent (wetland influent).
- Each cell effectively reduced TSS. Effluent TSS concentrations exceeded the monthly average NPDES permit limit of 30 mg/L during 3 months for cell 1 and 1 month for cell 2, but these occurred when vegetation was still sparse and water levels were not stabilized.
- Limited nitrification was occurring in each cell during most of the period because of the limited DO availability. Winter total nitrogen reductions and DO effluent concentrations were higher than summer total nitrogen reductions and DO effluent concentrations.
- Fecal coliform bacteria were effectively reduced in all three cells.
- Phosphorus reductions were small.

V. DESIGN CONSIDERATIONS FOR CONSTRUCTED WETLANDS

A. CONSTRUCTED WETLANDS CLASSIFICATION

There are two broad classifications for constructed wetlands: they are either a free water surface system (FWS) with shallow water depths, or a subsurface flow systems (SFS) which is also often referred to as a vegetated submerged bed system (VSB).[8,9] Only the design procedure for the FWS will be presented here.

A FWS system typically consists of basins or channels with a natural or constructed subsurface barrier of clay or impervious geotechnical material to prevent seepage, soil, or another suitable medium to

Table 10.16 Monthly Effluent Monitoring Results for Cells 1, 2, And 3, Benton, KY

Date	Substrate[b] temp. (°C)	Water temp. (°C)	Flow (mgd)	pH (s.w.)	DO (mg/l)	Alkalinity as CaCO$_3$ (mg/l)	Fecal coliforms (No./100 ml)	TSS (mg/l)	Total BOD$_5$ (mg/l)	Dissolved BOD$_5$ (mg/l)	Total P (mg/l)	Dissolved P (mg/l)	NH$_3$-N (mg/l)	NO$_3$-NO$_2$ (mg/l)	Organic N (mg/l)	TKN (mg/l)
						Cell 1										
Median[a]	15.5	16.9	0.1421	6.8	0.4	119	273	5	8	4	5.1	4.0	7.7	0.05	2.2	10.0
Average	15.0	15.1	0.1334	6.8	1.5	119		14	10	5	5.1	4.3	7.6	0.08	2.8	10.4
Maximum	22.8	21.9	0.3136	7.1	7.9	154	1,600	44	20	10	7.9	7.9	12.5	0.29	7.6	15.0
Minimum	1.1	3.4	0.0631	6.1	0.2	85	40	1	4	2	2.8	2.0	3.1	0.01	0.7	6.0
Removal %					84	−16	93	76	60	34	10	−2	−25	56	68	29
						Cell 2										
Median[a]	17.7	20.7	0.1406	7.0	3.3	123	29	11	9	5	4.4	4.3	5.3	0.17	2.4	7.9
Average	13.7	17.0	0.1569	7.1	3.6	125		18	11	5	4.7	4.3	5.6	0.25	3.4	9.0
Maximum	23.6	24.1	0.348	8.3	11.5	230	310	53	19	11	6.6	5.8	9.0	0.76	7.2	15.6
Minimum	2.8	4.5	0.0490	6.6	0.2	79	3	2	5	2	3.0	2.8	0.4	0.01	1.0	4.6
Removal %					62	−22	99	70	55	34	18	−2	7	−37	61	39
						Cell 3										
Median[a]	21.1	17.4	0.2088	7.1	0.6	174	57	3	5	3	4.9	5.3	10.7	0.05	1.8	15.0
Average	19.1	16.1	0.2656	7.0	1.1	69		6	8	3	4.9	4.1	10.4	0.80	1.9	12.3
Maximum	28.4	23.5	0.6476	7.4	4.1	233	360	29	72	9	7.3	6.1	15.0	2.60	4.2	16.8
Minimum	4.5	2.0	0.1183	6.3	0.1	91	10	1	2	1	2.5	0.01	2.4	0.01	0.3	3.4
Removal %					88	−65	98	90	68	61	14	2	−72	−340	78	16

[a] Geometric mean for fecal coliform.
[b] Substrate temperature measured within wetland cell.

Compiled from Reference 15.

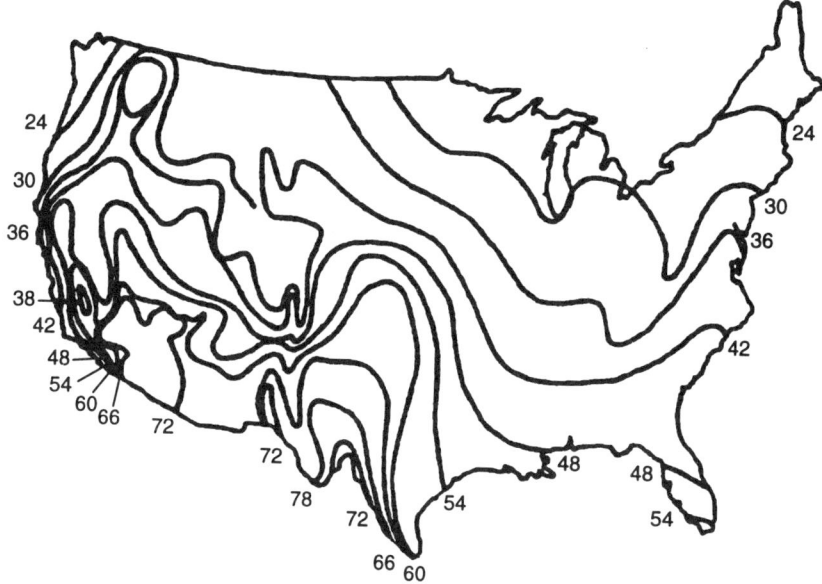

Figure 10.8 Map showing the approximate annual average lake evaporation rates (in inches) for the U.S.[8]

support the emergent vegetation, and which contain water at a relatively shallow depth flowing over the soil surface. The shallow water depth, low flow velocity, and presence of the plant stalks and litter regulate water flow and, especially in long, narrow channels, ensure plug-flow conditions.[9]

B. WATER BALANCE AND HYDROLOGIC FACTORS

A major factor in the design and performance of any constructed wetland system depends on several hydrologic factors such as precipitation, infiltration, evapotranspiration (ET), hydraulic loading rates, water depth, and hydraulic detention time.[8,9]

For a constructed wetland that has an impermeable barrier present, groundwater inflow and infiltration can be excluded.

A simple equation expressing the water balance for such a system is as follows:[9]

$$Q_i - Q_0 + P - ET = [dV/dt] \qquad [10.2.1]$$

where: Q_i = influent wastewater flow, volume/time
Q_0 = effluent wastewater flow, volume/time
P = precipitation, volume/time
ET = evapotranspiration, volume/time
V = volume of water
t = time

Information for precipitation and pan evaporation data are available for most areas in the U.S. from NOAA in Asheville, NC.[8] For instance, NOAA, as indicated earlier, reported that precipitation for Russellville, KY was 134.4 cm (52.91 in.) for 1990 and 122.4 cm (48.2 in.) for 1991.[2]

Figure 10.8 presents a general diagram of lake evaporation rates for the U.S. For wetlands that are continuously inundated, ET can generally be estimated as being equal to lake evaporation, or approximately 70 to 80% of pan evaporation values.[8]

If the wetland system operates at a relatively constant water depth

$$\left(\frac{dv}{dt} = 0\right)$$

then the effluent flow rate can be estimated by using Equation 10.2.1.

Hydraulic loading rates (HLR) vary for natural and constructed wetlands and of course vary with climate. To minimize vegetative changes and maximize treatment efficiencies, HLRs in natural wetlands should generally not exceed 1 to 2 cm/d. The maximum acceptable HLRs for constructed wetlands should generally not exceed 2.5 to 5 cm/d for FWS wetlands and 6 to 8 cm/d for VSB wetlands.[8]

Water depths in constructed wetlands (FWS) are related to the type of plants growing in the system. Water depths can vary from 0.15 to 1.0 m. Normal wetland design water depths are generally less than or equal to 0.5 m.[8]

The hydraulic residence time (t) for FWS constructed wetlands depends on the tolerance of the plant species in the system and, as mentioned earlier, the water depth.

The hydraulic residence or detention time can be calculated by dividing the wetland water volume by the average flow rate.

$$t = V/Q \qquad [10.2.2]$$

where: t = detention time, d
V = wetlands water volume, m^3
Q = average daily flow, m^3/d

The wetlands water volume is equal to its surface area times the water depth:

$$V = A_r d \qquad [10.2.3]$$

where: V = water volume, m^3
A_r = wetlands area, m^2
d = water depth, m

In a FWS wetland, a portion of the available volume is occupied by vegetation, so the actual detention time is a function of voids or porosity (e). Taking into account the various factors involved, the detention time can be calculated from the following equation:[8]

$$t = \frac{A_r d e}{Q} \qquad [10.2.4]$$

where: t = hydraulic residence time, days
A_r = wetland area, m^2
d = water depth, m
e = porosity or void fraction (as a decimal fraction, usually 0.75)
Q = flow rate, m^3/d

C. ORGANIC LOADING RATES

Organic loading rates (BOD$_5$) for FWS can vary from 1 kg of BOD$_5$/ha/d to 125 kg of BOD$_5$/ha/d and have removal rates that range from 49 to 95%.[8] A maximum BOD$_5$ loading rate of about 100 kg/ha/d has been recommended to help prevent the occurrence of mosquito populations.[8] An upper limit of about 110 kg of BOD$_5$/ha/d has been suggested by some authors.[10] Other authors have, however, recommended a maximum BOD$_5$ loading rate of 110 to 120 kg/ha/d for both FWS and VSB systems.[8]

D. NUTRIENT AND CONTAMINANT REMOVAL

While wetland systems exhibit seasonal variability they have achieved high removal rates for biochemical oxygen demand (BOD$_5$), suspended solids (SS), trace organics, nitrogen, and heavy metals. The reasons for these high treatment levels are not fully understood, however. Table 10.17 presents current thinking as to the removal mechanisms involved.[9,12]

The BOD$_5$ removal in a wetland has been described as a process that follows a first-order reaction. This, of course, is the same basic equation or model used to express BOD$_5$ degradation, the oxygen sag curve, and many other equations used to design wastewater treatment facilities.[9,16]

Table 10.17 Removal Mechanisms in Wetlands for the Contaminants in Wastewater

Mechanism	Contaminant effected[a]							Description
	Settleable solids	Colloidal solids	BOD	N	P	Heavy metals	Refractory organics	Bacteria and virus	
Physical									
Sedimentation	P[a]	S[a]	I[a]			I	I	I	Gravitational settling of solids (and constituent contaminants) in pond/marsh settings.
Filtration	S	S							Particulates filtered mechanically as water passes through substrate, root masses, or fish.
Adsorption		S							Interparticle attractive forces (van der Waals force).
Chemical									
Precipitation					P	P			Formation of or co-precipitation with insoluble compounds.
Adsorption					P	S			Adsorption on substrate and plant surfaces.
Decomposition						P		P	Decomposition or alteration of less stable compounds by phenomena such as UV irradiation, oxidation, and reduction.
Biological									
Bacterial metabolism[b]		P	P	P			P		Removal of colloidal solids and soluble organics by suspended, benthic, and plant-supported bacteria. Bacterial nitrification/denitrification.
Plant metabolism[b]				S			S	S	Uptake and metabolism of organics by plants. Root excretions may be toxic to organisms of enteric origin.
Plant adsorption				S	S	S	S		Under proper conditions, significant quantities of these contaminants will be taken up by plants.
Natural die-off								P	Natural decay of organisms in an unfavorable environment.

[a] P = primary effect; S = secondary effect; I = incremental effect (effect occurring incidental to removal of another contaminant).
[b] The term metabolism includes both biosynthesis and catabolic reactions.

Compiled from References 9 and 12.

$$[C_e/C_0] = \exp(-K_T t) \qquad [10.2.5]$$

where: C_e = effluent BOD$_5$, mg/L
C_0 = influent BOD$_5$, mg/L
K_T = temperature-dependent first-order reaction rate constant, d^{-1}
t = hydraulic residence time, d

The value of K_T for any given organic compound is temperature dependent. Because microorganisms are more active at higher temperatures, the value of K_T increases with increasing temperatures. The change in K_T can be approximated by the van't Hoff-Arrhenius model:[16]

$$K_T = K_{20}\theta^{T-20°} \qquad [10.2.6]$$

The value of θ can vary from 1.056 to 1.135 depending upon the temperature. At 20°C a θ value of 1.1 is often used for constructed wetlands design.[9] Combining the relationships in Equation 10.2.4 and Equation 10.2.5, and adding coefficients that have been developed in order to better define the important characteristics for BOD$_5$ removal in wetlands, results in Equation 10.2.7:[8-10]

$$C_e/C_0 = A \exp\left[-0.7 K_T (A_V)^{1.75} (A_r\, d\, e)/Q\right] \qquad [10.2.7]$$

where: A = fraction of BOD$_5$ not removed as settleable solids near headworks of the system (as decimal fraction)
A_V = specific surface area for microbial activity, m²/m³
A_r = wetlands area, m²
d = design depth of system, m
e = porosity of system (as a decimal fraction) (0.75)
Q = average hydraulic loading on the system, m³/d

Coefficients that have been developed for use in Equation 10.2.7 are as follows:[9]

A = 0.52
K_{20} = 0.0057 d^{-1}
A_V = 15.7 m²/m³
e = 0.75

Using these coefficients and taking the natural logarithm of both sides of Equation 10.2.7, as well as being sure to consider the proper negative or positive signs, gives Equation 10.2.8:[8-10]

$$A_r = \frac{Q(\ln C_0 - \ln C_e - 0.6539)}{65 \times K_T \times d} \qquad [10.2.8]$$

Note: if:

$$A = 0.52, \ln 0.52 = -0.6539$$

and

$$\left[-0.7 K_T (15.7\, m^2/m^3)^{1.75} A_r d\, (0.75)\right] =$$

$$\left[-0.7(123.82)(0.75) K_T A_r d\right] = -65 K_T A_r d$$

where: A_r = wetland area, m²
Q = flow, m³/d
C_0 = influent BOD$_5$, mg/L

C_e = effluent BOD$_5$, mg/L
K_T = temperature-dependent first-order rate constant, d^{-1}
d = water depth, m

This equation allows for an estimate of the wetland area required to achieve a given reduction of BOD$_5$ in a FWS constructed wetlands system.[8]

By combining Equations 10.2.4 and 10.2.8, Equation 10.2.9 can be developed in order to calculate detention times:

$$t = \frac{(\ln C_0 - \ln C_e) - 0.6539}{65 K_T} \qquad [10.2.9]$$

Porosity or the void fraction have already been included in both Equation 10.2.4 and Equation 10.2.8 and therefore Equation 10.2.9 does not take porosity or void fractions into account.

Example 10.2.1.
Design a wetland system with a free water surface (FWS) to produce an effluent with a BOD$_5$ of 10 mg/L or less. The influent BOD$_5$ is 110 mg/L. The mean summer temperature is 20°C (68°F), and the mean winter temperature is 4.4°C (40°F). The design flow is 1363 m³/d, the depth of flow is 0.3 m (fairly constant), and the wetlands slope is greater than 1%. For this particular area the precipitation averages 114 cm (45 in.) per year and the evapotranspiration is also about 114 cm (45 in.) per year. Determine the wetland area requirement as well as the hydraulic and organic loading rates and check if these values are in the appropriate ranges.

Solution

1. Determine the temperature-dependent first-order rate constants for summer and winter conditions using Equation 10.2.6:

    $K_T = K_{20} \theta^{T-20°C}$
    $K_{20} = 0.0057$ d^{-1}
    K_T at 20°C (summer)
    $K_T = 0.0057 (1.1)^{(20-20)}$
    $= 0.0057$ d^{-1}
    $K_T = 4.4°C$ (winter)
    $K_T = 0.0057 (1.1)^{(4.4-20)}$
    $= 0.0057 (1.1)^{-15.6}$
    $= 0.00129$ d^{-1}

2. Determine the wetlands area required to reduce the BOD$_5$ from 110 to 10 mg/L. For summer conditions (20°C temperatures) use Equation 10.2.8:

    $$A_r = \frac{Q(\ln Co - \ln Co - 0.6539)}{65 \cdot K_R \cdot d} \qquad [10.2.8]$$

    $$A_r = \frac{1363 \, m^3/d (\ln 110 - \ln 10 - 0.6539)}{65(0.0057 \, days)(0.3 \, m)}$$

    $$A_r = \frac{1363(4.7005 - 2.3025 - 0.6539)}{65(0.0057)(0.3)}$$

    $= 21,387$ m²

    $= 2.138$ ha

    $= 5.28$ ac.

3. Determine the detention time, summertime conditions, using Equation 10.2.9.

$$t = \frac{(\ln C_0 - \ln C_e) - 0.6539}{65 K_T}$$

$$t = \frac{(\ln 110 - \ln 10 - 0.6539)}{65 K_T}$$

$$= \frac{(4.7005 - 2.3025 - 0.6539)}{65(0.0057)}$$

$$= 4.7 \text{ d (porosity not included)} \qquad [10.2.9]$$

4. Determine the detention time, summertime conditions, when porosity or void fraction is considered. Use Equation 10.2.4.

$$t = \frac{A_r d e}{Q} \qquad [10.2.4]$$

$$= \frac{21{,}387 \text{ m}^2 (0.3 \text{ m})(0.75)}{1363 \text{ m}^3/\text{d}}$$

$$= 3.5 \text{ d}$$

5. Determine wetland area requirement, wintertime conditions, with a 4.4°C temperature using Equation 10.2.8.

$$A_r = \frac{1363(4.7005 - 2.3025 - 0.6539)}{65(0.00129)(0.3)}$$

$$A_r = 94{,}500 \text{ m}^2$$

$$= 9.45 \text{ ha}$$

$$= 23.35 \text{ ac.}$$

6. Determine the detention time for wintertime conditions using equation 10.2.9.

$$t = \frac{\ln 110 - \ln 10 - 0.6539}{65(0.00129)}$$

$$= 20.8 \text{ d (porosity not included)}$$

7. Determine the detention time for wintertime conditions when porosity or void fraction is considered; use Equation 10.2.4.

$$t = \frac{94{,}500 \text{ m}^2 (0.3 \text{ m})(0.75)}{1363 \text{ m}^3/\text{d}}$$

$$= 15.6 \text{ days}$$

Winter conditions obviously determine the size and detention times for these constructed wetlands. Using a safety factor of 1.5 gives a required area of 141,750 m² (14.18 ha = 35ac) and a detention time of 23 d (Equation 10.2.4).

8. The water balance for Example 10.2.1 can easily be determined by using Equation 10.2.1, especially with the precipitation and evapotranspiration being equal.
Equation 10.2.1:

$$Q_i - Q_0 + P + E = dv/dt \quad [10.2.1]$$

In this case the influent water flow equals the effluent water flow.

9. Determine the hydraulic loading rate (HLR) using equation 10.2.10.

$$HLR = \frac{Q}{A_r} \quad [10.2.10]$$

where: HLR = hydraulic loading rate, m/d
Q = flow, m³/d
A_r = wetlands area, m²

Then:

$$HLR = \frac{1363 \text{ m}^3/\text{d}}{141,750 \text{ m}^2}$$

$$= 0.0096 \text{ m/d}$$

$$= 1 \text{ cm/d}$$

This HLR is well below the acceptable HLR for constructed wetland which is generally not to exceed 2.5 to 5 cm/d.

10. Determine the organic loading rate (BOD_5) for this constructed wetland system using Equation 10.2.11.

$$BOD_5 \text{ organic loading rate} = \frac{Q \cdot C}{A_r} \quad [10.2.11]$$

where: Q = flow, m³/d
C = influent BOD_5 concentration, mg/L
A_r = wetland area, ha

Then: BOD_5 organic loading rate $= \dfrac{1363 \text{ m}^3/\text{d} \times 110 \text{ mg/L}}{14.18 \text{ ha}}$

$$= \frac{1363 \text{ m}^3/\text{d} \times 1000 \text{ L/m}^3 \times 110 \text{ mg/L}}{14.18 \text{ ha}}$$

$$= \frac{149,930,000 \text{ mg/d}}{14.18 \text{ ha}}$$

$$= \frac{149.93 \text{ kg/d}}{14.18 \text{ ha}}$$

$$= 10.57 \text{ kg/ha/d}$$

This organic loading rate is well below the suggested loading rate of 100 kg/ha/d. The configuration of the constructed wetlands depends upon the topography of the site; however, optimum treatment is obtained by using large surface areas and shallow water flow depths.

VI. BACKGROUND AND DEVELOPMENT OF THE LOGAN ALUMINUM WETLANDS PROJECT

A. PLANT LOCATION

The Logan Aluminum plant is located in Logan county in the central-south portion of the Pennyroyal (pronounced Pennyrile) region of Kentucky, adjacent to the State of Tennessee. Logan county has an area of 1665 km² (643 mi.²) and a 1990 population of 24,416, which increased 2.1% from 1980 to 1990. Russellville, the county seat, had a 1990 population of 7454 and showed a 0.9% decline in population from 1980 to 1990. The county is drained by the Gaspar and Mud rivers. The terrain in the Russellville area is undulating to rolling and interspersed with many sinkholes. This type of landscape is referred to as typical Karst topography.[17]

B. THE LOGAN ALUMINUM PLANT

In January 1981, the Anaconda Aluminum Company, a division of Atlantic Richfield Company., announced that they would build an aluminum processing plant in Logan County at an estimated cost of $400 million.[18] An estimated 1500 construction workers were to be employed while the plant was being built, and at that time it was estimated that between 400 and 500 permanent workers would be employed. The contractor for the complex was Daniels Construction Company of Greenville, SC. The contractor expected to complete construction on this project by May 1983.[18]

Anaconda indicated that the plant would meet all air emission standards and that the wastewater would be recycled in a special treatment facility.[18] In 1985 Logan Aluminum, Inc. was created for the purpose of managing and operating this Russellville facility. Logan Aluminum, Inc. is currently owned jointly by the Atlantic Richfield Company (Arco Aluminum) and the Aluminum Company of Canada (Alcan) on a 60/40 basis, respectively.[19]

In 1991, Logan Aluminum, Inc. started a 5-year, $345 million expansion. The plant, in 1991, employed 750 people.[20]

C. LOGAN ALUMINUM WASTEWATER TREATMENT

In 1989, Logan Aluminum, Inc. decided to investigate technology that would treat its process water while meeting the following objectives:[21]

- Zero discharge
- Cost effective
- Total reuse capability
- Improved and consistent quality
- Efficient
- Innovative

When the plant was initially constructed during the early 1980s, separate water supply and wastewater treatment facilities were installed to treat the process wastewater from all its processes and, in particular, the five main systems and the wastes created by the water supply systems:[21]

- Melt and casting
- Hot mill
- Cold rolling
- Coating line
- Cooling system

Each of these systems present their own unique and special treatment requirements, due to the nature of the process and the contaminants involved. The contaminants of major concern coming from the various processes include oils, greases, ammonia, fluorides, phosphates, and aluminum.

After a careful evaluation of the various technological options and alternatives available to meet the objectives, as well as special treatment requirements, it was decided to proceed with the development of a wetlands system.[19,21]

D. WETLANDS CONSTRUCTION

The wetlands construction started in July 1991 and was completed in October 1991, after moving some 230,000 m³ (300,000 cubic yards) of earth on the 16-ha (40-ac) site. To minimize earth movement, cut and fill construction techniques were employed. Each basin was lined with a compressed clay liner of at least 1.2 m (4 ft) thick to ensure no infiltration or leakage.

At the exits of each basin, controlled metering boxes were installed so accurate flow measurements could be made from basin to basin. The water flows from basin to basin through specially designed spillways. The spillways are lined with stone and provide aeration for the flowing water.

The system now consists of:[21]

- An overland flow area composed of two (2) separate units at a 4% slope and occupying 3 ha (7.4 ac.)
- Nine (9) basins of 12 ha (30 ac.)
- Berms, islands, and a buffer zone of 4 ha (10 ac)

Initially, some 144,000 seedlings of 13 different wetland species were planted. It is anticipated that other local plant species will grow in this area and increase the diversity of the plant system.[22] Plant species indigenous to the area were chosen for the initial planting, with water depth and plant availability acting as determining factors. The overland flow area was planted in reed canary grass and prairie grass, other grasses are to be added to ensure a good vegetative cover to inhibit erosion. Wool grass, rushes, coon tail, and pond weed were planted in the lagoon areas; cattails, bulrush, and giant reed were planted in the intermediate zones; and deep water duck potato and hardstem bulrush were also planted in the wetlands. Flowering wetland plants such as water lilies and aquatic iris have been planted to add aesthetic value to the site.[23]

The probability of wetlands becoming a breeding ground for mosquitoes is fairly great, however, purple martin birdhouses have been installed to lure the mosquito predators to the area. Also, the abundant population of dragon flies at this site appears to be the main mosquito control mechanism.

E. TREATMENT PROCESS

The wastewater treatment process starts with the wastewater being given primary treatment through the existing treatment systems, which includes pre-aeration, prechlorination, as well as settling and chemical treatment. The wastewater, after receiving primary treatment, is directed to the holding ponds (spray irrigation ponds) and is then pumped at a rate of 1360 m^3/d (360,000 gpd) through spray heads which are at the beginning of the overland flow area. The wastewater then flows in a sheet over this area. During this phase of the operation, aeration takes place causing an increase in the oxygen content, as well as removing chlorine and ammonia. Recent designs combining overland flow with FWS in the same system have shown promise as being effective in ammonia removal.[24] The water is in contact with the soil, providing electrostatic bonding sites which facilitate ion exchange/nonspecific adsorption/precipitation and complexation.[21]

The water then travels through the wetlands at the rate of 1360 m^3/d (360,000 gpd). A minimum detention period of 20 days is provided. The wastewater first accumulates in basin number one and is then allowed to spill into the subsequent basins on its minimum 20-day journey to the final basin.[21]

As the wastewater flows through the nine basins, oxygen is added, especially at the spillway points. Gravity causes the suspended solids to settle out of the wastewater, where they are biologically converted to nutrients for the plants. Physical filtration of the water is facilitated through the mass and root structure of the plants. The various mechanisms for removal of nutrients and contaminants in the wetlands is presented in Table 10.17. On the average, evapotranspiration (ET) is about 454 m^3/d (120,000 gpd) from the plants and basins, leaving approximately 908 m^3/d (240,000 gpd) available for reuse in the plant.[21]

After the water has passed through the eight basins, it is collected at the ninth and final basin for pumping back to a 7570 m^3 (2.0 mil gal) storage tank for ultimate reuse in the facility's processes.[21]

Figure 10.9 is a flow diagram showing the water sources, the various plant processes, the holding pond (SI ponds), the wetlands, and the recycling circuit.[19,21]

VII. KPDES WETLAND PERMIT

A wetlands system is considered a wastewater treatment facility by the regulatory agencies and therefore a Kentucky Pollutant Discharge Elimination System (KPDES) permit for effluent disposal was required. Although the Logan Aluminum facility does have a KPDES permit to discharge its wastewater to a nearby stream, it does not do so, as they have a zero discharge policy. Consequently, all the wastewater is reclaimed and reused.

The KPDES discharge permit covers six outfalls, however, only the effluent limitations and monitoring requirements for the potential discharges from the wetlands to Austin Creek is included here. See Table 10.18 for details regarding this outfall.[25]

With the Logan Aluminum plant having a zero discharge policy, at no time would they have been in violation of their permit. However, if by chance they did have to discharge water from the wetlands, all the water quality parameters are at levels well below the values presented in their permit. Table 10.19

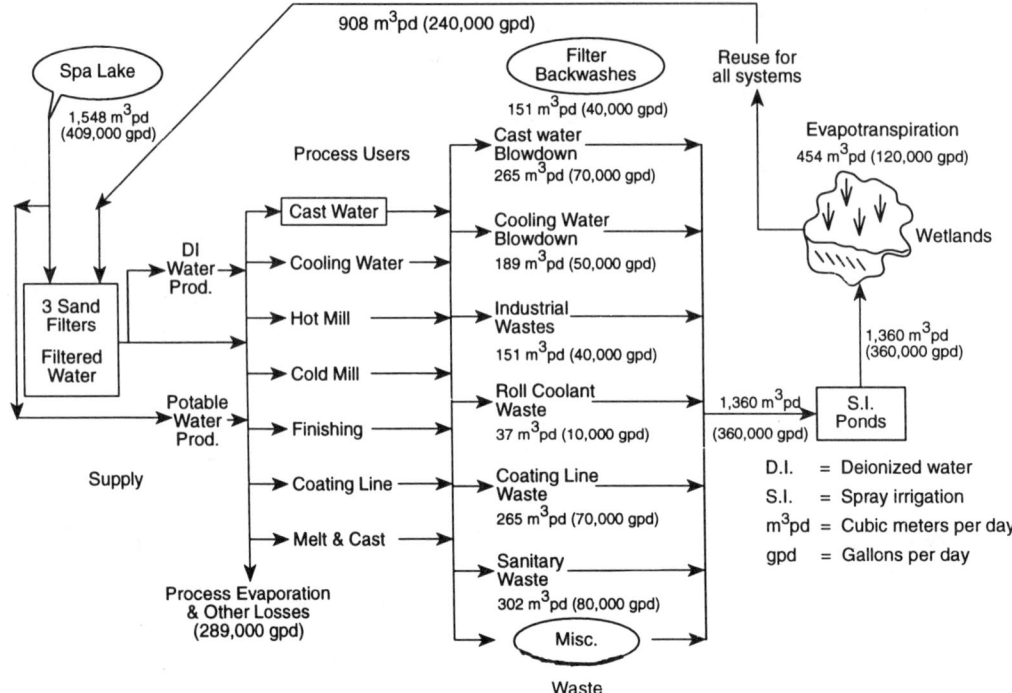

Figure 10.9 Flow diagram for the Logan Aluminum Inc. plant at Russelville, KY.[19,21]

gives the water quality values for the influent and effluent for the wetland system as well as their percent removal.[21] At no time does the average effluent quality exceed the figures listed in the permit.

VIII. COST OF CONSTRUCTION AND OPERATION OF WETLANDS

Many consultants use the figure of $10,000 per 0.4 ha (ac) to create a wetland site. This price does not include the cost of acquisition, which ranges from $1000 to $2000 per 0.4 ha (ac). In addition, maintenance of the created site and monitoring of its success generally add thousands of dollars to the cost of a project. Monitoring usually costs approximately 3% of the construction cost.[26]

Logan Aluminum, Inc. spent $1.6 million on their 16.2 ha (40 ac) constructed wetlands project and expect to recoup this cost within 3 years. It is estimated that the wetlands treatment unit reduces the operating costs by from $535,000 to $564,000 per year.[21,27]

IX. SUMMARY

A brief summary of the major highlights of this wetland project are presented here and come directly from Reference 21.

- 16.2 ha (40 ac) site
- 12.1 ha (30 ac) of water
- 52,990 m³ (14.0 mil gal) capacity
- Detention time 40 days
- 1363 m³ (360,000 gal) of wastewater treated per day
- Evaporation rate of 4.54 m³ (120,000 gal) per day
- 2.54 cm (1 in.) of rainfall will add 3785 m³ (1.0 mil gal)
- Nine different types of plants are present
- 144,000 Plants were planted
- Elevation drop from overland flow area to the lowest pond is 24.4 m (80 ft)
- Depths of water flow ranges from 0 m to 1.8 m (0 to 6 ft)
- Cost $1.6 million
- Operational savings from $535,000 to $564,000 annually

Table 10.18 Effluent Limitations and Monitoring Requirements for Logan Aluminum, Inc.

During the period beginning on the effective date of this permit and lasting through the term of this permit, the permittee is authorized to discharge from outfall(s) serial number(s): 004 — Plant process and cooling wastewaters intermittently discharged from wetlands to Austin Creek.

Such discharges shall be limited and monitored by the permittee as specified below:

EFFLUENT CHARACTERISTICS	DISCHARGE LIMITATIONS				MONITORING REQUIREMENTS	
	(lbs/day)		Other Units (mg/l)			
	Monthly Avg.	Daily Max.	Monthly Avg.	Daily Max.	Measurement Frequency	Sample Type
Flow, discharge (MGD)	—	—	—	—	1/Discharge	Instantaneous
Total Suspended Solids	—	—	46 mg/l	94 mg/l	1/Discharge	Grab
Chlorine, Total Residual	—	—	0.01 mg/l	0.019 mg/l	1/Discharge	Grab
Oil and Grease	—	—	24 mg/l	40 mg/l	1/Discharge	Grab
Chromium, Total Recoverable	—	—	0.1 mg/l	0.1 mg/l	1/Year	Grab
Zinc, Total Recoverable	—	—	0.191 mg/l	0.211 mg/l	1/Discharge	Grab
Aluminum, Total recoverable	—	—	—	—	1/Discharge	Grab
Cyanide, Amenable	—	—	0.005 mg/l	0.022 mg/l	1/Year	Grab
Temperature, °F	—	—	—	89°F	1/Discharge	Grab
Hardness (mg/l as $CaCO_4$)	—	—	—	—	1/Discharge	Grab
Dissolved Oxygen	—	—	5 mg/l (min.)	—	1/Discharge	Grab

The pH of the effluent shall not be less than 7.5 standard units nor greater than 9.0 standard units and shall be monitored 1/discharge by a grab sample.

There shall be no discharge of floating solids or visible foam in other than trace amounts.

Samples taken in compliance with the monitoring requirements specified above shall be taken at the following location(s): nearest accessible point after final treatment but prior to actual discharge to or mixing with the receiving waters.

Compiled from Reference 25.

Table 10.19 Influent and Effluent Water Quality for Wetlands

Parameter	Influent (mg/L)	Effluent (mg/L)	Percent removal
Suspended solids (TSS)	28	3	89
Ammonia	50	16	68
Fluoride	15.4	8.5	45
Aluminum	1.5	0.1	93
BOD_5	107	11	90
COD	346	100	71

Compiled from Reference 21.

CASE STUDY III
CLAYTON COUNTY — E. L. HUIE, JR. LAND APPLICATION FACILITY

I. INTRODUCTION

Georgia had an estimated population of 6,751,404 in 1991 and occupied 152,577 km² (58,910 mi²) with about 65% of this area being forested. In the south, Georgia forms part of the Atlantic coastal plain, divided from the Appalachians in the north by a piedmont plateau. Agricultural production in Georgia includes cotton, tobacco, peanuts, poultry, and livestock. Major industries in Georgia are the textile industry, agricultural processing, paper and pulp mills, and the production of transportation equipment.[1,2]

II. WEATHER AND METEOROLOGICAL DATA

Atlanta, the capital of Georgia, has a relatively mild climate with an average mean temperature of around 17°C (63°F), with daily average mean temperatures never falling below 0°C (32°F). The mean winter temperatures are around 8°C (47°F) with only one or two temporary light snows. The mean or average summer temperatures are around 21°C (70°F). Maritime tropical air masses dominate the climate in summer, while in other seasons continental polar air masses are not uncommon. Rainfall in and around Atlanta averages about 127 cm (50 in.) annually, while evapotranspiration is estimated at 89 cm/yr (35 in./yr).[3-5] The central part of Georgia is drier and averages about 112 cm (44 in.) of precipitation annually.[6,7]

III. WASTEWATER RECLAMATION AND REUSE IN GEORGIA

The Environmental Protection Division (EPD) of the Georgia Department of Natural Resources encourages the use of treated wastewater for land application as an environmental and economically viable alternative to discharging it to a receiving body of water. The EPD indicates that most wastewater land application systems in Georgia result in the growth and harvesting of crops ranging from sod farming and hay production to tree crops. Several golf courses in Georgia are now using treated wastewater for irrigation, and several other golf courses are now in the construction phase, with plans of utilizing treated wastewater for their irrigational needs. The EPD has drafted guidelines for *Design For Urban Water Reuse*.[8] These guidelines apply to golf courses, residential and commercial landscaping, parks, athletic fields, cemeteries, and roadway medians where there is unrestricted public access. Reclaimed water may also be used for fire protection and aesthetic purposes (landscape impoundments and fountains). In Georgia, as of 1992 there were approximately 151 land application systems permitted, for a total estimated flow of 330,000 m^3/d (87 mgd).[8]

Following are a few important excerpts from a draft copy of Georgia's clear, concise, guidelines for urban water reuse.[8]

WASTEWATER TREATMENT AND DISINFECTION[8]

General

Prior to urban reuse, the reclaimed wastewater must meet advanced treatment limits with a high level of disinfection.

Advanced Treatment

The treatment system shall include the following processes:
   Biological Oxidation/Clarification
   Coagulation/Filtration
   Disinfection
Treatment Criteria
   Process Control
      Turbidity ≤3 TU
   Treatment Criteria
      Total Suspended Solids (TSS) ≤5 mg/L
   Fecal Coliform ≤23 per 100 mL
   pH 6–9

MONITORING REQUIREMENTS

Turbidity shall be monitored continuously on the filtered water prior to disinfection for process control. Turbidity shall not exceed an average of 3 turbidity units. Reclaimed water exceeding 5 TU is to be considered reject water.

TSS and fecal coliform limits shall be monitored after disinfection and reported for permit compliance. The frequency of sampling for these parameters shall be determined at the time of permitting.

Fecal coliform organisms in the reclaimed water shall not exceed 23 per 100 mL, as determined from the geometric mean of bacteriological test results of the last 7 analyses which have been completed. The number shall not exceed 200 per 100 mL in any two consecutive samples.

The need to monitor groundwater and surface waters will be determined on a case-by-case basis at the time of permitting.

REJECT WATER STORAGE

Reclaimed water must meet the required treatment criteria before it is transported to the reuse area. Because of the need for continuous monitoring of water quality, turbidity monitoring prior to disinfection is required. If for any reason the turbidity limit is not met, the water must be rejected. An off-line system for storage of reject water shall be provided. At a minimum the capacity of this storage shall be equal to 3 days of flow at the average daily design flow of the treatment facility. Provisions for returning this reject water to the facility for further treatment shall be incorporated into the design.

SYSTEM RELIABILITY REQUIREMENTS

Biological Oxidation
    Equalization
    (for systems which have widely varying flows)
    Standby aeration equipment
    Accessible aeration equipment

Clarification
    Multiple units

Coagulation/Filtration

    Chemical feed facilities for coagulant, coagulant aids, and polyelectrolytes shall be provided. Such chemical feed facilities may be idle if turbidity limits are achieved without chemical addition.

    Multiple, multimedia filter units

Disinfection
    Automatic switchover feed system
    Standby disinfection source
    Multiple points of disinfection (before and after storage, prior to long transport lines, etc.)
    Consider multiple disinfection points

If ultraviolet light disinfection is used, multiple units must be provided.

Power Supply
    On site standby power source/separate feed line
    Automatic switchover

Alarms

  Alarms shall be installed to provide warning of:

    Loss of normal power supply
    Failure of pumping systems
    Failure of disinfection system
    Failure to meet turbidity limits

OPERATION REQUIREMENTS

Operation of reclaimed water systems must follow one of the following options:

1. The facility's chief operator and operator supervisor shall be class I. On site operation by a class II or higher operator 24 hours per day, 7 days per week. Class I operator is the highest level of expertise. Class II operator is the next level of expertise. Details as to operator classification and certification can be found in the Georgia Certification of Water and Wastewater Treatment Plant Operators Act.

2. The facility's chief operator and operator supervisor shall be a class I. On site operation by a class II or higher operator for a minimum of 6 hours per day, 7 days per week in conjunction with one or more of the following:
   a. Automatic diversion of reclaimed water that does not meet the turbidity criteria (reject water) to the reject water storage facility.
   b. Transport of acceptable reclaimed water to the reuse areas only during periods of operator presence.

Other sections in these guidelines include:[8]

- Storage of reclaimed water
- Limited wet weather discharges
- Access control and warning signs
- Portable water cross connections

IV. LAND TREATMENT DESIGN CONSIDERATIONS

A. BASIC LAND TREATMENT PROCESSES

There are three basic land treatment processes — these are slow rate, overland flow, and rapid infiltration. Design features for the three land treatment processes are presented in Table 10.20.[9] Site characteristics for the three processes are presented in Table 10.21[9] and the expected effluent quality for each process is presented in Table 10.22.[9] Only the slow rate (SR) land treatment system will be considered here. For extensive details regarding design procedures and operation of land treatment systems see References 3, 9, 10, 11, 12, and 13.

B. TYPES OF SLOW-RATE LAND TREATMENT SYSTEMS

There are two types of slow rate land treatment systems depending on the design objectives. A slow rate system is considered Type 1 when the principal objective is wastewater treatment and the amount of water applied is not controlled by the crop water requirement. Design of Type 1 systems is based on a limiting design parameter (LDP) of either the soil permeability or the allowable loading rate for a particular wastewater constituent such as nitrogen, BOD, or a trace metal. For municipal wastewater, the LDP is usually either the nitrogen loading or the soil permeability. Type 1 systems are designed to maximize the amount of applied wastewater per unit land area, thereby minimizing the required land area for the system. Type 1 systems are also referred to as slow infiltration systems.[3]

Type 2 SR systems, also known as crop irrigation systems, are designed to apply sufficient water to meet crop irrigation requirements. Water reuse for crop production is the primary objective; wastewater treatment is an additional objective. The irrigation requirement is the amount of water that must be

Table 10.20 Land Treatment Design Feature

Feature	Process		
	Slow rate	Overland flow	Rapid infiltration
Minimum pretreatment	Primary sedimentation[a]	Grit removal and comminution[b]	Primary sedimentation[b]
Annual loading rate, m/a	0.5–6	3–20	6–125
Field area for treatment, ha[c]	23–280	6–44	3–23
Water pathway	Evapotranspiration and percolation	Surface runoff and evaporation	Mainly percolation
Need for vegetation	Needed	Needed	Sometimes used to stabilize soil

[a] With restricted public access; crops not used for direct human consumption.
[b] With restricted public access.
[c] Field area for 3785 m^3/d (1 mgd) flow, not including area for roads or buffer zones.

Compiled from Reference 9.

Table 10.21 Site Characteristics for Land Treatment

Site characteristic	Process		
	Slow rate	Overland flow	Rapid infiltration
Soil depth, m	0.6	0.2	1.5
Soil permeability	Slow to moderately rapid (1.5 to 500 mm/h)	Slow (5 mm/h)	Rapid (50 mm/h)
Depth to groundwater, m	0.6–0.9	Not critical	3
Slope	20% Cultivated 40% Wooded	2–8%	5%
Storage	Usually needed	Usually needed	None

Compiled from Reference 9.

Table 10.22 Typical Expected Effluent Quality from Land Treatment Processes

Process	Average effluent quality, mg/L[a]				
	BOD	SS	Total N	Ammonia-N	Total P
Slow rate[b]	2	1	3	0.5	0.1
Overland flow[c]	10	10	5	4	4
Rapid infiltration[d]	5	2	10	0.5	1

[a] Quality expected with loading rates at the middle to low end of range given in Table 5.20.
[b] Percolation of primary or secondary effluent through 1.5 m (5 ft) of unsaturated soil.
[c] Runoff from a slope length of 30 to 37 m (100 to 120 ft) with comminuted and screened wastewater applied.
[d] Percolation of primary or secondary effluent through 4.6 m (15 ft) of unsaturated soil.

Compiled from Reference 9.

applied, in addition to precipitation, to meet the crop water needs for consumptive use and leaching. Design hydraulic loading rates are based on the crop irrigation requirement and the application efficiency of the distribution system. However, the allowable hydraulic loading rate based on nitrogen loading must also be checked to avoid excess nitrogen loading.[3]

Slow rate system design can be divided into two phases — preliminary and detailed design. The key steps in the preliminary design phase include:[3]

- Crop selection
- Distribution system selection
- Determination of hydraulic loading rates
- Determination of field area and buffer zone requirements
- Determination of storage volume requirements

Once these parameters are established, it is possible to develop planning-level cost estimates for economic comparison of alternatives. The design values determined during the detailed design phase include:

- Depth of applied water per irrigation
- Irrigation frequency
- Sizing of system components — pumps, distribution system, and drainage systems[3]

C. LAND AREA REQUIREMENTS

The total land area required for a SR system includes the cropped area, or field area, as well as land for pretreatment facilities, buffer zones, service roads, and storage reservoirs. The required field area can be determined from the design hydraulic loading rate according to the following equation:[3]

$$A_w = \frac{(Q)(365 \text{ d/a}) + \Delta V_s}{C(L_{W(P)})} \quad [10.3.1]$$

where: A_W = field area, ha
Q = average daily community wastewater flow (annual basis), m³/d
ΔV_S = net loss or gain stored wastewater volume from precipitation, evaporation, and seepage at storage pond, m³/a (a = annual basis)
C = constant, 100
$L_{W(P)}$ = design hydraulic loading rate, cm/a.

The first calculation of field area must be made without considering net gain or loss from storage. After the storage pond area is computed, the value of ΔV_S can be computed from precipitation and evaporation data. The field area must then be recalculated to account for ΔV_S.

D. DESIGN HYDRAULIC LOADING RATE — TYPE 1 SYSTEMS

The hydraulic loading rate is the volume of wastewater applied per unit area of land over a specified time period — typically weekly, monthly, or annually. The corresponding units of expression are cm/wk, cm/mo, and cm/a (in./wk, in./mo, and ft/yr).

The design hydraulic loading rate for Type 1 systems is the hydraulic loading rate calculated on the basis of the limiting design factor. For municipal wastewaters, the factors that must be considered are soil permeability and nitrogen loading. For industrial wastewaters, other factors such as organic loading, salt loading, or metals loading may require consideration.[3]

1. Monthly Hydraulic Loading Rate Based on Soil Permeability

The general water-balance equation, with rates based on a monthly time period, is used to determine the monthly hydraulic loading rate. The equation, with runoff of applied water assumed to be zero, is

$$L_{W(P)} = ET - P + W_p \quad [10.3.2]$$

where: $L_{W(P)}$ = wastewater hydraulic loading rate based on permeability, cm/mo
ET = design evapotranspiration rate, cm/mo
P = design precipitation rate, cm/mo
W_p = design percolation rate, cm/mo

For example, if the evapotranspiration rate for the month of January was 2.0 cm, the precipitation 4.2 cm, and the soil permeability (saturated hydraulic conductivity) 27.4 cm, the hydraulic loading rate in cm/mo would be as follows:

$L_{W(P)}$ = 2.0 − 4.2 + 27.4
= 25.2 cm/mo

These water balance calculations must be made for each month of the year, then the total annual water balance is calculated in order to provide numbers for determination of the yearly hydraulic loading rate.[3]

2. Hydraulic Loading Rate Based on Nitrogen Loading

If percolating water from an SR system enters a potable groundwater aquifer, then the system should be designed so that the concentration of nitrate-nitrogen in the receiving groundwater at the project boundary does not exceed 10 mg/L as NO_3–N. To meet this requirement, the allowable hydraulic loading ($L_{W(n)}$) must be estimated and compared to the previously calculated ($L_{W(P)}$). The lesser of the two values controls design. To estimate the allowable annual hydraulic loading rate based on nitrogen limits use the following equation:[3]

$$L_{W(n)} = \frac{(C_p)(P - ET) + (U)(10)}{(1-f)(C_n) - C_p} \quad [10.3.3]$$

where: $L_{W(n)}$ = allowable annual hydraulic loading rate based on nitrogen limits, cm/a
C_P = total nitrogen concentration in percolating water, mg/L
P = precipitation rate, cm/a
ET = evapotranspiration rate, cm/a
U = nitrogen uptake by crop, kg/ha · a
C_n = total nitrogen concentration in applied wastewater, mg/L (after losses in preapplication treatment)
f = fraction of applied total nitrogen removed by denitrification and volatilization

A similar procedure to that described for determination of the hydraulic loading rate can be used to calculate the nitrogen loading rate. Reference 3 provides details for this procedure, as well as information regarding values for the coefficients U and f.[3]

3. Storage Requirements

The procedure used to determine storage requirements is the same for Type 1 and Type 2 systems. In this procedure, an estimate of the storage volume requirement is first made by using a water balance computation. The final design storage volume is adjusted to account for net gain or loss from precipitation or evaporation. Some states prescribe a minimum storage volume (for example, 10 or 12 day storage). The designer should determine if such prescribed storage requirements exist.[3]

In Georgia, in some cases the minimum required storage period is 12 days. The storage volume can also be determined based on the month by month cumulative water balance calculations. The cumulative water balance storage requirements are based on the difference between the wastewater hydraulic loading rate and the wastewater available. The cumulative storage is then computed by adding the storage during one month to the accumulated storage from the previous month. The maximum cumulative storage is then multiplied by the field area to determine the required storage volume. See Reference 3 for a step-by-step procedure to estimate storage volume requirements.

Example 10.3.1

Determine the hydraulic loading rate, the land area requirement, minimum storage volume, and storage area requirements for a land treatment system utilizing a secondarily treated effluent with a flow of 73,808 m³/d (19.5 mgd). For this site the evaporation rate is 88.9 cm/yr (35 in./yr) and the precipitation rate is 127 cm/yr (50 in./yr). The soil is a sandy clay loam with a moderate permeability of 4.23 cm/wk (saturated soil conductivity 7 × 10⁻⁶ cm/s). Assume that the soil permeability, not the nitrogen loading rate, is the limiting design factor. The state regulations indicate a minimum storage period of 12 days. The storage pond is to have a depth of 3 m and is lined in order to prevent seepage.

Solution:

1. Determine the Hydraulic Loading Rate using Equation 10.3.2, but using yearly data rather than monthly data:

$$ET = 88.9 \text{ cm/yr}$$

$$P = 127 \text{ cm/yr}$$

$$W_p = 4.23 \text{ cm/wk} \times 52 \text{ wk/yr} = 220 \text{ cm/yr}$$

$$L_{W(P)} = ET - P + W_p \qquad [10.3.2]$$

$$= 88.9 \text{ cm/yr} - 127 \text{ cm/yr} + 220 \text{ cm/yr}$$

$$= 182 \text{ cm/yr}$$

$$= 3.5 \text{ cm/wk}$$

2. Determine the Field Area Required using Equation 10.3.1:

$$A_N = \frac{Q(365 \text{ d/a}) + \Delta V_S}{CL_{W(P)}} \qquad [10.3.1]$$

(At this point in the calculations $\Delta V_s = 0$)

$$= \frac{73{,}808 \text{ m}^3/\text{d}(365 \text{ d/a}) + 0}{100(182 \text{ cm/yr})}$$

$$= 1480 \text{ ha}$$

$$= 3657 \text{ ac.}$$

3. Determine the Storage Volume Required, based on a minimum storage period of 12 days and a flow of 73,808 m³/day (19.5 mgd). The storage volume required would be 73,808 m³/day × 12 days = 885,696 m³.
4. Determine the Storage Area Required. If the retention pond depth is to be 3 m (10 ft) the surface area required is

Surface area $= \dfrac{885{,}696 \text{ m}^3}{3 \text{ m}}$

$$= 295{,}232 \text{ m}^2$$
$$= 29.5 \text{ ha}$$
$$= 73 \text{ ac.}$$

5. Determine the Final Area Requirements including the ΔV_s term for a 29.5 ha (73 ac) storage pond:

ΔV_S = 127 cm/yr (precipitation) × 295,232 m² − 88.9 cm/yr (evapotranspiration) × 295,232 m²
$= 0.381$ m/yr (295,232 m²)
$= 112{,}483$ m³/yr (annually)

(seepage can be ignored as the pond is lined)

Substituting ΔV_s in Equation 10.3.1:

$$A_W = \frac{Q(365 \text{ d/a}) + \Delta V_S}{C(L_{W(P)})} \qquad [10.3.1]$$

$$A_W = \frac{(73{,}808 \text{ m}^3/\text{d})(365 \text{ d/a}) + 112{,}483 \text{ m}^3/\text{a}}{100(182 \text{ cm/yr})}$$

$$= 1486 \text{ ha}$$

$$= 3673 \text{ acres}$$

As indicated earlier, another approach to determining storage volume is to use the month by month cumulative water balance procedure. However, in this case the monthly data for precipitation and evapotranspiration are needed to carry out the required calculations.

V. CLAYTON COUNTY

Clayton County is located on the southern city limits of Atlanta. The population of the county in 1993 was estimated to be 210,000. By the year 2010 the population is projected to be 270,000.[14] The county in 1990 occupied a land area of 36,975 ha (142.65 mi²) with an additional water area of 425 ha (1.64 mi²).[15] Clayton County is a suburban area with some light industry but mainly commercial establishments.

A. CLAYTON COUNTY WATER AUTHORITY

In 1955 the Georgia State Legislature created the Clayton County Water Authority (CCWA) to provide water and sanitary sewer services to unincorporated and some incorporated areas in Clayton County.[14] The CCWA now has over a $30 million annual budget and serves over 37,000 customers. The CCWA

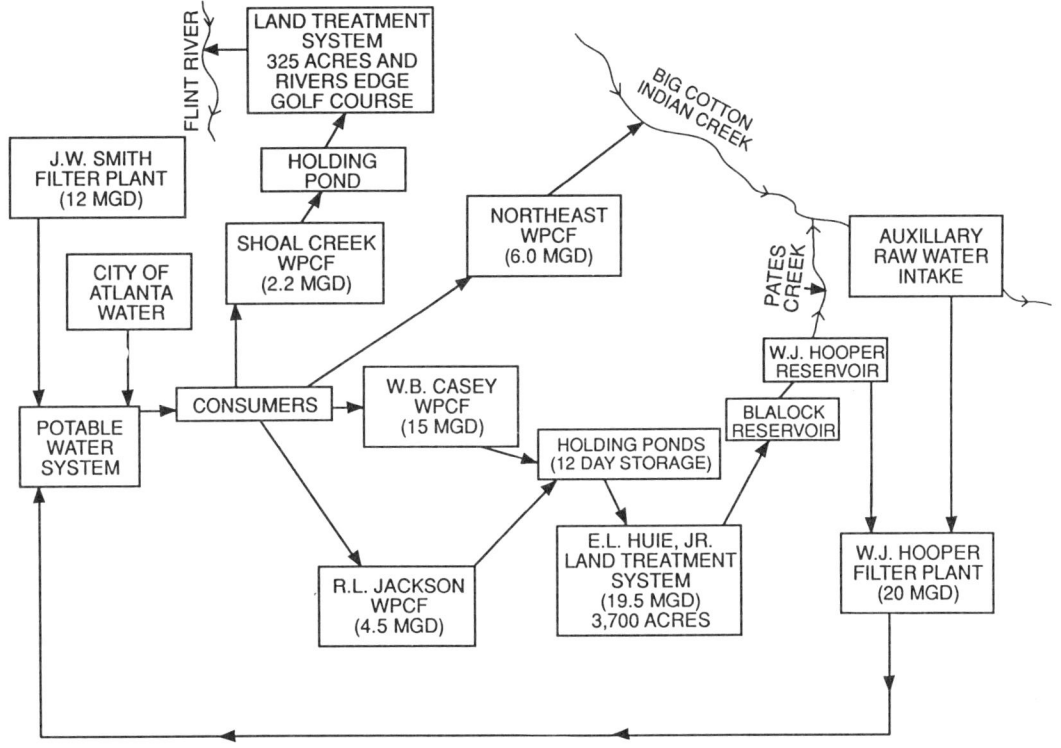

Figure 10.10 Flow schematic: Clayton County, GA wastewater recycling system.[18,19]

owns and maintains over 1862 ha (4600 ac) of land, 1588 km (987 mi) of water lines, and 1205 km (749 mi) of sewer lines.[16]

The CCWA operates the E.L. Huie, Jr. Land Application Facility. This slow-rate land treatment system uses secondarily treated wastewater for forest and golf course irrigation. The facility has two forest application sites. One site has 1497 ha (3700 ac) of forested area while the other site has a 131-ha (325 ac.) area.[17] The secondarily treated wastewater effluents from the W.B. Casey, Jr. and R.L. Jackson plants are pumped to a holding pond with a 12-day storage capacity of 885,690 m³ (234 mil gal). From the holding pond this water is sprayed on the 1497 ha (3700 ac) site, while the secondarily treated effluent from the Shoal Creek wastewater treatment facility goes first to a holding pond and is then sprayed on the second site as well as used for irrigational purposes at the Rivers Edge 18-hole public golf course.[18] Figure 10.10 presents a flow diagram for the system.[18,19]

The Clayton County land treatment system has eight primary fields. Seven of the fields are sprayed on a rotating daily basis each week. The eighth is held out of the rotation for maintenance and harvesting.

Each day, 6.5 cm (2.5 in.) of the treated effluent is sprayed on one of the seven fields. The water percolates through the soil and back to the raw water supply through Pates Creek.[16]

The CCWA has selected areas on the project open to the public as a wild life sanctuary. Bird watchers from the local Audubon Society have found 21 different varieties of shore birds at the land treatment site. Wild turkey and Canada geese abound, while the deer population has flourished.

Two raw water reservoirs have been constructed on the land application site — Lake Shamrock and the Edgar Blalock, Jr. Lake. These areas are now open to the public for recreational use, such as boating and fishing. The typical type of fish caught in these reservoirs include bass, bream, and catfish.

A community recreational area is now complete, with a paved boat ramp and a pedestrian bridge to man-made islands. A new community use facility has now been constructed on a hilltop overlooking Lake Shamrock and the Blalock Reservoirs. This facility is used for both private and public events such as training seminars.[16,18]

The land application site is used by the Soil Conservation Service, the Georgia Forestry Commission, and the University of Georgia for research projects.[20] One such research project now underway by Zoo Atlanta deals with trying to determine the color and style of bat-houses that would be most attractive to bats. Bats are known to eat night-flying insects. A single bat can consume up to about one-third its body weight in insects within a half hour. It is reported that each bat consumes about 1 g of insects per night.[21]

Another insect control program in operation at the holding ponds is the installation of purple martin birdhouses. Some purple martin birdhouses contain over 200 mated pairs of birds.[22] Purple martins are known to be voracious daytime insect eaters, and it is said that they eat their weight in mosquitoes every day.

The holding ponds have been stocked with pot-bellied top-water-feeding minnows *(Gambusia affinis)*. These minnows grow to a maximum length of approximately 7.6 cm (3 in.) and are avid feeders of mosquito larvae. The *Gambusia affinis* bear their young alive in successive broods numbering up to 50 or more throughout the spring and summer. They live equally well in brackish water, freshwater, or water with some pollution. These top-water minnows have been observed to eat up to 165 mosquito larvae in 1 day. The young minnows start to feed on the larvae immediately after they are born.[18,23]

In order for the land application system to be effective the trees must be healthy and fast growing. Initially it was planned to harvest 100 ac of trees on a rotating 16-year basis, however, experience has shown that a 32-year rotating basis is more appropriate.[18] The trees are harvested whole and fed into a chipper. These chips are used in both the composting and pelletization processes.[17]

Recently, the CCWA began irrigating the championship Rivers Edge 18-hole public golf course with treated wastewater. This golf course is adjacent to the 131-ha (325 ac.) land application site.[18,24]

B. BACKGROUND AND DEVELOPMENT OF THE CLAYTON COUNTY WASTEWATER TREATMENT PROJECT

In the early 1970s, Clayton County had several secondary treatment plants discharging to surface waters. Two of these, the Casey plant and the Jackson plant, discharged to the Flint River. In 1974 the mean annual flow in this river was only 0.34 m^3/s (12 ft^3/s) and wastewater effluent discharges averaged 0.14 m^3/s (5 ft^3/s). During the summer months, the wastewater effluent exceeded natural streamflow. The State of Georgia imposed new discharge standards and a limited flow allocation for these existing systems. Because the projected wastewater flows were significantly higher than the discharge allocation, it was necessary to limit growth in the county, provide very high levels of treatment to satisfy the discharge limits, or develop a new treatment method discharging to another watershed.[7]

Faced with this situation the CCWA undertook in 1974 a study in accordance with Section 201 of Public Law 92–500 of the Federal Water Pollution Control Act Amendments of 1972, to evaluate wastewater treatment alternatives to meet their stated water quality goals. Section 201 encourages the use of recycling technology and requires that land treatment be evaluated and compared with conventional treatment methods as part of the cost-effective analysis. At the time the study was initiated, the law required innovative and alternative technology (land treatment is specified as one such technology) to be selected if the cost did not exceed the cost-effective alternative by more than 15%.[4]

The major alternatives considered were providing high levels of advanced wastewater treatment (AWT) at the two existing activated sludge plants, or using land treatment at another location for wastewater management. Five potential land treatment sites were evaluated — all within 12 km (7.5 mi) of the two existing treatment systems, and all were forested to ensure that year-round operation would be possible. An extensive screening and evaluation process led to the selection of the largest contiguous site. It was 80% forested, had well-defined drainage networks, and would require displacing the fewest number of property owners. Detailed investigations on this site confirmed the suitability for land treatment. The cost comparisons for developing this site vs. AWT indicated that land treatment was the most cost-effective way to upgrade the two existing plants as compared to their abandonment and the construction of a new multiple-cell aerated lagoon.[7]

C. DESIGN CONSIDERATIONS

Determining the land area needed for the treatment system was a critical step in the design and was based on the limiting design parameter (LDP) approach.

The LDP is the parameter or wastewater constituent that requires the largest land area for acceptable performance. The LDP might be based on the ability of the soil to pass the design volume of water, or on some wastewater constituent such as nitrogen, phosphorus, organics, or metals. Experience with typical municipal effluents has shown that the LDP for this type of project is usually the hydraulic capacity of the soil or the ability to remove nitrogen, and thereby maintain drinking water levels for nitrate in the groundwater at the project boundary.[7]

Table 10.23 The Wastewater Characteristic Pumped from the Holdings Ponds

Parameter	Concentration[a] (mg/L)
pH	7.1
BOD_5	16
TSS	15
NO_3-N	0.7
Total P	1.2
Chloride	40
Iron	0.8
Copper	0.05
Zinc	0.09
Cadmium	<0.01
Chromium	<0.03
Nickel	0.06
Lead	0.02
Sodium	55
Calcium	4
Magnesium	2
Conductivity	400 μmhos

[a] All in mg/L except for pH and conductivity.

Compiled from Reference 18.

At Clayton County, the hydraulic capacity of the surface soils limited the design wastewater application rate to 6.4 cm/wk (2.5 in./wk). Thus, the design wastewater application would be 3.3 m/yr (10.8 ft/yr), and the average design daily hydraulic loading would be about 8 L/m²·d(0.2 gal/ft²·d). At this loading rate, the design projections indicated that nitrate-nitrogen in percolate entering Pates Creek or the groundwater would be well below the 10 mg/L limit required for drinking water.[7]

D. MAJOR COMPONENTS IN THE CLAYTON COUNTY LAND TREATMENT FACILITY

The major components in the Clayton County land treatment system include three wastewater treatment plants: the Casey Plant with a design capacity of 57,000 m³/d (15 mgd), the Jackson Plant with a design capacity of 17,000 m³/d (4.5 mgd), and the Shoal Creek Plant with a recently expanded capacity of 8300 m³/d (2.2 mgd).[18] The secondarily treated effluent is then pumped to holding ponds before being sprayed on the land application sites.

Table 10.23 presents the quality of the water that is being pumped from the holding ponds to the forested areas. The water quality characteristics for TSS and BOD_5 are well below the limits set in the state permit for this site. The holding ponds at the facility are of a multicell design with a combination storage of 12 days flow. The 12-day storage size was determined based on periods of nonirrigation resulting from adverse climatological conditions or mechanical problems.

Three buildings were constructed on the site: the pump station, the control building, and the service building. The pump station is located adjacent to the pond complex and supplies effluent to the various spray fields. The pump station contains three 1500-HP pumps for the main irrigation fields and a 60-HP pump for the test field. The control building houses the administrative offices for the Land Management Division of the CCWA, conference facilities, an office for the forester responsible for the forestry activities within the Land Management Division, and facilities for the operations staff, including a weather station and two computer systems. The service building is used by the maintenance staff of the Land Management Division, which is responsible for field maintenance and routine facilities maintenance.[20]

As indicated earlier, the land treatment system has eight primary fields, with seven fields being sprayed, on a rotating daily basis, with 6.4 cm (2.5 in.) of treated wastewater.[16] In the total irrigation system, there are approximately 442 km (275 mi) of buried pipeline ranging in size from 107 cm (42 in.) ductile iron pipe to 3.8 cm (1.5 in.) PVC, 11 motorized valves, 47 manual valves, and approximately 19,500 sprinklers.[18,20]

E. LAND TREATMENT OPERATION

The wastewater is sprinkled, in rotation, on the forested units in the land treatment site. Some of the applied wastewater percolates vertically and reaches the native groundwater table, but most of the applied wastewater infiltrates to a relatively shallow depth, percolates laterally through the soil, and emerges as surface or subflow in the site's drainage network, which eventually flows into Pates Creek. Operating the land treatment system has significantly increased flow in Pates Creek. During very dry summers up to 84% of the flow in this creek is due to the percolation from the land treatment site.[19] The dominant tree species on the site is loblolly pine (*Pinus taeda*) and there are also significant stands of mixed pine and hardwood trees. The hardwoods include Sweet Gum, Red Oak, and poplar. Open land was also planted with loblolly pine. Distribution piping was buried and wide access lanes were cleared at each sprinkler distribution row. Currently, all of the sprinklers in a given section are replaced on a regular schedule, and the removed units are taken to the shop for repair and rehabilitation and then used as replacement units elsewhere. This is more efficient than trying to find and repair individual malfunctioning sprinklers.[7]

The original management plan called for clear-cut harvesting for pulpwood of several blocks per year by a contractor on a 16-year rotation. The cleared areas would then be replanted with pine. This procedure was used for the first year of operations and then abandoned because of excessive erosion and related problems caused by heavy-duty harvesting equipment. As indicated previously, the 16-year rotation has now been changed to a 32-year rotation system.[7,18] Most forest land application systems use natural forests on which to spray their treated wastewater. The City of Edenton, NC uses a managed hardwood forest for its land application site. This system may have some advantages: following canopy closure, mowing, and other site access requirements could be dramatically reduced and thus operational and maintenance costs reduced.[25]

F. SYSTEM PERFORMANCE

The treated wastewater quality from the holding ponds according to Table 10.23 is clearly within the effluent limitation set out in the state land application system permit. The present flow to the land application sites in 1994 was 53,000 to 57,000 m^3/d (14 to 15 mgd) well below the design capacity of 73,800 m^3/d (19 mgd).

The system's performance is evaluated using monitoring wells on the application sites (see Section VI) as well as monitoring Pates Creek as it flows away from the site. Initially, nitrogen was of the greatest concerns, because nitrogen was the LDP. Nitrate contamination of drinking water sources had to be avoided. Total nitrogen concentrations in the groundwater and in Pates Creek at the project boundary were monitored because wastewater was applied beginning in 1979. Phosphorus was monitored because of its potential to limit the life of the site. Chlorides were monitored because they served as a tracer, confirming that wastewater percolate reached the sampling points.

Table 10.24 and Table 10.25 present the nitrogen, phosphorus, and chlorides concentrations in the groundwater at the site as well as nitrogen, phosphorus, and chloride concentrations in Pates Creek for 1978 and for 1984 to 1987.

The data indicate that chlorides increased from start-up until 1986 and then leveled off. The final concentration of chlorides is about one-half of that present in the applied wastewater, indicating that there is significant mixing and dilution with the surface water and groundwater.

The nitrogen and phosphorus data indicate excellent removal efficiency and, although the pattern for nitrogen is somewhat erratic, the concentration has always remained below the 10 mg/L target value. At the present nitrogen loading rate, this performance can be expected to continue indefinitely. Even if the chloride values in the groundwater and surface water were to double, a phenomenon that would indicate the presence of wastewater percolate without any dilution, the nitrogen concentration should only increase proportionally and still not exceed 5 mg/L. If the nitrogen loadings ever reach the design values, the percolate nitrogen reaching the groundwater and Pates Creek may approach, but still not exceed, the 10 mg/L target value.[7]

VI. EFFLUENT LIMITATIONS AND MONITORING REQUIREMENTS

The Environmental Protection Division of the Department of Natural Resources for the State of Georgia issued the Clayton County Water Authority a land application system permit in 1991.

Some of the most significant conditions in this permit as they apply to effluent standards for the discharge to the land application system are as follows:[26]

- *Biochemical Oxygen Demand (5-Day)*. The monthly average shall not exceed 50 mg/L.
- *Suspended Solids*. The monthly average shall not exceed 50 mg/L.

Table 10.24 Clayton County System Groundwater Quality

Year	Parameter, mg/L		
	Nitrogen	Phosphorus	Chloride
1979	0.69	0.03	1.0
1980	0.54	0.03	1.6
1981	0.63	0.04	2.3
1982	0.34	0.03	3.6
1983	0.44	0.02	10.8
1984	0.43	0.09	12.6
1985	0.15	0.01	15.4
1986	0.65	0.01	21.0
1987	0.18	0.04	21.5

Compiled from Reference 7.

Table 10.25 Pates Creek Water Quality

Year	Parameter, mg/L		
	Nitrogen	Phosphorus	Chloride
1978	0.65	0.05	1.0
1984	0.96	0.05	7.5
1985	1.15	0.26	13.3
1986	2.02	0.32	18.3
1987	1.04	0.17	18.9

Compiled from Reference 7.

(All analyses shall be made in accordance with the latest edition of *Standard Methods for the Examination of Water and Wastes, Methods for Chemical Analysis of Water and Wastes*, or other required methods.) Included in the permit are the monitoring parameters and monitoring schedule requirements for the discharge to spray fields, groundwater well monitoring, soil monitoring, and surface water monitoring.

The monitoring results indicating the concentrations for the various parameters are reported monthly to the Georgia Environmental Protection Division.[26]

- *Discharge to Sprayfields*

Parameter	Frequency
Flow	Daily (continuous recording measurements required)
Biochemical oxygen demand (5-Day)	Three/week
Suspended solids	Three/week
pH	Three/week
NO_3-N	One/week

- *Groundwater Well Monitoring*

Parameter	Frequency		
	Wells #13,14, 15, and 17	Wells #3,5,10, 11,18,19, and 20	Wells #1,8,12, 16,21, and 22
Depth to groundwater	One/month	One/quarterly	One/6 months
pH	One/month	One/quarterly	One/6 months
NO_3-N	One/month	One/quarterly	One/6 months
Fecal coliform bacteria	One/month	One/quarterly	One/6 months
Electrical conductivity	One/month	One/quarterly	One/6 months

Note: Where there are categorical and/or significant industrial discharges to the sewer system, the Permittee may be required to sample for additional parameters. These may include metals and organic compounds.

- *Soil Monitoring*

Parameter	Frequency
pH	One/year
Cation exchange capacity	If pH changes by one unit
Percent base saturation	If pH changes by one unit
Phosphorus adsorption	If pH changes by one unit

Note: Where there are categorical and/or significant industrial discharges to the sewer system, the Permittee may be required to sample for additional parameters. These may include heavy metals and organic compounds.

Soil samples should be collected from each major soil type.

- *Surface Water Monitoring.* Surface waters adjacent to or flowing through the Land Application Site should be monitored to detect any adverse impact. Samples for the following parameters should be collected from Pates Creek and Shamrock Lake:

Parameter	Frequency
Biochemical oxygen demand (5-Day)	One/quarterly
Suspended solids	One/quarterly
Dissolved oxygen	One/quarterly
pH	One/quarterly
Fecal coliform bacteria	One/quarterly

- *Groundwater.* Groundwater leaving the land application system boundaries must not exceed primary maximum contaminant levels for drinking water. If groundwater samples indicate contamination, the Permittee will be required, upon written notification by the EPD, to develop a plan which will insure that the primary maximum contaminant levels for drinking water are not exceeded. The plan will be implemented by the Permittee immediately upon EPD approval.[26]

VII. MANPOWER AND COSTS

The Clayton County Water Authority, which was established in 1955 and now operates the land treatment system, is composed of a 7-person board serving 5-year staggered terms. It has a general manager, 5 assistant managers, and 245 employees, with a budget in excess of $30 million.[14,18]

As of 1987 the staffing requirements for the land treatment operation, including the secondary treatment and sludge pelletizing operations and the land treatment component, totalled 61. The staff at the land treatment facility increased from 13 people when land treatment operations began in 1978 to 24 people in 1987. Staff was needed to harvest trees and to maintain and repair the sprinklers and appurtenant equipment.[7]

The actual capital costs for the Clayton County Land Treatment System was within 2% of the estimate. Table 10.26 presents a breakdown of these construction costs as well as the future salvage value for the land.[7]

This project was 75% funded by the EPA under their innovative and alternative (I/A) program. Up to 1984 some 1400 projects has been funded by this EPA program. In 1983, seven I/A projects were nominated for national recognition by the officers, member firms, and the awards committee of the American Consulting Engineers Council (ACEC) in their 1983 Engineering Excellence Award Program. Award-winning projects represented a cross section of I/A technologies. The Clayton County Water Authority project was selected as one of these seven I/A projects.[27]

Now, the EPA estimates that land application alternatives can cut construction costs by 25% and operating costs by as much as 50%.[16]

About 30% of the total capital costs for the Clayton County land application project were used to upgrade the existing activated sludge plants. If Clayton County had been required to build entirely new advanced activated sludge systems the total estimated capital costs would have been $37 million rather than the $26.73 million as shown in Table 10.26.[7] Also, if the county had constructed aerated lagoons as an alternative, the total estimated capital costs would have been $32 million in 1976 dollars or $71.2 million in 1994 dollars.[7,28]

Table 10.26 Clayton County System Capital Costs ($ × 10^6)

Upgrade Casey Plant		6.95
Upgrade Jackson Plant		1.42
	Sub total	8.37
Pump station and transmission line		1.70
Land treatment facility		
Storage ponds		0.46
Pump and distribution system		5.30
Structures, roads, and the like		0.60
Land costs (3000 ac)[a]		10.30
	Sub total	18.36
	Total	26.73
Future salvage value of the land		−(10.30)
	Total, with salvage	16.43

[a] ac × 0.4047 = ha.

Compiled from Reference 7.

Table 10.27 Clayton County System Total Annual Costs

Total direct costs		3,088,000
Depreciation/debt service		2,004,000
Indirect and administrative		1,595,000
Sale of sludge pellets		−(199,000)
	Total	6,488,000
Secondary treatment, $/1000 gal		0.85
Land treatment facility, $/1000 gal		0.41
	Total	1.26

Note: Based on 1987 dollars.
Total gallons treated = 5,146,500,000.

Compiled from Reference 7.

These figures indicate a total estimated capital cost saving of almost $16 million in 1976 dollars or $35.5 million in 1994 dollars. These calculations take into account the future land salvage value of $10.3 million (see Table 10.26).[7,28]

It is not easy to compare operational and maintenance (O&M) figures as these expenditures vary geographically as do the items included in the reported costs.

The 1987 operational and maintenance (O&M) costs for the Clayton County wastewater treatment facilities, including labor, chemicals, power, and other costs, were $0.11/m^3 ($0.40/1000 gal) for the activated sludge secondary treatment processes, and $0.05/m^3 ($0.20/1000 gal) for the land treatment system.[7] The operational and maintenance costs for the activated sludge secondary treatment systems in Clayton County are within the normal range, which can vary from $0.05 to $0.16/m^3 ($0.20 to $0.72/1000 gal) based on 1987 dollars.[7,29,30] However, the land treatment O&M costs are less than half that often reported for land treatment systems, which range from $0.06 to $0.21/m^3 ($0.22 to $0.81/1000 gal), again based on 1987 dollars.[30-33]

Table 10.27 presents the total annual costs in 1987 for the Clayton County wastewater treatment systems. The annual activated sludge secondary treatment cost was $0.22/m^3 ($0.85/1000 gal), while the annual cost for the land treatment system was $0.11/m^3 ($0.41/1000 gal).[7]

Again, the annual costs for the Clayton County activated sludge secondary treatment systems are in the normal range and vary from $0.11 to $0.37/m^3 ($0.43 to $1.41/1000 gal) based on 1987 dollars.[7,26,27]

However, the total annual O&M costs for the Clayton County land treatment system are clearly lower than the annual O&M costs reported for other land treatment systems. The total annual costs reported for other land treatment systems range from $0.09 to $0.36/m^3 ($0.34 to $1.35/1000 gal) based on 1987 dollars.[7,30-33]

A general rule of thumb is that total annual costs (which include capital costs) are about double the maintenance and operational costs.

VIII. CONCLUSIONS

The Clayton County GA Land Treatment Forest System has now been in operation for over 15 years. The secondarily treated effluent percolates through the soil and rock, removing pollutants on its way to Pates Creek. The water in Pates Creek then drains into Lake Shamrock, Lake Blalock, and then into the W.J. Hooper Reservoir which is a major source of the county's raw water supply system.

Trees grown on the land application site are commercially harvested for the production of pelletized fertilizer (Agri-plus 650) and also as fuel for heating the drying kilns.

The Clayton County land application site is used by the community for recreational purposes such as boating, fishing, and bird watching. A community use facility has recently been constructed on the site overlooking Shamrock and Blalock Reservoirs. This facility is available for both public and private use. A championship 18-hole golf course adjacent to the land application site is now irrigated with treated wastewater from this project.

Another important use of the land application site is for research projects being carried out by universities, governmental agencies, and other environmentally concerned institutions.

More detailed conclusions indicate that a hydraulic loading rate of 6.4 cm/week (2.5 in/d over one of the eight field sites) of secondarily treated wastewater has resulted in only a 20 mg/L increase in chloride concentrations in the groundwater and Pates Creek. The chloride concentrations leveled off in 1986 and now remains at the 20-mg/L level. It has been found that slopes as steep as 30% can be irrigated on a year-round basis with the treated wastewater. The water that percolates through the soil and rock to Pates Creek has increased the stream's flow, especially during low-flow periods.[4] At design flows, the wastewater represents approximately 84% of the water flowing into the water supply reservoirs during low-flow conditions, and approximately 33% during normal flow conditions.[19] Also, at design flows during low-flow conditions, wastewater could represent approximately 62% of the flow in Big Cotton Indian Creek at the auxiliary raw water intake.[19]

The capital costs savings in constructing the land application system has been estimated at $16 million in 1976 dollars ($35.5 million in 1994 dollars), when compared to other alternative treatment technologies.

The O&M costs were found to be about half the total annual costs when calculated on a per cubic meter or per 1000 gallon basis. The total annual costs are based on the annual operational and maintenance costs plus depreciation/debt service and indirect and administrative costs, minus the sale of sludge pellets.

The O&M charges for the activated sludge systems are in the normal range for such activities, however the O&M charges for the land application system are about half that reported for similar land treatment operations.

The Clayton County Water Authority has not only the responsibility for the management and operation of the wastewater treatment facilities and the accompanying land application system, but also the responsibility for the operation of the water supply system. As indicated, a large portion of the county's raw water supply is made up of water that either percolates through the soil and rock from the land treatment application site or from the tertiary treated wastewater flowing in Big Cotton Indian Creek.[18,19]

This indirect reuse of the treated wastewater includes political, administrative, legislative, regulatory, and jurisdictional responsibilities. Should a problem arise, the Clayton County Water Authority must review and evaluate the problem and then take the appropriate action, keeping in mind these many responsibilities.

OVERVIEW OF WASTEWATER RECLAMATION AND REUSE IN THE SULTANATE OF OMAN

I. INTRODUCTION

The last census conducted in the Sultanate of Oman was in December 1993, and indicates that the total population was 2,017,591, with 74% being Omani nationals and 26% being non-Omanis. The capital of the Sultanate of Oman is Muscat, with a total population of 622,506 as of December 1993.[1] This independent Sultanate occupies an area of 212,380 km^2 (82,000 mi^2) bounded on the east by the Gulf of Oman, on the south by the Arabian Sea, on the west by Yemen and Saudia Arabia, and on the north by the United Arab Emirates.[2]

The chief crops are dates, fruits, vegetables, wheat, and bananas, but there is also an abundance of sugarcane and cattle, including goats and sheep, produced mainly in the southwest part of the country. The major exports are petroleum products, with oil revenues accounting for more than half of the gross national product (GNP).[3]

II. WEATHER AND METEOROLOGICAL DATA FOR OMAN

The rainfall patterns, occurrence, and quantity within the Sultanate of Oman are varied. Regular rainy seasons are found only in the Dhofar Mountains in the southern part of the country and in the Hajar mountains in the northern part of the country. The interior regions (which make up around two-thirds of the country) receive less than 5 cm (2 in.) average rainfall annually. The Hajar Mountain region receives from 10 to 30 cm (4 to 12 in.) of rainfall annually and the Dhofar Region from 20 to 30 cm (8 to 12 in.) annually. Some of the rainfall in the Dhofar Region is due to the summer monsoons.

The summers in the coastal areas are hot and humid, while the winters are comparatively cool. In the interior, the climate is hot and dry in summer and somewhat cooler in winter. The average temperature in the Sultanate of Oman varies from 17.8 to 28.9°C (64.0 to 84.0°F). The hottest months are June and July, ranging from 30.7°C (87.2°F) in Saig to 46.1°C (115°F) in Fahoud. The coldest month is January with temperatures of 9.4°C (49°F) in Saig of 24°C (75°F) in Fahoud. The lowest temperature ever recorded in the Sultanate of Oman was –3.6°C (25.5°F) in Saig. Summer temperatures on the coast at Muscat often reach 36°C (97°F) while winters have lower temperatures of about 17°C (63°F).[3]

III. WATER RESOURCES IN OMAN

The total volume of water now being used in the Sultanate of Oman is estimated at 1.5 billion m^3 (396.3 billion gal) per year with more than 85% of this water going for agricultural irrigation. Recent hydrological studies estimate that the maximum increase in available water resources in the Sultanate of Oman over the next 30 years will not exceed 50% of the present available water resources.

Even with improvements in irrigation methods and with most of the drinking water coming from desalinization plants the maximum augmentation in water supplies will still only support an increase in cultivated (arable) land from the present 125,000 ha (308,900 ac) to 300,000 ha (741,000 ac) by the year 2020.[4] The sultanate has no permanent bodies of freshwater supplies.[3]

IV. WASTEWATER TREATMENT AND REUSE IN THE SULTANATE OF OMAN

Considerable information is available regarding the existing wastewater treatment plants in the Sultanate of Oman, as well as the use of the treated wastewater for irrigational purposes. For more particulars see References 5, 6, and 7.

Table 10.28 presents details regarding the existing wastewater treatment plants in the Sultanate of Oman as well as the volume of treated wastewater used for irrigational purposes. Table 10.28 presents the locations of the wastewater treatment plants, their design capacity, the actual flow rates as of 1993, and the average effluent quality for BOD$_5$, SS, and NH$_3$-N.[7]

Treated wastewater reuse is restricted to the irrigation of ornamental plants in the capital area. The stringent regulations introduced in 1986 for wastewater discharge and reuse have been revised in view of existing experience. These new regulations were established according to Ministerial Decision Number 145/93, addressing wastewater reuse and discharge. Reuse of wastewater was limited to irrigation of ornamental trees and shrubs, groundwater recharge in areas of no public exposure, and industrial closed circuit processes. Regulations have placed emphasis on the physical quality parameters of total dissolved solids, suspended solids, and electrical conductivity. The bulk of the regulations address chemical parameters. Bacteriological quality was evaluated by the faecal coliform bacteria and the viable nematode ova and cysts.[7,8,16]

In the Greater Matrah area (one of the important regions in the Muscat area) rainfall runoff (stormwater) is collected in a separate stormwater system and used for irrigational purposes.

The Darsait STP receives pretreated wastewater from the Al-Nahda hospital, domestic sewage from about 500 ha (1236 ac) of neighboring areas, as well as sewage from tanker trucks that service septic tanks in unsewered areas.[7,9]

The first phase in the operation of the Darsait STP was in 1977, and served a population of 10,000. In the second phase of operation in 1981, the Darsait plant increased its capacity to treat the wastewater

Table 10.28 Wastewater Treatment Plants in the Sultanate of Oman and the Volume of Treated Wastewater Used for Irrigational Purposes

Location	Design capacity (m³/d)	Plant flow rate (m³/d)	BOD (mg/L)	SS (mg/L)	Ammonia (N) (mg/L)	Reuse-irrigational water, (m³/d)
Darsait	10,800	11,500	10	10	1	8,000–11,500
AlAnsab	12,000	5,400	10	10	1	2,000–5,000
Shati AlQrum	1,350	800	10	10	1	800
Mabella	1,920	700	10	10		
AlKoud	1,200	700	20	30		
Bowshar	400	400	10	10	1	
AlAmerat	600	600	20	30		
Jibroo	70	100	10	10	1	50
AlAynt	60	100	10	10	1	
Total	28,400	20,300				10,850 to 17,350

from a population of 20,000. The plant was further expanded in 1986 to handle the wastewater coming from a population of 53,300. Now, the plant treats the wastewater coming from a population of around 90,000. The current wastewater flow to this plant is 11,500 m³/d (3.03 mgd) and is mainly domestic in origin.

The Darsait wastewater treatment facility provides preliminary, primary, secondary, and tertiary treatment for wastewater and is equipped with of screens, grit chambers, primary sedimentation tanks, activated sludge tanks, secondary sedimentation tanks, chlorination, and sand filtration.

The treated effluent is used to irrigate ornamental trees and shrubs. Buried drip irrigation lines with emitters and bubblers above the ground surface are used to distribute the treated wastewater.[7,10]

The treated effluent is also used for landscape irrigation along highway median and border strips, along main roads, on traffic circles in Muscat, as well as in some parks and recreational areas. Sludges at the Darsait plant are treated by anaerobic digestion and then dewatered before being used as fertilizer.[6] Other wastewater treatment plants in the Muscat area include: Al-Amerat, Al-Aynt, and five other facilities in the region. The treated wastewater from the Darsait and Ansab plants irrigate an estimated 160 ha (395 ac) using an application rate of 1 cm/d.[10]

The Ministry of Regional Municipalities and Environment in Muscat recommends that the treated wastewater be used for agricultural irrigation, especially for date palms and alfalfa which account for about 46% of the total cultivated crops in Oman.[4] Another recommended use for the treated wastewater is for recharging freshwater and brackish water aquifers, especially in areas where there is considerable withdrawal of groundwater for agricultural irrigation. It is also anticipated that recharging the groundwater aquifers by well injection could have an economic influence, as this reuse process would be less costly than sea water desalinization.[11]

V. WASTEWATER TREATMENT AND REUSE AT THE SULTAN QABOOS UNIVERSITY

The Sultan Qaboos University (SQU) generates about 3000 m³/d (0.8 mgd) of wastewater. SQU has a wastewater treatment facility that provides preliminary, primary, secondary, and tertiary treatment and includes comminution, screens, primary sedimentation, activated sludge, secondary sedimentation, rapid sand filtration, and chlorination. The treated effluent is then used for landscape irrigation on the university campus. The digested sludges are stabilized on sand drying beds and then used as fertilizer.[7,12]

VI. WASTEWATER QUALITY AND REUSE STANDARDS IN OMAN

The quality, standards, and regulations for wastewater reclamation and reuse in the Sultanate of Oman are set by the Ministry of Regional Municipalities and Environment. Details as to these standards and regulations will be found in References 8, 13, 14, 15, and 16. Excerpts from the most recent regulations dealing with wastewater reuse and discharge for the Sultanate of Oman are presented in Table 10.29 and Table 10.30.[16]

Table 10.29 Wastewater Reuse — Areas of Application of Standards A and B (See Table 10.30)

	A	B
Crops	Vegetables likely to be eaten raw Fruit likely to be eaten raw and within 2 weeks of any irrigation	Vegetables to be cooked or processed Fruit if no irrigation within 2 weeks of cropping Fodder, cereal and seed crops
Grass and Ornamental areas	Public parks, hotel Lawns recreational areas Areas with public access Lakes with public contact (except places which may be used for praying and hand washing)	Pastures Areas with no public access
Aquifer recharge	All controlled aquifer recharge	
Method of irrigation	Spray or any other method of aerial irrigation not permitted in areas with public access unless with timing control	
Any other re-use applications	Subject to the approval of the Ministry	

Compiled from Reference 16.

Table 10.30 Wastewater — Maximum Quality Limits

	Standards (See Table 10.29)	
	A	B
Parameter	(mg/L except where noted)	
Biochemical oxygen demand (BOD) (5 d @ 20°C)	15.000	20.000
Chemical oxygen demand (COD)	150.000	200.000
Suspended solids (SS)	15.000	30.000
Total dissolved solids (TDS)	1500.000	2000.000
Electrical conductivity (EC) (µS/cm)	2000.000	2700.000
Sodium absorption ratio (SAR)[a]	10.000	10.000
pH (within range), pH units	6–9.000	6–9.000
Aluminum (as Al)	5.000	5.000
Arsenic (as As)	0.100	0.100
Barium (as Ba)	1.000	2.000
Beryllium (as Be)	0.100	0.300
Boron (as B)	0.500	1.000
Cadmium (as Cd)	0.010	0.010
Chloride (as Cl)	650.000	650.000
Chromium (total as Cr)	0.050	0.050
Cobalt (as Co)	0.050	0.050
Copper (as Cu)	0.500	1.000
Cyanide (total as CN)	0.050	0.100
Fluoride (as F)	1.000	2.000
Iron (total as Fe)	1.000	5.000
Lead (as Pb)	0.100	0.200
Lithium (as Li)	0.070	0.070
Magnesium (as Mg)	150.000	150.000
Manganese (as Mn)	0.100	0.500
Mercury (as Hg)	0.001	0.001
Molybdenum (as Mo)	0.010	0.050
Nickel (as Ni)	0.100	0.100
Nitrogen: Ammoniacal (as N)	5.000	10.000
Nitrate (as NO_3)	50.000	50.000
Organic (Kjeldahl, as N)	5.000	10.000
Oil and grease (total extractable)	0.500	0.500

Table 10.30 (continued) Wastewater — Maximum Quality Limits

Parameter	Standards (See Table 10.29)	
	A	B
	(mg/L except where noted)	
Phenols (total)	0.001	0.002
Phosphorus (total as P)	30.000	30.000
Selenium (as Se)	0.020	0.002
Silver (as Ag)	0.010	0.010
Sodium (as Na)	200.000	300.000
Sulfate (as SO_4)	400.000	40.0000
Sulfide (total as S)	0.100	0.100
Vanadium (as V)	0.100	0.100
Zinc (as Zn)	5.000	5.000
Fecal coliform bacteria (per 100 mL)	200.000	1000.000
Viable nematode ova (/L)	<1.000	<1.000

a The effect of sodium on soil absorption.

VII. HOMEWORK PROBLEMS

1. Identify and explain the following acronyms: SWD, VSB, SFS, FWS, $CBOD_5$, TDH, AWT, and LDP.
2. What parameters are normally included in a NPDES permit regarding discharge limitation as they relate to reclaimed wastewater quality?
3. Indicate the processes normally used at a secondary wastewater treatment facility where the effluent is to be used for a beneficial purpose. Also indicate the purpose of each unit using these processes in their treatment operations.
4. What are the four principal cost components in a wastewater reclamation and reuse project?
5. What are at least five factors that influence the economics of developing a wastewater reclamation and reuse facility?
6. In the construction and operation of a wastewater reclamation and reuse project, what items fall under the capital cost construction category, and what items are included in the operational and maintenance costs?
7. A secondarily treated wastewater effluent with an average flow of 76,000 m³/d is to be chlorinated. The peak flow is 150% of the average flow and the effluent has a chlorine demand of 10 mg/L and a chlorine residual of 1.5 mg/L must be maintained after a 30-min contact time. Determine the volume of the chlorine contact basin and the average and peak chlorine dosages in kg/d.
8. A polymer feed system doses 2 mg/L to a 56,000 m³/d flow of treated wastewater in a flash mixing tank 5.5 × 3.0 × 3.0 m. What is the detention time in this tank, as well as the polymer dosage rate in kg/d?
9. Six Dynasand filters filter 34,000 m³/d of treated effluent. Each filter has four modules and each module has an area of 4.8 m². Calculate the average filtration rate. If the average filtration is increased to 0.43 m³/m²/min, what volume of treated wastewater could then be filtered each day?
10. How does the U.S. Federal Administration presently define wetlands? What are the two basic types of wetlands and what types of plants are found in each?
11. Indicate the various mechanisms involved in the removal of contaminants in the operation of a wetland treatment system.
12. What are the benefits provided by having healthy riparian areas adjacent to streams, rivers, or lakes?
13. Design a constructed wetland system using a free water surface (FWS) scheme that produces an effluent with a BOD_5 of 5 mg/L. The influent BOD_5 is 38 mg/L. The summer water temperature is 24.1°C and 4.5°C in winter. The design flow is 946 m³/d. The depth of flow is 0.45 m, and the wetlands slope is over 1%. The precipitation in this area is 135 cm/yr and the evapotranspiration is about the same. Calculate the wetland area requirement, as well as the hydraulic and organic loading rate, and check if these loading rates are in the appropriate range.
14. What natural control methods can be used to limit mosquitoes and other insect populations at land treatment storage facilities or wetland treatment systems?

15. What is the limiting total nitrogen concentration for wastewater that has percolated through the sand, soil, and rock at a wastewater land treatment site?
16. What are the common forms in which phosphorus is found in reclaimed wastewater planned for reuse?
17. A sprinkler system is selected for application of the wastewater for irrigation. The sprinklers are spaced in a rectangular grid pattern 10 m by 20 m, and each sprinkler nozzle discharges 2 L/s. What is the application rate in cm/hr? How many hours must the system be operated in a single area each week to satisfy the application rate of 6.5 cm/week?
18. Determine the hydraulic loading rate, the land area requirement, and storage volume required for a slow-rate land application system given the following information. The reclaimed water volume is 40,000 m^3/d, the precipitation is 80 cm/yr, the evaporation rate is 120 cm/yr (including uptake by the trees). The soil permeability is 0.75 cm/hr (percolation rate 6.3 cm/wk). The land treatment system can operate throughout the year. Governmental regulations require a 10-day storage period. The retention pond is to be 3 m deep. What storage volume is required? Compare the $L_{W(n)}$ with the $L_{W(p)}$ under the following conditions: the nitrogen concentration in the treated wastewater is 20 mg/L, the nitrogen uptake by the trees and plants on the site is 264 kg/ha/yr, the denitrification loss is 20%, and the nitrogen concentration in the percolate is 10 mg/L. Is $L_{W(n)}$ or $L_{W(p)}$ the limiting design parameter?
19. Compare the wastewater reclamation and reuse water quality standards for the Sultanate of Oman with those in the NPDES permits for the IRIS project at Boca Raton, FL, the Logan Aluminum Wetlands Project at Russellville, KY, and the E.L. Huie, Jr. Land Application Facility in Clayton County, GA.
20. Prepare a case study for any wastewater project listed in Chapter 11, Section IV. G. except for the E.L. Huie, Jr. Land Application Facility, Clayton County, GA, the IRIS project at Boca Raton, FL, and the Logan Aluminum Wetlands Project which have already been reported in this chapter. The case study should include the following:
    - Cover page
    - Name of reporter
    - Project name
    - Introduction
    - Reclaimed water production
    - Background of reuse project
    - Site selection
    - Design factors
    - Type of use
    - Volume of water reused
    - Physical, chemical, and biological monitoring
    - Effluent characteristics
    - Economics (costs involved)
    - Summary and conclusions
21. Review an article or publication that deals with a wastewater reclamation and reuse project. The review should be from two to three pages and present the significant elements involved in the project.

REFERENCES — CASE STUDY I

1. **York, D. W., and J. Cook,** Florida's Reuse Program Paves the Way, *Water Environ. Technol.,* 2, 12; 1990, 72.
2. FDER, Reuse of Reclaimed Water, Department of Environmental Regulation, Tallahassee, FL. 1991.
3. **Duplaix, N. and K. Fleming,** South Florida Water, Paying the Price, *National Geographic,* 178, 1, July 1990, 89.
4. **Chansler, J. M.,** The Future for Effluent Reuse, *Water Eng. Manage.,* May 1991, 31.
5. SFWMD, Annual Report, 1991, South Florida Water Management District, P.O. Box 24680, West Palm Beach, FL, 1991.
6. **York, D.** Reuse in Florida — An Update, *Fla. Water Resources J.,* August, 1991, 6.
7. **Kruger, K.,** Boca Homes Will be Able to Use Recycled Water, *Boca News,* December 12, 1991, 6.
8. CDM, *City of Boca Raton, Florida, Reclaimed Water System Master Plan,* CDM 6004–115-RT, Camp, Dresser & McKee, Inc., Fort Lauderdale, FL, September, 1990.

9. **Murphy, R. C. and J. C. Hancock,** Urban Reuse in the City of Boca Raton, Florida: The Master Plan. Presented at the WPCF 46th Annual Conference and Exposition, Toronto, October 7–10, 1991.
10. Glades Road Wastewater Treatment Expansion, 201 West Palmetto Park Road, Boca Raton, FL, 1989.
11. **Vidaurreta, E. J.,** O & M Superintendent, City of Boca Raton, 1501 W. Glades Road, Boca Raton, FL, NPDES Permit No. FL0026344, 1989.
12. **Chansler, J. M.,** Director of Public Utilities, City of Boca Raton, Personal Communication, September, 1993.
13. **Chansler, J. M.,** Director of Public Utilities, City of Boca Raton, Boca Raton, FL, FDER Permit No. DC 50–199935, 1992.
14. **Riley, S. and R. Hungate,** What Price Water Reuse, *Fla Water Resources J.,* August, 1991, 38.
15. **Davis, P. A., et al.,** A Comprehensive Review of Effluent Reuse Potential In Broward County, Florida, in *Proc. Urban Agricultural Water Reuse,* WEA and FPCA, Orlando, FL, 1992, 407.
16. **Castellanos, A.,** *Purgeable Organics Analysis,* Spectrum Laboratories, Inc., Fort Lauderdale, FL, 1991.

REFERENCES — CASE STUDY II

1. Anon., State of Kentucky's Environment: A Report of Progress and Problems, Environmental Quality Commission, Frankfort, 1992.
2. USDOC, National Oceanic & Atmospheric Administration, Climate data for Russellville, KY, NS CLIMAT., 157049, Mar. 9, 1993.
3. *The Encyclopedia Americana International Edition,* Grolier Inc., Danbury, CT, Vol. 16, 1986, 375.
4. **Wood, P. A. and K. Pidgeon,** Kentucky Department for Environmental Protection, Personal Communication, October, 1993.
5. **Lemonick, M. D.,** War Over the Wetlands, *Time,* August 26, 1991, 42.
6. **Mallory, B.,** Wetlands and a New Clean Water Act, *Water Environ. Technol.,* 5, 9, 1993, 11.
7. 23rd Annual Report of the Council on Environmental Quality, Washington, D.C., 1992, 235.
8. WPCF, Chapter 9, Wetlands Systems, *Natural Systems for Wastewater Treatment,* Manual of Practice FD-16, Water Pollution Control Federation, Alexandria, VA, 1990.
9. EPA, Design Manual — Constructed Wetlands and Aquatic Plant Systems for Municipal Wastewater Treatment EPA/625/1–88/022, Office of Research and Development, U.S. Environmental Protection Agency, Washington, D.C., 1988, chap. 4.
10. **Reed, S. C., E. J. Middlebrooks, and R. W. Crites,** *Natural Systems for Waste Management and Treatment,* McGraw-Hill, New York, 1988, chap. 6.
11. **De Boer, J., and K. D. Linstedt,** Advances In Water Reuse Applications, *Water Research,* U.K., 19, 11, 1985, 1455.
12. WPCF, *Water Reuse,* 2nd ed., Manual of Practice SM-3, Water Pollution Control Federation, Alexandria, VA, 1988, 159.
13. **Robert Adams,** Regional Supervisor for The Kentucky Cabinet For Environmental Protection, Bowling Green, 1993.
14. **Steiner, G. R., J. T. Watson, and D. A. Hammer,** Constructed Wetlands For Municipal Wastewater Treatment, Tennessee Valley Authority, Office of Natural Resources and Economic Development, Division of Air and Water Resources, Chattanooga, TN, March, 1988.
15. **Choate, K. D., G. R. Steiner, and J. T. Watson,** First Semi-Annual Monitoring Report Demonstration of Constructed Wetlands For Municipal Wastewater, March to December, 1988, TVA, WR, WQ, 89–5, Chattanooga, TN, 1989.
16. **Peavy, H. S., D. R. Rowe, and G. Tchobanoglous,** *Environmental Engineering,* McGraw-Hill, New York, 1985.
17. **Jillson, W. R.,** Geology and Mineral Resources of Kentucky, The Kentucky Geological Survey, Frankfort, KY, 1928, 223.
18. **Koeing, B.,** Anaconda to Logan, *Park City Daily News,* Bowling Green, KY, January 29, 1981, and March 31, 1982.
19. **Riley, J. M., and H. A. Wojnar,** A Case Study: Logan Aluminum, Russellville, KY, Problems and Solution Alternatives Associated with Zero Discharge and Re-Use of Treated Wastewater, KY-TN WPCA, 46 Annual Meeting, May, 1992.
20. Logan Aluminum Gets State Award, *The Courier-Journal,* Louisville, KY, 9/27/91, 4C.
21. **Zick, E. A., H. Wojnar, and P. Porter,** The Wetlands — Recycling Nature's Way — A Case Study, Unpublished Report, Logan Aluminum, Inc., Russellville, KY, 1993.
22. **Zick, E. A.,** Wetland Project An Economical Answer, *Plant Services,* February, 1992, 75.
23. **Clark, T.,** Wetlands Unit Purifies Recycles Wastewater, *Inside Environment,* 4, 4, 1992, 10.
24. **Reed, S. C., and D. S. Brown,** Constructed Wetland Design — The First Generation, *Water Environ. Res.,* 63(6), 1992, 776.
25. **Porter, P.,** Environmental Coordinator, Logan Aluminum, Inc., P.O. Box 300, Russellville, KY, Department for Environmental Protection, Frankfort, KY, NPDES No. KY0072630, Oct., 1991.
26. **Smith, S. E.,** Is the Wetland Program Falling Further into Disarray?, *Inside Environment,* August, 1992, 10.
27. Aluminum Plant Treats Waste With Wetland, *U.S. Water News,* 9, 3, 1992, 1.

REFERENCES — CASE STUDY III

1. *The World Almanac and Book of Facts 1994,* Funk & Wagnalls, Mahwah, NJ, 1994, 636.
2. *New Webster's Dictionary and the Thesaurus,* Lexicon Publications, Inc., Danbury, CT, 1993.
3. WPCF, *Natural Systems for Wastewater Treatment,* Manual of Practice FD-16, Water Pollution Control Federation, Alexandria, VA, 1990, chapter 4.
4. **Nutter, W. L.,** Forest Land Treatment of Wastewater in Clayton County, Georgia: A Case Study, *Proc. Forest Land Appl. Symp.,* College of Forest Resources, University of Washington, Seattle, 1985.
5. USDOC, National Oceanic & Atmospheric Administration, climate data for Atlanta, WSO AP, Georgia, NS, CLIMAT 090451, Mar. 9, 1993.
6. *The New Encyclopaedia Britannica,* Encyclopaedia Britannica, Inc., 15th ed., Chicago, Vol. 29, 1986.
7. **Reed, S., and R. Bastian,** Potable Water Via Land Treatment and AWT, *Water Environ. Technol.,* Alexandria, VA, 3, 8, 1991, 40.
8. **Earn, Ernest, V.,** Technology Transfer Coordinator, Environmental Protection Division, Georgia Dept. of Natural Resources, Atlanta, GA, personal communication, 1992.
9. WPCF, *Water Reuse,* 2nd ed., Manual of Practice SM-3, Water Pollution Control Federation, Alexandria, VA, 1989, 90, 153.
10. **Pettygrove, G. S., and T. Asano,** Ed., *Irrigation with Reclaimed Wastewater — A Guidance Manual,* Lewis Publishers, Chelsea, MI, 1985.
11. Metcalf & Eddy, Inc., *Wastewater Engineering: Treatment, Disposal, Reuse,* 3rd ed., McGraw-Hill, New York, 1991.
12. **Reed, S. C., E. J. Middlebrooks, and R. W. Crites,** *Natural Systems for Waste Management and Treatment,* McGraw-Hill, New York, 1988.
13. USEPA Process Design Manual: Land Treatment of Municipal Wastewater, EPA 62 5/1–81–013, EPA Center for Environmental Research Information, U.S. Environmental Protection Agency, Cincinnati, OH, 1981.
14. Northeast Clayton County Water Pollution Control Plant, Clayton County Water Authority, Morrow, GA, 1993.
15. DOC, Department of Commerce, Bureau of Census U.S.A. County Statistics, City County Data Base, Government Services Library, WKU, Bowling Green, KY, 1990.
16. E. L. Huie, Jr. Land Application Facility, Clayton County Water Authority, Morrow, GA, 1992.
17. Facts And Figures, Clayton County Water Authority, Morrow, GA, 1993.
18. Newman, Melvin, General Manager, Clayton County Water Authority, Morrow, GA, personal communication, 1994.
19. USEPA, Innovative and Alternative Technology Projects: 1986 Progress Report, Office of Municipal Pollution Control (WH-595), U.S. Environmental Protection Agency, Washington, D.C., 1986.
20. EPA — An Alternative Technology Application — Land Treatment Silvicuture — A Practical Approach, Clayton County Water Authority, Morrow, GA, 1993.
21. **Barbour, R. W., and W. H. Davis,** *Bats of America,* University Press of Kentucky, Lexington, 1969.
22. **Bull, J., and J. Farrand, Jr.,** *The Audubon Society Field Guide to North American Birds,* Alfred A. Knopf, Inc., New York, 1989.
23. **Ehlers, C. E., and E. W. Steel,** *Municipal And Rural Sanitation,* 5th ed., McGraw-Hill, New York, 1965, 317.
24. **Clark, M. J.,** personal communication, 1994.
25. **Rubin, A. R., D. Frederick, and R. Milosh,** Wastewater Irrigation Onto Managed Forest Lands, Water Environment Federation 66th Annual Conference & Exposition, Anaheim, CA, 1993.
26. Clayton County Water Authority, Land Application System Permit, No. GA02–008, Environmental Protection Division, Department of Natural Resources, State of Georgia, Atlanta, May 1991.
27. USEPA, Innovative And Alternative Technology Projects: A Progress Report, Office of Water, Programs Operations (WH-547), U.S. Environmental Protection Agency, Washington, D.C., 1984.
28. Construction Scoreboard, Market Trends, *Engineering News-Record, ENR,* Hightstown, NJ, 1976, 1977, 1987, 1994.
29. **Chansler, J. M.,** Director of Utilities, City of Boca Raton, Boca Raton, FL, personal communication, 1994.
30. **Culp, G., G. Wesner, R. Williams, and M. V. Hughes, Jr.,** *Wastewater Reuse And Recycling Technology,* Noyes Data Corp., Park Ridge, NJ, 1980.
31. **Davis, P. A., et al.,** A Comprehensive Review of Effluent Reuse Potential in Broward County, Florida, in Proc. Urban and Agricultural Water Reuse, WEA and FPCA, Orlando, FL, 1992, 407.
32. **Reed, S. C., and R. W. Crites,** *Handbook of Land Treatment Systems for Industrial and Municipal Wastes,* Noyes Data Corp., Park Ridge, NJ, 1984, 334.
33. Municipal Cost Indexes, *American City & County,* Atlanta, GA, 1977 to 1988 to 1994.

REFERENCES — SULTANATE OF OMAN

1. Results of the National Census, *Al-Watan Newspaper,* 23, December 28, 1, 1993, Muscat, Sultanate of Oman, (in Arabic).
2. *The Concise Columbia Encyclopedia,* J. S. Levy and A. Greenhall, Eds., Columbia University Press, New York, 1983, 622.

3. *The New Encyclopaedia Britannica,* Encyclopaedia Britannica, Inc., 15th ed., Chicago, Vol. 8, 1991, 945.
4. Ministry of Regional Municipalities and Environment, National Conservation Strategy: Environmental Protection and Natural Resources Conservation for Sustainable Development, Ministry of Regional Municipalities and Environment, Muscat, Sultanate of Oman: Vol. I, Synthesis and Policy Framework, 1992; Vol. II, Supporting Annexes, 1992.
5. **Al-Shaqsi, S.** Water Resources in Oman: The Present and Future, Paper presented at the National Seminar On Wastewater Reuse, organized by the Ministry of Health, Muscat, Sultanate of Oman, April 26–29, 1992.
6. **Palsson, L.,** Wastewater Treatment Practices in Oman, Paper presented at the National Seminar on Wastewater Reuse, organized by the Ministry of Health, Muscat, Sultanate of Oman, April 26–29, 1992.
7. **Abdel-Rahman, H. A., and Abdel-Magid, I. M.,** Water Conservation in Oman, *Water Int.,* 18(2), 95, 1993.
8. Ministry of Environment and Water Resources, Sultanate of Oman Law on the Conservation of the Environment and Prevention of Pollution: Regulations for Waste-Water Reuse and Discharge, Ministerial Decision 5/86 dated 17th May, Muscat, 1986.
9. Ministry of Environment and Water Resources, Sultanate of Oman Law on the Conservation of the Environment and Prevention of Pollution: Regulations for Septic Tanks and Holding Tanks, Ministerial Decision 5/86 dated 17th May, Muscat, 1986.
10. **Abdel-Rahman, H. A.** Appropriate Wastewater Irrigation Practices as Related to Crop Selection in Oman, Paper presented at the National Seminar on Wastewater Reuse, organized by the Ministry of Health, Muscat, Sultanate of Oman, April 26–29, 1992.
11. **Lootens, D. J.** Water Resource Management and Wastewater Reuse: A Case Study, Paper presented at the National Seminar on Wastewater Reuse, organized by the Ministry of Health, Muscat, Sultanate of Oman, April 26–29, 1992.
12. Sultan Qaboos University, Building and Planning Section, Muscat, Oman, personal communication, 1994.
13. Ministry of Commerce and Industry, Sultanate of Oman Standards: Drinking Water, No. 8, General Department of Standards and Weight, Muscat, Oman, 1978.
14. **Abdel-Magid, I. M., and El-Zawahry, A.,** Establishment of Water Quality Guidelines for the Sultanate of Oman, *Arab Water World J.,* September-October, 1992, 18.
15. **Abdel-Magid, I. M., and El-Zawahry, A.,** Preconditions and Requirements for Successful Environmental Policies in the Sultanate of Oman, the Sudan and Egypt, Conference on Preconditions and Requirements for Successful Environmental Policies in the Arab World, organized by the Earth and Environmental Science Department, Yarmouk University; the National Program for Environmental Awareness and Information; and Friedreich Naumann Stiftung, Irbid, Jordan, May, 1993.
16. Ministry of Regional Municipalities and Environment, Regulations For Wastewater Re-use And Discharge, Ministerial Decision 145/93 dated 13 June, Muscat, 1993.

Chapter 11

Wastewater Reclamation and Reuse Research

CONTENTS

I. Introduction 484
II. Sources of Information For Wastewater Reclamation and Reuse Research 495
    A. Literature Search 485
    B. Some Selected Informational Sources 485
        1. National Technical Information Service (NTIS) 485
        2. U.S. Environmental Protection Agency Public Information Center 487
        3. U.S. Environmental Protection Agency/VISITT Program 487
        4. U.S. Government Printing Office (GPO) 488
        5. The Federal Bulletin Board 489
        6. U.S. Department of the Interior (DOI) 489
        7. American Waterworks Association (AWWA) Waternet on CD-ROM 489
        8. Other Databases as Sources of Information For Wastewater Reclamation
           and Reuse 490
III. Potential Sources For Research Funds For Wastewater Reclamation and Reuse 491
    A. Introduction 491
    B. Major U.S. Governmental Agencies Involved in Water Resources Research 491
        1. U.S. Environmental Protection Agency 491
        2. U.S. Department of Defense 492
        3. U.S. Department of Energy 492
        4. U.S. Department of the Interior 493
        5. U.S. Department of Commerce 493
        6. U.S. Department of Health and Human Services 494
        7. U.S. Department of Agriculture 494
        8. The National Science Foundation 494
        9. The National Aeronautics and Space Administration 495
    C. Water Environment Research Foundation 495
    D. American Water Works Association Research Foundation 501
    E. The National Water Research Institute 503
IV. Wastewater Reclamation and Reuse Research 504
    A. Introduction 504
    B. Scope of the Wastewater Reclamation and Reuse Research Areas 504
        1. Wastewater Reuse — Physical Problems 505
        2. Wastewater Reuse — Chemical Problems 505
        3. Wastewater Reuse — Biological Problems 507
           a. Bacterial Pathogens 507
           b. Protozoan Pathogens 507

    c. Viral Pathogens 507
    d. Helminth Pathogens 508
  C. Societal Issues 508
  D. World Health Organization 508
  E. World Bank and World Health Organization 509
  F. Inventory For Wastewater Reclamation and Reuse 509
  G. Some Significant Wastewater Reclamation and Reuse Projects 513
 V. Homework Problems 513
References 514

I. INTRODUCTION

Research has been defined as a careful, systematic, patient scientific investigation in some field of knowledge aimed at uncovering and applying new facts, principles, techniques, and natural laws.[1,2]

A more expanded definition indicates that research is the primary activity of science and is a combination of theory and experimentation directed toward finding scientific explanations of phenomena. It is commonly classified into two types: pure research, involving theories with little apparent relevance to human concerns (invention); and applied research, concerned with finding solutions to problems of social importance, for instance in medicine and engineering (innovation). The two types are linked in that theories developed from pure research may eventually be found to be of great value to society.

Scientific research is most often funded by government and industry, and so a nation's wealth and priorities are likely to have a strong influence on the kind of work undertaken.[3,4]

In broad general terms, basic, pure, or fundamental research is carried out by scientists while applied research includes engineers, technologists, technicians, lawyers, economists, and health professionals. Currently, much of the research being done is by teams of specialists from many disciplines, and is often referred to as research and development (R & D).

Research and development not only includes the scientific investigation of theories, general principles, and observed phenomena, but also the practical application of this information including the experimental production and testing of models, devices, equipment, materials, and processes.[5]

Modern day R & D emerged from the demands made on military technology during World Wars I and II. Government-sponsored military R & D led to "spin-off" civilian industrial developments in areas such as aircraft engines, nuclear power, and computer electronics.

Government agencies are still the principal sponsors of R & D, contracting projects either to directly owned laboratories or to independent, company, or university laboratories. Research associations, which pool the R & D resources of an entire industry, have proven particularly effective.[6]

Not only can the research and development teams do basic and applied research, but they can also make evaluations of the efficiency and cost of a product.

Another term often used in the research area is operations research/management science (OR/MS). The terms operations research and management science can be used interchangeably as either OR/MS or MS/OR.[4,7]

MS/OR can be defined as the systematic study of a problem involving gathering data, building a mathematical model, experimenting with the model, predicting future operations, and getting the support of management for the use of the model.[7]

A slightly more extensive explanation of operations research is that it is a scientific approach to the analysis of many kinds of complex decision-making problems — economic, engineering, or environmental — as encountered by individuals and organizations of all types, be they profit making or nonprofit, private or governmental. Often the problem studied involves the design and/or the operation of systems or parts of systems. The aim is the evaluation of probable consequences of decision choices, usually under conditions requiring the allocation of scarce resources —funds, manpower, time, or raw materials. The objective is to improve the effectiveness of the system as a whole, with emphasis on the last three words.[8]

Operations research relies on six simple, methodological steps for its success:[9]

- Formulate the problem.
- Construct a model of the system.
- Select a solution technique.
- Obtain a solution to the problem.

- Establish controls over the system.
- Implement the solution.

Figure 11.1 presents a flow diagram of these six basic steps.[9]

OR/MS can be considered to be mainly an operational tool involved in formulating mathematical models used to solve problems, while R & D is mainly involved in industrial research projects to develop and evaluate the economies of the finished product. Ideally, a successful research program includes both the procedures and techniques in operations research/management science and research and development.

II. SOURCES OF INFORMATION FOR WASTEWATER RECLAMATION AND REUSE RESEARCH

A. LITERATURE SEARCH

One of the first steps in carrying out a research project is to do a critical review and evaluation of the accumulated body of knowledge for that topic.

There are many sources of information available to do a literature search and these include:

- Periodicals
- Published proceedings of conferences
- Research reports
- Patents
- Theses and dissertations
- Manufacturer's literature
- Handbooks
- Encyclopedias
- Reviews
- Abstracts
- Computer databases and databanks
- Textbooks

B. SOME SELECTED INFORMATIONAL SOURCES

Following are some excellent informational sources that provide extensive services in almost any scientific or technical research area.[10]

1. National Technical Information Service (NTIS)
5285 Port Royal Road, Springfield, VA 22101

Description of services. NTIS, an agency of the U.S. Department of Commerce, is the central source for the public sale of U.S.- and foreign government-sponsored research, development, engineering, and business reports. NTIS also provides information in electronic formats, including datafiles and software on magnetic tape and diskette. It also maintains a growing CD-ROM collection.

Approximately 70,000 new technical reports of completed research are added annually to the NTIS database. Anyone seeking the latest technical reports may search the NTIS Bibliographic Database online, using the services of vendors or organizations that maintain the database for public use. The entire database in machine readable form may be leased directly from NTIS.

NTIS is a unique agency supported by its customers. All costs such as salaries, production, acquisition, marketing, and postage are paid from sales income, not from tax-supported congressional appropriations.

The NTIS pilot project, FedWorldÖ, allows access to more than 100 computer bulletin board systems operated by the U.S. Government through the FedWorld GateWayÖ. These systems contain hundreds of useful datafiles, programs, databases, and information. You can browse and search a rapidly growing number of NTIS data files including a weekly listing of NTIS's 25 best sellers. In addition to gateway services, FedWorldÖ includes a bulletin information center, a library of computer files, conference areas where users can exchange information, online help, the ability to search and retrieve files and bulletins, and a teleconference area.

Primary focus. NTIS focuses on technical and business information from more than 200 government agencies, with a heavy emphasis on the publications of the Departments of Commerce, Defense, Energy, Health and Human Services, National Aeronautics and Space Administration (NASA), and the Environmental Protection Agency. NTIS provides archival service for all of its publications.[10]

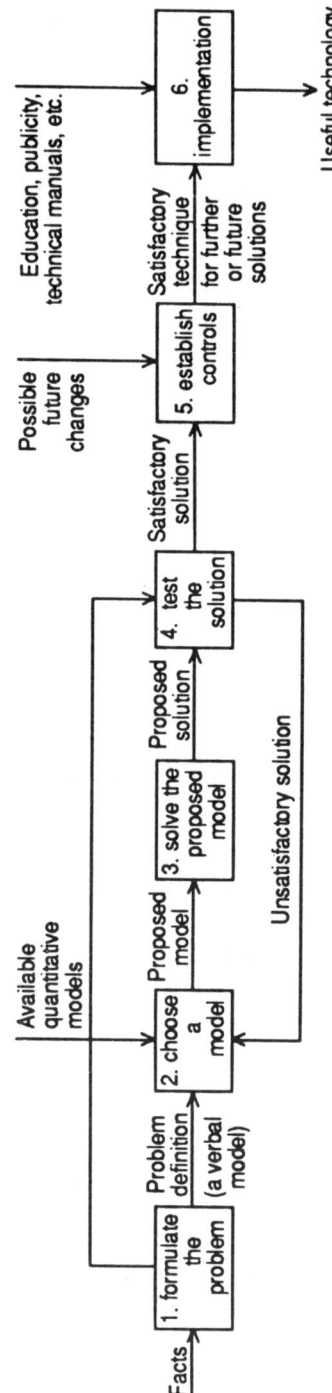

Figure 11.1 Operations research approach: the six basic rules of success. (From Parker, S. P., Ed., *Dictionary of Scientific and Technical Terms*, 3rd ed., McGraw-Hill, New York, 1984, 1362. With permission.)

An NTIS literature search can be made through the following Online Service Systems provided you have the necessary accession number and have determined the key words for the research topic.

Online Service Orders:

BRS: (800) 456–7248 or (800) 289–4377
CISTI, in Canada: (613) 993–1210
DATA-STAR: (800) 221–7754
DIALOG: (800) 334-2564
ESA/IRS, in Italy: (39/6) 941801 (Fax)
ORBIT: (800) 456–7248, in Virginia: (708) 442–0900
STN International: (800) 848–6533, in Ohio and Canada: (800) 848–6588.

There is a charge for this service.

2. U.S. Environmental Protection Agency Public Information Center

Publication No. GPO-055–000–00437–4
NTIS — PB93-170041

Description of Publication: EPA produces the ACCESS EPA directory to improve access to environmental information services provided by EPA and related public sector organizations. The directory is maintained by EPA's Office of Information Resources Management (OIRM) to support the Agency's mission. Because EPA believes that public access to environmental information fosters environmental awareness, the directory is made available through the Government Printing Office (GPO) and the National Technical Information Service (NTIS). The directory provides contact information and a description of services provided. ACCESS EPA includes information on documents, dockets, clearinghouses and hotlines, records, databases, models, EPA libraries, and state libraries.

Online access: ACCESS EPA is available online through the EPA Library Online Library System (OLS), and the GPO Federal Bulletin Board System (GPO BBS).

OLS access: for dial-in access, call (919) 549–0720. Use these transmission and format settings: 300–9600 baud rate, half duplex, even parity, 7 data bits per character, one Stop Bit.

To login: dial into the system. At the first system prompt, select OLS. At second menu, select A.
To Print: printing is only available through your telecommunications software.
To logout: Type Q or QUIT at the system prompt.

Access through Internet: Universities and institutions with Internet connection are encouraged to access OLS using Internet links to EPA. EPA's Internet address is EPAIBM.RTPNC.EPA.GOV. At the first menu, select "Public Access". At the second menu, select OLS. At third menu, select A.[10]

3. U.S. Environmental Protection Agency/VISITT Program

A fairly recent EPA database is the Vendor Information System for Innovative Treatment Technologies, VISITT.[11]

The Technology Innovation Office (TIO) of the U.S. Environmental Protection Agency (EPA) Office of Solid Waste and Emergency Response (OSWER) has developed the Vendor Information System for Innovative Treatment Technologies (VISITT) to provide current information on innovative treatment technology for the remediation of contaminated sites. VISITT contains technology information submitted by developers, manufacturers, and suppliers of innovative treatment technology equipment and services. This database provides a means for innovative technology vendors to make their products and capabilities known to state, federal, and private sector professionals.

VISITT Version 1.0 was first released in June, 1992 and the VISITT Version 2.0 in April, 1993. EPA now has available VISITT 3.0 which was released in September 1994.

VISITT is available on diskettes compatible with personal computers that use DOS operating systems.

Figure 11.2 shows the portion of VISITT 2.0 technologies represented by each major technology category. VISITT 2.0 contains data on 231 technologies offered by 141 companies.[12]

The technologies of most interest in VISITT in the wastewater reclamation and reuse field include the control of groundwater contamination as well as biological, chemical, and treatment technologies for contaminant removal.[11]

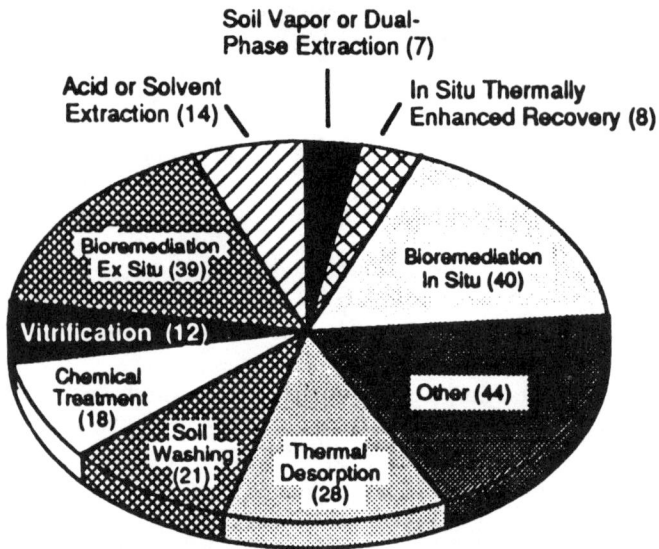

Figure 11.2 Major technology types.[12]

To register as a VISITT user, the following information is required:

- Name
- Organization
- Address
- Phone number
- Diskette size (3–1/2" or 5–1/4")
- Indicate if the request is to order the VISITT software or to register as a VISITT user only.

Or, mail the information to:

U.S. EPA/NCEPI
P.O. Box 42419
Cincinnati, OH 45242-0419

The VISITT system software requires an IBM-compatible computer running MS-DOS 3.3 or higher and at least 640K of RAM. You should have at least 4 megabytes of free disk space, and the file handles statement (FILES=n) for at least 45 should be included in your Config.sys file. Note that the VISITT installation program automatically changes the Config.sys configuration if the file handles statement is less than 45. For optimum performance, your environment space should be greater than 1024K.[11]

4. U.S. Government Printing Office (GPO)

Description of services. The mission of GPO is the production or procurement of printing for congress and the agencies of the federal government. GPO also disseminates information to the public through the Superintendent of Documents publications, sales, and depository library programs. Established as the official printer of the nation, GPO is part of the legislative branch of government.

Through its documents program, GPO disseminates what is possibly the largest volume of information in the world. The Superintendent of Documents offers approximately 12,000 titles to the public at any given time. These are sold principally by mail order and through government bookstores across the country.

Primary focus. GPO's primary mandate is to facilitate the printing of congressional work in an efficient and cost-effective manner. The *Congressional Record* and *Federal Register* are printed daily. Although it is often referred to as the "Nation's largest publisher", GPO neither initiates nor exercises control over the publications it sells. Virtually all government publications are issued by congress and the various agencies of the federal government. GPO prints or procures the printing of these publications and distributes them through its sales and/or depository library programs.

The Federal Bulletin Board, operated by GPO, enables Federal agencies to provide the public with immediate, self-service access to a variety of government information in electronic formats at reasonable rates.

5. The Federal Bulletin Board

The Federal Bulletin Board offers the public a way to obtain electronic documents in a rapid and cost-effective manner. If you have a personal computer, a modem, telecommunications software, and a telephone line, you can access The Federal Bulletin Board by calling (202) 512–1387. At any time of day or night, you can place an order for all GPO sales items or download files.

The following information from the Environmental Protection Agency is available on The Federal Bulletin Board: Access EPA, Information Resources; Code of Federal Regulations, Title 40; recent Federal Register documents; Federal Register Air Quality Designations; the Toxic Release Inventory; and "HyperVentilate", an EPA software program that assists in the determination of appropriate methods to ventilate contaminated soil (both Macintosh and DOS versions are available). The Federal Bulletin Board also carries a variety of government information from other federal agencies.[10]

Online service orders: The Federal Bulletin Board
(GPO_ORDR SIG)
DIALOG Information Retrieval System
(File Code 166, GPOPRF)

There is a charge for this service.

6. U.S. Department of the Interior (DOI)

The U.S. Department of the Interior (DOI), acting through the U.S. Geological Survey (USGS), provides financial support to some 54 state water resource research institutes located at land-grant universities in each of the 50 states and Puerto Rico, the District of Columbia, Guam, and the Virgin Islands.[13]

These water resource research institutes can be very helpful in providing information on publications dealing with wastewater reclamation and reuse through such database searchs as:[14]

- Selected water resources abstracts
- Enviroline
- Environmental bibliography

7. American Water Works Association (AWWA) Waternet on CD-ROM[15]

Some of the topics on the AWWA Waternet CD-ROM that have application in the wastewater reclamation and reuse field include:

- Drinking water treatment
- Lead and copper corrosion
- Reuse
- Ozonation
- Reverse osmosis
- Organics
- Water quality and analysis
- Sludge disposal
- Regulations and compliance
- Disinfection and byproducts
- Conservation
- Appropriate technology
- Alternative disinfectants
- Desalination
- Operation and maintenance
- Waterborne diseases
- Privatization
- Wastewater treatment

The minimum requirements for installing and running WATERNET are

- IBM PC (or compatible) with MS-DOS or PC-DOS, Version 3.0 or higher.
- 640K of RAM, with 500K of conventional RAM free.
- MS-DOS CD-ROM Extensions (MSCDEX) Version 3.1, 3.3, or higher.
- CD-ROM driver with ISO 9660 compatibility.
- Monochrome or color monitor.

Optimum performance can be achieved by upgrading the minimum requirements with the following:

- IBM PC (or compatible) with 80386 microprocessor or higher.
- MS-DOS or PC-DOS 5.00 or higher.
- 4MB RAM or more. Memory above 1MB configured as Extended or Expanded. (Extended is recommended.)
- Color VGA monitor.

8. Other Databases As Sources Of Information For Wastewater Reclamation And Reuse

Many sources are available to assist in a literature search that would be impossible to do by hand. The rapid development of electronic information and telecommunication systems have produced online databases. A recent directory, edited by Thomas F.P. Sullivan, *Directory of Environmental Information Sources,* lists over 140 databases of environmental information, many of which are available through one or more of the 15 different data services. Some databases can be accessed directly, some can be accessed only through a data service, and some can be accessed both ways.[16]

The price for either directly connecting to a database or for using a data service may include an initial fee, an annual fee, and a per-minute use fee.

Following are just three examples from the list of 140 databases that would clearly relate to the wastewater reclamation and reuse field.[16]

> Federal Research in Progress — FEDRIP[16]
> U.S. Department of Commerce
> National Technical Information Service
> Office of Data Base Services
> 5285 Port Royal Road
> Springfield, VA 22161
> 703/487-4807

FEDRIP contains descriptions of current and recently completed federally funded research projects in the physical sciences, engineering, and life sciences. All 297,828 records contain the project title, principal investigator, and the organizations sponsoring and conducting the research. Most records contain a description of the research, although the content of each depends on the sponsoring agency. One of the two subfiles is abridged and available worldwide. The other, available only in the U.S., is unabridged and includes U.S. Department of Energy (DOE) and National Aeronautics and Space Administration (NASA) records the abridged subfile doesn't. It is updated monthly and available on DIALOG and Knowledge Express.

> Innovative/Alternative Pollution Control Technology Facility File — IADB[16]
> U.S. Environmental Protection Agency
> Office of Water Program Operations
> 401 M Street, S.W.
> Washington, D.C. 20460
> 202/260-7277

This database combines information from the Grant Information and Control System (GICS), the Needs Survey (NEEDS), and the Financial Management System (FMS), with data on innovative and alternative wastewater treatment technologies entered by states and regions. It serves as a clearinghouse for municipalities and private developers considering using innovative/alternative technologies. For additional access information contact the University of West Virginia, National Small Flows Clearinghouse, at 800/624-8301.

Storage and Retrieval System of Water Quality Information[16]
U.S. EPA
401 M Street, S.W.
Washington, D.C. 20460
800/424-9067 or 202-260-7220

This database is a computerized utility maintained for the storage and retrieval of data pertaining to the quality of water (surface and groundwater) within the contiguous U.S. The STORET family includes several separate but related files: the Water Quality File, the BIOS Field Survey File, and the Daily Flow File. STORET contains more than 90 million individual observations of water quality parameters for both surface and groundwater, encompassing 800,000 sampling sites. STORET can be used by anyone who has access to the EPA central computer in Raleigh, NC. STORET may also be accessed through NAWDEX.

III. POTENTIAL SOURCES FOR RESEARCH FUNDS FOR WASTEWATER RECLAMATION AND REUSE

A. INTRODUCTION

The U.S. federal government funds just under half of all research and development (R & D) in the U.S. but performs only about 10% of it in its own laboratories. Industry conducts nearly 75% of the nation's R & D, and academia about 10%.[17]

The U.S. Congress has been asked to increase the federal research and development budget to $73 billion in Financial Year (FY) 1995. This is a 3% or $2 billion increase over the R & D federal budget for FY 1994.[18]

B. THE MAJOR U.S. GOVERNMENTAL AGENCIES INVOLVED IN WATER RESOURCES RESEARCH

Following are some of the major federal agencies involved in the water resources research area:

- Environmental Protection Agency (EPA)
- Department of Defense (DOD)
- Department of Energy (DOE)
- Department of Interior (DOI)
- Department of Commerce (DOC)
- Department of Health and Human Services (DHHS)
- Department of Agriculture (USDA)
- National Science Foundation (NSF)
- National Aeronautics and Space Administration (NASA)

1. U.S. Environmental Protection Agency (EPA)

The EPA is an independent agency in the U.S. Government. EPA was created in 1970 to provide coordinated and effective government action to protect and enhance our environment to the extent possible under the laws enacted by congress. Its mission is to control and abate pollution in the areas of air, water, solid waste, pesticides, radiation, and toxic substances, with an integrated coordinated attack cooperatively with state and local governments.

EPA engages in a variety of research, monitoring, standard setting, and enforcement activities and coordinates and supports research and antipollution activities by state and local governments, private and public groups, individuals, and educational institutions.[16]

To carry out this mandate, EPA has some 50 to 60 offices, administrations, and agencies involved in environmental programs.

The budget request for FY 1995 for EPA is up 8% over FY 1994 to $6.9 billion. Of that, $2.55 billion is for water-infrastructure funding, including $1.6 billion for wastewater state revolving loan funds and $700 million for a new (and yet to be authorized) drinking water state revolving loan fund. Another $150 million is being sought for clean water projects along the Mexican border in support of the North American Free Trade Agreement.[18]

In the area of research and development, EPA has at least eight offices, each with a mission relating to environmental research activities.[16]

EPA is the nation's primary environmental regulatory agency, with pollution prevention as one of its top priorities.[19] In 1993, EPA had a total civilian employment of 18,476.[20]

Recently, EPA initiated a new program entitled "Environmental Technology Initiative." The focus of this program is long-term research and pollution prevention by EPA, other federal agencies, and the private sector. The goal is to develop more advanced environmental systems and treatment techniques that can yield environmental benefits and increase exports of "green" technologies. This investment will aid in the transition away from a defense-oriented economy by stimulating the increased use of private sector R & D resources for environmental quality-related purposes.[21]

The initiative is funded at $36 million in FY 1994 and, in the President's plan, is to be funded at $80 million in FY 1995, with overall funding projected to be $1.8 billion over 9 years.[22]

The four objectives of this program are[23]

- Adapt EPA's policy, regulatory, and compliance framework to promote innovation.
- Strengthen the capacity of technology developers and users to succeed in environmental technology innovation.
- Strategically invest EPA funds in the development and commercialization of promising new technologies.
- Accelerate diffusion of innovative technologies at home and abroad.

2. U.S. Department of Defense (DOD)

The Department of Defense (DOD) is working with the private sector and other governmental agencies on environmental research and development.

In 1992, DOD established the Strategic Research and Development Program (SERDP) to address environmental matters of concern to DOD and the Department of Energy (DOE). The program identifies priorities in environmental research, technology, and information developed by the two departments for defense purposes. SERDP will update the DOD 5-year strategic environmental R & D plan and will provide government agencies and the private sector access to DOD environmental data. It also will identify energy technologies developed for national defense purposes that have environmentally sound, energy efficient applications for other DOD programs, for DOE and other government programs, and for industrial and commercial applications.[17]

Also an agency within the DOD, the U.S. Army Corps of Engineers (COE) is responsible for extensive civil works programs including R & D, planning, design, construction, operation, maintenance, and real estate activities related to rivers, harbors, and waterways. It also has significant responsibilities under laws for the preservation of navigable (most rivers and streams are defined as "navigable") waters and related natural resources such as wetlands.[16]

The COE and EPA regulate water issues such as wetlands filling and ocean dumping. The COE, along with the Bureau of Reclamation, oversees a vast system of levees, dams, and reservoirs for flood control, irrigation, hydropower, navigation, municipal and industrial use, fish and wildlife purposes, recreation, and environmental protection.[17]

For FY 1995, the U.S. Army Corps of Engineers' budget may be cut 14% below FY 1994, to $3.3 billion, precluding new construction and rehabilitation projects.[18]

3. U.S. Department of Energy (DOE)

The DOE was created to provide a framework for a comprehensive and balanced national energy plan through the coordination and administration of the various energy functions of the federal government while being fully responsive to the numerous related environmental laws and policies.[16]

In the DOE, there are at least five offices, administrations, or agencies involved in environmental programs; one such program relates to the Federal Water Pollution Control Act conducted by the Federal Energy Regulatory Commission (FERC). The FERC is an independent commission within the DOE which has all the regulatory responsibilities for interstate rates, construction, and operating regulations as mandated over the years by congress. Particular responsibilities include siting and construction of natural gas pipelines and hydroelectric facilities, all subject to the requirements of federal environmental laws such as the National Environmental Policy Act, Clean Air Act, and Federal Water Pollution Control Act.[16]

For FY 1995, it is proposed that the Department of Energy (DOE) environmental programs grow a marginal 3% to $6.52 billion, including $2.9 billion for waste management and $1.8 billion for environmental restoration.[18]

4. U.S. Department of the Interior (DOI)

The DOI has at least six bureaus, agencies, or services involved in environmental programs. Following are four of these organizations, and a brief description of their missions as taken from Reference 16.

- *Bureau of Land Management.* The Bureau has total management responsibility for 270 million acres of public lands and mineral management responsibility for an additional 370 million acres where mineral rights were reserved to the federal government. Land use plans are prepared with public involvement for the public lands to provide orderly use and development while maintaining and enhancing the environment. The Bureau is responsible for the survey of federal lands and maintains public land and mining claims records.[16]
- *Bureau of Reclamation.* The Bureau of Reclamation constructs, operates, and maintains public works for the storage, diversion, and development of waters for the reclamation of arid and semiarid lands in the Western states. Environmental information functions of the Bureau include basin-wide water resource studies, development of multi-use resource management plans, and the preparation of environmental impact statements.[16]
- *Fish and Wildlife Service.* The Service is responsible for conserving, protecting, and enhancing fish and wildlife and their habitats. Its activities include biological monitoring; surveillance of pesticides, heavy metals, and thermal pollution; studies of fish and wildlife populations; ecological studies; and environmental impact assessment of proposed hydroelectric dams, nuclear power sites, stream channelization, and dredge and fill operations.
- *Geological Survey (USGS).* The U.S. Geological Survey is responsible for identifying land, water, energy, and mineral resources; classifying federally owned lands for mineral and energy resources and water power potential; investigating natural hazards such as earthquakes, volcanoes, and landslides; and conducting a national mapping program. It prepares maps, collects and interprets geological data, performs fundamental and applied research, and publishes and disseminates the results in thousands of new maps and reports each year.[16] The USGS administers 54 State Water Resource Research Institutes located at land-grant universities in each of the 50 states and Puerto Rico, the District of Columbia, Guam, and the Virgin Islands (see Section II.B.6).

The Water Resources Research Act of 1984 authorized $10 million per fiscal year through fiscal 1989 for matching grants to the state institutes.[13]

Then, in 1990, PL 101–397 appropriated $10 million per fiscal year for this program through FY 1994, with provisions for matching grants. While $10 million was appropriated, only about $6 million was actually authorized on a matching basis of two nonfederal dollars for each federal dollar spent.

State institutes, along with other educational institutions, private foundations, and firms, individuals, and agencies of state or local governments, are eligible for matching grants.

Beyond FY 1994, the funds for this water resources research program are in question.[14]

The proposed federal budget for FY 1995 for the Department of the Interior (DOE) dropped 3.4% to $7.5 billion. The U.S. Geological Survey's federal program for water data collection and analysis could be cut 4% to $19 million, and the federal/state cooperative program for water data collection and analysis could be cut 2% to $58.1 million.

The Bureau of Reclamation could be cut $93 million (10%) to $782 million, continuing a 4-year trend of decreased funding for construction and increased funding for operations and maintenance.[18]

5. U.S. Department of Commerce (DOC)

National Oceanic and Atmospheric Administration (NOAA). The mission of NOAA is to explore, map, and chart the global ocean and its living resources; to manage, use, and conserve those resources; to describe, monitor, and predict conditions in the atmosphere, ocean, sun, and space environment; issue warnings against impending destructive natural events; develop beneficial methods of environmental modification; and assess the consequences of inadvertant environmental modification over several scales of time.

NOAA provides information to coastal zone states and assists them in dealing with the impact of offshore drilling for oil and gas, assists in licensing and deepwater ports, develops policy on proper ocean management and use along the U.S. coastline, and provides grants to institutions or marine research education and advisory services.

NOAA acquires, stores, and disseminates worldwide environmental data through a system of meteorological, oceanographic, geodetic, and seismological data centers. It includes the National Weather Service, National Ocean Survey, and National Marine Fisheries Service.[16]

6. U.S. Department of Health and Human Services (DHHS)

National Institute of Environmental Health Science (NIEHS). NIEHS conducts and supports fundamental research to define, measure, and understand the effects of chemical, biological, and physical factors in the environment on human health and well-being.[16]

7. U.S. Department of Agriculture (USDA)

Agricultural Research Service (ARS). The ARS administers fundamental and applied research to solve problems in animal and plant protection and production; the conservation and improvement of soil, water, and air; the processing, storage, and distribution of farm products, and human nutrition. The research applies to a wide range of goals, commodities, natural resources, fields of science, and environmental conditions. Research activities are carried out at 139 locations nationwide, much of it in cooperation with state universities and experimental stations. Management of the various research programs is the responsibility of eight area offices.[16]

8. The National Science Foundation (NSF)

NSF is an independent agency in the U.S. federal government that initiates and supports scientific research and development of science education programs.[16]

NSF has funded cooperative research centers/engineering research centers at major research universities, linking academia and industry. As of 1992, there were 23 such centers in operation at 21 universities in 16 states around the nation. The centers focus on environmental problems and technology solutions in fields such as microelectronics, telecommunications, biotechnology, energy resources and recovery, and design and manufacturing.[17]

An example of a recent NSF research plan is the Small Business Innovation Research (SBIR) program, NSF 93–18.

In the Biochemical Engineering section of this program a few of the technologies that relate to wastewater reclamation and reuse research are[24]

- Electrophoresis
- Selective precipitation
- Ultrafiltration
- Membrane technology
- Ion exchange techniques
- High pressure liquid chromatography
- Process monitoring and control

Also, in the Environmental Systems division of the Biological and Critical Systems section of this program, research proposals are sought in the application of engineering principles to ameliorate the impairment of land, air, and water resources by environmental pollutants. New knowledge is needed on the diffusion, dispersion, and interactions of pollutants; innovative water and wastewater treatment processes and systems; and engineering approaches to manage or eliminate discharges that degrade environmental quality. Proposals are not sought for research on the regulation of pollutants, improved monitoring for compliance with existing regulations, regulatory enforcement, or environmental policy issues. Techniques are sought to disinfect and decontaminate water; for strategies to control physical, chemical, and biological treatment processes; and for the management of products and by-products from treatment systems.

Areas of special interest within this subtopic include the following:

- Uses the ionizing radiation (ultraviolet, gamma, electron beam, and X-ray) for disinfection and decontamination of water, for treatment of solid, liquid, and gaseous emissions, and for destruction of environmentally hazardous substances.
- Technologies especially appropriate for investigations and/or engineering design of systems that address waste processing and management problems that are unique to U.S. coastal regions.[24]

For FY 1995, the proposed budget for the National Science Foundation (NSF), which supports most of the basic civil engineering research in the nation's universities, is slated for a 9.6% increase to $2.2 billion in its research activities budget. Overall, the NSF budget could climb 6% to $3.2 billion; the engineering directorate could rise 8.69% to $320 million.

9. The National Aeronautics And Space Administration (NASA)

NASA is an independent agency in the U.S. Government. NASA is responsible for the U.S. space exploration programs. A major necessity for this space exploration program is the provision of a high-quality water recovery system.

For crews aboard the informational space station, extended periods of stay will require a recycling of 98% of the water they take up with them.

One water purification system found suitable for space travel relies on multiple filters, an approach that won out over reverse osmosis in design competition. First, several filters remove hair and other particles. Then the water is heated to 121°C (250°F) for 20 minutes to kill microorganisms. Different types of carbonaceous filters remove hydrocarbons. Iodine is added to other filters to kill the remaining microorganisms. Ion-exchange resins remove metals.

The multifiltration subsystem is nearly 100% efficient, losing only a small amount of water when filters are changed every 30 days.

Another device, called the volatile removal assembly, removes low molecular weight compounds, such as fatty acids and ammonia, by creating larger molecules through biochemical, catalytic, and oxidation reactions.

The urine processor uses an energy-efficient process called vapor compression distillation. Urine recovery is about 90% efficient and will yield 6% of the total water supply, or about 1.4 kg (3 lb) daily per astronaut. Water is also removed from crewmembers' breath and evaporated perspiration by a condensing heat exchanger.

The Water Recovery Management Group in NASA is part of a $400 million project to provide a safe life support system for the international space station.[25,26]

C. WATER ENVIRONMENT RESEARCH FOUNDATION (WERF)

In 1990, the Water Pollution Control Federal Research Foundation (WPCFRF) was incorporated. When WPCF changed its name, the WPCERF then became known as the Water Environment Research Foundation (WERF).

The Water Environment Research Foundation (WERF) was established to advance science and technology for the benefit of the water quality profession and its customers. Funded through voluntary contributions, WERF manages research under four major *Thrust Areas:* Collection and Treatment Systems, Integrated Resources Management, Residuals Management, and Human Health and Environmental Effects. WERF seeks cost-effective, publicly acceptable, environmentally sound solutions to water pollution control problems. A 14-member Board of Directors composed of water quality professionals, volunteers from utilities, academia, consulting firms, and industry; a Utility Council composed of subscribing entities; an Advisory Council of industry leaders; and a Research Council of knowledgeable leaders in environmental sciences and engineering, are actively involved in applied and basic research program management.

While WERF manages the research and coordinates the parties involved, the actual work is carried out by individual organizations — primarily utilities, universities, industrial and commercial firms, and government laboratories. To ensure objectivity, an independent advisory committee of distinguished scientists and engineers helps select researchers, oversees the studies, and provides periodic review and advice. Benefits accrue in the form of services, technological advances, and information for direct application by the profession for its customers.[27,28]

Table 11.1 presents the ranking of each topic listed under the four Thrust Areas. Each subject is given a number indicating its rank out of the 51 topics, as well as indicating its priority within each of the four subgroups.[27]

WERF has also organized its research initiatives around four major research programs. These programs provide opportunities for the creation of new projects, enhancement and expansion of existing projects, and cooperative ventures with other organizations. The four programs are Solicited Research, Unsolicited Research, Targeted Collaborative Research, and Cooperative Research. All projects fall within one of the program areas.

WERF funded 28 research projects during 1991 and 1992. Eight of the Solicited Research projects initiated during this time period continued into 1993. The total value of the 28 research projects is estimated at over $11 million. Over the lifetime of these projects, WERF will provide $5,547,000 in funding. The remaining $5,527,000 will be contributed as in-kind services.

Table 11.1 Subscriber Priority Rankings of WERF Research Areas

Collection and Treatment Systems		Human Health and Environmental Effects	
4	Advanced Wastewater Treatment Processes	1	Water Quality Indicators
7	Disinfection	2	Transport and Fate of Toxic Contaminants
8	Pretreatment and Source Control	3	Bioassays and Measures of Toxicity
10	Biological Treatment Processes	5	Effluent and Residuals Disposal
21	Instrumentation and Process Controls	6	Risk Management and Assessment
25	Laboratory and Analytical Methods	16	Water Quality and Sediment Relationships
39	Planning, Operations & Design	22	Fresh Water Ecosystems
41	Industrial Treatment	23	Environmental Response and Crisis Management
42	Combined Sewer Overflows	30	Communicable Diseases
45	Collection Systems	32	Safety
50	Preliminary and Primary Treatment	36	Estuarine and Marine Systems
		44	Chronic Diseases
Integrated Resource Management		**Residuals Management**	
9	Watershed Management	15	Land Application
11	Surface Water Quality	18	System Design and Process Optimization
12	Modeling Pollutant Loads	24	Stabilization
13	Air Emissions	26	Biosolids
14	Storm Water Management	31	Bioremediation
17	Water Reclamation and Reuse	33	Thermal Processing
19	Waste Minimization	35	Dewatering
20	Urban Runoff Best Management Practices	43	Conditioning
27	Cross Media Effects	46	Thickening
28	Wetlands and Natural Systems	47	Incineration
29	Non Urban Runoff Best Management Practices	48	Anaerobic Digestion
34	Ground Water Resources and Recharge	49	Conveyance
37	Sampling and Monitoring Protocols	51	Aerobic Digestion
38	Environmental Auditing		
40	Nonstructural Pollution Controls		

Note: Listed in order of priority within each category; numbers indicate overall ranking.

From *Research Development Plan/1993–1997,* Water Environment Foundation, Alexandria, VA, 1993. With permission.

In 1993, WERF funded research included Solicited, Unsolicited, and Cooperative Research Programs. The Solicited Research Program consisted of 10 Requests For Proposals (RFPs) and three workshops. The three workshops involved topics listed under the heading of Human Health and Environmental Effects, and were conducted to advance the development of RFPs for release in 1994.

The distribution of $1,954,000 in 1993 among the four Thrust Areas for Solicited Research was as follows:

- Collection and Treatment Systems (CTS) 22%
- Integrated Resources Management (IRM) 35%
- Human Health and Environmental Effects (HHE) 24%
- Residuals Management (REM) 19%

This distribution will change when Unsolicited and Cooperative research proposals are selected for funding, increasing the total projected research expenditures to $2,454,000.[27]

Following are examples of WERF current research, cooperative research, and one 1993 research solicitation. Each example is from one of the Thrust Areas that relate directly or indirectly to wastewater reclamation and reuse research.

INTEGRATED RESOURCES MANAGEMENT
CURRENT RESEARCH[27]

Water Reuse Assessment

James Crook, Ph.D.
Camp Dresser & McKee Inc.
Cambridge, MA

Abstract:

Water reuse has become a necessary consideration for many communities as populations increase and the demands for additional water supplies continue to grow. The lack of adequate water supplies has made water reuse efforts desirable. However, when water reuse is considered, the cost of initiating reuse programs must be weighed against what is incurred in the development of new supplies. Along with considering costs and benefits, attention must also be given to possible adverse health impacts associated with the proposed end use. These issues have given rise to many difficult questions regarding the necessary degree of treatment, and how to reliably achieve that level of treatment to furnish the desired quality of water.

Much information has been presented over the years and published as proceedings of Reuse Symposia. However, this information needs to be consolidated and evaluated for the purpose of identifying areas where additional research is needed. Following the compilation of pertinent data, a workshop will be held for the purpose of identifying and ranking research needs.

Product:

The primary objective of this assessment project is to conduct a critical review of the status of existing information and valid scientific data relative to wastewater effluent reuse. Information and data shortfalls will be identified as a precursor to future specific research efforts.

FUNDING

	WERF Share	In-Kind Contribution	Total Cost
1992	$45,000	$20,000	$65,000

RESIDUALS MANAGEMENT
COOPERATIVE RESEARCH[27]

The Use of Treated Municipal Wastewater Effluents and Sludge in Production of Crops for Human Consumption
National Research Council

Abstract:

The conservation of natural resources and the emphasis on recycling and reusing materials and resources is an important issue. The application of treated effluents from municipal wastewater treatment plants to agricultural lands for irrigation is a prime example of reuse, particularly in arid agricultural areas. In these areas, higher quality water can be more appropriately assigned to other purposes. The agricultural use of treated sewage biosolids produced in municipal wastewater treatment offers yet another opportunity for recycling valuable organic matter and nutrients while improving soil and agricultural productivity.

This two-year study will focus on a number of issues associated with the use of treated effluents and sludges in the production of crops for human consumption. These include a general evaluation of technical issues, risk assessment, regulation, public education aspects, and potential barriers to implementation such as perception and liability issues. The study will include recommendations of the appropriate steps that can be taken to lower these barriers.

Product:

The study will be carried out by an NRC selected committee and will result in a final report to be published in early 1995.

FUNDING

	WERF Share	In-Kind Contribution	Total Cost
1992 - 1994	$25,000	$200,000	$225,000

HUMAN HEALTH AND ENVIRONMENTAL EFFECTS COOPERATIVE RESEARCH[27]

Collaborative National Study Using Molecular Techniques to Detect Hepatitis A Virus and Virulence Factor Genes in E. coli.

National Water Research Institute

The National Water Research Institute (NWRI) is conducting research regarding the discharge of indicator bacteria (*E. coli*) and pathogenic microorganisms into coastal aquatic environments. The multi-year project, funded at the level of $360,000, will use gene probe technology to assess microbial survivorship and recreational water quality. Foundation participation in this project involves a financial contribution of $10,000, and peer review of conducted research.

Sewage impact on our nation's coastal waters is of great concern to scientists, regulators, and the general public. Sewage discharge and storm water runoff has led to beach closures, which affects resident morale and tourism, and to the closure of valuable shellfish growing areas, which adversely impacts the nation's food supply. Coastal areas are closed based on high levels of coliform or enterococcus bacteria. Coliforms are poor indicators in the marine environment since many human pathogens, such as enteric viruses, persist longer in the marine environment than do the indicator organisms. In addition, indicator bacteria do not predict the presence or absence of indigenous marine pathogens, such as *Vibrio* species, and other pathogens, such as *Staphylococcus aureus*, that survive well in marine environments and may become part of the normal microbial flora.

Environmental monitoring of specific pathogens such as hepatitis A virus (HAV) and pathogenic *E. coli* has been hampered due to the lack of a sensitive and appropriate methodology. Advances in molecular biology, including gene probes and polymerase chain reactions, have recently been successfully applied to environmental water samples. This research will analyze sewage, stream water, storm drain water and coastal water for pathogens using the new molecular detection techniques.

Product:

The National Water Research institue does not require the preparation of a final report. However, investigators are encouraged to publish in peer reviewed journals as data is produced. Published articles are made available to Foundation Subscribers.

FUNDING

	WERF Share	In-Kind Contribution	Total Cost
1992	$10,000	$350,000	$360,000

COLLECTION AND TREATMENT SYSTEMS
1993 RESEARCH SOLICITATIONS[27]

Secondary Clarification Assessment

Rationale:

There is controversy among design and operating professionals over what constitutes the elements of good clarifier design (flocculation provisions, inlet and outlet conditions, sludge removal methods, etc.). The control of total suspended solids (TSS) is important because of water quality concerns and effluent requirements related to nonconventional pollutants associated with suspended solids. Examples of pollutants in suspended solids are nutrients, heavy metals and certain nonbiodegradable toxic organics.

Plants seeking low levels of phosphorus in effluents can remove soluble phosphorus to low levels through biological or chemical means, but either method concentrates phosphorus into the particulate fraction. Therefore, providing more efficient secondary sedimentation would be less expensive than providing tertiary filtration for reducing effluent phosphorus levels in many plants.

Objectives:

In 1987, EPA gathered together a group of experts to develop a research priority list for secondary clarification research. Since that time, research has progressed to answer some of the questions posed. For example, the American Society of Chemical Engineers Clarifier Research Technical Committee (CRTC) has developed a draft protocol for the conduct of full-scale clarifier tests. By the end of 1993, both rectangular and circular clarifiers will be tested.

Given the data already collected, the question remains as to which of the original research questions have been answered. An assessment of data collected since 1987, including CRTC work, should be done as well as an analytical evaluation to determine what fundamental design and operating parameters affect clarifier performance. Areas needing further research shall be addressed with the ultimate aim to be the development of rational design and operating parameters for secondary clarifiers of different types.

Benefits:

Municipalities and industries will have specific analysis and design guidance on available methods to improve existing clarifiers as well as for constructing new clarifiers. Research needs will be addressed. Measures for reductions of TSS in effluent will yield reductions in the levels of nutrients and toxicants in effluent. Disinfection methods sensitive to TSS in effluent will be rendered more efficient and economical.

PROPOSED FUNDING RANGE

1993 $45,000 - $60,000

Recently, the Chairman of WERF pointed out that, "Without research to study pollutants, their effects, and the technologies needed to control these effects, we cannot protect the environment or public health."[29]

He also indicated that there is a need for new research in order to maintain the water quality research momentum generated over the last 20 years. The Chairman also noted that research funding has fallen dramatically over the past 10 years. "Federal funding for practical wastewater treatment technology and process research has declined from $15.6 million annually to virtually nothing."[29]

D. AMERICAN WATER WORKS ASSOCIATION RESEARCH FOUNDATION (AWWARF)

The AWWA Research Foundation was created by the water supply community as its center for cooperative research and development. The foundation functions as a planning and management agency, awarding research contracts to water utilities, universities, engineering firms, and other organizations. The scientific and technical expertise of the foundation staff is enhanced by industry volunteers who serve on Project Advisory Committees and on other standing committees and councils. An extensive planning process involves many hundreds of water industry professionals in the important task of keeping the foundation's program responsive to the practical, operational needs of local utilities and to the general research and development needs of a progressive industry.

The foundation's research agenda embraces all aspects of water supply: resources, treatment and operations, distribution and storage, water quality and analysis, health effects, and economics and management. The ultimate purpose of the foundation's efforts is to assist local water purveyors in their efforts to deliver a reliable and economical supply of the highest possible quality drinking water.[30]

In 1993, the AWWARF funded 32 new projects with an investment of approximately $11.5 million. More than half this amount represented in-kind contributions from research contractors. The U.S. Congress supplemented this research by channeling $575,000 to the foundation in fiscal year 1993.[31]

In October, 1993 the AWWARF's technical advisory group, the Research Advisory Council, developed the following list of recommended projects for 1994:[32]

Water Resources

- Effectiveness of low-flush toilets in commercial settings
- The role of impaired quality water in water resources management
- Land use management practices in watersheds to improve water quality
- Predicting subsurface virus fate and transport

Treatment

- Arsenic removal and evaluation of arsenic residuals
- Biological stability in treatment and distribution systems
- Bench-scale membrane evaluation
- Enhanced coagulation for removal of particulates and microbial contaminants
- Reverse osmosis and nanofiltration for removing organic contaminants
- Clearwell baffle designs for regulatory compliance
- Ozone contactor design to minimize bromate formation
- Workshop on toxicity testing for membrane concentrates
- Workshop on tracer studies in treatment facilities

Distribution Systems

- Technologies to locate non-ferrous water pipes
- Innovative water main rehabilitation/renewal
- Distribution system water quality changes following implementation of corrosion control
- Guidance manual for installation of booster disinfection
- Identify and control odor-producing fungi in distribution systems
- Impact of wet-charge fire sprinkler systems on water quality
- Investigate sulfur compound joint failures
- Tank and storage reservoir water quality

Monitoring and Analysis

- Radon exposure while showering
- Simplified direct method for determining haloacetic acids
- Evaluation techniques for "total" NOM
- Incorporate a viability procedure for *Giardia* and *Cryptosporidium* into the ICR protocol
- Quantitative particle count methods
- Rapid method for *Giardia* and *Cryptosporidium* isolation and viability

Management and Administration

- Utility communications architecture demonstration
- Joint utility meter reading
- Performance benchmarks for water utilities
- Reliability of water treatment processes
- The value of drinking water system reliability
- Water treatment with ozone and energy efficiency
- Water utility data architecture
- The water utility of the future

Health Effects

- Gastrointestinal health effects due to the consumption of drinking water
- North American exposure to arsenic in drinking water
- Association of low birth weight, birth defects, and childhood cancer with disinfectants and their by-products
- Endemic waterborne disease risk for *Giardia* and *Cryptosporidium*

In January 1994 AWWARF's Board of Trustees selected 23 projects for 1994 funding. Requests for proposals (RFPs) were issued for 15 of these. The board approved $4 million to sponsor the solicited research projects, and $732,000 was appropriated for projects selected through the foundation's unsolicited project program.

The trustees selected the project topics from among the 40 high-priority research needs listed above. Unsolicited proposal funds were earmarked for innovative research ideas not included in the solicited projects.[33]

Following are several examples of AWWARF's RFPs that could directly or indirectly benefit the wastewater reclamation and reuse field; included are the maximum AWWARF funding amounts.[33]

- *Cyst and Oocyst Survival in Watersheds and Factors Affecting Inactivation (RFP 151)*. Evaluate the survival of *Cryptosporidium* oocysts and *Giardia* cysts exposed to different environmental conditions and determine subsequent effects on disinfection efficiency. Evaluate the effect of temperature, age, and physical stress on viability and susceptibility to disinfection — $270,000.
- *Enhanced and Optimized Coagulation for Removal of Particulates and Microbial Contaminants (RFP 155)*. Evaluate enhanced and optimized coagulation strategies for particle, cyst, and virus removal. Demonstrate treatment processes for removal credits for coagulation and sedimentation, and demonstrate the effects of an enhanced coagulation approach to particulate removals — $350,000.
- *Improved Methods for the Isolation and Characterization of Natural Organic Matter (NOM) and Identification of Physical-Chemical Properties Affecting Treatment (RFP 159)*. Develop improved methods to isolate and characterize NOM — $250,000.
- *Incorporation of a Vital Stain for* Giardia *and* Cryptosporidium *into the Immunofluorescence Assay (RFP 160)*. Incorporate a proven vital stain (viability testing) for *Giardia* cysts and *Cryptosporidium* oocysts into the immunofluorescence assay for drinking water — $100,000.
- *Quantitative Particle Count Method Development — Count Standardization and Sample Stability and Shipping Considerations (RFP 161)*. Evaluate particle count analytical standards for drinking water quality monitoring applications and develop a protocol that would allow the total number of particles measured by different instruments and utilities to be compared. This study would also develop a protocol for sample handling and shipping — $125,000.
- *Innovative and Rapid Methods for* Giardia *and* Cryptosporidium *Isolation, Detection, and Viability (RFP 162)*. Develop innovative and rapid methods for the isolation, detection, and viability testing of *Giardia* and *Cryptosporidium* — $225,000.

Following are two examples of AWWARF's research projects that relate to wastewater reclamation and reuse but for which RFPs were not issued:[33]

- Investigation of Reverse Osmosis and Nanofiltration for Removing Organic Contaminants and the Development of Piloting Procedures for Drinking Water Treatment. Develop a bench-scale and pilot-plant methodology for utilities and investigate and model pesticide and disinfection by-product precursor removal by diffusion-controlled membrane processes (reverse osmosis and nanofiltration). Cooperative project with KIWA (Dutch Research Association).
- Evaluation of Endemic Waterborne Disease Risk for *Giardia* and *Cryptosporidium*. Develop an epidemiological study design to quantify the waterborne risks of endemic giardiasis and cryptosporidiosis for general and special populations.

For details regarding research and research funding, the AWWA Research Foundation should be contacted.

E. THE NATIONAL WATER RESEARCH INSTITUTE (NWRI)

The NWRI was formed in 1991 and owes its creation to the encouragement and financial support of the Joan Irvine Smith and Athalie R. Clarke Foundation, as well as five southern California water and sanitation agencies: the Orange County Water District, the County Sanitation Districts of Orange County, the Municipal Water District of Orange County, the Irvine Ranch Water District, and the San Juan Basin Authority.[34]

The NWRI is independently governed by a board of directors consisting of one member from each of the contributing agencies.

The NWRI is unique in that its funding is a blend of both public and private sources. Today the Institute supports research projects throughout the U.S. which address both national and local water resource needs.[34]

The NWRI research priorities include:[34]

- Membrane research and development
- Health risk assessment
- Water quality improvement
- Water recycling
- Watershed management
- Public policy and institutional development

In 1992, in response to a request for proposals (RFPs), NWRI received 32 project proposals from 20 states and Canada. After reviewing and evaluating these proposals, the NWRI Research Advisory Board selected 12 finalists to make oral presentations. Six projects were then selected for funding, with a total investment of close to $2 million.[35]

Following is an example of one of these projects that relates directly to wastewater reclamation and reuse:[35]

Name of project: Organic Carbon Characterization of Advanced Treated Wastewater at Water Factory 21.
Project: Dr. Martin Reinhard, Stanford University
Value: NWRI $92,181 — Matching $92,181

> It is the intent of this research to develop a database which can be used to evaluate and compare the organic content of advanced wastewater treatment effluent and alternate potential drinking water sources such as groundwater, infiltrated secondary effluent, and surface waters, and to collect data and develop models of subsurface chemical processes that will aid in determining the optimal pretreatment for municipal wastewaters to be used for groundwater recharge.

The NWRI, also in 1992, initiated five cooperative joint venture research projects. The joint venture partners committed a total of $802,910 while NWRI committed $488,000, for a total program funding of $1,290,910.[35]

Following is an example of one of these projects that also relates directly to wastewater reclamation and reuse:[35]

Name of project: The Use of Treated Municipal Wastewater Effluent and Sludge in the Production of Crops for Human Consumption.
Project manager: National Academy of Science/National Research Council: Water Science and Technology Board
Value: NWRI $25,000 — NAS/WSTB $375,000

The study would focus on a number of issues associated with the use of treated effluent and sludge in the production of crops for human consumption. These would include a general evaluation of technical issues, risk assessment, regulation and public education aspects, and potential barriers to implementation such as preception and liability issues, including recommendations as to appropriate steps that can be taken to lower these barriers.

The National Water Research Institute (NWRI) announced that with the inclusion of its 1993 research program awards of $2.2 million, the total research program value is now more than $8 million.[36]

The centerpiece of the NWRI research program for the next 3 years is the pledge by the Joan Irvine Smith and Athalie R. Clarke Foundation to provide $1.25 million per year in support of the Institute's research programs. This vital multi-year base will enable the NWRI to expand its joint venture partnerships which have contributed in excess of $5 million in matching funds.

The current research program supports 20 projects in the areas of desalination, water quality, water reuse, health and regulations, and watershed management.

Major foundation blocks of this extensive program are the cooperative projects undertaken with the U.S. Environmental Protection Agency, Office of Drinking Water and Environmental Criteria and Assessment; the U.S. Bureau of Reclamation; the Electric Power Research Institute; the National Science Foundation; the National Research Council, Water Science and Technology Board; the American Water Works Association Research Foundation; and the Water Environment Research Foundation.

These cooperative projects have enabled each organization to leverage its often limited resources and thereby provide a more substantial level of support for all research projects.[36]

IV. WASTEWATER RECLAMATION AND REUSE RESEARCH

A. INTRODUCTION

A great deal is presently known about the use of reclaimed wastewater, yet much remains to be discovered, investigated, and resolved.

B. SCOPE OF THE WASTEWATER RECLAMATION AND REUSE RESEARCH AREAS

The scope of the areas involved in the wastewater reclamation area, both technical and societal, are presented in Table 11.2.[37]

To evaluate and to try to solve problems associated with these various areas in wastewater reclamation and reuse will require the skills and knowledge of many disciplines.[38]

The various modes of treatment involved in treating wastewater can be divided into three broad categories: physical, chemical, and biological processes. These three processes can be further categorized. This categorization is presented in Table 11.3.[39-42]

In many treatment operations, the physical, chemical, and biological processes overlap. For instance, in the adsorption process by activated carbon, both physical and chemical phenomena are involved.

While Table 11.2 and Table 11.3 present the scope, principal mode of treatment, and examples of each specific process involved, there are still many other areas that must be considered for investigation and research in the wastewater reclamation and reuse field. Following are a few of these areas:[43-46]

  Groundwater contamination
  Toxic contaminants
  Monitoring methods and instrumentation
  Biomonitoring
  Land and soil systems
  Removal of organohalogens
  Aquaculture
  Epidemiological studies
  Standards and regulations
  Environmental law

Table 11.2 Scope of the Wastewater Reclamation and Reuse Field

Research category	Levels of wastewater treatment technology	Primary modes of treatment
Technical	Preliminary (pretreatment)	Physical
		Chemical
	Primary	Physical
	Secondary	Physical
		Chemical
		Biological
Basic and applied	ªAdvanced wastewater treatment	Physical
		Chemical
		Biological
	ªTertiary (polishing process)	Physical
		Chemical
		Biological
	Disinfection	Physical
		Chemical
		Biological

Research category	Sociological and economical wastewater reuse issues
Societal	Public opinion
	Public acceptance — user concerns
	Public health concerns — health risks
	Economic feasibility — financing
	Markets
	Pricing structure
	Taxing policies (who pays)
	Benefits — cost effectiveness
	Safety
	Institutional policy issues
	Legal issues

ª Advanced wastewater treatment and tertiary treatment are often considered synonymous terms, however, when in doubt, refer to the definitions given in the publication or regulations.

Compiled from References 37 and 38.

1. Wastewater Reuse — Physical Problems

The most important physical and aggregate properties involved in wastewater treatment include:[47]

- Color
- Turbidity
- Odor
- Acidity
- Alkalinity
- Hardness
- Conductivity
- Salinity
- Solids (total dissolved solids)
- Temperature

For non-potable use of reclaimed wastewater, the physical parameters of primary concern are turbidity/solids, hardness (Ca^{2+}, Mg^{2+}), alkalinity, and conductivity. Many of the physical treatment processes presented in Table 11.3 can be used to control physical water quality parameters associated with wastewater reclamation and reuse. It is really a matter of cost that generally controls this situation.

2. Wastewater Reuse — Chemical Problems

The chemicals and chemical compounds encountered in the wastewater reclamation and reuse field can be divided into two broad groups, inorganic and organic.[48]

Table 11.3 Physical, Chemical, and Biological Categorization of Various Processes in Wastewater Treatment

Mode of treatment	Specific treatment processes involved
Physical	Residual removals
	Sedimentation
	Filtration
	Slow sand filters
	Rapid sand filters
	Upflow filters
	Dual-media filters
	Ultra filtration
	Microstraining
	Carbon adsorption
	Activated aluminum adsorption
	Distillation
	Aeration
	Centrifugation
	Reverse osmosis
	Ion-exchange
	Ultraviolet disinfection
Chemical (organic/inorganic)	Coagulation
	Flocculation
	Clarification
	Disinfection
	Chlorine
	Chlorine dioxide
	Ozonation
	Lime clarification
	Recarbonation
	Air-ammonia stripping
Biological	Trickling filters
	Activated sludge and modifications
	Aerated lagoons
	Stabilization ponds
	Intermittent sand filters
	Rotating biological contactors (RBC)

Compiled from References 39–42.

In the inorganic category, trace metals that have been given the most attention are arsenic, beryllium, cadmium, chromium, copper, lead, mercury, molybdenum, nickel, selenium, silver, and zinc.[48,49] The metals in this group that have been considered as potential carcinogens in test animals include arsenic, beryllium, cadmium, and nickel.[49]

Other metals and compounds of concern in reclaimed wastewater include sodium, calcium, magnesium, nitrates, phosphates, and chlorides. The sodium, calcium, and magnesium concentrations are of particular concern in the wastewater reclamation and reuse field as they relate to determining the sodium adsorption rate (SAR), which is a parameter that indicates the suitability of the reclaimed water for irrigational purposes.[50] Also, the hardness (Ca^{2+}, Mg^{2+}) of the reclaimed wastewater is important in urban reuse projects as this water can stain or leave a film on surfaces such as window panes or cars.

In the organic chemical category many, many compounds (over 1200) are associated with health effects. To obtain information on the carcinogenicity of these chemicals and compounds, the Integrated Risk Information (IRIS) database developed by EPA should be contacted.[52]

The classical methods used to measure organics in wastewater such as BOD_5, COD, TOC, organic nitrogen, and carbon-chloroform extraction have little or no relevance to the toxicological evaluation or health effects of organics in wastewater reclamation and reuse.[49]

A special group of organic compounds that are considered to be potential carcinogens for humans include the trihalomethane group (THMs) that can be formed in water disinfected with chlorine.[46] These organic compounds must be given special attention if the reclaimed wastewater is intended for potable use.

Table 11.4 Major Bacterial Pathogens Found in Raw Domestic Water and Also of Concern in Wastewater Reclamation and Reuse Projects

Shigella sonnei
Shigella dysenterae
Salmonella typhimurium
Campylobacter jejuni
Escherichia coli (enteropathogenic)
Vibrio cholerae
Vibrio comma
Legionella pneumophila

Compiled from References 54–56.

3. Wastewater Reuse — Biological Problems

The biological water quality problems involved in wastewater reclamation and reuse include exposure to pathogenic bacteria, viruses, protozoa, and helminths. These pathogenic organisms can be transmitted by direct contact, ingestion — such as by consumption of food irrigated with the reclaimed wastewater, drinking contaminated groundwater, or by inhaling spray aerosols.

Factors that can affect the transmission of these pathogenic microorganisms are the survival time of the pathogens, the minimal effective dose, as well as the susceptibility of the exposed population.

a. Bacterial Pathogens

It has recently been pointed out that since World War II, the presence in water of synthetic organic chemicals (SOCs) and disinfection by-products has concerned us, as it should. However, biological risks in our water supplies may be more critical today than chemical risks.[53] This too, would appear to be the case in the wastewater reclamation and reuse field, even though practically all reuse projects are for non-potable purposes.

Microorganisms of major concern in the wastewater reclamation and reuse field are presented in Tables 4.9 through 4.13 inclusive.

However, Table 11.4 presents major bacterial pathogens that are currently being given considerable research attention and are found in raw domestic water supplies as well as in wastewater intended for reuse.[54-56]

b. Protozoan Pathogens

The protozoan microorganisms that are presently at the top of the research list include *Cryptosporidium* (oocysts) and *Giardia lamblia* (cysts). This is the result of waterborne infections in Milwaukee, WI; New York, NY; Portland, OR; Washington, D.C. and Swinden, Oxfordshire and the Isle of Thanek, U.K.[54]

A recent AWWA conference dealing with "Preventing Waterborne Disease Outbreaks" listed the following unsettling facts:[57,58]

- In the U.S. during the last ten years, there have been at least five major (and scores of minor) waterborne disease outbreaks affecting nearly half a million people.
- In nearly 50% of the outbreaks, the cause is never determined.
- A recent study of 66 surface water systems found *Giardia* or *Cryptosporidium* in 32 of 83 finished water samples.
- *Giardia* and *Cryptosporidium* can be resistant to chlorine disinfection.
- Cryptosporidiosis can be a serious affliction as no cure is known, and recovery depends on the patient's own immune system.
- The only known effective means of removing *Cryptosporidium* from the water supply is filtration.
- Compliance with the Safe Drinking Water Regulations does not guarantee that treated water will be free of *Giardia* or *Cryptosporidium*.

These protozoan pathogens are of particular concern in the wastewater reclamation and reuse field due to their resistance to chlorine disinfection.

c. Viral Pathogens

Virus concentrations in raw sewage may be as high as about 100,000 infectious units per liter. If 99% of them are removed in a very well-run conventional treatment plant, the effluent will still contain 1000 virus particles per liter — a high number, since very few (perhaps even a single) infectious viruses are capable of inducing infection in humans.[59]

Some viruses that are presently being given attention in the wastewater research field include:[60-62]

- Infectious hepatitis A (HAV)
- Poliovirus type 1 and 2
- Echovirus
- Norwalk virus
- Human immunodeficiency virus (HIV) (retrovirus) (AIDS)
- Rotavirus

d. Helminth Pathogens

The World Health Organization has indicated that some helminths are serious pathogens in the transmission of disease when reclaimed wastewater is used, and considers the following helminths prime candidates for further research:[63]

Enterobius vermicularis (pinworm or threadworm)
Hymenolepis nana (dwarf tapeworm)
Ascaris lumbricoides (human roundworm)
Ancylostoma duodenale (hookworm)
Necator americanus (hookworm)
Trichuris trichiura (human whipworm)
Taenia saginata (tapeworm)
Taenia solium (tapeworm)

C. SOCIETAL ISSUES

In general, the technical research aspect of wastewater reclamation and reuse can be quantified, explained, evaluated, and solutions to the problems developed. The same can be said for some societal aspects of wastewater reclamation and reuse such as the economic, taxing, pricing, and cost-benefit issues. However, when it comes to public opinion, user concerns and public acceptance of the use of reclaimed wastewater along with emotional and culture attitudes must be considered. Considerable research work has been done in this area, but more needs to be done in order for governmental agencies to make decisions that involve public acceptance of wastewater reclamation and reuse projects.[64,65]

D. WORLD HEALTH ORGANIZATION

In 1989, the World Health Organization (WHO) sponsored a meeting of experts in the wastewater treatment and public health fields that resulted in a report entitled "Health Guidelines For the Use of Wastewater in Agriculture and Aquaculture," Technical Report #778.[63]

The following research needs were outlined for the wastewater reclamation and reuse area:

- Water quality assessment. A need exists to develop and improve a low-cost, reliable monitoring method for helminth egg detection.
- Wastewater treatment technologies. Technologies that should receive special attention include:
  - Waste stabilization pond technology.
  - Helminth egg removal systems such as biological filters, activated-sludge plants (with addition of maturation ponds which are entirely aerobic, and the use of sand filters).
  - Effectiveness of disinfection with particular reference to helminth egg inactivation.
  - Bacterial removal in stabilization ponds.
  - Treatments to inactivate pathogens in sludges, especially for use on a small scale.
- Irrigation technologies. There is a need to develop a nonclogging drip, trickle, and bubbler systems to facilitate "safe" application of reclaimed wastewater on crops.
- Epidemiology. Studies of wastewater reuse in irrigation in a variety of settings should be conducted, both where the microbiological quality guidelines are met and where they are not met. An urgent need for wastewater reuse projects is an evaluation of the improvement in the health of the agricultural workers and their children in situations where the new helminth egg guidelines are met. Studies are also needed to fill in the gaps in existing knowledge about excess morbidity and excess infections, especially the more virulent infections such as typhoid fever, as they relate to reclaimed wastewater reuse.
- Sociological and economical considerations. Research into the public's and users' attitudes to wastewater reuse is needed, both in areas where it is practiced and where it may be introduced. This may be done

by using case study analyses. The social and cultural factors affecting the acceptability and compliance with the health protection measures suggested (e.g., crop restriction) need to be evaluated in a variety of different cultural settings. Finally, research aimed at developing systematic methods of identifying the most cost-effective approach to health protection under any particular local conditions should be considered.[63]

- Aquaculture. There is an urgent need to conduct research into the microbiological and epidemiological aspects of the use of wastewater in aquaculture, so that guidelines may be proposed in the future with greater confidence. The health effects on both aquacultural workers and consumers of fish should be evaluated in a variety of sociocultural settings. Research is particularly needed in areas where the major trematode infections (e.g., clonorchiasis, schistosomiasis) do not occur and where bacterial infections are of great concern. Research on the impact of excreta use in aquaculture is also required.[63]

E. WORLD BANK AND WORLD HEALTH ORGANIZATION

In 1985, the World Bank and the World Health Organization held a meeting of environmental specialists and epidemiologists at Engelberg, Switzerland. This meeting dealt with the "Health Aspects Of Wastewater And Excreta Use In Agriculture And Aquaculture".[66]

Applied research priorities were developed for certain areas such as epidemiology, microbiology, sociology, and the technical aspects of using human waste in agriculture and aquaculture.

In the epidemiological area, it was indicated that an evaluation of the health risks associated with the application of excreta and wastewaters meeting certain guidelines (see Table 2.32) be compared to the health risks associated with the application of excreta and wastewater *not* meeting the suggested guidelines.

Epidemiological research was considered necessary in problem areas such as symptomatic diarrhea, enteric infections, intestinal nematode infections (particularly *Ascaris, Trichuris,* and the hookworms), typhoid fever, rotavirus diarrhea, and infectious hepatitis. In some situations, it was suggested that an examination of the risk of cholera, taeniasis, and trematode infections be evaluated.

In the microbial area, further research was considered necessary as to the appropriate treatment process needed for removal of microbial agents. A recent publication by one of the participants in "The Engelberg Report" deals with "The Development of Health Guidelines For Wastewater Reclamation", including epidemiological models, persistence of enteric pathogens, and stabilization ponds.[67]

The Engelberg Report considered that the study of the survival of intestinal nematode eggs was the first priority, followed by the survival of excreted bacteria. While the study of the survival of viruses was considered of lesser priority for the time being, it was felt that when good techniques for the isolation of rotavirus from wastewater becomes available, the monitoring of rotavirus should become a priority.

In the sociological area, the report considered that the following three aspects of wastewater reclamation and reuse needed attention.

1. Human behavior as a determining factor in disease transmission.
2. Prevention of the spread of diseases related to deep-rooted cultural factors which differ from society to society and which need to be considered in planning wastewater reclamation and reuse projects.
3. The successful utilization of new and innovative technologies in the wastewater field, with not only a technical evaluation but also the societal acceptability of the project.

F. INVENTORY FOR WASTEWATER RECLAMATION AND REUSE

While making a national wastewater reclamation and reuse inventory of projects may not seem to be research according to some, yet, by definition, the systematic collection and recording of facts accompanied by a careful, critical examination, interpretation, and analysis of these data in order to reach useful conclusions could, in most cases, be classified as applied research.

In any case, whether or not a national inventory of wastewater reclamation and reuse is applied research, no comprehensive national survey relating to wastewater reclamation and reuse has been made since 1975.[68] At that time, 536 wastewater reclamation and reuse projects were reported in the U.S. with an estimated wastewater reuse volume of 9.5×10^8 m^3/yr (2.5×10^{11} gal/yr).[69,70]

This information is clearly out of date; for instance, in 1975 California reported 283 wastewater reuse projects with a wastewater reuse volume of 2.37×10^8 m^3/yr (6.3×10^{10} gal/yr). In 1987, California reported 854 reclaimed water reuse projects with a volume of reclaimed water being 3.29×10^8 m^3/yr (8.7×10^{10} gal/yr) and this is expected to increase to 10.2×10^8 m^3/yr (2.7×10^{11} gal/yr) by the year 2000.[71,72]

Florida, in 1975, reported 6 wastewater reuse projects with a wastewater reuse volume of 9.1×10^6 m^3/yr (2.4×10^9 gal/yr), while in 1990 Florida reported 214 wastewater reuse projects with a reuse low of 4.4×10^8 m^3/yr (1.21×10^{11} gal/yr).[69,73]

The state agencies responsible for water and wastewater activities generally have information available regarding wastewater reclamation and reuse projects. However, there is a real need to collect this information in order to assist in planning wastewater reuse projects, including both capital and operational costs, standards and regulations, treatment processes, health aspects, as well as engineering decisions regarding the application of new technical and sociological knowledge.

Some of the basic information needed for a national wastewater reuse inventory include the state, location, name of permittee, county or parish, category costs, capital and operational costs, volume of wastewater storage, and wastewater quality.

Following is an example of California's municipal wastewater reclamation facility questionnaire.[71]

Names and telephone numbers have been deleted from the following California questionnaire as changes may have taken place since this information was gathered.

OFFICE OF WATER RECYCLING
CALIFORNIA STATE WATER RESOURCES CONTROL BOARD
Sacramento, CA

MUNICIPAL WASTEWATER RECLAMATION FACILITY QUESTIONNAIRE[71]

Instructions for Completing Questionaire

The purpose of this questionnaire is to collect information related to the types and quantities of use of reclaimed water in California and to the types of wastewater treatment provided for reclamation. The use of reclaimed water makes a significant contribution to the water supply of California. However, accurate data on wastewater reclamation facilities are not available because a detailed survey has not been done since 1978.

1. Please complete one questionnaire for each of your water reclamation facilities.
2. Please return questionnaire.
3. If another agency has the lead responsibility for managing the treatment, distribution, or sale of the reclaimed water and would be more capable of completing all or portions of the questionnaire, please complete as much as you can and forward the questionnaire to the other agency to complete remaining portions.
4. All information is to be based on 1987 calendar year.
5. Please include in the questionnaire all uses of reclaimed water, where the reclaimed water:
    a. Replaces a fresh water use.
    b. Augments a fresh water supply (for example: ground water recharge).
    c. Results in a useful product (for example: pasture, harvested crops, or recreational use of stream or lake that would not occur with natural water flows). For example, land disposal of effluent is considered a reclaimed water use if the land is used for grazing or growing a crop, even though this might not replace fresh water.
6. We are trying to distinguish whether the use of reclaimed water replaces fresh water or is primarily for disposal. Thus in part B of the questionnaire we ask for your opinion on whether fresh water would probably be used if reclaimed water were not available.
7. The following abbreviations are used in this questionnaire:

    ADWF: Average Dry Weather Flow
    MGD: Million Gallons per Day
    MG/yr: Million Gallons per Year
    $/acre-ft: Dollars per Acre Foot

8. Thank you for cooperating with this survey.

MUNICIPAL WASTEWATER RECLAMATION FACILITY QUESTIONNAIRE

A. RECLAIMED WATER PRODUCTION

1. Name and address of reclamation plant:

   _____

   _____

   _____

2. County: _____

3. Name and address of agency operating facility (if different from A.1): _____

   _____

4. Name and address of agency completing questionnaire (if different from A.3): _____

   _____

5. Name and title of person completing questionnaire:

   _____

   Telephone (_____) _____

6. Treatment Process (attach flow schematic if available):

   ____ Primary sedimentation   ____ Coagulation/Flocculation
   ____ Trickling Filter   ____ Filtration
   ____ Activated Sludge   ____ Other (explain): _____
   ____ Oxidation pond   _____
   ____ Disinfection   _____

7. Effluent disposal of reuse options available:

   ____ Discharge to fresh water

   ____ Discharge to marine/brackish water

   ____ Evaporation/percolation pond (effluent disposal, and not groundwater recharge, is primary purpose)

   ____ Planned groundwater recharge

   ____ Slow rate land disposal (for example: irrigation or overland flow where effluent disposal is the primary purpose)

   ____ Other reuse (explain): _____

   ____ Other (explain): _____

8. Is discharge to surface water prohibited during parts or all of the year?

   Yes ____ No ____

   If discharge is allowed all or part of the year, what general requirements on treatment or periods of discharge are there?

   ____ Secondary treatment

   ____ Nitrogen removal

   ____ Other (list): _____

9. 1987 Average Dry Weather Flow (ADWF): _____ MGD

10. Design Capacity of Treatment Plant (ADWF): _____ MGD

11. Amount of flow reused in 1987 (including groundwater recharge and productive land disposal):

    _____ Million Gallons per Year

12. Additional remarks: _____

    _____

B. RECLAIMED WATER USE

1. Did another agency have lead responsibility for the distribution or sale of the reclaimed water in 1987?

    Yes _____ No _____

    If yes, agency name and address: _____

    _____

    _____ Telephone No. (_____) _____

2. List Reclaimed Water Users in 1987 (for example: individual farmer, city, private golf course, etc.):

    Name of use area and address (approximate address is acceptable): _____

    _____

    Type of use (for example: in-plant use, landscape irrigation, cooling water, strawberry irrigation, cotton irrigation, recreation lake, etc.):

    _____

    Quantity: _____ MG/yr

    Entity managing use area: _____

    _____

    Phone No. (if known): (_____) _____

    Would fresh water probably be used if reclaimed water were not available?

    Yes _____ No _____

    If more than one reclaimed water user, please provide the same information for each user as that included above.

3. What is the price/fee structure for reclaimed water use? (check one or more):

    _____ User charged for reclaimed water

    Price range: _____
    (specify units, for example: $/acre-ft)

    _____ No monies exchanged for reclaimed water

    _____ User is paid to take reclaimed water

    Price range: _____
    (specify units, for example: $/acre-ft)

C. COMMENTS

Do you have plans for expanding water reuse? If so, briefly describe your plans with projected dates and quantities.

_____

_____

_____

Provide any comments that you think could be useful to other agencies or the state in planning for or regulating the use of reclaimed water. For example, you may comment on any positive or negative experiences you or users have had or with the application of state or county regulations. If any users have stopped the use of reclaimed water due to problems, such information would be very useful.

_____

_____

_____

_____

G. SOME SIGNIFICANT WASTEWATER RECLAMATION AND REUSE PROJECTS[38,74-77]

Bakersfield, CA 1929
Lubbock, TX 1938
Chanute, KS 1956
Whittiers Narrows, CA 1962
The City of Tallahassee, FL 1962
South Lake Tahoe Water Reclamation Plan, CA 1962
Tucson and Phoenix, AZ 1965
The Flushing Meadows Project, Phoenix, AZ 1967
Windhoek, Namibia, South Africa 1968
St. Petersburg Project, FL 1968
Muskegon County Wastewater Management System, MI 1968
Lancaster, CA 1974
Water Factory 21, Orange County Water District, CA 1976
Upper Occoguan Sewage Authority (UOSA), Fairfax County, VA 1978
E.L. Huie, Jr. Land Application Facility, Clayton County, GA 1979
Petromin Plant, Riyadh, Saudi Arabia 1979
Tahoe-Truckee Sanitation Agency Water Reclamation Plant, Reno, NV 1978
Dan Region Project, Tel Aviv, Israel 1986
APRICOT Project, Altamonte Springs, FL 1982
Potomac Estuary Experimental Water Treatment Plant, Washington, D.C. 1983
Potable Water Reuse Demonstration Plant, Denver, CO 1983
Potable Water Facility, San Diego, CA 1983
Fred Hervey Water Reclamation Plant, El Paso, TX 1985
Water Resource Recovery Pilot Plant, Tampa, FL 1986
Mexico City, Mexico 1988
WATER CONSERV II, Orlando, FL 1988
IRIS, Boca Raton, FL 1990
Logan Aluminum Wetlands Project

V. HOMEWORK PROBLEMS

1. Define:
   a. Research
   b. Research and Development
   c. Operations Research
   d. Epidemiology
2. What are the methodological steps involved in conducting an operations research project?
3. What equipment is needed to access an online electronic database?
4. According to the Council On Environmental Quality (CEQ), what percentage of the total funds spent on R & D is contributed by the U.S. Government, and what percentage of this research is conducted by academia?
5. What Federal U.S. Governmental agency is the nation's primary environmental regulator?
6. Define:
   MS-DOS
   PC-DOS
   RAM
   CD-ROM
   Modem
   Megabyte
   Baud Rate
7. Which of the trace metals are of major concern in the wastewater reclamation and reuse field and why?
8. What major bacteria, protozoa, viruses, and helminths are to be found in reclaimed wastewater and why are they now being given top priority in the research field?

9. List the physical, chemical, and biological characteristics of treated wastewater that require attention if this reclaimed wastewater is to be used for beneficial purposes.
10. Outline the general treatment technologies available in order to make wastewater suitable for nonpotable reuse purposes.
11. What are the four top (priority) research topics the WREF subscribers have identified?
12. Prepare a research proposal for a significant topic in the wastewater reclamation and reuse field. The proposal should include the following:
    Cover Page
    Title of Proposed Research
    Principal Investigator
    Project Duration
    Project Rationale
    Project Objectives
    Literature Evaluation & References
    Research Plan (Various Phases)
    Mathematical and Computational Analyses (Data Management)
    Proposed Application, Benefits or Relevance of the Results
    Budget: Personnel
            Travel
            Equipment
            Supplies
            Overhead
    Biographical Sketch

REFERENCES

1. **Parker, S. P., Ed.,** *Dictionary of Scientific and Technical Terms,* 3rd ed., McGraw-Hill, New York, 1984, 1362.
2. **Neufeldt, V., and D. B. Guralnik, Eds.,** *Webster's New World Dictionary,* 3rd College ed., Simon and Schuster, New York, 1986, 1141.
3. **Elliott, S. P., et al., Eds.,** *Webster's New World Encyclopedia,* Prentice Hall, New York, 1992, 942.
4. **Chacko, G. K.,** *Operations Research/Management Science,* McGraw-Hill, New York, 1993, 11, 166.
5. **Lee, C. C.,** *Environmental Engineering Dictionary,* 2nd ed., Government Institutes, Inc., Rockville, MD, 1992.
6. *The New Encyclopaedia Britannica,* Vol. 9, Micropaedia, 15th ed., University of Chicago Press, Chicago, 1989.
7. **Levin, R. I., et al.,** *Quantitative Approach To Management,* 7th ed., McGraw-Hill, New York, 1989, 21.
8. **Daellenbach, H. G., J. H. George, and D. C. McNickle,** *Introduction to Operations Research Techniques,* 2nd ed., Allyn & Bacon, Newton, MA, 1983.
9. *Encyclopedia of Science and Technology,* 7th ed., McGraw-Hill, New York, 1992, 385.
10. **Harris, K.,** Underground Injection Control Officer, Region IV, E.P.A., Database Information Pathfinder, and EPA National Online Library, Research Triangle Park, personal communication, 1994.
11. EPA, VISITT Vendor Information System for Innovative Treatment Technologies, User Manual (VISITT Version 2.0), EPA 542-R-93-001, U.S. Environmental Protection Agency, Washington, D.C., 1993.
12. EPA, VISITT Vendor Information System for Innovative Treatment Technologies, EPA 542-N-93-004, U.S. Environmental Protection Agency, Cincinnati, OH, 1993.
13. **Goldfarb, W.,** *Water Law,* Lewis Publishers, Chelsea, MI, 1989, 100.
14. **Huffsey, R.,** Personal communication, University of Kentucky, Water Resources Research Institute, Lexington, 1994.
15. **Byerly, G. P.,** AWWA Reorganizes to Better Serve Membership, *AWWA Mainstream,* 37, 12, 1993.
16. **Sullivan, T. F. P., Ed.,** *Directory of Environmental Information Sources,* 4th ed., Government Institutes, Inc., Rockville, MA, 1992, 10, 209.
17. Council on Environmental Quality: 23rd Annual Report of the Council on Environmental Quality, Washington, D.C., 1992.
18. **Dinges, C.,** Clinton Administration's Investment Budget For FY 1995, *Civil Engineering,* 64, 4, 1994, 128.
19. Council on Environmental Quality, 21st Annual Report of the Council on Environmental Quality, Washington, D.C., 1990, 113.
20. *The World Almanac 1994,* Funk & Wagnalls, Mahwah, NJ, 1993, 139.
21. US EPA, Environmental Technology Initiative: FY 1994 Program Plan, EPA 543-K-93-003, U.S. Environmental Protection Agency, Washington, D.C., 1994.
22. *Federal Register,* EPA's Technology Innovation Strategy, 59–19, Friday 28, 1994, 4068.
23. US EPA, Technology Innovation Strategy, Strategy Committee, Innovative Technology Council, U.S. Environmental Protection Agency, Washington, D.C., 1994.

24. National Science Foundation (NSF 93–18), Small Business Innovation Research (SBIR), NSF 93–18, Science & Technology Information System, Washington, D.C., 1993, 39.
25. **Coghlan, A.,** Water in Space Goes Round and Round Again, *New Scientist,* May 11, 1991, 24.
26. **Asker, J. A.,** Space Station Water Recycling Test Will Mark Milestone Life Support, *Aviation Week and Space Technology,* 135, 7, 1991, 54.
27. *Research Development Plan/1993–1997,* Water Environment Foundation, Alexandria, VA, 1993.
28. Research Notes, *Water Environ. Technol.,* Alexandria, VA, 5, 7, 1993, 84.
29. **Barnes, G. D.,** Importance of Water Quality Research Stressed by WEF, WERF in Congressional Testimony, *Highlights,* Water Environment Federation, Alexandria, VA, 30, 11, 1993.
30. AWWA, *Research Projects: A Summary, 1993/94,* American Water Works Association Research Foundation (AWWARF), Denver, CO, 1993.
31. Board of Trustees and Staff, AWWA Research Foundation, *Mainstream,* 38, 3, 1994, 10.
32. *Drinking Water Research,* AWWA Research Foundation, Denver, CO, 3, 6, 1993, 6.
33. AWWA, Research Foundation Approves $4.7 million for 1994 Projects, *AWWA Mainstream,* 38, 2, 1994, 1.
34. NWRI, *Meeting the Ever-Growing Demand for Water,* Brochure, National Water Research Institute, Fountain Valley, CA, 1993.
35. NWRI, *Six New Research Projects For 1992,* National Water Research Institute, Fountain Valley, CA, 1, 3, 1992, 1.
36. NWRI, *NWRI Research Program Reaches $8 Million,* National Water Research Institute, Fountain Valley, CA, 2, 1, 1993, 1.
37. **Haas, C. N.,** The State of Water Environment Research, *Water Environ. Res.,* 64, 5, 1992, 659.
38. **Lieuwen, A.,** *Effluent Use In The Phoenix And Tucson Metropolitan Areas,* Arizona Water Resources Research Center, University of Arizona, Phoenix, 1990.
39. **Argo, D. G.,** Water Reuse: Where Are We Headed?, *Environ. Sci. Technol.,* 19, 3, 1985, 208.
40. **Cook, J.,** Water Reuse in California, *J. AWWA,* July, 1985, 60.
41. **Odendaal, P. E.,** Wastewater Reclamation Technologies And Monitoring Techniques, *Water Sci. Technol.,* 24, 9, 1991, 173.
42. **Donovan, J. F. and J. E. Bates,** Guidelines For Water Reuse, Camp Dresser and McKee Inc., Publication No. EPA-600/8–80–936, Municipal Environmental Research Laboratory, U.S. Environmental Protection Agency, Cincinnati, OH, 1980, 95.
43. **Amy, G., et al.,** Fate of Chlorination Byproducts and Nitrogen Species During Effluent Recharge and Soil Aquifer Treatment (SAT), *Water Environ. Res.,* 65, 6, 1993, 726.
44. **Rogers, S. E., D. L. Peterson, and W. C. Lauer,** Organic Contaminants Removal For Potable Reuse, *Journal WPCF,* 59, 7, 1987, 722.
45. **Carry, C. W., et al.,** An Update on Wastewater Reclamation Research And Development In Los Angeles County, *Water Sci. Technol.,* 21, 1989, 409.
46. **Bauman, L. C., and M. K. Stenstrom,** Removal of Organohalogens and Organohalogen Precursors in Reclaimed Wastewater. I, *Water Res.,* 24, 8, 1990, 949.
47. APHA, *Standard Methods for Examination of Water and Wastewater,* 18th ed., American Public Health Association, Washington, D.C., 1992, 2–1.
48. **Sheikh, B., et al.,** Monterey Wastewater Reclamation Study for Agriculture, *Research Journal WPCF,* 62, 3, 1990, 216.
49. WPCF, *Water Reuse,* 2nd ed., Manual of Practice SM-3, Water Pollution Control Federation, Alexandria, VA, 1989, 201.
50. **Asano, T. and G. S. Pettgrove,** Using Reclaimed Municipal Wastewater for Irrigation, *Calif. Agri.,* 44, 3–4, 1987, 15.
51. **Chansler, J. M.,** Personal communication, 1994.
52. USEPA, Integrated Risk Information System (IRIS), U.S. Environmental Protection Agency, Cincinnati, OH.
53. **Okun, D.,** A World View of Water Supply, *J. Am. Water Works Assoc.,* 86, 3, 1994, 19.
54. **Hrudey, S. E., E. J. Hrudey, and N. J. Low,** Health Effects Associated with Waste Treatment, Disposal and Reuse, *Water Environmental Research,* 64, 4, 1992, 593.
55. **Moore, A. C., et al.,** Waterborne Disease in the United States 1991 and 1992, *J. Am. Water Works Assoc.,* 86, 2, 1994, 81.
56. **Crook, James, Camp Dresser and McKee Inc.,** *Water Science and Technology,* IAWPC, Great Britain, 24, 9, 1991, 109.
57. AWWA, New Satellite Teleconference, Preventing Waterborne Disease Outbreaks, Conference, April 8, 1994.
58. **Rose, J. B., C. P. Gerba, and W. Jakubowski,** Survey of Potable Water Supplies for *Cryptosporidium* and *Giardia, Environ. Sci. Technol,* 25, 1991, 1393.
59. **Schaefer, W. J.,** Health Aspects of Reuse of Treated Wastewater for Irrigation, World Health Organization Eastern Mediterranean Region, Intercountry Seminar on Wastewater Reuse, Manama, Bahrain, 1984.
60. **Kindzierski, W. B., R. E. Rogers, and N. J. Low,** Health Effects Associated with Wastewater Treatment, Disposal and Reuse, *Water Environment Research,* 65, 4, 1993, 599.
61. **Smith, R. G. and M. R. Walker,** Water Reclamation and Reuse, *Water Environ. Res.,* 64, 4, 1992, 402.

62. **Yanko, W. A.**, Analysis of 10 Years of Virus Monitoring Data from Los Angeles County Treatment Plants Meeting California Wastewater Reclamation Criteria, *Water Environ. Res.,* 65, 3, 1993, 22.
63. WHO, Health Guidelines for the Use of Wastewater in Agriculture and Aquaculture, Report of a Scientific Group, Technical Report Series 778, World Health Organization, Geneva, 1989.
64. **Bruvold, W. H.**, Obtaining Public Support for Reuse Water, *Journal AWWA,* July, 1985, 72.
65. **Bruvold, W. H.**, Public Opinion on Water Reuse Options, *Journal WPCF,* 60, 1, 1988, 45.
66. WHO, The Engelberg Report, Health Aspects of Wastewater and Excreta Use in Agriculture and Aquaculture, Sponsored by the World Bank and World Health Organization, Engelberg, Switzerland, 1985.
67. **Shuval, H. I.**, The Development of Health Guidelines for Wastewater Reclamation, *Water Sci. Technol.,* 24, 7, 1991, 149.
68. **Metcalf and Eddy, Inc., revised by G. Tchobanoglous and F. L. Burton,** *Wastewater Engineering Treatment, Disposal and Reuse,* 3rd ed., McGraw-Hill, New York, 1991, 1138.
69. **Culp, G., G. Wesner, R. Williams, and M. V. Hughes, Jr.,** *Wastewater Reuse and Recycling Technology,* Noyes Data Corporation, Park Ridge, NJ, 1980, 54, 58, 60.
70. **Asano, T. and G. Tchobanoglous,** The Role of Wastewater Reclamation and Reuse in the USA, *Water Sci. Technol. (Great Britain),* 23, 1991, 2049.
71. *California Municipal Wastewater Reclamation in 1987,* California State Water Resources Control Board, Sacramento, CA, 1990.
72. **Billings, C. H.**, Environmental Wastes Control Digest, *Public Works,* June, 1992, 96.
73. **Chansler, J. M.**, The Future for Effluent Reuse, *Water/Eng. Manage.,* May 138, 4, 1991, 31.
74. **Hamann, C. L. and B. McEwen,** Potable Water Reuse, *Water Environ. Technol.,* January, 1991, 75.
75. **Culp/Wesner/Culp,** Water Reuse And Recycling, Volume 2, Evaluation of Treatment Technology, OWRT/RV-79/2, U.S. Dept. of the Interior, Washington, D.C., 1979.
76. **WPCF,** *Water Reuse,* 2nd ed., Manual of Practice SM-3, Water Pollution Control Federation, Alexandria, VA, 1989.
77. **Rowe, D. R., K. Al-Dhowlia, and A. Whitehead,** *Reuse of Riyadh Treated Wastewater,* Project 18/1402, King Saud University, Riyadh, Saudi Arabia, 1988.

Appendix

Table 1   Periodic Table of Elements.
Table 2   Saturation Values of Dissolved Oxygen in Water at Various Temperatures.
Table 3   Physical Properties of Water.
Table 4   Most Probable Number (MPN) — 3 Tubes Used for Each 10 mL, 1.0 mL, and 0.1 mL Sample.
Table 5   Most Probable Number (MPN) — 5 Tubes Used for Each 10 mL, 1.0 mL, and 0.1 mL Sample.
Table 6   Conversion Factors.
Table 7   Fundamental SI Units (Système International d'Unités) or the Metric System.
Table 8   Formulated SI Units with Commemorative Names.
Table 9   Prefixes for SI Units.
Table 10   Greek Characters.
Table 11   Formula Weights and Equivalent Weights of Ions Found in Water.
Table 12   Abbreviations For Selected Units of Measurement
Table 13   Computer Program for the Freundlich Isotherm
References

PERIODIC TABLE OF THE ELEMENTS

Periodic table image omitted - contains standard element data including atomic numbers, symbols, atomic weights, electron configurations, and oxidation states for all elements, arranged with IUPAC group numbering 1-18 and previous IUPAC/CAS group labels (IA-VIIIA, IB-VIIIB).

KEY TO CHART:
- Atomic Number → 50
- Oxidation States → +2, +4
- Symbol → Sn
- 1989 Atomic Weight → 118.71
- Electron Configuration → 18-18-4

The new IUPAC format numbers the groups from 1 to 18. The previous IUPAC numbering system and the system used by Chemical Abstracts Service (CAS) are also shown. For radioactive elements that do not occur in nature, the mass number of the most stable isotope is given in parentheses.

REFERENCES

1. G. J. Leigh, Editor, *Nomenclature of Inorganic Chemistry*, Blackwells Scientific Publications, Oxford, 1990.
2. *Chemical and Engineering News*, 63(5), 27, 1985.

Table 2 Saturation Values of Dissolved Oxygen in Water Exposed to Water-Saturated Air Containing 20.9% Oxygen Under a Pressure of 760 mmHg [1013 mbar]

Temp. (°C)	Chloride concentration in water (g/m³ chloride)					Difference per 100 mg chloride	Vapour pressure [mmHg]
	0	5000	1000	15000	20000		
	Dissolved oxygen [g/m³]						
0	14.62	13.79	13.0	12.1	11.3	0.0165	5
1	14.23	13.41	12.6	11.8	11.0	0.0160	5
2	13.84	13.05	12.3	11.5	10.8	0.0154	5
3	13.48	12.72	12.0	11.2	10.5	0.0149	6
4	13.13	12.41	11.7	11.0	10.3	0.0144	6
5	12.80	12.09	11.4	10.7	10.0	0.0140	7
6	12.48	11.79	11.1	10.5	9.8	0.0135	7
7	12.17	11.51	10.9	10.2	9.6	0.0130	8
8	11.87	11.24	10.6	10.0	9.4	0.0125	8
9	11.59	10.97	10.4	9.8	9.2	0.0121	9
10	11.33	10.73	10.1	9.6	9.0	0.0118	9
11	11.08	10.49	9.9	9.4	8.8	0.0114	10
12	10.83	10.28	9.7	9.2	8.6	0.0110	11
13	10.60	10.05	9.5	9.0	8.5	0.0107	11
14	10.37	9.85	9.3	8.8	8.3	0.0104	12
15	10.15	9.65	9.1	8.6	8.1	0.0100	13
16	9.95	9.46	9.0	8.5	8.0	0.0098	14
17	9.74	9.26	8.8	8.3	7.8	0.0095	15
18	9.54	9.07	8.6	8.2	7.7	0.0092	16
19	9.35	8.89	8.5	8.0	7.6	0.0089	17
20	9.17	8.73	8.3	7.9	7.4	0.0088	18
21	8.99	8.57	8.1	7.7	7.3	0.0086	19
22	8.83	8.42	8.0	7.6	7.1	0.0084	20
23	8.68	8.27	7.9	7.4	7.0	0.0083	21
24	8.53	8.12	7.7	7.3	6.9	0.0083	22
25	8.38	7.96	7.6	7.2	6.7	0.0082	24
26	8.22	7.81	7.4	7.0	6.6	0.0080	25
27	8.07	7.67	7.3	6.9	6.5	0.0079	27
28	7.92	7.53	7.1	6.8	6.4	0.0078	28
29	7.77	7.39	7.0	6.6	6.3	0.0076	30
30	7.63	7.25	6.9	6.5	6.1	0.0075	32

Saturation at a barometric pressure other than 760 mmHg, c', is related to the corresponding tabulated values, c_s by Equation 3.35 Chapter 3, section V.G.

From Whipple, G. G. and Whipple, M. C., *J. Amer. Chem. Soc.*, 33, 1911, 362. With permission.

Table 3 Physical Properties of Water

Temperature (°C)	Temperature (°F)	$\rho \times 10^3$ Density at 1 atmosphere (kg/m³)	$\mu \times 10^3$ Dynamic viscosity (N · s/m²)	$\mu \times 10^6$ Kinematic viscosity (m²/s)	$\sigma \times 10^{-3}$ Surface tension against air (N/m)	Vapor pressure (mmHg)
5	41.0	0.999965	1.5188	1.5189	74.92	6.543
6	42.8	0.999941	1.4726	1.4727	74.78	7.013
7	44.6	0.999902	1.4288	1.4289	74.64	7.513
8	46.4	.999849	1.3872	1.3874	74.50	8.045
9	48.2	.999781	1.3476	1.3479	74.36	8.609
10	50.0	.999700	1.3097	1.3101	74.22	9.209
11	51.8	.999605	1.2735	1.2740	74.07	9.844
12	53.6	.999498	1.2390	1.2396	73.93	10.518
13	55.4	.999377	1.2061	1.2069	73.78	11.231
14	57.2	.999244	1.1748	1.1757	73.64	11.987
15	59.0	.999099	1.1447	1.1457	73.49	12.788
16	60.8	.998943	1.1156	1.1168	73.34	13.634
17	62.6	.998774	1.0875	1.0889	73.19	14.530
18	64.4	.998595	1.0603	1.0618	73.05	15.477
19	66.2	.998405	1.0340	1.0357	72.90	16.477
20	68.0	.998203	1.0087	1.0105	72.75	17.535
21	69.8	.997992	0.9843	0.9863	72.59	18.650
22	71.6	.997770	.9608	.9629	72.44	19.827
23	73.4	.997538	.9380	.9403	72.28	21.068
24	75.2	.997296	.9161	.9186	72.13	22.377
25	77.0	.997044	.8949	.8976	71.97	23.756
26	78.8	.996783	.8746	.8774	71.82	25.209
27	80.6	.996512	.8551	.8581	71.66	26.739
28	82.4	.996232	.8363	.8395	71.50	28.349
29	84.2	.995944	.8181	.8214	71.35	30.043
30	86.0	.995646	.8004	.8039	71.18	31.824
31	87.8	.995340	.7834	.7871	71.02	33.695
32	89.6	.995025	.7670	.7708	70.86	35.663
33	91.4	.994702	.7511	.7551	70.70	37.729
34	93.2	.994371	.7357	.7399	70.53	39.898
35	95.0	.99403	.7208	.7251	70.38	42.175
36	96.8	.99368	.7064	.7109	70.21	44.563
37	98.6	.99333	.6925	.6971	70.05	47.067
38	100.4	.99296	.6791	.6839	69.88	49.692

Van der Leeden, F., Troise, F. L., and Todd, D. K., *The Water Encyclopedia,* 2nd ed., Lewis Publishers, Chelsea, MI, 1990, 794. With permission.

Table 4 MPN Index and 95% Confidence Limits for Various Combinations of Positive and Negative Results When Three 10-mL Portions, Three 1-mL Portions, and Three 0.1-mL Portions are Used

Number of tubes giving positive reaction out of			MPN index per 100 mL	95% Confidence limits	
3 of 10 mL each	3 of 1 mL each	3 of 0.1 mL each		Lower	Upper
0	0	1	3	<0.5	9
0	1	0	3	<0.5	13
1	0	0	4	<0.5	20
1	0	1	7	1	21
1	1	0	7	1	23
1	1	1	11	3	36
1	2	0	11	3	36
2	0	0	9	1	36
2	0	1	14	3	37
2	1	0	15	3	44
2	1	1	20	7	89
2	2	0	21	4	47
2	2	1	28	10	150
3	0	0	23	4	120
3	0	1	39	7	130
3	0	2	64	15	380
3	1	0	43	7	210
3	1	1	75	14	230
3	1	2	120	30	380
3	2	0	93	15	380
3	2	1	150	30	440
3	2	2	210	35	470
3	3	0	240	36	1300
3	3	1	460	71	2400
3	3	2	1100	150	4800

From *Standard Methods For the Examination of Water and Wastewater,* 17th ed., American Public Health Association, Washington, D.C., 1978. With permission.

Table 5 MPN Index and 95% Confidence Limits for Various Combinations of Positive Results When Five Tubes Are Used per Dilution (10 mL, 1.0 mL, 0.1 mL)

Combination of positives	MPN index/ 100 mL	95% Confidence limits		Combination of positives	MPN index/ 100 mL	95% Confidence limits	
		Lower	Upper			Lower	Upper
0-0-0	<2	—	—	4-2-0	22	9.0	56
0-0-1	2	1.0	10	4-2-1	26	12	65
0-1-0	2	1.0	10	4-3-0	27	12	67
0-2-0	4	1.0	13	4-3-1	33	15	77
				4-4-0	34	16	80
1-0-0	2	1.0	11	5-0-0	23	9.0	86
1-0-1	4	1.0	15	5-0-1	30	10	110
1-1-0	4	1.0	15	5-0-2	40	20	140
1-1-1	6	2.0	18	5-1-0	30	10	120
1-2-0	6	2.0	18	5-1-1	50	20	150
				5-1-2	60	30	180
2-0-0	4	1.0	17	5-2-0	50	20	170
2-0-1	7	2.0	20	5-2-1	70	30	210
2-1-0	7	2.0	21	5-2-2	90	40	250
2-1-1	9	3.0	24	5-3-0	80	30	250
2-2-0	9	3.0	25	5-3-1	110	40	300
2-3-0	12	5.0	29	5-3-2	140	60	360
3-0-0	8	3.0	24	5-3-3	170	80	410
3-0-1	11	4.0	29	5-4-0	130	50	390
3-1-0	11	4.0	29	5-4-1	170	70	480
3-1-1	14	6.0	35	5-4-2	220	100	580
3-2-0	14	6.0	35	5-4-3	280	120	690
3-2-1	17	7.0	40	5-4-4	350	160	820
4-0-0	13	5.0	38	5-5-0	240	100	940
4-0-1	17	7.0	45	5-5-1	300	100	1300
4-1-0	17	7.0	46	5-5-2	500	200	2000
4-1-1	21	9.0	55	5-5-3	900	300	2900
4-1-2	26	12	63	5-5-4	1600	600	5300
				5-5-5	≥1600	—	—

From *Standard Methods For the Examination of Water and Wastewater,* 17th ed., American Public Health Association, Washington, D.C., 1978. With permission.

Table 6 Conversion Factors

Multiply	by	to obtain
Area		
Acre	0.4047	ha
Acre	43560	ft^2
Acre	4047	m^2
cm^2	0.155	in^2
ft^2	0.0929	m^2
hectare, ha	2.471	acre
ha	10^4	m^2
in^2	6.452	cm^2
km^2	0.3861	mile^2
m^2	10.76	ft^2
mm^2	0.00155	in^2
Concentration		
mg/L	8.345	lb/million U.S. gal
ppm	1	mg/L
Density		
g/cm^3	1000	kg/m^3
g/cm^3	1	kg/L
g/cm^3	62.43	lb/ft^3
g/cm^3	10.022	lb/gal (Br.)
g/cm^3	8.345	lb/gal (U.S.)
kg/m^3	0.001	g/cm^3
kg/m^3	0.001	kg/L
kg/m^3	0.6242	lb/ft^3
Flow rate		
ft^3/sec	448.8	gal/min
ft^3/sec	28.32	L/sec
ft^3/sec (cusec)	0.02832	m^3/s
ft^3/sec	0.6462	M gal/d
gal/min	0.00223	ft^3/sec
gal/min	0.0631	L/sec
L/sec	15.85	gal/min
M gal/d	1.547	ft^3/sec
m^3/hr	4.4	gal/min
m^3/sec	35.31	ft^3/sec
Length		
ft	30.48	cm
in	2.54	cm
km	0.6214	mile
km	3280.8	ft
m	3.281	ft
m	39.37	in
m	1.094	yard
mile	5280	ft
mile	1.6093	km
mm	0.03937	in
yard	0.914	m
Mass		
g	2.205·10^{-3}	lb
kg	2.205	lb
lb	0.4536	kg
lb	16	ounce
ton	2240	lb
tonne, t	1.102	ton (2000 lb)

Table 6 (continued) Conversion Factors

Multiply	by	to obtain
Power		
Btu	252	cal
Btu	778.2	ft-lb
Btu	$3.93 \cdot 10^{-4}$	HP-hr
Btu	1055	J
Btu	$2.93 \cdot 10^{-4}$	kW-hr
HP	0.7457	kW
Pressure		
atm	33.93	ft water
atm	29.92	in Hg
atm	$1.033 \cdot 10^4$	kg/m^2
atm	760	mm Hg
atm	10.33	m water
atm	$1.013 \cdot 10^5$	N/m^2
bar	10^5	N/m^2
cm water	98.06	N/m^2
in water	1.8665	mm Hg
in Hg	0.49116	lb/in^2
in Hg	25.4	mm Hg
in Hg	3386	N/m^2
k Pa	0.145	lb/in^2, psi
lb/in^2	0.0703	kg/cm^2
lb/in^2	6895	N/m^2
mm Hg	13.595	kg/m^2
mm Hg	0.01934	lb/in^2
mm Hg	133.3	N/m^2
mm Hg	1	torr
torr	133.3	N/m^2
Temperature		
Celsius	(9C/5) + 32	F
Fahrenheit (F)	5(F − 32)/9	Celsius (C)
Kelvin	C + 237.16	K
Rankine	F + 459.67	F
Velocity		
cm/sec	0.03281	ft/sec
cm/sec	0.6	m/min
m/sec	196.8	ft/min
m/sec	3.281	ft/sec
ft/min	0.508	cm/sec
ft/sec	30.48	cm/sec
ft/sec	1.097	km/hr
mile/hr	1.609	km/hr
poise (g/cm.sec)	0.1	$N.sec/m^2$
Viscosity		
centipoise	0.01	g/cm.sec
centistoke	0.01	cm^2/sec
stoke	10^{-4}	m^2/sec
Volume		
ft^3	6.229	gal (Br.)
ft^3	7.481	gal (U.S.)
ft^3	28.316	L
ft^3	0.02832	m^3
gal (Br.)	0.1605	ft^3
gal (U.S.)	0.1337	ft^3

Table 6 (continued) Conversion Factors

Multiply	by	to obtain
gal (U.S.)	0.833	gal (Br.)
gal	3.785	L
in^3	16.39	cm^3
L	0.03532	ft^3
L	0.22	gal (Br.)
L	0.2642	gal (U.S.)
L	0.001	m^3
m^3	35.314	ft^3
m^3	1000	L

Table 7 Fundamental SI Units (Système International d'Unités) or the Metric System

Quantity	Unit	Symbol
length	meter	m
mass	kilogram	Kg
time	second	s
thermodynamic temperature	Kelvin	K
electric current	ampere	A
amount of substance	mole	mol
plane angle	radian	rad
solid angle	steradian	sr

Table 8 Formulated SI Units with Commemorative Names

Quantity	Name	Symbol	Units
conductance	siemens	S	$s^3 \cdot A^2 / Kg \cdot m^2$ or A/V
electromotive force	volt	V	$Kg \cdot m^2 / s^3 \cdot A$ or W/A
energy	joule	J	$Kg \cdot m^2 / s^2$ or $N \cdot m$
force	newton	N	$Kg \cdot m / s^2$
power	watt	W	$Kg \cdot m^2 / s^3$ or J/s
pressure	pascal	Pa	$Kg / m \cdot s^2$ or N/m^2
radioactivity	becquerel	Bq	dps[a]

[a] dps — Disintegrations of atoms per second.

Table 9 Prefixes for SI Units

Prefix	Symbol	Multiplier	Synonymous
tera	T	10^{12}	trillion
giga	G	10^9	billion
mega	M	10^6	million
kilo	k	10^3	thousand
hecto	h	10^2	hundred
deka	da	10^1	ten
deci	d	10^{-1}	tenth part
centi	c	10^{-2}	hundredth part
milli	m	10^{-3}	thousandth part
micro	μ	10^{-6}	millionth part
nano	n	10^{-9}	billionth part
pico	P	10^{-12}	trillionth part

Table 10 Greek Characters

A	α	Alpha	N	ν	Nu
B	β	Beta	Ξ	ξ	Xi
Γ	γ	Gamma	O	o	Omicron
Δ	δ	Delta	Π	π	Pi
E	ε	Epsilon	P	ρ	Rho
Z	ζ	Zeta	Σ	σ	Sigma
H	η	Eta	T	τ	Tau
Θ	θ	Theta	Y	υ	Upsilon
I	ι	Iota	Φ	φ	Phi
K	κ	Kappa	X	χ	Chi
Λ	λ	Lambda	Ψ	ψ	Psi
M	μ	Mu	Ω	ω	Omega

Table 11 Formula Weights and Equivalent Weights of Ions Found in Water
[Weights expressed to three significant figures]

Ion	Formula weight	Equivalent weight	Ion	Formula weight	Equivalent weight
Al^{3+}	27.0	9.00	Fe^{3+}	55.8	18.6
Ba^{++}	137.	68.7	Pb^{++}	207.	104.
HCO_3^-	61.0	61.0	Li^+	6.94	6.94
Br^-	79.9	79.9	Mg^{++}	24.3	12.2
Ca^{++}	40.1	20.0	Mn^{++}	54.9	27.5
$CO_3^=$	60.0	30.0	Mn^{4+}	54.9	13.7
Cl^-	35.5	35.5	NO_3^-	62.0	62.0
Cr^{4+}	52.0	8.67	PO_4^{3-}	95.0	31.7
Cu^{++}	63.6	31.8	K^+	39.1	39.1
F^-	19.0	19.0	Na^+	23.0	23.0
H^+	1.01	1.01	Sr^{++}	87.6	43.8
OH^-	17.0	17.0	$SO_4^=$	96.1	48.0
I^-	127.	127.	$S^=$	32.1	16.0
Fe^{++}	55.8	27.9	Zn^{++}	65.4	32.7

Van der Leeden, F., Troise, F. L., and Todd, D. K., *The Water Encyclopedia,* 2nd ed., Lewis Publishers, Chelsea, MI, 1990, 794. With permission.

Table 12 Abbreviations for Selected Units of Measurement

Symbol	Description	SI Unit
AW	atomic weight	g
BCR	benefit-cost ratio	%
BOD_5	5th day biochemical oxygen demand	mg/L or l
BOD_c	carbonaceous oxygen demand	mg/L or l
BOD_1	ultimate biochemical oxygen demand	mg/L or l
BOD_n	nitrogenous oxygen demand	mg/L or l
Bq	becquerel	[a]dps or counts/s
Ci	curie	[a]dps or counts/s
COD	chemical oxygen demand	mg/L or l
CPI	consumer price index	dimensionless
DWF	dry weather flow	L/d
EC	electrical conductivity or specific conductance	μmohs/cm
°C	degrees Celsius	°C
F/M	food-to-microorganism ratio	d^{-1}
HRL	annual hydraulic loading rate	cm/yr.
Le	effluent BOD	mg/L or l
Li	influent BOD	mg/L or l
Lo	ultimate BOD	mg/L or l
Lt	amount of first stage BOD	mg/L or l
MLSS	mixed liquor suspended solids	mg/L or l
MPN	most probable number	coliform/100 ml
NTU	nephelometric turbidity unit	NTU
PE	population equivalent	dimensionless
pH	negative logarithm of the hydrogen ion concentration	mol/L or l
Gy	gray	m^2/s^2 or J/Kg
J	joule	$Kg \cdot m^2/s^2$ or $N \cdot m$
N	newton	$Kg \cdot m/s^2$
Pa	pascal	N/m^2
Q	discharge flow rate	m^3/s
RAD	radiation absorbed dose	m^2/s^2 or J/Kg
RBE	relative biological effects	dimensionless
REM	roentgen equivalent man	ergs/g
R	roentgen	esu/g of air
Re	reynolds number	dimensionless
S	siemens	$s^3 \cdot A^2/Kg \cdot m^2$
SA	sludge age	d
SAR	sodium adsorption ratio	dimensionless
SDI	sludge density index	g/ml
Sv	sievert	ergs/g
SS	suspended solids content	mg/L or l
SVI	sludge volume index	mL/g
t	time	s
v	velocity	m/s
V	volume	m^3
W	weight	Kg
y	year	d
ZP	zeta potential	volt
ρ	density	Kg/m^3
γ	gamma ray	Ci or Bq
α	alpha particle	Ci or Bq
β	beta particle	Ci or Bq
μ	dynamic, absolute viscosity	$N \cdot s/m^2$
ν	kinematic viscosity	m^2/s
Φ	particle shape factor	dimensionless

[a] dps — Disintegrations per second.

Table 13 Computer Program Freundlich Isotherm[6] (see Chapter 5, Section IV.B.2)

```
*****************************************************
DIM M(50), C(50), X(50), Q(50): Z = 2.302585093#
   CLS
   INPUT "Name of Adsorption material:   "; N$
   INPUT "Name of Contaminant material:   "; C$
   INPUT "Number of measurements:   "; T
   INPUT "Control value (mg / L):   "; CO
   FOR I = 1 TO T: PRINT
   PRINT "Measurement #"; I; "; mg / L; of; "; N$; " "; : INPUT M(I)
   INPUT "mg/L of Continement remaining:   "; C(I)
      X(I) = CO - C(I)
      Q(I) = X(I) / M(I)
      NEXT I
   PRINT : PRINT
   INPUT "Press ENTER to continue "; ent$
   CLS
   Print data
   PRINT TAB(1); " [M]"; TAB(16); " [C]"; TAB(29); " [X]"
   PRINT "milli-"; TAB(11); "Concentration"; TAB(28); "Amount"; TAB(43); " X/M"
   PRINT "grams"; TAB(13); "remaining"; TAB(27); "adsorbed"; TAB(43); "mg/g"
   PRINT "-------"; TAB(11); "----------------"; TAB(27); "----------"; TAB(43);
   FOR I = 1 TO T
   PRINT M(I), C(I), X(I), Q(I)
   NEXT I
   PRINT : PRINT "Control value used was "; CO; " mg/L"
   PRINT : PRINT
   INPUT "Press ENTER to continue "; ent$
   CLS
   PRINT TAB(1); "C"; TAB(10); "log C"; TAB(20); "log X"; TAB(30); "log M"; TAB(
   FOR I = 1 TO T
   C1 = LOG(C(I)) / Z: REM  ln (C(I))
   C2 = C1 ^ 2
      IF X(I) = 0 THEN C3 = 0: GOTO step2
   C3 = LOG(X(I)) / Z: REM  ln (X(I))
step2:
   C4 = LOG(M(I)) / Z
   C5 = C3 - C4
   C6 = C1 * C5
   A1 = A1 + C1: A2 = A2 + C5: A3 = A3 + C2: A4 = A4 + C6
   PRINT C(I); TAB(8); C1; TAB(20); C3; TAB(30); C4; TAB(40); C5; TAB(50); C2; T
   NEXT I: PRINT : PRINT
   PRINT "sum of log(C)'s:     "; A1
   PRINT "sum of log(X/M)'s:   "; A2
   PRINT "sum of (log C)^2's:  "; A3
   PRINT "sum of (log C * log[X/M]):    "; A4: PRINT : PRINT
   D = (T * A3) - (A1 ^ 2)
   D1 = (A2 * A3) - (A4 * A1)
   D2 = (T * A4) - (A2 * A1)
   S = D2 / D: I = D1 / D
   S = ABS(S)
   PRINT "Slope     ="; S; " = 1/n"
   I = 10 ^ I
   PRINT "Intercept = "; I
   V = I * (CO ^ S)
   PRINT "Ultimate sorption capacity of absorbent = "; V; "mg/g"
   PRINT : PRINT "*****************************************************"
END
```

Table 13 (continued) Computer Program Freundlich Isotherm[6] (see Chapter 5, Section IV.B.2)

```
       Computer Program Freundlich  Isotherm - Example

Name of Adsorption material:   ? DIRAB SOIL
Name of Contaminant material:  ? ZINC
Number of measurements:    ? 5
Control value (mg / L):    ? 1.65

Measurement # 1 ; mg / L; of; DIRAB SOIL ? 250
mg/L of Contaminent remaining:   ? 0.9

Measurement # 2 ; mg / L; of; DIRAB SOIL ? 500
mg/L of Contaminent remaining:   ? 0.47

Measurement #·3 ; mg / L; of; DIRAB SOIL ? 1000
mg/L of Contaminent remaining:   ? 0.30

Measurement # 4 ; mg / L; of; DIRAB SOIL ? 2000
mg/L of Contaminent remaining:   ? 0.15

Measurement # 5 ; mg / L; of; DIRAB SOIL ? 4000
mg/L of Contaminent remaining:   ? 0.12


Press ENTER to continue ?
```

[M] milligrams	[C] Concentration remaining	[X] Amount adsorbed	X/M mg/g
250	.9	.75	.003
500	.47	1.18	.00236
1000	.3	1.35	.00135
2000	.15	1.5	.00075
4000	.12	1.53	.0003825

Control value used was 1.65 mg/L

Press ENTER to continue ?

Table 13 (continued) Computer Program Freundlich Isotherm[6] (see Chapter 5, Section IV.B.2)

C	log C	log X	log M	log(X/M)	log(C^2)	logC*(log[X/M])
.9	-.0457575	-.1249387	2.39794	-2.522879	2.093749E-03	.1154406
.47	-.3279021	7.188199E-02				
			2.69897	-2.627088	.1075198	.8614278
.3	-.5228787	.1303337	3	-2.869666	.2734022	1.500487
.15	-.8239087	.1760913	3.30103	-3.124939	.6788255	2.574664
.12	-.9208187	.1846914	3.60206	-3.417368	.8479072	3.146777

```
sum of log(C)'s:         -2.641266
sum of log(X/M)'s:       -14.56194
sum of (log C)^2's:        1.909748
sum of (log C * log[X/M]):   8.198797


Slope     = .9842839
Intercept = 4.051007E-03
Ultimate sorption capacity of absorbent =  6.631763E-03 mg/g

***************************************************************

C:\BOOK>
```

REFERENCES

1. **Lide, D. R. and Frederikse, H. P. R.,** *CRC Handbook of Chemistry and Physics,* 74th Edition, CRC Press, Boca Raton, FL, 1993–1994, inside front cover.
2. **Whipple, G. C. and Whipple, M. C.,** Solubility of oxygen in sea water, *J. Amer. Chem. Soc.,* 33, 1911, 362.
3. **van der Leeden, F., F. L. Troise, and D. K. Todd,** *The Water Encyclopedia,* 2nd ed., Lewis Publishers, Chelsea, MI, 1990, 774.
4. **USPHA,** *Standard Methods For the Examination of Water and Wastewater,* 17th ed., American Public Health Association, Washington, D.C., 1989 (pp 881, table 808:v).
5. **van der Leeden, F., F. L. Troise, and D. K. Todd,** *The Water Encyclopedia,* 2nd ed., Lewis Publishers, Chelsea, MI, 1990, 779.
6. **Chansler, J. M., W. G. Lloyd, and D. R. Rowe,** Soil Adsorption of Zinc According to the Freundlich Isotherm, publication pending, 1995.

Index

A

Abbreviations, 527
Absolute ownership rights, 397
Accidents, 363
Accuracy, 333
Acidity, 76
Acquisition, of water supply system, 302–303
Activated carbon, 29, 34
Activated sludge process, 8, 124, 288–289
   trace metals, 30
Adenovirus, infective dose, 281
Administrative bodies, 397
Administrative penalties, 415, 419–420
Adsorption, 176, 194–197
   boron, 28
   carbon, 29, 34
   isotherms, 194, 196
   sodium adsorption ratio (SAR), 24–27, 356
Advanced treatment methods (tertiary), 124–126, 131–132, 174–251, 468
Aedes aegypt, 113, 115, 134
Aerated rotating biological contactors, 8
Aeration, 8, 245–247
Aerobic digestion, 226–227
Agencies, 397–398
   funding by, 491–495
Agricultural irrigation, 5, 23–31, 427
   case study, 459–474
   design for disease vector control, 142–150
   guidelines, 24–25, 417
   Israel, 6
   Mexico, 6
   removal of nematode eggs, 283
   Sultanate of Oman, 475–476
   WHO criteria, 51
Agriculture, uses of water in, 2
Air stripping, 34
Air unit risk, 275

Alarms, 376–378
Aldicarb, 138
Algae, 89, 98–99. *See also* Biofouling
Algicides, 99
Alkalinity, 75–77
   maximum contaminant level, 9
   sampling, 332, 334
Alkyl benzene sulfonates (ABS), 9
Allethrin, 138
Allocation of costs, 317–318
Alpha particles, 68–69
Alpha spectrometry systems, 365–366
Alternative technologies
   EPA funding, 403–405
   VISITT Program, 487–488
Aluminosilicates, 218
Aluminum, water quality criteria for cooling water, 36
American Dye Manufacturers Institute, 335
American Water Works Association (AWWA) database, 489–490
American Water Works Association Research Foundation (AWWARF), 501–503
Ames mutagenicity test, 268
Amino acids. *See* Proteins
Ammonia, 81–82, 167
   analysis, 345–347
   maximum contaminant level, 9, 416
   reaction with chlorine, 221–223
   water quality
      cooling water, 36
      fishing, 39
Amoeba, 51
Amortized capital, 302
Amperometric titration method, 344
Amphoteric ion exchangers, 218
Anaconda Aluminum Company, 456
Anaerobic decomposition, 83, 225–226, 232
Analysis of samples, 333–365
Ancylostoma, 122

Animals
- in hazard identification studies, 267–268
- as vectors of disease, 118, 134–135

Anions
- detergents, 83
- in ion exchange, 218
- sewage water, 6
- toxicity, 82

Ankylostoma duodenale, 111
Annual Book of ASTM Standards, 327
Annual reports, 380, 436
Anopheles, 112
Apparent color, 65, 335
Appendix, 517–530
Applied microbiology, 88–89
Aquaculture, 51, 126
Area under probability curves, 389–390
Argentometric method, 341
Arid regions, 5–11
Army Corps of Engineers, 397
Artificial radioactivity, 68
Asbestos, 47
Ascaris, 51, 112–113, 115, 122
- incubation period, 116
- survival time, 114, 285

Ascomycetes, 98
Atlantic Richfield Company, 456
Atomic absorption spectrophotometry, 356–358
Australia, 5–6
Automated ferricyanide method, 341
AWWA (American Water Works Association) database, 489–490
AWWARF (American Water Works Association Research Foundation), 501–503

B

Back-flow prevention, 19
Backup systems, 375
Backwashing, 186–187, 189
- hydraulics, 191–193

Bacteria, 96–98, 119–120. *See also Escherichia coli*
- shapes of, 96
- survival times, 121
- in wastewater, 281–282, 507

Bancroftian filariasis, 112, 134
Basidiomycetes, 98
Batch type culture system, 92–93
BATEA (best available technology economically achievable), 402
BCR (benefit-cost ratio), 315–316
Becquerel units, 363
Beer's law, 62–63

Benefit-cost ratio (BCR), 315–316
Benefits, 315–316, 320
Best available technology economically achievable (BATEA), 402, 408
Best practicable control technology currently available (BPCTCA), 402
Best practicable waste treatment technology (BPWTT), 402
Beta particles, 69–70
Bias, 333
Bibliography
- for Palmer-Bowlus flumes, 371
- for radionuclides, 363
- for sampling methods, 327

Bilharziasis, 110
Biocarbonate, agricultural irrigation, 24
Biodegradation, 177, 246–247
Biofouling, 35, 98
- in agricultural irrigation units, 23
- in osmotic filtration, 212

Biological contactors, 8, 30
Biological control agents, 139–140, 467–468
Biological filtration, 124, 177
- redundant facilities for, 377

Biological oxygen demand. *See* BOD (biological oxygen demand)
Biostimulants, 82. *See also* Ammonia; Nitrogen; Phosphates; Phosphorous
- wetland removal efficiency, 442–444, 450–455

Black fly, 132, 134
- environmental management, 155–156

Boca Raton, Florida, case study, 426–438
BOD (biological oxygen demand), 80, 83
- for biodegradation, 246–247
- constituents of, 352
- constructed wetlands, 442–444, 450
- effluent levels, 6, 11, 448
- kinetics, 83–88
- maximum contaminant level, 8–9, 416
- and population equivalent, 171
- sampling and analysis, 351–352
- water quality criteria for cooling water, 36

Boilers, 428
Boron, 23, 132
- in agricultural irrigation, 24, 27–29
- in reclaimed wastewater, 127–128
- sampling and analysis, 340–341

BPCTCA (best practicable control technology currently available), 402
BPT (best practicable technology), 402
BPWTT (best practicable waste treatment technology), 402
Breakpoint chlorination, 222
Brunauer isotherm, 197
Budget estimates, 301

Buffer zones, 463
By-passing, 375–376

C

Cadmium, 23, 30–31. *See also* Trace metals
Calcium, in agricultural irrigation, 25–27
Calcium carbonate, 79, 200, 212
Calcium phosphate, 200
Calcium silicate, 200
Calibration, of electromagnetic flowmeters, 373
Campulobacter jejune, 110
Campylobacter jejuni, 122
Canal design, and disease vector control, 142–150
Cancer, 129, 277, 279–280. *See also* Carcinogenicity
Capillary suction time, 237–238
Capital investment, 316
Capital recovery factor, 307
Carbamate compounds, 138
Carbohydrates, 82–83
Carbon, 29, 34, 354
Carbonaceous biological oxygen demand, 352
Carbon analyzer, 354
Carbon dioxide, 167
  water quality standards for fishing, 39
Carbon tetrachloride, 267
Carcinogenicity, 40, 274–279, 411
  of radioactive constituents, 129
Carman-Kozeny equation, 179
Carmine method, 340
Case studies, 425–478
Cations
  causing water hardness, 77
  detergents, 83
  exchangeable. *See* Calcium; Magnesium; Sodium
  in ion exchange, 218
  sewage water, 6
CD (Criteria Documents), 17–18, 413
Cellular biology, 89
  division and growth, 90–92
  inorganic substances needed, 95
  organization, 91
Cellulose, 82–83
Centrifugation, 229–231, 238–242
  helminth egg analysis, 362
Cercariae, 111
CERCLA, 398, 405–406
Certification of operators, 19
Chagas' disease. *See* Trypanosomiasis
Chain of custody, 332–333
Chelation, in ion exchange, 218
Chemical characteristics, 73–88, 469, 505–506
Chemical coagulation, 29

Chemical filtration, 176–177
Chemical oxygen demand. *See* COD (chemical oxygen demand)
Chemicals of concern, 263
Chemosynthesis, 96, 98
Chezy's formula, 144, 305
Chick's law, 224–225
Chlorides
  in agricultural irrigation, 23–24
  maximum contaminant level, 9, 416
  sampling and analysis, 340–342
  sources of, 81
Chlorinated hydrocarbons. *See* Organochlorine compounds; Pesticides
Chlorination, 220–225
  calculation of pathogen removal, 288–291
  kinetics, 224–225
Chlorine
  cylinders, 378
  in disinfection, 125, 220–221, 291, 432
  dosages, 225
  free
    maximum contaminant level, 9
    recommended limits for agricultural irrigation, 32
    water quality criteria for cooling water, 36
  residuals, 224, 343
    for bactericidal disinfection, 222
    maximum contaminant level, 9
  sampling and analysis, 332, 342–344
  scales, 378
  standby supply, 378
Chlorine contact basins, 432
Cholera, 109, 122, 285
Chromates, 82. *See also* Anions
Chromium, 30–31. *See also* Trace metals
Chronic daily intake, 267
Chronic reference dose, 269–274
Cipolletti weir, 369–370
City of Boca Raton, Florida, case study, 426–438
Clayton County, Georgia, case study, 459–474
Clayton County Water Authority, 467–469, 472
Cleanup regulations, 405–406
Clean Water Act Amendments, 1991–1992 pending, 408
Clean Water Act Amendments of 1981, 398, 406–407
Clean Water Act of 1977, 381, 398, 403–405, 421
Clogging, of filters, 183–186
Clonorchis, 122
Closed reflux methods, 353
Coagulation, 377–378
Cocoa Beach, Florida, 428

COD (chemical oxygen demand), 80, 85, 88
  maximum contaminant level, 9, 416
  sampling and analysis, 353–354
  water quality
    cooling water, 36
    recreational reuse, 38
Coefficient of variation, 388–389
Coliform
  analysis, 358–361
  effluent levels, 11, 448
  fecal, maximum contaminant level, 8
  sampling, 332
  survival time, 285
  total, 17
    analysis, 360–361
    and livestock drinking water, 38–39
    maximum contaminant level, 9, 416
    water quality criteria, 38
Colilert test system, 361
Collection, of samples. *See* Sampling
Colloidal solids, 67
Color, 64–65, 334–335
Colorimetric methods, 340, 344–345, 353, 356
Common law, 396
Communities, 151–153, 155–156
Community awareness. *See* Public awareness
Community Right-to-Know Act, 406
Comparison of costs, 318–320
Complete mixed culture system, 94–96
Comprehensive Environmental Response, Compensation and Liability Act of 1980 (CERCLA), 398, 405–406
Conductivity. *See* Electrical conductivity
Confidence limits, intervals, and levels, 389
Congressional Record, 488
CONSERV II, 428
Constructed wetlands, 441, 444–459
Construction costs, example parameters, 308
Construction grants, 406–407
Consumer Price Index, 319
Contaminant identification, 263
Contaminants, in Safe Drinking Water Act, 17–18
Contingencies, in cost estimates, 302
Continuous culture systems, 93–94
Conversion factors, 523–525
Copper, 30–31, 220. *See also* Trace metals
Corrosion, 35
Cost allocation, 317–318
Cost benefit analysis, 315–317
Cost comparisons, 317–320
Cost-effectiveness analysis, 316
Costs. *See also* Estimating costs
  activated sludge system, 472
  constructed wetland, 458
  Florida wastewater project, 433–435, 437
  land treatment system, 472–474
  treatment for potable use, 34
Council on Environmental Quality, 400, 402
Coxsackie virus
  infective dose, 281
  survival time, 285
Cracking-time test, 233
Criminal law, 418–419
Criteria. *See* Water quality, criteria
Criteria Documents (CDs), 17–18, 413
Crops
  Boron tolerance, 28, 128
  in manufacturing processes, 132
  selection for land treatment system, 463
Cross-connection control, 19, 378–379
Crustaceans, 99–100
Cryptosporidium, 110
Culex pipens, 112, 134
Culture systems, 92–96
Curcumin method, 340–341
Curie units, 70, 363
Curtis Stanton Energy Center, 428
Cyanides, 39, 82. *See also* Anions

D

Darcy's head loss equation, 179
Darcy's law for flow, 233
Data, precision and accuracy, 333
Databases, for research information, 485, 487–491
Data collection, 263
Data evaluation, 263–264
Death phase, 96
Dechlorination, 212
Definite estimates, 301
Denitrification, 131
Density, 68, 71
Deposition, in a filter, 183–186
Depreciation, 302
Dermal exposure, 36–37
Desalination, 197–218. *See also* Distillation
Design
  canals, 142–150
  constructed wetlands, 445–447, 449–455
  disease vector control through, 142–150
  principles for reliability, 375
  wastewater treatment facility, 430, 462–463
Design for Urban Water Reuse, 460
Detection limit, 356
Detergents, 83, 355
Dewatering, of sludge, 226–242
  factors influencing, 231–232
DHHS (U.S. Department of Health and Human Services), 494
Diarrhea, 110, 122, 134, 155

Die-off factors, of pathogens, 120
Digestion, of sludge, 225–226
Dilution, 242
Direct costs, 302
Direct injection, for groundwater recharge, 31
Directory of Environmental Information Sources, 490
Disbenefits, 316
Discharge limitations, 382, 459
  zero, 402
Disease. *See* Health effects
Disinfectants
  drinking water guidelines, 44–46
  risks of by-products, 40–41
Disinfection, 19, 34, 124, 220–225
  expected removal of microorganisms, 288
  redundant facilities, 378
  types, 220
Dispersion model, 277–278
Dissociation constant, 75
Dissolved gases, 80
Dissolved oxygen, 79–80
  influent and effluent, 9, 448
  sampling and analysis, 332, 344–345
  saturation values, 519
  water quality
    fishing, 39, 245
    recreational reuse, 38
Dissolved solids, 23–25, 67
  and conductivity, 66
  recommended limits for agricultural irrigation, 32
  sampling and analysis, 336
  in seawater, 198
  water quality
    cooling water, 36
    fishing, 39
    livestock and wildlife, 40
Distillation, 197
  double, 202
  multistage, 202
  principles of, 198–202
  solar, 205–207
  unit layout, 203
Distribution, of water supply, 302–303, 312–313, 463
Distributions, statistical, 387–391
DOC (U.S. Department of Commerce), 493
DOE (U.S. Department of Energy), 363, 492
DOI (U.S. Department of the Interior), 4, 489, 493
DOJ (U.S. Department of Justice), 419
Domestic wastewater, 167
  volume calculations, 169–170
Dose-response assessment, 262, 268–279
Dose-response curves, 268–269, 271

for MCLG development, 17
projected by models, 285–286
Double distillation, 202
DPD ferrous titrimetic method, 344
Dracontiasis, environmental management, 155
Drinking water
  chronic daily intake, 267
  constituents giving rise to complaints, 46
  disease reduction with potable, 154
  ingestion, 270
  from reuse water, 22, 33–35, 63
  standards, 21, 403, 413–415, 416
    radionuclides, 364
    Sultanate of Oman, 41, 48–49
  WHO guidelines, 39–41, 42–47, 414, 416, 421
Drinking Water Criteria Documents, 17–18, 413
Drying beds, 34, 227–228
Dry weather flow (DWF), 169–170
Dwarf tape worm, 11
Dynamic viscosity, 71
Dynasand filters, 432
Dysentery, 51, 109, 115, 122

E

Easements, cost estimates, 306–307
EC. *See* Electrical conductivity (EC)
Echovirus, infective dose, 281
Economics, 299–320
Effluent guidelines, 6, 11, 411
  Florida wastewater treatment plant, 430, 433
  Georgia land treatment, 470–472
  Kentucky constructed wetland, 448, 459
  Kentucky land treatment, 463
Effluent polishing. *See* Tertiary treatment
EIS (environmental impact statements), 400
Electrical conductivity (EC), 24–25, 66, 217
  and livestock, 38
  measuring, 337–338
Electrodialysis, 198, 213, 216–218
Electromagnetic flowmeters, 373–375
Elements, in bacteria, 97
Elephantiasis, 112
Elutriation, 231
Emergency Planning and Community Right-to-Know Act (EPCRA), 406
Emergency response, 405
Emmett isotherm, 197
Enforcement, 415, 418–420
Engelberg report, 51–52, 361, 509
English common law, 396, 418
Entamoeba histolytica, 11, 109–110, 122, 123
  incubation period, 116
  survival times, 114, 121, 295
Enterococci, fecal coliform analysis, 358–361

Enteroviruses, survival time, 285
Environmental fate and transport, 265
Environmental impact assessment, 158
Environmental impact statements (EIS), 400
Environmental management, for vector control, 140–141, 150–151, 155–156
Environmental microbiology, 88–90
Environmental reuse, 22
  aquatic, 38
  wildlife, 38–40
Enzymes, 89–90
EPA. See U.S. Environmental Protection Agency
Epidemiology
  for hazard identification, 267–268
  lack of pathogen studies, 40
  for MCLG development, 17
Equipment, for sampling, 329, 331
Escalation, 318
Escherichia coli, 110, 122
  destruction with chlorination, 289
  drinking water guidelines, 42
  fecal coliform analysis, 358–361
  infective dose, 282
Esters of alcohol, 82
Estimating costs, 301–302
  distribution system, 312–313
  force mains, 310–311
  pumping station, 311–312
  storage, 313–314
  wastewater reclamation and reuse, 308–314
  water supply system, 302–308
Euler diagram, 300–301
Evaporation, 3, 169–170, 449
Evapotranspiration, 457
Exponential phase, of growth, 92
Exposure assessment, 262, 264–268, 279
  calculating chemical intake, 270
  quantification, 265, 267, 279
Exposure pathways. See Routes of exposure
Exposure setting, 264–265
Extended aeration, 8
Extrapolation, in risk modeling, 285–287
Exxon Corporation, 420
Exxon Shipping, 420

F

FAO (Food and Agricultural Organization), 28
Fats, oils, and grease, 354–355
Fatty acids, 82, 354–355
Feces. See also Coliform
  pathogens excreted in, 122
Federal Bulletin Board, 489
Federal Register, 488
Federal Research in Progress (FEDRIP) database, 491
Federal Water Pollution Control Act Amendments of 1961, 398–400
Federal Water Pollution Control Act of 1948, 397–399
Federal Water Pollution Control Act of 1956, 398–399, 421
Federal Water Pollution Control Act of 1972, 381, 398, 401–402, 419–421, 468
FEDRIP database, 491
Felony, 419
Fertilizer, 6, 226
Field techniques. See Sampling
Filariasis, 155
Filter aids, 232
Filter leaf test, 237–238
Filter press, 230
Filtration, 174–194
  clogging, 183–186
  hydraulics, 179–183
  multimedia, 178, 188
  operation and control, 188–191
  osmotic, 207–213
  pressure, 186, 188, 229
  rate, 189
  redundant facilities, 378
  sand, 125, 177–178, 186–194, 432
  of sludge, 233–238
  traveling bridge, 432
  trickling. See Trickling filter process
  up-flow, 186, 188
  vacuum, 34, 229–230
Financing, 317–320. See also funding
Fish and fishing, 39, 245
Flame emission photometry, 356–357
Flatworms, 120
Florida Administrative Code, 380–381
Florida case study, 426–438
Florida Department of Environmental Regulation, 375, 379, 433
Flow
  Darcy's law, 233
  dry weather, 169–170
  in filtration, 5, 186, 188
  land treatment system, 474
  measurements, 365–375
  prevention of back-flow, 19
  standards, 21
Flow diagrams
  treatment technologies, 131
  wastewater reclamation facility, 432
  wastewater treatment facility, 429, 458
Flowmeters, 373–375
Fluidized bed carbon reactivation furnace, 34
Fluids, types of, 72
Flumes, 370–372

Fluoride, 108
Foaming, 83
FOG (fats, oils, and grease), 354–355
Food, water content of, 2
Food and Agricultural Organization (FAO), 28
Force mains, estimating costs, 310–311
Forest irrigation, 467, 474
Formulated Si Units, 525
Fouling, 183–186
   biofouling, 23, 35, 98, 212
   by oil and grease, 82
   in osmotic filtration, 212
Free water surface wetlands, 441, 447, 449–455
Freezing, 197–198
Frequency, of sampling, 328–329
Freundlich isotherm, 196–197
   computer program for, 528–530
Fruits, 28, 128
Fundamental SI Units, 525
Funding, 317–320, 408–409
   land treatment system, 472–474
   restoration of the environment, 420
   wastewater reclamation and reuse, 491–504
Fungi, 89, 97–98
Fungi-imperfecti, 98

G

Gallatin Report, 397
Gambiense, 112
Gamma radiation, 70, 365–366
Gases, 167
   from anaerobic decomposition, 83
   dissolved, 80
   ideal, 207
Gas proportional counters, 365
Gaussian distribution, 387–391
Gavage tests, 271
General Petroleum and Minerals Organization, 10
Generation time, 95, 96
Genetic vector control, 140
Georgia case study, 459–474
Georgia Department of Natural Resources, 460
Germanium diodes, 365
Giardia, 17, 109, 122–123
   in farm workers, 11
   helminth as indicator for, 51
   self-reinfection, 110
   survival time, 285
GICS database, 491
Glycerol, 82
Golf course irrigation, 428, 467
Governmental agencies, 397–398
GPO (U.S. Government Printing Office), 488–489
Grant Information and Control System (GICS), 491
Grass filtration, 5
Gray units, 363–364
Grease, 82
   analysis, 354–355
   maximum contaminant level, 9
Greek characters, 526
Gross National Product Deflator, 319
Groundwater
   legal aspects, 396–397, 403
   percent of water resources, 4
   quality in Clayton County, Georgia, 471
   well monitoring, 471
Groundwater recharge, 5, 22, 31, 33
   water quality criteria, 33, 417
Growth, of cells and cell cultures, 90–96
Guidelines for Drinking-Water Quality, 16, 41, 414, 421
Guidelines for Water Reuse, 375
Guinea worm, 111, 115, 120
Gypsum, 200

H

Habitats, of disease vectors, 133, 135–136
Hardness, of water, 76–79
Hazard identification, 262, 267–268
Hazard index, 274
Hazardous waste regulations, 403, 405–406
Head
   on a flume, 370–371
   loss, 179–182, 370
   total, 304–305
   on a weir, 366–367
Health effects
   of drinking water contaminants, 47
   hazard identification, 267–268
   livestock, 5–6
   of reclaimed water, 107–158
   relative risks of untreated wastewater, 119
   transmission routes of disease, 113–114, 116–118
Health Effects Assessment Summary Tables (HEAST), 271
Heat treatment, 124, 220
Heavy metals
   in agricultural irrigation, 23, 32
   drinking water guidelines, 42–43
   in pesticides, 139
   in reclaimed wastewater, 126–127
   removal by sedimentation, 131
   sampling and analysis, 355–359
   toxicity, 40, 82

fishing, 39
livestock and wildlife, 40
Helminths, 100, 111–113, 120–122, 282–283
analysis for eggs, 361–362
candidates for research, 508
in feces, 122
as indicator species, 51
survival times, 121
Hepatitis, 51, 109, 115, 122
survival time, 285
Herbicides. *See* Pesticides
Hippocrates, 108
Historical background
connection between water quality and disease, 108
drinking water standards, 414
public health protection, 409
water quality regulations, 420–421
Holding time, of samples, 332
Hookworm, 11, 51, 111–113, 115
found in feces, 122
survival time, 285
Human body, water content, 1–2
Human exposure studies, 17
Human habitation
factors effecting sanitation, 153
modifications for vector control, 151–152, 155–156
Hydraulic loading rates, 450, 463–466
Hydraulic radius, 305
Hydraulic residence time, 450
Hydraulics, in filtration, 179–183
Hydrogen sulfide, 64, 82, 167
Hydrologic cycle, 2
Hydrologic factors, of constructed wetlands, 449–450
Hydrosphere, 2–3
Hyphae, 98
Hypochlorite ion, 343
Hypochlorous acid, 343

I

IADB database, 490
Illness. *See* Health effects
Impact analysis, 316
Incinerators, 428
In-City Reclamation Irrigation System (IRIS), 428
Incubation periods, of disease organisms, 116
India, 11
Indian (Native American) water rights, 397
Indirect costs, 302, 309
Indirect methods, Boron, 340
Induced radioactivity, 68
Inductively coupled plasma method, 340, 357–359
Industrial reuse, 5, 22, 35–36
Israel, 6
uses of water in, 2, 35
Industrial wastewater, 167
Infiltration, for groundwater recharge, 31, 33
Inflation, 319
Information sources, 485–491
Ingestion, 28, 128, 276. *See also* Crops; Drinking water
Inhalation, 130, 276
of radioactive constituents, 128
Innovative and Alternative Technology program, 403–405
Pollution Control Technology Facility File, 490
VISITT Program, 487–488
Insecticides. *See* Pesticides
Insects
biological management, 467–468
environmental management, 155–156
vector relationship with water, 133
as vectors of disease, 112–113, 118, 134
Intake
acceptable daily, 270
calculating chemical, 270
chronic daily, 267
Integrated Risk Information System (IRIS), 271, 506
Interest, 302
International Commission on Radiological Protection, 363
Intestinal infections. *See* Parasites
Inventory of projects, 509–513
Iodine, in disinfection, 220
Iodometric methods, 344–345
Ion chromatographic method, 341
Ion exchange, 11, 34, 132, 218–220
Ionization constant, 342
IRIS (In-City Reclamation Irrigation System), 428
IRIS (Integrated Risk Information System), 271, 506
Irrigation. *See* Agricultural irrigation; Landscape irrigation
Israel, 6, 31

J

Jackson Candle Turbidimeters, 62
Judicial penalties, 418

K

Kentucky case study, 438–459
Kentucky Division of Water, 444

Kentucky Pollutant Discharge Elimination System (KPDES), 382, 444, 457–458
Kinematic viscosity, 71
Kjeldahl method, 347
KPDES (Kentucky Pollutant Discharge Elimination System), 382, 444, 457–458

L

Labeling samples, 331
Labor, 302
Laboratory methods, 333–365
Lagooning, 5–6
Lag phase, 91–92
Lakes. *See* Surface water
Lambert's law, 62–63
Land filtration, 5
Landscape irrigation, 5, 22, 417, 428–429, 476
Land treatment systems, 405, 462–466, 469–470
   operation and maintenance costs, 473–474
Langmuir isotherm, 197
Leaching requirement, 25
Lead, 30–31. *See also* Trace metals
Leak detection, 19
Legal aspects
   sampling and monitoring, 379–382
      chain of custody, 332–333
   wastewater reclamation and reuse, 395–421
Legal systems, 396–397, 418
Legislation. *See* Regulations
Leishmaniasis, 134, 136
Leptospira, 109, 116, 285
Leukemia, 129
Liability. *See* Legal aspects
Light
   absorption of, 62–63
   effect on disease vectors, 133
Lime treatment, 34
Limiting design parameter, 468
Linear alkylate sulfonate, 355
Litigation, 415, 418–420
Livestock, 5–6, 38–40
Local regulations, 411–413
LOEL (lowest-observed-effect level), 272–273
Logan Aluminum, Inc., case study, 446–459
Lowest-observed-effect level (LOEL), 272–273
Low lift pump station, 431
Loxahatchee River environmental control district, 428

M

Magnesium, 25–27
Magnesium hydroxide, 200
Maintenance. *See* Operation and maintenance costs

Malaria, 113–114, 134, 135
   environmental management, 155
   habitat, 136
Malathion, 138
Manning-Kutter formula, 144–145
Mannitol potentiometric method, 340
Manual for the Certification of Laboratories Analyzing Drinking Water, 361
Manual - Guidelines for Water Reuse, 375
Marshes, 142
Materials, cost of, 302
Maturation ponds, 34
Maximum contaminant goal level (MCGL), 16–18, 414, 416
Maximum contaminant levels (MCLs), 9, 335, 403, 413–416
McKlay Bay Refuse-to-Energy Facility, Florida, 428
MCL. *See* Maximum contaminant levels (MCLs)
Mean, 388
MEB (multiple effect boiling distillation), 198, 202, 204
Media, of filters, 178
Median, 388
Membrane electrode method, 344
Membrane filter technique, 359–361
Membranes
   for electrodialysis, 213
   for osmotic filtration, 211–212
Mercuric nitrate method, 341–342
Mercury, 30–31, 332. *See also* Trace metals
Mesophiles, 95
Metals. *See also* Heavy metals; Trace metals
   removal by land filtration, 6
   Saudi Arabian Standard, 9
   soil sorption, 10
   water quality criteria for cooling water, 36
Methane, 167
Methemoglobinemia, 128
Methods. *See* Analysis; Sampling
Methods for Chemical Analysis of Water and Wastes, 327
Methods for Collection and Analysis of Water Samples for Dissolved Minerals and Gases, 327
Metric units, 523–525
Mexico, 6
MFD (multistage flash distillation), 198, 202
Microbiology, 88–100
Misdemeanor, 419
Mode, 388
Modeling, 264
   carcinogenic risk, 279–280
   for chemical exposure, 284–288
   of infectious agents, 287–288

slope factor calculation, 275
Moisture content, of sludges, 72–73
Molality, 73, 208
Molarity, 73–74
Molds, 89–90, 97–98
Molybdenum, 23. *See also* Trace metals
Monitoring
   analysis, 333–365
   flow measurements, 365–375
   legal aspects, 379–382
   reliability, 375–379
   sampling, 327–333
Monitoring programs, 19, 430, 433–434, 459–461
Mosquitos, 112, 132, 157
   biological management, 467–468
   environmental management, 155–156
   problems in impoundments, 142
Most probable number (MPN) index, 521–522
Multiple effect boiling distillation (MEB), 198, 202, 204
Multiple-tube fermentation, 38–39, 359
Multistage flash distillation (MFD), 198, 202
Murphy's Law, 375
Mutagenicity, 268, 411
Mycelium, 98

N

Naples, Florida, 428
National Academy of Sciences, 262
National Aeronautics and Space Administration (NASA), 495
National Contingency Plan, 406
National Council on Radiation Protection and Measurements, 363
National Environmental Policy Act of 1969 (NEPA), 398, 400–401
National Pollutant Discharge Elimination System (NPDES), 381–382, 401–402, 407
   sample permit requirements, 409–411, 430–431
National Priorities List, 406
National Science Foundation (NSF), 494
National Technical Information Service (NTIS), 485, 487
National Water Research Institute (NWRI), 503–504
National Wildlife Federation, 444
Native American water rights, 397
Natural radioactivity, 68
Natural wetlands, 441
Navigable waters, 401, 419. *See also* Refuse Act; Rivers and Harbors Act of 1899
Necator, 111, 122

Nematodes. *See* Roundworms
NEPA, 398, 400–401
Nephelometric Turbidity Units (NTUs), 6
Newtonian fluids, 72
Newton's drag coefficient, 182
Nickel, 30–31. *See also* Trace metals
Nitrates, 81, 128
   maximum contaminant level, 9
   sampling and analysis, 332, 345–347
Nitrification-denitrification process, 8
Nitrite, 81
Nitrogen, 81
   analysis, 345–347
   gas, 167
   guidelines for agricultural irrigation, 24
   and hydraulic loading rate, 464–465
   water quality criteria for livestock and wildlife, 40
   wetland removal efficiency, 442
Nitrogen cycle, 346
Nitrogenous biological oxygen demand, 352
NOEL (no-observed-effect level), 272–273
Noncancer hazard quotient, 274
Noncarcinogenic response, 269, 272
Nonionic detergents, 83
Nonpoint source pollution, 402
No-observed-effect level (NOEL), 272–273
Normal distribution, 387–391
NPDES (National Pollutant Discharge Elimination System), 381–382, 401–402, 407
NTUs (Nephelometric Turbidity Units), 6
Nuclear Regulatory Commission, 363
Nutrients, 82
   wetland removal efficiency, 442–444, 450–455
NWRI (National Water Research Institute), 503–504

O

Oak Ridge National Laboratory, 271
Odors, 64, 82, 245
Oil, 82, 420–421
   analysis, 354–355
Onchocerciasis, 113, 134
   environmental management, 155
Open reflux method, 353
Operation and maintenance costs, 302–303, 305, 307
   example parameters, 308–309
   land treatment system, 473–474
Operations research/management science (OR/MS), 484–486
Optimum force main diameter, 310
Oral route, 276

Order of magnitude estimates, 301
Oregon Department of Environmental Quality, 375–379
Organic constituents
    drinking water guidelines, 42–43, 46
    effect on disease vectors, 132
    in pesticides, 139
    reaction with chlorine, 221–222
    in reclaimed wastewater, 127
    sampling and analysis, 332, 351–355
Organic loading rate, 450
Organochlorine compounds, 137
Organophosphate compounds, 138
Orlando, Florida, wetlands, 428
OR/MS, 484–486
Orthophosphate, 349
Osmotic filtration, 207–213
Overland flow, Australia, 5
Oxygen, dissolved. *See* Dissolved oxygen
Oxygen demand, 80
Oxygen depletion, 246–247
Oxygen renewal, 245–247
Ozonation, 34, 291–292
Ozone, in disinfection, 220

P

Palmer-Bowlus flumes, 371–372
Parasites, 39, 51, 111–114. *See also* individual parasite names
Paratyphoid, 109
Parshall flumes, 370–371
Partition method, 355
Pathogens. *See* Bacteria; Helminths; Parasites; Protozoa; Viruses
Pathways. *See* Routes of exposure
Peaking factor, 170
Penalties, 415, 418–420
Percolation, for groundwater recharge, 31
Periodic table of elements, 518
Permeability, of soil, 464
Permethrin, 138
Permits, 381–382, 397. *See also* NPDES
Personnel, for sampling, 331
Pesticides, 130, 137–139
    analysis, 355
    classes of, 138–139
    in reclaimed wastewater, 127
    sampling, 332
    water quality
        drinking water, 43–44
        fishing, 39
        livestock and wildlife, 40
pH, 75
    analysis, 347–349
    calculating, 27
    and color, 65
    and enzyme reactions, 90
    guidelines for agricultural irrigation, 24
    and heavy metal uptake, 126
    influent and effluent, 9, 448
    lime treatment, 34
    recommended limits for agricultural irrigation, 32
    relationship between ions, 347
    sampling, 332
    sludge, 232
    water quality
        cooling water, 36
        fishing, 39
        livestock and wildlife, 40
        recreational reuse, 38
Pharmacokinetics, 17, 268
Phenolic compounds, 83, 127
    water quality standards for fishing, 39
pH meter, 334, 348
Phosphates
    analysis, 349
    ecological impact, 349
    maximum contaminant level, 9
    water quality criteria for cooling water, 36
Phosphorous, wetland removal efficiency, 442
Photosynthesis, 2, 98, 245
Phycomycetes, 98
Physical characteristics, 60–68, 505
    sampling, 332, 334–339
    water, 520
Pinworm, 11
Pipe installation, cost estimates, 303–304, 310–311
Planning process, for reuse program, 300
Plants
    effect on disease vectors, 133
    uptake of metals, 6, 23, 126
Plasmodium, 113
Poise, 71
Poliovirus, 122
    infective dose, 281
    survival time, 114, 285
Pollutants
    toxic, 401, 405–406, 408
    wetland removal efficiency, 442
Pollution
    effects of wastewater discharge, 245
    nonpoint source, 402
Polymer feed system, 431–432
Polymers
    ion exchange, 218
    for osmotic filtration, 211
    sludge treatment, 227
Polypeptides. *See* Proteins

Population
  exposed, 265, 268
  Florida, 426–427
  Georgia, 459
  Kentucky, 438
Population equivalent, 171, 310
Porosity, of filter bed, 180–182
Positive net-present value, 316
Potable quality reuse water, 22, 33–35, 63
  water quality data, 35
Potassium permanganate, 220
Potency factor (slope), 274–279
Potentiometric method, 341
POTWs. *See* Publicly owned treatment works (POTWs)
Power, 305, 375
  boilers, 428
  refuse-to-energy facilities, 428
  standby, 376–377
Power plant accidents, 363
Precipitation (solids removal), 176, 212
Precision, 333
Preliminary treatment. *See* Primary treatment
Preservation of samples, 331–332
Pressure
  in distillation units, 202
  osmotic, 207–208
  in Venturi meters, 371–372
Pressure filtration, 34, 229–230
Pretreatment guidelines, 411–412
Primary treatment, 123–124, 126, 130–131, 173
  percent of pathogen removal, 289
  redundant facilities, 377
Prince William Sound, 420
Prior appropriation rule, 396–397
Probability curves, 389–390
Project APRICOT, 428
Propoxur, 138
Proteins, 82–83, 232
Protozoa, 99, 123, 282, 506
  infective dose, 283
Psychrophiles, 95
Public awareness, 20, 154, 157–158, 291–295.
  See also Risk communication
Public health. *See* Health effects
Public laws
  PL 80–845, 397–399
  PL 84–660, 398–399, 421
  PL 87–88, 398–400
  PL 89–234, 398, 400
  PL 91–224, 398, 400–401
  PL 92–500, 381, 398, 401–402, 411, 419–421, 468
  PL 93–523, 17, 18, 363–364, 398, 402–403
  PL 94–580, 398, 403
  PL 95–217, 381, 398, 403–405, 421
  PL 97–117, 398, 406–407
  PL 100–4, 398, 407–408, 411–412
  42-U.S.C. 9601–9657, 398, 405–406
Publicly owned treatment works (POTWs), 35
  funding, 409
  regulations, 402, 407, 411–413, 420
Pueblo water rights, 397
Pump systems, 431–432
  cost estimates, 303–306, 311–312
Pyrethrin, 137
Pyrethroids, 138

Q

Quantity, 170–171
Questionnaire, for projects inventory, 509–513

R

Radioactive constituents
  analysis, 363–366
  drinking water guidelines, 46
  in reclaimed wastewater, 128–130
Radioactivity, 68–70, 363–364
  common radioactive isotopes, 71
  regulatory agencies, 363
Radionuclides
  bibliography, 363
  drinking water standards, 364
  water quality criteria for livestock and wildlife, 40
  WHO guidelines, 364
Rad units, 363–364
Range, 388
Raoult's law, 208
Rapid sand filters, 177–178, 186–193, 432
Rats, as disease vector, 114, 116, 118, 134
RBE (relative biological effect), 364
RCRA (Resource Conservation and Recovery Act of 1976), 398, 403
Reaeration, oxygen renewal, 245–247
Reagents for analysis
  chlorides, 341
  COD, 353
  coliform, 360
  dissolved oxygen, 344
  pH, 348
  sulfates, 350
Reasonable use rule, 397
Recarbonation, 34
Reclamation projects. *See* Wastewater reclamation and reuse projects
Recordkeeping, 331
Recreational reuse, 5, 22, 36–38, 474
  swimming pool diseases, 37
  WHO criteria, 50

Rectangular contracted weir, 367–368
Rectangular suppressed weir, 369
Redundancy, 375, 377
Reference dose (RfD), 269–274
Refrigeration of samples, 332
Refuse Act of 1899, 397, 419–421
Regulations
   federal, 398–408
   local, 411–413
   sampling and monitoring, 381–382, 430
   state, 20–22, 375–382, 408–411
Regulatory agencies, 397–398
Relative biological effect (RBE), 364
Relative error, 389
Relative standard deviation, 388–389
Reliability, 375–379, 461
Removal efficiency, of pathogens, 126
Rem units, 363-364
Replicates, 329–330
Research, 484. *See also* Funding
   for hazard identification, 267
   wastewater reclamation and reuse, 483–513
Research and development (R & D), 484
Resins, for ion exchange, 218
Resmethrin, 138
Resource Conservation and Recovery Act (RCRA) of 1976, 398, 403
Resources, informational, 485–491
Reuse. *See* Wastewater reclamation and reuse
Reuse of Reclaimed Water and Land Application, 375, 379
Reverse osmosis, 11, 34, 124, 197, 209–210
Reverse osmosis plants, 212–215
Revolving loan fund, 408–409
Reynold's number, 180
Rheological properties, 70–73
Rickettsia alkari, 116
Riparian regulations, 396, 440
Risk, 262, 279–280. *See also* Exposure assessment; Risk assessment; Toxicity assessment
   lifetime, 275
Risk assessment, 17, 261–295. *See also* Health effects
Risk characterization, 262, 279–288
Risk communication, 262, 280, 293. *See also* Public awareness
Risk management, 262
Risk perception, 34
   drinking water complaints, 46
Rivers. *See* Surface water
Rivers and Harbors Act of 1899, 397–399, 419
Roentgen units, 363
Roman law, 396, 418
Rotating biological contactors, 8, 30

Rotavirus, 51, 122
Rotifers, 99
Roundworms, 11, 113, 115, 120
Routes of exposure, 18, 129–130, 265–266, 268
Runoff, 402

S

Safe Drinking Water Act of 1974, 17–18, 363–364, 398, 402–403
Safe Drinking Water Act of 1988, 413–414
Sag curves, 247–251
Salinity, 23–25, 66–67
   desalination, 197–218
   effect on disease vectors, 132
   and livestock, 38
Salmonella, 115, 122, 281
   incubation period, 116
   infective dose, 282
   survival time, 114, 285
Salts
   build-up, 25
   sampling and analysis, 340–342
Sampling, 263, 327–333
   bibliography, 327
   40CFR Part 136, Chapter 1, Subchapter D, 327
   legal aspects, 379–382
Sand filtration, 125, 177–178, 186–193, 432
Sandfly, environmental management, 155–156
Sandfly fever, 114
SARA (Superfund Amendments and Reauthorization Act), 405–406
SAR (sodium adsorption ratio), 24–27, 356
Saturation values, of dissolved oxygen, 519
Saudi Arabia, 6–10, 30
Scabies, 115
Scale formation, 35, 200, 202
   in osmotic filtration, 212
Schistosoma, 51, 110–111, 120, 122
   environmental factors, 154
   incubation period, 116
Schistosomiasis, 110, 115, 134–135
   environmental management, 155
   habitat, 136
Scintillation detectors, 365–366
Scoping activities, 262–264
Secondary treatment, 124, 126, 131, 173. *See also* Activated sludge process; Rotating biological contactors; Trickling filter process
   redundant facilities, 377
Sedimentation, 123–124, 130–131, 176
   expected removal of microorganisms, 288
   helminth egg analysis, 362
   redundant facilities, 377
   using centrifugation, 229–231

Sediments. *See* Turbidity
Selection criteria, 314–317
Selective chelating groups, 218
Semidetailed estimates, 301
Settleable solids, 67, 130
    maximum contaminant level, 9
    sampling and analysis, 337
    water quality for recreational reuse, 38
Sewage treatment units, 172–173, 175
Shape factor, 180–182
Shearing forces, 71–72
Shearing strength, 232
Shellfish, 99–100
Shigella, 109–110, 122
    incubation period, 116
    infective dose, 282
    survival time, 114, 285
Sievert units, 363–364
Sieving, 175
Silver, 47, 220
Single-effect still, 201
Site description, 264–265
    constructed wetlands, 445–446
    land treatment, 463
Site location, 328
Sleeping sickness, 114, 134
Slope factor, 274–279
Slow-rate land treatment systems, 462–464, 467
Slow sand filters, 177–178, 193–195
Sludge, 167. *See also* Vacuum filtration
    activated, 8, 30, 124, 288–289
    centrifugation, 229–231, 238–242
    dewatering, 131–132, 226–242
    digestion, 225–226
    dilution, 242
    disposal into natural waters, 242–251
    drying beds, 34, 227–228
    elutriation, 231
    filtration, 229–230, 233–238
    moisture content, 72–73
    protein content, 82
    regulations, 407
    Sultanate of Oman standards, 54
    treatment and disposal, 173, 225–251, 429
    units of analysis, 74
Snails, 154, 156
Snow, John, 108
Societal issues, 508
    in cost benefit analysis, 316
    factors effecting sanitation, 153
Sodium
    agricultural irrigation, 23, 25–27
    guidelines, 24
Sodium adsorption ratio (SAR), 356
    guidelines for agricultural irrigation, 24–27
Sodium azide method, 344
Sodium iodide, 365–366
Soil
    monitoring, 472
    permeability, 464
    sorption of metals, 10
Solar distillation, 205–207
Solids content, 67–68, 335–337. *See also* Dissolved solids; Settleable solids; Suspended solids; Volatile solids
Solid waste regulations, 403
Solubility product constant, 74
Solvents
    of oil and grease, 82
    osmosis of, 207–211
    water as, 1
Sources
    of contaminants, 265, 267
    of pollution, 402
    of wastewater, 16
South Africa, 10, 33
Soxhlet extraction method, 355
Spanish law, 396–397
Specific conductance. *See* Electrical conductivity
Specific ion exchangers, 218
Specific resistance
    apparatus, 233–235
    testing, 235–237
Spectrophotometry, 335, 356–357
Spills, 405, 420–421
Sprinkler land treatment, 405
SRF (State Revolving Loan Fund), 408–409
St. Petersburg dual distribution system, 428
Stabilization ponds, 124, 288–289
Standard deviation, 387
Standard error of the mean, 388
Standard Methods for the Examination of Water and Wastewater, 327, 334, 382
Standards
    for color, 335
    drinking water, 21, 402–403, 413–415
    fishing, 39, 245
    flow, 21
    radionuclides, 364
    for reclaimed water flows, 21
    Sultonate of Oman, 41, 48–49, 52–54, 476–477
    water quality, 16, 18–20, 41
Starch, 82–83
State regulations, 20–22, 408–411
    reliability, 375–379
    sampling and monitoring, 379–382
State Revolving Loan Fund (SRF), 408–409
Stationary growth phase, 96
Statistics, 387–391

Stills, 201–202
Stoke's frictional forces, 230
Storage and Retrieval System of Water Quality Information, 491
Storage facilities, 431–432
   cost estimates, 303, 307, 313–314
Storage impoundments, 141–143, 288
   rejection of water for, 461
   volume, 463, 465–466
Stormwater runoff, 168, 402, 411
Straining, 175–176
Streams. *See* Surface water
Streptococcus
   fecal coliform analysis, 358–361
   infective dose, 282
   survival time, 114
Strickler's formula, 144–145
Subsurface flow system, 441
Sugars, 82–83
Sulfates, 200
   analysis, 349–351
   ecological impact, 349
Sulfur cycle, 350
Sultanate of Oman
   standards, 41, 48–49, 52–54, 476–477
   wastewater reclamation, 474–478
Sultan Qaboos University, 476
Supercolloidal solids, 67
Superfund Amendments and Reauthorization Act (SARA), 398, 405–406
Surface application technique, 405
Surface spreading, for groundwater recharge, 31
Surface water
   disposal of sludge, 242–251
   monitoring, 472
   oxygen depletion and renewal, 245–247
   percent of water resources, 4
   riparian regulations, 396, 440
Surfactants, 83, 355
Surveys, opinion on wastewater reuse, 294–295
Survival factors, of pathogens, 114, 120–121, 283–285
Suspended solids, 23, 67, 471
   effluent levels, 11, 448
   maximum contaminant level, 8–9
   and population equivalent, 171
   sampling and analysis, 336
   water quality
      cooling water, 36
      recreational reuse, 38
   wetland removal efficiency, 442–443
Swimmer's itch, 110
Swimming pool diseases, 37, 110
Syringaldazine (FACTS) method, 344

T

Taenia, 6, 122, 124
   survival time, 285
Tagging samples, 331
Tallahassee spray irrigation system, 428
Tapeworm. *See Taenia*
Taste, 63–64
Teller isotherm, 197
Temperature, 60–61
   and culture generation time, 95
   effect on disease vectors, 133
   and enzyme reactions, 89
   of heat treatment, 124
   influent and effluent, 9, 448
   measuring, 338
   water quality criteria for recreational reuse, 38
Tennessee Valley Authority, 444
Teratogenicity, 411
Tertiary treatment methods, 124–126, 131–132, 174–251
Thermophiles, 95
THM. *See* Trihalomethanes (THM)
Threshold dose, 269, 271
Time-dependent fluids, 72
Time-independent fluids, 72
Tin, 47
Title III of SARA, 406
Total amperometric chlorine residual, 290
Total organic carbon, analysis, 354
Toxic Chemical Release Inventory (TRI) database, 406
Toxicity, 82. *See also* Health effects
   chemicals in reclaimed wastewater, 125–128
   metals, 127
   radioactive constituents, 128–129
Toxicity assessment, 267–279
Toxicity studies, for MCLG development, 17
Trace metals. *See also* Cadmium; Chromium; Copper; Lead; Mercury; Molybdenum; Nickel; Zinc
   agricultural irrigation, 24, 29–32
   drinking water guidelines, 42–43
   removal by secondary treatment, 131
   sampling and analysis, 332, 355–359
   toxicity, 82
   water quality
      cooling water, 36
      livestock and wildlife, 40
Trachoma, 115
Tragelaphus, 112
Transmission routes, of disease, 113–114, 116–118
Trapezoidal weir, 369–370
Traveling bridge filter, 432

Treatment plants, cost estimate, 302, 304
Treatment technologies
   flow sheet, 131
   for pathogen removal, 123–125
   reasons for, 171, 173
   wastewater reclamation and reuse, 174–251
Trichloroethylene, 284–285
Trichuris, 51, 122
Trickling filter process, 8, 124
   percent of pathogen removal, 289
   for potable reuse, 34
   trace metals, 30
Trihalomethanes (THM), 221, 225, 506
   sampling, 332, 343
Tristimulus filter method, 335
TRI (Toxic Chemical Release Inventory) database, 406
True color, 65, 335
Trypanosomiasis, 112, 115, 134–135
   environmental management, 155
   habitat, 136
TSS. *See* Suspended solids
Tularemia, 109
Turbidimeters, 62
Turbidity, 61–63
   maximum contaminant level, 403
   Nephelometric Turbidity Units, 6
   sampling and analysis, 338–340
   water quality
      cooling water, 36
      recreational reuse, 38
t values, 330
Tyndall effect, 61–62
Typhoid fever, 51, 109, 115, 122

U

Ultraviolet disinfection treatment, 220
Uncertainty factor, 273, 275
United Arab Emirates (U.A.E.), 10–11
United Nations, 28
Unit processes, 172–173, 175
Units, conversion factors, 523–525
Universal gas constant, 207
Urban reuse, nonpotable, 22
U.S. Army Corps of Engineers, 397
U.S. Constitution, 396
U.S. Department of Agriculture (USDA), 494
U.S. Department of Commerce (DOC), 493
U.S. Department of Defense (DOD), 363, 492
U.S. Department of Energy (DOE), 363, 492
U.S. Department of Health and Human Services (DHHS), 494
U.S. Department of Justice (DOJ), 419
U.S. Department of the Interior (DOI), 4, 489, 493
U.S. Environmental Protection Agency
   enforcement actions, 419
   establishment of criteria, 16
   funding by, 491–492
   Public Information Center, 487
   referrals to DOJ, 418–419
   reliability requirements, 375–376
   risk assessment database, 271
   VISITT Program, 487–488
   wetlands demonstration project, 444
U.S. Government Printing Office (GPO), 488–489
U.S. Gross Domestic Product Deflator, 319
USDA (U.S. Department of Agriculture), 494
User charges, 435

V

Vacuum filtration, 34, 229–230
Van Leeuwenhoek, Anton, 108
Vapor compression distillation, 198
Variance, 388
Vectors of disease
   animals, 114, 116, 118, 134–135
   control, 135, 137–158
   ecological factors, 132–133
   insects, 112–113, 118, 134
   preferred habitats, 133, 135–136
Vegetables
   Boron tolerance, 28
   and Boron uptake, 128
   cost savings using reclaimed water, 6
   water content, 2
Vegetated submerged bed wetlands, 441
Venn diagram, 300–301
Venturi meters, 371–373
Vibrio, 109–110, 122
   incubation period, 116
   infective dose, 282
   survival time, 114
Viruses, 51, 100, 123
   destruction with chlorination, 290
   destruction with ozonation, 292
   in human drinking water, 39
   infective dose, 281
   in livestock drinking water, 39
   survival times, 121
   in wastewater, 281, 507–508
Viscometer, 70
Viscosity, 70–72
   kinematic, 182
VISITT (Vendor Information System for Innovative Treatment Technologies), 487–488
V-notch weir, 367–368

Volatile organic chemicals, 17
Volatile solids, 67, 337
Volume of sample, 331

W

Wastewater
  biological and bacteriological characteristics, 88–100
  chemical characteristics, 73–88, 469, 505–506
  estimating unit costs, 310
  health effects of reclaimed, 107–158
  in a metropolitan system, 411
  physical characteristics, 60–68, 332, 334–339, 469, 505
  quantity generated, 170–171
  sources and classification, 167–170
  units of analysis, 74
Wastewater reclamation and reuse, 22, 108–109
  cost savings, 6
  economics, 6, 299–320
  estimating costs, 308–314
  example parameters, 308, 461–462
  fertilizer, 6
  funding, 491–504
  legal aspects, 395–421
  monitoring, 327–382
  pathogens in reclaimed water, 117–123, 126
  public opinion survey, 294–295
  research, 483–513
  risk assessment, 280–284
  standards for drinking water, 21
  standards for water flows, 21
  toxic chemicals, 125–128
  treatment technologies, 165–251, 457
Wastewater reclamation and reuse projects
  California, 4–5
  case studies, 426–459
  Florida, 428–438
  Georgia, 460–474
  inventory of projects, 509–513
  Kentucky, 438–459
  Saudi Arabia, 6–10, 30
  Sultanate of Oman, 475–478
Wastewater treatment plants, 380, 429–438, 468–470
  costs, 433–435, 437
Water
  biological and bacteriological characteristics, 88–100
  chemical characteristics, 73–88
  conductivity, 66
  dissociation constant, 75
  human consumption statistics, 2–4, 428, 438, 460
  importance of, 1–2
  physical characteristics, 60–68, 332, 334–339, 520
  reaction with chlorine, 221
  in sludges, 72–73
  velocity and disease vector control, 146
  weights of ions in, 526
Water balance, of wetlands, 449–450
Water content
  in fruits and vegetables, 2
  in human body, 1–2
Water demand, 427
Water Environmental Research Foundation (WERF), 495–501
Water mains, estimating costs, 310–311
Waternet database, 489–490
Water Pollution Control Act, 381
Water quality
  criteria, 15–18, 413
    industrial cooling water, 35–36
    for livestock and wildlife, 40
    recreational reuse, 38
    toxic pollutants, 408
    trace metals, 30
    WHO for wastewater reuse, 50–51
  and disease, 109–117
  guidelines, WHO. See Drinking water, WHO guidelines
  standards, Sultanate of Oman, 41, 48–49, 476–477
Water Quality Act of 1965, 398, 400
Water Quality Act of 1987, 398, 407–408
Water resources
  earth, 2
  Florida, 427–428
  Kentucky, 438–439
  Saudi Arabia, 7
  Sultanate of Oman, 475
  U.S., 2–5
Water rights, 396–397
Water supply systems
  estimating costs, 302–308
  example parameters, 308
Water unit risk, 275
Weight-of-evidence, 274–276
Weirs, 365–370
WERF (Water Environmental Research Foundation), 495–501
Wetland removal efficiency, 442–444, 450–455
Wetland removal mechanisms, 451
Wetlands, 427–428, 436–459
  constructed, 441, 444–459
  natural, 441–442
Whipworm, 11
WHO. See World Health Organization (WHO)
Wild life sanctuaries, 467

Winkler method (iodometric), 344–345
Winters Doctrine, 397
World Bank, 509
World Bank's Manufacturers Unit Value Index, 319
World Health Organization (WHO), 16, 20, 508–509
  criteria for wastewater reuse, 50–51
  definition of health, 132
  guidelines for drinking water, 39–47, 414, 416, 421
  guidelines for radioactive constituents, 364

Worms. *See* Helminths
Wuchereria bancroftii, 112

Y

Yellow fever, 113–115
Yersinia, 122

Z

Zero discharge, 402
Zinc, 30–31. *See also* Trace metals